Microbial Biochemistry

G.N. Cohen

Microbial Biochemistry

Third Edition

Springer

G.N. Cohen
Institut Pasteur
Paris
France

Institut de Génomique
Commissariat à l'Energie Atomique
Evry
France

ISBN 978-94-017-7626-4 ISBN 978-94-017-8908-0 (eBook)
DOI 10.1007/978-94-017-8908-0
Springer Dordrecht Heidelberg New York London

Printed on acid-free paper

Springer is part of Springer Science+Business Media (www.springer.com)

Foreword

This book originates from almost 60 years of living in the company of microorganisms, mainly with *Escherichia coli*. My scientific life has taken place almost exclusively at the Institute Pasteur in Paris, where many concepts of modern molecular biology were born or developed.

The present work emphasizes the interest of microbial physiology, biochemistry and genetics. It takes into account the considerable advances which have been made in the field in the last 30 years by the introduction of gene cloning and sequencing and by the exponential development of physical methods such as X-ray crystallography of proteins.

The younger generation of biochemists is legitimately interested in the problems raised by differentiation and development in higher organisms, and also in neurosciences. It is however my feeling that the study of prokaryotes will remain for a long time the best introduction to general biology.

A particular emphasis has been given to particular systems which have been extensively studied from historical, physiological, enzymological, structural, genetic and evolutionary points of view: I present my apologies to those who may find that this choice is too personal and reflects too much my personal interest in subjects in which I have either a personal contribution or where important results have been obtained by some of my best friends.

I am grateful to the Philippe Foundation for the help it has given to me and to many of my students for many years.

I dedicate this book to the memory of my wife, Louisette Cohen, for her patience and help, not only while this book was written, but also during our 68 years of common and happy life.

This work is a tribute to the memory of my beloved colleagues, my mentor Jacques Monod, and the late Harold Amos, Dean B. Cowie, Michael Doudoroff, François Jacob, Ben Nisman, Earl R. Stadtman, Roger Y. Stanier, Germaine Stanier, Huguette and Kissel Szulmajster and Elie Wollman.

Paris, France Georges N. Cohen

Abbreviations

aceA	Isocitrate lyase
aceB	Malate synthase A
aceK	Isocitrate dehydrogenase kinase/phosphatase
acnA	Aconitase A
acnB	Aconitase B
ACP	Acyl carrier protein
ADP	Adenosine diphosphate
ADPG	Adenosinediphosphoglucose
AICAR	5-aminoimidazole-4-carboxami-do-1-b-ribofuranosyl-5′-phosphate
AIR	Aminoimidazole ribonucleotide
AK	Aspartokinase
ALA	d-aminolevulinic acid
alr	Alanine racemase
AMP	Adenosine monophosphate
APS	Adenosine-5′-phosphosulfate
arcA	Global regulatory network
AS	Anthranilate synthetase
ATCase	Aspartate transcarbamylase
ATP	Adenosine diphosphate
BCCP	Biotin carboxyl carrier protein
CAIR	5-amino-4-imidazole carboxylate ribonucleotide
cAMP	Cyclic AMP
CAP	Catabolite activator protein
cDNA	Complementary DNA
CDP	Cytidine diphosphate
CDRP	Anthranilate deoxyribulotide
CH2THF	5–10 methylene tetrahydrofolate
citA	Citrate synthase (*B. subtilis*)
citC	Isocitrate dehydrogenase (*B. subtilis*)
citH	Malate dehydrogenase (*B. subtilis*)
citR	Citrate synthase regulator (*B. subtilis*)

citZ	Citrate synthase (*B. subtilis*)
CMP	Cytidine monophosphate
CoA	Coenzyme A
CRM	Cross-reacting material
CTP	Cytidine triphosphate
dADP	Deoxyadenosine diphosphate
DAHP	3-deoxy-D-arabino-heptulosonate-7-phosphate
DAP	Diaminopimelate
DAPA	7–8 diaminopelargonate
dATP	Deoxyadenosine triphosphate
dCDP	Deoxycytidine diphosphate
dCMP	Deoxycytidine monophosphate
dCTP	Deoxycytidine triphosphate
ddlB	D-alanyl-D-alanine ligase
DEAE	Dimethylaminoethyl
dGDP	Deoxyguanosine diphosphate
dGTP	Deoxyguanosine triphosphate
DHNA	1, 4-dihydroxy-2-naphthoic acid
DHQ	Dehydroquinate
DMK	Demethylmenaquinone
DNA	Deoxyribonucleic acid
dNDP	Deoxynucleoside diphosphate
dNTP	Deoxynucleoside triphosphate
dTDP	Deoxythymidine diphosphate
dTMP	Deoxythymidine monophosphate
dTTP	Deoxythymidine triphosphate
dUDP	Deoxyuridine diphosphate
dUMP	Deoxyuridine monophosphate
E4P	Erythrose-4-phosphate
eda	2-keto-3-deoxy-6-phosphogluconate aldolase
edd	6-phosphogluconate dehydrase
eno	Enolase
F	Female
FAD	Flavin adenine dinucleotide
FAICAR	5′-phosphoribosyl-4-carboxamide-5-formamidoimidazole
fbaA	Fructose 1, 6-bisphosphate aldolase
fbp	Fructose bisphosphatase
FeMoco	Iron-molybdenum cofactor
FGAM	Formylglycinamidine ribonucleotide
FGAR	Formylglycinamide ribonucleotide
fmd	Molybdenum-containing formylmethanofuran dehydrogenase
FMN	Flavin mononucleotide
FPP	Farnesylpyrophosphate
ftr	Formylmethanofuran tetrahydro-methanopterinformyltransferase

fumA	Fumarase A
fumB	Fumarase B
fumC	Fumarase C
fur	Global regulatory network
fwd	Tungsten-containing formylmethanofuran dehydrogenase
gapA	Glyceraldehyde 3-phosphate dehydrogenase
gapB	4-phosphoerythronate dehydrogenase
GAR	Glycinamide ribonucleotide
GAT	Glutamine amidotransferase
gcd	Glucose dehydrogenase
GDP	Guanosine diphosphate
glcB	Malate synthase induced by glycolate
glmU	Glc-NAc-1-phosphate uridylyltransferase
gltA	Citrate synthase (*E. coli*)
GMP	Guanosine monophosphate
gnd	Phosphogluconate dehydrogenase
gpm	Phosphoglyceromutase
GS	Glutamine synthetase
GSH	Glutathione
GSSG	Oxidized glutathione
GTP	Guanosine triphosphate
HDH	Homoserine dehydrogenase
Hfr	High frequency of recombination
hisM	Membrane-bound component of the histidine ABC transporter
hisP	ATP-binding membrane-bound protein of the histidine ABC transporter
hisQ	Membrane-bound component of the histidine ABC transporter
HPr	Heat-stable protein of the PTS system
iclR	Transcriptional repressor of the *ace* operon
idh	Isocitrate dehydrogenase
IGP	Indoleglycerol phosphate or imidazoleglycerol phosphate (cf. context)
IMP	Inosinic acid
IPTG	Isopropylthiogalactoside
7-KAP	7-keto-8-aminopelargonate
Lac	Lactose
LAO	Lysine-arginine-ornithine permeation system
Lrp	Leucine-responsive regulatory protein
MBP	Maltose-binding protein
mch	Methenyltetrahydromethanopterin cyclohydrolase
mdh	Malate dehydrogenase
mer	Methylenetetrahydromethanopterin dehydrogenase
MK	Menaquinone
mraY	Lipid I synthesis
mRNA	Messenger RNA

mtrEDCBAFGH	Methylreductase operon
murA	UDP-GlcNAc-enolpyruvoyltransferase
murB	UDP-GlcNac-enolpyruvate reductase
murC	Alanine addition
murD	Glutamate addition
murE	Meso-diaminopimelate addition
murF	D-alanyl D-alanine addition
murG	Lipid II synthesis
murI	Glutamate racemase
^{5}N-methyl-H_4PtGlu	^{5}N-methyltetrahydropteroylglutamate
NAD^+	Nicotinamide adenine dinucleotide
NADH	Reduced form of NAD^+
$NADP^+$	Nicotinamide adenine dinucleotide phosphate
NADPH	Reduced form of $NADP^+$
NDP	Nucleoside diphosphate
NEM	N-ethylmaleimide
NMR	Nuclear magnetic resonance
NTP	Nucleoside triphosphate
OMHMB	2-octaprenyl-3-methyl-5-hydroxy-6-methoxy-1, 4-benzoquinone
OMP	Orotidine-5′-phosphate
ONPF	*o*-nitrophenylfucoside
ONPG	*o*-nitrophenylgalactoside
ORF	Open reading frame
OSB	*o*-succinylbenzoate
OTCase	Ornithine transcarbamylase
p-AB	*p*-aminobenzoic acid
pabA	4-amino-4-deoxychorismate synthase, Gln amidotransferase subunit
pabB	4-amino-4-deoxychorismate synthase, amination subunit
pabC	Elimination of pyruvate from 4-amino-4-deoxychorismate
p-ABG	*p*-aminobenzoylglutamic acid
PALA	N-phosphonacetyl-L-aspartate
PAPS	3′-phosphoadenosine-5′-phosphosulfate
pdxb	4-phosphoerythronate dehydrogenase
PEP	Phosphoenolpyruvate
pfkA	Phosphofructokinase 1
pfkB	Phosphofructokinase 2
pgi	Glucose-6-phosphate isomerase
pgk	Phosphoglycerate kinase
pgl	6-phosphogluconolactonase
Pi	Inorganic phosphate
PII	Protein interacting with glutamine synthetase adenylyltransferase

PLP	Pyridoxal phosphate
PPi	Pyrophosphate
PQQ	Pyrroloquinoline quinone
PRPP	5-phosphoribosyl-1-pyrophosphate
PR-PRFAR	Phosphoribosyl-PR-formimino-5-aminoimidazole-4-carboxamide ribonucleotide
PRT	Anthranilate phosphoribosyltransferase
PTS	Phosphotransferase system
pykA	Pyruvate kinase activated by AMP
pykF	Pyruvate kinase activated by fructose 1, 6-bisphosphate
RNA	Ribonucleic acid
rpe	Phosphopentose isomerase
rpi	Ribose phosphate isomerase
SAH	S-adenosylhomocysteine
SAICAR	5-amino-4-imidazole-N-succinylcarboxamide ribonucleotide
SAM	S-adenosylmethionine
sdhA	FAD subunit of succinate dehydrogenase
sdhB	Non-heme iron subunit of succinate hydrogenase
sdhC	Membrane anchor of succinate dehydrogenase, cytochrome b661-5
sdhD	Membrane anchor of succinate dehydrogenase
soxRS	Global regulatory network
sucA	Dehydrogenase of the ketoglutarate dehydrogenase complex
sucB	Succinyltransferase of the ketoglutarate dehydrogenase complex
sucC	Succinyl CoA synthetase beta subunit
sucD	Succinyl CoA synthetase alpha subunit
tal	Transaldolase
TDG	Thiodigalactoside
tkt	Transketolase
TMG	Thiomethylgalactoside
TPEG	Thiophenylethylgalactoside
TPG	Thiophenylgalactoside
tpi	Triose phosphate isomerase
TPP	Thiamine pyrophosphate
TQQ	Tryptophyl tryptophanquinone
tRNA	Transfer RNA
UDP	Uridine diphosphate
UDPG	Uridine diphosphoglucose
UMP	Uridine monophosphate
UR	Uridylyl-removing enzyme
UTase	Uridylyltransferase
UTP	Uridine triphosphate
XMP	Xanthylic acid
zwf	Glucose 6-phosphate dehydrogenase

Contents

Introduction

The ability of cells to multiply, leading to a net increase in mass, is due to a network of chemical reactions which can be classified as anabolic. Their study forms a chapter in biochemistry to which the name biosynthesis can be given. Biosynthetic reactions require energy; this is provided by another set of chemical reactions which are called catabolic.

A study of cellular metabolism must therefore concern itself with the reactions which produce energy and with the reactions of biosynthesis. This distinction, useful in a didactic way, must not obscure the fact that many intermediates involved in the classical degradation processes, glycolysis and the tricarboxylic acid cycle, are branch points from which purely biosynthetic pathways arise. The degradation sequences are therefore not only important in so far as they provide energy in the form of ATP, but also as they provide the carbon atoms which are necessary for the synthesis of cellular constituents.

Furthermore, if we consider the growth of a bacterium such as *Escherichia coli* on succinate as sole source of carbon, it is evident that this organism must be able to carry out the reactions of glycolysis in the reverse direction in order to obtain, for example, glucose 6-phosphate, which when transformed to erythrose 4-phosphate is required in the biosynthesis of the aromatic amino acids. In this instance, the glycolytic reactions have a purely biosynthetic role. The term "amphibolic" has been introduced to describe such reactions which function in both catabolism and anabolism.

Let us consider the bacterium *E. coli*: it can grow exponentially on a mineral medium containing a usable carbon source. Table 1 gives the composition of one such medium and Table 2 is a very incomplete list of carbon sources for this bacterium. Such a cell will first of all have to be brought into contact with the carbon source of the medium. We shall see that this is generally achieved not by simple diffusion but by means of proteins localized in the cytoplasmic membrane, which are responsible for the ingress of metabolites into the intracellular space.

Phosphate is in large excess, in order to buffer the medium against pH variations. Metals such as zinc, molybdenum cobalt, etc. found in proteins and certain vitamins exist as trace impurities in the commercial salts used.

Table 1 Composition of a minimal medium for the growth of *Escherichia coli*

$KH2PO4$	13.6 g
$(NH4)2SO4$	2.0 g
$MgSO4$, $7H2O$	0.2 g
$FeSO4$, $7H2O$	0.0005 g
Bring to pH7.0 with KOH	
$H2O$ to 1 l	

Table 2 Some carbon sources used by *Escherichia coli*

Monosaccharides	Disaccharides	Acids	Polyols
Glucose	Maltose	Acetate	Glycerol
Fructose	Lactose	Succinate	Mannitol
Mannose			Dulcitol
Galactose			Sorbitol
Arabinose			
Xylose			
Rhamnose			

Chapter 1
Bacterial Growth

The Lag Phase

When cells are inoculated from an agar slant into a synthetic medium, a certain time elapses before cells grow at a constant rate. This time is called the lag phase during which no growth is first detected and then cells experience an accelerating growth rate before the constant rate is established.

Most authors agree that the lag phase is the expression of phenomena which are not linked to the growth proper, and which are due to contingent reasons:

(a) after growth has stopped, the number of ribosomes per cell diminishes and protein synthesis cannot resume at a convenient rate until new ribosomes have been produced.
(b) in a culture where a part of the population is no longer viable, the dead cells continue to diffuse light. Since the bacterial mass is often measured by nephelometry rather than by viable count, a certain time will be necessary before a change in optical density becomes significant. Actually, the lag phase determined by viable count is shorter than the one determined by nephelometry.
(c) a transfer from a certain medium to a different one may create a lag time if the utilization, say, of the carbon source of the second medium, requires the synthesis of one or more inducible enzymes, or if growth on the first medium has caused the repression of some enzymes which must be synthesized in the second one.
(d) finally, if exponentially growing bacteria are transferred in a fresh medium identical to the initial one, all other conditions being equal, one observes no lag time.

G.N. Cohen, *Microbial Biochemistry*,
DOI 10.1007/978-94-017-8908-0_1,

The Exponential Phase

When no essential nutrient is limiting, the culture grows at a constant rate, and the growth rate is proportional to the culture density. The growth curve is therefore an exponential. After n generations, the population of an exponential culture equal to x_0 at zero time becomes

$$x = x_0 2^n.$$

If μ is the number of divisions per unit time, x the density at time t and x_0 the density at time t_0, one obtains:

$$x = x_0 2^{\mu t}$$

from which the value of μ is extracted:

$$\mu = \frac{\log x - \log x_0}{(t - t_0)\log 2}.$$

Consequently, if the size of the population is known at two times in the exponential phase, the value of μ, the growth rate, can be derived. In practice, the time unit chosen is the hour and μ is the number of divisions per hour.

The best graphic representation of exponential growth is obtained by putting as ordinates the optical density of the culture and in abscissas the time at which the measures were made. The simple inspection of the straight line obtained on semi-logarithmic paper gives the value of μ. A simple interpolation gives the time necessary to obtain a doubling of the population.

If an essential nutrient disappears during growth, if oxygen becomes limiting, or if the medium becomes either too acidic or too basic, the growth rate decreases and finally reaches zero. This phase is ill-defined: its characteristics and duration depend on the cause which has limited growth. The number of viable cells and the bacterial mass cease to increase almost simultaneously. On the other hand, the accumulation of toxic substances can inhibit cell division (the number of viable cells remaining constant) but in this case, the optical density of the cultures may continue to increase.

Linear Growth

When certain amino acid analogues are added to a culture of *E. coli* growing exponentially, growth does not stop altogether, but the exponential growth is transformed into arithmetic (linear) growth, i.e., where the increase in mass becomes directly proportional to time, according to

$$x - x_0 = kt.$$

Table 1.1 Total growth and yield of *E. coli* cultures in function of the concentration of glucose in the synthetic medium

Glucose (mg/l)	Total growth (mg dry wt/l)	Yield	Deviation from the mean value (%)
200	47.0	0.235	+0.9
180	42.4	0.236	+1.3
160	37.4	0.234	+0.43
140	33.0	0.236	+1.3
120	28.0	0.234	+0.43
90	20.8	0.230	−1.3
70	16.0	0.228	−2.1
50	11.5	0.230	−1.3
25	5.4	0.258	+10.7
		Mean yield: 0.233	

Everything happens as if an essential component ceases to be synthesized, at least under its active form, as a result, growth becomes directly proportional to the amount of the essential component present before the addition of the analogue.

Microscopic examination of bacteria grown under these conditions show long filaments due to an anomaly in cell division. After a certain increase in cell mass, growth stops altogether, but in general the bacteria remain viable and division resumes upon removal of the analogue.

The Yield of Growth

It is defined by the following ratio:

$$\frac{\text{Total growth}\left(\text{mg dry weight per unit volume}\right)}{\text{Carbon source concentration}\left(\text{mg per unit volume}\right)}.$$

Examination of Table 1.1, taken from Monod's classical work, shows that total growth is directly proportional to the initial concentration of the carbon source in the medium; in other terms, the growth yield is independent of the concentration of the limiting nutrient, within the limits of the concentrations studied. It has been obviously verified that all the glucose was used up in this experiment.

Careful experiments have shown that growth yield is totally independent of growth rate, when the variations of growth rate are caused by oxygen limitation, for example. Figure 1.1, taken from the same work by Monod, shows the growth of two cultures of *E. coli* on the same synthetic medium. One culture is shaken; the other is not. In the second, oxygen becomes limiting and growth is slower. However, the two cultures reach the same maximum.

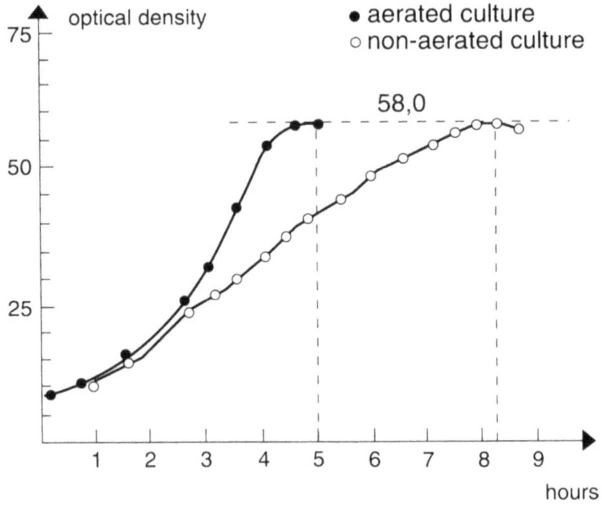

Fig. 1.1 Growth of two *E. coli* cultures growing under two different aeration conditions

The growth yield measured as a function of different carbon sources is highly variable and seems roughly to reflect the different ATP yields characteristic of each catabolic pathway. However, if the growth yield is calculated as a function of the amount of ATP synthesized, it is roughly constant and equal to 10 g of dry matter per "mole" of energy rich bond.

This view has been seriously challenged by Marr in a 1991 review. Marr provides evidence that the growth rate of *E. coli* is set by the flux of a precursor metabolite and of the monomers derived from it rather than by the flux of ATP or, with the possible exception of the maximal growth rate, the rate of protein synthesis. The physiology of *E. coli* at low growth rates is clearly different from that at moderate to high growth rates. Although one can account for the rate of protein synthesis only in terms of rate of peptide elongation, it seems likely that initiation of translation is modulated to prevent the depletion of pools of amino acids and aminoacyl-tRNAs. Modulation of initiation might be one of a set of controls which confer resistance to starvation. Marr speculates that the signal for regulation of protein synthesis is the ratio of uncharged tRNA to charged aminoacyl-tRNA, that this signal controls the concentration of guanosine tetraphosphate, which itself controls the transcription of the ribosomal RNA genes. The protein synthesizing system is an important but not singular example of synthesis on demand which is also characteristic of peptidoglycan, phospholipids, and even DNA. Unfortunately, in no case is the mechanism of such modulation well understood.

Variation of the Growth Rate at Limiting Carbon Source Concentrations

The growth rate is constant only above a certain concentration of the carbon source. Below that concentration, it decreases considerably. The curves describing the variation of the growth rate of *E. coli* in a synthetic medium in function of the concentration of glucose, mannitol or lactose are hyperbolas expressing the relation

$$\mu = \mu_0 \frac{C}{C_1 + C},$$

μ being the growth rate, μ_0 the maximal growth rate, C the concentration of the limiting nutrient. The ratio $C/C_1 + C$ being equal to 1/2 when $C = C_1$. Constant C1 is therefore the concentration of the carbon source for which the growth rate is half of the maximum rate. The above equation has a form identical to a Langmuir isotherm. What limits the growth rate is the existence of a saturable system limiting the utilization of the carbon source. This system has been shown to be the active transport of the carbon source (see Chap. 4).

Continuous Growth: The Chemostat

Under the usual batch conditions of growth, the exponential phase can last only during a limited number of generations, a nutrient or the aeration of the medium becoming limiting thereafter. It can be prolonged by successive transfers of the bacteria into fresh medium. Monod, and Novick and Szilard have independently introduced the more practical method of continuous cultures which, in addition, has proven extremely interesting both conceptually and experimentally.

The principle is the following: bacterial growth takes place in a vessel B connected to a reservoir R containing fresh medium which can be continuously delivered at the desired rate; on the other hand, vessel B is equipped with an exit tubing placed in such a way that a volume of culture equal to the volume of medium delivered from the reservoir be withdrawn. A non limiting aeration and an efficient stirring are necessary in vessel B. If the bacteria having grown under these conditions are required, they can be instantaneously taken from vessel P, and cooled so as to prevent further growth (Fig. 1.2).

How does N, the bacterial population in the growing vessel B, vary as a function of time? By derivation of the exponential growth equation,

$$N_t = N_0 e^{-\mu t}$$

one obtains the instantaneous growth rate:

$$\frac{\mathrm{d}N}{\mathrm{d}t} = \mu N.$$

Let V be the volume of the medium in the growth flask and v the volume delivered from the reservoir per unit time (equal to the volume withdrawn from flask B per unit time). The rate of removal of cells is given by

$$\frac{\mathrm{d}N}{\mathrm{d}t} = -N \frac{v}{V_t}$$

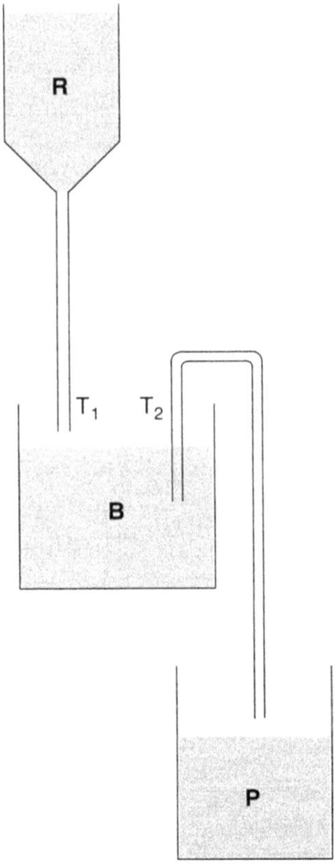

Fig. 1.2 The chemostat, an apparatus for continuous culture. The actual chemostats are much more elaborate. The figure only explains the principle on which they are built

Let us call $\frac{v}{V_t} = D$, the dilution rate.

The net population change in vessel B is given by the algebraic sum of these two functions:

$$\frac{dN}{dt} = \mu N - ND = N(\mu - D).$$

We have previously noted that, for a given microorganism, μ cannot exceed a maximal value μ_0 under experimental conditions where the value of μ depends exclusively on the concentration of a limiting factor. If the dilution rate D becomes higher than μ, the expression dN/dt becomes negative and the population in the growth flasks is gradually depleted. If D is smaller than μ, dn/dt becomes positive and the size of the population increases in vessel B. The increase however cannot be indefinite and will continue up to a moment when a nutrient of the reservoir becomes limiting. Then μ decreases and when it becomes equal to D, the exponentially growing culture of vessel B attains a constant size.

We are thus in presence of an autoregulated system, the chemostat, where the growth rate will adjust automatically to its equilibrium value and remain constant indefinitely. The only way, all other conditions being equal, to interfere with the value of μ is to increase or decrease the rate of admission of the medium from the reservoir and in consequence to act on the dilution rate.

When the equilibrium is reached, after one generation, we can write

$$N = 2N_0 \mathrm{e}^{\mu t},$$

from which the following ensues:

$$e^{\mu t} = 2 \text{ and } \mu t = \ln 2 = 0.69.$$

Since at equilibrium,

$$\mu N = -N \frac{v}{Vt}$$

then,

$$\mu t = \frac{v}{V}$$

and the generation time is defined by the value $v/V = 0.69$, i.e., when a volume equal to 0.69 V has circulated through the growth flask.

Advantages of the Continuous Exponential Culture

The theory and the practice of the simplified chemostat we just described show that it is possible to obtain a culture maintained in the exponential phase by continuous dilution calculated in order that the growth of the microorganisms is exactly compensated. In principle, such a culture grows indefinitely, at a constant rate, under constant conditions. As the growth rate can be varied at will, one can study its specific effects on the cell size and chemical composition. These are very marked effects: the DNA, RNA and protein content vary quite significantly (Table 1.2).

Using auxotrophic mutants, the essential metabolite required as the result of the mutation can be made easily limiting with the chemostat and the effect of this limitation on the synthesis of more specific components can then be studied, as we shall see in later chapters.

Table 1.2 Relative contents of macromolecules in *E. coli* grown at different growth rates

		DNA	RNA	Protein
Growth rate μ	Cells/g dry weight	mg/g dry weight		
2.4	$1.3 \cdot 10^{12}$	30	310	670
1.2	$3.1 \cdot 10^{12}$	35	220	740
0.6	$4.8 \cdot 10^{12}$	37	180	780
0.2	$6.3 \cdot 10^{12}$	40	120	830

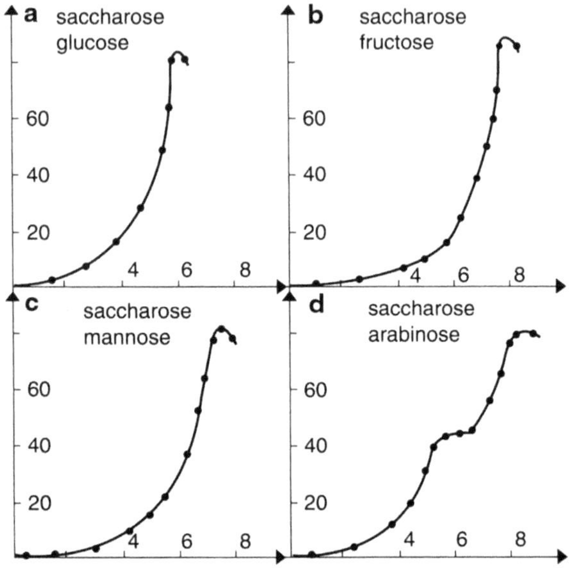

Fig. 1.3 Growth curves of cultures of B. Subtilis grown on the following mixtures: saccharose and glucose (**a**), saccharose and fructose (**b**), saccharose and mannose (**c**) and saccharose and arabinose (**d**). The abscissa reflects time in hours and the ordinate reflects the optical density

Diauxic Growth

In 1941, while studying the growth of cultures of *B. subtilis* in a synthetic medium containing mixtures of carbohydrates, Monod observed a phenomenon to which he gave the name "*diauxie*". Whereas in certain mixtures, the growth curve is normal (with a growth rate value intermediate between those obtained with the two isolated carbohydrates), certain other mixtures gave rise to a curve which could be decomposed in two distinct exponential phases separated by a lag phase.

If mixtures are studied where one carbohydrate, e.g., saccharose, is always present and the other is varied, the results show that certain mixtures grow in the usual way (Figs. 1.3a–c and 1.4b), whereas in certain other mixtures, the diauxie is clearly visible (Figs. 1.3d and 1.4a, c, d).

If the sugars which, associated to saccharose, give rise or not to a diauxie, are grouped into two lists, one obtains.

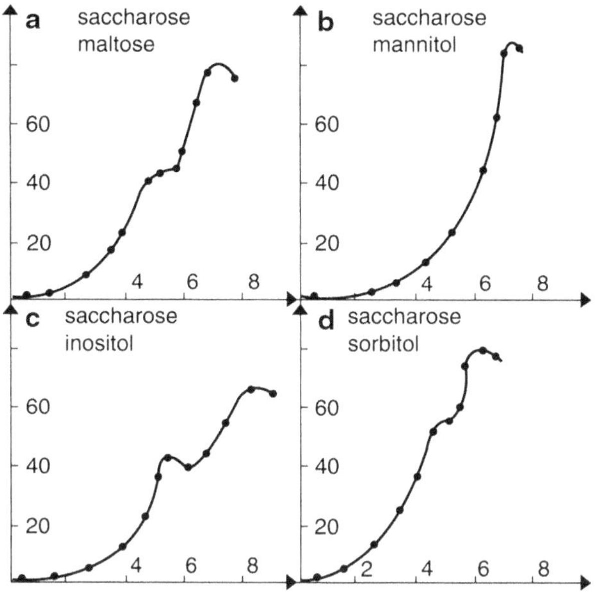

Fig. 1.4 Growth curves of cultures of B. Subtilis grown on the following mixtures: saccharose and maltose (**a**), saccharose and manitol (**b**), saccharose and inositol (**c**) and saccharose and sorbitol (**d**). The abscissa reflects time in hours and the ordinate reflects the optical density

"Non diauxic" sugars with saccharose	"Diauxic" sugars with saccharose
A	B
Glucose	Sorbitol
Fructose	Inositol
Mannose	Arabinose
Mannitol	Maltose
Saccharose	Dextrin

No chemical similarity is specific of one or the other group. Monod made all the possible binary combinations with any sugar of the first list with any sugar of the second: he always observed a diauxic growth; such a curve was never observed with two sugars belonging to the same list.

A similar phenomenon is observed with *E. coli* or *Salmonella typhimurium* where the two lists are as follows:

A	B
Glucose	Arabinose
Mannose	Xylose
Fructose	Rhamnose
Mannitol	Sorbitol
	Dulcitol
	Maltose
	Lactose

The lists are similar to the ones obtained with *B. subtilis*, the few differences observed being due to the fact that enterobacteria do not utilize saccharose or inositol and that other sugars are not utilized by *B. subtilis*.

An analysis of the phenomenon reveals that, in both types of bacteria, the first growth phase is always associated with the total utilization of a sugar from list A. Sugar B is never, even partially utilized during the first growth cycle. The cultures appear unable to metabolize simultaneously the constituents of the A-B couple. We shall return later to the mechanism of diauxie (Chap. 16). It is useful to discuss its nature at this point. Monod has dismissed, by a set of convincing experiments, the improbable hypothesis that the second cycle is due to the selection of mutants able to metabolize B sugars.

Another hypothesis to explain diauxie is based on observations made by Emile Duclaux and reported in his Traité de Microbiologie (1899).

He observed that in *Aspergillus niger*, certain enzyme activities were always present whereas the appearance of other enzymes required the presence of their substrates in the growth medium. It was only in 1930, that this phenomenon was carefully studied in Karström's doctoral thesis. We shall give an example of the numerous experiments performed by Karström, which all show that the type of "adaptation" suspected by Duclaux is a direct temporary answer of the cell to the constituents of the growth medium, which disappears when the microorganism is grown in the absence of the stimulating agent: a xylose-fermenting strain of *Aerobacter aerogenes* is grown on whey and the washed suspension of the organism added to xylose in presence of chalk; no fermentation occurs in 15 h. A source of nitrogen is then added: fermentation sets in only two hours later. If the organism had been grown in xylose broth and treated as before, fermentation sets in immediately, while the fermentation of glucose, as opposed to xylose, occurred whether glucose was present in the growth medium or not. The glucose-fermenting enzyme, being apparently a constant constituent of the cell was called "constitutive" by Karström, whilst he called the xylose-fermenting enzyme "adaptive".

Monod was aware of Karstrom's thesis and hypothesized that the conclusions of the Finnish scientist could plausibly explain the diauxic growth curves; bacteria would always be equipped with the enzymes necessary for degrading sugars A. The enzymes responsible for the degradations of sugars B would appear only after a beforehand adaptation, slow enough to explain the two successive cycles characteristic of diauxie. In addition, sugars A should inhibit the adaptation to sugars B.

Quorum Sensing and Quorum Quenching

Quorum sensing is a system of stimulus and response correlated to population density. Many species of bacteria use quorum sensing to coordinate gene expression according to the density of their local population. Quorum sensing can occur within a single bacterial species as well as between diverse species, and can regulate a host of different processes, in essence, serving as a simple indicator of population density or the diffusion rate of the cell's immediate environment. A variety of different molecules can be used as signals. Common classes of signaling molecules are oligopeptides in Gram-positive bacteria, N-acylhomoserine lactones (AHL) in Gram-negative bacteria, and a family of autoinducers known as autoinducer 2 (AI-2) in both Gram-negative and Gram-positive bacteria. The term "autoinduction" was first coined in 1970, when it was observed that the bioluminescent marine bacterium *Vibrio fischeri* produced a luminescent enzyme (luciferase) only when cultures had reached a threshold population density. At low cell concentrations, *V. fischeri* did not express the luciferase gene.

However, once the cultures had reached exponential growth phase, the luciferase gene was rapidly activated. This phenomenon was termed "autoinduction" because it involved a molecule (autoinducer) that accumulated in a growth medium and induced the synthesis of components of the luminescence system. Subsequent research revealed that the actual autoinducer used by *V. fischeri* is an acylated homoserine lactone (AHL) signaling molecule.

Bacteria have a receptor that can specifically detect the signaling molecule (inducer). When the inducer binds the receptor, it activates transcription of certain genes, including those for inducer synthesis. There is a low likelihood of a bacterium detecting its own secreted inducer. Thus, in order for gene transcription to be activated, the cell must encounter signaling molecules secreted by other cells in its environment. When only a few other bacteria of the same kind are in the vicinity, diffusion reduces the concentration of the inducer in the surrounding medium to almost zero, so the bacteria produce little inducer. However, as the population grows, the concentration of the inducer passes a threshold, causing more inducer to be synthesized. This forms a positive feedback loop, and the receptor becomes fully activated. Activation of the receptor induces the up-regulation of other specific genes, causing all of the cells to begin transcription at approximately the same time. This coordinated behavior of bacterial cells can be useful in a variety of situations. For instance, the bioluminescent produced by *Vibrio fischeri* would not be visible if it were produced by a single cell. By using quorum sensing to limit the production of luciferase to situations when cell populations are large, *V. fischeri* cells are able to avoid wasting energy on the production of a useless product.

Three-dimensional structures of proteins involved in quorum sensing were first published in 2001, when the crystal structures of three LuxS orthodox were determined by X-ray crystallography. In 2002, the crystal structure of the receptor LuxP of *Vibrio harveyi* with its inducer AI-2 (which is one of the few biomolecules containing boron) bound to it was also determined. Many bacterial species, including *E. coli*, an enteric bacterium and model organism for Gram-negative bacteria, produce AI-2. A comparative genomic and phylogenetic analysis of 138 genomes of bacteria, archaea, and eukaryotes found that the LuxS enzyme required for AI-2 synthesis is widespread in bacteria, while the periplasmic binding protein LuxP is present only in *Vibrio* strains.

Quorum quenching is the process of preventing quorum sensing by disrupting the signaling. This may be achieved by degrading the signaling molecule. A well-studied quorum quenching bacterium has been isolated and its AHL degradation kinetic has been studied by using rapid resolution liquid chromatography.

Selected References

Bacterial Growth: Diauxie

Monod J (1958) Recherches sur la croissance des cultures bactériennes Thèse de1942, 2ème édn, Hermann, Paris

Linear Growth

Cohen GN, Munier R (1959) Biochim Biophys Acta 31:347–356

Continuous Growth: The Chemostat

Maaloe O, Kjeldgaard NO (1966) Control of macromolecular synthesis. W. A. Benjamin, New York/Amsterdam

Marr AJ (1991) Microbiol Rev 55:316–333

Monod J (1950) Influence of growth rate on cellular constituents. Ann Inst Pasteur 79:390–410

Adaptive (Inducible) Enzymes: Prehistory

Karström H (1930) Ueber die Enzymbildung in Bakterien and über einige physiologische Eigenschaften der untersuchten Bakterienarten, Thesis, Helsingfors

Quorum Sensing and Quorum Quenching

Miller MB, Bassler BL (2001) Ann Rev Microbiol 55:165–199

Ng WL, Bassler BL (2009) Ann Rev Genet 43:197–222

Chapter 2
The Outer Membrane of Gram-Negative Bacteria and the Cytoplasmic Membrane

The Outer Membrane of Gram-Negative Bacteria

The major permeability barrier in any membrane is the lipid bilayer structure, and its barrier property is inversely correlated with its fluidity. Bacteria cannot make this membrane much less fluid or it will start to interfere with the normal functions of the membrane proteins, so some bacteria have constructed an additional structure that surrounds the cell outside the cytoplasmic membrane. An example of this are Gram-negative bacteria, such as *E. coli*, which surround themselves with a second outer membrane which functions as an effective barrier.

It was actually shown by electron microscopy that the Gram-negative bacteria are covered by a membrane layer outside the peptidoglycan layer. This outer membrane (OM) should not be confused with the cytoplasmic or inner membrane. The two membranes differ in their buoyant densities, and the OM can be isolated from bacterial lysates by sucrose equilibrium density centrifugation.

The outer leaflet of the outer membrane bilayer is composed of an unusual lipid, lipopolysaccharide (LPS), rather than the usual glycerophospholipid found in most other biological membranes. LPS is composed of three parts: a proximal hydrophobic lipid A region, a core oligosaccharide region connecting a distal O-antigen polysaccharide region to lipid A. This distal region protrudes in the medium.

During lipid A synthesis (Raetz pathway), the acyl carrier protein shuttles acyl intermediates linked to its prosthetic 4′-phosphopantetheine group among four acyltransferases, including LpxD. The crystal structures of three forms of the *Escherichia Coli* acyl carrier protein engaging LpxD, which represent stalled substrate and liberated products along the reaction coordinate, show the intricate interactions at the interface that optimally position acyl carrier protein for acyl delivery and that directly involve the pantetheinyl group. Conformational differences among the stalled acyl carrier proteins provide the molecular basis for the association–dissociation process. An unanticipated conformational shift of 4′-phosphopantetheine groups within the LpxD catalytic chamber shows an unprecedented role of the acyl carrier protein in product release.

G.N. Cohen, *Microbial Biochemistry*,
DOI 10.1007/978-94-017-8908-0_2,

All the fatty acid chains present in LPS are saturated which significantly reduces the fluidity. Also, the LPS molecule contains six or seven covalently linked fatty acid chains, in contrast to glycerophospholipids that contains only two fatty acid residues.

Hydrophobic probe molecules have been shown to partition poorly into the hydrophobic portion of LPS and to permeate across the outer membrane bilayer at about one-fiftieth to one-hundredth of the rate seen in the usual lipid bilayers. The vast majority of clinically important antibiotics and chemotherapeutic agents show some hydrophobicity that allows them to diffuse across the membrane. The LPS-containing asymmetric bilayer of the bacterial outer membrane serves as an efficient barrier against rapid penetration by these lipophilic antibiotics and chemotherapeutic agents.

Bacteria with this barrier must develop methods to bring in nutrients from their surroundings, however. Apart from other systems which will be developed in Chap. 4, the outer membrane contains for this purpose porins, a special class of proteins, which produce non-specific aqueous diffusion channels across the membrane. The properties of the porin channels exclude antibiotics crossing them by having a very small diameter (7 by 10 Angström in their most constricted portion) which slows down or completely stops antibiotic influx, and by lining the channel with charged amino acid residues which orient the water molecules in a fixed orientation. These charged residues make the influx of lipophilic molecules difficult because the energetically favorable orientation of the water would be disturbed.

Pseudomonas aeruginosa PAO1 is a paradigm for outer membrane impermeability and antibiotic resistance. It is one of the most notorious Gram-negative pathogens and is responsible for around 10 % of hospital-acquired infections. *P. aeruginosa* is intrinsically resistant to many antibiotics, which is partly due to a lack of large channel porins such as *E. coli* OmpF/C. Instead of porins, *P. aeruginosa* has approximately 30 substrate-specific channels. Nineteen of those belong to the outer membrane carboxylate channel (Occ) family, making Occ proteins responsible for uptake of the majority of small molecules, including antibiotics, in *P. aeruginosa*. The archetype of the family, OccD1 (formerly OprD), is known to be an uptake channel for basic amino acids and clinically important carbapenem antibiotics. The recent structural and biochemical characterization of the majority of the Occ protein family has revealed that the two Occ subfamilies (OccD and OccK) have very different substrate specificities. OccD family members transport positively charged amino acids, and OccK proteins have a preference for cyclic compounds with a net negative charge, such as benzoate. The requirement of substrates for all family members is that they possess a carboxyl group. Taken together, the data showed how bacteria containing Occ proteins are able to effectively take up a range of substrates while still maintaining a highly effective outer membrane barrier.

Gram-positive mycobacteria have also been found to have developed an outer leaflet to protect themselves from drugs. The mycobacterial barrier also consists of an unusual lipid, mycolic acid. Mycolic acid contains more than 70 carbon atoms with only a few double bonds. Also, where LPS had six or seven fatty acids joined to a single head group, here, hundreds of mycolic acid residues are covalently linked to a common head group, an arabinogalactan polysaccharide, which in turn is covalently linked to the underlying peptidoglycan structure. Nutrients diffuse into the cells through mycobacterial porin, which is present in very small amounts and allows only very slow diffusion of small molecules through its channel. Antibiotics are severely retarded in entering the bacteria by the low permeability combination of the porin channels and the lipid matrix, which allows some resistance.

The Cytoplasmic Membrane

The cytoplasmic membrane, also called cell membrane or plasma membrane, is about 7 nm thick. It lies internal to the cell wall and encloses the cytoplasm of the bacterium.

Like all biological membranes in nature, the bacterial cytoplasmic membrane is composed of phospholipid and protein molecules. In electron micrographs, it appears as two dark bands separated by a light band and is actually a fluid phospholipid bilayer imbedded with proteins. With the exception of the mycoplasmas (the only bacteria that lack a cell wall), prokaryotic membranes lack sterols. Many bacteria, however, do contain sterol-like molecules called hopanoids. Like the sterols found in eukaryotic cell membranes, the hopanoids most likely stabilize the bacterial cytoplasmic membrane.

The phospholipid bilayer is arranged so that its polar ends (the phosphate and glycerol portion of the phospholipids) form the outermost and innermost surface of the membrane while its hydrophobic ends (the fatty acid portions of the phospholipids) which are insoluble in water, make up the inner core of the membrane.

The cytoplasmic membrane is a selectively permeable membrane that determines what goes in and out of the organism. All cells must take in and retain all the various chemicals needed for metabolism. Water, carbon dioxide and oxygen, and lipid-soluble molecules simply diffuse across the phospholipid bilayer, diffusion being powered by the potential energy of a concentration gradient and does not require the expenditure of metabolic energy. All other molecules require carrier molecules to transport them through the membrane. Mechanisms by which materials move across the cytoplasmic membrane will be examined in Chap. 4.

A number of other functions are associated with the cytoplasmic membrane: in addition of being the site of peptidoglycan synthesis, both in the growing cell wall and in the transverse septum that divides the bacterium during bacterial division and the site of phospholipid and some protein synthesis required for the production of more cytoplasmic membrane, it is the site of energy production through the electron transport system for bacteria with aerobic and anaerobic respiration and photosynthesis for bacteria converting light energy into chemical energy.

Energy Generation

Many cells use respiratory processes to obtain their energy. During respiration, organic or inorganic compounds that contain high energy electrons are broken down, releasing those electrons to do work. These electrons find their way to the membrane where they are passed down a series of electron carriers. During this operation, protons are transported outside the cell. The outside of the membrane becomes positively charged; the inside becomes negatively charged.

This proton gradient energizes the membrane, much like a battery is charged. The energy can then be used to do work directly, a process known as the proton motive force, or can be channeled into a special protein known as ATP synthase. ATP synthase can convert ADP to ATP, the ATP doing the work.

Unlike eukaryotic cells, where the aerobic generation of ATP is made possible by the passage of protons through the membrane of mitochondria, the site of oxidative phosphorylation in bacteria is the inner cytoplasmic membrane within which the major proteins involved in the translocation of hydrogen ions (flavoproteins, quinones, iron-sulfur proteins, cytochromes) are inserted. Their combined action leads to lower concentrations of hydrogen ions in the cytoplasm compared to the external medium as well as differences in the electrical charges across the inner membrane. Together they create the proton motive force, which provides the energy for a number of processes that occur on the cytoplasmic membrane, such as the active transport of food molecules and certain inorganic ions into the cytoplasm and the generation of ATP from ADP by ATP synthase. In some bacteria, such as *Halobacterium halobium*, the sunlight energy is converted into proton gradients that serve as energy source. When oxygen is scarce, *H. halobium* switches to a photosynthetic mode and synthesizes large amounts of a protein, bacteriorhodopsin, a light driven proton pump which converts light into a proton-motive force which is used to synthesize ATP. Light-driven proton-pumping rhodopsins are widely distributed among microorganisms. A new class of these proteins has been described in the marine flavobacterium *Krokinobacter eikastus*, which harbors two rhodopsins, one being a typical proton pump, while the second pumps sodium ions outward, but converts to a classical proton pump when presented with potassium chloride or salts of larger cations.

ATP Synthase

The ATP synthase enzymes have been remarkably conserved through evolution. The bacterial enzymes are essentially the same in structure and function as those from mitochondria of animals, plants and fungi, and the chloroplasts of plants. The early ancestry of the enzyme is seen in the fact that the Archaea have an enzyme which is clearly closely related, but has significant differences from the Eubacterial branch. The H^+-ATP-ase found in vacuoles of the eukaryote cell cytoplasm is similar to the archaeal enzyme, and is thought to reflect the origin from an archaeal ancestor.

In most systems, the ATP synthase sits in the membrane (the "coupling" membrane), and catalyses the synthesis of ATP from ADP and phosphate driven by a flux of protons across the membrane down the proton gradient generated by electron transfer. The flux goes from the protochemically positive (P) side (high proton electrochemical potential) to the protochemically negative (N) side. The reaction catalyzed by ATP synthase is fully reversible, so ATP hydrolysis generates a proton gradient by a reversal of this flux. In some bacteria, the main function is to operate in the ATP hydrolysis direction, using ATP generated by fermentative metabolism to provide a proton gradient to drive substrate accumulation, and maintain ionic balance.

$$\mathrm{ADP + Pi + nH^{+}P \Leftrightarrow ATP + nH^{+}N}$$

In bacteria, the P side is the outside (the periplasm in gram negative bacteria), the N side the cytoplasm in chloroplasts.

Subunit Composition of ATP Synthase

F-type ATP synthases are the major supplier of chemically bound energy in the form of ATP in all living cells. These enzymes use a transmembrane electrochemical ion gradient as an energy source to convert ADP and inorganic phosphate to ATP. The simplest system is that from *E. coli*. The ATP synthase can be dissociated into two fractions by relatively mild salt treatments.

A soluble portion, the F1 ATPase, contains five subunits, in a stoichiometry of $3\alpha:3\beta:1\gamma:1\delta:1\in$. Three substrate binding sites are in the β-subunits. Additional adenine nucleotide binding sites in the α-subunits are regulatory. The F1 portion catalyzes ATP hydrolysis, but not ATP synthesis.

Dissociation of the F1 ATPase from the membranes of bacteria or organelles leaves behind a membrane embedded portion called Fo. This consists (in *E. coli*) of three subunits a, b and c, with relative stoichiometries of 1:2:9–12. The c-subunit is very hydrophobic, and forms a helix turn helix structure which spans the membrane twice, with a hydrophilic loop on the side of attachment of F1. There is a conserved acidic residue half-way across the membrane in the C-terminal helix.

The c-rings of ATP synthases consist of individual c-subunits, all of which harbor a conserved motif of repetitive glycine residues (G×G×G×G) important for tight transmembrane α-helix packing. The c-ring stoichiometry determines the number of ions transferred during enzyme operation and has a direct impact on the ion-to-ATP ratio, a cornerstone parameter of cell bioenergetics. In the extreme alkaliphile *Bacillus pseudofirmus*, the glycine motif is replaced by an alanine repeat (A × A × A × A). A structural study on two mutants with alanine-to-glycine changes using atomic force microscopy and X-ray crystallography, showed that mutants form smaller c12 rings compared with the wild-type c13. The molar growth yields of *B. pseudofirmus* cells on malate further revealed that the c12 mutants have a considerably reduced capacity to grow on limiting malate at high pH. These results demonstrate that the mutant ATP synthases with either c12 or c13 can support ATP synthesis, and also underscore the critical importance of an alanine motif with c13 ring stoichiometry for optimal growth at pH >10. The data indicate a direct connection between the precisely adapted ATP synthase c-ring stoichiometry and its ion-to-ATP ratio on cell physiology, and also demonstrate the bioenergetic challenges and evolutionary adaptation strategies of extremophiles.

Both complexes can function as separate and independent rotary motors, fueled by either H+ or Na+ (Fo) or ATP (F1). Structurally, they are connected by a static peripheral (δb2) and a rotating central (γε) stalk. Ion translocation through the Fo complex induces torque, which leads to rotation of the enzyme's rotor (γεcn). The central stalk γ-subunit rotates inside the F1 (αβ)3 headpiece and elicits sequential conformational changes in the three catalytic β-subunits, finally leading to ATP synthesis. In the reverse direction, along the same operation principle, the enzyme can act as an ATP hydrolysis-driven pump.

After dissociation, the membranes are permeable to protons. The proton leak can be stopped by addition of inhibitors, which are also inhibitors of ATP synthesis in the functional complex. Two "classical" inhibtors are commonly used. Oligomycin binds at the interface between Fo and F1; dicyclohexylcarbodiimide (DCCD) binds covalently to the conserved acidic residue in the c-subunit of Fo. One DCCD per ATPase is sufficient to block turn-over, suggesting a cooperative mechanism. Single-molecule analysis showed that a single modification by bound DCDD in the multiple *c* subunit complex significantly inhibits F0F1. Bound DCCD induces steric hindrance of the *c* subunit and hence blocks F0F1 rotation (see below). The action of these inhibitors indicates that the proton permeability of Fo is a part of its functional mechanism.

The proton leak can be plugged, and a functional ATP synthase can be reconstituted, by adding back the F1 portion to membranes containing the Fo portion.

A model of ATP synthase, derived from image averaging and cryo-electron microscopy, showing a second stalk is presented in Figs. 2.1 and 2.2.

The structure of the soluble (F1) portion of the ATP synthase from beef heart mitochondria, as well as the structure of the complete F1-Fo ATP synthase from yeast mitochondria have been solved by X-ray crystallography by John Walker and his associates.

The ATP synthase operates through a mechanism in which the three active sites undergo a change in binding affinity for the reactants of the ATP-ase reaction, ATP, ADP and phosphate, as originally predicted by Paul Boyer. The change in affinity accompanies a change in the position of the γ-subunit relative to the *a*, *b*-ring, which involves a rotation of the one relative to the other. In the direction of ATP synthesis, the rotation is driven by a flux of H^+ down the proton gradient, through a coupling between the γ-subunit, and the c-subunit of Fo. This rotation has now been demonstrated experimentally.

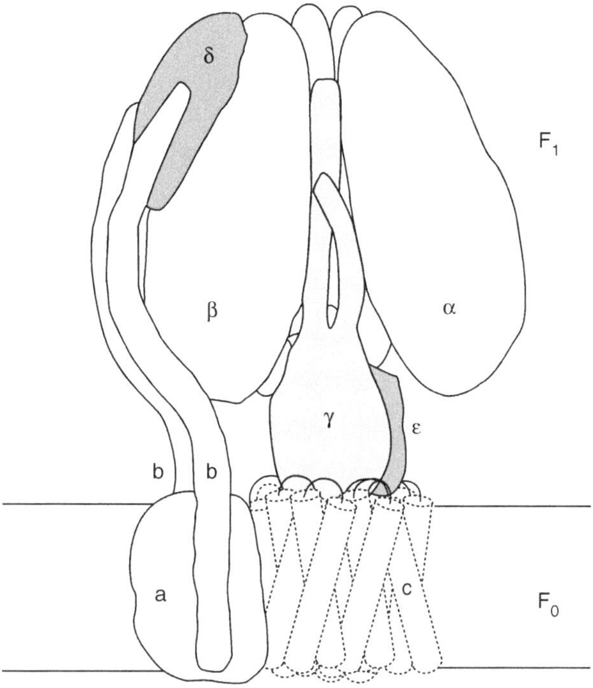

Fig. 2.1 Model of ATP synthase. Fo is embedded in the membrane (From Jungeet al. (1997))

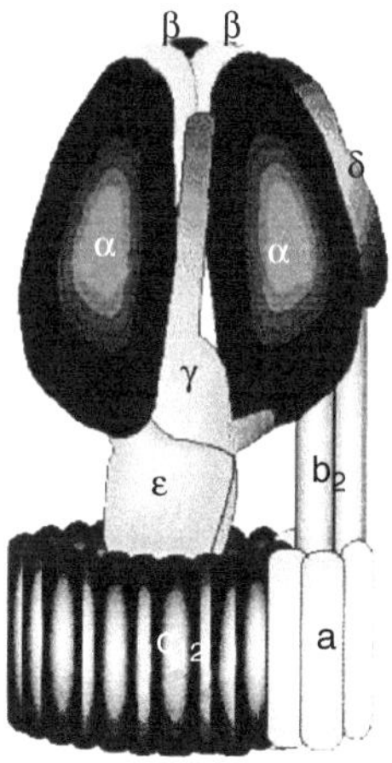

Fig. 2.2 Model of ATP-synthase. As in Fig. 2.1, seen from another angle

Thus, the F_1-ATPase (F_1) is an ATP-driven rotary motor wherein the γ subunit rotates against the surrounding $\alpha_3\beta_3$ stator ring. The three catalytic sites of F_1 reside on the interface of the α and β subunits of the $\alpha_3\beta_3$ ring. While the catalytic residues predominantly reside on the β subunit, the α subunit has 1 catalytically critical arginine, termed the arginine finger. The role of the arginine finger has been studied by analyzing the rotation of a mutant F_1 with a lysine substitution of the arginine finger. The mutant showed a 350-fold longer catalytic pause than the wild-type; this pause was further lengthened by the slowly hydrolyzed ATP analog ATPγS. On the other hand, the mutant F_1 showed highly unidirectional rotation with a coupling ratio of 3 ATPs/turn, the same as wild-type, suggesting that cooperative torque generation by the three β subunits was not impaired. The principal role of the arginine finger is not to mediate co-operativity among the catalytic sites, but to enhance the rate of the ATP cleavage by stabilizing the transition state of ATP hydrolysis. Lysine substitution also caused frequent pauses because of severe ADP inhibition, and a slight decrease in ATP-binding rate.

The peripheral stator stalk of *Escherichia coli* ATP synthase contains two *b* subunits. Using disulfide bond formation, one *b* subunit was cross-linked to *a*, α, and δ and the other to β. The *b* subunits adopt distinct positions within the stator to generate stability. The different positions imply different roles in counteracting the torque generated by the rotor.

ATP Synthesis in Archaea

Archaea are a heterogeneous group of microorganisms that often thrive under harsh environmental conditions such as high temperatures, extreme pHs and high salinity. As other living cells, they use chemiosmotic mechanisms along with substrate level phosphorylation to conserve energy in the form of ATP. Because some archaea are rooted close to the origin in the tree of life, these unusual mechanisms are considered to have developed very early in the history of life and, therefore, may represent on of the first energy-conserving mechanisms. A key component in cellular bioenergetics is the ATP synthase. The enzyme from archaea represents a new class of ATPases, the A1A0 ATP synthases. They are composed of two domains that function as a pair of rotary motors connected by a central and peripheral stalk(s). The structure of the chemically-driven motor (A1) was solved by small-angle X-ray scattering in solution, and the structure of the first A1A0 ATP synthases was obtained recently by single particle analyses. These studies revealed novel structural features such as a second peripheral stalk and a collar-like structure. In addition, the membrane-embedded electrically-driven motor (A0) is very different in archaea with a sometimes novel, exceptional subunit composition and coupling stoichiometries that may reflect the differences in energy-conserving mechanisms as well as adaptation to temperatures at or above 100 °C.

Photosynthetic cells also have a membrane system. Here light excites electrons and the electrons are again passed down through a series of electron carriers, a proton motive force is generated and ATP is synthesized. All the photosynthetic machinery is situated in the membrane. The cytoplasmic membrane contains the bases of flagella used in motility.

Selected References

Outer Membrane

Eren E, Parkin J, Adelanwa A, Cheneke B, Movileanu L, Khalid S, van den Berg B (2013) Toward understanding the outer membrane uptake of small molecules by Pseudomonas aeruginosa, J Biol Chem 288:12042–12053

LPS and Lipid A

Bartling CM, Raetz CRH (2008) Biochemistry 47:5290–5302

Masoudi A, Raetz CRH, Zhou P, Pemble CW (2014) Nature 505:422–426

Ion Pumps

Inoue K, Ono H, Abe-Yoshizumi R, Yoshizawa S, Ito H, Kogure K, Kandori H (2013) Nat Commun 4, Article number 1678

Kaback HR (1970) Transport Ann. Rev. Biochemistry 39:561–598

Stoeckhenius W, Oesterhelt D (1975) Biophys J 15:955–962

ATP Synthase

Abrahams JP, Leslie AG, Lutter R, Walker JE (1994) Nature 370:621–628

Brandt K, Maiwald S, Herkenhoff-Hesselmann B, Gnirss K, Greie J-C, Dunn SD, Deckers-Hebestreit G (2013) J Biol Chem 288:24465–24479

Hakulinen JK, Klyszejko AL, Hoffmann J, Eckhardt-Strelau L, Brutschy B, Vonck J, Meier T (2012) Proc Natl Acad Sci 109:2050–2056

Junge W, Lill H, Engelbrecht S (1997) TIBS 22:420–423

Komoriya Y, Ariga T, Iino R, Imamura H, Okuno D, Noji H (2012) J Biol Chem 287:15134–15142

Noji H, Ryohei Y, Masasuke Y, Kinosita K Jr. (1997) Nature, 386:299–302

Preiss L, Klyszejko AL, Hicks DB, Liu J, Fackelmayer OJ, Yildiz O, Krulwich TA, Meier T (2013) Proc Natl Acad Sci USA 110:7874–7879

Toei M, Noji H (2013) J Biol Chem 288:25717–25726

Chapter 3
Peptidoglycan Synthesis and Cell Division

General Structure

Peptidoglycan (or murein) is a continuous covalent macromolecular structure found on the outside of the cytoplasmic membrane of almost all bacteria and exclusively in these organisms. In particular it is absent in Archaea.

Its function is essentially to maintain the cell integrity by withstanding the internal osmotic pressure, to maintain a certain cell shape. It is involved in the cell division process.

The peptidoglycan of *E. coli* is composed of N-acetylglucosamine(GlcNAc), N-acetylmuramic acid (MurNAc, which is GlcNAc substituted at C-3 with D-lactic ether), L-alanine, D-glutamic acid, *meso*-di-aminopimelic acid and D-alanine, all in equimolar amounts except for D-alanine, the amount of which may vary. These elements form the basic repeating unit of the peptidoglycan. The two sugars are linked together by $\beta - 1,4$ glycosidic bonds. Attached by an amide linkage to the carboxyl group of muramic acid one finds a short peptide, L-alanyl-D-isoglutamyl-L-diaminopimelyl-D-alanine. This composition is shared by nearly all Gram-negative and by a few Gram-positive rods.

Within the peptidoglycan layer, the two sugars alternate with each other in chains. Adjacent tetrapeptides are linked by an oligopeptide bridge to form a crosslink between two chains. These crosslinks make the peptidoglycan layer a giant molecule.

The synthesis of peptidoglycan can be divided in two stages. The first stage is the assembly of a disaccharide-peptide monomer unit by enzymes located at the inner surface of the cytoplasmic membrane. The second stage involves the polymerization of this monomer unit on the outer surface of the cytoplasmic membrane and the binding of the nascent peptidoglycan to the preexisting peptidoglycan sacculus located in the periplasm.

G.N. Cohen, *Microbial Biochemistry*,
DOI 10.1007/978-94-017-8908-0_3,

Assembly of the Peptidoglycan Unit

Formation of UDP-MurNAc

This compound arises in two steps from UDP-Glc-NAc, the initial precursor of the pathway. UDP-Glc-NAc is, synthesized by Glc-Nac-1-phosphate uridylyltransferase (*glmU*),

$$\text{GlcNac-1-P} + \text{UTP} \rightarrow \text{UDP-GlcNAc}.$$

Enolpyruvate is added to position 3 of the GlcNAc residue by UDP-GlcNAc-enolpyruvoyltransferase (*murA*),

$$\text{UDP-GlcNAc} + \text{phosphoenolpyruvate} \rightarrow \text{UDP-GlcNAC-enolpyruvate}.$$

The covalent MurA-phosphoenolpyruvate adduct was captured in different reaction states. The covalent adduct primes phosphoenolpyruvate for catalysis and enables feedback inhibition by UDP-*N*-acetylmuramic acid, the product of MurB. In the second step, the enolpyruvoyl moiety is reduced to D-lactate by UDP-GlcNAc-enolpyruvate reductase (murB),

$$\begin{aligned}&\text{UDP-N-acetyl-3-(1-carboxyvinyl)-D-glucosamine} + \\ &\quad \text{NADPH} \Leftrightarrow \text{UDP-N-acetylmuramate} + \text{NADP}^+.\end{aligned}$$

Peptide Addition

UDPMurNAc peptides are formed by the sequential addition of L-alanine, D-glutamic acid, *meso*-diaminopimelic acid and D-alanyl-D-alanine on the D-lactyl group of UDPMurNAc. These reactions, coded respectively by the *murC*, *murD*, *murE* and *murF* genes, are catalyzed by Mg^{++}-ATP dependent enzymes:

$$\begin{aligned}&\text{UDP-N-acetylmuramate} + \text{L-Alanine} + \text{ATP} \Leftrightarrow \\ &\quad \text{UDP-N-acetylmuramoyl-L-alanine} + \text{Pi} + \text{ADP}\end{aligned}$$

$$\begin{aligned}&\text{UDP-N-acetylmuramoyl-L-alanine} + \text{D-Glutamate} + \\ &\text{ATP} \Leftrightarrow \text{UDP-N-acetylmuramoyl-L-alanyl-D-glutamate} + \\ &\text{Pi} + \text{ADP}\end{aligned}$$

$$\begin{aligned}&\text{UDP-N-acetylmuramoyl-L-alanyl-D-glutamate} + \\ &\textit{meso}\text{-2,6-diaminopimelate} + \text{ATP} \Leftrightarrow \\ &\text{UDP-N-acetylmuramoyl-L-alanyl-D-gamma-} \\ &\text{glutamyl-}\textit{meso}\text{-2,6-diaminopimelate} + \text{Pi} + \text{ADP}\end{aligned}$$

$$\begin{aligned}&\text{UDP - N - acetylmuramoyl - L - alanyl - D - glutamyl - }\textit{meso}\text{ -}\\&\text{2,6 - diaminopimelate + D - Alanyl - D - alanine + ATP} \Leftrightarrow\\&\text{UDP - N - acetylmuramoyl - L - alanyl - D - glutamyl - }\textit{meso}\text{ - 2,6 -}\\&\text{diaminopimeloyl - D - alanyl - D - alanine + Pi + ADP}\end{aligned}$$

MurC is responsible of the addition of the first residue (L-alanine) onto the nucleotide precursor UDP-MurNAc. Since phosphorylation of proteins by Ser/Thr protein kinases has recently emerged as a major physiological mechanism of regulation in prokaryotes, the hypothesis of a phosphorylation-dependent mechanism of regulation of the MurC activity was investigated in *Corynebacterium glutamicum*. Actually, MurC was phosphorylated *in vitro* by the PknA protein kinase. An analysis of the phosphoamino acid content indicated that phosphorylation exclusively occurred on threonine residues. The catalytic activity of MurC was impaired following mutation of these threonine residues. The finding that phosphorylation is correlated with a decrease in MurC enzymatic activity could have significant consequences in the regulation of peptidoglycan biosynthesis.

Three auxiliary enzymes are necessary to perform the above reactions, an alanine and a glutamate racemase as well as a D-alanyl-D-alanine ligase:

$$\begin{aligned}&\text{L - Alanine} \Leftrightarrow \text{D - Alanine}(\textit{alr})\\&\text{L - Glutamate} \Leftrightarrow \text{D - Glutamate}(\textit{murl})\\&\text{D - Alanine + D - Alanine + ATP} \Leftrightarrow\\&\text{D - Alanyl - D - alanine + Pi + ADP}(\textit{ddlB})\end{aligned}$$

All the *E. coli* genes involved in the assembly of the peptidoglycan are cytoplasmic. They have been identified, cloned and sequenced; they all map at 2 min on the chromosome. The three auxiliary enzymes do not belong to the 2 min cluster. All the corresponding enzymes have been investigated to some extent.

Methionine and *meso*-DAP are generated by pathways that diverge from aspartate semialdehyde (see Chapter 22). Enzymes of the methionine pathway possess an intriguing latent capability for recruitment to function in replacing *meso*-DAP. The latter is a critical amino acid component of cross-linking the disaccharide pentapeptide. In mutants lacking the ability to synthesize *meso*-DAP, a potential for substituting sulfur containing amino acids (*meso*-lanthionine, L-cystathionine, and L-*allo*cystathionine) exists. Surprisingly, such a remodeled peptidoglycan, which must possess gross stereochemical perturbations of configuration, does not cause loss of viability. The endogenous production of the "non physiological" sulfur amino acids requires the overexpression of *met B* and inactivation of *met C*. Cystathionine synthase (Met B) exhibits broad substrate specificity and is able to convert L-cysteine to lanthionine. Cystathionase (Met C) normally converts cystathionine to homocysteine during methionine biosynthesis and is highly efficient in utilization of L-lanthionine and L-allocystathionine as substrates as well. Thus, the scavenging capabilities of wild type Met C mask the potential of the cell to incorporate sulfur-containing amino acids into the peptidoglycan in place of *meso*-DAP. It is worth mentioning that in the cell wall of *Fusobacterium nucleatum*, *meso*-lanthionine has been found to replace *meso*-diaminopimelate. In this species, it is not constructed *in situ* but incorporated as a free metabolite into peptidoglycan precursors, just as *meso*-DAP in wild-type *E. coli*.

The Membrane Steps

A translocase catalyzes the transfer of UDP-N-acetyl-muramoyl-L-alanyl-D-glutamyl-6-carboxy-L-lysyl-D-alanyl-D-alanine to the C55 membrane acceptor undecaprenylphosphate and the compound undecaprenyl-diphospho-N-acetylmuramoyl-L-alanyl-D-glutamyl-*meso*-2,6-diaminopimeloyl-D-alanyl-D-alanine, named lipid I for simplicity, is obtained:

$$\text{UDP-N-acetylmuramoyl-L-alanyl-D-glutamyl-}\textit{meso}\text{-2,6-diaminopimeloyl-D-alanyl-D-alanine} + \text{Undecaprenyl phosphate} \Leftrightarrow \text{Undecaprenyl-diphospho-N-acetylmuramoyl-L-alanyl-D-glutamyl-}\textit{meso}\text{-2,6-diaminopimeloyl-D-alanyl-D-alanine} + \text{UMP}\left(\textit{mraY}\right)$$

The crystal structure of MraY from *Aquifex aeolicus* at 3.3 Å resolution allows for visualization of its overall architecture, locates Mg2+ within the active site, and provides a structural basis of catalysis for this class of enzymes. A transferase adds GlcNAc to lipid I, yielding lipid II, which is the substrate of the polymerization reaction:

$$\text{Undecaprenyl-diphospho-N-acetylmuramoyl-L-alanyl-D-glutamyl-}\textit{meso}\text{-2,6-diaminopimeloyl-D-alanyl-D-alanine} + \text{UDP-N-acetyl-D-glucosamine} \Leftrightarrow \text{Undecaprenyl-diphospho-N-acetylmuramoyl-}\left(\text{N-acetylglucosamine}\right)\text{-L-alanyl-D-glutamyl-}\textit{meso}\text{-2,6-diaminopimeloyl-D-alanyl-D-alanine}\left(\text{lipid II}\right) + \text{UDP}\left(\textit{murG}\right)$$

The synthesis of the lipid carrier undecaprenyl phosphate (C_{55}-P) requires the dephosphorylation of its precursor, undecaprenyl pyrophosphate (C_{55}-P-PP). The latter lipid is synthesized *de novo* in the cytosol and is also regenerated after its release from C_{55}-PP-linked glycans in the periplasm. In *Escherichia coli* the dephosphorylation of C_{55}-PP was shown to involve four integral membrane proteins, BacA, and three members of the type 2 phosphatidic acid phosphatase family, PgpB, YbjG, and YeiU. The PgpB protein was purified to homogeneity. This enzyme was shown to catalyze the dephosphorylation of C_{55}-PP with a relatively low efficiency when compared to diacylglycerol pyrophosphate and farnesyl pyrophosphate (C_{15}-PP) lipid substrates.

Assembly of the Murein Sacculus

Lipid II is transferred to the outer surface of the cytoplasmic membrane and is directly used for the assembly of the murein sacculus. The assembly of murein, a molecule which is associated with a number of other macromolecules of the cell envelope, must be coordinated with the various steps of the cell cycle, such as elongation and septation. This highly complex process is not easily accessible to biochemical investigation. The general structure of murein from *E. coli* is that of a polymer in which the linear glycan chains have repeating disaccharide peptide units in which the peptide cross bridges are directly established between two monomer units.

The assembly of the glycan chains presumably occurs in the periplasm by a transglycosylation reaction and supposes a release of undecaprenylpyrophosphate which is dephosphorylated, the undecaprenylphosphate is then available for the synthesis of lipids I and II.

The main crosslinkage is between the penultimate D-alanine residue of one subunit and the diaminopimeloyl residue of another subunit and is the result of a transpeptidation reaction.

Penicillin Sensitivity

The antibiotic penicillin kills only actively dividing cells by inhibiting the synthesis of murein. Both Gram-positive and Gram-negative cells possess in their cytoplasmic membrane a set of proteins which are involved in the late steps of murein synthesis and are the targets of the antibiotic. These proteins, called PBPs (for penicillin-binding proteins) bind penicillin covalently. *E. coli* possesses at least seven PBPs which have been identified, cloned and sequenced. Four of them are essential for the formation of the murein sacculus, while the others are non essential. When lipid II is incubated with an essential PBP, an insoluble murein is formed. Analysis of this material indicates that both transglycosylation and transpeptidation have taken place, the PBPs being therefore bi-functional enzymes. The *pbp*B gene, coding for PBP3, belongs to the above mentioned 2 min gene cluster, which in addition to the genes involved in murein synthesis, contains genes involved in the cell division process.

Cell Division

Cytokinesis and cell division, in almost all bacteria, uses a cytoskeletal element called the Z ring, which consists of filaments of polymerized FtsZ protein. Cell division occurs through the formation of this protein ring (division ring) at the site of division, with FtsZ being its main component. FtsZ is the prokaryotic ortholog of eukaryotic tubulin; FtsZ also has GTPase activity and the ability to polymerize *in vitro*.

The Z ring provides a scaffold to recruit at least ten additional auxiliary Fts proteins required for cytokinesis, many of which link the Z ring to synthesis of a new cell wall that accompanies formation of the septum.

In vitro, FtsZ assembles into short, one-stranded protofilaments, averaging 30 subunits and 125 nm in length. In the bacterial cell, these are further assembled into a long, thin filamentous structure attached to the inner bacterial membrane. Normally this filament forms a single Z ring at the center of the cell.

The self-assembly of the tubulin homologue FtsZ at the mid-cell is a critical step in bacterial cell division. Dynamic light scattering spectroscopy is a new method to study the polymerization kinetics of FtsZ in solution. Analysis of the data obtained by this method indicates that the FtsZ polymers are remarkably monodisperse in length, independent of the concentrations of GTP, GDP, and FtsZ monomers. Measurements of the diffusion coefficient of the polymers demonstrate that their length is remarkably stable until the free GTP is consumed. The mean size of the FtsZ polymers within this interval of stable length is estimated to be between 9 and 18 monomers. The rates of FtsZ polymerization and depolymerization are likely influenced by the concentration of GDP, as the repeated addition of GTP to FtsZ increased the rate of polymerization and slowed down depolymerization. Increasing the FtsZ concentration did not change the size of FtsZ polymers; however, it increased the rate of the depolymerization reaction by depleting free GTP. Transmission electron microscopy shows that FtsZ forms linear polymers in solutions which rapidly convert to large bundles upon contact with surfaces in time scales as short as several seconds.

In the dividing bacterial cell, a Z ring is present long before constriction begins. This suggests that one of the roles for the additional cell division proteins is regulatory – to hold the Z ring in check until the appropriate signal for constriction is received. Spatial regulation ensures that the Z ring forms at the cell center.

Escherichia coli cell division is effected by a large assembly of proteins called the divisome, of which a subcomplex consisting of three bitopic inner membrane proteins, FtsQ, FtsB, and FtsL, is an essential part. These three proteins, hypothesized to link cytoplasmic to periplasmic events during cell division, contain large periplasmic domains that are of major importance for function and complex formation. The essential nature of this subcomplex, its low abundance, and its multiple interactions with key divisome components in the relatively accessible periplasm make it an attractive target for the development of protein-protein interaction inhibitors. Although the crystal structure of the periplasmic domain of FtsQ has been solved, the structure of the FtsQBL complex was unknown, with only very crude indications of the interactions in this complex. An *in vivo* site-specific photo cross-linking was used to probe the surface of the FtsQ periplasmic domain for its interaction interfaces with FtsB and FtsL. An interaction hot spot for FtsB was identified around residue Ser-250 in the C-terminal region of FtsQ and a membrane-proximal interaction region for both proteins around residue Lys-59. Sequence alignment revealed a consensus motif overlapping with the C-terminal interaction hot spot, underlining the importance of this region in FtsQ. The identification of contact sites in the FtsQBL complex will guide future development of interaction inhibitors that block cell division.

In E coli, an important component of spatial regulation is a rapid oscillator that shuttles an inhibitor of Z ring assembly between the ends of the cell. The oscillator consists of an adenosine triphosphatase (ATPase) (MinD) and an activator (MinE). The dynamic behavior of these proteins results from the ATP-dependent accumulation of MinD on the membrane followed by its recruitment of MinE. MinE stimulates the ATPase, releasing the proteins from the membrane.

The Min proteins oscillate many times between the cell poles during the time course of a cell division cycle, resulting in a time-averaged concentration of the inhibitor that is lowest at mid-cell, thus selecting the cell center as the division site. This dynamic pattern has been proposed to arise by self-organization of these proteins, and several models have suggested a reaction-diffusion type mechanism. Actually, Min proteins spontaneously formed planar surface waves on a flat membrane *in vitro*. The formation and maintenance of these patterns, which extended for hundreds of micrometers, required adenosine 5′-triphosphate (ATP), and they persisted for hours. A reaction-diffusion model of the MinD and MinE dynamics that accounts for this observation and for the in vivo oscillations has been recently proposed.

The spatial and temporal control of filamenting temperature sensitive mutant Z (FtsZ) Z-ring formation is crucial for proper cell division in bacteria. In *Escherichia coli*, the synthetic lethal with a defective Min system, SlmA protein helps mediate nucleoid occlusion, which prevents chromosome fragmentation by binding FtsZ and inhibiting Z-ring formation over the nucleoid. However, to perform its function, SlmA must be bound to the nucleoid. To deduce the basis for this chromosomal requirement, biochemical, cellular, and structural studies have been performed. Strikingly, structures show that SlmA dramatically distorts DNA, allowing it to bind as an orientated dimer-of-dimers. Biochemical data indicate that SlmA dimer-of-dimers can spread along the DNA. Combined structural and biochemical data suggest that this DNA-activated SlmA oligomerization would prevent FtsZ protofilament propagation and bundling. Bioinformatic analyses localize SlmA DNA sites near membrane-tethered chromosomal regions, and cellular studies show that SlmA inhibits FtsZ reservoirs from forming membrane-tethered Z rings. These combined data indicate that SlmA DNA helps block Z-ring formation over chromosomal DNA by forming higher-order protein-nucleic acid complexes that disable FtsZ filaments from coalescing into proper structures needed for Z-ring creation.

Though conserved in most bacteria, the auxiliary Fts and the Min proteins are not present in Archaea that lack a cell wall. In most Archaea, only FtsZ is found, raising the question of whether it is sufficient to constrict an organism which lacks a rigid cell wall. Whereas the two bacterial lineages (Gram+and Gram−) display fundamental differences in cell envelope structure, a central differential characteristic between the two main archaeal branches concerns the cell division machinery: euryarchaea utilize an FtsZ-dependent mode of division, similar to that of bacteria, whereas the crenarchaeal cell division machinery appears to be different: the crenarchaeal genome encodes homologs of the eukaryotic endosomal sorting system components Vps4(an ATPase) and ESCRT (endosomal sorting complex required for transport). For example, in *Sulfolobus acidocaldarius* ESCRT and Vps4 homologs undergo regulation of their expression during the cell cycle. The proteins interact and are specifically localized to the mid-cell during cell division. Overexpression of a catalytically inactive mutant of Vps4 in *Sulfolobus* resulted in the accumulation of enlarged cells, indicative of failed cell division. Thus, the crenarchaeal ESCRT system plays a key role in cell division.

Selected References

Chung BC, Zhao J, Gillespie RA, Kwon D-Y, Guan Z, Hong J, Zhou P, Lee S-Y (2013) Science 341:1012–1016

Fiuza M, Canova MJ, Patin D, Letek M, Zanella-Cleon I, Becchi M, Mateos LM, Mengin-Lecreulx D, Molle V, Gil JA (2008) J Biol Chem 283:36553–36563

Park JT (1996) The murein sacculus. In: Frederick N (ed) *Escherichia coli* and *Salmonella*, 2nd edn. The ASM Press, Washington, DC, pp 48–57

Richaud C, MenginLecreulx D, Pochet S, Johnson EJ, Cohen GN, Marlière P (1993) J Biol Chem 268:26827–26835

Touze T, Blanot D, Mengin-Lecreulx D (2008) J Biol Chem 283:16573–16583

van den Berg van Saparoea HB, Glas M, Vernooij IG, Bitter W, den Blaauwen T, Luirink J (2013) J Biol Chem 288:24340–24350

van Heijenoort J (1996) Murein synthesis. In: Frederick N (ed) *Escherichia coli* and *Salmonella*, 2nd edn. The ASM Press, Washington, DC, pp 1025–1034

Zhu J-Y, Yang Y, Han H, Betzi S, Olesen SH, Marsilio F, Schonbrunn E (2012) J Biol Chem 287:12657–12667

Cell Division

Hou S, Wieczorek SA, Kaminski TS, Ziebacz N, Tabaka M, Sorto NA, Foss MH, Shaw JT, Thanbichler M, Weibel DB, Nieznanski K, Holyst R, Garstecki P (2012) J Biol Chem 287:23878–23886

Lindås A-C, Karlsson EA, Lindgren MT, Ettema TJG, Bernander R (2008) Proc Natl Acad Sci USA 105:18942–18946

Loose M, Fischer-Friedrich E, Ries J, Kruse K, Schwille P (2008) Science 320:789–792

Lutkenhaus J, Addinall SG (1997) Annu Rev Biochem 66:93–116

Osawa M, Anderson DE, Erickson HP (2008) Science 320:792–794

Samson RY, Obita T, Freund SM, Williams RL, Bell SD (2008) Science 322:1710–1713

Tonthat NM, Milam SL, Chinnam N, Whitfill T, Margolin W, Schumacher MA (2013) Proc Natl Acad Sci USA 110:10586–10591

Chapter 4
Cellular Permeability

Any object that enters or leaves a cell, whether a nutrient, must penetrate one or more enclosing membranes. The importance of transport activity can be appreciated from the non-trivial metabolic costs of pumping molecules across cell membranes, which are estimated to consume 10–60 % of the ATP requirements of bacteria, depending on conditions.

Gram-negative bacteria have a complex cell surface, consisting of three layers: an outer membrane, composed of protein, phospholipid, and lipopolysaccharide, the cell wall proper or peptidoglycan and an inner, or cytoplasmic membrane, composed of protein and phospholipid. Nutrients, therefore, have to pass through this rather formidable architectural and protective structure. The outer membrane constitutes an important barrier against a variety of assaults, such as enzymatic and detergent attack in the digestive tract, and antibiotic entry. Limited permeability to small solutes is available through the outer membrane by way of proteinaceous channels, constituted of a special class of proteins which can be substrate-non specific (the proteins in this case are named porins) or substrate-specific. The cell wall proper is commonly regarded as an entirely permeable layer, conferring rigidity to the cell while forming a widely open network through which nutrient diffusion occurs readily. The cytoplasmic membrane, on the other hand, is impermeable to almost every solute unless a special transport system is provided.

The selective permeation of certain molecular species across certain tissues, or into certain cells, has been recognized since the beginning of the century as a phenomenon of fundamental importance in animal physiology. Up to 1955–1957, the situation was quite different in the field of microbiology. Although the importance of recognizing and studying selective permeability effects had been frequently emphasized, the available evidence appeared ambiguous, and the very concept of selective permeation was looked upon with suspicion by many microbiologists who believed that, in the absence of direct proof, it served mostly as a verbal "explanation" of otherwise obscure results.

In the second part of the 1950s, however, definite proof was obtained of the existence in bacteria of stereospecific permeation systems, functionally specialized and distinct from metabolic enzymes. We now know that the entry of most of the

G.N. Cohen, *Microbial Biochemistry*,
DOI 10.1007/978-94-017-8908-0_4,

metabolically accessible organic and inorganic nutrilites into any given type of bacterial cell is in fact mediated by such specific permeation systems. Many of these systems have been characterized as composed of one or more proteins, and defined by a combination of highly characteristic properties. The generic name "permeases" has been suggested for these systems. Although this designation has been criticized, it has the overwhelming advantage that its general meaning and scope are immediately understood.

Accumulation, Crypticity, and Selective Permeability

That the entry of organic substrates into bacterial cells may be mediated by more or less selective permeation systems has been suggested primarily by two kinds of observations concerning respectively: (a) the capacity of certain cells to accumulate certain nutrilites; (b) the state of "crypticity" of certain cells toward certain substrates, i.e., their incapacity to metabolize a given substrate, even though they possess the relevant enzyme system.

Let us see why both the accumulation and the crypticity phenomena were strongly suggestive, yet inconclusive, as evidence of selective permeation systems.

The classical work of Gale on the uptake of amino acids by staphylococcal cells posed the problem of the accumulation mechanism as early as 1947. Gale and his associates found that these cells, grown on casein hydrolyzate, contain large amounts of glutamic acid, lysine, and other amino acids which could be extracted by water from crushed, but not from intact, cells. These observations appeared to indicate that the cells were highly impermeable to amino acids. If this were true, then the entry of amino acids could not occur by simple diffusion, which is by definition a reversible process: it had to be mediated by some special, unidirectional transfer mechanism. This conclusion was also suggested by the fact that glutamic acid enters the cells only in the presence of glucose. However, lysine, which is accumulated to a similar extent as glutamic acid, and is equally retained by intact cells, does not require glucose for its entry. An alternative mechanism therefore has to be considered, namely that the amino acids are retained within staphylococcal cells by intermolecular forces, for instance by some kind of intracellular receptors. If so, no permeable barrier, nor any permeation mechanism, needs to be assumed to account for the accumulation. These two alternatives must both be considered and weighed against each other, whenever attempting to interpret the mechanism of accumulation of a compound by a cell. A choice between them is always difficult; all the more so since they are not mutually exclusive: proving a contribution to the accumulation process by one of these mechanisms does not in itself disprove contribution by the other.

The paradoxical finding that enzymes active against a given substrate may, in some cases, be extracted from cells which, when intact, are inert toward the same substrate, has been noted many times and diversely interpreted by puzzled microbiologists. There is of course no paradox when the "cryptic" state of the cells concerns a whole class of chemical compounds, since there is no difficulty in assuming that the solubility and/or electrical properties of a class of compounds may forbid their passage through the cell membrane. The phosphorylated metabolites (nucleotides, hexose phosphates) provide classical examples.

The paradox arises when interpretations in terms of non-specific forces or properties become inadequate; that is to say, when crypticity is highly stereospecific. The study of the metabolism of disaccharides by yeast has provided some of the earliest cases of specific crypticity. For instance, intact baker's yeast does not ferment maltose, although autolyzates of the same yeast does contain maltase (α-glucosidase). Analogous observations have been made with other yeasts for cellobiose and cellobiase (β-glucosidase), sucrose and sucrase, etc.

Similarly, it was observed in 1939 that dried preparations of a strain of *E. coli* which did not ferment lactose possessed lactase (β-galactosidase).

An essential point is that the cells which are cryptic towards a given carbohydrate nevertheless behave as a rule quite normally towards other carbohydrates. For instance, an *E. coli* cell which is cryptic towards lactose metabolizes glucose, maltose and other carbohydrates at a high rate. Now, if crypticity to a particular carbohydrate is attributed to the impermeability of the cell membrane, then the membrane must be impermeable to all compounds presenting similar solubility properties and molecular weight, that is to say to virtually all the carbohydrates. Therefore, those carbohydrates that do enter the cell and are metabolize at a high rate must use some highly specific stratagem for getting through the barrier.

The actual demonstration and identification of a specific permeation system, as distinct from other similar systems and from intracellular metabolic enzymes has rested upon a sort of operational isolation *in vivo*, which has required a combination of several approaches, and we shall now examine it in some detail.

β-Galactoside Permease

Before introducing the system, it should be recalled that *E. coli* metabolizes lactose and other galactosides via the inducible β-galactosidase system.

Analogs of β-galactosides where the oxygen atom of the glycosidic linkage is replaced by sulfur (Fig. 4.1) are not split by β-galactosidase, nor are they used by *E. coli* as a source of energy, carbon or sulfur.

Accumulation in Induced Cells: Kinetics and Specificity

When a suspension of *E. coli*, previously induced by growth in the presence of a galactoside, is shaken for a few minutes with an ^{35}S-labeled thiogalactoside, and the cells are rapidly separated from the suspending fluid by membrane filtration, they are found to retain an amount of radioactivity corresponding to an intracellular

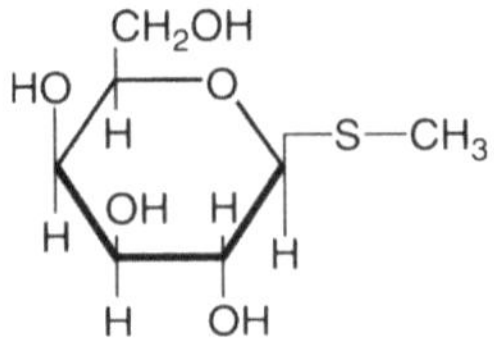

Fig. 4.1 A thiogalactoside, methyl-β-D- thiogalactoside (TMG)

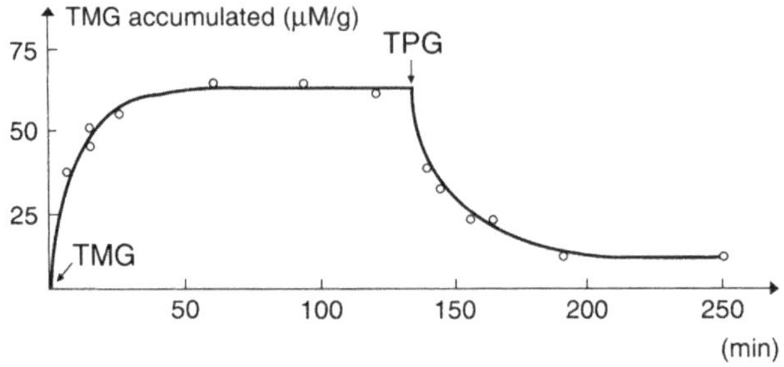

Fig. 4.2 Accumulation of methyl-β-D-galactoside (*TMG*) by induced bacteria. Non-radioactive β-D-phenylthiogalactoside (*TPG*) was added at the point shown by an *arrow*. The experiment was carried out at 0 °C to enable the kinetics to be followed (From G. N. Cohen and J. Monod (1957) Bacteriol Rev 21:169–194. Used with permission of Bacteriological Reviews)

concentration of galactoside which may exceed by 100-fold or more its concentration in the external medium. Non-induced cells, i.e., cells grown in the absence of a galactoside do not accumulate any significant amount of radioactivity.

The accumulated radioactivity is quantitatively extracted by boiling water. Chromatographic analysis of extracts show a major spot, which by all criteria corresponds to the free unchanged thiogalactoside. A minor compound accumulates if the incubation is prolonged. It corresponds to 6-acetylthiogalactoside, and its formation may be disregarded in discussing the kinetics of accumulation.

The accumulation is reversible: when the uptake of galactoside is followed as a function of time, a stable maximum is seen to be reached gradually (within 5–20 min at 34 °C, depending on the galactoside used). If at this point, an unlabeled galactoside is added to the medium at a suitable concentration, the radioactivity flows out of the cell (Fig. 4.2).

The amount of intracellular galactoside at equilibrium, in presence of increasing external concentrations of galactoside, follows quite accurately an adsorption isotherm (Fig. 4.3).

Calling the external concentration of galactoside G_{ex}, and the amount taken up by the cells at equilibrium ${}^{eq}G_{in}$, one may write

$$ {}^{eq}G_{in} = Y \frac{G_{ex}}{G_{ex} + K} \tag{4.1} $$

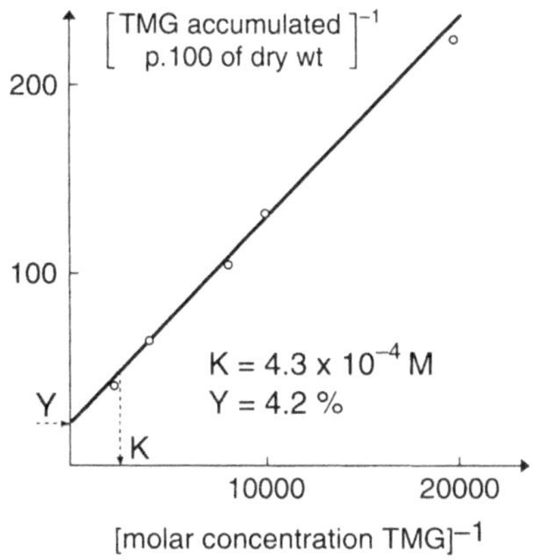

Fig. 4.3 Accumulation of radioactive thiogalactoside as a function of external concentration. Y and K are the constants of Eq. (4.1)

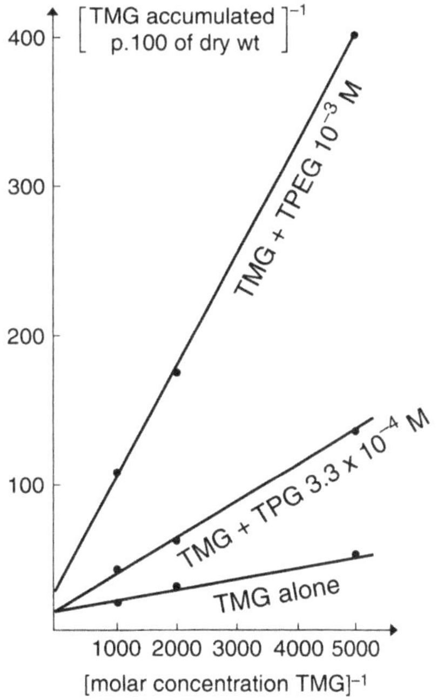

Fig. 4.4 Competitive displacement of radioactive thiomethylgalactoside by thiophenylethylgalactoside (*TPEG*) and thiophenylgalactoside (*TPG*), in *E. coli*

K is the dissociation constant of the "bacterium-galactoside complex" and Y is another constant, called capacity, which expresses the maximal amount of galactoside which the cells take up at saturating concentration.

The displacement of a labeled galactoside by another, unlabeled, galactoside also follows quite accurately the classical laws of competition for a common site (Fig. 4.4).

This allows the determination of specific affinity constants for any competitive compound. The specificity of the system proves very strict: only those compounds which possess an unsubstituted galactosidic residue (in either α or β linkage) present detectable affinity for the competition site. Glucosides or other carbohydrates, even though they may differ from galactosides only by the position of a single hydroxyl, do not compete with the galactosides. Moreover, all the effective competitors which have been tested have also proved to be accumulated within the cells. The affinity constant for each can then be determined either directly, by measurement of accumulation, or indirectly by displacement of another galactoside. The two values agree reasonably well.

These results show that the accumulation of galactosides within induced cells is due to, and limited by, stereospecific sites able to form a reversible complex with α- and β-galactosides.

A stoichiometric model according to which galactosides are accumulated within the cells in combination with specific receptor sites is dismissed mainly by an argument of common sense: in induced cells, the level of galactoside accumulation may be very high, and actually exceed 5 % of the dry weight of the cells. If the intracellular galactoside was adsorbed onto stereospecific sites (presumably associated with cellular proteins) there would have to exist, in highly induced cells, one such site for each fraction of cellular protein of molecular weight 2,000, an assumption which seems quite unreasonable.

The second interpretation (the catalytic or permease model) assigns to the specific "permease" sites the role of catalyzing the accumulation of the galactosides into the cell, rather than serving as their final acceptors. In order to account for the properties of the system, one is then led to the following scheme:

$$G_{ex} \xrightarrow[\text{(entry)}]{(y)} G_{in} \xrightarrow[\text{(exit)}]{} G_{ex}$$

according to which the intracellular galactoside (G_{in}) is a steady-state intermediate between an entry reaction, catalyzed by the stereospecific sites, and an independent exit reaction. The entry reaction involves the transitory formation of a specific complex between the sites and the galactoside and should follow the kinetics of enzyme reactions. The exit reaction is assumed not to involve the sites, and its rate is to be proportional to the amount of intracellular galactoside G_{in}. According to these assumptions, the rate of increase of the intracellular galactoside is given by:

$$\frac{dG_{in}}{dt} = y\frac{G_{ex}}{G_{ex}+K} - cG_{in} \tag{4.2}$$

If $\frac{y}{c} = Y$, Eq. (4.2) reduces to Eq. (4.1) for equilibrium conditions, when $\frac{dG_{in}}{dt} = 0$. The constant K again corresponds to the dissociation constant of the galactoside-site complex, while the capacity Y is now the ratio of the permease activity y to the exit rate constant c. Insofar as the latter remains constant, the level of intracellular galactoside is proportional to the activity of the permease.

The kinetics of intracellular galactoside accumulation also prove incompatible with the stoichiometric model while it is in good agreement with the permease model. But the most decisive reason for adopting the permease interpretation is the evidence that the cells genetically or otherwise devoid of permease are specifically cryptic towards galactosides. This evidence will be reviewed later in this chapter.

The permease model assigns the binding essentially to a high degree of impermeability of the cell membrane (or other osmotic membrane) toward carbohydrates. The permease sites catalyzing the entry must then be assumed to be associated with the osmotic barrier itself.

The following experiment depends on the fact that protoplasts of *E. coli* retain the permeability properties of intact cells while their resistance to osmotic pressure is very greatly reduced. Differences in osmotic pressure between the intra- and extracellular phases may result from either a decrease of the external osmolality or an increase of the internal pressure resulting, for example, from intracellular accumulation of a compound in an osmotically active state, i.e., in a free state. The bursting of protoplasts is easily measured as a decrease of the optical density of the suspension, so that protoplasts can be used as sensitive indicators of variation of their own osmotic pressure. By comparing the extent of lysis which occurs when the internal pressure increases as a result of permease activity with that caused by a known decrease in external pressure, one can determine approximately the intracellular concentration of free permease substrate at the time of lysis.

Using this experimental principle, the lysis caused by accumulation of various galactosides in *E. coli* protoplasts has been studied. The strain used was a mutant possessing an inducible galactoside permease and devoid of galactosidase, i.e., able to accumulate, but unable to hydrolyse lactose or other galactosides. It was observed that protoplasts prepared from induced cells (i.e., cells grown in presence of an inducer of galactoside permease) and suspended in 0.1 M phosphate buffer underwent lysis in the presence of 10^{-2} M lactose, while protoplasts from uninduced cells were insensitive to the addition of lactose. Comparisons with lysis provoked by decreasing the molarity of the suspending buffer showed that the induced protoplasts had accumulated free lactose to the extent of about 22 % of their dry weight, a figure closely approximating direct estimations made on intact cells. The addition of a rapidly metabolizable carbohydrate, namely glucose, did not result in any significant lysis.

These experiments demonstrate that the bulk, if not the totality, of the substrates of galactoside permease are accumulated within the cells in a free form. Therefore the accumulating mechanism must be catalytic, and the retention is due to a permeability barrier, not to binding or other intermolecular forces.

The Induced Synthesis of Galactoside Permease

The fact that galactoside permease is an inducible system has been of particular value for its study and characterization. As mentioned above, the system is active only in cells previously grown in the presence of a compound possessing a

Table 4.1 Induction of galactoside permease and β-galactosidase by various thiogalactosides. The specific activities are respectively μmoles TMG concentrated/g dry weight and nmoles ONPG hydrolyzed per min and per mg dry weight. The cultures were on synthetic medium, in presence of 10^{-3} M inducer. Permease was measured with 10^{-3} M TMG

Inducer	Permease	β-galactosidase
None	<2	2
Methyl-β-D-thiogalactoside	176	8,500
Propyl-β-D-thiogalactoside	181	8,500
Isopropyl-β-D-thiogalactoside	124	9,500
Hexyl-β-D-thiogalactoside	14	8.5
Phenyl-β-D-thiogalactoside	5	2
Benzyl-β-D-thiogalactoside	10	16
Phenylethyl-β-D-thiogalactoside	50	110
Galactoside-β-D-thiogalactoside	10	4

free unsubstituted galactosidic residue. No other carbohydrates show any inductive activity. Not even all galactosides are inducers. The specificity of induction can best be studied using thiogalactosides which are not hydrolyzed or "transgalactosylated" in the cells. The specificity pattern of induction is strikingly parallel to that of β-galactosidase, although there are some minor differences which might be significant (Table 4.1) in the relative inducing activity of different compounds.

Since all galactosides are concentrated by the permease, all inducers are also "substrates" of the system. However, certain substances known to be actively concentrated (phenyl-β-D-thiogalactoside; thio-di-β-galactoside) show little or no inducing activity.

The induction is effective only under conditions allowing the synthesis of protein; it is blocked by chloramphenicol or in the absence of a required amino acid. The kinetics of the induced synthesis show that the increase in capacity corresponds to a *de novo* synthesis of protein.

Functional Significance of Galactoside Permease. Specific Crypticity

If the permease model is correct, i.e., if (a) the cells are virtually impermeable to galactosides, except for the specific activity of the permease, (b) the permease is distinct from galactosidase, and (c) galactosidase is strictly intracellular, one should expect at least two phenotypes among *E. coli* mutants unable to metabolize galactosides, one type corresponding to the loss of the permease, the other to the loss of the galactosidase. Actually, the two predicted types have been found.

Cells of the first mutant type ("absolute-negative"), grown in presence of a suitable inducer (thiomethylgalactoside), accumulate normal amounts of galactoside, but they form no detectable trace of β-galactosidase. These organisms, called

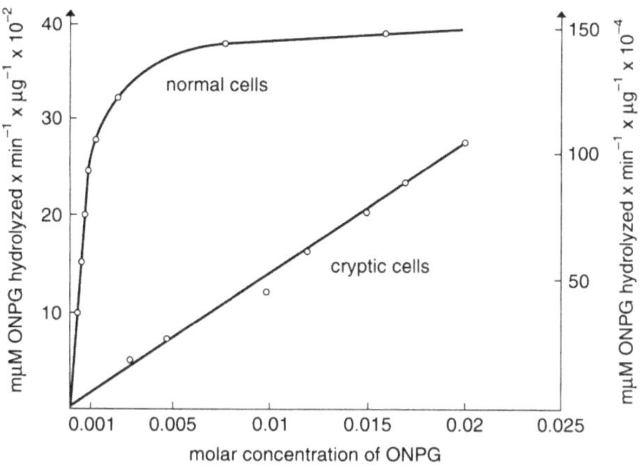

Fig. 4.5 *In vivo* hydrolysis of ortho-nitrophenyl-β-D-galactoside (*ONPG*) by whole cells of *E. coli*. *Upper curve*: wild type. *Lower curve*: cryptic (y^-) mutant. Ordinates on *left* apply to *upper curve*, on *right* to *lower curve*

thereafter z^-, accumulate in particular large amounts of lactose (up to 20 % dry weight), while the induced normal cells do not accumulate lactose, which is split and metabolized as soon as it is taken up.

Cells of the second mutant type (cryptics or y^-), grown in presence of sufficiently high concentrations of thiomethylgalactoside, form normal amounts of galactosidase (as revealed by extraction), but none or only traces of permease. So long as they are physiologically intact, these cells hydrolyze galactosides at a much slower rate than normal cells possessing equal amounts of galactosidase. Moreover, the rate of hydrolysis is a linear function of galactoside concentration, instead of being hyperbolic, as in the wild type (Fig. 4.5).

This indicates that the rate of hydrolysis in these cells is limited by a diffusion process rather than by a catalyst. The organisms are, in particular, almost inert toward lactose, while their metabolic behavior towards other carbohydrates is normal. In other words, these organisms are specifically cryptic towards galactosides.

These properties of the cryptic mutants are immediately explained if β-galactosidase is effectively inside a highly permeable barrier which the galactosides can cross only by forming a complex with the permease. The existence of the two mutant types proves that permease and β-galactosidase are genetically and functionally distinct and that they form a metabolic sequence *in vivo*.

The study of the hydrolysis of galactosides *in vivo* in wild-type organisms, also brings out the functional role of the permease. For instance, the *in vivo* hydrolysis of true galactosides is inhibited by thiogalactosides, allowing a determination of the affinities of the inhibitors for the total system. The pattern of affinities which is thus disclosed is quite different from that of β-galactosidase studied *in vitro*, but is very close to the pattern of affinities of the permease showing that, in normal induced wild-type, the permease, rather than the galactosidase, is the limiting factor for the hydrolysis of galactosides and that *in vivo*, it controls the communication between the galactosidase and the external medium.

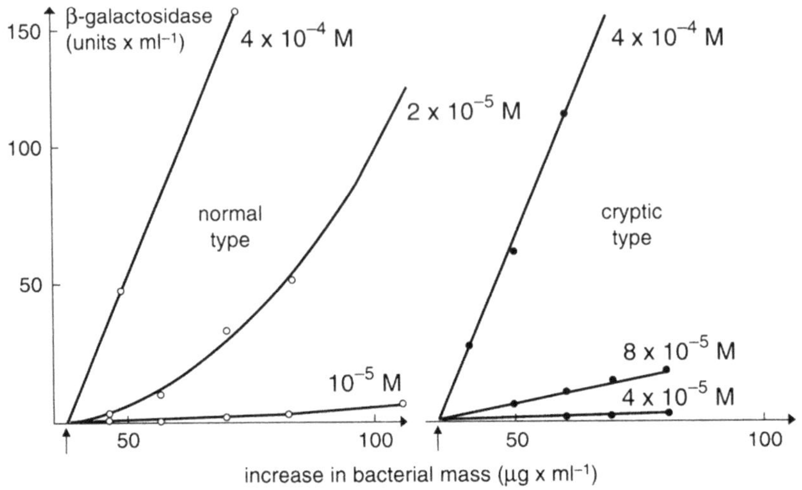

Fig. 4.6 Kinetics of induced β-galactosidase synthesis in normal *E. coli* and in the cryptic mutant type. Inducer: Isopropylthiogalactoside at the concentrations indicated. Carbon source: succinate

Functional Relationships of Permease: Induction

Galactosides are not only substrates or specific inhibitors of permease and galactosidase; they are also inducers of the two systems. Therefore, by controlling the intracellular concentration of inducers, the permease should control the kinetics of induction of β-galactosidase and of its own induction.

This is borne out by the experiments summarized in Fig. 4.6. The cryptic mutants, devoid of permease, show linear kinetics of induction at all concentrations of inducer. In contrast, with the normal cells, at low concentrations of inducer, the induction of the permease-galactosidase system is autocatalytic, since permease is a system which concentrates its own inducer.

Genetic Relationships of Galactosidase and Galactoside Permease

Although most mutations affect specifically either permease or galactosidase, giving rise to the "cryptic" and "absolute negatives" mentioned previously, some appear to suppress both simultaneously. They are deletions of the *lac* locus of the bacterial chromosome which comprises both the *z* and the *y* cistron. But another mutant type deserves special mention. This is the constitutive in which both the permease and β-galactosidase are formed in the absence of inducer. The mutation to constitutivity may occur in "cryptic" organisms devoid of permease. These "constitutive cryptics" which form very large amounts of β-galactosidase without external inducer but are unable to metabolize lactose, may in turn mutate to a permease-positive type able to grow on lactose. When this occurs, the permease is also constitutive. Therefore, a single step mutation controls the constitutive versus inducible character of both galactosidase and permease. It should be added that this pleiotropic constitutive mutation is closely linked to the loci coding specifically for the structure of galactosidase and permease. We shall have to examine further the relationships between the inducible and constitutive states of this locus when we examine the mechanisms of induction.

Galactoside Permease as Protein

At least *in vivo*, the functioning of the permease depends on the integrity of sulfhydryl groups; it is inhibited by *p*-mercuribenzoate, the inhibition being partially suppressed by cysteine and reduced glutathione. The thiogalactosides, permease substrates, protect against the inactivation, the ones with the greater affinity being the best protectors. Fox and Kennedy have taken advantage of this property of the permease in the following manner: intact cells are treated with N-ethylmaleimide (NEM), a reagent specific of the sulfhydryl groups which it covalently modifies, in the presence of thiodigalactoside (TDG), a high affinity substrate of permease. After several centrifugations and washings to eliminate the excess NEM and TDG, the cells are treated with radioactive NEM in the absence of protecting TDG. This, in principle, should allow the specific labeling of permease. One can indeed extract from the membrane fraction (with detergents) a radioactive protein which is the product of the *y* gene: the synthesis of this protein is inducible and it is distinct from β-galactosidase. The protein is firmly linked to the membrane, from which it cannot be extracted with the usual aqueous buffers, but is extractible with buffers containing ionic or non-ionic detergents.

Circular dichroism studies of the pure protein and examination of the hydropathic profile of its sequence have allowed the proposal of a model of its secondary structure: the polypeptide possesses a short hydrophilic amino-terminal sequence on the internal face of the membrane followed by twelve transmembrane hydrophobic domains, each of them being an α-helix, connected by more hydrophilic segments, and then a hydrophilic 17 amino acid C-terminal end, again on the internal face of the membrane (Fig. 4.7). Evidence supporting the general features of the model has been obtained from laser Raman spectroscopy, limited proteolysis, immunological studies and chemical modification. Exclusive support for the 12-helix motif has been obtained from an extensive series of lac permease-alkaline phosphatase (lac Y-pho A) fusions.

Mechanistic studies of the galactoside permease have greatly benefited from the utilization by Kaback and his colleagues of membrane vesicles since the beginning of the 1970s to this day. The vesicles are prepared by osmotic lysis of *E. coli*

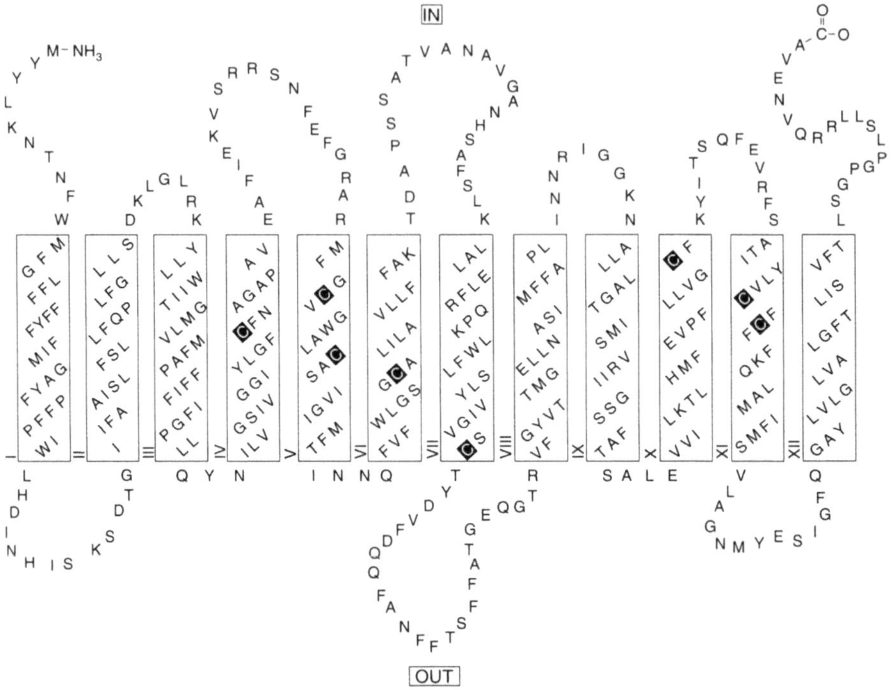

Fig. 4.7 Model for the secondary structure of the β-galactoside permease based on a variety of data. The cysteine residues are highlighted in boldface characters. One-letter code (From P. R. van Iwaarden, J. C. Pastore, W. N. Konings and H. R. Kaback, with permission of Biochemistry)

spheroplasts under well-defined conditions; they are generally closed right-side out sacs of a diameter from 0.1 to 0.5 μm delimited by a double-layer membrane 70 Å thick. They are devoid of cytoplasmic proteins and of cofactors, and are unable to synthesize ATP by oxidative phosphorylation. However, in the presence of a suitable energy source, they are able to accumulate sugars, amino acids and metals at rates comparable to those observed with intact cells. Vesicles obtained from bacteria unable to synthesize phosphatidylethanolamine are unable to accumulate galactosides and as a result carry out only facilitated diffusion. The accumulation strictly obeys the same genetic determinism as the bacteria from which they have been prepared. For example, vesicles prepared from a y^- strain do not accumulate thiogalactosides. Respiration is indispensable for transport which is not accompanied by the phosphorylation of substrate.

If a pH gradient is imposed, the most acid part being outside the vesicle, accumulation of lactose takes place within the vesicle. A membrane potential, negative inside, generated by a gradient of potassium ions also causes the accumulation of lactose. These results have revealed that the force leading to lactose accumulation is the protonmotive force through the membrane. Reconstituted vesicles containing only β-galactoside permease inserted in a phosphatidylcholine bilayer are perfectly able to transport lactose, the entry of each molecule of the disaccharide being coupled with the simultaneous movement of a proton from outside to the interior of the artificial vesicle; in addition to its galactoside binding site, the permease actually contains a specific site for the binding of protons. Under the physiological conditions, the proton gradient necessary for active transport is generated by the electron flux originating from donors like NADH+. We are thus in the presence of a further example of the unifying concept proposed by Peter Mitchell. Several sugars, amino acids and succinate are also transported via this mechanism of proton symport.

Kaback and his colleagues have further studied the mechanism of β-galactoside/H+symport by site-directed mutagenesis of *y*, the lactose permease gene.

We shall try to summarize the main results of these studies. The studies of Fox and Kennedy, reported above, suggested the involvement of an essential sulfhydryl in or near the binding galactoside site. Subsequently, the substrate-protectable cysteine residue was identified as Cys148, predicted to be in putative helix V. However, when this cysteine is converted to a glycine or to a serine, the permease is still active, although it is still inactivated by NEM; this inactivation is not protected by TDG. Therefore, although Cys148 is important for substrate protection against NEM inactivation, it does not play a direct role in the mechanism of lactose/H+symport. The seven other cysteines have been subjected individually to site-directed mutagenesis and only Cys154 is important for the symport although its mutants bind galactosides normally. Furthermore, in 1991, Kaback has obtained a lactose permease where all the eight cysteines have been replaced, and which catalyzes active lactose transport, in a manner that it is resistant to inactivation by NEM, as well as efflux and equilibrium exchange. Subsequently, individual putative intramembrane charged residues were replaced by neutral residues. Some of these replacements abolished lactose transport. The analysis of double replacements provided strong evidence that the putative transmembrane helix VII lies next to helices X and XI in the tertiary structure of the lactose permease.

Other mutants have been obtained where individual tyrosine, proline, tryptophan, arginine, histidine or glutamic acid residues have been changed. The results show that disulfide-sulfhydryl interchange is not an integral part of the mechanism, that the pathways for H+and lactose translocation may share common residues, that the association of ligand probably involves large-scale environmental changes with small secondary structure alterations, and that the pathway for lactose may involve "binding sites" that include hydrophilic segments 2, 8 and 12 on the periplasmic surface, residues within putative helices IX and X and hydrophilic segments 5 and 11 on the cytoplasmic surface. Furthermore, the data in their entirety suggest possible arrangements of the 12 transmembrane helices in the permease, whereby key regions of the protein, containing residues vital for various aspects of function, are within reasonable proximity. One possible arrangement is shown in Fig 4.8. although others can be postulated.

An interesting approach to model building deserves to be somewhat detailed: co-expression of the *lacY* gene encoding either the N-terminal or the C-terminal halves of the permease, or the first two putative helices and the remainder of the

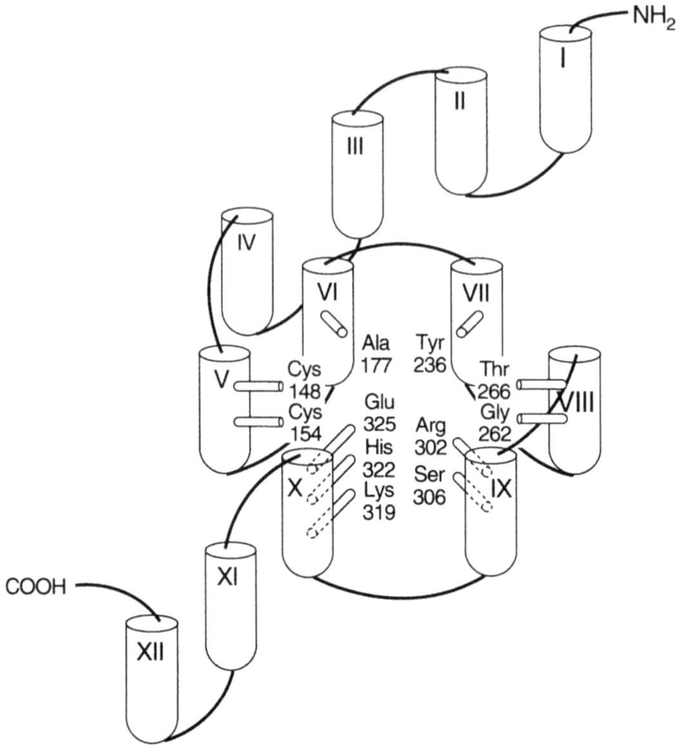

Fig. 4.8 Hypothetical model for the tertiary structure of the β-galactoside permease. The model depicts 12 cylindrical transmembrane segments, in an α-helical conformation, connected by hydrophilic segments (From R. J. Brooker, with permission of Research in Microbiology)

protein, leads to membrane insertion of the peptides and formation of functional complexes. No transport activity is observed when gene fragments are expressed individually. These experiments have been extended to new constructions based on the hypothetical model, and have provided evidence for a physical interaction between independently expressed pairs of complementing peptides, by co-immunoprecipitation with an anti C-terminal antibody. In addition, since helices I and II are independently inserted (they are separated in the model by an external loop) and form a stable complex, there is a strong argument against the hypothesis that the N-terminus inserts into the membrane as a helical hairpin.

Few side chains in the *E. coli* β- galactoside/H + symporter-galactoside permease are irreplaceable for an alternating access mechanism in which sugar binding induces closing of the cytoplasmic cavity and reciprocal opening of a periplasmic cavity. In this study, each irreplaceable residue was mutated individually, and galactoside-induced opening or closing of periplasmic or cytoplasmic cavities was probed by site-directed alkylation. As expected, mutation of Glu126 (helix IV) or Arg144 (helix V), which are essential for sugar binding, completely block sugar-induced periplasmic opening. Remarkably, however, replacement of Glu269 (helix

VIII), His322 (helix X), or Tyr236 (helix VII) causes spontaneous opening of the periplasmic cavity in the absence of sugar and decreased closing of the cytoplasmic cavity in the presence of galactoside. In contrast, mutation of Arg302 (helix IX) or Glu325 (helix X) has no such effect, and sugar binding induces normal opening and closing of periplasmic and cytoplasmic cavities. Possibly, Glu269, His322, and Tyr236 act in concert to coordinate opening and closing of the cavities by binding water, which also acts as a cofactor in H + translocation. Mutation of the triad results in loss of the bound water, which destabilizes LacY, and the cavities open and close in an uncoordinated manner. Thus, the triad mutants exhibit poor affinity for sugar, and galactoside/H + symport is abolished as well.

Lac permease has been crystallized with and without thiodigalactoside. Studies of its tridimensional structure confirms most of the above predictions.

Periplasmic Binding Proteins and ATP Binding Cassettes

In order to survive, bacteria use a battery of different transport systems. Their versatility in transporting a wide range of nutrients is exemplified by the set of transport systems that require several dozen binding proteins found in the periplasmic space of Gram-negative, and more recently of Gram-positive bacteria.

Nossal and Heppel have observed 35 years ago that a certain number of proteins are liberated from the bacterial cell envelope by an osmotic shock in the cold after treatment by a variety of agents. Among these proteins are some whose unique function seems to be the binding of small molecular weight substances, such as sugars, amino acids or ions. The ease with which these proteins are released suggests that they are located near the cell surface, in the periplasm, a space between the cell wall and the cytoplasmic membrane. In many cases, it has been shown that cells having been subjected to an osmotic shock have a reduced capacity to accumulate various molecules. The property of the periplasmic proteins to restore a normal accumulation and the fact that they present association constants for their specific ligands in the order of the constants of intact cells, suggest that these proteins play an essential role in transport. Further arguments come from genetics: mutants of *E. coli* unable to accumulate sulfate are devoid of the corresponding periplasmic protein; the same is true of mutants respectively unable to accumulate arabinose or galactose.

In addition to those devoid of a sulfate-binding periplasmic protein, there are other mutants unable to accumulate sulfate which possess a normal binding protein; one has to admit that the mutation has affected another component of the system in those mutants; ten times more sulfate is needed to reach the half-saturation of the transport system than required for the periplasmic binding protein. Similar mutants, affected in their specific permeation, but with a normal periplasmic protein, have been described in the case of maltose and arabinose.

Notwithstanding the differences in size, amino acid composition and sequence, and to ligand specificity, the tridimensional structure of the periplasmic proteins that have been analyzed to high resolution share the following features: they have overall similar tertiary structures, consisting of two similar lobes connected by a deep cleft; the ligand is bound and sequestered within the cleft; hinge motion between the two lobes allows access to and from the ligand binding site; finally, molecular recognition and binding of the different ligands are achieved principally by hydrogen-bonding interactions.

The sulfate-, arabinose-, maltose- and galactose-binding proteins have been crystallized as well as the ones binding phosphate, valine-leucine-isoleucine, lysine-arginine-ornithine (LAO) (at 2.7 Å resolution) and histidine (at 1.89 Å resolution). The structure of the LAO binding protein showed that the molecule has a bilobal structure and that its topological structure, although different from other amino acid-binding proteins, is similar to that of the sulfate- and maltose-binding proteins. High sequence homology between LAO and the histidine-binding protein, which share the same membrane-bound protein complex (see below) allow for the definition of the functional regions responsible for the ligand binding and for the interaction with the membrane complex. Furthermore, the high degree of sequence similarity suggests that they are likely to have evolved from a tandem gene duplication event and to function similarly.

The crystal structure of the arabinose-binding protein has been refined at 1.7 Å resolution, which allows a quantitative description of the interaction between the protein and arabinose. The refined structure of the complex of the protein with the sugar clearly shows an almost enclosed sugar molecule, the bound sugar not being able to leave without a conformation change. This change allows essential residues of the protein to be properly oriented for binding. A hinge-bending motion between the two domains of the protein results in closure of the structure around the ligand (Venus fly trap hypothesis) and shields it from water. The ligand-induced conformational change is believed to be crucial to the function of the periplasmic proteins in transport. It generates the appropriate stereochemistry for the specific interaction of the liganded protein with membrane-bound protein components, thus initiating the translocation process.

This brings us to the actual passage of the ligand (which is still in the periplasm) across the cytoplasmic membrane. In the case of histidine and maltose transport, it has been shown that the membrane potential is both insufficient and unnecessary to energize periplasmic transport.

The ABC (ATP binding cassette) transporter is one of the active transport systems of the cell, which is widespread in archaea, eubacteria, and eukaryotes. It is also known as the periplasmic binding protein-dependent transport system in Gram-negative bacteria and the binding-lipoprotein-dependent transport system in Gram-positive bacteria. It is responsible for the ATP-powered translocation of many substrates across membranes. The transporter shows a common global organization with three types of molecular components. Typically, it consists of two integral membrane proteins each having six transmembrane segments, two peripheral membrane proteins that bind and hydrolyze ATP, and the above mentioned periplasmic (or lipoprotein) substrate-binding protein. The ATP-binding protein component is the most conserved, the membrane protein component is somewhat less conserved, and the substrate-binding protein component is most divergent in terms of the sequence similarity. The ABC transporters form the largest group of paralogous genes in bacterial and archaeal genomes and the genes for the three components frequently form an operon.

Importers and exporters represent the ABC transporters. ABC transporters include nucleotide binding domains, transmembrane spanning domains and the periplasmic solute binding proteins. In the case of exporters, the latter are absent. Also inherent to the ABC transporters is the conserved organizational nature of the genes involved.

Incontrovertible evidence that ATP is the direct energy source has been obtained by proteoliposome reconstitution using purified periplasmic proteins and bacterial phospholipids. The reconstitution of the histidine permease utilized pure histidine-binding periplasmic protein (coded by *hisJ*), about 80 % pure complex comprising hisQ and hisM, two hydrophobic membrane-bound components and hisP, an ATP-binding membrane-bound protein, *E. coli* phospholipids and internally trapped ATP. The reconstituted system has an efficiency comparable to that in reconstituted right-side out vesicles or whole cells. It shows that the periplasmic protein alone is insufficient to explain the transport. Periplasmic proteins are subject upon binding of their specific ligands to conformation changes leading to an affinity change and to the delivery of the ligand to the external part of the cytoplasmic membrane where other hydrophobic membrane-bound components of the system would affect the actual translocation. Evidence that these components (as shown for HisQ and HisM) are accessible on both the inner and outer membrane with the carboxyl termini to the inside surface of the cytoplasmic membrane and their amino termini in the periplasmic space has been obtained. There are indications that both proteins span the membrane five times. ATP binds to one of these components and is hydrolyzed concomitantly with transport.

Maltose transport across the periplasmic membrane of *E. coli* is dependent upon the presence of the periplasmic binding protein MBP, the product of the *malE* gene. The products of the *malF*, *malG* and *malK* genes form a membrane associated complex that catalyzes the hydrolysis of ATP to provide energy for the transport event. Mutants have been isolated that have gained the ability to grow on maltose in the absence of MBP. After reconstitution of the complex into proteoliposomes, measurements of the ATPase activity of mutant and wild type complexes in the absence and presence of MBP have been carried out: they reveal that the wild type complex hydrolyzes ATP rapidly only when both maltose and MBP are present. In contrast, the mutant complexes have gained the ability to hydrolyze ATP in the absence of maltose and MBP. The rate of hydrolysis in different complexes is directly proportional to the growth rate (on maltose as the carbon source) of the different mutants from which they were derived, indicating that constitutive ATP hydrolysis and presumably the cyclic conformational changes of the complex produce maltose transport in the absence of MBP. These results suggest that ATP hydrolysis is not directly coupled to ligand transport even in the wild type cells and that one important function of MBP is to transmit a transmembrane signal through the transmembrane proteins MalF and MalG to the MalK protein on the other side of the membrane, so that ATP hydrolysis can occur.

The *Escherichia coli* maltose transport system selectively imports malto-oligosaccharides into the cell as nutrients. Through crystal structures of MBP-MalFGK2 captured in two different conformational states it has been found that the substrate specificity is conveyed by both the periplasmic binding protein MBP and the ATP-binding cassette transporter MalFGK2. These structures show that the periplasmic binding site (formed by MBP and MalG) interacts with only four glucosyl units from the reducing end of the polymer, and that the transmembrane-binding site (in MalFGK2) binds only three glucosyl units from the opposite, nonreducing end. These structures essentially lead to a single concept: that transport selectivity can be explained through the polarity of substrate binding to the two-component (MBP + transporter) system.

MBP is the first binding protein involved in transport in which the structures of the liganded and the unliganded formed have been determined and refined at high resolution. The unliganded structure (open form) shows a rigid-body "hinge-bending" between two globular domains by approximately 35°, relative to the maltose-bound structure (closed form), opening the sugar binding groove located between the two domains.

Phosphotransferases: The PTS System

In 1962, it was shown that α-methylglucoside is partially accumulated within *E. coli* cells as a phosphate ester. Four years later, it was observed that the ester can be detected only during one minute, which suggested a specific mechanism of dephosphorylation, since in longer experiments the α-methylglucoside is found unchanged in the intracellular space.

Due to the efforts of Roseman and his colleagues who have analyzed this system during more than 35 years, a phosphotransferase was isolated in 1964 from several bacterial species which catalyzes the following sequence of reactions:

$$\text{Phosphoenolpyruvate} + \text{HPr} \xrightarrow{\text{I, Mg}^{++}} \text{HPr-P} + \text{pyruvate} \tag{4.3}$$

$$\text{HPr-P} + \text{sugar} \xrightarrow{\text{II, III}} \text{sugar - phosphate} + \text{HPr} \tag{4.4}$$

The sum of these two reactions leads to:

$$\text{Phosphoenolpyruvate} + \text{sugar} \xrightarrow[\text{Mg}^{++},\ \text{HPr}]{\text{I, II, III}} \text{sugar - phosphate} + \text{pyruvate}\,(1+2)$$

The unusual feature of this PTS system is that phosphoenolpyruvate, rather than ATP, is the phosphoryl donor.

Crude extracts of *E. coli* contain a protein kinase that phosphorylates Enzyme I at a histidine residue. The activity of this ATP-dependent kinase depends on the presence of NAD^+ (or $NADP^+$). NADH reverses the activation by NAD^+, and thus the kinase level depends on the NAD^+/NADH ratio.

Enzyme I catalyzes the transfer of the phosphoryl group of phosphoenolpyruvate to a specific histidine residue (N5-atom of His15) of HPr which is a small thermostable protein. The phosphohistidine is an unstable energy-rich compound which transfers its phosphoryl group to "Enzyme III" (a misnomer, Enzyme III is actually a phosphocarrier protein like HPr), a peripheral membrane protein that interacts with enzyme II, the actual channel integral protein, which is responsible for the final step, the transfer of the phosphoryl group to the sugar that is being translocated. A functional enzyme complex has been obtained *in vitro* from the four purified proteins and an extensive study of the mechanism involving several critical amino acid residues has being carried out, involving among others the use of site-directed mutagenesis. In particular, it appears that Arg17 of HPr is partly responsible for the instability of the phosphorylated HPr.

In *E. coli*, the phosphotransferase system is active on glucose (and α-methylglucoside), fructose, mannose and mannitol (In *Staphylococcus aureus* and *Bacillus subtilis*, similar systems are respectively responsible for lactose and sucrose transport). Whereas HPr and Enzyme I are constitutive and soluble and participate in the transport of all sugars handled by the PTS, enzymes II and III are specific for particular sugars and inducible: there are different enzymes II and III for glucose, mannose and fructose. The three-dimensional structure of the enzyme III specific for glucose (IIIGlc) has very recently been solved. The molecule is a β-sheet sandwich with six antiparallel strands on either side. The active site is a shallow extremely hydrophobic depression near the center of one face, and is a good candidate for interaction with HPr. Mutants unable to synthesize a normal HPr or Enzyme I loose the capacity to utilize numerous carbohydrates while mutants of Enzymes II or III are unable to transport only a particular sugar.

Since the PTS is the mode of transport and of introduction of many sugars into the metabolic network, the question arises as to whether it is general. The answer is negative. Among others, the accumulation of β-galactosides is inhibited by uncoupling agents such as azide and dinitrophenol, which are without effect in the accumulation of α-methylglucoside by the PTS system. Moreover, membrane vesicles from a mutant devoid of enzyme I normally accumulate β-galactosides by the system linked to respiration.

In addition to its translocation function, the PTS has important regulatory functions. It regulates the uptake of its sugar substrates by controlling adenylate cyclase and inhibiting the uptake of inducers by the non PTS-permeases. This aspect, which is in full development, is discussed at length in a 1990 review by Roseman and associates referred to in the references at the end of this chapter.

TRAP Transporters

Until recently, extracytoplasmic solute receptor (ESR)-dependent uptake systems were invariably found to possess a conserved ATP-binding protein (the ATP-binding cassette protein or ABC protein), which couples ATP hydrolysis to the translocation of the solute across the cytoplasmic membrane. While it is clear that this class of ABC transporter is ubiquitous in prokaryotes, it is now firmly established that other, unrelated types of membrane transport systems exist which also have ESR components. These systems have been designated tripartite ATP-independent periplasmic (TRAP) transporters, and they form a distinct class of ESR-dependent secondary transporters where the driving force for solute accumulation is an electrochemical ion gradient and not ATP hydrolysis. Currently, the most well characterized TRAP transporter at the functional and molecular level is the high-affinity C4-dicarboxylate transport (Dct) system from *Rhodobacter capsulatus*. This transporter consists of three essential proteins; a periplasmic binding receptor (DctP) and small (DctQ) and large (DctM) integral proteins, which span the membrane 4 and 12 times, respectively. Homologues of the *R. capsulatus* DctPQM proteins are present in a diverse range of prokaryotes, both bacteria and archaea, but not in eukaryotes. The deduced structures and possible functions of these homologous systems have been described. In addition to the DctP family, other types of ESRs can be associated with TRAP transporters. Sometimes, the 4-TM(DctQ) and 12-TM(DctM) integral transmembrane proteins(TM) are fused.

In both the ABC and TRAP systems, a specific periplasmic protein from the ESR family is often involved for the recruitment of the solute and its presentation to the membrane complex. The TRAP family is widespread in prokaryotes, as predicted from sequence analysis of bacterial genomes. However, the physiological role of few of them has been elucidated since ligands for ESRs of TRAP transporters have only been evidenced for C4-dicarboxylate, ectoine, glutamate, xylulose, α-keto acids and sialic acid. The best characterized TRAP transporters at functional and molecular levels are the high-affinity C4-dicarboxylate transport system (dctPQM) from *Rhodobacter capsulatus*, the α-keto acid transporter from *Rhodobacter sphaeroides* and the sialic acid transporter (SiaPQM) from *Haemophilus influenzae*. In the latter, the structure of the periplasmic subunit (SiaP) was solved very recently at high resolution, revealing, among others, an overall topology similar to ABC ESR proteins

In *Rhodobacter sphaeroides*, TakP is an ESR that binds α-keto acids *in vitro*. Its high-resolution crystal structure of its unliganded form and as a complex with sodium pyruvate has been determined. The results show a limited "Venus flytrap" conformational change induced by substrate binding. In the liganded structure, a cation (most probably a sodium ion) is present and plays a key role in the association of the pyruvate to the protein. The structure of the binding pocket gives a rationale for the relative affinities of various ligands that were tested from a fluorescence assay. The protein appears to be dimeric in solution and in the crystals, with a helix-swapping structure largely participating in the dimer formation. A 30Å long water channel buried at the dimer interface connects the two ligand binding cavities of the dimer. The concerted recruitment by TakP of the substrate group with a cation could represent a first step in the coupled transport of both partners, providing the driving force for solute import. Furthermore, the unexpected dimeric structure of TakP suggests a molecular mechanism of solute uptake by the dimeric ESR via a channel that connects the binding sites of the two monomers

Another TRAP family, the TAXI family has been defined by its structural characteristics. Substrates for this transporter family are not fully characterized but, besides C4 dicarboxylates, may include mannitol and other compounds.

A Few Well-Identified Cases of Specific Cellular Permeability

Amino Acid Permeases

At the same time the initial work on β-galactoside permease was done, the same research team discovered specific permeases for amino acids in *E. coli*. A single permease catalyzes the intracellular concentration of the branched-chain amino acids valine, leucine and isoleucine; another is responsible for the entry into the intracellular space of methionine and of its structural analogue, norleucine. Later on, a third permease was described for the aromatic amino acids. Other systems were found responsible for the entry and accumulation of histidine, arginine, proline, etc.

Mutants devoid of one or the other amino acid permease have been selected by layering the wild-type cells on plates containing a toxic structural analogue of the amino acid. These analogues are toxic because they are incorporated in proteins by replacing their natural counterparts at their normal positions in the protein sequences, thus leading most of the times to inactive enzymes. Some resistant colonies appear on the plates. Among these are found some which resisted the toxic action of the analogue, because they are devoid of the permease responsible both for the entry of the natural amino acid and the analogue. Thus, bacteria devoid of the arginine permease have been isolated among the resistants to canavanine (an arginine analogue), bacteria unable to concentrate glycine using D-serine as the analogue, and other lacking the histidine permease among the triazole alanine-resistant bacteria. With the use of this and other selection methods, practically all the specific permeases for the amino acids have been identified and the genes governing their synthesis have been localized on the *E. coli* or the *S. typhimurium* chromosome. Membrane vesicles, identical to those used for the study of the galactoside permease have been shown to concentrate proline, whereas membranes from bacteria without proline permease are unable to do so. Proline is thus concentrated by a system similar to the galactoside permease. In contrast, the protein responsible for the transport of valine, leucine and isoleucine involves an ATP binding cassette similar to the one involved in histidine transport.

Peptide Permeases

The growth of *E. coli* auxotrophs requiring valine, leucine or isoleucine is inhibited competitively by either of the two amino acids not required for growth. It has been shown conclusively that this competition is at the level of the common permease for the three amino acids, which has an association constant of the same order for the three compounds. However, if peptides containing the amino acids are used as competitors, they exert only a slight inhibitory effect when they are used at a concentration 100 times higher than the ones where the free amino acids exert a total inhibition of growth. Thus, peptides escape the amino acid permease, and since they can be hydrolyzed at a slow rate and provide the free amino acids when they have entered the cells, there must be some specific mechanism for their entry. The isolation of mutants unable to transport a given amino acid, but growing on peptides containing this amino acid has confirmed the independance of the accumulation mechanisms.

There exists an immense number of peptide structures, 400 for the dipeptides, 8,000 for the tripeptides, etc… This precludes the existence of highly specific transport systems. The availability of radioactive peptides has brought an answer to this problem.

The existence of two independent peptide permeases has been shown in *E. coli*, the first responsible for the transport of dipeptides, the second for oligopeptides. Dipeptides can be transported to a limited extent by the oligopeptide permease, but the reciprocal is not true. The two systems require a free amino terminal group, but only the dipeptide permease activity depends on a free carboxyl terminal group.

Oligopeptide permease mutants are available and have been isolated in the following way: triornithine is toxic for *E. coli* and mutants resistant to it can be isolated. The resistance is due to the lack of oligopeptide permease and therefore to the incapacity to bring the toxic tripeptide inside the cells. Auxotrophs carrying this mutation cannot grow on oligopeptides, but grow perfectly on free amino acids or dipeptides (Fig. 4.9).

A single transport system exists for all oligopeptides, as shown by Fig. 4.10 which illustrates the competition, measured by the inhibition of the growth rate, of the utilization of trilysine or tetraglycine by trityrosine, in the corresponding lysine and glycine auxotrophs.

The peptide permeases belong to the multicomponent periplasmic binding dependent transport, linked to ATP hydrolysis (ATP binding cassette).

Porins

The cytoplasmic membrane of Gram-negative and most other bacteria is supported on the outside by the peptidoglycan layer, which gives the bacteria their characteristic shape and protect them from osmotic lysis. This peptidoglycan consists of a network of amino sugars (N-acetylglucosaminyl-N-acetylmuramyl dimers) forming

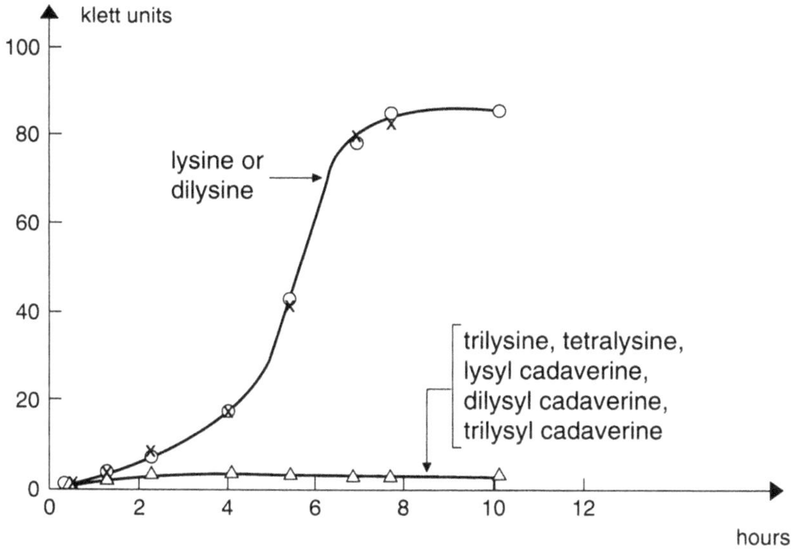

Fig. 4.9 Growth of an *E. coli* lysine auxotroph resistant to triornithine. Each culture contains either lysine or an equivalent concentration of lysine residues (From JW Payne and C Gilvarg (1971) Adv Encymol 35:187–244)

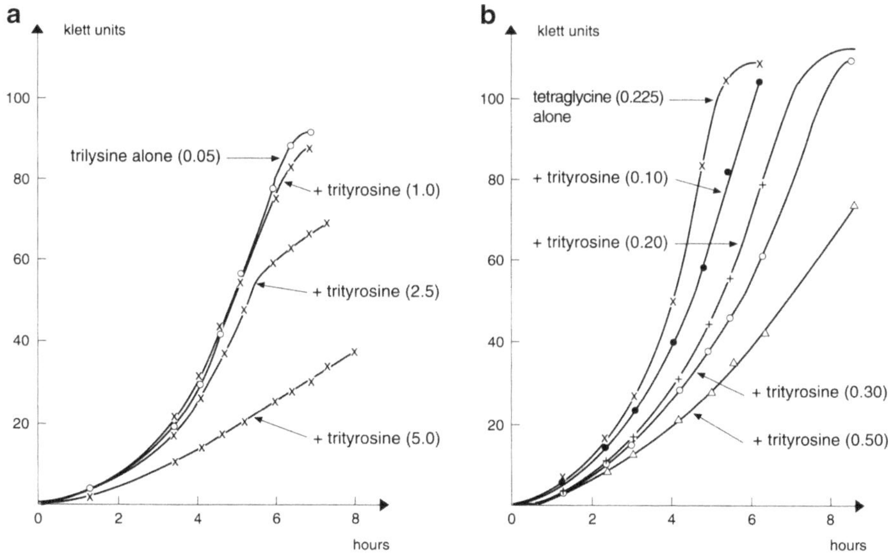

Fig. 4.10 Effect of trityrosine on the growth of: (**a**) a lysine auxotroph (**b**) a glycine auxotroph. Trityrosine has been selected because of its absolute lack of structural relationship with either lysine or glycine. Trityrosine does not affect the growth rate of the mutants growing on free lysine or glycine. Numbers refer to concentrations of oligopeptides in micromoles per milliliter (From JW Payne and C Gilvarg (1971) Adv Encymol 35:187–244)

long linear strands covalently linked together through their muramyl residues by short peptides. Gram-negative bacteria are covered by the outer membrane, an additional membranous structure present outside the cytoplasmic membrane and the peptidoglycan layer. The major function of this outer membrane is to act as a permeability barrier excluding many of the noxious molecules that exist in the external medium. Gram-positive bacteria, which are exposed to many peptidoglycan-lytic enzymes they find in their environment, elaborate many different types of peptidoglycan apparently to resist hydrolysis by such enzymes; in contrast, practically all Gram-negative bacteria produce a peptidoglycan of identical structure, presumably because they successfully exclude such enzymes. The components of the outer membrane are bound to the peptidoglycan in such a way to form a tight network.

The Gram-negative bacteria must nevertheless exchange nutrients and waste products with the environment. Much of this flux through the outer membrane takes place through protein channels. These include a) porins, which allow non-specific and spontaneous diffusion of small solutes b) specific channels, which contain specific ligand-binding sites within the channel and c) high affinity receptors which carry out the energy-coupled translocation of large nutrient molecules that exist in very low concentrations outside (eg. vitamin B12, Fe^{+++}-complexes).

Porins, also called "peptidoglycan-associated proteins" or "matrix proteins", are very important for Gram-negative bacteria because all hydrophilic nutrients must pass through them to cross the outer membrane. They are diffusion channels often tightly bound to the peptidoglycan layer. Porins from many species of bacteria have been isolated and characterized. Each species contains a very limited number of porins; for example, according to the strain examined. *E. coli* contains only one (OmpF) or two (OmpF and OmpC) porins. The OmpF protein is produced at low osmolarity, while the OmpC protein is produced to a greater extent in high osmolarity. Their expression depends on two genes, *ompR*, coding for a cytoplasmic repressor which binds to the promoter regions of both the *ompF* and the *ompC* genes, and on *envZ*, coding for an inner transmembrane protein, EnvZ, which functions as an osmolality sensor located in the cell envelope. EnvZ is a histidine kinase which can phosphorylate the OmpR protein. The phosphorylated OmpR exhibits an increased binding affinity to the *ompF* and *ompC* promoters. OmpF and OmpC exist as tight trimeric complexes which are not dissociated by sodium dodecylsulfate unless the protein is denatured by heating.

Most porins form wide channels filled with water in the outer membrane, which allows for the diffusion of different types of solutes according to their molecular weight, the larger exclusion limit being 5,000 Da for the porin of *Pseudomonas aeruginosa*.

Apart from the non-specific porins which form channels that permit general diffusion of hydrophilic molecules below a certain size, and thus are responsible for the exclusion limit of the outer membrane, there exists a class of substrate-specific porins that act as facilitated diffusion channels for specific substrates by virtue of having a substrate-specific binding site in their channel. To date only a few substrate-specific porins have been defined. Some examples: growth of *E. coli* on maltose is accompanied by the synthesis of the LamB protein which facilitates the permeation of maltose and maltodextrins across the outer membrane; the nucleoside-specific porin Tsx of *E. coli* (coded by *tsx*); and in *Pseudomonas aeruginosa*, a basic amino acid-specific porin (OprD), a phosphate-specific porin (OprB) and a polyphosphate-specific porin (OprO).

Small, hydrophilic compounds such as β-lactams diffuse across the outer membrane of Gram-negative bacteria through porin channels, which were originally thought to be nonspecific channels devoid of any specificity. However, since the discovery of an ampicillin-binding site within the OmpF channel in 2002, much attention has been focused on the potential specificity of the channel, where the binding site was assumed either to facilitate or to retard the penetration of β-lactams. Since the earlier studies on porin permeability were done without the knowledge of the contribution of multidrug efflux pumps in the overall flux process across the cell envelope, a study has been carried out where both the porin permeability and active efflux of ampicillin and benzylpenicillin have been studied: the influx occurs apparently by spontaneous passive diffusion without any indication of specific binding within the concentration range relevant to the antibiotic action of these drugs, and the higher permeability for ampicillin is totally as expected from the gross property of this drug as a zwitterionic compound. It is concluded that the original nonspecific model is still valid and that it is misleading to overemphasize the role of specific binding sites.

Some porins are inducible: phosphate limitation leads to the induction of the OprP protein of *Pseudomonas*, which allows the passage of phosphate. The induction of OprO requires cells to be in the stationary growth phase.

The primary structure of many porins has been determined either by direct sequence of the protein or deduced from the DNA sequence of their genes. Electron diffraction studies and Fourier transform infrared studies have shown that the polypeptide chain of porins crosses the membrane more than a dozen times. The β-structures are predominant. Whereas the function of the general diffusion porins is understood by the pore characteristics, chemical modification and genetic mutations, our knowledge of their structure and of their selectivity has been clarified when the crystal structures of a porin from *Rhodobacter capsulatus* and of *E. coli* porins OmpF and PhoE have been determined by X-ray analysis. They reveal a homotrimeric structure, each of the three subunits consisting of a 16-stranded antiparallel-β-barrel containing a pore. A long loop inside the barrel contributes to a constriction of the channel where the charge distribution affects the ion selectivity. The structures of OmpF and PhoE are compatible at the molecular level with the functional characteristics of the two porins and with their alterations by known mutations.

Functions of porins OmpF, OmpC and PhoE were analyzed at various pHs. Preliminary results from lipid bilayer membranes and liposome swelling assays indicated that porins have at least two open-channel configurations with a small and a large size *in vitro*. The small channels were stabilized at low pH while the larger channels were detected under basic conditions. The size switch occurs over a very narrow range near neutral pH, and the two major open-channel configurations responded differently to variations in voltage (applied in order to measure the electrical conductance across the lipid bilayer). The results obtained by modification of histidine 21 of OmpC and OmpF from *E. coli* with diethylpyrocarbonate (a specific histidine reagent) suggests that this residue is involved in the pH-induced change in channel size.

Porins have been extracted from the mitochondrial outer membrane of organisms as diverse as *S. cerevisiae*, *N. crassa*, human B-lymphocytes, *Paramecium* and *Dictyostelium discoideum*. Their sequences show no particular homology to those of bacteria, although the function of porins as channels are similar to those of the mitochondrial porins.

Iron Uptake

The outer membrane of Gram-negative bacteria constitutes a permeability barrier that protects the cell from exterior hazards, but also complicates the uptake of nutrients. In the case of iron, the challenge is even greater, because of the scarcity of this indispensable element in the cell's surroundings. To solve this dilemma, bacteria have evolved sophisticated mechanisms whereby the concerted actions of receptor, transporter and energy-transducing proteins ensure that there is a sufficient supply of iron-containing compounds, such as siderophores.

Iron is one of the most important nutrients of bacteria, because of its essential metabolic role. However, iron is scarcely available under physiological conditions; first, because of its propensity to form insoluble complexes, and second, as a result of the existence of numerous iron-binding proteins that the host itself uses to store and transport iron. Bacteria have evolved a wide range of strategies to overcome iron shortage and to ensure sufficient uptake. One of these relies on the synthesis and excretion of siderophores, or small compounds that either bind free iron or sequester it from iron-binding proteins in the bacterial environment. In Gram-negative bacteria, recovery of iron-loaded siderophores involves a sophisticated uptake mechanism. At the outer-membrane level, high-affinity receptors capture siderophores and mediate their translocation into the periplasmic space. This is powered by the Ton complex, which resides in cytoplasmic membrane and is coupled to the proton-motive force. Siderophores are subsequently transported across the periplasm and the cytoplasmic membrane by periplasmic-binding proteins and ATP-dependent membrane transporters. In recent years important advancements have been made in the characterization of the siderophore uptake system, owing to the combination of structural, biophysical, biochemical and genetic approaches. Among these, it is worth noting the determination of the atomic structure of several outer-membrane receptors and the use of spectroscopic techniques to monitor the uptake process *in vivo*.

However, fundamental questions concerning almost every aspect of the uptake process remain unclear and are the subject of continued debate between researchers.

The structure and biosynthesis of enterochelin, also called enterobactin, a siderophore found widely among the *Enterobacteriaceae* will be found in Chap. 31. In addition, some strains of *E. coli* and *S. typhimurium* synthesize the hydroxamate siderophore alrobactin. Fungi produce siderophores such as rhodotorulic acid, and non-ferrated forms of ferrichrome and coprogen.

Conclusion

At least five different systems playing a role in the passive or active transport of small solutes in bacteria have been identified: porins linked to the outer membrane, periplasmic proteins linked to cytoplasmic membrane proteins and where the transport has been shown in many cases to be linked to ATP hydrolysis (ATP-Binding Cassette, ABC superfamily, TRAP transporters), membrane carriers linked to respiration (e.g. β-galactoside permease), and the phosphotransferase (PTS) system where vectorial translocation is linked to phosphorylation of the substrate.

Comparison of amino acid sequences show strong conservation not only among bacterial transporters but also among them and many transporters of animal cells. A unifying design among membrane transporters is gradually emerging: common structural motifs and evolutionary origins among transporters with various energy-coupling mechanisms suggest that many transporters contain a module forming a transmembrane channel through which the solute may pass with size or/and steric constraints. Energy coupling mechanisms can then be viewed as secondary features added to the fundamental translocation units.

Selected References

Coupling of energy to transmembrane solute translocation in bacteria, 6th Forum in Microbiology (1990) Res Microbiol 141:281–395

β-Galactoside Permease

Abramson J, Smirnova I, Kasho V, Verner G, Kaback HR, Iwata S (2003) Science 301:610–615
Cohen GN, Monod J (1957) Bacteriol Rev 21:169–194
Kaback HR (1986) Annu Rev Biophys Biophys Chem 15:279–319
Zhou Y, Jiang X, Kaback HR (2012) Proc Natl Acad Sci USA 109:12438–12442

Amino Acid Permeases

Cohen GN, Rickenberg HV (1956) Ann Inst Pasteur 91:693–720

Periplasmic Proteins and ATP-Binding Cassettes

Biemans-Oldehinkel E, Doeven MK, Poolman B (2006) FEBS Lett 580:1023–1035
Davidson AL, Shuman HA, Nikaido H (1992) Proc Natl Acad Sci USA 89:2360–2364
Ferro-Luzzi Ames G (1986) Annu Rev Biochem 55:397–425
Higgins CF (1992) Annu Rev Cell Biol 8:67–113
Quiocho FA (1986) Ann Rev Biochem 55:287–315
Rees DC, Johnson E, Lewinsohn O (2009) Nat Rev Mol Cell Biol 10:218–227
Saurin W, Dassa E (1994) Protein Sci 3:325–344
Yao N, Trakhanov S, Quiocho FA (1994) Biochemistry 33:4769–4779

Maltose Transport

Oldham ML, Chen S, Chen J (2013) Proc Natl Acad Sci USA 110:18132–18137

Phosphotransferase System

Meadow ND, Fox DK, Roseman S (1990) Annu Rev Biochem 59:497–542
Postma PW, Lengeler JW, Jacobson GR (1993) Microbiol Rev 57:543–594
Van Nuland NAM, Boelens R, Scheek RM, Robillard GT (1995) J Mol Biol 246:180–193

Peptide Permeaes

Payne JW, Gilvarg C (1971) Adv Enzymol 35:187–244

TRAP Transporters

Gonin S, Arnoux P, Pierru B, Lavergne J, Alonso B, Sabaty M, Pignol D (2007) BMC Struct Biol 7:11

Porins

Benz R (1988) Annu Rev Microbiol 42:359–393
Kojima S, Nikaido H (2013) Proc Natl Acad Sci USA 110:2629–2634
Nikaido H (1992) Mol Microbiol 6:419–423
Nikaido H (1994) J Biol Chem 269:3905–3908
Nikaido H, Saier MH Jr (1992) Science 258:936–942

Chapter 5
Allosteric Enzymes

One of the most remarkable properties of the cell is its capacity to coordinate efficiently its different biochemical activities. A steady state is thus maintained between the different catabolic processes and the myriad of synthetic reactions necessary for the reproduction of the cell.

The last 50 years have seen a striking advance in our knowledge of numerous regulatory processes, and in our understanding of their role in the cellular economy.

The present chapter will be confined to an examination of the properties common to many enzymes whose activity is subject to metabolic regulation. These controls will be examined in some detail when certain regulated functions will be encountered.

Experiments carried out with *E. coli* some 60 years ago at the Department of Terrestrial Magnetism (!) of the Carnegie Institution of Washington showed clearly that each end product of a biosynthetic sequence is part of the system responsible for its synthesis. Thus, the de novo synthesis of different amino acids from glucose is halted when these amino acids are added to the growth medium. The mechanisms which underlie the control by end products became clear when it was shown that each of these essential metabolites possesses the property of repressing the synthesis of one or several of the enzymes belonging to its biosynthetic pathway and also of inhibiting the activity of the early enzymes (often the first enzyme) in the pathway.

These two control mechanisms are distinct, as shown by the existence of mutants with one or the other controls affected. The two mechanisms nevertheless often occur in the regulation of the same metabolic pathway.

Which are the enzymes subject to control? Every highly complex metabolic system contains some reactions which produce intermediates occurring at branch points. The branches may include biosynthetic pathways leading to different end products, or mixtures of degradative and biosynthetic pathways. In this entangled network of reactions, certain enzyme activities occur at strategic points, where their regulation becomes important for the preservation of the delicate equilibrium required to integrate the different metabolic functions. During evolution, natural selection has retained those enzymatic species with structures best suited as targets

G.N. Cohen, *Microbial Biochemistry*,
DOI 10.1007/978-94-017-8908-0_5, © Springer Science+Business Media B.V. 2014

for activations or inhibitions endowed with a regulatory signification. The existence of control mechanisms is one of the principal basis on which most of teleological arguments are built. The natural selection of these systems can make every natural phenomenon appear useful.

Since many metabolic functions are largely reversible, mass action might have a significant share in regulation. However, most pathways involve one or several irreversible steps which could not be controlled by mass action. Moreover, it is a general observation that the intracellular concentration of most intermediary metabolites is vanishingly small, indicating that mass action only plays a limited role and suggesting that other mechanisms must intervene in metabolic regulation.

The examination of a large number of enzymes subject to control of their activity reveals certain common characteristic, structural as well as kinetic, which are not generally found for other enzymes. The most fundamental of these characteristics is the ability of this particular class of enzymes to be activated or inhibited by metabolites other than their substrates or analogs. Often, there is no structural similarity between "effectors" (activitors or inhibitors) and substrates. A few examples will be sufficient to substantiate this:

L-isoleucine is a specific inhibitor and L-valine a specific activator of *E. coli* L-threonine dehydratase (also called L-threonine deaminase), the first enzyme specifically involved ("committed") in the synthesis of isoleucine. It could be argued that these compounds are analogs of L-threonine because they share a carboxyl and amino group in the same configuration. However, threonine dehydratase is not inhibited by any other naturally occurring amino acid other than L-isoleucine. Many other cases of strict specificity together with the absence of a structural relationship between effector and substrate will be encountered throughout this book. We shall quote here some examples in order to make a general rule. Let us take the case of aspartate transcarbamylase, the first enzyme in the biosynthesis of pyrimidines; in *E. coli*, this enzyme is inhibited by cytidine triphosphate and activated by adenosine triphosphate. Examination of the structural formulas of the substrates, carbamyl phosphate and aspartate, and of the inhibitor (Fig. 5.1) is sufficient to show that there is no similarity between them.

Likewise, in the bacterium *Rhodopseudomonas spheroides*, hemin inhibits the synthesis of δ-aminolevulinic acid from glycine and succinyl CoA. This acid is the compound from which the specific porphyrin synthesis diverges from the common catabolic path (Fig. 5.2).

The very fact that the effector is not a steric analogue of the substrate has led to the name *allosteric effector* being given to it, and the name *allosteric sites* to sites on the enzyme to which the effectors specifically bind. By extension, the enzymes subject to control by these effectors are called *allosteric proteins* or *allosteric enzymes*.

One of the most frequent characteristics of the allosteric enzymes is the atypical relationship between their activity and the concentration of substrate and effector concentration. For the majority of enzymes, this relationship is described by a rectangular hyperbola, representing graphically the Henri-Michaelis equation (Fig. 5.3).

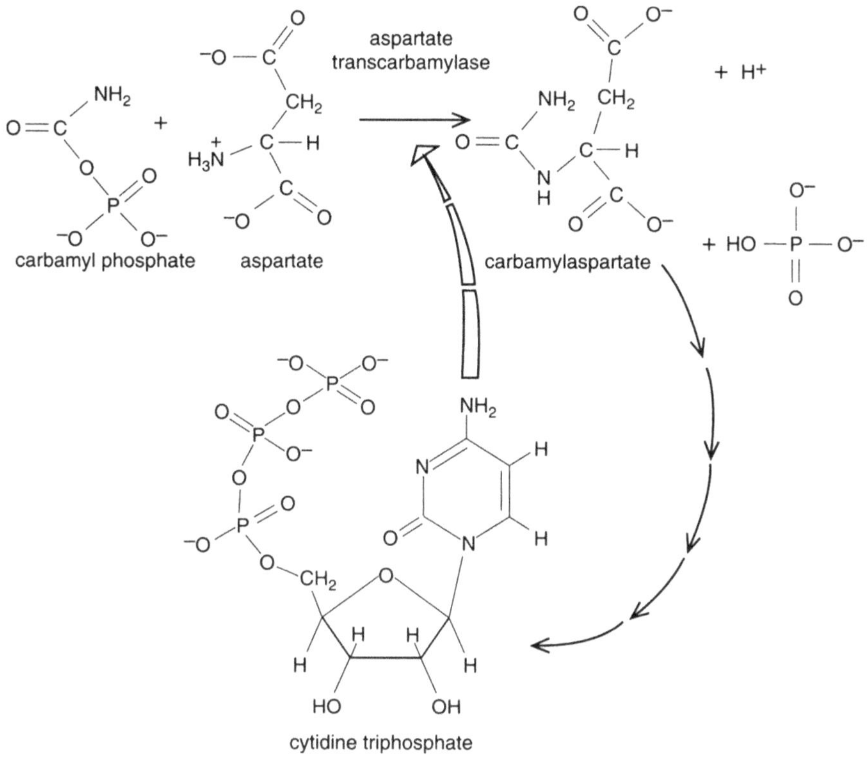

Fig. 5.1 Feedback inhibition in the synthesis of pyrimidines by *E. coli*. Cytidine triphosphate is an allosteric inhibitor of aspartate transcarbamylase

In the case of many enzymes subject to allosteric control, this relationship is represented by a sigmoid curve (Fig. 5.4).

Such a curve indicates that at least two molecules of substrate react with the enzyme and that the binding of one molecule of substrate in some way makes the binding of the second easier. Expressed differently, there is a cooperative effect in binding more than one substrate molecule to the enzyme. One often finds similar cooperative interactions in the binding of allosteric effectors, which suggests that allosteric enzymes contain more than one allosteric site per molecule (Fig. 5.5). These interactions must reflect some fundamental property. Whatever the mechanism, they have a physiological significance: a sigmoidal response of the enzymatic activity to increasing concentrations of substrate or effector is in practice a "threshold" effect. At concentrations lower than the threshold, the enzymatic activity is effectively unaltered by changes in substrate or effector concentration, but on crossing the threshold, important activity changes are obtained by only slight changes in concentration. These threshold effects make the adjustment of enzymatic activity very sensitive to minimal changes in concentration within a narrow range.

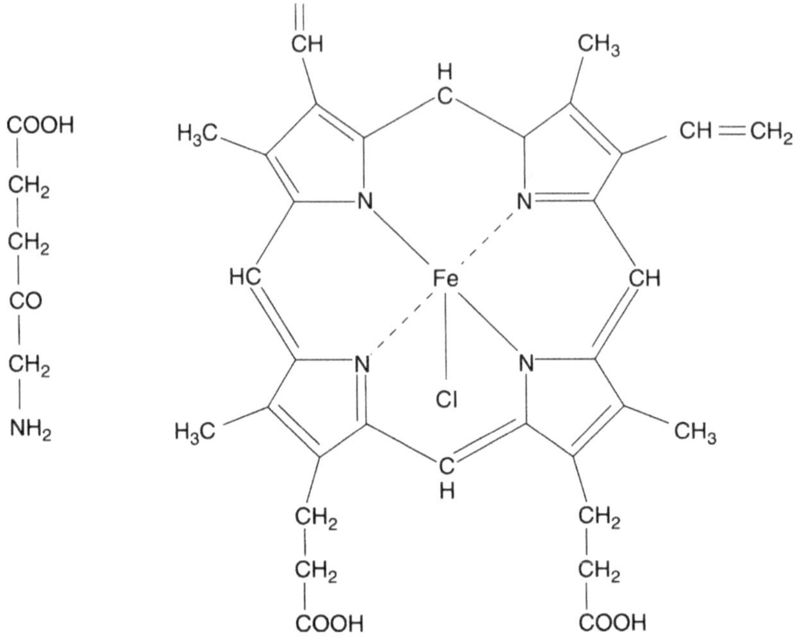

Fig. 5.2 δ-aminolevulinic acid (ALA) is a precursor of hemin. Hemin inhibits the synthesis of ALA from simpler precursors. Examination of the structural formulas above shows that, even with the best will, no structural similarities can be found between the two compounds

Fig. 5.3 Hyperbolic relationship between the concentration of sucrose and its rate of hydrolysis by sucrase

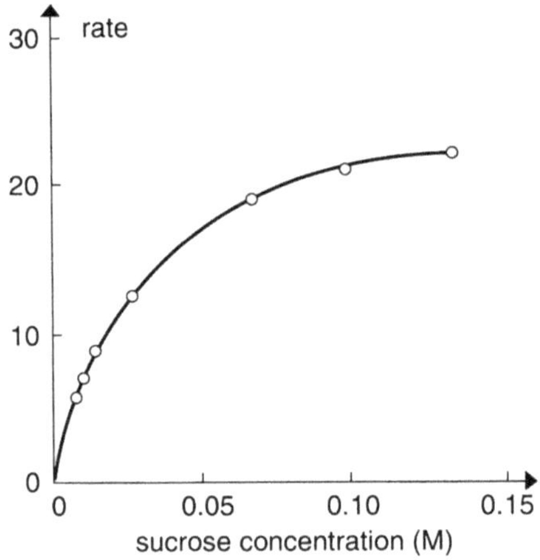

Allosteric Inhibition and Activation

Classical kinetic analysis of the enzyme inhibitions caused by allosteric effectors show that they are reversible, and can be competitive, non-competitive or of the mixed type, with kinetics quite often obeying very complex laws. When inhibition

Fig. 5.4 Sigmoid relationship between threonine concentration and the rate of its deamination by *E.coli* threonine deaminase (From J.-P. Changeux, with permission of the Société Française de Biochimie et de Biologie Moléculaire)

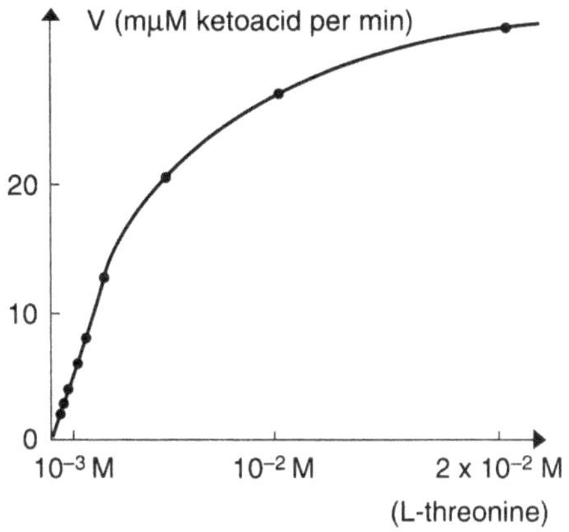

Fig. 5.5 Sigmoid relationship between the concentration of isoleucine (allosteric inhibitor) and the activity of biosynthetic threonine deaminase of *E. coli* (From J.-P. Changeux with permission of the Société Française de Biochimie et de Biologie Moléculaire)

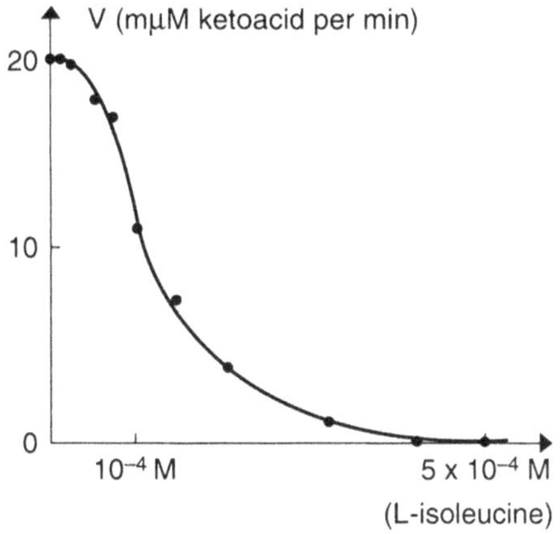

is non-competitive or mixed, the inhibitor binds to sites distinct from the catalytic site. Competitive inhibition is more difficult to explain. In the case of "classical" enzymes, competitive inhibitions are obtained with steric analogs of the substrate, and it is reasonable to assume that the substrate and the inhibitor compete with each other for occupation of the catalytic site. In some of the cases of interest here, the absence of structural similarity between the highly specific allosteric inhibitors and the substrates of the enzymes which they inhibit seems to exclude the possibility of such a competition, since the catalytic sites are usually characterized by a strict degree of specificity. In fact, allosteric effectors have been found to bind to sites separate from the catalytic sites; numerous allosteric enzymes can be made by mutation or by various treatments insensitive (desensitized) to their allosteric effectors without their catalytic activity being affected. In one case at least, it has been possible to physically separate the different subunits of an allosteric enzyme subject to competitive inhibition and to show that one kind of subunit carries the catalytic site, and the other the binding site for the allosteric effector. This proves unequivocally that the two types of sites reside on different parts of the protein and that the apparent competitive inhibition does not result from a direct interaction with the

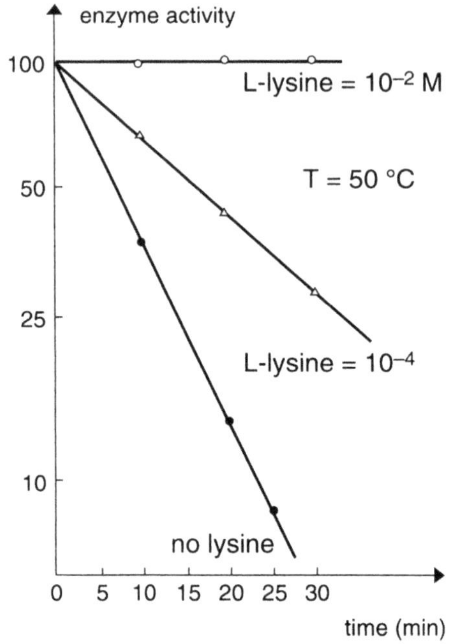

Fig. 5.6 Protection of *E. coli* aspartokinase III against inactivation by heat. The inactivation can be totally prevented by L-lysine, a specificallosteric inhibitor of this aspartokinase (From P. Truffa-Bachi and G. N. Cohen, with permission of Biochimica et Biophysica Acta)

catalytic site. Rather, the inevitable conclusion is that the binding of the inhibitor causes some conformational change in the enzyme which results in a diminished affinity for the substrate at the catalytic site.

Some allosteric effectors cause conformational changes which increase the affinity for the substrate. In the case of the enzymes showing a sigmoidal relationship between their activity and substrate concentration, the addition of these allosteric activators change the sigmoid relationship to a hyperbolic one. In a number of cases, it is found that allosteric activators permit a modulation of the enzymatic activity to the advantage of the cell, as in the case of the threshold effects described above for allosteric inhibitors.

A number of allosteric enzymes are protected against thermal inactivation by their allosteric effectors (Fig. 5.6). Such effects are attributable to an increased stability of the protein due to the conformational changes which accompany the reversible binding of allosteric ligands to their specific sites.

Polymeric Nature of Allosteric Enzymes. The model of Monod, Wyman and Changeux

All carefully studied allosteric enzymes are made up of subunits. Reversible associations and dissociations are found with some of them, which may or may not be accompanied by inactivation or a number of changes in susceptibility to allosteric effectors. Although this absence of a general rule would indicate that association and dissociation are not inherent in the allosteric effects, it is possible that such an absence indicates that the strength of the binding non-covalent interactions between subunits differs from enzyme to enzyme.

On the other hand, the polymeric nature of allosteric enzymes has been a decisive element in shaping the ideas about the mechanism of allosteric regulation. One of the models which has in most cases withstood the test of time has been proposed by Monod, Wyman and Changeux in 1965 and can be described by the following hypotheses:

1. Allosteric proteins are oligomers whose protomers are associated in such a way that they all occupy equivalent positions. This implies that the molecule possesses at least one axis of symmetry.
2. On each protomer, there exists one and only one site capable of forming a stereospecific complex with a given ligand. In other words, the symmetry of each group of stereospecific sites is the same as the symmetry of the molecule.
3. The conformation of each protomer is constrained by its association with the other protomers.
4. There are at least two states (R and T) in equilibrium in which allosteric oligomers can exist. These states differ in the distribution and /or energy of the inter-protomer bonds, and accordingly in the conformational constraints imposed on the protomers.
5. As a result, the affinity of one or several of the stereospecific sites for the corresponding ligand is modified when there is a transition from one state to the other.
6. During these transitions, the molecular symmetry and the symmetry of the conformational constraints are conserved.

Let us analyze the consequences of this model: in the absence of substrates and allosteric effectors, the distribution of molecular species among the different states will be a function of the free energies of the formation of these states, which are in turn determined by the strength of the interactions between protomers of the different states. In the case where a given ligand has a greater affinity for one of the conformational states, the presence of this ligand at low concentration will result in the preferential binding of a single ligand molecule to the protomer in the state for which it has the greater affinity. This will bring about a displacement of the equilibrium in favor of that state, and will thus facilitate the subsequent binding of additional ligand molecules, because of the simultaneous formation of more than one reactive site (the exact number will depend on the number of identical protomers in the oligomer). It is evident that the sigmoid relationship between activity and substrate (or effector) concentration is perfectly explained by this model. Monod, Wyman and Changeux have elaborated a mathematical treatment of the model which accounts for a good many allosteric enzymes.

Let us analyze the interactions of an ideal allosteric enzyme with a single ligand F endowed with differential affinity toward the two accessible states. In the absence of ligand, the two states, symbolized as R_0 and T_0 are assumed to be in equilibrium.

Let L be the equilibrium constant for the $R_0 \Leftrightarrow T_0$ transition. In order to differentiate it from the dissociation constants of the ligand, the authors called it the "allosteric constant". Let K_R and K_T be the microscopic dissociation constants of a ligand F

bound to a stereospecific site, in the R and T states, respectively. By reason of symmetry and because the binding of any one ligand molecule is assumed to be intrinsically independent of the binding of any other, these microscopic dissociation constants are the same for all homologous sites in each of the two states. Assuming n protomers (and therefore n homologous sites) and using the notation R_0, R_1, R_2, … R_n; T_0, T_1, T_2, … T_n to designate the complexes involving 0, 1, 2, … n molecules of ligand, the successive equilibriums can be written as follows:

$$\begin{array}{lcl} R_0 & \rightarrow & T_0 \\ R_0 + F \rightarrow R_1 & & T_0 + F \rightarrow T_1 \\ R_1 + F \rightarrow R_2 & & T_1 + F \rightarrow T_2 \\ \dots\dots\dots & & \dots\dots\dots \\ R_{n-1} + F \rightarrow R_n & & T_{n-1} + F \rightarrow T_n \end{array}$$

Taking into account the probability factors for the dissociations of the R_1, R_2, … R_n and T_1, T_2, … T_n complexes, we may write the following equilibrium equations:

$$T_0 = L\, R_0$$

$$\begin{array}{ll} R_1 = R_0 n \dfrac{F}{K_R} & T_1 = T_0 n \dfrac{F}{K_T} \\ R_2 = R_1 \dfrac{n-1}{2} \dfrac{F}{K_R} & T_2 = T_1 \dfrac{n-1}{2} \dfrac{F}{K_T} \\ \dots\dots\dots & \dots\dots\dots \\ R_n = R_{n-1} \dfrac{1}{n} \dfrac{F}{K_R} & T_n = T_{n-1} \dfrac{1}{n} \dfrac{F}{K_T} \end{array}$$

Let us now define two functions corresponding respectively to:

(a) the fraction of protein in the R state:

$$\bar{R} = \frac{R_0 + R_1 + R_2 + \dots + R_N}{(R_0 + R_1 + R_2 + \dots + R_n) + (T_0 + T_1 + T_2 + \dots + T_n)}$$

(b) the fraction of sites actually bound by the ligand:

$$\bar{Y}_F = \frac{(R_1 + 2R_2 + \dots + nR_n) + (T_1 + 2T_2 + \dots + nT_n)}{n(R_0 + R_1 + R_2 + \dots + R_n) + (T_0 + T_1 + T_2 + \dots + T_n)}$$

Using the equilibrium equations, and setting

$$\frac{F}{K_R} = \alpha \text{ and } \frac{K_R}{K_T} = c$$

we have, for the "function of state" $\bar{R}$:

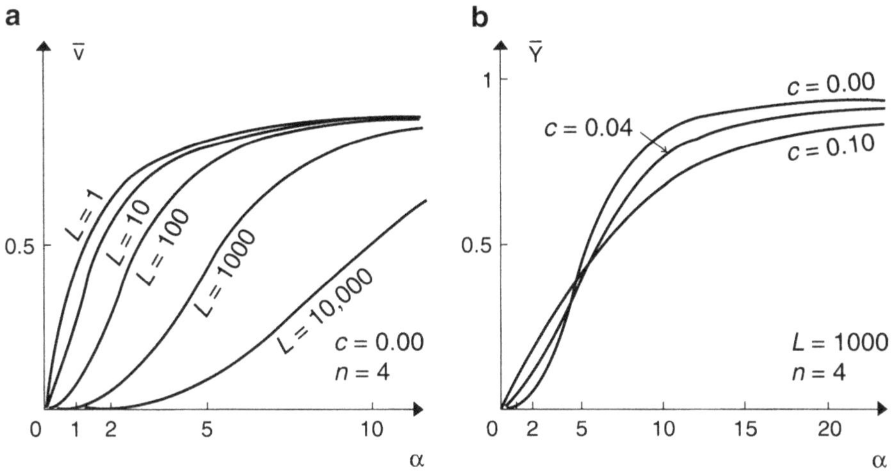

Fig. 5.7 (**a, b**) Theoretical curves of the saturation function Y_F function drawn to various values of the constants L and c, with $n=4$, i.e. for a tetramer (From J. Monod, J. Wyman and J-P. Changeux, with permission of the Journal of Molecular Biology)

$$\bar{R} = \frac{(1+\alpha)^n}{L(1+c\alpha)^n + (1+\alpha)^n}$$

and for the "saturation function" $\bar{Y}_F$:

$$\bar{Y}_F = \frac{Lc\alpha\,(1+c\alpha)^{n-1} + -(1+\alpha)^{n-1}}{L(1+c\alpha)^n + (1+\alpha)^n}.$$

In Fig. 5.7a, b, theoretical curves of the Y_F function have been drawn, corresponding to different values of the constants L and c. In such graphs the cooperative effect of the ligand, predicted by the symmetry properties of the model, is expressed by the curvature of the lower part of the curves. The graphs illustrate the fact that the "cooperativity" of the ligand depends on the values of L and c. The cooperativity is more marked when the allosteric constant is large (i.e. when the $R_0 \Leftrightarrow T_0$ equilibrium is strongly in favor of T_0) and when the ratio of the microscopic dissociation constants ($c=K_R/K_T$) is small.

It should be noted that when $c=1$ (i.e., when the affinity of the ligand towards both states is the same) and also when L is negligibly small, the Y_F function simplifies to:

$$\bar{Y}_F = \frac{\alpha}{1+\alpha} - = \frac{F}{K_R + F}$$

that is, to the Michaelis-Henri equation.

The model therefore accounts for the homotropic cooperative effects which are invariably found in allosteric proteins.

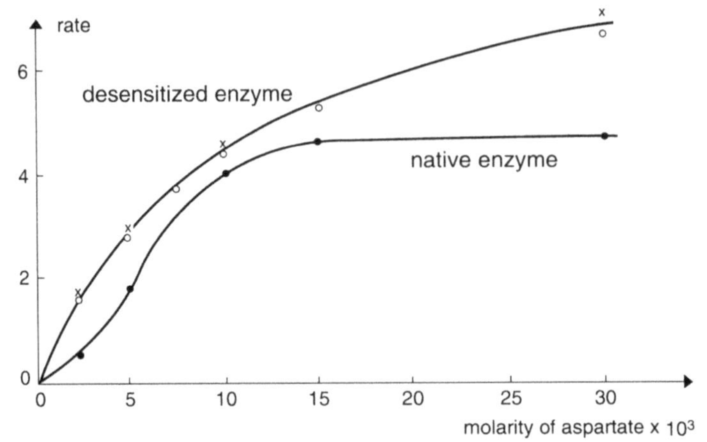

Fig. 5.8 Kinetic differences between *E. coli* aspartate transcarbamylase and the same enzyme desensitized by heat or by a mercurial. Whereas the rate is a sigmoid function of the substrate concentration in the first case, the kinetics are "normalized" in both the cases of desensitization

The properties of the model with respect to heterotropic interactions between different allosteric ligands has also been analyzed by Monod, Wyman and Changeux and results in an equation expressing that the heterotropic action of an allosteric ligand upon the saturation function for another allosteric ligand should be to modify the homotropic interactions of the latter.

The main merit of the model is that it is suitable for experimental analysis almost fifty years after its inception, now that a number of allosteric enzymes have been crystallized and that their tridimensional structure can often be analyzed in the liganded and unliganded state.

However, there is no a priori reason for believing that all enzymes subject to allosteric control have become so by the same mechanism. Each case must be examined individually.

It is useful to describe two classes of allosteric effects:

(a) homotropic effects, or interactions between identical ligands;
(b) heterotropic effects, or interactions between different ligands.

Homotropic effects are present in most allosteric proteins for at least one type of ligand (substrate, inhibitor or activator). While heterotropic effects may be either cooperative or antagonistic, homotropic effects are always cooperative. Experimental conditions which change heterotropic interactions often change homotropic ones (e.g., the desensitization of threonine dehydratase or aspartate transcarbamylase is invariably accompanied by the disappearance of cooperative effects between substrate molecules); in other words, desensitization of these enzymes is connected with a normalization of the kinetics to the Michaelis type (Fig. 5.8).

Two classes of allosteric enzymes can be defined:

1. Those in which the weaker substrate affinity predominates in the absence of substrate. Since the other state (*R* state) has the greater affinity for substrate, the

equilibrium is displaced in its favor when substrate is added. The affinity of the whole population of enzyme molecules for substrate thus increases progressively as more molecules of substrate are bound. In contrast, an allosteric inhibitor displaces the equilibrium in the direction of the state (*T* state) with the weaker affinity for the substrate.

Enzymes in this category are called allosteric enzymes of type K. Both the substrate and the effector are allosteric ligands which affect the equilibrium between the two enzyme states, *R* and *T*. It is this change in the equilibrium which is defined as the allosteric transition. Cooperative homotropic effects are found for both the substrate and effector. The presence of the effector modifies K_m, the apparent affinity of the enzyme for its substrate, and conversely, the substrate concentration modifies the affinity of the enzyme for its effector.

2. A second class where the two states of the allosteric enzyme have the same apparent affinity for the substrate. In this case, there is no effect of substrate or effector on the combination of the other ligand with the enzyme. Consequently, the effector is able to affect the enzymatic reaction only if the two states differ in their catalytic activity. Depending on whether the effector has a greater affinity for the active *R* state or inactive (or less active) *T* state, it will behave as an activator or an inhibitor. From the fact that it influences only V_{max} in the reaction and has no influence on the value of *Km*, enzymes of this second class have been called enzymes of type *V*. Here the substrate is not an allosteric ligand and cooperative interactions are not found between substrate molecules but only between effector molecules.

In practice, pure *K* systems will behave as enzymes inhibited competitively. The Henri-Michaelis curve will appear sigmoid rather than hyperbolic. Another useful representation is derived from an equation given by Hill in 1910. Letting V_{max} be the velocity of the enzyme when saturated with substrate, v the velocity at any non-saturating concentration and *S*, the substrate concentration, a plot of log $v/(V_{max}-v)$ as ordinate and log *S* as abscissa will give a curve with a linear portion of slope *n*. The number *n* (or Hill coefficient) is a measure of the interaction between the different substrate-binding stereospecific sites of the enzyme molecule. If the substrate causes an allosteric transition, *n* is always greater than unity.

In pure *V* systems, variation of enzyme activity with substrate concentration obeys classical Michaelis kinetics: *n* is therefore equal to 1. But if the enzymatic activity is plotted against inhibitor concentration *I*, a sigmoid relationship is obtained. Plotted as log $v(V_0-v)$, where V_0 is the non-inhibited activity and v the activity measured with a certain concentration of inhibitor, against log *I*, a curve is obtained which has a linear portion of slope n' which is a measure of the interaction between different inhibitor-specific sites of the enzyme molecule.

An Alternative Model

Whereas the model presented above proposes that the subunits of a homopolymeric enzyme change shape in a concerted manner to preserve the symmetry of the entire molecule as it is displaces the equilibrium between one conformation *T* and a second conformation *R* under the influence of a ligand, an alternative model postulates that each subunit changes shape by binding the ligand, so that changes in one subunit lead to distorsions in the shape and/or interactions of other subunits of the protein. This alternative model predicts that the first ligand to bind, instead of lowering the threshold for the binding of subsequent ligands, could make it more difficult. The Monod-Wyman-Changeux model allows no such alternative, which explains the numerous cases of negative cooperativity reported in the literature.

Throughout this book, we shall see that the activity of many proteins, including metabolic enzymes, molecular machines, and ion channels, is often regulated by conformational changes that are induced or stabilized by ligand binding. In cases of multimeric proteins, such allosteric regulation has often been described by the concerted Monod–Wyman–Changeux and sequential Koshland–Némethy–Filmer classic models of cooperativity. Despite the important functional implications of the mechanism of cooperativity, it has been impossible in many cases to distinguish between these various allosteric models using ensemble measurements of ligand binding in bulk protein solutions. Structural mass spectrometry (MS) offers a way to break this impasse by providing the full distribution of ligand-bound states of a protein complex. Given this distribution, it is possible to determine all the binding constants of a ligand to a highly multimeric cooperative system, and thereby inferring the underlying its allosteric mechanism. The dissection of allosteric mechanisms relies on advances in MS—which provide the required resolution of ligand-bound states—and in data analysis. This approach has been validated using the well-characterized *Escherichia coli* chaperone GroEL (see Chap. 19), a double-heptameric ring containing 14 ATP binding sites, which has become a paradigm for molecular machines. The values of the 14 binding constants of ATP to GroEL were determined, and the ATP-loading pathway of the chaperone was characterized. The methodology and analyses presented are directly applicable to numerous other cooperative systems and are expected to promote further research on allosteric systems.

Probing Allostery Through DNA

Allostery is well documented for proteins but less recognized for DNA-protein interactions. Specific binding of a protein on DNA can be substantially stabilized or destabilized by the nearby binding of another protein. The free energy of the ternary complex oscillates as a function of the separation between the two proteins with a periodicity of ~10 base pairs, the helical pitch of B-form DNA, and a decay length of ~15 base pairs. The binding affinity of a protein near a DNA hairpin is similarly dependent on their separation, which, together with molecular dynamics simulations, suggests that deformation of the double-helical structure is the origin of DNA allostery. The physiological relevance of this phenomenon is illustrated by its effect on gene expression in live bacteria.

Conclusion

By their indirect nature, allosteric interactions do not depend on the structure or on the particular reactivity of ligands, but solely on the structural properties of the protein which acts as a relay. Growth, metabolism and division require, in addition to the functioning of the main metabolic pathways responsible for the obtention of energy and of chemical building blocks, that the activity of the different metabolic pathways be precisely coordinated by a network of appropriate specific interactions. The creation and the development of such networks would have been impossible if they depended on strict direct interactions on the protein surface; such interactions would have been limited by the chemical structure, the reactivity or the absence of reactivity of the metabolites for which the existence of an interaction would have proven physiologically significant. The invention of indirect allosteric interactions depending on the structure of the protein, that is in last analysis on the genetic code, would free molecular evolution from these limitations.

Selected References

Dyachenko A, Gruber R, Shimon L, Horovitz A, Michal S (2013) Proc Natl Acad Sci USA 110:7235–7239

Kim S, Broströmer E, Xing D, Jin J, Chong S, Ge H, Wang S, Gu C, Yang L, Gao YQ, Su XD, Sun Y, Xie XS (2013) Science 339:816–819

Koshland DE Jr, Hamadani K (2002) J Biol Chem 277:46841–46844

Koshland DE Jr, Nemethy G, Filmer D (1966) Biochemistry 5:365–385

Monod J, Wyman J, Changeux J-P (1965) J Mol Biol 12:88–118

Perutz MF (1989) Quart Rev Biophys 22:1229–1236

Chapter 6
Glycolysis, Gluconeogenesis and Glycogen and Cellulose Synthesis

Glycogen Degradation

The bacterial glycogen phosphorylase is a pyridoxal phosphate enzyme that transform glycogen to glucose-1-phosphate, which is then transformed to glucose-6-phosphate, a normal intermediate of glycolysis.

Glycogen is a large branched polymer of glucose residues mostly linked by α-1,4 glycosidic bonds. However, branches at about every ten residues are created by α-1,6 bonds.

Carl and Gerty Cori showed in 1947 that glycogen is cleaved by orthophosphate to yield a new kind of phosphorylated sugar (the Cori ester) which they identified as glucose-1-phosphate. The Coris also isolated and crystallized glycogen phosphorylase, the enzyme that catalyzes the reaction.

$$\text{Glycogen}_{(\text{n units})} + \text{Pi} \rightarrow \text{glucose-1-phosphate} + \text{glycogen}_{(\text{n}-1\,\text{units})}$$

Phosphorylase catalyzes the sequential removal of glycosyl residues from the non-reducing end of glycogen (the end with a free 4-OH group).

The reaction is readily reversible *in vitro*. However, the phosphorolytic cleavage of glycogen is energetically advantageous; a hydrolytic cleavage would lead to glucose which would have to be phosphorylated to enter glycolysis and could diffuse out of the cell, whereas the ionized glucose-1-phosphate cannot.

Glycolysis

Glycolysis or the Embden-Meyerhof pathway is the main pathway for anaerobic degradation of carbohydrates found in most groups of organisms (Fig. 6.1). For each mole of glucose consumed, 36 kcal of energy is released. The organism obtains a net yield of 2 moles ATP per mole of glucose. The starting material is glycogen

G.N. Cohen, *Microbial Biochemistry*,
DOI 10.1007/978-94-017-8908-0_6, © Springer Science+Business Media B.V. 2014

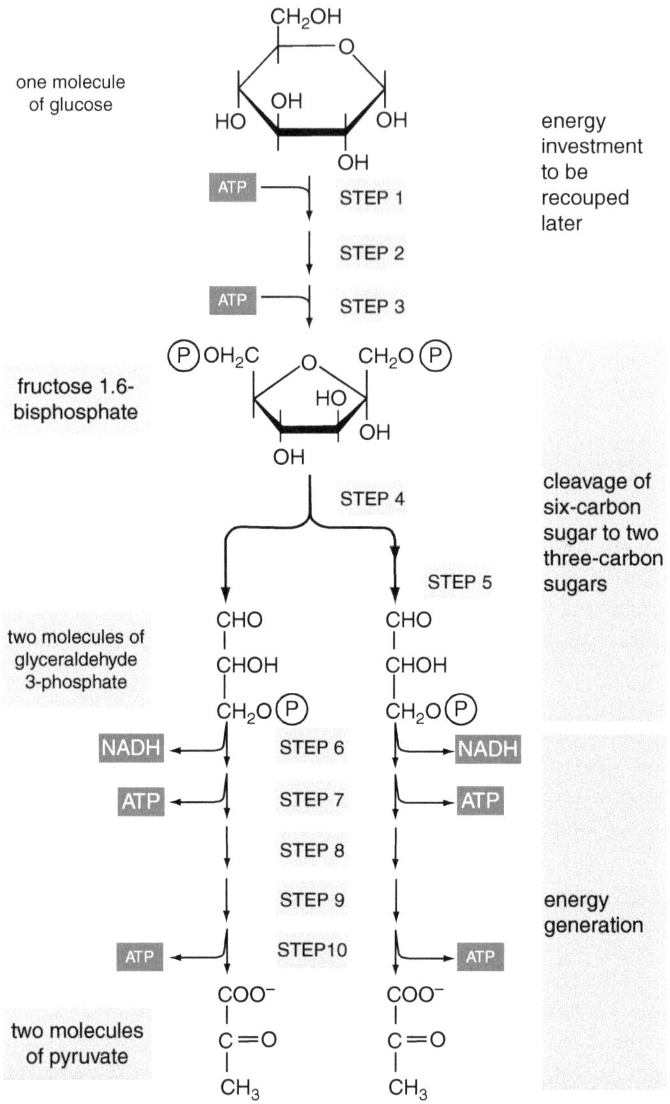

Fig. 6.1 Individual enzymes of glycolysis

(in bacteria and animals) or starch (in plants), which is hydrolyzed to glucose-1-phosphate or glucose monomers. Glycolysis can be divided into four phases:

(a) the formation of two molecules of triose phosphate (glyceraldehyde-3-phosphate and dihydroxyacetone-phosphate) from one molecule of hexose. Two molecules of ATP are consumed in this phase.
(b) dehydrogenation of the triose phosphates to 2-phosphoglycerate. In this process, NAD^+ is reduced to NADH. One molecule of ATP is generated per triose phosphate, which makes up for the ATP invested in the first phase.

(c) conversion of 2-phosphoglycerate to pyruvate via phosphoenol-pyruvate. Another ATP is generated here for each molecule of pyruvate.
(d) the reduction of pyruvate to regenerate NAD^+. Whereas in muscle, the pyruvate is converted to lactate, in yeast it is reductively decarboxylated to ethanol.

Hexokinase

In eukaryotes, glucose enters most cells through specific transport proteins and has one principal fate: it is phosphorylated by ATP to form glucose-6-phosphate. This transfer is catalyzed by hexokinase, which transfers the phosphoryl group of ATP to a variety of six-carbon sugars (hexoses), such as glucose and mannose or D-glucosamine. Hexokinase, like all other kinases, requires Mg^{++} for activity, which is replaceable by Mn^{++}. The divalent metal ion forms a complex with ATP. Hexokinase specifically binds to porin, a protein of the outer mitochondrial membrane which permits ADP and sugars to penetrate this membrane. Thus, although hexokinase is a soluble enzyme, it may therefore be associated with the mitochondrion under physiological conditions. This notion is supported by observations that mitochondrion-associated hexokinase has a higher *Km* for MgATP and is more susceptible to product inhibition by glucose-6-phosphate. X-Ray studies of hexokinase crystals grown in the presence or absence of glucose indicate that when glucose is bound, one lobe of the molecule rotates through 12° to close the substrate binding cleft. In this conformation, water is excluded from the cleft. The conformation change also occurs in an aqueous environment. The conformational change is required for activity; analogs of glucose that are too bulky to allow for this conformational change are not substrates, although they may be inhibitors.

In *E. coli*, glucose-6-phosphate is not formed under the action of a hexokinase, but is generated by the PTS system, as we have seen in the Chapter on cellular permeability.

Glucose-6-Phosphate Isomerase

The next step in glycolysis is the isomerization of glucose-6-phosphate to fructose-6-phosphate. The six-membered pyranose ring is converted to the five-membered ring of fructose-6-P converting an aldose to a ketose.

pgi mutants grow slowly on glucose using the pentose-P pathway (see chap. 7) with oxidative formation of pentose-P from glucose-6-P and with non-oxidative reactions yielding fructose-6-P and glyceraldehyde-3-P. Glucose isomerase is used industrially for fructose production to increase the sweetness of drinks and to increase the plasticity of paper.

The structure and the function of an enzyme from *Actinoplanes missouriensis* catalyzing the isomerization of the non-phosphorylated glucose into fructose has been studied by X-ray cristallography and site-directed mutagenesis after cloning and overexpression in *E. coli.*

Phosphomannose isomerase is used in both sugar catabolism (growth on mannose) and polysaccharide synthesis. Mutants lacking this isomerase are blocked for growth on mannose and are defective in mannose containing polysaccharides.

Phosphofructokinase

A second phosphorylation follows the isomerization step. Fructose-6-P is phosphorylated by ATP to fructose-1,6-bisphosphate. This reaction is catalyzed by phosphofructokinase (PFK), an allosteric enzyme which is the most important control element in the glycolytic pathway.

There are two main types of phosphofructokinases, the allosteric ATP-dependent phosphofructokinases and the pyrophosphate (PPi)-dependent phosphofructokinases, both being the rate-limiting enzymes of glycolysis. Typical ATP-dependent homotetrameric PFKs have been described in *Bacillus stearothermophilus* and *E. coli*, and their tridimensional structure determined.

Pyrophosphate-dependent PFKs are found in plants and in the protists *Giarda lamblia, Toxoplasma gondii, Trichomonas vaginalis* and *Naegleria fowleri*. Their difference with the ATP-dependent PFKs, especially with the human enzyme, makes them an important drug target. Current therapy for infections caused by these pathogens is inadequate, especially for children, pregnant women and immune compromised people. For the purpose of designing drugs to treat parasitic infections, a model of PPi-PFK from *E. histolytica* has been constructed based on the three-dimensional structure of the ATP-PFK from *B. stearothermophilus*. The predicted drug-enzyme interactions suggests that two bisphosphonates would be competitive inhibitors of pyrophosphate. Despite a low-level of overall sequence identity between these two groups of kinases, similarities in active-site residues permit a convincing amino acid alignment. Employing recent protein sequence and site-directed mutagenesis data along with the known three-dimensional coordinates of *E. coli* ATP-dependent PFK, a model of the active site of PPi-dependent PFK was proposed. In addition to provide compatible placement of residues shown to be important by earlier mutagenesis studies, the model predicted an important role for two arginyl residues that are conserved in all known PPi-PFK sequences. An alignment of all PFK of the two types shows a conserved methionine residue that appears from the crystal structure of *E. coli* PFK to be interacting with fructose-6-P. Very conservative substitutions for this methionine with leucine or isoleucine by site-directed mutagenesis of *E. coli* ATP-PFK and *Entamoeba histolytica* PPi-PFK produced profound decreases either in the apparent affinity for fructose-6-P or in maximal velocity, or both. Methionine provides a highly specific interaction with fructose-6-P for binding and for transition state stabilization.

Regulation of Phosphofructokinase in Bacteria

Phosphofructokinase (ATP/PFK) from *B. stearothermophilus* shows cooperative kinetics with respect to the substrate fructose-6-P, allosteric activation by ADP and inhibition by phosphoenolpyruvate. The three ligand binding sites have been located by crystallography. Two of these form the active site and bind the substrate fructose-6-P and ATP. The third binding site binds both allosteric activator and inhibitor. These structural studies show that the substrate fructose-6-P is bound between two subunits of the tetramer and allosteric effectors between another pair of subunits. The substrate cooperativity and allosteric control are mediated and explained by these ligand bridges between subunits. Analysis of the similar crystal structure of the complex of phosphofructokinase from *E. coli* with its reaction products as well as the study of mutations at the interface between subunits have allowed the identification of residues involved in the subunit interactions.

In addition to the most studied phosphofructokinase 1 (coded by *pfkA*), the *Escherichia coli* K12 strain contains an additional ATP phosphofructokinase: whereas phosphofructokinase 1, behaves as an allosteric enzyme, phosphofructokinase 2 (coded by *pfkB*) presents the features of a Michaelian enzyme. In fact, phosphofructokinase 2 also presents some regulatory properties *in vitro*: at high concentrations, ATP is an inhibitor of phosphofructokinase 2 and it provokes the tetramerization of the dimeric native enzyme. The binding of the two substrates to phosphofructokinase 2 is sequential and ordered as for phosphofructokinase 1, but in the former case fructose-6-phosphate is the first substrate to be bound and ADP the first product to be released. Each dimer of phosphofructokinase 2 binds two molecules of fructose-6-phosphate but only one molecule of the product fructose-1,6-bisphosphate. This non allosteric protein is a tetramer of a slightly higher molecular weight than phosphofructokinase 1. It does not show cooperative interactions with fructose-6-P, inhibition by phosphoenolpyruvate or activation by ADP. Unlike ATP-PFK1, it is somewhat sensitive to inhibition by fructose-1,6-bisphosphate and can use tagatose-6-P as a substrate. Although both phosphofructokinases of *E. coli* K12 present some regulatory properties *in vitro*, the mechanism of regulation of the activity of the two enzymes is strikingly different. It can be asked whether or not these mechanisms operate *in vivo*.

The FruR regulator of *E. coli* controls the initiation of transcription of several operons encoding a variety of proteins involved in carbon and energy metabolism. FruR target sites showing a high degree of symmetry were detected in 13 genes or operons: *pfkA* is one of them and the expression of the corresponding enzyme is negatively regulated by FruR.

Fructose 1,6-Bisphosphate Aldolase

It is a tetrameric metal-dependent class II aldolase (coded in *E. coli* by *fbaA*) which reversibly cleaves fructose-1,6-bisphosphate into the two triose phosphates, dihydroxyacetone phosphate and D-glyceraldehyde phosphate. The equilibrium concentrations lie at 89 % fructose bisphosphate and 11 % triose phosphate. However, it is displaced by the next reactions of glycolysis. The enzyme catalyzes the condensation of a number of other aldehydes with dihydroxyacetone phosphate and can also cleave fructose-1-P. The yeast enzyme is inhibited by cysteine. Fe^{++}, Zn^{++} and Co^{++} ions lead to reactivation.

Triose Phosphate Isomerase

As for the preceding enzyme, this one is not subject to metabolic regulation. Triosephosphate isomerase (*tpi*) is an ubiquitously expressed enzyme that catalyses the interconversion of dihydroxyacetone phosphate (DHAP) and glyceraldehyde-3-phosphate in the energy-generating glycolytic pathway, through an enediol intermediate. Inherited defects in the gene are characterized biochemically by markedly reduced enzyme activity in all tissues resulting in a metabolic block in glycolysis, resulting in an accumulation of DHAP, particularly in red cells. Clinical TPI deficiency is a rare autosomal recessive multisystem disorder characterized by non-spherocytic hemolytic anemia, recurrent infections, cardiomyopathy, severe and fatal neuromuscular dysfunctions. After more than 30 years since TPI deficiency was first described, the disease still remains without effective therapy.

Glyceraldehyde-3-Phosphate Dehydrogenase (GAPDH)

The preceding steps have transformed one molecule of glucose into two molecules of glyceraldehyde-3-phosphate. No energy has yet been extracted. On the contrary two molecules of ATP have been invested. A series of steps that harvest the energy contained in glyceraldehyde-3-phosphate will now be studied. The initial reaction in this conversion is the conversion of this compound into 1,3-bisphosphoglycerate (1,3-BPG) catalyzed by GAPDH (*gapA*). A high potential phosphorylated compound is generated in this oxidation-reduction reaction. The aldehyde group at C-1 is converted into an acylphosphate. The mechanism of this reaction has been extensively studied.

Phosphoglycerate Kinase

This enzyme (*pgk*) catalyzes the transfer of the phosphoryl group from the acylphosphate of 1,3-BPG to ADP and generates ATP and 3-phosphoglycerate.

Phosphoglyceromutase

Mutants for this enzyme (*gpm*) have not been reported yet.

Phosphoglycerate mutase exists in two unrelated forms. Vertebrates have only a 2,3-bisphosphoglycerate-dependent enzyme (dPGM), whilst higher plants have only the cofactor-independent enzyme (iPGM). Certain eubacteria possess genes encoding both enzymes, and their respective metabolic roles and activities are unclear. The two PGMs coexist in *E. coli*. Both are expressed at high levels, but dPGM has a 10-fold higher specific activity than iPGM. Differential inhibition by vanadate was observed. The presence of an integral manganese ion in iPGM was confirmed by EPR spectroscopy. The active conformation of the dimeric cofactor-dependent phosphoglycerate mutase (dPGM) from *E. coli* has been elucidated by crystallographic methods to a resolution of 1.25 Å. The active site residue His (10), central in the catalytic mechanism of dPGM, is present as a phosphohistidine.

Sequence analysis of archaeal genomes did not find PGMs of either kind, but identified a new family of proteins, distantly related to iPGMs. These predicted archaeal PGMs from *Pyrococcus furiosus* and *Methanococcus jannaschii* have been functionally produced in *E. coli*, and characterization of the purified proteins has confirmed that they are iPGMs. Analysis of the available microbial genomes indicates that this new type of iPGM is widely distributed among archaea and also encoded in several bacteria. In addition, as has been demonstrated in certain bacteria, some archaea appear to possess an alternative, cofactor-dependent PGM.

2,3-bisphosphoglycerate (2,3-BPG) is a controller of oxygen transport in erythrocytes. It binds to deoxyhemoglobin and lowers its oxygen affinity by a factor of 26, which is essential in enabling hemoglobin to unload oxygen in tissue capillaries. Red blood cells have a high concentration of 2,3-BPG, typically 4 mM, which is in the same order of magnitude as the hemoglobin concentration, in contrast with most other cells that have only trace amounts. The synthesis and degradation of 2,3-BPG are a short detour from the main glycolytic pathway. 2,3-BPG must have been present long before it was recruited by red cells to control the oxygen affinity of hemoglobin!

Enolase

An enol is formed by the dehydration of 2-phosphoglycerate. This dehydration reaction, catalyzed by enolase (*eno*), elevates the group transfer potential of the phosphoryl group. Enolase is a phosphoprotein in enterobacteria.

Pyruvate Kinase

This enzyme catalyzes the third irreversible step in glycolysis and controls the outflow from this pathway (the two other controlling enzymes are hexokinase and phosphofructokinase). This final step yields pyruvate and ATP. Pyruvate is a central metabolic intermediate that can be oxidized further or used as a building block. In enterobacteria, there are two pyruvate kinases. They differ kinetically, Pyk-1 (*pykF*) being activated by fructose-1,6-bisphosphate and Pyk 2 (*pykA*) being activated by AMP. They also differ structurally and in their expression. Double mutants are impaired on sugars other than substrates of the phosphotransferase system.

L-Glucose Catabolic Pathway

L-Glucose, the enantiomer of D-glucose, was believed not to be utilized by any organisms. However, an L-glucose-utilizing bacterium, *Paracoccus sp.* 43P, was recently isolated from soil by enrichment cultivation in a minimal medium containing L-glucose as the sole carbon source. In cell-free extracts from this bacterium, NAD^+-dependent L-glucose dehydrogenase was detected as having sole activity toward L-glucose. This enzyme, LgdA, was purified, and the *lgdA* gene was found to be located in a cluster of putative inositol catabolic genes. L-Gluconate dehydrogenase activity was also detected in cell-free extracts and represents the reaction product of LgdA activity toward L-glucose. Enzyme purification and gene cloning revealed that the corresponding gene resides in a nine-gene cluster, the *lgn* cluster, which may participate in aldonate incorporation and assimilation. Kinetic and reaction product analysis of each gene product in the cluster indicated that they sequentially metabolize L-gluconate to glycolytic intermediates, D-glyceraldehyde-3-phosphate, and pyruvate through reactions of C-5 epimerization.

Gluconeogenesis

It has been known for a long time that the resynthesis of glucose from pyruvate and lactate is an important physiological process. However, progress has been hindered for a long period because of the well-entrenched belief that all the reactions of glycolysis were reversible and that gluconeogenesis was achieved by the same enzymes. This concept did not take into account that three reactions of the glycolytic pathway are highly exothermic and constitute important energy barriers to the reversal of glycolysis under physiological conditions. These three reactions are respectively catalyzed by hexokinase, phosphofructokinase and pyruvate kinase.

We know by now that gluconeogenesis makes use of all the reversible reactions of glycolysis except for four systems which substitute for the three impossible reactions which are strongly endergonic in the direction of glucose synthesis. These four reactions are the pyruvate carboxylase, the synthesis of phosphoenolpyruvate by a reaction different from the one catalyzed by pyruvate kinase, the formation of fructose-6-phosphate by fructose-1,6-bisphosphate and the hydrolysis of glucose-6-phosphate by glucose-6-phosphatase (Fig. 6.2).

Fructose Bisphosphatase in Microorganisms

Mutants deficient in fructose bisphosphatase have been obtained in *E. coli* (*fbp*). They cannot grow on acetate, glycerol and succinate, demonstrating the importance of bisphosphatase for neoglucogenesis; these mutants have an absolute requirement

Fig. 6.2 The reactions specific of gluconeogenesis are indicated by *reverse arrows*

Glucose ← Glucose 6-phosphate ⇔ Fructose 6-phosphate ← Fructose 1,6-bisphosphate ⇔ Glyceraldehyde 3-phosphate ⇔ 3-Phosphoglycerate ⇔ 2-Phosphoglycerate ⇔ Phosphoenolpyruvate ← Oxaloacetate ← Pyruvate

for hexoses, for they have no other means of synthesizing ribose-5-phosphate, for example, which is indispensable for the synthesis of nucleic acids, histidine and tryptophan; or erythrose-4-phosphate required in the biosynthesis of the aromatic nucleus. Wild type organisms possess higher levels of bisphosphatase when they are grown on glycerol than on glucose, for they must synthesize glucose.

Glycogen Synthesis

Glycogen Synthase

The reaction catalyzed by glycogen phosphorylase is reversible and glycogen could be synthesized by it from glucose-1-phosphate in the presence of a branching enzyme. However, the thermodynamics of the phosphorylase reaction do not permit synthesis at physiological substrate concentrations and proceeds *in vivo* in the direction of glycogen degradation.

It was shown in 1957 by Leloir and his colleagues in Buenos Aires that glycogen is synthesized by a totally different mechanism, also used in the synthesis of starch and cellulose.

A glucosyl residue is transferred not from glucose-1-P but from a nucleoside diphosphate glucose to a primer whose length may vary, the primer being itself made up of a linear chain of glucosyl residues. The equation below shows the type of reaction common to the synthesis of the three polysaccharides.

$$\text{XDPG} + (\text{glucose})_n \rightarrow \text{XDP} + (\text{glucose})_{n+1}$$

In the case of glycogen synthesis in mammals, the glucose donor is uridine diphosphoglucose (UDPG). For starch and bacterial glycogen, the donor is adenosine diphosphoglucose (ADPG).

The acceptor, $(\text{glucose})_n$, can be an oligosaccharide but the best acceptor is glycogen itself.

$$(\text{Glycogen})_n + \text{UDP glucose} \rightarrow (\text{Glycogen})_{n+1} + \text{UDP}$$

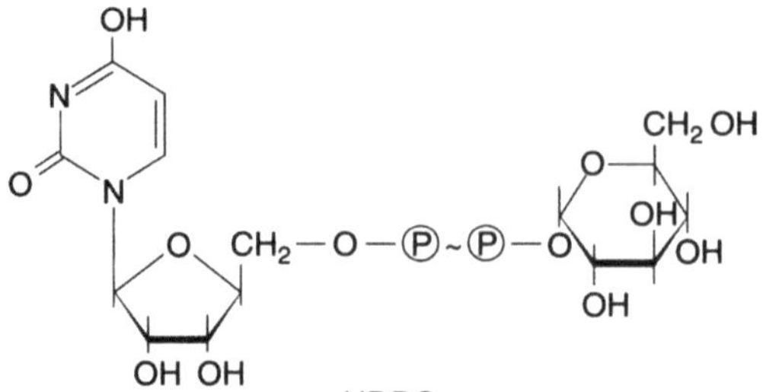

UDPG

UDPG is synthesized from glucose-1-phosphate and uridine triphosphate by a specific enzyme, UDPG pyrophosphorylase.

$$\text{UTP} + \text{glucose-1-phosphate} \rightarrow \text{UDPG} + \text{Ppi}$$

Pyrophosphate, coming from the two outer phosphates of UTP, is rapidly hydrolyzed by a pyrophosphatase, which makes the synthesis of UDPG practically irreversible and drives the reaction in the direction of UDPG synthesis.

Glycogen synthase catalyzes only the synthesis of a linear polysaccharide made of α-1,4 linkages. A branching enzyme breaks an α-1,4 link and forms an α-1,6 link about every seven residues on the average.

In the last analysis, it is glucose1-phosphate which is the origin of glycogen; now in gluconeogenesis the last intermediate before glucose is glucose-6-phosphate: we may remember that glucose-1-phosphate and glucose-6-phosphate are readily interconvertible by phosphoglucomutase.

Control of Glycogen Biosynthesis

The synthesis of glycogen in prokaryotes is regulated at the level of ADP-glucose pyrophosphorylase, an allosteric enzyme that is inhibited by AMP, ADP and inorganic phosphate, and activated by intermediates in carbohydrate dissimilation such as pyruvate, fructose-6-phosphate and fructose-1,6-bisphosphate. With respect alike to its mechanism and to its site of regulation, glycogen synthesis in prokaryotes resembles starch synthesis in algae and higher plants. It differs from glycogen synthesis in yeast and mammals, in which the substrate for the synthase is UDP-glucose, and allosteric regulation occurs at the level of glycogen synthase, not ADP-glucose pyrophosphorylase.

Branching Enzyme

Glycogen synthase catalyzes only the synthesis of α-1,4 linkages. Another enzyme forms the α-1,6-linkages that make glycogen a branched polymer. Branching is important, because it increases the solubility of glycogen. Furthermore, branching creates a large number of terminal residues, the sites of action of glycogen phosphorylase and synthase. Thus, branching increases the rate of glycogen synthesis and degradation.

Branching occurs after a number of glucosyl residues are joined in α–1,4 linkage by glycogen synthase. A branch is created by the breaking of α–1,4 link and the formation of an α-1,6 link: this reaction is different from debranching. A block of residues, typically seven in number, is transferred to a more interior site.

The branching enzyme is quite exacting. The block of seven or so residues must include the non-reducing terminus and come from a chain at least eleven residues long. In addition, the new branch point must be at least four residues away from a preexisting one.

Cellulose Synthesis

Cellulose is a linear extracellular polysaccharide. This unbranched polymer residues of glucose residues joined by β-1,4 likages is synthesized by membrane-embedded glycosyltransferases that processively polymerize UDP-activated glucose. Polymer synthesis is coupled to membrane translocation through a channel formed by the cellulose synthase. Although eukaryotic cellulose synthases function in macromolecular complexes containing several different enzyme isoforms, prokaryotic synthases associate with additional subunits to bridge the periplasm and the outer membrane. In bacteria, cellulose synthesis and translocation is catalyzed by the inner membrane-associated bacterial cellulose synthase BcsA and BcsB subunits. Robust *in vitro* cellulose synthesis has been reconstituted from purified BcsA and BcsB proteins from *Rhodobacter sphaeroides*. Although BcsA is the catalytically active subunit, the membrane-anchored BcsB subunit is essential for catalysis. The purified BcsA-BcsB complex produces cellulose chains of a degree of polymerization in the range 200–300. Catalytic activity critically depends on the presence of the allosteric activator cyclic-di-GMP, but is independent of lipid-linked reactants. These data reveal feedback inhibition of cellulose synthase by UDP but not by the accumulating cellulose polymer and highlight the strict substrate specificity of cellulose synthase for UDP-glucose. A truncation analysis of BcsB localizes the region required for activity of BcsA within its C-terminal membrane-associated domain.

Selected References

Boyer PD (ed.) (1972) The enzymes, 3rd edition, vol. 5 through 10, 17 and 18. Academic Press (New York)

Omadjela O, Narahari A, Strumillo J, Melida H, Mazur O, Bulone V, Zimmer J (2013) Proc Natl Acad Sci USA 110:17856–17861

Preiss J, Romeo T (1989) Adv Microbiol Physiol 30:183–238

Chapter 7
The Pentose Phosphate and Entner-Doudoroff Pathways

The Pentose Phosphate Pathway

The pentose phosphate pathway of carbohydrate metabolism is an oxidative pathway, in contrast to the Embden-Meyerhof pathway of glycolysis. In this pathway, glucose-6-phosphate, derived from glucose by phosphorylation can be totally degraded to carbon dioxide, accompanied by the reduction of $NADP^+$ to NADPH.

The importance of the pathway lies in the production not only of reduced NADPH which is required for many biosyntheses but also of pentoses required for the synthesis of nucleic acid components. By the action of a transhydrogenase system, or NADPH cytochrome c reductase, the NADPH can be reoxidized to produce energy (36 molecules of ATP per molecule of glucose), but operation of the pathway for the sole purpose of energy production is unusual.

The pathway consists (1) of an oxidative phase resulting in the production of NADPH, CO_2 and D-ribulose-5-phosphate and (2) of a nonoxidative phase, consisting of a fairly complex series of sugar interconversions.

The Enzymes of the Oxidative Phase

Glucose-6-Phosphate Dehydrogenase

Glucose-6-phosphate dehydrogenase (G6PDH, *zwf*) (older name: Zwischenferment) was discovered in 1931 by Otto Warburg. In 1934, Warburg and Christian discovered $NADP^+$ as a result of their investigation of the oxidation of glucose-6-phosphate by erythrocytes. The product of G6PDH is 6-phosphogluconolactone. The enzyme occurs widely in plants and animals and has been shown to be formed from inactive subunits. It is a homotetrameric enzyme with a molecular weight ranging from 206,000 in *Neurospora crassa* to 240,000 in erythrocytes. Its dimers are held

G.N. Cohen, *Microbial Biochemistry*,

DOI 10.1007/978-94-017-8908-0_7,

together by $NADP^+$. Recombinant human glucose-6-phosphate dehydrogenase has been crystallized and its structure solved by molecular replacement.

6-Phosphogluconolactonase

The product of G6PDH is 6-phosphoglucono-δ-lactone, which is an intramolecular ester between the C-1 carboxyl group and the C-5 hydroxyl group. The next reaction is catalyzed by 6-phosphogluconolactonase (*pgl*) and 6-phosphogluconate is obtained.

6-Phosphogluconate Dehydrogenase (Decarboxylating)

This six-carbon sugar is then oxidatively decarboxylated by 6-phosphogluconate dehydrogenase (*gnd*) to yield ribulose-5-phosphate.

In *E. coli*, the *gnd* gene has been the subject of detailed studies, including specialized transducing phages, cell-free enzyme synthesis and comparisons of sequence and expression.

The enzyme from *Schizosaccharomyces pombe* differs from the enzyme from other sources in that it is tetrameric instead of dimeric.

Crystals of a mammalian and of a protozoan enzyme have been obtained and structures determined. An X-ray diffraction study of the enzyme from *Lactococcus lactis* has been published.

Ribose Phosphate Isomerase

The final step in the synthesis of ribose-5-phosphate is the isomerization of ribulose-5-phosphate by ribose phosphate isomerase (*rpi*). This reaction, leading to the sugar present in nucleotides, is similar to the glucose-6-phosphate → fructose-6-phosphate and to the dihydroxyacetone phosphate → glyceraldehyde-3-phosphate reactions of glycolysis. All the three ketose-aldose isomerizations proceed through an enediol intermediate.

The Enzymes of the Non Oxidative Phase

The preceding reactions yield two NADPH and one ribose-5-P for each glucose-6-P oxidized. Many cells need however NADPH for reductive biosyntheses much more than ribose-5-P for incorporation into nucleotides. In these cases, ribose-5-P is converted into glyceraldehyde-3-P and fructose-6-P by transketolase and transaldolase.

These enzymes create a reversible link between the pentose pathway and glycolysis by catalyzing the following three reactions:

$$C5+C5 \xrightarrow{\text{transketolase}} C3+C7$$

$$C7+C3 \xrightarrow{\text{transaldolase}} C4+C6$$

$$C5+C4 \xrightarrow{\text{transketolase}} C3+C6$$

The net result of these reactions is the formation of two hexoses and one triose from three pentoses.

Transketolase transfers a two-carbon unit, whereas transaldolase transfers a three-carbon unit. The sugar which donates the 2 or 3 carbon unit is always a ketose, whereas the acceptor is always an aldose.

The first of these three reactions is the formation of glyceraldehyde-3-phosphate and sedoheptulose-7-phosphate from two pentoses.

The donor of the two-carbon unit is xylulose-5-phosphate, an epimer of ribulose-5-phosphate formed from the latter by a phosphopentose isomerase (*rpe*).

Glyceraldehyde-3-P and sedoheptulose-7-P then react to form fructose-6-P and erythrose-4-P.

In the third reaction, transketolase catalyzes the synthesis of fructose-6-P and glyceraldehyde-3-P from erythrose-4-P and xylulose 5-P.

The sum of these reactions is

$$\text{2-xylulose-5-P} + \text{ribose-5-P} \rightarrow \text{2-fructose-6-P} + \text{Glyceraldehyde-3-P}$$

Since xylulose 5-phosphate can be formed from ribose 5-phosphate, the net reaction from ribose 5-phosphate is

$$\text{3-ribose-5-P} \rightarrow \text{2-fructose-6-P} + \text{glyceraldehyde-3-P}$$

Thus, ribose-5-P formed in excess can be completely converted into glycolytic intermediates. It is evident that the carbon skeleton of sugars can be extensively rearranged according to the physiological needs.

Transketolase

This enzyme (*tkt*) requires thiamine diphosphate (TPP) and a divalent metal ion for its activity. TPP is tightly bound to the enzyme. The mechanism of catalysis involves the transfer of an activated aldehyde unit to the thiazole ring of TPP. More than 30 full-length sequences are known. The three dimensional structure of the enzyme has been determined at 2.5 Å resolution.

Transaldolase

In contrast to transketolase, transaldolase (*tal*) does not contain a prosthetic group. Rather, a Schiff base is formed between the carboxyl group of the ketose substrate and the ∈-amino group of a lysine residue at the active site of the enzyme. The Schiff base becomes protonated, the bond between C-3 and C-4 is split and an aldose is released.

Ribulose-5-Phosphate-3-Epimerase

This enzyme catalyzes the interconversion of ribulose-5-P and xylulose-5-P in the Calvin cycle and in the nonoxidative pentose pathway. The enzyme from potato chloroplasts, a homohexamer was expressed in *E. coli*, isolated and crystallized. Its tridimensional structure was determined at 2.3 Å resolution. The fold of the subunit is an (aβ) 8 barrel. The structure has allowed to produce a model for the mechanism of the enzyme. An *rpe* mutant of *E. coli*, lacking the enzyme, is unable to utilize single pentose sugars, indicating that the gene supplies the only ribulose-5-P epimerase activity. Figure 7.1 summarizes the pentose phosphate pathway.

Regulation of the Pentose Phosphate Pathway

The rate of the pentose phosphate pathway is controlled by the level of $NADP^+$. Glucose-6-phosphate dehydrogenase catalyzes an essentially irreversible reaction which is in fact rate-limiting under physiological conditions and serves as the control site. The most important regulatory factor is the intracellular concentration of $NADP^+$. NADPH competes with $NADP^+$ in binding to the enzyme.

The flow of glucose-6-phosphate depends on the need for NADPH, ribose-5-phosphate and ATP. For example, rapidly dividing cells need ribose-5-P for the synthesis of nucleic acids. In contrast, large amounts of NADPH are utilized by adipose tissue in the reductive synthesis of fatty acids from acetyl CoA, as in the mammary gland during lactation.

The Entner-Doudoroff Pathway

In many oxidative bacteria (aerobic Pseudomonads, *Caulobacter* and *Azotobacter* groups, *Agrobacterium, Rhizobium* and *Spirillum*), and facultative anaerobes able to ferment sugars (*Zymomonas*), sugar metabolism occurs primarily via

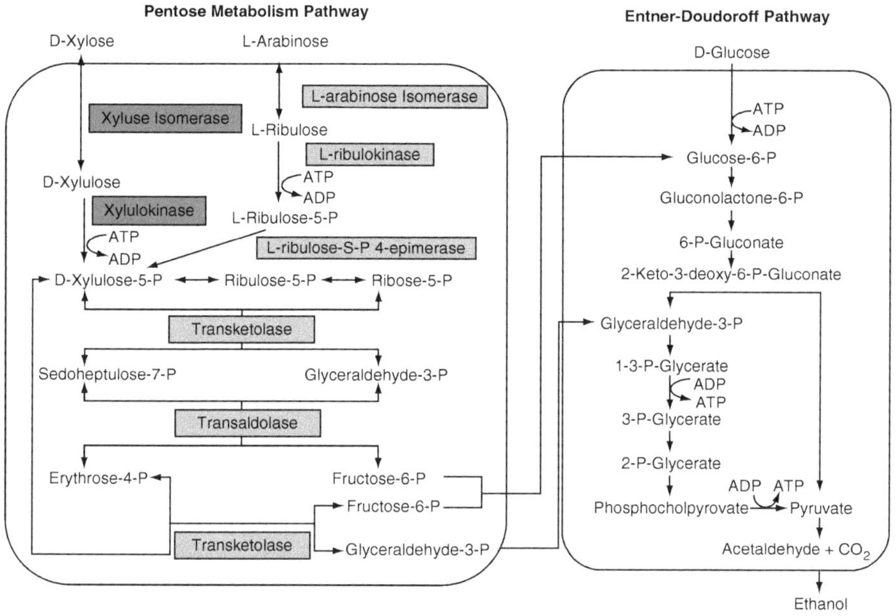

Fig. 7.1 The pentose metabolism and Entner-Doudoroff pathways

6-phosphogluconate and its cleavage by the two Entner-Doudoroff enzymes; 6-phosphogluconate dehydrase (*edd* in *E. coli*) and 2-keto-3-deoxy-6-phosphogluconate aldolase (*eda* in *E. coli*) giving pyruvate and glyceraldehyde-3-P, which is metabolized by enzymes of the Embden-Meyerhof pathway to yield a second molecule of pyruvate. The net yield of the metabolism of one molecule of glucose through the Entner-Doudoroff pathway is one molecule of ATP and two molecules of NADPH.

The pathway comes in various forms. Enteric bacteria use it for the inducible metabolism of gluconate (glucose oxidized in the C-1 carbon atom), which is converted to 6-phosphogluconate by gluconokinase. The aldolase is also involved in catabolism of glucuronate and galacturonate and possibly glyoxylate. The *eda* gene is immediately downstream in an operon from *edd*, with the major gluconate-induced transcript probably governed by *gntR*, which is also involved in the regulation of expression of gluconate transport and of gluconokinase.

Although under the usual laboratory conditions the metabolism of glucose itself in *E. coli* does not occur via the Entner-Doudoroff pathway, a glucose dehydrogenase (*gcd*) apoenzyme which requires the co-factor pyrroloquinoline quinone (PQQ) is widely present. PQQ addition permits metabolism of glucose via gluconate in mutants otherwise blocked and formation of gluconate in wild type. Oxidation in the periplasm is coupled with to ubiquinone reduction and is energy conserving.

Selected References

The Pentose Phosphate Pathway

Horecker BL (1972) In: Kornberg A, Cornudella L, Horecker BL, Oro J (eds) Reflections on biochemistry. Pergamon Press, New York, pp 65–72

The Entner-Doudoroff Pathway

Entner N, Doudoroff M (1952) J Biol Chem 196:853–862

Chapter 8
The Tricarboxylic Acid Cycle and the Glyoxylate Bypass

The oxidative decarboxylation of pyruvate to form acetyl CoA, which occurs in the mitochondrial matrix in eukaryotes, is the link between glycolysis and the tricarboxylic (or citric) acid cycle.

This irreversible funneling of the product of glycolysis into the tricarboxylic acid cycle is catalyzed by the pyruvate dehydrogenase complex.

The Origin of Acetyl CoA: The Pyruvate Dehydrogenase Complex

The pyruvate dehydrogenase complex, an organized assembly of three kinds of enzymes, catalyzes the oxidative decarboxylation of pyruvate in the formation of acetyl CoA.

The net reaction catalyzed is

$$\text{Pyruvate} + \text{CoA} + \text{NAD}^{+} \rightarrow \text{acetyl CoA} + \text{CO}_2 + \text{NADH}$$

The mechanism of this reaction is more complex than might be suggested by its stoechiometry. Thiamine pyrophosphate (TPP), lipoamide and FAD serve as catalytic cofactors in addition to CoA and NAD^+, the stoechiometric cofactors.

There are four steps in the conversion of pyruvate into acetyl CoA.

First, pyruvate is decarboxylated after it combines with TPP. This reaction is catalyzed by E1 of the multienzyme complex.

Second, the hydroxyethyl group attached to TPP is oxidized to form an acetyl group and concomitantly transferred to lipoamide. The oxidant in this reaction is the disulfide bridge of lipoamide, which is converted into the sulfhydryl group. This reaction, also catalyzed by the pyruvate dehydrogenase component E1, yields acetyllipoamide and TPP.

G.N. Cohen, *Microbial Biochemistry*,
DOI 10.1007/978-94-017-8908-0_8,

The 2.4 Å crystal structure of the homodimeric E1 from *Staphylococcus aureus* has been determined. The structures of the *S. aureus* and *E. coli* E1 domains are very similar. The orientation of the subdomains relative to each other, however, is different.

$$\text{Pyruvate} + \text{TPP} \longrightarrow \text{hydroxyethyl-TPP} + \text{CO2}$$

$$\downarrow \text{lipoamide}$$

$$\text{TPP} + \text{HS} - (\text{CH}_2)_2 - \underset{\substack{| \\ \text{S} - \text{CO} - \text{CH}_3}}{\text{CH}} - (\text{CH}_2)_4 - \text{COOH}$$

6-acetyllipoamide

Third, the acetyl group is transferred from acetyllipoamide to CoA to form acetyl CoA. Dihydrolipoyl transacetylase (E2) catalyzes this reaction, where the energy-rich thioester bond is preserved as the acetyl group is transferred to CoA.

$$6\text{-S-acetyllipoamide} + \text{CoA} \rightarrow \text{acetyl CoA} + \text{dihydrolipoamide}$$

Deductions from nuclear magnetic resonance studies provide a clear indication that the N-terminal region of the E1 interacts with the E2 peripheral subunit binding domain and that this interaction precedes reductive acetylation. These results provide the first structural support to the notion that the N-terminal region of the E1 component of bacterial pyruvate dehydrogenase complexes is responsible for binding the E2 component.

Fourth, the oxidized form of lipoamide is regenerated by dihydrolipoyl dehydrogenase (E3). Two electrons are transferred to an FAD prosthetic group of the enzyme and then to NAD^+. The unusual electron transfer potential of FAD bound to this protein enables it to transfer electrons to NAD^+.

$$\text{Dihydrolipoamide} + \text{NAD}^+ \rightarrow \text{oxidized lipoamide} + \text{NADH} + \text{H}^+$$

The activated intermediates are tightly bound to the complex.

In all these reactions, lipoic acid exists in the form of lipoamide, its carboxyl group being linked through an amide group to an $\in$ -lysine group of the E2 protein. This attachment provides a highly mobile flexible arm of 14 Å enabling the lipoyl moiety of a transacetylase subunit to interact with the thiamine pyrophosphate unit of an adjacent pyruvate dehydrogenase E1 subunit and with the flavin unit of an adjacent dihydrolipoyl dehydrogenase E3. Furthermore, the acetyl group can be shuttled from one lipoyl unit to another.

Kinetic, spectroscopic, and structural analysis tested the hypothesis that a chain of residues connecting the 4′-aminopyrimidine N1′ atoms of thiamin diphosphates (ThDPs) in the two active centers of the *Escherichia coli* pyruvate dehydrogenase

Table 8.1 Composition of the pyruvate dehydrogenase complex

Component	Chain numbers
E1	24
E2	24
E3	12

complex E1 component provides a signal transduction pathway. Substitution of the three acidic residues (Glu^{571}, Glu^{235}, and Glu^{237}) and Arg^{606} resulted in impaired binding of the second ThDP, once the first active center was filled, suggesting a pathway for communication between the two ThDPs.

The enzyme complex from *E. coli* has been studied extensively. It has a mass of about 5×10^6 Da and contains 60 polypeptide chains (see Table 8.1). A polyhedral structure of about 30 nm in diameter is evident in electron micrographs. The transacetylase polypeptide chains (E2) form the core of the complex, whereas the pyruvate dehydrogenase and lipoyl dehydrogenase units are bound to the outside of the transacetylase core.

Overview of the Tricarboxylic Acid (TCA) Cycle

The tricarboxylic acid (TCA) cycle is also called the citric acid cycle or Krebs cycle (from the name of the biochemist who established it). It is the final common catabolic pathway for the oxidation of fuel molecules. Two carbons enter the citric acid cycle as acetyl CoA and two carbons leave as CO_2. In the course of the cycle, four oxidation-reduction reactions take place to yield reduction potential in the form of three molecules of NADH and one molecule of FADH2. A high energy phosphate bond GTP is also formed.

A four carbon compound (oxaloacetate) condenses with a two-carbon acetyl unit to yield a six-carbon tricarboxylic acid (citric acid). An isomer of citrate is then oxidatively decarboxylated. The resulting five-carbon compound (α-ketoglutarate) is oxidatively decarboxylated to yield a four-carbon compound (succinate). Oxaloacetate is then regenerated from succinate.

Two carbon atoms enter the cycle as an acetyl unit and two carbon atoms leave the cycle in the form of two molecules of CO_2. An acetyl group is more reduced than CO_2, and so oxidation-reduction reactions must take place in the cycle. In fact, there are four such reactions: three hydride ions (hence six electrons) are transferred to three NAD^+ molecules, whereas one pair of hydrogen atoms (two electrons) is transferred to a flavin adenine dinucleotide (FAD) molecule. These electron carriers yield nine molecules of ATP when they are oxidized by oxygen in the electron transport chain. In addition, one high-energy phosphate bond is formed in each round of the citric acid cycle itself.

Origin of the Oxaloacetate

In bacteria, phosphoenolpyruvate carboxykinase is responsible for the fixation of CO_2.

$$\underset{\displaystyle OPO_3H_2}{HOOC\!-\!\overset{|}{C}}\!=\!CH_2 + CO_2 + ADP \rightarrow HOOC\!-\!CO\!-\!CH_2\!-\!COOH + ATP$$

There are nine steps in the cycle, which is illustrated in Fig. 8.1.

The net reaction of the cycle is

$$\text{Acetyl CoA} + 3\ \text{NAD}^+ + \text{FAD} + \text{NDP} + \text{Pi} + 2\text{H}_2\text{0}$$
$$\downarrow$$
$$2\ \text{CO}_2 + 3\ \text{NADH} + \text{FADH}_2 + \text{NTP} + 2\text{H}^+ + \text{CoA}$$

In summary the two carbon atoms of acetate become the carbon atoms of CO_2.

The NADH and $FADH_2$ formed are oxidized by the electron transport chain. The transfer of electrons from these carriers to O_2, the ultimate electron acceptor leads to the pumping of protons. The proton-motive force then powers the generation of ATP, the stoechiometry being about 2.5 ATP per NADH and 1.5 ATP per FADH2 (see the section on ATP synthase, Chap. 2).

Molecular oxygen does not participate directly. The cycle operates under aerobic conditions only because NAD and FAD must be regenerated by the transfer of electrons to molecular oxygen.

Citrate Synthase

The keto group of oxaloacetate reacts with acetyl CoA in an aldol condensation:

$$\text{oxaloacetate} + \text{acetyl-CoA} + \text{H}_2\text{O} \rightarrow \text{citrate} + \text{CoA} + \text{H}^+$$

The tridimensional structures of the enzymes from *E. coli*, from the thermophilic archaeon *Thermoplasma acidophilum* and from an Antarctic psychrotolerant bacterium have been determined. These structural studies have shown that citrate synthase adopts open and closed forms by a hinged domain motion to bind substrates and release products and to perform the catalytic condensation reaction.

The importance of this reaction lies in the fact that it allows the introduction of carbohydrate, lipid and protein catabolism into the cycle for complete oxidation.

The reaction catalyzed differs from the majority of reactions involving CoA derivatives in that it uses the methyl group of acetate in the condensation and not the carboxyl group.

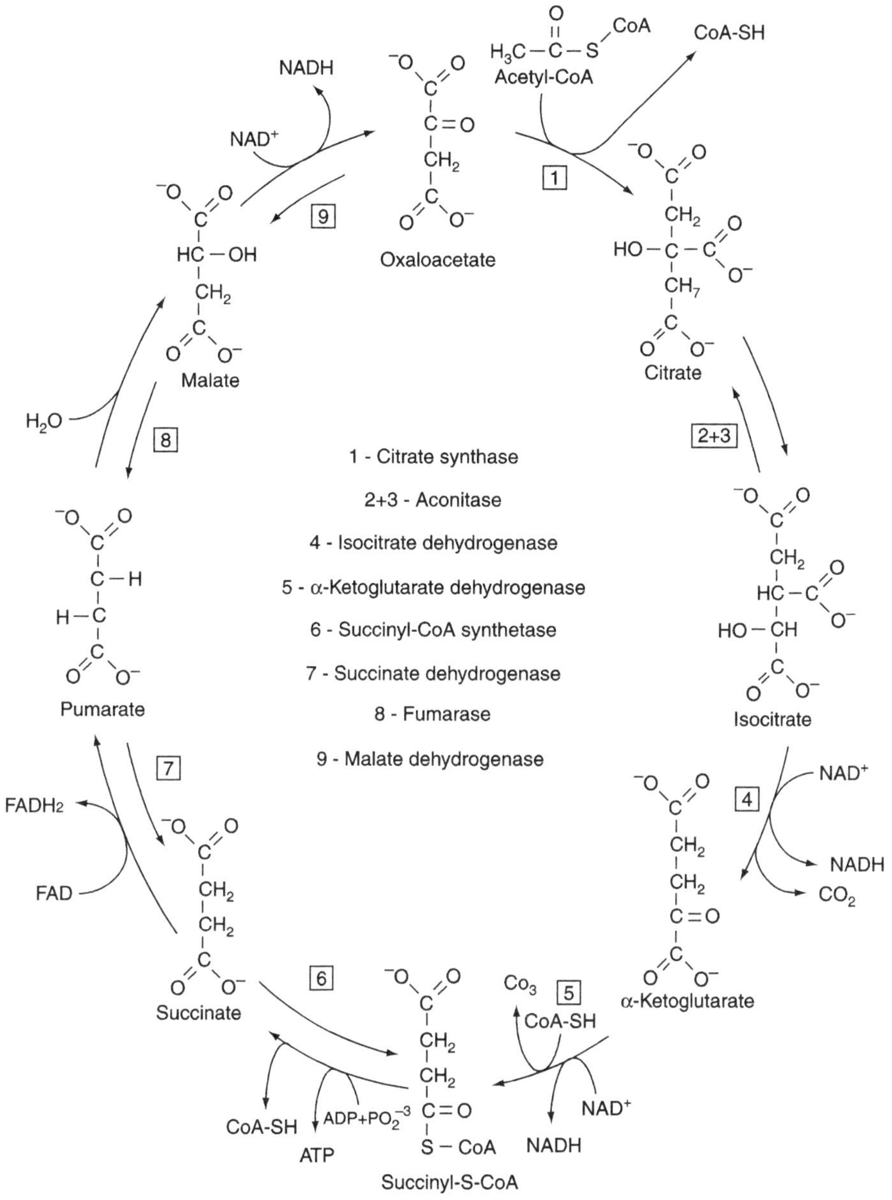

Fig. 8.1 The tricarboxylic acid cycle

The reaction can be decomposed into two parts: the true condensation and the thioester hydrolysis. The overall reaction is practically irreversible ($\Delta F = -8$ kcal).

The enzyme from *E. coli* behaves like a trimer of dimeric subunits, the dimer being the basic catalytic unit, but the hexamer is required for allosteric regulatory sensitivity. By contrast, citrate synthase from animals, plants and some bacteria is a simple dimer.

Citrate synthase is an important control point of the cycle. ATP is an inhibitor. Its effect is to increase the *Km* for acetyl CoA. Thus, as the level of ATP increases, less of the enzyme is saturated with acetyl CoA and so less citrate is formed.

Enzyme synthesis is subject to catabolite repression, is repressed by glucose and anaerobiosis, induced by acetate and oxygen. When acetate is the carbon source, citrate synthase is rate-limiting for the TCA cycle.

The complete sequence of the *gltA* gene encoding citrate synthase in *E. coli* has been determined. K^+ions are a non-allosteric activator, acetyl CoA an allosteric activator, α-ketoglutarate and ATP are competitive inhibitors; NADH, oxaloacetic acid and NAD^+ are allosteric inhibitors. The primary physiological regulators of enzyme activity are NADH and α-ketoglutarate.

Two distinct *Bacillus subtilis* genes (*citA* and *citZ*) were found to encode citrate synthase isozymes that catalyze the first step of the citric acid cycle.

The *citA* gene was cloned by genetic complementation of an *E. coli* citrate synthase mutant strain and present in a monocistronic transcriptional unit. A divergently transcribed gene, *citR*, could encode a protein with strong similarity to the bacterial LysR family of regulatory proteins. A null mutation in *citA* had little effect on citrate synthase enzyme activity or sporulation.

The residual citrate synthase activity was purified from a *citA* null mutant strain, and the partial amino acid sequence for the purified protein (CitZ) was determined. The *citZ* gene proved to be the first gene in a tricistronic cluster that also included *citC* (coding for isocitrate dehydrogenase) and *citH* (coding for malate dehydrogenase). A mutation in *citZ* caused a substantial loss of citrate synthase enzyme activity, glutamate auxotrophy, and a defect in sporulation.

Aconitase (From Citrate to Isocitrate Through cis-*Aconitate)*

Citrate must be isomerized into isocitrate to enable the six-carbon unit to undergo oxidative decarboxylation. The isomerization of citrate is accomplished by a dehydration step followed by a hydration step. The result is an interchange of an H and an OH. The enzyme catalyzing both steps is called aconitase because *cis*-aconitate is an intermediate.

Aconitase contains iron that is not bound to heme. Rather, its four iron atoms are complexed to four inorganic sulfides and three cysteine sulfur atoms.

A parenthesis is needed at this point concerning iron-sulfur proteins; they are defined as proteins carrying iron-sulfur clusters in which the iron is at least partially coordinated by sulfur. Iron-sulfur clusters are prosthetic groups commonly found in various proteins that participate in oxidation-reduction reactions and catalysis. The components and molecular mechanisms involved in the assembly of the Fe/S clusters have been identified only partially. In most instances, the iron is bound to sulfur(s) from cysteine residues in the protein and also to inorganic sulfurs in the

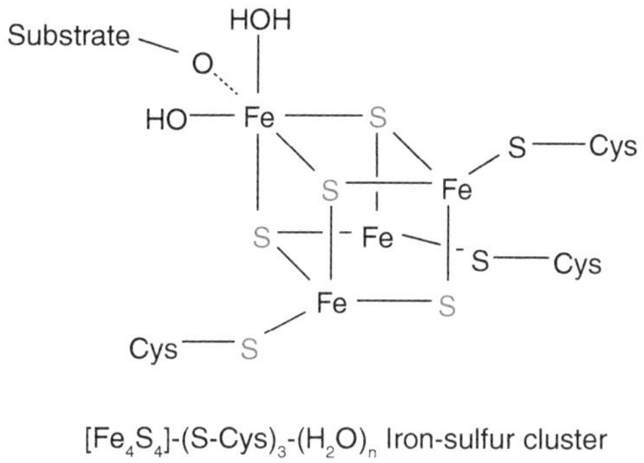

Fig. 8.2 The iron-sulfur cluster of aconitase

iron-sulfur centers. The chief role of the iron-sulfur cluster is to facilitate electron transfer, while in many proteins, it contributes to the catalytic function. The major forms of the Fe-S clusters are (1) the [2Fe-2S] cluster, where the two iron atoms are bridged to one another by two inorganic sulfur atoms and ligated to four cysteines from the protein backbone: (2) the [4Fe-4S] cluster where the four iron atoms are ligated to four cysteines and form a cubic structure with four inorganic sulfur atoms; (3) the [3Fe-4S] cluster where a single iron of the preceding cluster is missing.

This Fe-S cluster binds citrate and participates in dehydrating and rehydrating the bound substrate.

Aconitase has an active $[Fe_4S_4]^{2+}$ cluster, which may convert to an inactive $[Fe_3S_4]^{+}$ form. Three Cys residues have been shown to be ligands of the $[Fe_4S_4]$ centre. In the active state, the labile iron ion of the $[Fe_4S_4]$ cluster is not coordinated by Cys but by H_2O (Fig. 8.2).

By contrast with the majority of iron-sulfur proteins that function as electron carriers, the Fe-S cluster of aconitase reacts directly with an enzyme substrate.

Since crystallization of aconitase with the substrates citrate and *cis*-aconitate has not been possible because the enzyme turns over, resulting in the selection of the enzyme with isocitrate bound into the crystal lattice, the crystal structures of the enzyme in a complex with inhibitor analogs of these two substrates have been solved. The structure with nitrocitrate bound provides a model for citrate binding. The structure with *trans-aconitate* bound provides a model for *cis*-aconitate binding in two ways: Fe_4 of the [4Fe-4S] cluster is five-coordinate and the carbon at the C-β position is trigonal. These results allow the model for the reaction mechanism to be extended to all three natural substrates of aconitase. The results support a model in which citrate and isocitrate form similar chelate structures related by 180° rotation about the C–α–C–β bond while the intermediate *cis*-aconitate binds in either of two ways (citrate mode or isocitrate mode). In both inhibitor complexes a H_2O molecule is also bound to Fe_4. In the structure with nitrocitrate bound, partial occupancy of sulfate in the active site is observed accompanied by hydroxyl binding to Fe_4. Comparison of the structures with isocitrate, *trans*-aconitate, nitrocitrate and sulfate bound reveals preferred orientations for the three types of oxygens ligated to Fe_4 (carboxyl, hydroxyl and H_2O) supporting the proposed roles for two histidine and one aspartic residues in the catalytic mechanism.

The nucleotide sequence of *acn*, the aconitase gene has been determined. Its amino acid sequence shows similarity with mitochondrial aconitases, isopropylmalate isomerases but is most similar to the human iron-responsive-element-binding protein.

In addition to AcnA, *E. coli* contains a second genetically distinct aconitase, named aconitase B. It is coded for by the *acnB* gene. Aconitase B is also an iron-sulfur protein. Its unstable activity appears to be regulated differently than aconitase A. The *acnA* gene resembles other citric acid cycle genes in being subject to CRP-mediated catabolite repression and ArcA-mediated anaerobic repression and activation by the SoxRS oxidative stress regulatory system, and by the ferric uptake regulator (Fur), thus belonging to at least four global regulatory networks, *crp, arcA, fur* and *soxRS*. In contrast, the aconitase B activity decreased after exposure to oxidative stress and was less affected by anaerobiosis. Comparable studies with the fumarase genes (*fumA, B and C*) indicated that *fumA* (encoding the unstable aerobic iron-sulfur-containing fumarase) is activated by the ferric uptake regulator (Fur) and *fumC* (encoding the stable fumarase) is activated by the SoxRS oxidative stress regulatory system.

Isocitrate Dehydrogenase (IDH)

This enzyme acts on the CH-OH groups as a donor with NAD^+ or $NADP^+$ as an electron acceptor. It catalyzes the following reaction:

$$\text{Isocitrate} + \text{NAD}^+ \left(\text{or NADP}^+\right) \Leftrightarrow \alpha\text{ - ketoglutarate} + \text{CO}_2 + \text{NADH}\left(\text{or NADPH}\right)$$

The intermediate in this reaction is oxalosuccinate, an unstable β-keto acid. While bound to the enzyme, oxalosuccinate loses CO_2 to form α-ketoglutarate. There are marked differences in the properties of enzymes from different sources.

In *E. coli*, the cold-sensitive enzyme is not an allosteric protein as isocitrate dehydrogenases from other sources are. Phosphorylation of the enzyme, on a serine residue, by isocitrate dehydrogenase kinase/phosphatase inactivates it, and dephosphorylation by the phosphatase reactivates it.

Phosphorylation affects the binding of NADP. The modulation of this key enzyme activity enables *E. coli* to make rapid shifts between the tricarboxylic acid cycle and the glyoxalate bypass pathways. The state of phophorylation of isocitrate dehydrogenase determines its activity. The arginine-specific reagent phenylglyoxal inactivates the active dephosphorylated form of *E. coli* isocitrate dehydrogenase rapidly in a pseudo-first-order process. Both $NADP^+$ and NADPH protect the enzyme against inactivation. Phenylglyoxal appeared to react with one arginine residue per subunit, and the extent of the reaction is proportional to the extent of the inactivation. In contrast, the phosphorylated form of isocitrate dehydrogenase does not react detectably with phenylglyoxal. The data indicate that the coenzyme-binding site of isocitrate dehydrogenase contains a reactive arginine residue that is protected by phosphorylation, and are consistent with the hypothesis that phosphorylation of the enzyme occurs close to or at its active site.

The structure of the NADP-dependent isocitrate dehydrogenase from *E. coli* has been solved. Its structure with isocitrate, nicotinamide adenine dinucleotide phosphate, and calcium has also been solved at 2.5 Å resolution and compared by difference mapping against previously determined enzymatic complexes. Calcium replaces magnesium in the binding of the metal-substrate chelate complex, resulting in a substantially reduced turnover rate.

The structure shows the following: (i) A complete, structurally ordered ternary complex (enzyme, isocitrate, $NADP^+$, and $Ca2^+$) is observed in the active site, with the nicotinamide ring of $NADP^+$ exhibiting a specific salt bridge with isocitrate. The binding of the cofactor nicotinamide ring is dependent on this interaction. (ii) Isocitrate is bound by the enzyme with the same interactions as those found for the magnesium/substrate binary complex, but the entire molecule is shifted in the active site by approximately 1 Å in order to accommodate the larger metal species and to interact with the nicotinamide ring. The distances from isocitrate to the bound calcium are substantially longer than those previously found with magnesium. (iii) $NADP^+$ has a novel binding site and conformation as compared to previously solved dehydrogenases. (iv) The orientation and interactions of the nicotinamide ring with the substrate are consistent with the stereospecificity of the enzyme-catalyzed reaction. The structure of the same enzyme with bound α-ketoglutarate, Ca^{++}, and NADPH was solved at 2.7 Å resolution. The α-ketoglutarate binds in the active site at the same position and orientation as isocitrate, with a difference between the two bound molecules of about 0.8 Å. The Ca^{++} metal is coordinated by α-ketoglutarate, three conserved aspartate residues, and a pair of water molecules. The largest motion in the active site relative to the isocitrate enzyme complex is observed for tyrosine 160, which originally forms a hydrogen bond with the labile carboxyl group of isocitrate and moves to form a new hydrogen bond with Asp 307 in the complex with α-ketoglutarate. This triggers a number of significant movements among several short loops and adjoining secondary structural elements in the enzyme, most of which participate in dimer stabilization and formation of the active site cleft. These rearrangements are similar to the ligand-binding-induced movements observed in globins and insulin and serve as a model for an enzymatic mechanism which involves local shifts of secondary structural elements during turnover, rather than large-scale domain closures or loop transitions induced by substrate binding such as those observed in hexokinase or triosephosphate isomerase.

Yeast cells contain three isozymes of isocitrate dehydrogenase: mitochondrial NAD^+- and $NADP^+$-specific enzymes and a cytosolic $NADP^+$-specific enzyme. Independent metabolic functions of these enzymes in *S. cerevisiae* were examined by analyses of expression and of phenotypes displayed by mutants containing all possible combinations of isozyme gene disruptions. All three isocitrate dehydrogenases are expressed at high levels with growth on non-fermentable carbon sources, whereas the mitochondrial NADP-specific enzyme constitutes the major cellular activity with growth on glucose. Distinct growth phenotypes are observed for mutants expressing a single isozyme, and expression of at least one isozyme is necessary for glutamate-independent growth. The NADP-specific tricarboxylic acid cycle isocitrate dehydrogenase from *E. coli* was expressed in mitochondrial and cytosolic compartments of the yeast disruptions. The bacterial enzyme is competent for restoration of NADP-specific functions in either compartment but does not compensate for function of the yeast NAD-specific tricarboxylic acid cycle enzyme.

In yeast, NAD-dependent isocitrate dehydrogenase is an allosterically regulated enzyme composed of four Idh1p subunits and four Idh2p subunits.

E. coli *Isocitrate Dehydrogenase Kinase-Phosphatase*

This bifunctional enzyme, coded by the *aceK* gene, catalyzes the following reactions.

$$\text{ATP} + \text{isocitrate dehydrogenase}$$
$$\Uparrow\Downarrow$$
$$\text{ADP} + \text{isocitrate - dehydrogenase - P}$$

$$H_2O + \text{isocitrate - dehydrogenase - P}$$
$$\downarrow$$
$$\text{Pi} + \text{isocitrate dehydrogenase}$$

Photo-affinity labelling shows that *E. coli* isocitrate dehydrogenase kinase/phosphatase contains a single ATP-binding site.

Isocitrate dehydrogenase kinase catalyzes the phosphorylation of homogeneous active isocitrate dehydrogenase with a stoichiometry of just under one phosphate group incorporated per subunit. This almost completely inactivates the dehydrogenase. There is a good correlation between phosphorylation and inactivation. Analysis of a partial acid hydrolysate of phosphorylated isocitrate dehydrogenase shows that the only phosphoamino acid present is phosphoserine. Isocitrate dehydrogenase phosphatase catalyzes the release of ^{32}P from ^{32}P-phosphorylated isocitrate dehydrogenase; it requires either ADP or ATP for activity. In the presence of ADP, or ATP plus an inhibitor of the kinase, the phosphatase catalyzes full reactivation of isocitrate dehydrogenase and there is a good correlation between reactivation and the release of phosphate. In the presence of ATP alone the phosphatase catalyzes the release of ^{32}P from phosphorylated isocitrate dehydrogenase but the activity of the dehydrogenase remains low, indicating that the kinase and phosphatase are active simultaneously in these conditions. The active and inactive forms of isocitrate dehydrogenase can be resolved by non-denaturing gel electrophoresis; the two forms of the enzyme are interconverted by phosphorylation and dephosphorylation *in vitro*. The extent of the interconversion correlates well with the changes in isocitrate dehydrogenase activity.

The flow of isocitrate through the glyoxylate bypass (see below) in *E. coli* is regulated via the phosphorylation-dephosphorylation of isocitrate dehydrogenase mediated by this bifunctional enzyme. The *aceK* gene coding for this enzyme is part of the polycistronic *ace* operon, which also includes the *aceB* and *aceA* genes coding, respectively, for malate synthase and isocitrate lyase, the two glyoxylate bypass enzymes. The nucleotide sequence of the *aceK* gene and its 5′-flanking region has been determined. It codes for a 66,528-Da protein. The 5′ flanking region presents an unusual intercistronic structural pattern consisting of two consecutive long dyad symmetries, almost identical in sequence, which can yield very stable stem-loop units. These structures are probably responsible for the drastic downshifting in expression observed in acetate-grown bacteria between the *aceK* gene and the *aceA* gene located immediately upstream in the *ace* operon.

The effectors controlling the two activities belong to two distinct classes that differ in mechanism and in the locations of their binding sites. NADPH and isocitrate are representative members of one of these effector classes. NADPH inhibits both IDH kinase and IDH phosphatase, whereas isocitrate inhibits only IDH kinase. Isocitrate can "activate" IDH phosphatase by reversing product inhibition by dephospho-IDH. Mutations in *icd*, which encodes IDH, has parallel effects on the binding of these ligands to the IDH active site and on their effects on IDH kinase and phosphatase, indicating that these ligands regulate IDH kinase/phosphatase through the IDH active site. Kinetic analyses suggests that isocitrate and NADPH prevent formation of the complex between IDH kinase/phosphatase and its protein substrate.

α-Ketoglutarate Dehydrogenase Complex

The oxidative decarboxylation of α-ketoglutarate closely resembles that of pyruvate.

$$\alpha\text{-ketoglutarate} + \text{CoA} + \text{NAD}^+ \rightarrow \text{succinyl CoA} + \text{CO}_2 + \text{NADH}$$

The same cofactors are participants, TPP, lipoamide, CoA, FAD and NAD^+. The reaction is catalyzed by an enzyme complex structurally similar to the pyruvate dehydrogenase complex. It contains the three kinds of enzymes, an α-ketoglutarate complex (E′1), a transsuccinylase (E′2) and a dihydrolipoyl dehydrogenase (E3). E′1 and E′2 are different from E1 and E2 of the pyruvate dehydrogenase complex, but the dihydrolipoyl dehydrogenase part of the complexes is the same protein, E3. Although the catalytic cycles of E1 from the pyruvate dehydrogenase (E1p) and branched chain α-ketoacid dehydrogenase (E1b) complexes have been elucidated, little structural information is available on E1o, the first component of the α-ketoglutarate dehydrogenase complex, despite the central role of this complex at the branching point between the TCA cycle and glutamate metabolism. Structural evidence that MsKGD, the E1o from *Mycobacterium smegmatis*, shows two conformations of the post-decarboxylation intermediate, each one associated with a distinct enzyme state. An overall picture of the catalytic cycle, reconstructed by either crystallographic snapshots or modelling, showed that the conformational change leading the enzyme from the initial (early) to the late state, although not required for decarboxylation, plays an essential role in catalysis, and possibly in the regulation of mycobacterial E1o.

Succinyl CoA Synthetase

The succinyl thioester of CoA has an energy-rich bond. The ΔGo of succinyl CoA is about 8 kcal/mol, which is comparable with that of ATP. The cleavage of the thioester bond of succinyl CoA is coupled to the phosphorylation of a nucleoside diphosphate.

$$\text{Succinyl CoA} + \text{Pi} + \text{NDP} \rightarrow \text{succinate} + \text{NTP} + \text{CoA}$$
$$\left(\text{N denotes adenosine or guanosine}\right)$$

This readily reversible reaction (ΔGo = –0.8 kcal/mol) is catalyzed by succinyl CoA synthetase via a phosphorylated histidine intermediate.

It is the only step in the cycle that directly yields a high-energy phosphate bond. (ATP and GTP are used as a phosphoryl donor in protein synthesis and GTP in signal transduction processes).

The yeast succinyl CoA synthetase was shown to utilize ATP but not GTP for succinyl CoA synthesis.

There are two classes of this enzyme. The enzyme complex found in the pig heart acts as a dimer, whereas others, like the complex in *E. coli* acts as a tetramer. Genes for the 2 subunits are coordinately expressed, so α and β subunits are produced in equimolar proportions.

There is one ATP binding site and one phosphorylation histidine per monomer, but in the dimer, an alternating mode utilizes only one at a time.

A molecular model of the enzyme from *E. coli*, crystallized in the presence of CoA, has been refined against data collected to 2.3 Å resolution. The copies of the αβ-dimer are similar, each having one active site where the phosphorylated histidine residue and the thiol group of CoA are found. CoA is bound in an extended conformation to the nucleotide-binding motif in the N-terminal domain of the α subunit. The phosphoryl group of the phosphorylated histidine residue is positioned at the amino termini of two alpha-helices, one from the C-terminal domain of the α-subunit and the other from the C-terminal domain of the β-subunit. These two domains have similar topologies, despite having only 14 % sequence identity. By analogy to other nucleotide-binding proteins, the binding site for the nucleotide may reside in the N-terminal domain of the β-subunit. If this is so, the catalytic histidine residue would have to move about 35 Å to react with the nucleotide.

Two potential nucleotide-binding sites were predicted in the β-subunit, and have been differentiated by photoaffinity labeling with 8-N3-ATP and by site-directed mutagenesis. It was demonstrated that 8-N3-ATP is a suitable analogue for probing the nucleotide-binding site of succinyl CoA synthetase. Two tryptic peptides from the N-terminal domain of the β-subunit were labeled with 8-N3-ATP. These correspond to residues belonging to two regions lying along one side of an ATP-grasp fold. A mutant protein with changes on the opposite side of the fold was unable to be phosphorylated using ATP or GTP, but could be phosphorylated by succinyl CoA and Pi. From extensive further studies, it was concluded that the nucleotide-binding site is located in the N-terminal domain of the β-subunit. This implies that there are two active sites approximately 35 Å apart, and that the loop comprising the active site actually moves between them during catalysis. The genes encoding both subunits of the succinyl-CoA synthetase of *E. coli* have been identified as distal genes of the *suc* operon, which also encodes the dehydrogenase (*sucA*) and succinyltransferase (*sucB*) components of the α-ketogutarate dehydrogenase complex; *sucC* and *sucD*, correspond to the β and α subunits of succinyl-CoA synthetase, respectively. The genes are located together with the citrate synthase (*gltA*) and succinate dehydrogenase (*sdh*) genes, in a cluster of nine citric acid cycle genes: *gltA-sdhCDAB-sucABCD*. Further enzymological studies indicated that expression of succinyl CoA synthetase is coordinately regulated with that of α-ketoglutarate dehydrogenase.

Pure preparations of the denatured α and β subunits of succinyl CoA synthetase of *E. coli* have been obtained. Activity (50–60 %) can be recovered following renaturation of an equimolar mixture of the two subunits. The substrate ATP is required for reconstitution of activity. Knowing that ATP phosphorylates the α subunit in the native enzyme, these data suggest that phosphorylation of this subunit is necessary for correct assembly. The enzyme from *E. coli* has been crystallized and the primary structure of the two *E. coli* subunits has been deduced from the nucleotide sequence.

Concentrations of GDP, which are expected to bind to the catalytic site and inhibit the autophosphorylation of succinyl CoA synthetase when NTP is used as a substrate, were found to increase the level of phosphoenzyme formed. The ability of GDP to do so is dependent upon the presence of a protein distinct from succinyl CoA synthetase. This effector protein could be separated from the enzyme by ammonium sulfate fractionation. Reconstitution experiments show that the protein inhibits succinyl CoA synthetase, that the inhibition is relieved by GDP, and that the inhibitor recognizes both *E. coli* and eukaryotic forms of the synthetase. The inhibitor is itself regulated by the conditions used to grow the bacteria and in a manner that appears distinct from that of succinyl CoA synthetase.

Succinate Dehydrogenase

Escherichia coli

This essential component of the cycle catalyzes the conversion of succinate to fumarate

$$\text{succinate} + \text{FAD} \Leftrightarrow \text{fumarate} + \text{FADH}_2$$

with transfer of the reducing equivalents to ubiquinone. It is induced by aerobic growth on non-fermentable carbon sources and is subject to catabolic repression by glucose.

The succinate dehydrogenase of *E. coli* was found by immunochemical resolution and biophysical characterization to be a 4-subunit enzyme complex.

The *sdh* genes for this enzyme are encoded in an operon and are expressed as four polypeptides. The two larger catalytic subunits are polar and contain respectively a covalently bound flavin adenine dinucleotide (SdhA) and non-heme iron (SdhB) whereas the two smaller membrane-embedded proteins (SdhC and SdhD) are highly hydrophobic and anchor the larger subunits to the membrane, and also participate in electron transport (SdhC is cytochrome b_{661-5}). Another protein, SdhE, was required for growth on succinate as a sole carbon source and for the function, but not stability, of succinate dehydrogenase. SdhE interacts with the flavoprotein SdhA, directly bound the flavin adenine dinucleotide co-factor, and is required for the flavinylation of SdhA.

The large subunits, which have been cloned and sequenced are related in sequence to the two proteins encoded by the anaerobic fumarate reductase, suggesting a common evolutionary origin (the flavoprotein and the iron-sulfur protein of fumarate reductase are also anchored to the membrane by a pair of very hydrophobic membrane proteins unrelated to SdhC and SdhD).

Two global transcriptional regulators, Fnr and Arc, are involved in the regulation of anaerobic and aerobic metabolism. The role of multiple ArcA recognition sites in anaerobic regulation of succinate dehydrogenase (*sdhCDAB*) gene expression in *E. coli* was investigated. This expression is negatively regulated by the *arcA* and *fnr* gene products during anaerobic cell growth conditions. The controlled synthesis of succinate dehydrogenase, the sole membrane-bound enzyme of the tricarboxylic acid cycle allows optimal participation in the aerobic electron transport pathway for the generation of energy via oxidative phosphorylation reactions. Four distinct and independent ArcA binding sites exist in the *sdhC* promoter region. ArcA sites, designated sites 1 and 2, are centered at −205 bp and −119 bp upstream of the *sdhC* promoter, respectively, whereas ArcA site 3 overlaps the −35 and −10 regions of the *sdhC* promoter. A fourth ArcA site is centred at +257 bp downstream of the *sdhC* promoter. They are bound with differing affinity by ArcA and ArcA phosphate. The *in vivo* studies, in combination with the *in vitro* studies, indicate that ArcA site 3 is necessary and sufficient for the ArcA-dependent repression of *sdhC* gene expression, while the DNA region containing ArcA site 2 contributes to maximal gene expression. The DNA-containing ArcA sites 1 and 4 provide minor roles in the ArcA regulation of *sdhC* expression. Lastly, the Fnr-dependent control of *sdhCDAB* gene expression was shown to occur independently of the ArcA and to require DNA sequences near the start of *sdhC* transcription.

The mechanism underlying the glucose repression is not yet clear, because the involvement of the general catabolite regulators such as CRP and CRA has been dismissed. The results of genetic analyses demonstrate that a *ptsG* gene product (see Chap. 4 on bacterial permeability), a component of the major glucose transporter, acts as a crucial mediator in glucose repression.

Yeast

The succinate dehydrogenase (SDH) of *S. cerevisiae* is composed of four non-identical subunits encoded by the nuclear genes SDH1, SDH2, SDH3, and SDH4. The hydrophilic subunits, Sdh1p and Sdh2p, comprise the catalytic domain involved in succinate oxidation. They are anchored to the inner mitochondrial membrane by two small, hydrophobic subunits, Sdh3p and Sdh4p, which are required for electron transfer and ubiquinone reduction.

Comparison of the deduced primary sequence of the yeast Sdh4p subunit to Sdh4p subunits from other yeast species reveals the presence of an unusual 25–30 amino acid carboxyl-terminal extension following the last predicted transmembrane domain. The extension is predicted to be on the cytoplasmic side of the inner mitochondrial membrane. Truncation results reveal that the carboxyl-terminal extension is necessary for respiration and growth on nonfermentable carbon sources, for ubiquinone reduction, and enzyme stability. They suggest that the extension and more specifically, residues 128–135 are involved in the formation of a ubiquinone binding site. The findings support a two-ubiquinone binding site model for the *S. cerevisiae* SDH. The Sdh3p amino acid residues involved in ubiquinone binding have been identified.

Fumarase

The reaction catalyzed by fumarase, or fumarate hydratase, is the following:

$$\text{Fumarate} + \text{H20} \Leftrightarrow \text{malate}$$

Isotope effects have been used to analyze the mechanism of fumarase. The absolute stereochemistry of fumarate-to-malate transformation as catalyzed by the enzyme fumarase has been determined via neutron diffraction.

Two different types of fumarase were found in sonic extracts of *E. coli*; one required Fe-S for the enzyme activity, and the other did not. When the cells were grown without aeration, the Fe-S-independent enzyme occupied over 80 % of the overall fumarase activity. Highly purified Fe-S-independent enzyme was suggested to be composed of four subunits (Mr = 48 kDa) by SDS-polyacrylamide gel electrophoresis and gel filtration. Amino acid and N-terminal sequence analyses supported the possibility that the enzyme is a product of the *fumC* gene (FumC). In aerobically grown cells, however, the content of FumC was low and the Fe-S-dependent fumarase occupied over 80 % of the overall activity. The Fe-S-dependent enzyme appeared to be labile and the activity was rapidly lost during purification. Although the spontaneous inactivation was previously ascribed to thermal lability, the activity could be restored by anaerobic incubation with ferrous ions and SH-compounds.

A more thorough analysis shows that there are definitely two distinct classes of fumarase in *E. coli*, comprising the products of the three fumarase genes, *fumA* (FumA), *fumB* (FumB) and *fumC* (FumC). The Class I enzymes include FumA, FumB, and the immunologically related fumarase of *Euglena gracilis*. These are characteristically thermolabile dimeric enzymes containing identical subunits of Mr 60,000. FumA and FumB are differentially regulated enzymes that function in the citric acid cycle (FumA) or to provide fumarate as an anaerobic electron acceptor (FumB, the product of the FNR-regulated *fumB* gene), and their affinities for fumarate and L-malate are consistent with these roles. The Class II enzymes include FumC, and the fumarases of *Bacillus subtilis, S. cerevisiae* and mammalian sources. They are thermostable tetrameric enzymes containing identical subunits Mr 48,000–50,000. The Class II fumarases share a high degree of sequence identity with each other (approx. 60 %) and with aspartase (approx. 38 %) and argininosuccinase (approx. 15 %), and it would appear that these are all members of a family of structurally related enzymes. It is also suggested that the Class I enzymes may belong to a wider family of iron-dependent carboxylic acid hydrolyases that includes maleate dehydratase and aconitase. Apart from one region containing a Gly-Ser-X-X-Met-X-X-Lys-X-Asn consensus sequence, no significant homology was detected between the Class I and Class II fumarases.

The cloning, mapping, and expression of the fumarase genes of *E. coli* and *B. subtilis* has been achieved.

Definite evidence that fumarase A from *E. coli* is an iron-sulfur cluster containing enzyme has been obtained by spectroscopic methods. In addition, it catalyzes the isomerization of enol and keto oxaloacetic acid apparently at the same active site as the fumarase reaction. Fumarase A and aconitase, two enzymes with 4Fe-4S clusters that bind a linear 4-carbon dicarboxylic acid moiety in the *trans* conformation during their normal hydrolyase reaction, do catalyze this isomerization.

Expression of the *fumA* and *fumC* genes was lowest during anaerobic cell growth, in support of the proposed roles of FumA and FumC as aerobic fumarases. Transcription of the *fumC* gene was shown to be complex: it was dependent on both the *fumA* and *fumC* promoters. Anaerobic expression from the *fumA* promoter was derepressed in both an *arcA* and a *fnr* mutant, while expression from the *fumC* promoter was derepressed in only the *arcA* strain. The *fumA* promoter was also shown to be catabolite controlled, whereas the *fumC* promoter was relatively unaffected by the type of carbon used for cell growth. Cellular iron limitation stimulated *fumC* but not *fumA* expression. Superoxide radicals also caused increased *fumC* gene expression; *fumA* expression was unaffected. Both the superoxide control and the iron control of *fumC* expression required the SoxR regulatory protein. These studies suggest different physiological roles for the FumA and FumC fumarases. The iron-containing FumA fumarase is the more abundant enzyme under most conditions of aerobic cell growth except when iron is limiting; FumC, which lacks iron, is a member of the soxRS regulon; this regulon does not respond to changes in O_2-concentration but perhaps does respond to some consequence of a decrease in the ratio of NADPH to $NADP^+$. The modulation of the fumarases of *E. coli* in response to oxidative stress was further studied by the same group: The [4Fe-4S]-containing fumarases A + B are susceptible to oxidative inactivation, while fumarase C, which is not an iron-sulfur protein, is induced under oxidative conditions.

The crystal structure of the tetrameric enzyme, fumarase C from *E. coli*, has been determined to a resolution of 2.0 Å. The polypeptide conformation is composed of three domains. The central domain, D2, is a unique five-helix bundle. The association of the D2 domains results in a tetramer which has a core of 20 alpha-helices. The other two domains, D1 and D3, cap the helical bundle on opposite ends giving both the single subunit and the tetramer a dumbbell-like appearance.

In *S. cerevisiae*, the single translation product of the FUM1 gene (fumarase) is processed in mitochondria before being distributed between the cytosol and mitochondria.

Cloning of the *S. cerevisiae* FUM1 gene downstream of the strong GAL10 promoter resulted in inducible overexpression of fumarase. The overproducing strain exhibited efficient bioconversion of fumaric acid to L-malic acid, much higher than parameters known for industrial bacterial strains.

Malate Dehydrogenase

This enzyme is responsible for the regeneration of oxaloacetic acid and a new turn of the cycle. It catalyzes the reaction:

$$\text{malate} + \text{NAD}^{+} \Leftrightarrow \text{oxaloacetic acid} + \text{NADH} + \text{H}^{+}$$

Malate dehydrogenases are widely distributed and alignment of the amino acid sequences show that the enzyme has diverged into two main phylogenetic groups. Multiple amino acid sequence alignments of malate dehydrogenases also show that there is a low degree of primary structural similarity, apart from that in several positions crucial for nucleotide binding, catalysis, and the subunit interface. The 3-dimensional structures of several malate dehydrogenases are similar, despite their low amino acid sequence identity. The coenzyme specificity of malate dehydrogenase may be modulated by substitution of a single residue, as can the substrate specificity. The mechanism of catalysis of malate dehydrogenase is similar to that of lactate dehydrogenase, an enzyme with which it shares a similar 3-dimensional structure. Substitution of a single amino acid residue of a lactate dehydrogenase changes the enzyme specificity to that of a malate dehydrogenase, but a similar substitution in a malate dehydrogenase resulted in relaxation of the high degree of specificity for oxaloacetate. Knowledge of the 3-dimensional structures of malate and lactate dehydrogenases allows the redesign of enzymes by rational rather than random mutation and may have important commercial implications.

The malate dehydrogenase gene of *E. coli*, which is susceptible to catabolite and anaerobic repression, has been cloned and the sequence of the *mdh* gene determined. All information necessary for expression of the *mdh* structural gene was mapped upstream of the gene within a 1.3 kbp restriction fragment. Compared with the untransformed wild type, transformations with a vector containing this fragment, gave up to 40-fold more malate dehydrogenase activity in both *E. coli* wild type and *mdh* mutant recipients. Catabolite repression was not affected in the transformants. A possible CRP binding site in the promoter region of the *mdh* gene provides evidence for a co-regulation with the *fumA* gene, the structural gene of fumarase, which is also subject to catabolite repression.

The *mdh* gene appears to be highly regulated to adapt to changing conditions of aerobic and anaerobic cell growth with various types of carbon substrates.

A *B. subtilis* gene for malate dehydrogenase (*citH*) was found downstream of genes for citrate synthase and isocitrate dehydrogenase. Disruption of *citH* caused partial auxotrophy for aspartate and a requirement for aspartate during sporulation. In the absence of aspartate, *citH* mutant cells were blocked at a late stage of spore formation.

Methanobacterium thermoautotrophicum was found to contain two malate dehydrogenases, which were partially purified and characterized. One was specific for NAD^+ and catalyzed the dehydrogenation of malate at approximately one-third of the rate of oxalacetate reduction, and the other could equally well use NAD^+ and $NADP^+$ as coenzyme and catalyzed essentially only the reduction of oxalacetate. Via the N-terminal amino acid sequences, the encoding genes were identified in the genome of *M. thermoautotrophicum*. Comparison of the deduced amino acid sequences revealed that the two malate dehydrogenases are phylogenetically only distantly related. The NAD^+-specific malate dehydrogenase showed high sequence similarity to L-malate dehydrogenase from *Methanothermus fervidus*, and the $NAD(P)^+$-using malate dehydrogenase showed high sequence similarity to L-lactate dehydrogenase from *Thermotoga maritima* and L-malate dehydrogenase from *Bacillus subtilis*.

The structures of the malate dehydrogenases of the wild-type and of a Glu → Arg mutant of *Halobacter marismortui* -an extreme halophilic archaeon- were determined to a high resolution. They highlight a variety of novel protein-solvent features involved in halophilic adaptation. The tetramer appears to be stabilized by ordered water molecule networks and inter-subunit complex salt bridges "locked" in by bound solvent chloride and sodium ions. The analysis of the crystal structures showed that halophilic adaptation is not aimed uniquely at "protecting" the enzyme from the extreme salt conditions, as may have been expected, but, on the contrary, consists of mechanisms that harness the high ionic concentration in the environment.

The cytosolic yeast isozyme of malate dehydrogenase, MDH2, was shown to be subject to rapid inactivation and proteolysis following the addition of glucose to yeast cultures growing on nonfermentable carbon sources. MDH2 is actually phosphorylated during the process of glucose-induced degradation. A truncated active form of MDH2 lacking the first 12 residues of the amino terminus is resistant to glucose-induced degradation and is not subject to phosphorylation.

Organization of the Enzymes of the Tricarboxylic Acid Cycle

Sonic oscillation of mitochondria usually leads to the release of a number of tricarboxylic acid cycle enzymes. These enzymes have, therefore, been referred to as soluble matrix enzymes. Gentle sonic or osmotic disruption can be used to obtain a mitochondrial preparation where these enzymes appear to be organized in a large complex of proteins. Using citrate synthase as a marker for these enzymes, it is shown that the proposed complex is easily sedimented at 32,000×g in 30 min. The exposed citrate synthase in these complexes can be inhibited by its antibody, indicating that the enzymes are not merely entrapped in substrate-permeable vesicles. The effects of pH, temperature, ionic strength, and several metabolites on the ability to obtain the sedimentable citrate synthase have been tested. These studies indicate that the complex is stable at conditions presumed to exist *in situ*. Electron microscopic studies show that gentle sonic oscillation gives rise to an efflux of mitochondrial matrix contents which tend to remain attached to the original membranes. The sedimentable fraction also contained four other presumably soluble tricarboxylic acid cycle enzymes: aconitase, NAD^+-isocitrate dehydrogenase, fumarase, and malate dehydrogenase.

The enzymes of the tricarboxylic acid cycle in mitochondria are proposed to form a supramolecular complex, in which there is channeling of intermediates between enzyme active sites. While interactions have been demonstrated *in vitro* between most of the sequential tricarboxylic acid cycle enzymes, no direct evidence had been obtained *in vivo* for such interactions. In the *S. cerevisiae* gene encoding the tricarboxylic acid cycle enzyme citrate synthase Cit1p, an "assembly mutation," i.e. a mutation that causes a tricarboxylic acid cycle deficiency without affecting the citrate synthase activity has been isolated. A 15-amino acid peptide from wild type Cit1p encompassing the mutation point inhibits the tricarboxylic acid cycle in a dominant manner; the inhibitory phenotype is overcome by a co-overexpression of Mdh1p, the mitochondrial malate dehydrogenase. These data provide the first direct *in vivo* evidence of interaction between two sequential tricarboxylic acid cycle enzymes, Cit1p and Mdh1p, and indicate that the characterization of assembly mutations by the reversible transdominant inhibition method may be a powerful way to study multienzyme complexes in their physiological context.

A general analysis of the regulation of the citric acid cycle is hampered by the intimate interplay believed to exist between the various surrounding pathways. Two main regulatory mechanisms are thought to determine the flux through the cycle: (1) regulation of individual cycle enzymes, and (2) reversible complex formation between various enzymes of the cycle and related pathways. The latter mechanism allows a cell to maintain a high flux of substrates with a moderate number of intermediates, and offers a means of metabolite channeling.

The enzymes which are responsible for catalyzing sequential reactions in several metabolic pathways have been proposed to be highly organized in supramolecular complexes termed *metabolons*. However, the *in situ* existence of these weak complexes is difficult to demonstrate because many of them are dissociated during isolation due to dilution effects. Consequently, the metabolon concept is subject to controversy.

A multienzyme complex of tricarboxylic acid cycle enzymes, catalysing the consecutive reactions from fumarate to α-ketoglutarate, has been identified in extracts of *Pseudomonas aeruginosa* prepared by gentle osmotic lysis of the cells. Its study supports the idea of a 'metabolon' in this organism.

The Tricarboxylic Acid Cycle Is a Source of Biosynthetic Precursors

In the overview, it was stressed that the cycle was the major degradative pathway for the generation of ATP. The citric acid cycle also provides intermediates for biosyntheses. From oxaloacetate, aspartate is produced; from α-ketoglutarate, glutamate and glutamine are produced, and the majority of the carbon atoms in porphyrins come from succinyl CoA. The cycle intermediates must be replenished if any are drawn off for biosyntheses. The cycle will cease to operate unless new oxaloacetate is formed, because acetyl CoA cannot enter the cycle unless it condenses with oxaloacetate. Even though oxaloacetate is used catalytically, a minimal level must be maintained to allow the cycle to function.

The Anaplerotic Glyoxylic Pathway Bypass

The glyoxylate bypass is essential for growth on carbon sources such as acetate or fatty acids because this pathway allows the net conversion of acetyl CoA to metabolic intermediates. Strains lacking this pathway fail to grow on these carbon sources since acetate carbon entering the tricarboxylic acid cycle is quantitatively lost as CO_2 and thus there is no means to replenish the dicarboxylic acids consumed in amino acid biosynthesis.

During growth on acetate, about 75 % of the IDH is converted to the inactive phosphorylated form. This inhibition of IDH slows the tricarboxylic acid cycle and thus forces isocitrate through the bypass.

The first insight into the regulation of IDH came in the early 1970s when it was reported that the activity of this enzyme was increased when *E. coli* made the transition between growth on acetate and growth on preferred carbon sources such as glucose or pyruvate. Increased activity did not simply result from induction of IDH expression since it occurred even in the presence of protein synthesis inhibitors. Addition of glucose to a culture growing on acetate produced a metabolic crossover at IDH: the cellular level of isocitrate went down while that of α-ketoglutarate increased, suggesting that IDH had been activated during the transition. However, the mechanism responsible for regulation of IDH was not reported until 1979, when it was demonstrated that IDH activity was controlled by phosphorylation (see above). Growth of *E. coli* on acetate requires operation of the anaplerotic sequence known as the glyoxylate bypass (Fig. 8.3).

In this pathway three different enzymes are activated: malate synthase, isocitrate lyase and isocitrate dehydrogenase kinase/phosphatase which are encoded by genes *aceB, aceA* and *aceK*, respectively. These three genes are clustered, in that order, in the same acetate (*ace*) operon whose expression is under the transcriptional control of the *iclR* gene located downstream from *aceK*. The operon is expressed from a single promoter during growth on acetate. The IclR repressor has been cloned overproduced, then purified to homogeneity in a one-step procedure by cation exchange

Fig. 8.3 The anaplerotic glyoxylic acid cycle

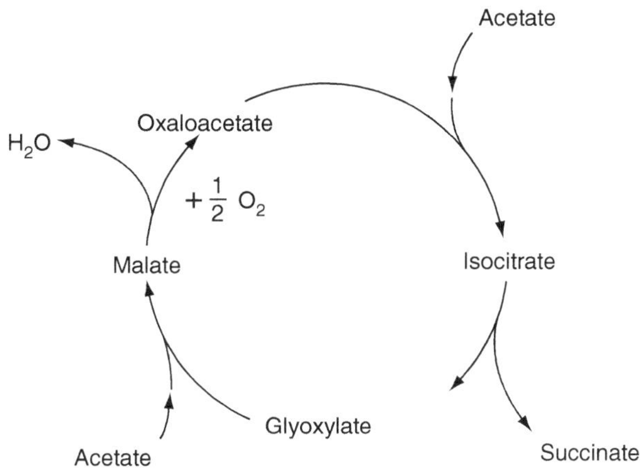

chromatography after ammonium sulfate fractionation. Its specific interaction with the operator/promoter region of the *ace* operon has been analyzed by gel retardation and DNase I footprinting experiments. The IclR repressor has been shown to recognize a 35 bp palindromic sequence, which largely overlaps the −35 recognition site of RNA polymerase. Moreover, the formation of the complex between IclR and the operator/promoter region has been found to be impaired by phosphoenolpyruvate but to be insensitive to acetate. Translation of *iclR* sequence revealed a protein with Mr 29,741 preceded by a potential Shine-Dalgarno ribosome-binding site. The deduced amino acid sequence includes a region at the amino terminus that may form a helix-turn-helix motif, a structure found in many DNA-binding domains. A binding site for integration host factor (IHF) was identified upstream of the *aceBAK* promoter. Under inducing conditions, IHF activates *aceB: lacZ* expression by opposing IclR repression. In contrast, IHF has little effect on *aceB: lacZ* expression under repressing conditions. The ability of IHF to relieve repression under inducing but not repressing conditions allows this protein to amplify the induction of *aceBAK*.

The response of *aceBAK* to fatty acids is mediated, in part, by FadR, initially identified as a repressor of the genes encoding fatty acid degradation.

The AceB protein is called malate synthase A (A for induction by acetate) to distinguish it from a second malate synthase, malate synthase G (G for induction by glycolate), encoded by the *glcB* gene, which functions in growth on glycolate or glyoxylate as sole carbon sources. The two malate synthases are distinguished by different stability and inhibition patterns. The enzymology of malate synthase A from *E. coli* is little explored, although its deduced amino acid sequence is known. It does not appear to have been purified to homogeneity and its subunit structure has not been reported. The enzyme from *Corynebacterium glutamicum* is a monomer and has been somewhat better characterized. In contrast, isocitrate lyase, a tetrameric protein has been purified and the active site and mechanism have been studied in some detail. Active isocitrate lyase is phosphorylated on a histidine residue and this modification seems essential for activity. The tetrameric enzyme from *E. coli* has been crystallized and its structure has been determined at 2.1 Å resolution. Comparison with the structure of the prokaryotic isocitrate lyase from *Aspergillus nidulans* reveals a different domain structure following the deletion of approximately 100 residues from the larger eukaryotic enzyme. Despite this, the active sites of the prokaryotic and eukaryotic enzymes are very closely related, including the apparent disorder of two equivalent segments of the protein that are known to be involved in a conformational change as part of the enzyme's catalytic cycle.

One role of the IDH phosphorylation cycle is to regulate the branch point between the glyoxylate bypass and the tricarboxylic acid cycle, during steady-state growth on acetate or fatty acids. Phosphorylation of IDH diverts some of the flux through the TCA cycle to the glyoxylate bypass. The immediate effect of phosphorylation is to inhibit IDH activity. The resulting increase in the level of isocitrate increases the velocity of isocitrate lyase, the first enzyme of the bypass. Mutant strains which are deficient in IDH kinase fail to grow on acetate, suggesting that phosphorylation of IDH is required for use of the glyoxylate bypass. IDH kinase/phosphatase also controls the glyoxylate bypass during transitions between preferred carbon sources (e.g., glucose or pyruvate) where the bypass is unnecessary, to a culture on acetate.

Selected References

The Pyruvate Dehydrogenase Complex

Green JD, Perham RN, Ullrich SJ, Appella E (1992) J Biol Chem 267:23484–23488
Nemeria NS, Arjunan P, Chandrasekhar K, Mossad M, Tittmann K, Furey W, Jordan F (2010) J Biol Chem 285:11197–11209
Oberholzer AE, Schneider P, Siebold C, Baumann U, Erni B (2009) J Biol Chem 284: 33169–33176
Song J, Park Y-H, Nemeria NS, Kale S, Kakalis L, Jordan F (2010) J Biol Chem 285:4680–4694
Teplyakov A, Lim K, Zhu PP, Kapadia G, Chen CC, Schwartz J, Howard A, Reddy PT, Peterkofsky A, Herzberg O (2006) Proc Natl Acad Sci USA 103:16218–16223

The Tricarboxylic Acid Cycle

Baldwin JF, Krebs H (1981) Nature 291:381–382
LaPorte DC (1993) J Cell Biochem 51:14–18

The Anaplerotic Glyoxylic Bypass

Kornberg HL (1966) Biochem J 99:1–11

Chapter 9
ATP-Generating Processes: Respiration and Fermentation

The Biological Oxidation of Organic Metabolites Is the Removal of Electrons

Respiration

Respiration is a process in which electrons are transferred sequentially through a series of membrane-bound protein carriers, the electron transport chain. Electrons are removed from membrane carriers by reducing some terminal electron acceptor such as oxygen (aerobic respiration) or nitrogen, sulfate, or carbon dioxide (anaerobic respiration). This process occurs in mitochondria in most eukaryotic cells, or in the cell membrane of prokaryotic cells.

Electron transport begins when electron carriers such as reduced nicotinamide adenine dinucleotide (NADH) release electrons (typically in the form of hydrogen atoms) to membrane-bound electron carriers,

Protons are translocated across the cell membrane, from the cytoplasm to the periplasmic space just outside the membrane. As protons accumulate outside the membrane, hydroxyl ions accumulate inside the membrane and combine with electrons and H+ ions to produce water.

As electrons flow through the electron transport chain, protons are simultaneously translocated across the membrane. As more electrons flow, more protons accumulate just outside the membrane, resulting in a substantial proton gradient. This gradient can be used for a variety of purposes, such as ATP synthesis, transport, or flagellar motion.

Proton gradients store energy, both because of charge separation and concentration differential, and protons would rapidly cross the membrane to restore equilibrium if allowed. The cell membrane is impermeable to protons, except through protein complexes called ATP synthases, containing a proton channel that allows re-entry of protons.

G.N. Cohen, *Microbial Biochemistry*,
DOI 10.1007/978-94-017-8908-0_9,

Energy released by their passage is coupled to synthesis of ATP from ADP and phosphate (Pi).

$$ADP + Pi \rightarrow ATP$$

This process is often called oxidative phosphorylation, since oxygen is a frequent electron acceptor, but is more accurately called chemiosmotic phosphorylation, since phosphorylation is coupled to the discharge of a chemiosmotic gradient.

Respiration can thus be defined as an ATP-generating process in which either organic or inorganic compounds serve as electron donors and inorganic compounds serve as electron acceptors. Usually, the ultimate electron acceptor is molecular oxygen. However, in the anaerobic respiration, a special class of respiratory processes characteristic of some bacteria, inorganic compounds other than oxygen, such as sulfates, nitrates and carbonates serve as ultimate electron acceptors.

NADH:ubiquinone oxidoreductase I (NDH-1) is an NADH dehydrogenase that catalyzes the transfer of electrons from NADH to the quinone pool in the cytoplasmic membrane and is able to generate a proton electrochemical gradient. It is part of both the aerobic and anaerobic respiratory chain of the cell. The study of this enzyme is of great interest, because it is considered to be a structurally minimal form of a proton-pumping NADH:ubiquinone oxidoreductase and serves as a model for the more complex mitochondrial enzyme of eukaryotes. The *Escherichia coli* enzyme, considered to be the minimal form of complex I, consists of 14 subunits. It contains flavin mononucleotide and FeS center prosthetic groups.

The number of protons pumped across the membrane is as yet unknown. Plausible mechanisms of electron transfer and its coupling to proton translocation have been deduced from the crystal structure of the peripheral arm of the *Thermus thermophilus* enzyme. Electron transfer from NADH via FMN to the iron-sulfur centers has been measured in real time.

The purified enzyme can be separated into three components: a soluble fragment composed of the NuoE, F and G subunits which catalyzes the oxidation of NADH, representing the electron input part of the enzyme ; an amphipathic connecting fragment composed of the NuoB, CD and I subunits; and a hydrophobic membrane fragment composed of the NuoA, H, J, K, L, M and N subunits. The soluble subunits contain all iron-sulfur clusters and the FMN cofactor; the redox properties of those cofactors have been studied. Results from crosslinking analysis suggest that the ubiquinone-binding site of the enzyme is located on the membrane subunit NuoM,but it has also been modeled to the interface between NuoB and NuoCD based on its location in the *T. thermophilus* enzyme. The NuoJ, NuoK, NuoM and NuoN subunits are implicated in the ability to generate an electrochemical gradient.

NDH-1 consists of three subcomplexes referred to as the peripheral, connecting, and membrane fragments (Fig. 9.1).

The 14 *nuo* genes encode the subunits of the type I (energy-conserving) NADH dehydrogenase) These genes are arranged as a single, large operon that is expressed from a complex promoter region upstream of *nuoA*. The subunit composition of

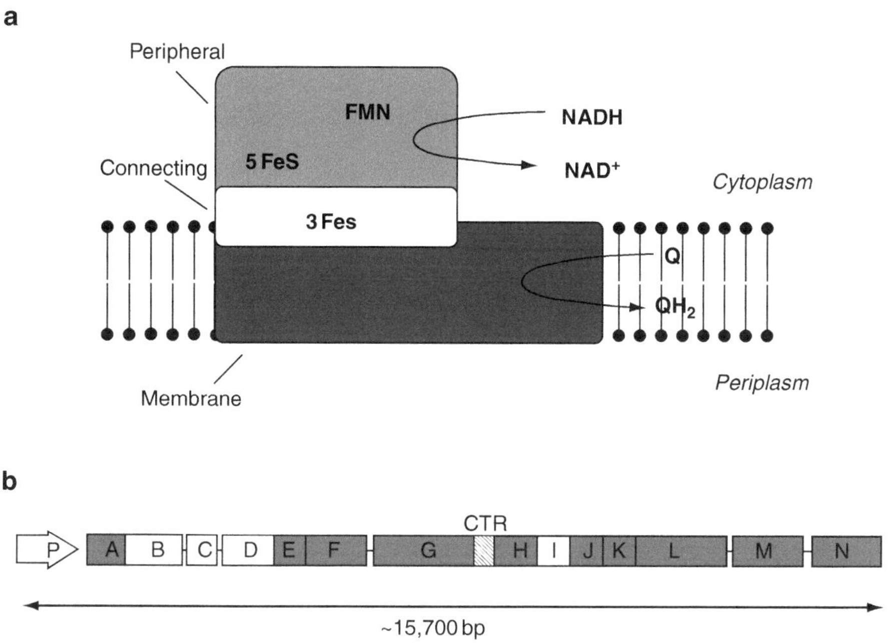

Fig. 9.1 Schematic of E. coli complex I and the corresponding *nuo* locus. (**a**) *E. coli* complex I is comprised of three distinct fragments: the peripheral (*light gray*), connecting (*white*), and membrane (*dark gray*). The peripheral fragment (NDF) is comprised of *NuoE, -F*, and *-G* and exhibits NADH dehydrogenase activity that oxidizes NADH to NAD^+; the connecting fragment is comprised of *NuoB, -C, -D,* and *-I*; and the membrane fragment is comprised of *NuoA, -H*, and *-J* to *-N* and catalyzes ubiquinone (*Q*) to its reduced form (*QH2*). FMN, flavin mononucleotide. (**b**) The *E. coli nuo* locus encodes the 14 Nuo subunits that constitute complex I. The 5′ half of the locus contains a promoter (*nuoP*) and the majority of genes that encode the peripheral and connecting subunits (*light gray* and *white*, respectively). The 3′ half of the locus contains the majority of the genes encoding the membrane subunits (*dark gray*). The 3′ end of *nuoG* encodes a C-Terminal region (*CTR*) of the *NuoG* subunit (hatched) (With permission of the American Society of Microbiology)

these three fragments correlates approximately with the organization of the 14 structural genes (*nuoA* to *nuoN*) of the *nuo* (for NADH:ubiquinone oxidoreductase) locus an organization that is conserved in several other bacteria.

Quinones are lipophilic molecules dissolved within the lipid bilayer of the cytoplasmic membrane. The *E. coli* membrane contains three types of quinones which all have an octaprenyl side chain (C40). These are a benzoquinone, ubiquinone, and two naphthoquinones, menaquinone and dimethylmenaquinone. Ubiquinone-8 is predominant during aerobic growth but is replaced by menaquinone-8 upon the transition from aerobiosis to anaerobiosis.

Modern α-proteobacteria are thought to be closely related to the ancient symbiont of eukaryotes, an ancestor of mitochondria. Respiratory complex I from α-proteobacteria and mitochondria is well conserved at the level of the 14 "core" *nuo* subunits, consistent with that notion. Mitochondrial complex I contains the core subunits, present in all species, and up to 31 "supernumerary" subunits, generally thought to have originated only within eukaryotic lineages. However, the full protein composition of an α-proteobacterial complex I had not been established until the first purification and characterization of complex I from the α-proteobacterium *Paracoccus denitrificans*. Single particle electron microscopy shows that the complex has a well-defined L-shape. Unexpectedly, in addition to the 14 core subunits, the enzyme also contains homologues of three supernumerary mitochondrial subunits. This finding suggests that evolution of complex I via addition of supernumerary or "accessory" subunits started before the original endosymbiotic event that led to the creation of the eukaryotic cell. It also provides further confirmation that α-proteobacteria are the closest extant relatives of mitochondria.

Three-dimensional reconstruction and 2-D crystals of the NDH-1 complex based on cryo-electron microscopy showed an L-shaped form with an integral membrane and a peripheral arm. A model of the spatial arrangement of the subunits and the possible functional mechanism of proton pumping has been proposed. Under low ionic strength conditions, the complex appears to adopt a horseshoe-like conformation. NDH-1 is primarily an electrogenic proton pump which may have secondary Na+/H+ antiport activity. It produces reactive oxygen species, mainly in the form of H2O2, at the NADH dehydrogenase active site, involving the FMN cofactor. The rate of O2 reduction is dependent on the NAD^+/NADH ratio.

NDH-1 is one of two distinct NADH dehydrogenases. In *E.coli* a second enzyme, NDH-2 (encoded by *ndh*), utilizes NADH exclusively, while NDH-1 can utilize both NADH and NADPH. NADH:ubiquinone oxidoreductase II (NDH-2) is a NADH dehydrogenase that catalyzes the transfer of electrons from NADH to the quinone pool in the cytoplasmic membrane. It is thus part of the aerobic respiratory chain of the cell, and its primary function may be maintenance of the [NADH]/[NAD^+] balance of the cell. In contrast to NDH-1 (encoded by the 14 *nuo* genes) the electron flow from NADH to ubiquinone does not generate an electrochemical gradient. NDH-2 is a strongly membrane-associated protein, which copurifies with phospholipids and contains an FAD cofactor and a thiolate-bound Cu(I) ion. NDH-2 can generate superoxide radicals and hydrogen peroxide by autooxidation of the FAD cofactor when the enzyme is overproduced or in the absence of quinones. However, under regular growth conditions, it is not the primary source of intracellular hydrogen peroxide.

Fermentation

Fermentation can be defined as an ATP-generating process in which organic compounds serve both as electron donors (becoming oxidized) and electron acceptors (becoming reduced). The compounds performing these two functions are usually two different metabolites derived from a single fermentable substrate (such as a sugar). The substrate gives rise to a mixture of end products, some more oxidized, some more reduced. The average oxidation level of the end products is identical to that of the substrate.

Pyridine nucleotides (NAD or NADP) reduced in one step of the process are subsequently oxidized in another. This general principle is illustrated by two fermentations, alcoholic fermentation (typical of the anaerobic metabolism of yeasts) and the homolactic fermentation (typical of the metabolism of certain lactic acid bacteria in the presence of air). Both are slight modifications of the Embden-Meyerhof pathway, whereby the two molecules of NADH oxidized are reduced in reactions involving the subsequent metabolism of pyruvate.

Carbohydrates are the principal substrates of fermentation. Among the bacteria, some compounds belonging to other chemical classes can also be fermented; organic acids, amino acids, purines and pyrimidines.

Some of these fermentations are described below.

Acetone-butanol Fermentation

The acetone butanol fermentation is one of the oldest fermentation known. The fermentation is based on culturing various strains of Clostridia in carbohydrate rich media under anaerobic conditions to yield butanol and acetone. *Clostridium acetobutylicum* is the organism of choice in the production of these organic solvents. These fermentations were out of favor till very recently because of the availability of acetone and butanol from the petroleum industry.

Today there is considerable amount of interest in these fermentations. However, the concentration of end products in these fermentations is quite small and the fermentations are a type of mixed fermentation yielding a mixture of compounds such as butyric acid, butanol, acetone etc. Attempts to increase yields by use of genetically altered strains or change in fermentation conditions have been partially successful.

The Stickland Reaction

Coupled deamination between two amino acids acting as donor and acceptor of hydrogen respectively is the principal chemical reaction by means of which *Clostridium sporogenes* and other anaerobic bacteria obtain their energy when growing on amino acids as sole source of carbon and nitrogen.

Stickland observed that, in the presence of suspensions of *Clostridium sporogenes*, certain naturally occurring L-amino acids reduced methylene blue and benzylviologen, thus showing them to be hydrogen donors. Among these were alanine, valine, leucine, isoleucine. Other amino acids were found to reoxidize reduced phenosafranin and benzylviologen in the presence of suspensions of the same organism, that is to be hydrogen acceptors. Examples of such amino acids were glycine, proline and ornithine. Stickland then showed that the interaction of a donor and acceptor amino acid in the presence of a bacterial suspension results in a mutual deamination as determined by ammonia formation.

In the case of the (alanine + glycine) and (alanine + proline) couples, the overall reactions are the following

$$\text{Alanine} + 2\text{glycine} + 2H_2O \rightarrow 3\text{acetate} + 3NH_3 + CO_2$$
$$\text{Alanine} + 2\text{proline} + 2H_2O \rightarrow 1\text{acetate} + 2\text{-}\delta - \text{aminovalerate} + CO_2$$

Ornithine Fermentation

L-ornithine, as a single substrate, is both an electron donor and acceptor and metabolized by some anaerobic bacteria in a way similar to the Stickland reaction: it is oxidized to acetate, alanine, and ammonia (oxidative pathway) and reduced to 5-aminovalerate through the formation of proline (reductive pathway). The oxidative degradation pathway, starts with the conversion of L-ornithine to the D isomer by ornithine racemase. D-Ornithine is next converted to 2,4-diaminopentanoate through the action of D-ornithine aminomutase), an adenosylcobalamine and pyridoxal phosphate dependent enzyme.

$$\underset{\text{D-ornithine}}{H_2N-(CH_2)_3-CH(NH_2)-COOH} \rightarrow \underset{\text{2,4-diaminopentanoate}}{CH_3-CH(NH_2)-CH_2-CH(NH_2)-COOH}$$

2,4-diaminopentanoate then undergoes a NAD + or NADP$^+$ dependent oxidative deamination by a specific dehydrogenase), leading to 2-amino-4-ketopentanoate.

$$CH_3-CH(NH_2)-CH_2-CH(NH_2)-COOH \rightarrow \underset{\text{2-amino-4-ketopentanoate}}{CH_3-CO-CH_2-CH(NH_2)COOH}$$

This compound is metabolized by a PLP-dependent thiolase (an α2β2 heterotetramer)through a thiolytic cleavage with coenzyme A (CoA) to form acetyl-CoA and D-alanine. So far,this thiolase has no significant homology with any known enzyme family.

Glycine and Proline Degradation

These two amino acids serve as hydrogen acceptors in the Stickland reaction. The knowledge of *Cl.sticklandii* and *Eubacterium aminophilum* genomes allows to attribute the proteins involved in glycine degradation to corresponding genes (see paragraph on the glycine reductase complex in Chap. 30). Interestingly, in *Cl.sticklandii*, a glycine riboswitch is localized upstream these clustered genes and cooperatively binds glycine to regulate the expression of the downstream genes. The energy-rich compound acetyl CoA is the product of glycine degradation.

In addition to the above systems, glycine is degraded to CO2 and NH3 by the four proteins of the glycine cleavage system: The P protein (*gcvP*) catalyzes the decarboxylation of glycine to CO2 and an aminomethyl group ; the H protein (*gcvH*) which contains a covalently bound lipoic acid prosthetic group serves as an electron sink and as a carrier of the aminomethyl group ; The T protein (*gcvT*) catalyzes the transfer of the C1 unit from H protein to tetrahydrofolate and the release of ammonia. The L protein (*lpd*), common to the lipoamide dehydrogenase of the pyruvate and α-ketoglutarate dehydrogenase complexes reoxidizes the reduced lipoic acid from the H protein to the disulfide form.

L-proline is first transformed to D-proline by proline racemase. D-proline is then degraded in *Cl.sticklandii* by D-proline reductase, consisting of three subunits, alpha, beta and gamma respectively. The alpha and beta subunits are the products of the cleavage of a preprotein.. The product of the degradation is 5-aminopentanoate (δ-aminovalerate), which is not further degraded by *Cl.sticklandii*, since no genes coding for proteins catalyzing its catabolism, similar to those detected in *Cl.aminovalericum* have been detected.

Threonine Degradation

Clostridium sticklandii can degrade threonine either using the catabolic threonine dehydratase (*tdcB*)

$$CH_3-CHOH-CH(NH_2)-COOH \rightarrow \underset{\alpha-\text{ketobutyrate}}{CH_3-CH_2-CO-COOH} + NH_3$$

or via the oxidoreductase pathway, using successively threonine dehydrogenase (*tdh*) and 2-amino-3-ketobutyrate coenzyme A ligase (*kbl*).

$$CH_3-CHOH-CH\left(NH_2\right)-COOH + NAD^+ \rightarrow \underset{\text{2-amino-3-ketobutyrate}}{CH_3-CO-CH\left(NH_2\right)-COOH} + NADH + H^+$$

The latter compound is cleaved to acetyl CoA and glycine, which can be further degraded (see preceding paragraph).

$$CH_3-CO-CH(NH_2)-COOH + CoA \rightarrow \underset{\text{Acetyl CoA}}{CH_3-CO-S-CoA} + \underset{\text{Glycine}}{H_2N-CH_2-COOH}$$

A third threonine degrading enzyme is also present in *Cl.sticklandii*, threonine aldolase, which converts threonine to glycine and acetaldehyde.

$$HOOC\text{-}(CH_2)_2\text{-}CH(NH_2)\text{-}COOH \xrightarrow[1]{} HOOC\text{-}CH(CH_3)\text{-}CH(NH_2)\text{-}COOH$$

Glutamate Methylaspartate

$$\xrightarrow[2]{} HOOC\text{-}CH(CH_3)CH\text{-}COOH \xrightarrow[3]{} HOOC\text{-}C(OH)CH_3)\text{-}CH_2\text{-}COOH$$

Mesaconate Citramalate

$$\xrightarrow[4]{} CH_3\text{-}CO\text{-}COOH + CH_3\text{-}COOH$$

Pyruvate Acetate

Fig. 9.2 Methylaspartate pathway of glutamate fermentation. *1*: coenzyme B12-dependent glutamate mutase; *2* :methylaspartate ammonia lyase; *3* : citramalate dehydratase; *4*: citramalate lyase

$$HOOC\text{-}(CH_2)_2\text{-}CH(NH_2)\text{-}COOH \xrightarrow[1]{} HOOC\text{-}(CH_2)_2\text{-}CO\text{-}COOH \xrightarrow[2]{}$$

Glutamate 2-oxoglutarate

$$HOOC\text{-}(CH_2)_2\text{-}C(OH)\text{-}COOH \xrightarrow[3]{} HOOC\text{-}(CH_2)_2\text{-}CO\text{-}SCoA \xrightarrow[4]{}$$

2-hydroxyglutarate 2-hydroxyglutarylCoA

$$HOOC\text{-}CH_2\text{-}CH{=}CH\text{-}CO\text{-}SCoA \xrightarrow[5]{} CH_3\text{-}CH{=}CH\text{-}COSCoA$$

Glutaconyl CoA Crotonyl CoA

Fig. 9.3 Hydroxyglutarate pathway of glutamate fermentation. *1*: glutamate dehydrogenase; *2*: 2-hydroxyglutarate dehydrogenase; *3*: gluconate/2-hydroxyglutarate CoA decarboxylase; *4*: 2-hydroxyglutaryl CoA dehydratase; *5*: glutaconyl CoA decarboxylase

Glutamate Degradation

Two pathways have been involved in the anaerobic fermentation of glutamate to acetate, butyrate, carbon dioxide and ammonia : the methylaspartate and the hydroxyglutarate pathway. Enzymes specific for each pathway were assayed in crude extracts of many organisms. The methylaspartate pathway appears to be used only by species of Clostridia.

Glutamate is first degraded to ammonia, acetate and pyruvate. Pyruvate is then oxidatively decarboxylated to acetyl-CoA by pyruvate-ferredoxin oxidoreductase. The reduced ferredoxin is oxidized by acetyl-CoA and protons leading to butyrate and hydrogen. The intermediates of this degradation of glutamate are methylaspartate, mesaconate, and citramalate (Fig. 9.2)

Some bacteria ferment glutamate by the hydroxyglutarate pathway (Fig.9.3)

Crotonyl-CoA is then reduced to butyryl-CoA, which reacts with acetoacetate to form butyrate and acetoacetyl-CoA. The latter is converted to acetate via acetyl-CoA and acetyl phosphate (*atoA* and *atoD*).

Lysine Degradation

In the early 1950s, it was demonstrated that lysine was decomposed to acetate, butyrate and ammonia by *Clostridium sticklandii*. This catabolic pathway comprises ten distinctive reactions. The two first reactions are aminomutases. The catalysis requires the participation of a 5′-deoxyadenosyl radical which is provided either by S-adenosylmethionine (SAM) or by coenzyme B12 (adenosylcobalamin). The first aminomutase is a homohexamer, coded by *kamA*.It contains one pyridoxal phosphate and one Zn++ or Co++ per subunit and 3(4Fe-4S) clusters per hexamer. SAM is not tightly bound to the enzyme but is required as a cofactor. The second aminomutase,a coenzyme B12-dependent enzyme, is composed of two different subunits,α (57 kDa) coded by *kamD* and β (29 kDa) coded by *kamE*, forming a heterotetramer α2β2. Both enzymes have similar mechanisms with respect to the amlno group migration, but differ in formation of the 5′-deoxyadenosyl radical which originates from SAM in the first case and from coenzyme B12 in the second.

The product of the second aminomutase is 3,5-diaminohexanoate, which undergoes an oxidative deamination to 3-keto-5-aminohexanoate under the influence of a dehydrogenase (*kdd*).

$$\underset{\beta\text{-lysine}}{H_2N-(CH_2)_4-CH(NH_2)-COOH} \rightarrow \underset{3,5\text{-diaminohexanoate}}{CH_3-CH(NH_2)-CH_2-CH(NH_2)-CH_2-COOH}$$

$$\underset{3,5\text{-diaminohexanoate}}{CH_3-CH(NH_2)-CH_2-CH(NH_2)-CH_2-COOH} \rightarrow \underset{3\text{-keto-5-aminohexanoate}}{CH_3-CH(NH_2)-CH_2-CO-CH_2-COOH}$$

This compound is further converted into 3-aminobutyryl-CoA and acetoacetate (in the presence of acetyl-CoA). This reaction, catalyzed by 3-keto-5-aminohexanoate cleavage enzyme and encoded by *kce* presents a new type of β-ketoacid cleavage and acetoacetate synthesis.

$$\underset{3\text{-keto-5-aminohexanoate}}{CH_3-CH(NH_2)-CH_2-CO-CH_2-COOH} \rightarrow \underset{3\text{-aminobutyryl CoA}}{CH_3-CH(NH_2)-CH_2-CO-S-CoA}$$

Crystals of this enzyme with and without substrate have been obtained and their study has allowed to propose a tentative catalytic mechanism. Although the enzyme shows the ubiquitous triose phosphate isomerase (TIM) barrel fold and a Zn(2+) cation reminiscent of metal-dependent class II aldolases, results based on a combination of x-ray snapshots and molecular modeling point to an unprecedented mechanism that proceeds through deprotonation of the 3-keto-5-aminohexanoate substrate, nucleophilic addition onto an incoming acetyl-CoA, intramolecular transfer of the CoA moiety, and final retro-Claisen reaction leading to acetoacetate and 3-aminobutyryl-CoA. This model also accounts for earlier observations showing the origin of carbon atoms in the products, as well as the absence of detection of any covalent acyl-enzyme intermediate.

In the next step, 3-aminobutyryl-CoA is deaminated to crotonyl-CoA by an ammonia lyase (*kal*).

$$\underset{\text{3 aminobutyryl CoA}}{CH_3-CH(NH_2)-CH_2-CO-S-CoA} \rightarrow \underset{\text{Crotonyl CoA}}{CH_3-CHCH-CO-C-CoA} + NH_3$$

Arginine Fermentation

Analysis of several genomes from aerobic and anaerobic bacteria reveals that arginine can be degraded via the so-called ADI pathway, implicating the successive action of arginine dihydrolase, catabolic ornithine carbamoyl transferase and carbamate kinase, the latter yielding one mole of ATP per mole of arginine degraded:

$$\text{L}-\text{Arginine}+H_2O \rightarrow \text{L}-\text{citrulline}+NH_3\ (arcA)$$
$$\text{L}-\text{citrulline}+\text{Pi} \rightarrow \text{carbamoyl phosphate}+\text{L}-\text{ornithine}\ (arcB)$$
$$\text{Carbamoyl phosphate}+\text{ADP} \rightarrow \text{ATP}+NH_3+CO_2\ (arcC)$$

The assignation of the above three genes in *Clostridium sticklandii* has been validated by comparison to the authentic proteins from *Streptococcus pyogenes*, *Bacillus licheniformis* and *Enterococcus faecium*, respectively.

Arginine degradation may also involve the Roc system, since a putative arginase (*rocF*) and genes corresponding to the RocB, RocC and RocR proteins are present in the genome of *Cl.sticklandii*. These proteins code for the transport of arginine, for a regulatory protein and for arginase

$$\text{L}-\text{arginine}+H_2O \rightarrow \text{L}-\text{ornithine}+\text{urea}$$

Organisms possessing the genes coding for urease and its accessory proteins can then degrade urea to CO2 and NH3

$$H_2N-CO-NH_2+H_2O \rightarrow CO_2+2NH_3$$

Methionine Degradation

Many bacteria catabolize L-methionine through α-ketobutyrate by three main pathways:

(a) conversion of methionine to cystathionine through S-adenosylmethionine and homocysteine, and then to cysteine, α-ketobutyrate and ammonia

(b) deamination to α-keto-γ-methylthiobutyrate and the subsequent dethiomethylation to α-ketobutyrate
(c) simultaneous deamination and dethiomethylation to α-ketobutyrate by L-methionine γ-lyase

The latter enzyme is pyridoxal phosphate dependent. It catalyzes the direct conversion of L-methionine into ammonia, α-ketobutyrate and methylmercaptan:

$$\underset{\text{Methionine}}{H_3C-S-(CH_2)_2-CH(NH_2)-COOH} \rightarrow \underset{\alpha-\text{ketobutyrate}}{CH_3-CH_2-CO-COOH} + NH_3 + CH_3SH$$

It is present in various bacteria, such as several pseudomonads and clostridia and is induced by methionine. Alpha-ketobutyrate, a main product of methionine catabolism, is converted to propionyl coenzyme A by the pyruvate dehydogenase complex or to α-aceto-α-hydroxybutyrate with pyruvate by α-acetolactate synthase, an enzyme of isoleucine biosynthesis (see Chap. 29).

In addition to *mdeA*, which codes for the methionine γ-lyase, the *mde* operon contains *mdeB*, encoding a homodimeric protein, homologous to the E1 component of the pyruvate dehydrogenase complex, with high specificity for α- ketobutyrate rather than for pyruvate. An *mdr* gene product, Mdr, a member of the leucine-responsive regulatory family (Lrp, see Chap. 26), acts as an essential positive regulatory protein allowing the expression of the *mdeAB* operon. The three genes have been cloned and sequenced in *Pseudomonas putida.*

D-selenocystine and D-cysteine Degradation

In addition to L-cysteine desulfurase, involved in the construction of Fe-S clusters, *Cl.sticklandii* possesses also a gene coding tor D-cysteine desulfhydrase (*dcyD*), homologous to the authentic corresponding gene in *E.coli*. This enzyme, producing H2S, ammonia and pyruvate could be a defense mechanism against D-cysteine. Actually, D-cysteine inhibits the growth of *E.coli* by inhibiting threonine deaminase, a key enzyme in the valine-leucine-isoleucine biosynthetic pathway. The simultaneous addition of these three amino acids antagonizes the growth inhibition by D – cysteine.

Twenty years ago, an enzyme was found in *Cl.sticklandii* that catalyzed the degradation of selenocystine in pyruvate, ammonia and elemental selenium. The gene for this D-selenocystine α,β-lyase has not been identified yet. The enzymatic properties this enzyme resemble those of *E. coli* D-cysteine desulfhydrase (see above). It behaves as a pyridoxal phosphate-containing enzyme and, similarly to the *E. coli*, D-cysteine desulfhydrase catalyzes elimination reactions with D-selenocystine as well as D-cysteine and D-cystine as substrates. Because of these similarities, it could therefore be suspected that D-cysteine desulfhydrase and D-selenocystine α,β -lyase are the same enzyme. However, immunochemical analysis shows that, in the case of *E.coli*, the two enzymes are different.

Selected References

NADH-ubiquinone Oxidoreductases

Archer CD, Elliott T (1995) J Bacteriol 177:2335–2342
Baranova EA, Holt PJ, Sazanov LA (2007) J Mol Biol 366:140–154
Calhoun MW, Gennis RB (1993) J Bacteriol 175:3013–3019
Esterházy D, King MS, Yakovlev G, Hirst J (2008) Biochemistry 47:3964–3967
Falk-Krzesinski HJ, Wolfe AJ (1998) J Bacteriol 180:1174–1184
Juarez O, Nilges MJ, Gillespie P, Cotton J, Barquera B (2008) J Biol Chem 283:33162–33167
Meio AM, Bandeira TM, Teixeira M (2004) Microbiol Mol Biol Rev 68:603–616
Stolpe S, Friedrich T (2004) J Biol Chem 279:18377–18383
Torres-Bacete J, Sinha PK, Matsuno-Yagi A, Yagi T (2011) J Biol Chem 286:34007–34014
Yip C, Harbour M, Jayawardena K, Fearnley IM, Sazanov LA (2011) J Biol Chem 286:5023–5033

From Quinones to Oxygen

Borisov VB, Verkhovsky MI (2009) Module 2.3.7., Oxygen as acceptor. In: Böck A, Curtiss R III, Kaper JB, Karp PD, Neidhardt FC, Nyström T, Slauch JM, Squires CL, Ussery D (eds) EcoSal – *Escherichia coli* and Salmonella: cellular and molecular biology. ASM Press, Washington, DC, http://www.ecosal.org
Gennis RB, Stewart V (1996) In: *Escherichia coli and Salmonella typhimurium* – cellular and molecular biology, Curtiss III R, Ingraham JL, Low ECKB, Magasanik B, Reznikoff WS, Riley M, Schaechter M, Umbarger HE (eds.), vol 1, 2nd edn., pp 217–261, American Society for Microbiology, (Washington DC)

The Stickland Reaction

Nisman B (1954) Bact Revs 18:16–42
Stickland LH (1934) Biochem J (London) 28:288–297

Arginine and Ornithine Degradation

Fonknechten N, Perret A, Perchat N, Tricot S, Lechaplais C, Vallenet D, Vergne C, Zaparucha A, Le Paslier D, Weissenbach J, Salanoubat M (2009) J Bacteriol 191:3162–3167
Gardan R, Rapoport G, Debarbouille M (1995) J Mol Biol 249:843–856
Schimke RT, Berlin CM, Sweeney EW, Carroll WR (1966) J Biol Chem 241:2228–2236
Wolthers KR, Rigby SE, Scrutton NS (2008) J Biol Chem 283:4615–34625
Wolthers KR, Levy C, Scrutton NS, Leys D (2012) J Biol Chem 285:13942–13950

Threonine Degradation

Schmidt A, Sivaraman J, Li Y, Larocque R, Barbosa JA, Smith C, Matte A, Schrag JD, Cygler M (2001) Biochemistry 40:5151–5160
Simanshu DK, Savithri HS, Murthy MR (2006) J Biol Chem 281:39630–39641

Wagner M, Andreesen JR (1995) Arch Microbiol 163:286–290

Glycine Degradation

Bednarski B, Andreesen JR, Pich A (2001) Eur J Biochem 268:3538–3544
Graentzdoerffer A, Pich A, Andreesen JR (2001) Arch Microbiol 175:8–18

Proline Degradation

Kabisch UC, Graentzdoerffer A, Schierhorn A, Ruecknagel KP, Andreesen JR, Pich A (1999) J Biol Chem 274:8445–8454
Rudnick G, Abeles RH (1975) Biochemistry 14:4515–4522

Glutamate Degradation

Buckel W, Barker HA (1974) J Bacteriol 117:1248–1260

Lysine Degradation

Bellinzoni M, Bastard K, Perret A, Zaparucha A, Perchat N, Vergne C, Wagner T, de Melo Minardi RC, Artiguenave F, Cohen GN, Weissenbach J, Salanoubat M, Alzari P (2012) J Biol Chem 286:27399–27405
Kreimeyer A, Perret A, Lechaplais C, Vallenet D, Médigue C, Salanoubat M, Weissenbach J (2007) J Biol Chem 282:7191–7197
Moss M, Frey PA (1987) J Biol Chem 262:14859–14862
Stadtman T (1973) Adv Enzymol Relat Areas Mol Biol 38:413–448

Methionine γ-lyase

Inoue H, Inagaki K, Eriguchi S-I, Tamura T, Esaki N, Soda K, Tanaka H (1997) J Bacteriol 179:3956–3962
Wiesendanger S, Nisman B (1953) C R Acad Sci 237:764–765

D-selenocystine and D-cysteine Degradation

Esaki N, Seraneeprakarn V, Tanaka H, Soda K (1988) J Bacteriol 170:751–756
Nagasawa T, Ishii T, Kumagai H, Yamada H (1985) Eur J Biochem 153:541–551
Soutourina J, Blanquet S, Plateau P (2001) J Biol Chem 276:40864–40872

Chapter 10
Biosynthesis of Lipids

Biosynthesis of Short Chain Fatty Acids

During the late 1930s, Barker isolated an anaerobic bacterium which he called *Clostridium kluyveri*. This organism is able to grow on a mixture of acetate and ethanol and produces *n*-butyrate and *n*-caproate. From this organism, Barker and Stadtman obtained cell-free extracts which could carry the reactions leading to the C4 and C6 acids. Ethanol is oxidized to acetic acid in two steps, respectively, catalyzed by alcohol dehydrogenase and an acetaldehyde dehydrogenase depending on the presence of coenzyme A:

$$CH_3-CH_2OH + NAD^+ \rightarrow CH_3-CHO + NADH + H^+$$
$$CH_3-CHO + CoASH + NAD^+ \rightarrow CH_3-CO-SCoA + NADH + H^+$$

One notes that the reactive species acetyl CoA is formed, rather than acetic acid. A β-ketothiolase condenses then two molecules of acetyl CoA by joining the methyl group of one molecule with the carboxyl group of the other (head to tail condensation):

$$2\,CH_3-CO\text{–}SCoA \rightarrow \underset{\text{Acetoacetyl CoA}}{CH_3-CO-CH_2-CO-SCoA} + CoASH$$

Acetoacetyl CoA is reduced by β-hydroxybutyryl CoA dehydrogenase:

$$CH_3-CO-CH_2-COSCoA + NADH + H^+ \rightarrow \underset{\beta\text{-hydroxybutyryl CoA}}{CH_3-CHOH-CH_2-COSCoA} + NAD^+$$

This compound is then dehydrated by a specific crotonase yielding crotonyl CoA, CH_3–CH=CH–COSCoA which is reduced to butyryl CoA CH_3CH_2–CH_2–COSCoA by butyryl CoA dehydrogenase.

G.N. Cohen, *Microbial Biochemistry*,
DOI 10.1007/978-94-017-8908-0_10,

The synthesis of caproate occurs by an analogous series of reactions starting from the condensation of acetyl CoA with butyryl CoA giving rise to β-ketocaproyl CoA. If *Cl. kluyveri* is grown on ethanol and propionate, one obtains the odd-numbered acids, *n*-pentanoic and *n*-heptanoic acids by the same type of reactions.

Biosynthesis of Long-chain Fatty Acids

Synthesis of acetyl CoA

In addition to the synthesis generated by the pyruvate dehydrogenase complex, bacteria and plants can synthesize acetyl CoA from acetate and CoA by an ATP-driven reaction catalyzed by acetyl CoA synthetase.

$$CH_3COOH + CoA + ATP \rightarrow Acetyl\,CoA + AMP + PPi$$

Animals synthesize acetyl CoA through the pyruvate dehydrogenase complex.

Synthesis of Malonyl CoA

After having observed in 1957 that bicarbonate is required for fatty acid synthesis, Wakil identified malonyl CoA as the first intermediate in the synthesis of long-chain (C10 to C18) fatty acids. He discovered that its synthesis is catalyzed by a biotin-containing enzyme, the biotin been covalently linked to an ε-amino group belonging to a lysine residue. This enzyme, acetyl CoA carboxylase, catalyzes a carboxylation of acetyl CoA in the presence of ATP:

$$CH_3-COSCoA + CO_2 + ATP \rightarrow \underset{\text{Malonyl CoA}}{HOOC-CH_2-COSCoA} + ADP + Pi$$

In yeast, acetyl CoA carboxylase is a tetramer of a multifunctional protein encodes by the FAS3 gene.

E. coli acetyl CoA carboxylase has been resolved into three subunits, each of which catalyze a partial reaction. Biotin is linked covalently to a small 16,688 Da protein (deduced from the DNA sequence), called "biotin carboxyl carrier protein" (BCCP).

The carboxylation of the biotin is due to another subunit of the enzyme, biotin carboxylase. The DNA sequence of the carboxylase corresponds to a protein of a molecular weight of 49,000 (449 amino acid residues). The gene encoding BCCP is located upstream of the biotin carboxylase gene and the two genes are transcribed as a single mRNA species that contains an unusually long untranslated leader preceding the BCCP gene. The last subunit is a transcarboxylase responsible for the transfer of the activated CO_2 of carboxybiotin to acetyl CoA. The length and flexibility of the arm linking biotin to its carrier allow the activated carboxyl group to move from one subunit to the other.

The reaction catalyzed by acetyl CoA carboxylase is the rate-limiting step in fatty acid synthesis. The bacterial enzyme is allosterically inhibited by palmityl CoA, the end product of the synthesis.

The mammalian enzyme is activated by citrate which is not a substrate for the reaction. In the enzyme from adipose tissue and rat liver, this important activation is accompanied by a conformational change in the protein: its sedimentation coefficient which is normally 18.8 S becomes 43 S when it is incubated and centrifuged with citrate under the same conditions as are necessary for activation. The rate of activation of the enzyme is concentration dependent and there are grounds for thinking that activation and aggregation are two facets of the same phenomenon. Citrate accumulates when NAD-dependent isocitrate dehydrogenase is non-functional due to the lack of AMP; besides the effect of citrate on the carboxylase, the exclusive functioning of the other isocitrate dehydrogenase, the one requiring NADP, increases the amount of NADPH which becomes available for fatty acid synthesis. We should note that the addition of citrate to tissue slices causes an increase in fatty acid synthesis and a decrease in sterol synthesis, both of which start from acetyl CoA.

Mammalian acetyl CoA carboxylase is also regulated covalently by phosphorylation which inactivates it and dephosphorylation which reactivates it.

From Malonyl CoA to Palmitate

It is known that the main compound synthesized from acetate by extracts purified from pigeon liver is palmitic acid, the saturated fatty acid with a linear chain of 16 carbon atoms. The overall equation representing the stoichiometry of its synthesis is the following:

$$\text{Acetyl CoA} + 7\,\text{malonyl CoA} + 14\ \text{NADPH} + 14\ \text{H}^+$$
$$\downarrow$$
$$\text{CH}_3-(\text{CH}_2)_{14}-\text{COOH} + 7\,\text{CO}_2 + 8\ \text{CoASH} + 14\ \text{NADP}^+ + 6\ \text{H}_2\text{O}$$

palmitate

The carbon atoms of acetyl CoA are incorporated into the methyl group and the last methylene group of palmitic acid, whilst the carbon atoms of malonyl CoA appear as carbon atoms 1–14. Acetyl CoA can be substituted in this reaction by the butyryl, hexanoyl or tetradecanoyl CoA derivatives, a fact that led to the belief that these compounds were probable intermediates in the synthesis of palmitate. We shall see that this is not so.

Work on the synthesis of palmitate has been mainly carried out with three systems, isolated respectively from *E. coli*, the yeast *S. cerevisiae* and chicken liver. It has been shown in *E. coli* that the product of the condensation of acetyl CoA and malonyl CoA is an acetoacetyl derivative bound to a heatstable protein fraction; this protein has been called acyl carrier protein (ACP).

ACP is a polypeptide of 77 residues in which a 4′-phosphopantetheine residue (see the section on pantothenic acid and CoA synthesis) is linked through a phosphodiester bond to a serine residue. ACP can be considered as a giant prosthetic group, a macro CoA.

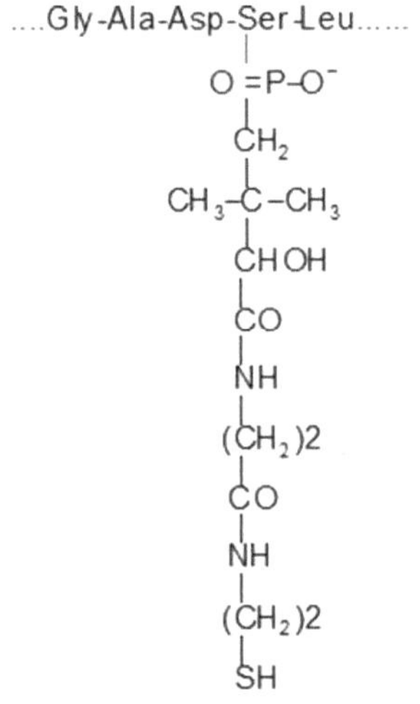

The acyl carrier protein

It is to the SH-group of phosphopantetheine that, during the synthesis, the acetyl, malonyl, and acyl groups of intermediate size are covalently linked.

The gene that encodes the acyl carrier protein from *E.coli* (*acpS*) lies within a cluster of known fatty acid biosynthetic genes.

The reactions that lead to butyryl-ACP are the following:

$$\text{Malonyl CoA} + \text{ACP} - \text{SH} = \text{Malonyl} - \text{S} - \text{ACP} + \text{CoASH}$$
$$\text{Acetyl CoA} + \text{ACP} - \text{SH} = \text{Acetyl} - \text{S} - \text{ACP} + \text{CoASH}$$
$$\text{Acetyl} - \text{S} - \text{ACP} + \text{Malonyl} - \text{S} - \text{ACP} = \text{Acetoacetyl} - \text{S} - \text{ACP} + \text{ACP} - \text{SH} + \text{CO}_2$$
$$\text{Acetoacetyl} - \text{S} - \text{ACP} + \text{NADPH} + \text{H}^+ = \beta - \text{hydroxybutyryl} - \text{S} - \text{ACP} + \text{NADP}^+$$
$$\beta - \text{hydroxybutyryl} - \text{S} - \text{ACP} = \text{Crotonyl} - \text{S} - \text{ACP} + \text{H}_2\text{O}$$
$$\text{Crotonyl} - \text{S} - \text{ACP} + \text{NADPH} + \text{H}^+ = \text{Butyryl} - \text{S} - \text{ACP} + \text{NADP}^+$$

Once butyryl-ACP is formed, a secound round of synthesis takes place where butyryl-ACP condenses with malonyl-ACP to form a β-ketoacyl-ACP with six carbon atoms; the successive reactions lead to caproyl-ACP which is the substrate for a new elongation cycle. This continues until palmityl-ACP with 16 carbon atoms is obtained. This compound is not a substrate for the condensation enzyme: it is hydrolyzed and liberates ACP and free palmitate. Thus coenzyme A esters are not involved in this synthesis after the first two steps which form the ACP derivatives of acetate and malonate. The next four reactions previously demonstrated to occur with CoA reflect a lack of absolute specificity in the enzymes whose normal substrates are the ACP-bound derivatives.

The enzymes which constitute the bacterial system dissociate when the cells are disrupted.

In autotrophic members of the Sulfolobales (Crenarchaeota, see Chap. 12) , the product of the initial acetyl-CoA carboxylation with CO2, malonyl-CoA, is further reduced to malonic semialdehyde by an NADPH-dependent malonyl-CoA reductase. Structural analysis of *Sulfolobus tokodaii* malonyl-CoA reductase in the substrate-free state and in complex with $NADP^+$and CoA revealed an unexpected reaction cycle in which $NADP^+$and CoA successively occupy identical binding sites. Both coenzymes are pressed into an S-shaped, nearly superimposable structure imposed by a fixed and preformed binding site. The template-governed cofactor shaping implicates the same binding site for the 3′- and 2′-ribose phosphate group of CoA and $NADP^+$, respectively, but a different one for the common ADP part: the β-phosphate of CoA aligns with the α-phosphate of $NADP^+$. Evolution from an $NADP^+$to a bispecific $NADP^+$and CoA binding site involves many amino acid exchanges within a complex process by which constraints of the CoA structure also influence $NADP^+$binding. Based on the paralogous aspartate-β-semialdehyde dehydrogenase (see Chap. 25) structurally characterized with a covalent Cys-aspartyl adduct, a malonyl/succinyl group can be reliably modeled into malonyl-CoA reductase and discussed regarding its binding mode, the malonyl/succinyl specificity, and the catalyzed reaction.

The fatty acid synthetases of eukaryotes, in contrast to the bacterial ones, are well defined multifunctional proteins. The one from yeast has a mass of 2,200 kDa and appears under the electron microscope as an ellipsoid having axes of 250 and 210 Å. It consists of only two multifunctional polypeptide chains, α and β. The enzyme has an $\alpha6\beta6$ structure. The α chain (206 kD) carries the condensation enzyme, β-ketoacylreductase activities and ACP, while the β chain (229 kDa) carries the acyl transacylase, the malonyl transacylase, the β-hydroxyacyl dehydratase and the enoyl reductase activities. It is the genetic study of yeast mutants which has demonstrated that the genes coding for the seven activities belonged to only two complementation groups. This led to the reexamination of the previous belief that these proteins were distinct objects associated by non covalent bonds. The yeast synthetase has been examined by negative-stain and cryo electron microscopy and a model has been proposed where the six fatty acid synthesizing centers are composed of two complementary half-α subunits and a β subunit, an arrangement having all the partial activities of the multifunctional enzyme required for fatty acid synthesis.

The fatty acid synthetase from animals (230 kD) is a dimer of identical subunits organized in a head to tail manner, such as the cysteine of the active group of the β-ketoacylreductase (peripheral – SH group) of one subunit be very near the sulfhydryl group of the ACP of the other subunit. Each subunit is organized in three globular domains linked by flexible hinges as demonstrated by limited proteolysis experiments. Domain I plays the role of a condensation unit, containing acetyl and malonyl transacylases and the condensation enzyme. Domain II plays the role of a reduction unit and contains ACP, β-ketoacylreductase, dehydratase and enoylreductase. Domain III is responsible for the thioesterase activity which liberates the palmitic acid.

The elongation begins on subunit A with the condensation of the acetyl unit, fixed on a peripheral sulfhydryl with a malonyl residue fixed on ACP. A molecule of CO_2 is then liberated and an acetoacetyl-S-phosphopantetheinyl unit is formed on ACP, liberating the peripheral SH of the active site of the condensation enzyme. The acetoacetyl group is transferred on domain II of subunit B and is subject to the three-step reduction to butyryl-ACP. This C-4 unit is then transferred from the SH of ACP to the SH of the condensation enzyme of subunit A. The enzyme is now ready for a second round. When the growing chain has reached a length of 16 carbon atoms, the palmitoyl residue still hooked upon ACP becomes the substrate of the esterase of the opposite chain. This remarkable translocation mechanism is reminiscent of that of the elongation of the nascent polypeptide chain on ribosomes. The flexibility and the maximal length of 20 Å of the articulated arm constituted by phosphopantetheine are essential for the functioning of this elaborated multifunctional enzyme. It is not necessary for the subunits to undergo important rearrangements to interact with the substrates: the latter reside on a long and flexible arm which can reach the active sites of all the enzymes involved in the overall reaction, without dilution of the intermediates which are sequestered and protected from competing reactions.

In eukaryotes, fatty acids are synthesized in the cytoplasm whereas acetyl CoA is synthesized from pyruvate in the mitochondria. Acetyl CoA must therefore be transferred; this is done by condensation of acetyl CoA with oxaloacetate to yield citrate. Whereas mitochondria are impermeable to acetyl CoA, they are not to citrate. Once citrate is in the cytoplasm, it is cleaved by citrate lyase:

$$\text{Citrate} + \text{ATP} + \text{CoA} + H_2O \rightarrow \text{acetylCoA} + \text{ADP} + \text{oxaloacetate} + \text{Pi}$$

Thus, the transfer is made at the expense of an ATP molecule.

Fatty acids longer than palmitate are synthesized by enzymes located at the internal side of the membrane of the endoplasmic reticulum, which add sequentially two-carbon atom units to the carboxyls of saturated and unsaturated acids. The same microsomal system introduces the double bonds in the long chain fatty acids, at least until carbon atom 9. Mammals do not possess enzymes allowing the introduction of double bonds beyond that limit; so linoleic and linolenic acid, which possess double bonds respectively at C-12, and C-12 and C-15, are essential fatty acids which must be present in the diet (Fig. 10.1).

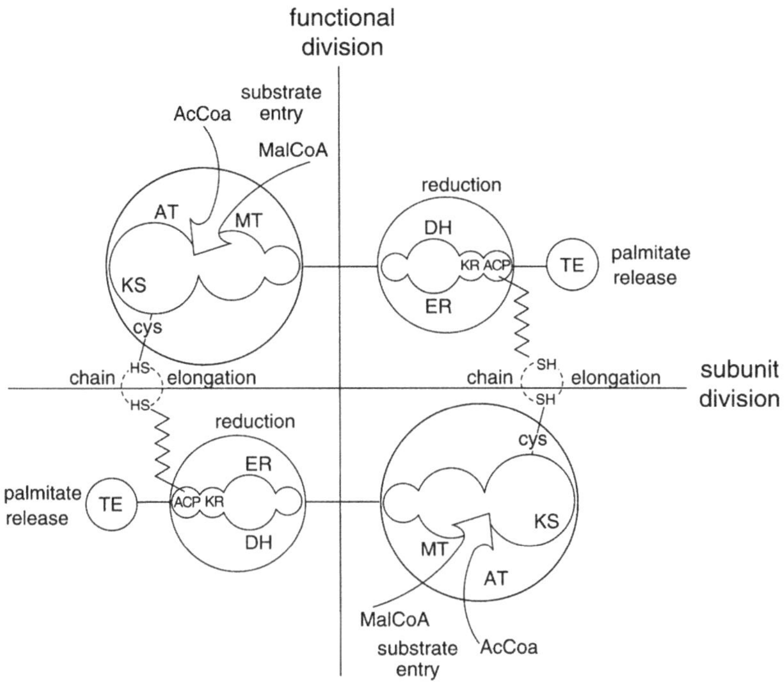

Fig. 10.1 Map of the chicken fatty acid synthetase. The sketch is based on the results of limited proteolysis experiments and on evidence from assays of catalytic activities, and binding of substrates, of specific inhibitors or antibodies. Two subunits are drawn in head-to-tail arrangement (subunit division) so that two sites of palmitate synthesis are constructed (functional division). The abbreviations for enzyme activities are *AT* acetyl transacylase, *MT* malonyl transacylase, *KS* β-ketoacyl synthetase, *KR* β-ketoacyl reductase, *DH* dehydratase, *ER* enoyl reductase, *TE* thioesterase, *ACP* acyl carrier protein. The wavy line represents the 4′-phosphopantetheine prosthetic group (From S. J. Wakil, J. K. Stoops and V. C. Joshi, with permission of the Annual Review of Biochemistry)

Regulation of Yeast Fatty Acid Synthesis at the Genetic Level

The two unlinked genes FAS1 and FAS2, coding respectively for the β and α subunit of yeast fatty acid synthetase are mapped to yeast chromosomes XI and XVI, respectively. The synthesis of the heteromultimeric fatty acid synthetase requires the expression of FAS1 and FAS2 at comparable rates. This is achieved by the existence of an upstream activation site (analogous to the enhancers) having the same sequence in both cases TYTTCACATGY (Y coding for any pyrimidine), functioning in any orientation. A protein trans-acting factor was identified which binds to this sequence. It is suspected that additional *cis-* and *trans-* acting elements are involved in the expression of the yeast fatty acid synthetase genes. The above consensus sequence is also present in the upstream regions of many other yeast genes involved in phospholipid biosynthesis and a general control mechanism affecting most components of yeast membrane biogenesis has been suggested, implying the involvement of common *trans*-acting factors.

Each of the two yeast genes FAS1 and FAS2, encoding respectively the α and the β subunit, were individually deleted by one-step gene disruption. The deletion of FAS2 does not influence the transcription of FAS1, whereas the transcription of FAS2 is significantly reduced in the FAS1-deleted strain. This suggests an activating role of the β subunit on the expression of FAS2, or a repression of FAS2 by an excess of its own product. On the other hand, differential proteolytic degradations remove the excess of either subunit, at least under starvation conditions. Thus, a combination of both regulation of expression and proteolysis of free FAS polypeptides may explain the equimolar amounts of α and β in yeast wild-type cells.

The gene FAS3, coding for yeast acetyl CoA carboxylase seems also to be coordinately regulated with FAS1 and FAS2.

Regulation of Fatty Acid Synthesis in Bacteria

The multifunctional fatty acid synthetase from *Brevibacterium ammoniogenes* is composed of a single polypeptide of 3104 residues; the seven active centers of it have been identified at different locations of the polypeptide and the overall structure is formally consistent with a head to tail fusion of the two yeast genes FAS1 and FAS2.

In *E. coli, fabA* codes for β-hydroxydecanoylthioester dehydrase, the introduction of the double bond of unsaturated fatty acids occurring at the C10 level. The FadR protein of *E. coli*, which acts as a repressor of fatty acid degradation, acts as an activator of the *fabA* gene: it binds at the −40 region of the major *fabA* promoter. The transcription of *fabA* is markedly decreased when fatty acids are added to the cultures, and acyl CoA esters specifically release the FadR protein from its DNA binding site. It has been shown that acyl CoA esters bind to the FadR protein.

Biosynthesis of Triglycerides

A direct acylation of glycerol has never been observed. A particulate fraction and a soluble fraction from *Clostridium butyricum* have been obtained, which together can acylate glycerol-3-phosphate in the presence of ACP and palmityl CoA. The same acylation is obtained directly starting from palmityl ACP. The product of the reaction is lysophosphatidic acid:

Lysophosphatidic acid	Phosphatidic acid
$CH_2O{-}CO{-}R$	$CH_2{-}O{-}CO{-}R$
|	|
$CHOH$	$CH{-}O{-}CO{-}R$
|	|
$CH_2{-}O{-}PO_3H_2$	$CH_2{-}O{-}PO_3H_2$

In addition to lysophosphatidic acid, phosphatidic acid and monopalmitine resulting from the dephosphorylation of lysophosphatidic acid are obtained. On the other hand, the dephosphorylation of phosphatidic acid leads to an αβ diglyceride which is the common precursor of triglycerides and choline; actually, an enzyme from chicken liver synthesizes triglycerides from diglycerides and acyl CoA:

$$R-CO-SCoA + \alpha\beta\text{-diglyceride} \rightarrow \text{triglyceride} + CoA-SH$$

Biosynthesis of Phosphoglycerides

An intermediate compound, cytidine diacylglycerol, is formed by reaction between cytidine triphosphate and phosphatidic acid, with the formation of pyrophosphate, the hydrolysis of which drives the reaction;

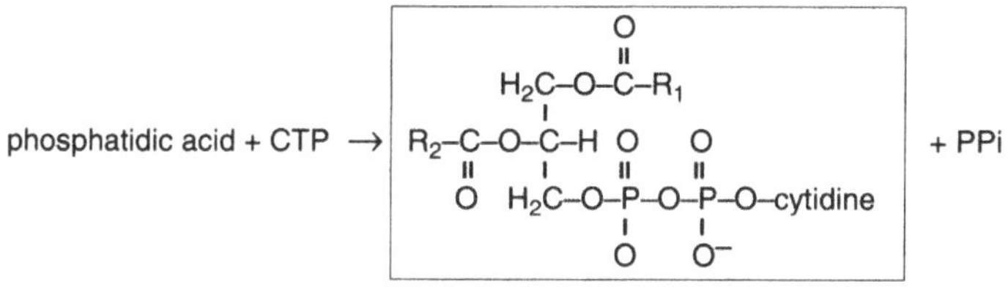

cytidine diacylglycerol

The activated phosphatidyl unit reacts with a polar alcohol while CMP is liberated. In eukaryotes, if the alcohol is inositol, phosphatidylinositol is formed. Further phosphorylations by specific kinases lead to polyphosphorylated forms of phosphatidylinositol, whose hydrolysis yields two intracellular messengers, diacylglycerol and inositol 1, 4, 5-triphosphate which elicit a great variety of physiological responses in many tissues. If the alcohol is serine, phosphatidylserine is formed.

In bacteria, phosphatidylserine is decarboxylated by a pyridoxal phosphate enzyme to phosphatidyl ethanolamine:

$$H_2N-CH_2-CH_2-OR \qquad R = \text{phosphatidyl residue}$$

In yeast, this phosphatidylserine decarboxylase is catalyzed by two isofunctional enzymes, one of which is a protein of 500 residues with 28–43 % identity to the *E. coli* and mammalian enzyme.

Phosphatidyl ethanolamine is the substrate of three sucessive methylations to yield phosphatidylcholine, with S-adenosylmethionine as the methyl donor:

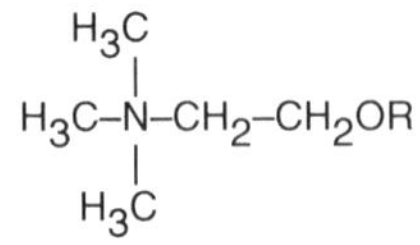

phosphatidyl choline

In mammals, which obtain choline from their diet, choline is directly phosphorylated by ATP to phosphorylcholine which then reacts with CTP to yield CDP-choline and pyrophosphate. CDP-choline is transferred to a diacylglycerol to finally give phosphatidylcholine.

Choline and trimethylamine (TMA) are small molecules that play central roles in biological processes throughout all kingdoms of life. These ubiquitous metabolites are linked through a single biochemical transformation, the conversion of choline to TMA by anaerobic microorganisms. This metabolic activity, which contributes to methanogenesis and human disease, has been known for over a century but has eluded genetic and biochemical characterization. A gene cluster responsible for anaerobic choline degradation within the genome of a sulfate-reducing bacterium has been now found. Its function has been verified using both a genetic knockout strategy and heterologous expression in *Escherichia coli*. Bioinformatics and electron paramagnetic resonance spectroscopy revealed the involvement of a C–N bond cleaving glycyl radical enzyme in TMA production, which is unprecedented chemistry for this enzyme family.

Cyclopropane Fatty Acid Synthase (CFA Synthase)

Cyclopropane fatty acids have been long known to occur in the lipids of many eubacteria although physiological and regulatory aspects of their formation has been almost exclusively been studied in *E. coli*. Their synthesis proceeds by the transfer of a methylene group from the activated methyl of S-adenosylmethionine to the *cis* double bond of an unsaturated fatty acid chain, resulting in the replacement of the double bond with a methylene bridge. The enzyme catalyzing the reaction is found in the cytoplasm, the substrate being not the free unsaturated fatty acid but rather phospholipids containing it. The enzyme has thus one soluble substrate, S-adenosylmethionine, and an insoluble one, the phospholipid bilayer. The double bond has to be positioned 9–11 carbon atoms away from the ester bond linking the fatty acid to the phospholipid backbone and is thus located within the hydrophobic core of the bilayer structures of the phospholipid vesicles or the membranes of intact bacteria.

The enzyme binds only to vesicles of phospholipids which contain either unsaturated or cyclopropane fatty acid moieties. CFA synthase is active on phosphatidylglycerol, phosphatidylethanolamine, and cardiolipin, the major phospholipids of *E. coli*, and also has some activity on phosphatidylcholine. The enzyme is equally active on phospholipid vesicles in the ordered or the disordered states of the lipid phase transition. Studies with a reagent that reacts only with the phosphatidylethanolamine molecules of the outer leaflet of a phospholipid bilayer indicate that CFA synthase reacts with phosphatidylethanolamine molecules of both the outer and the inner leaflets of phospholipid vesicles. The cloning of the gene coding for the cyclopropane fatty acid synthase (*cfa*), its sequence, its overexpression and the purification of the protein in milligram quantities has clarified the mechanisms of the production and of the activity this fascinating enzyme. The deduced amino acid sequence predicts a polypeptide of 382 residues that lacks long hydrophobic segments. The CFA synthase sequence has no significant similarity to known proteins except for sequences found in other enzymes that utilize S-adenosyl-L-methionine.

In *Pseudomonas denitrificans*, the cyclopropane fatty acid, methylene hexadecanoic acid, constitutes from 1 % to upwards of 30 % of the total lipid fatty acids. The amount of this component varied along with the levels of cyclopropane fatty acid synthase. The effect of the carbon source and oxygen tension on cyclopropane fatty acid synthase activity was studied. A correlation between the state of reduction of respiratory components and the accumulation of methylenehexadecanoic acid by *Ps. denitrificans* has been found: the uniqueness of succinate respiration in promoting the synthesis of cyclopropane synthase under limited oxygen conditions was assigned to the high degree of oxidation of respiratory components observed under this condition. The effect of ppGpp on *E. coli* cyclopropane fatty acid synthesis is mediated through the RpoS sigma factor: Strains of *E. coli* carrying mutations at the *relA* locus are deficient in cyclopropane fatty acid synthesis. The RelA protein catalyzes the synthesis of guanosine-3′, 5′-bisdiphosphate (ppGpp); therefore, ppGpp could be a putative direct regulator of CFA synthesis. The effect of RelA on CFA synthesis is indirect. In vitro and in vivo experiments show no direct interaction between ppGpp and CFA synthase activity. The relA effect is due to ppGpp-engendered stimulation of the synthesis of the alternative sigma factor, RpoS, which is required for function of one of the two promoters responsible for expression of CFA synthase.

Selected References

Short Chain Fatty Acid Synthesis

Barker HA (1956) Bacterial fermentations, Ciba lectures in microbial biochemistry. John Wiley, New York, 95pp

Fatty Acid Synthesis and Its Regulation

Demmer U, Warkentin E, Srivastava A, Kockelkorn D, Potter M, Marx A, Fuchs G, Ermler U (2013) J Biol Chem 288:6363–6370

Schweitzer E (1980) In Bisswanger H, Schmincke-Ott E (eds) Multifunctional proteins. John Wiley, New York, pp 197–215

Singh N, Wakil SJ, Stoops JK (1985) Biochemistry 24:6598–6602
Wakil SJ, Stoops JK, Joshi VC (1983) Annu Rev Biochem 52:537–579

Trimethylamine Production

Craciun S, Balskus EP (2012) Proc Natl Acad Sci USA109:21307–21312

Cyclopropane Fatty Acid Synthetase

Chang YY, Eichel J, Cronan JE Jr (2000) J Bacteriol 182:4288–4294
Eichel J, Chang YY, Riesenberg D, Cronan JE Jr (1999) J Bacteriol 181:572–576
Wang AY, Grogan DW, Cronan JE Jr (1992) Biochemistry 31:11020–11028

Chapter 11
Iron-Sulfur Proteins

Iron-sulfur proteins are proteins in which non-haem iron is coordinated with cysteine sulfur and usually also with inorganic sulfur. They are divided into three major categories: redoxins; "simple iron-sulfur proteins", containing only iron-sulfur clusters; and "complex iron-sulfur proteins", containing additional active redox centres such as flavin, molybdenum or haem.

Iron-Sulfur Clusters

Fe–S clusters were discovered by Helmut Beinert in the early 1960s by purifying enzymes with characteristic electron paramagnetic resonance signals. Some of the first Fe–S proteins to be discovered include plant and bacterial ferredoxins and respiratory complexes of bacteria and mitochondria

Iron–sulphur (Fe–S) clusters have long been recognized as essential and versatile cofactors of proteins involved in catalysis, electron transport and sensing of ambient conditions. Protein-bound Fe-S clusters (Fe-S) are polynuclear combinations of iron and sulfur atoms. In these complexes, Fe ions are bound to the polypeptide by heteroatoms of amino acid side chains (such as sulfur from cysteine). Yet alternative ligands of iron(Asp, Arg, Ser, CO, CN^-) are known, particularly in more complex Fe–S clusters. Clusters are linked to one another through sulfide bridges, resulting in materials with unique spectroscopic and chemical properties. Several different structural folds have been recognized to coordinate simple Fe–S clusters. Nevertheless, it has remained difficult to predict the presence of Fe–S clusters from protein sequences.

Iron-sulfur (FeS) clusters are important cofactors for numerous proteins involved in electron transfer, in redox and non-redox catalysis, in gene regulation, and as sensors of oxygen and iron. These functions depend on the various FeS cluster prosthetic groups, the most common being [2Fe-2S] and [4Fe-4S]. Figure 11.1 depicts the structure of the different iron-sulfur clusters

G.N. Cohen, *Microbial Biochemistry*,
DOI 10.1007/978-94-017-8908-0_11,

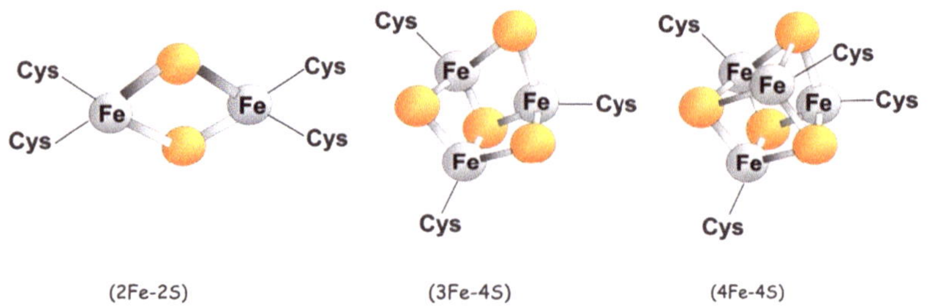

Fig. 11.1 The three most frequently occurring Fe-S clusters. The orange balls illustrate the sulfur atoms. Cys indicates the thiolates from the protein cysteine residues that coordinate the clusters (With permission of Dr. Marc Fontecave and the American Society of Microbiology)

In almost all Fe-S proteins, the Fe centers are tetrahedral and the terminal ligands are thiolate sulfur centers from cysteinyl residues. The sulfide groups are either two- or three-coordinated. Three distinct kinds of Fe-S clusters with these features are most common.

2Fe-2S Clusters

The simplest polymetallic system, the [Fe_2S_2] cluster, is constituted by two iron ions bridged by two sulfide ions and coordinated by four cysteinyl ligands (in Fe_2S_2 ferredoxins) or by two cysteines and two histidines (in Rieske proteins). The oxidized proteins contain two Fe^{3+} ions, whereas the reduced proteins contain one Fe^{3+} and one Fe^{2+} ion. These species exist in two oxidation states, $(FeIII)_2$ and FeIIIFeII.

4Fe-4S Clusters

A common motif features a four iron ions and four sulfide ions placed at the vertices of a cubane-type structure. The Fe centers are typically further coordinated by cysteinyl ligands. In the bacterial ferredoxins, the pair of oxidation states are [Fe3+, $3Fe^{2+}$] ($Fe_4S_4^{+}$) and [$2Fe^{3+}$, $2Fe^{2+}$] ($Fe_4S_4^{2+}$). The potentials for this redox couple range from −0.3 to −0.7 V. The two families of 4Fe-4S clusters share the $Fe_4S_4^{2+}$ oxidation state. Some 4Fe-4S clusters bind substrates and are thus classified as enzymes. In aconitase, the Fe-S cluster binds aconitate at the one Fe centre that lacks a thiolate ligand. The cluster does not undergo redox, but serves as a Lewis acid catalyst to convert aconitate to isocitrate (see Chap. 8). In the radical-SAM enzymes, the cluster binds and reduces S-adenosylmethionine to generate a radical, which is involved in many biosyntheses.

3Fe-4S Clusters

Proteins are also known to contain [Fe_3S_4] centres, which feature one iron less than the more common [Fe_4S_4] cores. Three sulfide ions bridge two iron ions each, while the fourth sulfide bridges three iron ions. Their formal oxidation states may vary from $[Fe_3S_4]^+$ (all-Fe^{3+} form) to $[Fe_3S_4]^{2-}$ (all-Fe^{2+} form). In a number of iron-sulfur proteins, the [Fe_4S_4] cluster can be reversibly converted by oxidation and loss of one iron ion to a [Fe_3S_4] cluster. For example, the inactive form of aconitase possesses an [Fe_3S_4] and is activated by addition of Fe^{2+} and reductant.

Other Fe-S Clusters

More complex polymetallic systems are common. Examples include both the 8Fe and the 7Fe clusters in nitrogenase. Carbon monooxide dehydrogenase and the [FeFe]-hydrogenase also feature unusual Fe-S clusters.

Biosynthesis of Fe-S Clusters

In the late 1960s, chemists and biochemists devised chemical reconstitution protocols to assemble Fe–S clusters into apoproteins in vitro, leading to the view that these cofactors can assemble spontaneously on proteins. However, genetic, biochemical and cell-biological studies in the 1990s provided ample evidence that the maturation of Fe–S proteins in living cells is a catalysed process rather than a spontaneous one. In striking contrast to the chemical simplicity of Fe–S clusters, their biosynthesis in vivo appears to be a rather complex and coordinated reaction. In the past decade, numerous biogenesis components were identified, and the first insights into the mechanisms of biogenesis were obtained.

The biogenesis of iron sulfur clusters has been studied most extensively in *E. coli* and *Azotobacte vinelandii* and in the yeast *Saccharomyces cerevisiae.* Three different biosynthetic systems have been identified, namely the *nif*, *suf*, and *isc* systems. The *nif* system is responsible for the clusters in the enzyme nitrogenase. The *suf* and *isc* systems are more general. Several proteins constitute the biosynthetic machinery via the *isc* pathway. The process occurs in two major steps: 1)the Fe/S cluster is assembled on a scaffold protein followed by transfer of the preformed cluster to the recipient proteins. The first step of this process occurs in the cytoplasm of prokaryotic organisms. FeS cluster assembly is a complex process involving the mobilisation of Fe and S atoms from storage sources, their assembly into [Fe-S] form, their transport to specific cellular locations, and their transfer to recipient apoproteins.

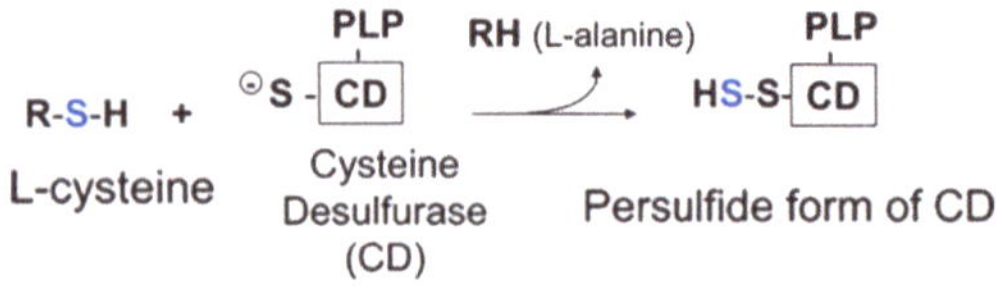

Fig. 11.2 The ISC system

Two FeS assembly machineries have been identified, which are capable of synthesizing all types of [Fe-S] clusters: ISC (iron-sulfur cluster) and SUF (sulfur assimilation). The ISC system is conserved in eubacteria and has broad specificity, targeting general FeS proteins. In *E.coli*, it is encoded by the *isc* operon (*iscRSUA-hscBA-fdx-iscX*). IscS is a cysteine desulfurase, which obtains S from cysteine (converting it to alanine) and serves as an S donor for FeS cluster assembly (Fig. 11.2).

IscU and IscA act as scaffolds to accept S and Fe atoms, assembling clusters and transferring them to recipient apoproteins. HscA is a molecular chaperone and HscB is a co-chaperone. Fdx is a [2Fe-2S]-type ferredoxin. IscR is a transcription factor that regulates expression of the *isc* operon. IscX (also known as YfhF) appears to interact with IscS and may function as an Fe donor during cluster assembly.

The SUF system is an alternative pathway to the ISC system that is thought to operate under iron starvation and oxidative stress. It is found in eubacteria and archaea. The SUF system is encoded by the *suf* operon (*suf*ABCDSE), and the six encoded proteins are arranged into two complexes (SufSE and SufBCD) and one protein (SufA). SufS is a pyridoxal-phosphate (PLP) protein displaying cysteine desulfurase activity. SufE acts as a scaffold protein that accepts S from SufS and donates it to SufA. SufC is an ATPase with an unorthodox ATP-binding cassette (ABC)-like component. No specific functions have been assigned to SufB and SufD. SufA is homologous to IscA acting as a scaffold protein in which Fe and S atoms are assembled into [FeS] cluster forms, which can then easily be transferred to apoproteins targets. The SufBCD complex contributes to the assembly or repair of oxygen-labile iron-sulfur clusters under oxidative stress.

As stated above, three multiprotein systems are known for iron-sulfur (Fe-S) cluster biogenesis in prokaryotes and eukaryotes as follows: the NIF (nitrogen fixation), the ISC (iron-sulfur cluster), and the SUF (mobilization of sulfur) systems. In all three, cysteine is the physiological sulfur source, and the sulfur is transferred from cysteine desulfurase through a persulfidic intermediate to a scaffold protein. However, the biochemical nature of the sulfur source for Fe-S cluster assembly in archaea is unknown, and many archaea lack homologs of cysteine desulfurases. *Methanococcus maripaludis* is a methanogenic archaeon that contains a high amount of protein-bound Fe-S clusters (45 nmol/mg protein). Cysteine in this archaeon is synthesized primarily via the tRNA-dependent pathway. When a cysteine auxotroph was grown with 34S-labeled sulfide and unlabeled cysteine, <8 % of the cysteine, >92 % of the methionine, and >87 % of the sulfur in the Fe-S clusters in proteins were labeled, suggesting that the sulfur in methionine and Fe-S clusters was derived predominantly from exogenous sulfide instead of cysteine. Therefore, this investigation challenges the concept that cysteine is always the sulfur source for Fe-S cluster biosynthesis in vivo and suggests that Fe-S clusters are derived from sulfide in those organisms, which live in sulfide-rich habitats.

Iron-Sulfur Proteins

In most iron-sulfur proteins, the clusters function as electron-transfer groups, but in others they have other functions, such as catalysis of hydratase/dehydratase reactions, maintenance of protein structure, or regulation of activity.

The most common function of Fe–S clusters, electron transfer, is based on the propensity of Fe to formally switch between oxidative states +2 and +3. Within proteins, Fe–S clusters can adopt redox potentials from −500 mV to +300 mV. Thus, Fe–S clusters can serve as excellent donors and acceptors of electrons in a variety of biological reactions. Examples are bacterial respiratory complexes, photosystem I, ferredoxins and hydrogenases. Another well-studied function of Fe–S clusters is in enzyme catalysis, the classic example being aconitase (see Chap. 8), in which a non-protein-coordinated Fe at one edge of a [4Fe–4S] cluster serves as a Lewis acid to assist H2O abstraction from citrate (the substrate), which is converted to isocitrate. Other special cases are the radical S-adenosyl-L-methionine (SAM) enzymes biotin synthase and lipoate synthase, which bind two Fe–S clusters each. It is believed that one of these clusters is disassembled during the formation of the products biotin and lipoic acid, respectively, thus serving as a sulphur donor.. However, in many cases the precise role of the Fe–S cluster is still unclear, and it is therefore possible that in some proteins the Fe–S cluster simply plays a structural role. This may be true for the recently discovered Fe–S clusters in adenosine triphosphate (ATP)-dependent DNA helicases involved in nucleotide excision repair. A third general role of Fe–S clusters is in sensing environmental or intracellular conditions to regulate gene expression Examples are the bacterial transcription factors FNR, IscR and SoxR, which sense O_2, Fe–S clusters and superoxide/NO, respectively. The switch between the activating and repressed states depends on the presence or absence of an Fe–S cluster (FNR or IscR) or the redox state of the [2Fe–2S] cluster present in SoxR.

Table 11.1 lists iron-sulfur proteins that have been unambiguously characterized (by iron and sulfur content analysis and/or by various spectroscopic methods) in Enterobacteriaceae. According to Fontecave, *E. coli* could contain as much as 200 different iron-sulfur proteins.

Table 11.1 List of characterized Fe-S cluster proteins in *E. coli*

Function (no. of proteins)	Protein(s)[a] (gene[s])
Electron transfer (52)	Ferredoxins (*fdx, bfd, yfaE*)
	Succinate dehydrogenase (*sdhB*)
	Sulfite reductase (*cysI*)
	Nitrite reductase (*nirB*)
	Formate-dependent nitrite reductase (*nrfC*)
	2,4-Dienoyl-CoA reductase (*fadH*)
	DMSO reductase (*dmsA, dmsB*)
	SoxR reductase (*rseC, rsxB, rsxC*)
	Formate dehydrogenase N (*fdnG, fdnH, fdhD*)
	Formate dehydrogenase H (*fdhF*)
	Formate dehydrogenase O (*fdoG, fdoH*)
	Ferrichrome reductase (*fhuF*)
	Fumarate reductase (*frdB*)
	Glutamate synthase (*gltB, gltD*)
	HCP (*hcp*)
	HCP reductase (*hcr*)
	Periplasmic nitrate reductase (*napA, napF, napG, napH*)
	Respiratory nitrate reductase 1 (*narG, narH*)
	Respiratory nitrate reductase 2 (*nary, narZ*)
	NADH-quinone oxidoreductase (*nuoB, nuoE, nuoF, nuoG, nuoI*)
	Anaerobic glycerol phosphate dehydrogenase (*glpB*)
	Hydrogenase 1 (*hyaA*)
	Hydrogenase 2 (*hybO, hybA*)
	Hydrogenase 3 (*hycB, hycF, hycG*)
	Hydrogenase 4 (*hyfA, hyfH, hyfI, hyfR*)
	Xanthine dehydrogenase (*xdhC*)
	Rieske proteins (*yeaW, hcaC*)
Metallosite assembly (7)	Scaffold proteins (*iscU, nfuA*)
	A-type transporters (*iscA, sufA, erpA*)
	Mocofactor biosynthesis protein (*moaA*)
	SufB (*sufB*)
Nucleic acid binding (7)	
Gene expression regulation (4)	Fumarate and nitrate reduction regulator (*fnr*)
	Isc regulator (*iscR*)
	Redox-sensitive transcriptional activator (*soxR*)
	Transcriptional activator (*yeiL*)
DNA repair (2)	Adenine glycosylase (*mutY*)
	Endonuclease III (*nth*)
RNA modification (1)	RNA 5-methyluridine methyltransferase (*rumA*)
Redox catalysis (12)	
Radical-SAM catalysis (10)	Biotin synthase (*bioB*)
	Coproporphyrinogen synthase (*hemN*)
	Lipoate synthase (*lipA*)
	Thiazole biosynthesis protein (*thiH*)
	RNA thiomethyl transferase (*miaB*)
	Anaerobic ribonucleotide reductase-activating component (*nrdG*)
	Pyruvate-formate lyase-activating component (*pflA*)
	Sulfatase maturase (*aslB or atsB*)
	Methylthio transferase (*rimO*)

Function (no. of proteins)	Protein(s)[a] (gene[s])
Other (2)	4-Hydroxy-3-methylbut-2-en-1-yl diphosphate synthase (*ispG*)
	4-Hydroxy-3-methylbut-2-enyl diphosphate reductase (*ispH*)
Non-redox catalysis (10)	Aconitases (*acnA, acnB*)
	Fumarases (*fumA, fumB*)
	Dihydroxyacid dehydratase (*ilvD*)
	Isopropylmalate dehydratase (*leuC*)
	Quinolinate synthase (*nadA*)
	Serine-threonine deaminase (*sdaA, sdaB*)
	Serine dehydratase (*tdcG*)

Table reproduced with permission of Dr. Marc Fontecave and the American Society of Microbiology
[a]*CoA* coenzvme A, *DMSO* dimethvl sulfoxide, *HCP* hvbrid cluster protein.

Selected References

Barras F, Loiseau L, Py B (2005) Adv Microb Physiol 50:41–101
Fontecave M (2006) Nat Chem Biol 2:171–174
Johnson DC, Dean DR, Smith AD, Johnson MK (2005) Annu Rev Biochem 74:247–281
Lill R (2009) Nature 460:831–838
Liu Y, Sieprawska-Lupa M, Whitman WB, White RH (2010) J Biol Chem 285:31923–31929
Roche B, Aussel L, Ezraty B, Mandin P, Py B, Barras F (2013) Biochim Biophys Acta 1827: 455–469

Chapter 12
The Archaea

The domain Archaea was not recognized as a major domain of life until quite recently. Until the twentieth century, most biologists considered all living things to be classifiable as either a plant or an animal. But in the 1950s and 1960s, most biologists came to the realization that this system failed to accommodate the fungi, protists, and bacteria. By the 1970s, a system of Five Kingdoms had come to be accepted as the model by which all living things could be classified. At a more fundamental level, a distinction was made between the prokaryotic bacteria and the four eukaryotic kingdoms (plants, animals, fungi, and protists). The distinction recognizes the common traits that eukaryotic organisms share, such as nuclei, cytoskeletons, and internal membranes.

In the late 1970s, Carl Woese and his colleagues were studying relationships among the prokaryotes. Because ribosomes are so critically important is the functioning of living things, they are not prone to rapid evolution. A major change in the sequence of ribosomal sequences can render the ribosome unable to fulfill its duties of building new proteins for the cell. This slow rate of molecular evolution made the ribosome sequence a good choice for unlocking the secrets of bacterial evolution. Using the sequences of 16S RNA in a wide diversity of bacteria, they found that there were two distinctly different groups. Those "bacteria" that lived at high temperatures or produced methane clustered together as a group well away from the usual bacteria and the eukaryotes. Because of this vast difference in genetic makeup, Woese proposed that life be divided into three domains: Eukaryota, Eubacteria, and Archaebacteria. He later decided that the term Archaebacteria was a misnomer, and shortened it to Archaea, to emphasize the fact that they are not bacteria. They have also been called Extremophiles in recognition of the extreme environments in which they have been found.

Archaea include inhabitants of some of the most extreme environments on the planet. Some live near rift vents in the deep sea at temperatures well over 100 °C. Others live in hot springs or in extremely alkaline or acid waters. They live in the anoxic muds of marshes and at the bottom of the ocean, and even thrive in petroleum deposits deep underground. However, archaea are not restricted to extreme environments; new research is showing that they are also quite abundant in the

G.N. Cohen, *Microbial Biochemistry*,
DOI 10.1007/978-94-017-8908-0_12,

plankton of the open sea. They have been found thriving inside the digestive tracts of cows, termites, and marine life where they produce methane. The methanogens were already known to be chemical oddities in the microbial world since they had cell walls different from all the other known bacteria.

The different types of extremophiles

One can distinguish

(a) the thermophiles, who live at high temperatures,
(b) the hyperthermophiles, who live at really high temperatures (present record is 113 °C)
(c) the psychrophiles, who grow best in the cold (one in the Antarctic grows best at 4 °C)
(d) the acidophiles, who live at low pH (as low as pH 1 and who die at pH 7)
(e) the alkaliphiles, who thrive at a high pH.

Most of the more than 250 named species that have been discovered so far have been placed in three groups, according to the sequence of their 16S RNAs: the Euryarcheota, the Crenarcheota and the Korarcheota.

There are three main groups of Euryarcheota, the Methanogens, the Halophiles and the Thermoacidophiles. The methanogens will be studied in the next chapter. As their name suggests, the thermoacidophiles thrive in hot and acid environments. They are found in such places as acidic sulfur springs (e.g., in Yellowstone National Park) and undersea vents ("smokers"). The halophiles can survive the dessicating effects of extremely saline waters. Among them, one finds *Halobacterium*, a well-studied archaeon. The light-sensitive pigment bacteriorhodopsin gives *Halobacterium* its purple color and provides it with chemical energy. Bacteriorhodopsin pumps protons to the outside of the membrane. When these protons flow back, they are used in the synthesis of ATP, which is the energy source of the cell. This protein is chemically very similar to the light-detecting pigment rhodopsin, found in the vertebrate retina.

The first members of Crenarcheota to be discovered like it really hot and so are called hyperthermophiles. One, *Pyrolobus fumaris*, lives at 113 °C. Many live in acidic sulfur springs at a pH as low as 1 (the equivalent of dilute sulfuric acid). These use hydrogen as a source of electrons to reduce sulfur in order to get the energy they need to synthesize their food (from CO_2). Other members of this group seem to make up a large portion of the plankton in cool, marine waters. As yet, none of these has been isolated and cultivated in the laboratory.

The Korarcheota are only known from environmental 16S RNA sequences.

Very recently, the cultivation of a new nanosized hyperthermophilic archaeon from a submarine hot vent has been reported. This archaeon cannot be attached to one of the three aforementioned grouops groups and therefore must represent an unknown phylum (which has been named Nanoarchaeota) and species, which has been named *Nanoarchaeum equitans*. Cells *of N. equitans* are spherical, and only about 400 nm in diameter. They grow attached to the surface of a specific archaeal host, a new member of the genus *Ignicoccus*. The distribution of the Nanoarchaeota is so far unknown. Owing to their unusual 16S ribosomal RNA sequence, members remained undetectable by commonly used ecological studies based on the polymerase chain reaction. *N. equitans* harbors the smallest archaeal genome, only 0.5 megabases in size. This organism should provide insight into the evolution of thermophily, of tiny genomes and of interspecies communication.

During the past few years, diverse groups of uncultivated mesophilic archaea have been discovered and affiliated with the Crenarchaeota. Based on the first genome sequence of a crenarchaeote, Crenarchaeum symbiosum, it was deduced that these mesophilic archaea are different from hyperthermophilic Crenarchaeota and branch deeper than was previously assumed. The results indicate that C. symbiosum and its relatives are not Crenarchaeota, but should be considered as another archaeal phylum, the Thaumarchaeota (from the Greek θαυμα, meaning wonder). An eukaryotic-like DNA topoisomerase is present in the recently sequenced genomes of two archaea of this newly proposed phylum. Phylogenetic analyses suggest that this topoisomerase was present in the last common ancestor of Archaea and Eucarya.

Chemical Characteristics of Archaea

The most striking chemical differences between Archaea and other living things lie in their cell membrane. There are four fundamental differences between the archaeal membrane and those of all other cells: (1) chirality of glycerol, (2) ether linkage, (3) isoprenoid chains, and (4) branching of side chains.

1. Chirality of glycerol: The basic unit from which cell membranes are built is the phospholipid. The glycerol used to make archaeal phospholipids is a stereoisomer of the glycerol used to build bacterial and eukaryotic membranes. While bacteria and eukaryotes have D-glycerol in their membranes, archaeans have L-glycerol in theirs. This is more than a geometric difference. Chemical components of the cell have to be built by enzymes, and the "handedness" (chirality) of the molecule is determined by the shape of those enzymes. A cell that builds one form will not be able to build the other form.
2. Ether linkage: When side chains are added to the glycerol, most organisms bind them together using an ester linkage By contrast, archaeal side chains are bound using an ether linkage. This gives the resulting phospholipid different chemical proerties from the membrane lipids of other organisms.
3. Isoprenoid chains: The side chains in the phospholipids of bacteria and eukaryotes are fatty acids, chains of usually 16 to 18 carbon atoms. Archaea do not use fatty acids to build their membrane phospholipids. Instead, they have side chains of 20 carbon atoms built from isoprene (see Chap. 33 for the biosynthesis of isoprenoids).
4. Branching of side chains: Not only are the side chains of achaeal membranes built from different components, but the chains themselves have a different physical structure. Because isoprene is used to build the side chains, there are side branches off the main chain. The fatty acids of bacteria and eukaryotes do

not have these side branches (the best they can manage is a slight bend in the middle), and this creates some interesting properties in archaeal membranes. For example, the isoprene side chains can be joined together. This can mean that the two side chains of a single phospholipid can join together, or they can be joined to side chains of another phospholipid on the other side of the membrane. No other group of organisms can form such transmembrane phospholipids. Another interesting property of the side branches is their ability to form carbon rings of five carbon atoms. Such rings are thought to provide structural stability to the membrane, since they seem to be more common among species that live at high temperatures. They may work in the same way that cholesterol does in eukaryotic cells to stabilize membranes.

There are other molecular species which differ between bacteria and Archaea. For example, the sequence of all their ribosomal proteins is homologous to the corresponding sequences of eukaryotes rather to those of bacteria. The same is true of some of their transfer RNAs. The cell walls of bacteria contain the chemical peptidoglycan. Archaeal cell walls do not contain this compound, though some species contain a similar one. Likewise, archaea do not produce walls of cellulose (as do plants) or chitin (as do fungi).

Archaea have a self-assembling proteinaceous surface (S-) layer as the primary and outermost boundary of their cell envelopes. The S-layer maintains structural rigidity, protects the organism from adverse environmental elements, and yet provides access to all essential nutrients. The crystal structure of one of the two "homologous" tandem polypeptide repeats that comprise the *Methanosarcina acetivorans* S-layer protein has been determined. The S-layer model reveals a negatively charged molecular sieve that presents both a charge and size barrier to restrict access to the cell periplasmic-like space. The β-sandwich folds of the S-layer protein are structurally homologous to eukaryotic virus envelope proteins, suggesting that Archaea and viruses have arrived at a common solution for protective envelope structures. These results provide insight into the evolutionary origins of primitive cell envelope structures, of which the S-layer is considered to be among the most primitive.

The archaea have a curious mix of traits characteristic of the other prokaryotes (the bacteria) as well as traits found in eukaryotes.

The eukaryotic traits include the DNA replication machinery, the presence of histones, nucleosome-like structures, the transcription machinery (structure of RNA polymerase, existence of a TFIIB like protein and of a TATA-binding protein), the translation machinery (eukaryotic-like initiation and elongation factors, eukaryotic-like ribosomal proteins, inhibition by diphtheria toxin). The prokaryotic traits include the single, circular chromosome, the organization of genes in operons, the absence of introns, the bacterial-type membrane transport, many metabolic processes (energy production, nitrogen fixation, polysaccharide synthesis).

We could conclude from this that many traits found in the bacteria first appeared in the ancestors of all the present-day groups. The split leading to the archaea and the eukaryotes would have occurred after the bacteria had gone their own way. However, the acquisition by eukaryotes of mitochondria (probably from an ancestor of today's rickettsias) and chloroplasts (from cyanobacteria) occurred after their line had diverged from the archaea.

Their ability to live in extreme environments, their autotrophism; that is, their ability to make food using materials (H_2, S, CO_2) in the earth's crust have suggested that the archaea may be the little-changed descendants of the first forms of life on earth.

Archaea: Fossil Record

Archaea and Bacteria cells may be of similar sizes and shapes, so the shape of a microbial fossil does not help in determining its origin. Instead of physical features, micropaleontologists rely on molecular fossils that are found in just one group of organisms, that are not prone to chemical decay, or decay into predictable and recognizable secondary chemicals.

In the case of the Archaea, there is a very good candidate to preserve as a molecular fossil from the cell membrane, namely the isoprenoid structures, because they are unique to Archaea, are not as prone to decomposition at high temperatures. Molecular fossils of Archaea in the form of isoprenoid residues were first reported in 1979 from the Messel oil shale of Germany. These are Miocene desposits whose geologic history is well known. Material from the shale was dissolved and analyzed using a combination of chromatography and mass spectrometry. The fingerprint of the Messel shale included isoprene compounds identical to those found in some Archaea. Based on the geologic history of the Messel area, thermophiles and halophiles are not likely to have ever lived there, so the most likely candidates are archaeal methanogens. Since their discovery in the Messel shales, isoprene compounds indicative of ancient Archaea have been found in numerous other localities including Mesozoic, Paleozoic, and Precambrian sediments. Their chemical traces have even been found in sediments from the Isua district of west Greenland, the oldest known sediments on Earth at about 3.8 billion years old. This means that the Archaea (and life in general) appeared on Earth within one billion years of the evolution of the planet, and at a time when conditions were still quite inhospitable for life. The atmosphere of the young Earth was rich in ammonia and methane, and was probably very hot. Such conditions, toxic to plants and animals, are not toxic for Archaea, which may thus represent remnants of once-thriving communities that dominated the world when it was young.

Economic Importance of the Archaea

Because they have enzymes that can function at high temperatures, considerable effort is being made to exploit the archaea for commercial processes such as providing enzymes to be added to detergents (maintain their activity at high temperatures and pH), obtaining an enzyme converting corn starch into dextrins, obtaining Taq polymerase, a DNA polymerase isolated from *Thermus aquaticus*, a denizen of a Yellowstone hot spring, used for PCR (the polymerase chain reaction). Archaea may also be enlisted to aid in cleaning up contaminated sites, e.g., petroleum spills.

Selected References

Arbing MA, Chan S, Shin A, Phan T, Ahn CJ, Rohlin L, Gunsalus RP (2012) Proc Natl Acad Sci USA 109:11812–11817

Brochier-Armanet C, Boussau B, Gribaldo S, Forterre P (2008a) Nat Rev Microbiol 6:245–252

Brochier-Armanet C, Gribaldo S, Forterre P (2008b) Biol Direct 3:54

Fox GE, Magrum LJ, Balch WE, Wolfe RS, Woese CR (1977) Proc Natl Acad Sci USA 74:4537–4541

Kates M, Kushner DJ, Matheson AT (1993) The biochemistry of archaea (Archaebacteria). Elsevier Science, Amsterdam

Woese CR, Fox GE (1977) Proc Natl Acad Sci USA 74:5088–5090

Chapter 13
Methanogens and Methylotrophs

The Italian physicist Volta must be credited for observation that the bubbles arising at the surface of Lago Maggiore were inflammable (1776). Thirty years later, this combustible air was identified as methane (CH_4), whose microbial origin was established only in 1868.

Methanogens, the methane generators are strict anaerobes, which thrive in mud deposits, in certain trees growing in the vicinity of rivers and lakes, in poorly drained soils. Methane is also produced in the garbage dumping grounds; in the intestinal tract of all animals including humans (therefore in sewers), in the sludge of sewage treatment plants, in the digestive tract of termites. One main source of methane is the paunch (the rumen), the first stomach of ruminants. The gas from rumen consists in 40 % methane and 60 % CO_2. The CO_2 originates from the catabolism of the organic substances present in the ruminant diet. It is converted to methane according to the overall equation

$$CO_2 + 4H_2 \rightarrow CH_4 + 2H_2O + \text{energy}\left(\text{in the form of ATP}\right)$$

The hydrogen necessary for the reduction of CO_2 is produced through the fermentation of organic substances by other bacterial species present in the rumen.

Methylotrophs form a very diverse group of microorganisms utilizing compounds containing a single carbon atom in a more reduced state than CO_2, or compounds containing two or more methyl groups not directly linked to one other (e.g. dimethylether, CH_3–0–CH_3). They assimilate the carbon at the level of oxidation of formaldehyde. Methanotrophs are a subgroup of methylotrophs which can utilize methane as the sole source of carbon and energy.

Due to their role in the chemistry of atmosphere and to their impact in biotechnology, the ecological studies on methylotrophs are important. For example, the destruction of trichloroethylene, of vinyl chloride, of chloroform and of toxic or carcinogenic compounds present in the underground waters could use in situ bioremediation using the indigenous methylotrophic bacteria. It is thus important to stimulate the growth of the species able to degrade these chemical compounds,

DOI 10.1007/978-94-017-8908-0_13,

which entails the monitoring of the activity, the biomass and other characteristics of the microbes growing in these waters. Therefore, oligonucleotide probes have been developed in order to detect the specific groups of methylotrophs.

The distinguishing property of these organisms is the ability to derive both carbon and energy from the same compound. By far the most abundant C1 compound is the gas methane, which occurs in coal and oil deposits and is continuously produced in enormous quantities by the anaerobic methanogenic bacteria present in the rumen of cattle and sheep, in waste disposal plants, decomposition swamps and lake sediments. Methanol is formed during the degradation of pectins and other natural substances containing methyl esters or ethers. Methylated amines (mono- di- and trimethylamine) and their oxides are present in animal and plant tissues.

Methanogens and Methanogenesis

Methanogenesis is a biological process that produces approximately one billion tons of methane per year.

Nature has invented a unique metabolic system during which the step producing the energy necessary for the maintenance of the methanogenic species is associated with the reduction of a methyl group to methane (see above overall equation).

All methanogens share this biochemlcal reaction, although the electrons necessary for the reduction may arise, according to the species, from hydrogen, formate, methanol, methylamines or acetate, and in some rare cases, from ethanol or propanol.

The phylogenetic diversity of methanogens is evident from their dispersal throughout the Euryarcheota kingdom of the domain Archaea. Most methanogens are capable to reduce a single substrate, CO_2, to methane. Only one order of methanogenic Archaea, the Methanosarcinales, has evolved the ability to reduce other compounds to methane. As a result, this highly successful group is found in a number of different environments. For example, Methanosarcina barkeri can produce methane autotrophically by reducing CO_2, acetotrophically by cleavage of acetate to methane and CO_2,or methylotrophically by the dismutation of methanol, methylated thiols or methyated amines to methane and CO_2.

To study the biochemistry of methanogens, procedures for mass culture of these organisms that yield kilogram quantities of cells have been developed. Anoxic methods of fractionation of enzymes and coenzymes by FPLC and HPLC procedures as well as the use of anaerobic chambers are routinely used.

Reduction of CO_2

Most methanogens are able to reduce CO_2 to methane via the formyl, methenyl, methylene and methyl stages; the structures and the roles of six new coenzymes have been defined during this process (Fig. 13.1). The reduction of CO_2 occurs

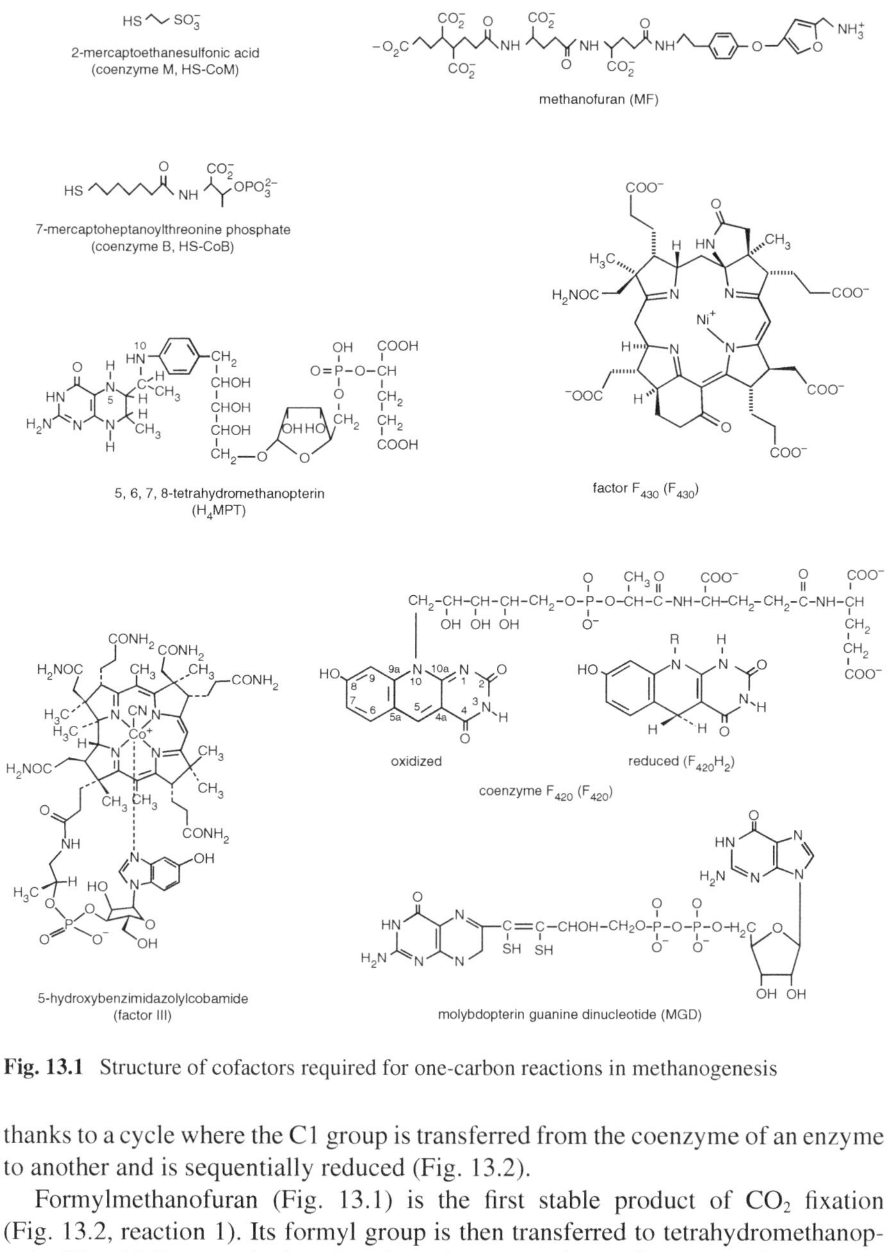

Fig. 13.1 Structure of cofactors required for one-carbon reactions in methanogenesis

thanks to a cycle where the C1 group is transferred from the coenzyme of an enzyme to another and is sequentially reduced (Fig. 13.2).

Formylmethanofuran (Fig. 13.1) is the first stable product of CO_2 fixation (Fig. 13.2, reaction 1). Its formyl group is then transferred to tetrahydromethanopterin (Fig. 13.1), a pterin found so far only among the methanogens. The enzyme catalyzing this reaction is formylmethanofuran: tetrahydromethanopterin formyltransferase (Fig. 13.2, reaction 2). The formyl group is then reduced to a methenyl group by 5,10-methenyltetrahydromethanopterin cyclohydrolase (Fig. 13.2, reaction 3). Reduced deazaflavin (also called coenzyme 420) (Fig. 13.1) provides the electrons required to reduce the double bond of the methenyl group, which becomes

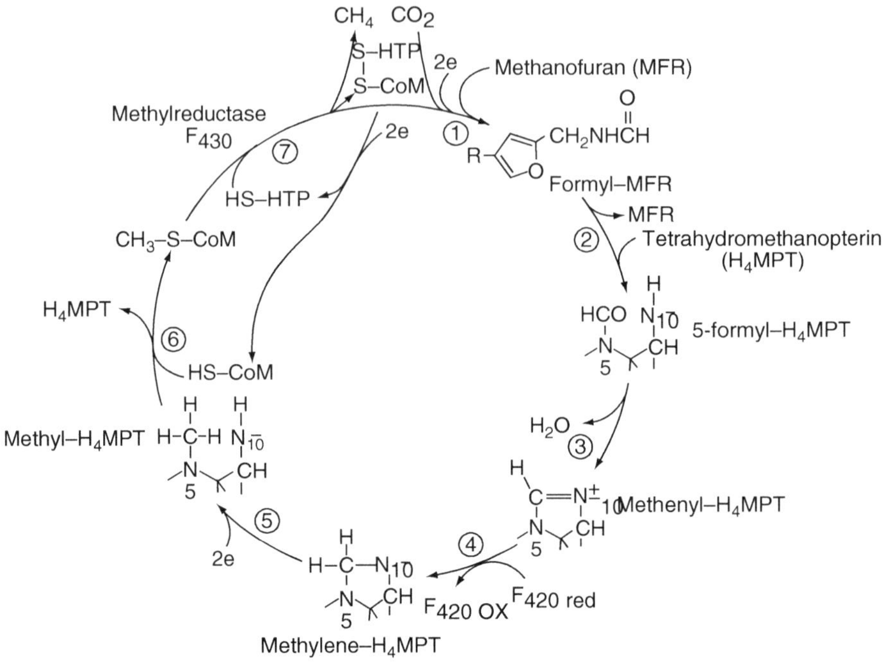

Fig. 13.2 The methanogenic pathway

a methylene. This reaction is catalyzed by methylenetetrahydromethanopterin: coenzyme F_{420} oxidoreductase (Fig. 13.2, reaction 4). Most reactions have been documented by homogeneous enzyme preparations. If the methylene group is chemically reduced by borohydride, 5-methyl-tetrahydromethanopterin is obtained, and its methyl group is reduced to methane by cell extracts. Reaction 7 (Fig. 13.2, reaction 7) schematizes the methylcoenzyme M methylreductase, the details of which will be found later in this chapter. This last reaction is involved in the activation of the first reaction (synthesis of formyltetrahydrofuran) and thus completes the cycle.

Formylmethanofuran Dehydrogenase

This enzyme catalyzes the first reaction of the cycle. Methanofuran (for short) is a long linear molecule with a hydrophilic tetracarboxylic part at one of its extremities and a furan ring substituted by a primary amine at the other (Fig. 13.1, reaction 1). In *Methanobacterium thermoautotrophicum*, methanofuran is 4 (N-(4, 5, 7-tricarboxyheptanoyl-γ-glutamyl-γ-L-glutamyl)p-(β-aminoethyl) phenoxymethyl) p(aminomethyl) furane. In *Methanosarcina barkeri* glutamyl units are found instead of the tetracarboxylic structure.

The enzyme catalyzes the reversible conversion of CO_2 and methanofuran to formylmethanofuran, the formyl group being linked to the primary amine of the furan cycle. CO_2 and not bicarbonate is the substrate of the enzyme.

Methanobacterium wolfei and *M. thermoautotrophicum* as well as other methanogens possess two isoenzymes, one containing tungsten (Fwd) and the other molybdenum (Fmd). The primary structures of the four subunits (FwdABCD) of the tungsten-containing enzyme from *M. thermoautotrophicum* have been determined by cloning and sequencing of the corresponding genes. FwdB contains sequence motifs characteristic of enzymes containing a molybdopter-in-dinucleotide, indicating that this subunit carries the active site. FwdA, FwdC and FwdD show no significant similarity with proteins found in the databases. A Northern blot analysis demonstrates that the 4 *fwd* genes form a transcriptional unit with 3 genes *fwdE, fwdF* and *fwdG*. Two proteins have been inferred from *fwdE* and *fwdG*, respectively, of 17.8 and 8.6 kDa. Each contains [4Fe-4S] binding motifs fwdF encodes a 38.6 kDa protein containing 8 such motifs, suggesting that the gene product is a polyferredoxin. The seven *fwd* genes have been expressed in *E. coli*, yielding proteins of the expected size. The *fwd* operon is located in a region of the genome coding for molybdenum enzymes and for proteins involved in the synthesis of molybdopterins.

The group of "molybdate genes" comprises three open phases (*fmdECB*). The *fmdB* gene codes for the subunit binding molybdopterin guanine dinucleotide and carries the active site. It is thus functionally equivalent to FwdB. The 5′ terminal part of *fmdC* codes for a protein similar to FwdC, the C-terminal part of which is similar to FwdD. FmdC is thus functionally equivalent to FwdC+FwdD. The *fmd* operon does not contain a *fmdA* gene coding for a FmdA subunit. Since the FmdA protein has the same apparent molecular weight and the same N-terminal sequence as FwdA, one supposes that FmdA et FwdA are encoded by *fwdA* belonging to the *fwd* operon. Actually, it has been found that *M. thermoautotrophicum* cells grown in the presence of tungstate or molybdate transcribe the *fwdHFGDACB* operon whereas the *fmd ECB* gene cluster is transcribed only by cells grown in the presence of molybdate.

The molybdenum enzyme from *M. barkeri* contains about 30 molecules of non-heme iron and 30 molecules of labile sulfur. The FmdB subunit, which carries the molybdenum active site binds one [4Fe-4S] cluster *fmdF* codes for a protein with 4 tandemly repeated motifs; this protein is predicted to be a polyferredoxin containing up to 32 iron atoms in eight [4Fe-4S] clusters. The fact that FmdF forms a stable complex with the other subunits of formylmethanofuran dehydrogenase is in favor of a function of polyferredoxin during catalysis. A *malE-fmdF* fusion has been constructed and expressed in *E. coli* and has led to the production of the polyferredoxin apoprotein in preparative amounts.

Formylmethanofuran: Tetrahydromethanopterin Formyltransferase

Formylmethanofuran: tetrahydromethanopterin formyltransferase (FTR) from *M. thermoautotrophicum* has been cloned and its sequence determined. The active enzyme has been synthesized under the control of the *lac* promoter. The *ftr* gene codes for an acidic protein (MW=31,401) the sequence of which presents no homology with the sequences present in the databases, including those having pterin substrates.

FTR transfers the formyl group of formylmethanofuran to tetrahydromethanopterin of known structure, which is the central transporter in methanogenesis (Fig. 13.1, reaction 2) to which the C1 structure will remain successively linked as the formyl, methenyl, methylene and methyl derivatives.

Methenyltetrahydromethanopterin Cyclohydrolase

This enzyme has been obtained in the homogeneous state from *M. thermoautotrophicum*. Its subunit has a molecular weight of 41 kDa. Since the native protein has been found by exclusion chromatography to have a molecular weight of 82 kDa, it is thought to be a homodimer. The corresponding *mch* gene has been cloned, sequenced and overexpressed in *E. coli*: the deduced sequence of the subunit indicates a molecular weight of 37 kDa.

Mch has been crystallized and its structure determined at a resolution of 2 Å. It shows a new type of α/β fold with two domains separated by a pocket formed by a long conserved sequence. One phosphate ion is in the pocket whereas another one is adjacent to the pocket. The latter is displaced by the phosphate residue of the substrate, formyl-H(4)MPT according to a hypothetical model.

5, 10-methylenetetrahydromethanopterin Dehydrogenase

Coenzyme F_{420} is the phosphodiester N-(N-lactyl–L-glutamyl)-L-glutamic acid of 7, 8-didemethyl-8-hydroxy-5-deazariboflavin -5′ phosphate. F_{420} is the first described deazaflavin.

In most methanogens, two hydrogenase systems can catalyze the reduction of Coenzyme 420 by H_2.

1. The first type of hydrogenases catalyzes the reversible conversion of N5, N10-methylene-H4MPT to N5, N10-methenyl-H4MPT. These hydrogenases do not contain nickel, nor iron or iron-sulfur-clusters and are considered to be metal-free. They catalyze the reaction via a special reaction mechanism.
2. The second type of hydrogenases catalyzes the splitting of H_2 and reduces a deazaflavine, the cofactor F_{420}. Therefore these enzymes are called F_{420}-reducing hydrogenases. The cofactor F_{420} is an important electron carrier in methanogens, but it is not unique to Archaea. The F_{420}-reducing hydrogenases also react with artificial dyes such as benzylviologen. Typically these hydrogenases are composed of three subunits. The large subunit (or subunit A) contains the primary reaction centre. The thiol-groups of four cysteines are ligands of the bimetallic

active site. Two of these cysteines are conserved in the amino-terminal part, the other two in the carboxy-terminal part of the subunit. The conserved motif - Asp - Pro - Cys - X - X - Cys - X - X - His - is found in the carboxyterminal region of the large subunit. The molecular hydrogen is split and the electrons are transferred to the third subunit (subunit B) via [FeS]-clusters, which are localized in the small subunit (or subunit G). This subunit probably contains one flavin and transfers the electrons to the cofactor F_{420}.

In most methanogens, two hydrogenase systems can catalyze the reduction of Coenzyme 420 by H_2: (1) a nickel Fe-S flavoprotein composed of three different subunits and (2) the N(5), N(10)-methylenetetrahydromethanopterin dehydrogenase, composed of a H_2-forming methylenetetrahydromethanopterin dehydrogenase and a F_{420}-dependent methylenetetrahydromethanopterin dehydrogenase, both metal-free and without an apparent prosthetic factor. In nickel-limited chemostat cultures of *M. thermoautotrophicum*, the specific activity of the first system is negligible whereas that of H_2-forming methylenetetrahydromethanopterin dehydrogenase is six times higher, and that of F_{420}-dependent methylenetetrahydromethanopterin dehydrogenase is four times higher than in cells where nickel is not limiting. These experiments suggest that, when nickel is limiting, the reduction of F_{420} by H_2 is catalyzed by the second system.

The methylenetetrahydromethanopterin dehydrogenase from *M. thermoautotrophicum* has been purified to homogeneity. It is a homohexamer (MW of the subunit: 36 kDa). Its coding gene *mer* codes for an acid polypeptide of 321 residues. The enzyme catalyzes the reversible oxidation of 5, 10-methylene-5, 6, 7, 8-tetrahydromethanopterin to the methenyl derivative. There is a strict requirement for the electron carrier F_{420} as co-substrate.

5, 10-methylenetetrahydromethanopterin F_{420} Oxidoreductase

5, 10-methylenetetrahydromethanopterin oxidoreductase from has been purified to homogeneity. It catalyzes the reduction of the 5,10-methylene derivative to 5-methyltetrahydromethanopterin (Fig. 13.2, reaction 5). Here again, the electron carrier F_{420} is a specific co-substrate.

The reaction is reversible; however, the reduction of the methylene derivative is thermodynamically favored. The enzyme is a monomer (MW = 35,000). It contains neither flavin nor Fe-S clusters.

The Methylreductase. Methyl Coenzyme M Reductase

The coupling of CH_3-S-CoM methylreductase with the reduction of CO_2 to CH_4 has been discovered by R.P. Gunsalus in 1980; in the presence of CH_3-S-CoM, the rate of reduction of CO_2 to CH_4 is increased 30-fold and the amount of methane produced 11-fold. The effect can be reinitiated by further additions of CH_3-S-CoM. If $^{14}CO_2$ is used, the synthesis of ^{14}C formyl-methanofuran can be detected only if CH_3-S-CoM is added. The transient intermediate has been identified: it is the mixed disulfide CoM-S-S-HTP (Fig. 13.2, reactions 6 and 7) which is the product of the methylreductase reaction. In the absence of CH_3-S-CoM, the mixed disulfide CoM-S-S-HTP stimulates the reduction of CO_2 42-fold, activates the biosynthesis of formylmethanofurane biosynthesis and is reduced to HS-CoM and HS-HTP by cell extracts.

Coenzyme M (HS-CoM) (Fig. 13.1) is mercaptoethane sulfonic acid, its methylated form (CH_3-S-CoM) being the substrate of the methylcoenzyme M methylreductase system (Fig. 13.2, reaction 7). This system is a complex of several protein fractions and cofactors. It has been first resolved in three fractions called A, B and C, according to their elution order during an anion exchange chromatography. A and C are protein fractions whereas B is 7-mercaptoheptanoylthreonine phosphate (HS-HTP), a small dialyzable molecule which the organic acid chemists have synthesized. Fraction A consists of three proteins A1, A2 et A3 and FAD (Furthermore, ATP is required for the enzyme activity, whereas coenzyme F_{420} and vitamin B12 stimulate, but are not absolute requirements). The methylcoenzyme M reductase system is extremely sensitive to oxygen.

The methylation of coenzyme M is catalyzed by a methyltransferase composed of several subunits subunits,one of which is MtsA (the methylcobalamin:coenzyme M methyltransferase proper) another (MtsB) being homologous to a class of corrinoid proteins involved in methanogenesis.

Component C of *M. thermoautotrophicum*, the methylreductase proper, is stable in the presence of oxygen. Its molecular weight is of about 300,000 and consists of three different subunits of respectively 68,000, 45,000, and 38,500 kDa, corresponding to an $\alpha 2\beta 2\gamma$ stoichiometry.

The genes of the three subunits of *Methanococcus voltae, Methanococcus vannielii*, and *Methanosarcina barkeri* have been cloned and sequenced. They are part of an operon which contains in addition two open reading frames, Orf1 and Orf2 which codes for two polypeptides (16 and 21 kDa). Component C amounts to 10 % of the cell proteins and is highly conserved in all methanogens. It presents a characteristic yellow color with a peak at 420 nm and a shoulder at 440 nm. The isolated chromophore, factor F_{430}, is a tetrahydrocorphin, a highly reduced porphinoid, which chelates a nickel atom and which is the first biological Ni-tetrapyrrole structure to have been isolated (Fig. 13.1). The displacement from 421 to 430 nm is representative of the interactions between the co-factor and component C (reconstitution studies show that a molecule of F_{430} is bound to each of the two large subunits of Component C).

Extended x-ray absorption fine structure spectroscopy (EXAFS) and Raman resonance spectroscopy reveal structure differences between two free forms of F_{430}. The first is a hexacoordinated structure with long Ni-N bonds (2.1 Å) and a plane cycle. The nickel atom of the second form is tetracoordinated with shorter Ni-N bonds (1.9 Å) which suggests a bent form of the corphine nucleus.

The structure of bound F_{430} differs from the two free forms: the nickel atom is hexacoordinated and under the form of "EPR silent" Ni(II) The reduction of free F_{430} is difficult and occurs under non physiological conditions. In addition to the two molecules of F_{430}, component C binds also two molecules of coenzyme M which are incorporated in the CH_3-S-CoM form only when the methylreductase functions. Finally, component C binds also two molecules of HS-HTP.

As noted above, the reduction of CH_3-S-CoM requires in addition to component C other protein fractions (A1, A2 and A3). A2 is a colorless monomer (MW = 59,000) with no known enzymatic activity.

Component A1 is a complex fraction containing several unidentified proteins A1 has often been considered as the hydrogenase of the methylreductase system because it co-migrates on a phenyl-Sepharose column with the hydrogenase which specifically reduces F_{420}. However, all the efforts to replace H_2 by reduced F_{420} have met no success. Component A3 is extremely oxygen-sensitive.

Methylcoenzyme M methylreductase requires the presence of 7-mercaptoheptanoylthreonine phosphate. The latter is part of a cytoplasmic factor. A hydrolytic fragment of this cytoplasmic factor has been characterized as uridine 5′-O-(2-acetamido)-2-deoxy-β-mannopyranuronosic acid substituted by 2-acetamido-2-deoxy-α-glucopyranosyl diphosphate, by mass spectrometry and by high resolution NMR. It is thought that this UDP- disaccharide serves as an anchor for 7-mercaptoheptanoyl threonine phosphate at the active site of the methylreductase complex.

Two essential observations have been made: (1) HS-HTP is the electron donor for the reduction of CH_3-S-CoM (2) The SH-HTP/ HS-CoM heterodisulfide has been identified as the other product of the reduction of CH_3-S-CoM to methane.

$$CH_3-S-CoM+HS\text{-}HTP \rightarrow CH_4+CoM-S-S-HTP$$

Simplification of the Methylreductase System

Homogeneous preparations of component C from *M. thermoautotrophicum* catalyze the reduction of CH_3-S-CoM by dithiothreitol. This reaction requires catalytic amounts of HS-HTP and is stimulated by vitamin B_{12}. No other protein component nor ATP are required. The EPR spectrum of these preparations have a peak corresponding to the Ni atom of F_{430} in the Ni(I) or Ni(III) state. The hyperfine study of the signal shows an interaction of the four nitrogen atoms of the tetrahydrocorphine ring. Furthermore, a correlation has been established between the intensity of the EPR signal and the specific activity of the component C preparations. The component C yielding the EPR signal is the active form Ca of the enzyme and contains Ni (I) in its prosthetic group, the inactive component C being EPR silent. Ca alone can demethylate CH_3-S-CoM in the presence of HS-HTP, which demonstrates unequivocally that it is the methylreductase per se.

A model of the methylreductase has been proposed, in which a role for each of the enzymes and the coenzymes of the system is given.

Reaction 1 describes the conversion of Ci into Ca by reduction of Ni (II) to Ni (I), the components A2, A3 and ATP being involved and the necessary electrons being provided by H_2 or by titanium (III) citrate. ATP is supposedly required to induce a conformational change of Ci necessary for the reduction of Ni (II).

Reaction II is the methylreductase proper, catalyzed by Ca.

During reaction III, the heterodisulfide CoM-S-S-HTP is reduced, either chemically into HS-CoM et HS-HTP by titanium (III) citrate or dithiothreitol in presence of B_{12}, or enzymatically by a reductase linked to a hydrogenase (reaction 7 of Fig. 13.1).

The heptanoyl chain of HS-HTP is very specific. The hexanoyl and octanoyl analogues are not substrates, but strong inhibitors of the methylreductase.

Structure of the Methylreductase

The methylreductase contains eight different subunits which have all been cloned and sequenced.

Sodium dodecylsulfate electrophoresis of the purified complex has shown the presence of eight polypeptides of respective molecular weights of 34 (MtrH), 28 (MtrE), 24 (MtrC), 23 (MtrA), 21 (MtrD), 13 (MtrG), 12.5 (MtrB) and 12 kDa (MtrF).

Cloning and sequencing of the corresponding genes has revealed the presence of eight *mtr* genes organized in a 4.9-kbp segment in the order *mtrED-CBAFGH.*

A Northern blot analysis reveals a transcript of 5 kbp Since mtrE and mtrH DNA probes hybridize with the transcript, the eight genes form an operon.

This composition is to be compared with the tridimensional structure of the methylreductase which, as noted above, is a 300 kDa protein hexameric protein. It has been crystallized in the inactive EPR silent state. Two molecules of the coenzyme F_{430} Niporphinoid are sandwiched between subunits α, α', and γ on one hand and α', α, and γ' on the other, forming two identical active sites. The substrate methylcoenzyme M can reach each site by a narrow channel which closes after the second substrate HS-HTP (coenzyme B) is fixed.

Thanks to another structure crystallized with the heterodisulfide, a reaction mechanism is proposed which involves a free radical intermediate and an organonickel compound.

The crystal structure of methyl-coenzyme M reductase from *M. thermoautotrophicum* has revealed the presence of five amino acids post translationally modified in the subunit, near the active site. Four of them are C- N- et S-methylations, among which 2-methylglutamine and 5-methylarginine had never been encountered before. These modifications have been confirmed by mass spectrometry of chymotryptic peptides. When methyl-coenzyme M reductase was purified from cells grown in presence of L-[methyl-^{3}D] methionine, it was shown that the methyl group of the modified amino acids originates from the methyl group of methionine.

Source of the Energy Needed for the Growth of Methanogens

The reduction of CH_3-S-CoM by H_2 is an exergonic process ($\Delta G° = -85$ kJ) and is the source of energy necessary for ATP synthesis. In *M. barkeri*, methanogenesis from methanol and H_2 or from acetate is linked to ATP synthesis by a proton-motive force, indicating a chimioosmotic mechanism of ATP synthesis. In *M. thermoautotrophicum*, the C component is localized in the vicinity of the membrane only if the cells have been grown with limiting concentrations of nickel. By electron microscopy, it is found that component C associates with the cytoplasmic membrane forming an organite, which has been called methanoreductosome. This association is compatible with the chimioosmotic model.

Although six new coenzymes have been discovered and several enzymes have been purified to homogeneity, there are major gaps in our understanding of methanogenesis. The future goal is to use genetics to further study its biochemistry.

Biosynthesis of Some Cofactors Involved in Methanogenesis

Methanogenesis, the anaerobic production of methane from CO_2 or simple carbon compounds, requires specific organic coenzymes.

Methanofuran

The biosynthesis of the 2-(aminomethyl)-4-(hydroxymethyl)furan (F1) and of the 1,3,4,6-hexane tricarboxylic (TCA) subunits of methanofuran has been studied using gas chromatography and mass spectrometry to follow stable isotope incorporation into these subunits. It was rconcluded that F1 is generated from the condensation of dihydroxyacetone phosphate with pyruvate. The resulting dihydroxy-substituted tetrahydrofuran after elimination of 2 mol of water would produce the phosphate ester of 2-carboxy-4-(hydroxymethyl)furan. Reduction of the carboxylic acid to an aldehyde and subsequent transamination would produce the phosphate ester of F1. It was first thought that TCA was produced by condensation of α-ketoglutarate and malonate to form 1,1,2,4-butanetetracarboxylic acid, which was further condensed with a second molecule of malonate. However, this mechanism has been questioned.

Methanopterin

The biosynthesis of methanopterin (MPT) is similar to that of folic acid (see Chapter). The pterin ring arises from a ring expansion of guanosine triphosphate and the *p*-aminobenzoyl part from shikimic acid.

The pathway for the biosynthesis of methanopterin in *Methanosarcina thermophila* has been shown to proceed by the following series of reactions. First, 5-phospho-alpha-D-ribosyl diphosphate (PRPP) and 4-aminobenzoic acid condense together to produce 4-(beta-D-ribofuranosyl)aminobenzene 5′-phosphate, which then reacts with 6-hydroxymethyl-7,8-H_2pterin pyrophosphate to produce 7,8-H_2pterin-6-ylmethyl-4-(beta-D-ribofuranosyl)aminobenzene 5′-phosphate. This compound is then reduced to 7,8-H_2pterin-6-ylmethyl-l-(4-aminophenyl)-1-deoxy-D-ribitol 5′-phosphate in a reaction stimulated by the addition of FMN or factor F(420). Dephosphorylation of the latter compound leads to 7,8-H_2pterin-6-ylmethyl-1-(4-aminophenyl)-1-deoxy-D-ribitol which condenses with another molecule of PRPP to form 7,8-H_2pterin-6-ylmethyl-1-(4-aminophenyl)-1-deoxy-5-[1-alpha -D-ribofuranosyl 5-phosphate]-D-ribitol. This molecule, in the presence of ATP, condenses with 2-hydroxyglutaric acid to form demethylated H_2methanopterin, a known precursor to methanopterin. The occurrence of this pathway was confirmed by the chemical and/or biochemical synthesis of most of the proposed intermediates, by the detection of these intermediates in cell-free extracts, and by the measurement of their conversion to demethylated methanopterin and/or other intermediates in the pathway. Details of this complex series of reactions can be found in the references at the end of this chapter.

Coenzyme M

Methanogenic archaea begin the production of this essential cofactor by sulfonating phosphoenolpyruvate to form 2-phospho-3-sulfolactate. After dephosphorylation, this precursor is oxidized, decarboxylated and then reductively thiolated to form Coenzyme M.

ComA catalyzes the first step in coenzyme M biosynthesis, a Michael addition of sulfite to phosphoenolpyruvate over a broad range of temperature and pH conditions. This enzyme has no significant sequence similarity to previously characterized enzymes; however, its Mg^{++} dependent enzyme reaction mechanism may be analogous to one proposed for enolase.

A thermostable Mg-dependent phosphosulfolactate phosphohydrolase(ComB) catalyzing the second step in CoM biosynthesis, was identified in the hyperthermophilic euryarchaeon *Methanococcus jannaschii*. ComB has a narrow substrate specificity and an amino acid sequence dissimilar to any biochemically characterized protein. Like other phosphatases, it functions via covalent phosphoenzyme intermediates, Homologs of *comB* are identified in all available cyanobacterial genome sequences and in genomes from phylogenetically diverse bacteria and archaea; most of these organisms lack homologs of other Coenzyme M biosynthetic genes. The broad and disparate distribution of *comB* homologs suggests that the gene has been recruited frequently into new metabolic pathways.

The decarboxylation of sulfopyruvic acid to sulfoacetaldehyde. is the fourth step in the biosynthesis of coenzyme M. The *Methanococcus. jannaschii* sulfopyruvate decarboxylase is a heterododecamer composed of six alpha and six beta subunits. The two subunits are designated ComD and ComE.

The figure below shows the compounds successively involved in Coenzyme M biosynthesis.

CH_2=C-COOH
|
PO_3H_2

phosphoenolpyruvate

H_2O_3S-CH(OPO_3H_2)
|
COOH

L-sulfolactate phosphate

H_2O_3S -CHOH-COOH

L-sulfolactate

H_2O_3S -CO-COOH

sulfopyruvate

H_2O_3S -CHO

sulfoacetaldehyde

H_2O_3S -CH_2-S-CH_2-CH(NH_2)-COOH

sulfoethylcysteine

H_2O_3S-CH_2-CH_2SH

Coenzyme M

7-mercaptoheptanoylthreoninephosphate(Coenzyme B)

Gas chromatographic-mass spectrometric analysis on the distribution of alpha-ketodicarboxylic acids in various bacteria determined that alpha-ketoglutarate and alpha-ketoadipate are widely distributed in all the bacteria examined, whereas alpha-ketopimelate and alpha-ketosuberate are found only in the methanogenic archaebacteria. Labeling experiments with stable isotopes indicated that each of these acids arises from alpha-ketoglutarate by repeated alpha-ketoacid chain elongation. The final product in this series of reactions, alpha-ketosuberate, serves in the methanogenic bacteria as the biosynthetic precursor to the 7-mercaptoheptanoic acid portion of 7-mercaptoheptanoylthreonine phosphate,

$$\mathrm{HOOC-(CH_2)5-CO-COOH}$$
$$\alpha-\text{ketosuberate}$$

Two proteins from the methanogen *Methanocaldococcus jannaschii* formed the first homoaconitase shown to catalyze both the dehydration of (*R*)-homocitrate to form *cis*-homoaconitate, and its hydration is shown to produce homoisocitrate. A combination of the homoaconitase with the *M. jannaschii* homoisocitrate dehydrogenase catalyzed all of the isomerization and oxidative decarboxylation reactions required to form 2-oxoadipate, 2-oxopimelate, and 2-oxosuberate, completing three iterations of the 2-oxoacid elongation pathway.

The α-ketosuberate then undergoes a nonoxidative decarboxylation to form 7-oxoheptanoic acid

$$\mathrm{CH_3-(CH_2)4-CO-COOH}$$
$$7\text{-oxoheptanoate}$$

Hydrogen sulfide was found to supply the sulfur for the production of the 7-mercaptoheptanoic acid in a reaction that was shown to obtain its reducing equivalents from hydrogen via an F420-dependent hydrogenase

$$\mathrm{HS-(CH_2)6-COOH}$$
$$\text{7-Mercaptoheptanoate}$$

The biochemical mechanism for the formation of the amide bond in N-(7-mercaptoheptanoyl)-L-threonine phosphate has been studied by measuring the incorporation of tritiated threonine into N-(7-mercaptoheptanoyl)-L-threonine (HS-HT) by cell extracts of *Methanosarcina thermophila* incubated with different precursors. Data indicate that HS-HT is produced by the phosphorylation of mercaptoheptanoate with ATP. The phosphorylated compound then reacts with L-threonine to produce HS-HT.

Biosynthesis of Coenzyme F420

The hydride carrier coenzyme F(420) contains the unusual chromophore 7,8-didemethyl-8-hydroxy-5-deazariboflavin (FO). FO synthase, composed of two subunits (CofG and CofH) catalyzes the radical-mediated transfer of the hydroxybenzyl group from 4-hydroxyphenylpyruvate (HPP) to 5-amino-6-ribitylamino-2,4(1H,3H)-pyrimidinedione to form 7,8-didemethyl-8-hydroxy-5-deazariboflavin (FO).

The next step is catalyzed by LPPG:FO 2-phospho-L-lactate transferase(CofD) which transfers the 2-phospholactate moiety from lactyl diphospho-(5')guanosine (LPPG) to 7,8-didemethyl-8-hydroxy-5-deazariboflavin (FO) with the formation of the L-lactyl phosphodiester of 7,8-didemethyl-8-hydroxy-5-deazariboflavin (F420-0) and GMP. 2-phospholactate is a natural product that was chemically identified in *Methanobacterium thermoautotrophicum, M. thermophila,* and *M. jannaschii.* CofD has no recognized sequence similarity to any previously characterized enzyme.

The addition of the glutamyl residues is performed by two different ligases,catalyzing respectively the GTP-dependent successive addition of multiple gamma-linked L-glutamates to the L-lactyl phosphodiester of 7,8-didemethyl-8-hydroxy-5-deazariboflavin (F420-0) to form polyglutamate F420-2 (F420-0:gamma-glutamyl ligase, CofE) and the ATP-dependent addition of one L-glutamate molecule to gamma-F420-2, producing the mature coenzyme alpha-F420-3. (Gamma-F420-2:alpha-L-glutamate ligase, CofF).

Biosynthesis of Factor F_{430}

The biosynthetic route leading to the formation of coenzyme F_{430} has not been elucidated completely. It has been shown that it is produced from uroporphyrinogen-III (which in methanogens is synthesized from L-glutamate), through the intermediate precorrin-2 (see Chap. 37). A plausible scheme for the biosynthesis of coenzyme F_{430} from precorrin-2 has been proposed and is illustrated here. However, all of these intermediates, except the seco-precursor, are considered hypothetical at the moment.

$$\text{Precorrin2} \rightarrow \text{pyrrocorphinate} \rightarrow \text{dihydrocorphinate} \rightarrow \text{tetrahydrocorphinate} \\ \rightarrow F_{430}\ \text{seco} - \text{precursor} \rightarrow F_{430}$$

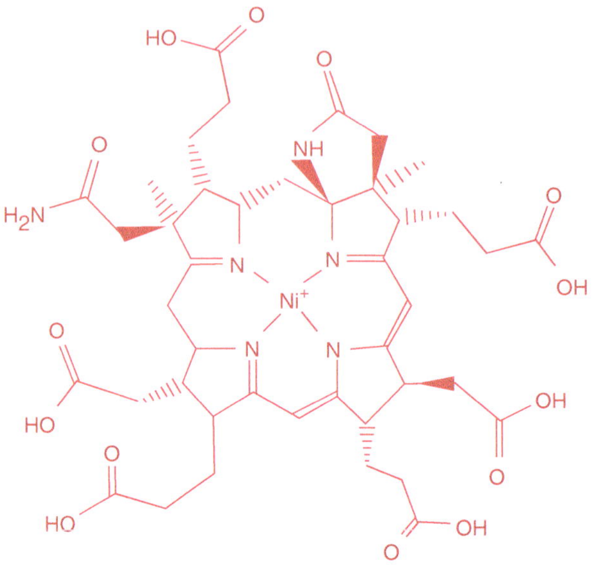

F_{430} seco-precursor

Biosynthesis of Factor III

Factor III is obtained from cobalt siderohydrochlorin (see Chap. 37)

$$\begin{aligned}&S-adenosyl-L-methionine+cobalt-sirohydrochlorin\\&\rightarrow S-adenosyl-L-homocysteine+cobalt-factor\ III+H^{+}\end{aligned}$$

The Aerobic Formation of Methane

Methane is produced in significant quantities by aerobic marine organisms. These bacteria apparently catalyze the formation of methane through the cleavage of the highly unreactive carbon-phosporus bond in methylphosphonate(MPn). The ocean-dwelling *Nitrosopumillus maritimus* catalyzes the biosythesis of MPn from 2-hydroxyethylphosphonate and the bacterial C-P lyase complex converts MPn to methane. PhnJ from the C-P lyase complex catalyzes a C-P cleavage reaction of ribose-1-phosphonate-5-phosphate to methane and ribose -1,2- cyclic -phosphate-5-phosphate. This reaction requires a redox active Fe-S cluster and S-adenosylmethionine.

Methylotrophs

Obligate methylotrophs can all grow on methane. The two other only substrates allowing their growth are dimethylether and methanol. Facultative methylotrophs can grow at the expense of methanol or methylamines, but not of methane. Other substrates allowing them to grow are formate and a small number of simple C2 or C4 compounds.

Methanotrophs

Methane is abundant in the atmosphere. Its concentration increases every year by 1–3 %, reaching presently 2 parts per million. Because it absorbs infra-red radiation more efficiently than CO_2, it may contribute significantly to the temperature increase of our planet. Toundra and humid environments which produce methane when they are soaked can absorb it when they are dry. This phenomenon may result in a reduction of the atmospheric methane concentration. Little is known on the types and amounts of methanotrophs in these environments, which consume 10^{13} g of methane per year.

Methanotrophic bacteria contain an enzyme, methane monooxygenase (MMO) which allows them to grow on methane as the sole carbon source. It is the first of the four enzymes required for the transformation of methane to CO_2. It exists under two forms, soluble (sMMO) or particular, membrane-associated (pMMO). Most methanotrophs possess both forms, although certain species are devoid of the soluble form.

PMMO converts methane to methanol bu using molecular oxygen as cosubstrate. Its activity requires the presence of copper ions, which can be detected by paramagnetic resonance. It oxidizes also butane and shorter hydrocarbons. Its expression is associated with intracytoplasmic membrane structures. Little is known on its structure and mechanism. It resists purification and it is only recently that a gene cluster *pmoCAB* has been cloned and sequenced from *Methylosinus trichosporium* and from a strain of *Methylocystis*. These genes are down-regulated if copper is deficient, under conditions where it is the soluble sMMO which is expressed.

The soluble enzyme from four organisms has been well characterized. In three of them, *Methylococcus capsulatus, Methylosinus trichosporium* and *Methylosinus sporium*, three protein components are necessary for the activity, an hydroxylase, a monomeric reductase homologous to bacterial ferredoxins and a regulatory protein.

The hydroxylase and the reductase from a strain of *Methylomonas* have been purified. The hydroxylase (240 kDa) is composed of three subunits of respectively 56, 43 and 27 kDa and is an (αβγ) 2 protein. It contains 3.78 mol Fe/mol. The α subunit contains a domain bis-μ-hydroxo similar to that of the B2 subunit of *E. coli* ribonucleoside diphosphate reductase; the catalysis seems here also to use a free radical mechanism. FAD is the cofactor of the monomeric reductase (42 kDa) which has a (2Fe-2S) center Finally, the partially purified regulatory protein (16 kDa) contains no metal or cofactor, and increases the activity of the hydroxylase-reductase system by a factor 40.

Metabolism of Methyl Compounds

The steps of total oxidation of methane and methanol seem simple at first view:

$$CH_4 \rightarrow CH_3OH \rightarrow HCHO \rightarrow HCOOH \rightarrow CO_2$$

The enzymology of these reactions is however complex. Experiments performed with $^{18}O2$ have shown that the oxygen atom of methanol derives from molecular oxygen and not from water. The initial step involves an oxygenation. The fact that dimethylether CH_3-O-CH_3 is a substrate for all the methane oxidizing bacteria had first suggested that it was the primary product of the oxidation. Actually certain Gram-positive bacteria can utilize diethylether as growth substrate, according to

$$C_2H_5 - O - C_2H_5 + O_2 + NADH + H^+ \rightarrow C_2H_5OH + CH_3CHO + NAD^+ + H_2O$$

An analogous oxidation of dimethylether would produce methanol and formaldehyde, but this reaction has never been found in methylotrophs.

In contrast to the oxidation of other primary alcohols, catalyzed by NAD^+ dependent dehydrogenases, the oxidation of methanol is catalyzed by an enzyme with unusual characteristics, methanol dehydrogenase.

Methanol Dehydrogenase (MDH)

This periplasmic enzyme is a quinoprotein having pyrroloquinoline quinone (PQQ) as cofactor. The structure of the tetrameric MDH from *Methylobacterium extorquens* has been established. The tridimensional structure and the gene sequence of the enzyme from *Methylophilus sp* have also been determined. The enzyme requires Ca^{2+} for its activity and utilizes cytochrome *c1* as electron acceptor. PQQ is maintained in the right position by a coplanar tryptophan residue and by a disulfide bridge between two adjacent cysteine residues. The enzyme is a heterotetramer H_2L_2, the molecular weights of the subunits being 60 and 9 kDa respectively. The active site contains a tightly bound Ca^{++} ion coordinated to residues of the protein and to PQQ by its 7-carboxy group, by the nitrogen atom of its cycle and by the oxygen atom of a carbonyl group.

Anaerobic Oxidation of Methane

A methane-oxidizing consortium has been isolated from the anoxic sediments of a freshwater canal which was subject to agricultural run-off with a high nitrogen load. The sediment was saturated with methane and contained nitrate at a concentration of 1 mM. The culture obtained grew very slowly on methane and nitrate as the sole energy sources. It consisted mainly of two different types of microorganisms, a bacterium (80 %) and an archaeon (10 %).

CO_2 and N_2 are produced by this consortium according to the following equation

$$5CH_4 + 8NO_3^- + 8H^+ \rightarrow 5CO_2 + 4N_2 + 14H_2O$$

The only other known anaerobic methane-oxidizing process is coupled to sulfate reduction, mediated by a consortium of methane consuming archaea and sulfate-reducing proteobacteria. The nitrate-dependent anaerobic oxidation of methane in the sediment is predicted to occur close to the oxic-anoxic interface, where in the oxic phase ammonia is oxidized to nitrate by nitrifying organisms. Nitrate then diffuses into the anoxic sediments saturated with methane, which is produced from the degradation of plant materials

Recently, it has been shown that birnessite, a lamellar manganese oxide (MnO_6),and ferrihydrite (an ubiquitous iron oxyhydroxide) can serve as excellent electron acceptors

In the anaerobic oxidation of methane in marine sediments, in the absence of nitrate or sulfate. Large amounts these oxides are provided to oceans from rivers, indicating that manganese- and iron-dependent anaerobic oxidation of methane have the potential to be globally important.

Anaerobic oxidation of methane (AOM) is critical for controlling the flux of methane from anoxic environments. AOM coupled to iron, manganese and sulphate reduction have been demonstrated in consortia containing anaerobic methanotrophic (ANME) archaea. More recently it has been shown that the bacterium *Candidatus 'Methylomirabilis oxyfera'* can couple AOM to nitrite reduction through an intra-aerobic methane oxidation pathway.

Methylamine Dehydrogenase

Methylamine can be used as carbon and nitrogen source by two groups of bacteria, differing by the way methylamine is cleaved. The first group, illustrated by *M. extorquens*, cleaves directly methylamine to formaldehyde and ammonia

$$CH_3NH_2 + H_20 + \text{acceptor}$$
$$\left(HCHO + NH_3 + \text{reduced acceptor}\right)$$

The members of the second group, among which pseudomonads are found, exhibit a complex mode of cleavage, involving the formation of N-methylglutamate which is oxidized regenerating glutamate and yielding formaldehyde.

The primary structure of a certain number of methylamine dehydrogenases has been determined by partial protein sequencing or deduced from the sequencing of the corresponding genes. The species studied have been several pseudomonads, *Paracoccus denitrificans, Thiobacillus versutus* and *M. extorquens*. The most striking aspect of this heterotetramer L_2H_2 is the diversity of the structures of the α subunits and the homology of all the β subunits. The latter is a very compact structure due to the presence of 5 disulfide bridges and of tryptophyltryptophan quinone (TQQ) which derives from a covalent bond between two tryptophan residues belonging to the polypeptide chain.

Carbon Assimilation by Methylotrophs

Methylotrophs synthesize all their carbon containing constituents from C-1 compounds. Radioactive tracer experiments have shown that the great majority of this carbon originates from formaldehyde and not from CO_2.

Two cyclic pathways exist in methylotrophs. The first is the ribose phosphate pathway, analogous to the dark phase of the photosynthesis cycle, the key reaction being the addition of formaldehyde to ribose 5-phosphate, producing the phosphorylated ketohexose, allulose 6-phosphate, which is epimerized to fructose 6-phosphate, then degrades by the classical glycolysis cycle. Part of the fructose-6-phosphate serves to regenerate ribose 5-phosphate, which can then accept new molecules of formaldehyde. The second, called the serine pathway, involves a totally different series of cyclic reactions. A C1-unit derived from formaldehyde is transferred to tetrahydrofolic acid (THF) and from there to glycine resulting in the formation of serine:

$$\text{C1-THF} + \text{glycine} \rightarrow \text{THF} + \text{serine}$$

Serine is then converted by a series of reactions to phosphoglycerate, a part of which is assimilated by conversion to triose phosphate, and a part serving to regenerate glycine, the primary C1 acceptor. Glycine regeneration is complex and implies the following series of transformations:

$$\text{Phosphoglycerate} \rightarrow \text{phosphoenolpyruvate} \rightarrow \text{malate}$$
$$\text{Malate} \rightarrow \text{glyoxylate} + \text{acetyl CoA}$$

The transformation of phosphoenolpyruvate to malate implies the fixation of one molecule of CO_2.

Figures 13.3 and 13.4 summarize the two pathways of formaldehyde assimilation.

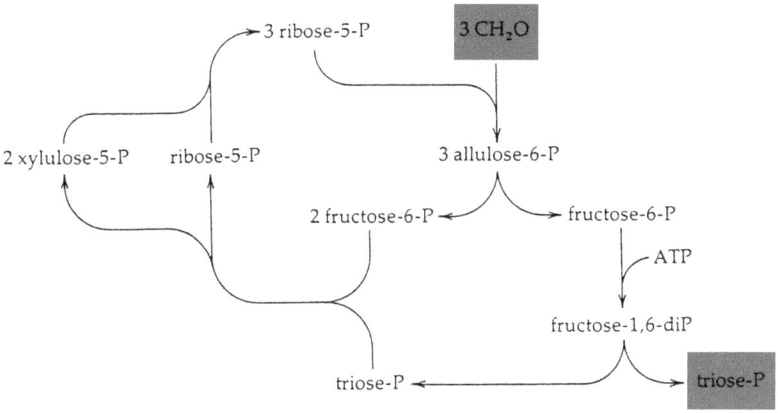

Fig. 13.3 The ribose pathway used for the assimilation of formaldehyde by some methylotrophs

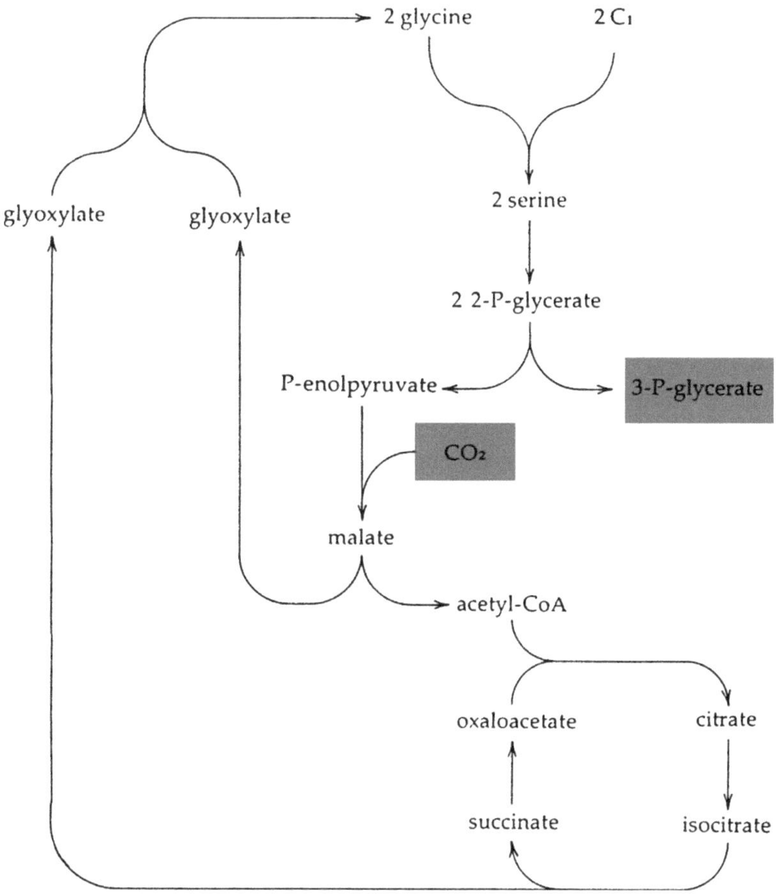

Fig. 13.4 The Serine pathway used for the assimilation of formaldehyde by some methylotrophs

Carboxydotrophs

These bacteria are characterized by the chemolitho-autotrophic utilization of CO, the carbon monoxide, as the sole source of carbon and energy under aerobic or denitrifying conditions. Their habitats are soil and water and natural enrichment conditions are found in the soil covering charcoal installations.

These organisms belong to about 20 phylogenetically distinct groups. In addition to carbon monoxide, most strains can utilize H_2+CO_2, or in a limited manner, organic substrates. *Streptomyces thermoautotrophicus* is a notable exception: it grows exclusively on CO or H_2+CO_2, and fixes atmospheric nitrogen.

The reaction catalyzed by carboxydotrophs is the following:

$$CO + H_2O + \text{acceptor} \rightarrow CO_2 + \text{reduced acceptor}$$

It requires the presence of several proteins which are either absent in other microorganisms, or present in different combinations. In *Pseudomonas carboxydovorans* they are (a) A CO dehydrogenase (CODH) which is a molybdenum-containing flavoprotein and possesses a Fe-S cluster for the catabolic oxidation of CO. CODH is composed of three subunits; (b) a membrane-linked hydrogenase necessary for the catabolic utilization of H_2; (c) an electron acceptor cytochrome b_{561} anchored to the membrane; (d) a cytochrome b_{563}, serving as a terminal oxidase and (e) a ribulosebisphosphate carboxylase and a phosphoribulokinase, key enzymes for fixing CO_2 and bringing its carbon into the pentose phosphate cycle.

The genetics of the system is less advanced than its biochemistry. Several carboxydotrophs carry from one to three relatively large plasmids. In *P. carboxydovorans*, the structural genes coding for the three subunits of CODH, *coxL, coxM* and *coxS* reside on a 123kbp plasmid. The structural genes of ribulosebisphosphate carboxylase and phosphoribulokinase are duplicated, one copy being on the plasmid and the other on the chromosome. The genes coding for the two cytochromes (*pac* and *tox*) are carried by the chromosome. In bacteria lacking the plasmids, all the genes are chromosomal.

The study of the CO dehydrogenases of carboxydotrophs has led to the characterization of molybdopterin dinucleotides as the organic components of the molybdenum cofactor, and more specifically of molybdopterin cytosine dinucleotide in the case of *P. carboxydovorans*.

A group of anaerobic thermophilic bacteria growing on carbon monoxide and producing hydrogen and carbon dioxide according to the same equation

$$CO + H_2O \rightarrow CO_2 + H_2$$

has been isolated from samples of diverse thermal habitats: terrestrial hot springs of Kamchatka, Yellowstone, Kuril and Kermadec Islands and Baikal region, deep-sea hot vents of West and East Pacific and Atlantic oceans. At present they are represented by six bacterial genera (all new) and one archaeal. These organisms were found to be neutrophiles or alkaliphiles, moderate, extreme or hyperthermophiles. CO was found to be either the only or preferable substrate and molecular hydrogen to be the only reduced product during the growth on CO. CO-utilizing bacteria of genera *Carboxydothermus, Carboxydocella, Carboxydotropha*, were found to be obligate lithotrophs, while *Carboxydobrachium* was capable also of organotrophic growth. From the samples of Yellowstone National Park, a novel genus *Carboxydosinus* was isolated, capable of ferric iron reduction coupled with CO oxidation. All these bacteria are Gram-positive but do not belong to a single phylogenetic group. The representative of Archaea isolated from the deep-sea hot vents of East Pacific Rise was found to belong to the genus *Thermococcus*.

Carbon monoxide is regarded as the common component of volcanic gases both in terrestrial and deep-sea environments. Being the energy source of litho-autotrophic prokaryotes, carbon monoxide might support the primary biomass production in volcanic habitats.

The above data actually show that diverse thermal environments are inhabited by thermophilic CO-oxidizing anaerobic prokaryotes.

Selected References

Methanogenesis

Mukhopadhyay B, Johnson EF, Wolfe RS (1999) Appl Environ Microbiol 65:5059–5065

Wasserfallen A, Wolfe RS (1994) Proc R Swed Acad Sci Nobel Symp 1991:145–155

Wolfe RS (1993) An historical overview of methanogenesis. In: Ferry JG (ed) Methanogenesis, ecology, physiology, biochemistry, and genetics. Chapman and Hall, New York, pp 1–32

Biosynthesis of the Methanogenic Cofactors

White RH (2001) Vitam Horm 61:299–337

Biosynthesis of Methanofuran

Gorkovenko A, Roberts MF, White RH (1994) Appl Environ Microbiol 60:1249–1253

White RH (1988) Biochemistry 27:4415–4420

Biosynthesis of Methanopterin

White RH (1996) Biochemistry 35:3447–3456

Biosynthesis of Coenzyme M

Graham DE, Graupner M, Xu H, White RH (2001) Eur J Biochem 268:5176–5188
Graupner M, Xu H, White RH (2000) J Bacteriol 182:4862–4867
Wise EL, Graham DE, White RH, Rayment I (2003) J Biol Chem 278:45858–45863

Biosynthesis of 7-mercaptoheptanoylthreoninephosphate (Coenzyme B)

Drevland RM, Jia Y, Palmer DRJ, Graham DE (2008) J Biol Chem 283:28888–28896
Howell DM, Harich K, Xu H, White RH (1988) Biochemistry 37:10108–10117
Solow B, White RH (1997) Arch Biochem Biophys 345:299–304

Biosynthesis of Coenzyme F420

Forouhar F, Abashidze M, Xu H, Grochowski LL, Seetharaman J, Hussain M, Kuzin A, Chen Y, Zhou W, Xiao R, Acton TB, Montelione GT, Galinier A, White RH, Tong L (2008) J Biol Chem 283:11832–11840
Graham DE, Graupner M, White RH (2003) Arch Microbiol 180:455–464
Graupner M, White RH (2001) Biochemistry 40:10859–10872
Graupner M, Xu H, White RH (2002) Biochemistry 41:3754–3761
Graupner M, Xu H, White RH (2003) Biochemistry 42:9771–9778

Biosynthesis of Factor F430

Pfaltz A, Kobelt A, Hüster R, Thauer RK (1987) Eur J Biochem 170:459–467

Anaerobic Oxidation of Methane

Beal EJ, House CH, Orphan VJ (2009) Science 325:184–187
Haroon MF, Hu S, Shi Y, Imelfort M, Keller J, Hugenholtz P, Yuan Z, Tyson GW (2013) Nature 500:567–570
Raghoebarsing AA, Pol A, van de Pas-Schoonen KT, Smolders AJ, Ettwig KF, Rijpstra WI, Schouten S, Damsté JS, Op den Camp HJ, Jetten MS, Strous M (2006) Nature 440:918–921

Aerobic Formation of Methane

Kamat SB, Williams HJ, Dangott LJ, Chakrabarti M, Raushel FM (2013) Nature 497:132–136

Carboxydotrophs

Cypionka H, Meyer O (1983) Arch Microbiol 135:293–298
Ferry JG (1995) Annu Rev Biochem 49:305–333
Henstra AM, Sipma J, Rinzema A, Stams AJM (2007) Curr Opin Biotechnol 18:200–206

Chapter 14
Enzyme Induction in Catabolic Systems

Enzymatic adaptation, as it had been studied by Karström can be defined as the induced appearance of specific enzymes after a preliminary "adaptation" of microorganisms on their substrates. The term of enzymatic adaptation has been replaced by enzyme induction.

The Specificity of Induction

The relationships between the specificity of the induced enzymes and the specificity of induction present a capital importance for the interpretation of the phenomenon. Two levels of specificity can be defined for an enzyme: (a) specificity of action, defined by the enzyme activity on different substrates; (b) specificity of binding, defined by the affinity of the enzyme for substances forming a specific complex with it, whether these substances are substrates or not. These enzyme ligands comprise substrates and inhibitors.

What are the relationships between the inducing properties of a given substance and its substrate or ligand properties with regard to the enzyme it induces? The hypotheses belong to three classes:

Functional hypothesis: The synthesis of the enzyme depends on its activity. The inducer acts as a substrate. The corollary of this hypothesis is that action and induction specificities are identical.

Equilibrium hypothesis: The synthesis of the enzyme is limited by the dynamic equilibrium:

$$\text{Precursors} \rightleftharpoons \text{Enzyme}$$

This equilibrium is perturbed when the enzyme is bound to a ligand

$$\text{Enzyme} + \text{L} \rightarrow \text{Enzyme} - \text{L}$$

G.N. Cohen, *Microbial Biochemistry*,
DOI 10.1007/978-94-017-8908-0_14, © Springer Science+Business Media B.V. 2014

and more enzyme is produced. This hypothesis is based on the identity between the induction and binding specificities.

Template hypothesis: The inducer plays an "organizing" role not by combination with the enzyme itself, but by its action on a synthesizing system. This hypothesis does not presuppose the identity between substrate, inducer or ligand. Monod, Cohen-Bazire and Cohn have compared the properties of a series of sugar derivatives as inducers, substrates or ligands of β-galactosidase, enzyme whose physiological function is the hydrolysis of the disaccharide lactose into glucose and galactose. Table 14.1 highlights their main results, which show that the specificity of combination of the purified enzyme is limited to the substances possessing the β-galactoside or the derived α-L-arabinoside configuration. Any substitution, inversion or oxidation causes the complete or almost complete loss of affinity; α-galactosides are neither substrates or ligands of the enzyme. Substitution of a sulfur atom in place of the natural oxygen atom of the galactoside link leaves the affinity almost intact whereas activity toward the analogue is entirely lost. As a result, for example, thiophenyl-galactoside is a powerful competitive inhibitor of β-galactosidase. If one considers the inducing properties of these substances, the following conclusions can be drawn:

1. The inducing power is independent of the substrate property. This is best demonstrated by the inducing capacity of α-methylgalactoside and of melibiose. In addition, thiomethyl, thioethyl, and thioisopropylgalactosides which are among the best known inducers of β-galactosidase are absolutely not hydrolyzed by this enzyme. Such non substrate inducer molecules have been called gratuitous inducers.
2. The inducing power is independent from the ligand property: there is no systematic correlation between the values of the affinity and the inducing power (Table 14.1). For example, 2-deoxygalactose which retains a significant affinity has no inducing power. Phenyl-β-D-thiogalactoside (TPG), endowed with a considerable affinity is not an inducer. However, it certainly penetrates the cells since it inhibits competitively β-galactosidase in vivo. In this respect, it is interesting to note that TPG which inhibits the growth of *E. coli* on lactose is devoid of any intrinsic toxic effect when maltose is the carbon source.
3. The inducing power is associated with the presence of an intact α- or β-galactoside link. The case of raffinose shows that the presence of this link is a necessary but not sufficient condition.

These results are incompatible with either the functional or the equilibrium hypothesis, nor with any other hypothesis relying on a specific combination between the inducer and the enzyme as a sine qua non condition for induction to occur. We shall later reexamine the template hypothesis in its modern version.

De novo Synthesis of β-galactosidase

The students of induction who were thinking in terms of mechanisms were confronted with the following problem: does enzyme induction correspond to the appearance of a new molecular species synthesized de novo from amino acids, or to

Table 14.1 Comparison of the inducing, combination and substrate specificities of *E. coli* β-galactosidase

Compound	Inducer[a]	Assayed as Ligand[b]	Substrate[c]
Galactose	420	30	+
β-D-galactosides			
methyl-	2800	10	+
n-butyl-	2800	400	+
o-nitrophenyl-(ONPG)	1060	1000	+
β-naphthyl-	42	200	+
4-glucose-(*lactose*)	2500	100	+
mannose-	2400	10	+
Thiophenyl-	0	700	0
α-D-galactosides			
Methyl-	140	0	0
6-glucose(*melibiose*)	2400	0	0
6-sucrose-(*raffinose*)	0	0	0
Carbon 2, 3, 4, and 6 substitutions of β-methyl-galactoside			
2-methyl-	20	0	0
2, 6-dimethyl-	0	0	0
3, 4-dimethyl-	0	0	0
2, 4, 6-trimethyl-	0	0	0
Reduced derivatives			
2-deoxygalactose	0	10	–
1, 2-dideoxygalactose	0	0	–
6-deoxygalactose	0	0	–
Oxidation at carbon 6			
D-galacturonic acid	0	0	–
and its methylester	0	0	–
Suppression of carbon 6			
L-arabinose	0	0	0
methyl-β-L-arabinose	0	–	0
o-nitrophenyl L-α-arabinose	0	50	+
Sugars without galactoside cycle			
D-xylose, D-tagatose, D-mannose, D-glucose, mathyl-β-D-glucoside, maltose, cellobiose	0	0	– or 0

[a]Activity after growth in the presence of 10^{-3} inducer, expressed as μmoles ONPG hydrolyzed/min/mg dry weight

[b]Determined by measure of the competitive inhibition of the hydrolysis of ONPG, taken as reference substrate, and to which the value 1,000 is arbitrarily attributed

[c]+ and 0 mean respectively that the compound is hydrolyzed or not hydrolyzed; – = not tested

a type of molecular reorganization which cannot be regarded as synthesis? For example, the conversion of trypsinogen to trypsin cannot be considered as the synthesis of a protein, although it implicates the appearance of a new protein species. Thus, the interpretation of induced biosynthesis must rely on an estimation of synthesis independent of activity measurements.

Two research teams have established beyond doubt that β-galactosidase is synthesized de novo from amino acids. Hogness, Cohn and Monod have grown *E. coli*

Fig. 14.1 The *left rectangle* represents the total bacterial proteins, labeled by ^{35}S, synthesized in the absence of inducer. *Right*, after induction in a non-radioactive growth medium, β-galactosidase appears as the sole non radioactive protein (recognized by specific antibodies) (From D.S. Hogness, M. Cohn and J. Monod, with permission of Biochimica et Biophysica Acta)

in presence of radioactive sulfate during several generations in order to obtain uniformally labeled bacteria. The cells are then washed to eliminate the excess radioactivity and transferred in a medium containing non radioactive sulfate and a gratuitous inducer of β-galactosidase. After several generations, the cells are collected and disrupted and the β-galactosidase purified from the extract, the last purification step consisting in a specific precipitation by anti-β-galactosidase antibodies. The radioactivity of the isolated enzyme is of about 4 % of that of the enzyme obtained by induction in the presence of radioactive sulfate (of specific radioactivity identical to that used in the non induced culture). This maximum of 4 % is observed whether the mass increase during induction is of 10 or 100 %. The authors conclude that if any precursors of β-galactosidase existed before induction, they must be at a very low concentration. Spiegelman and Rotman have performed essentially the same experiment, using ^{14}C-lactate as a carbon source for growth. After having grown in this medium, the bacteria are induced in a non radioactive medium, the enzyme purified by electrophoresis on a starch gel and precipitated by the specific antibody. Here again β-galactosidase possesses less than 3 % of the specific radioactivity of the other proteins present in the induced culture.

The results of these experiments schematized in Fig. 14.1, indicate that the contribution of preexisting proteins or of other preexisting carbon compounds is minimal if any. They suggest therefore that the enzyme is synthesized from amino acids which in turn derive from the carbon source.

It is important to note that, had there been an important proteolysis of cell proteins, yielding free amino acids, β-galactosidase would have been radioactive. The fact that it does contain minimal amounts of radioactivity whatever the duration of induction suggests that there is no significant protein turnover during exponential growth.

Constitutive Mutants

If we consider Karström's separation of the sugar degradation enzymes into "constitutive" or "adaptative", we observe that it is purely empirical and does not implicate differences in the mechanism of synthesis of the enzymes belonging to the two classes. The main reason to refuse to consider different mechanisms has been that the the same enzyme can be constitutive in one strain and inducible in another strain of the same species, or even partly constitutive; in the latter case, it is synthesized, albeit at a lower rate, even in the absence of inducer.

It has been unambigously demonstrated that the penicillinase constitutively produced by *Bacillus subtilis* is identical to the enzyme produced -in much larger amounts- when penicillin is present in the culture medium.

Mutations have been obtained, which transform an inducible system into a constitutive one. These changes correspond to a single mutational event which does not affect the structure of the enzyme, but only its induciible or constitutive character: the inducible *E. coli* β-galactosidase has the same molecular weight, the same kinetic parameters, the same degree of activation by ions, the same pattern of heat inactivation and the same immunochemical specificity as the enzyme extracted from a constitutive mutant.

The isolation of constitutive mutants involves selection methods which vary from one case to the other and are based on a selective advantage conferred to the constitutive mutants; here are some examples of selection of mutants constitutive for the synthesis of β-galactosidase:

(a) thiophenyl-β-D-galactoside (TPG) is an inhibitor of β-galactosidase, but mainly a powerful inhibitor of its induction. If a mutagenized population is layered on a solid medium containing lactose as the sole carbon source and TPG at a convenient concentration, the inducible bacteria will not be induced and will not grow; the only colonies which will appear arise from constitutive bacteria.
(b) phenyl-β-D-galactoside is a substrate of β-galactosidase, but is devoid of inducing activity. If it is incorporated as the sole carbon source into a solid medium, only constitutive bacteria will be able to grow and to give rise to colonies.
(c) finally, a more general method is based on the difference in growth rates of the inducible wild type and the constitutive mutants in different media.

Pleiotropy of the Constitutive Mutants

During the study of β-galactoside permease, two types of structural mutations had been observed, *y*-mutations which affect the structure of the permease and *z-mutations*, which occur within the structural gene of β-galactosidase.

When the permease activity of the β-galactosidase constitutive mutants was assayed, it was observed with a great surprise that it was also constitutuve, i.e., it was present without a previous culture in presence of an inducing galactoside.

One recalls that, during the same study, when the bacteria having accumulated radioactive thiomethylgalactoside, a minor spot was identified with 6-acetylthiomethylgalactoside. This compound is the product of an enzyme normally inducible in the wild type of *E. coli*, which catalyzes the following reaction:

$$\text{Acetyl CoA} + \text{thiogalactoside} \rightarrow \text{6-acetylthiogalactoside} + \text{CoA}$$

The physiological function of thiogalactoside transacetylase is unknown to this day. What interests us however here, is the fact that its synthesis is constitutive in the mutants constitutive for the galactosidase and the permease. We are thus confronted with a new type of mutation, a pleiotropic mutation, which affects several characters simultaneously. At this point, it is important to emphasize that it is not the structures of the three proteins which are affected by the pleiotropic mutation, which determines only the inducible or constitutive character of the synthesis of these catalysts. We shall call i^+ the wild type inducible allele, and i^- the mutated constitutive allele. We shall call the *i* gene a regulatory gene by contrast with the *z, y, ac* genes which are structural genes whose base sequence codes for the corresponding amino acid sequence.

Pleiotropic constitutive mutants have been found to affect the regulation of the synthesis of several enzymes involved in the degradation of arabinose, galactose, maltose, etc…

The Genetic Control and the Cytoplasmic Expression of Inducibility in the Synthesis of β-galactosidase in *E. coli*. The Lac Repressor

Any hypothesis on the mechanism of induction must provide an interpretation of the differences between the inducible and constitutive systems. Since specific point mutations can transform a typically inducible system into a fully constitutive one, an analysis of the genetic structure and of the biochemical effects of such mutations should lead to such an interpretation.

If, for reasons of simplicity, one does not consider the transacetylase, the following eight genetic combinations can be obtained:

$i^+z^+y^+$, wild type, inducible for galactosidase and permease.

$i^+z^+y^-$, cryptic mutant (devoid of permease), but possessing a normally inducible galactosidase.

$i^+z^-y^+$, mutant inducible for permease, not producing active β-galactosidase. It has been shown in many z^- mutants, that an inactive protein is synthesized, with affinity for the *E. coli* anti-β-galactosidase antibodies (CRM protein, for cross-reacting material). In this class, the synthesis of CRM is inducible.

$i^+z^-y^-$, mutant devoid of the two activities, but where the synthesis of CRM is inducible.

$i^-z^+y^+$, constitutive mutant synthesizing permease and galactosidase without inducer added.

$i^-z^+y^-$, constitutive cryptic mutant, devoid of permease.

$i^-z^-y^+$, mutant constitutive for the syntheses of permease and CRM.

$i^-z^-y^-$, mutant without permease and constitutive for CRM.

The genetic structure of the *lac* region which commands the utilization of lactose has to be known for the comprehension of the capital experiment described later.

(a) the recombination frequency between *z* and *y* mutations is very low, of the order of one hundredth of that observed between the *leu* (1 min on the circular chromosome) and *gal* (17 min) markers.
(b) if in the cross $y^+i^-z^-$ x $y^-i^+z^+$, the y^+z^+ recombinants are selected by their ability to grow on lactose-containing plates (on which none of the parents strains can grow), one observes that i^+ remains associated with z^+ in 85 % of the cases.
(c) the frequency of cotransduction of *i* and *z* (when z^+ is selected for) is higher than 90 %, whereas the frequency of cotransduction of *i* with *y* is only of 70 %.

These studies, and more detailed ones, show that the *z, y, ac*, and *i* genes are all grouped in the same region and that their topological relationships can be represented as shown in the scheme below

gal — *i* — *z* — *y* — *ac* — *leu, thr*

It should be recalled that conjugation in *E. coli* involves the injection of the chromosome of a male Hfr (high frequency of recombination) into a receptor F^- (female) strain, which generally results in the formation of an incomplete zygote or merozygote. The recombination between the female chromosome and the segment of the injected male chromosome does not start before 60–90 min after injection; furthermore, segregation of recombinants from the merozygotes takes place only after several hours, which leaves ample time to study the interactions between the two genomes, their expression in the cytoplasm and their dominance relationships. It is equally important for the interpretation of the results that the contribution of the Hfr partner to the merozygotes is strictly chromosomic, whereas the F^- bacteria have both a cytoplasmic and chromosomal contribution. In order to be able to draw a conclusion of the experiments on the expression of the *z, y* and *i* factors, it is necessary to find conditions where the merozygotes, but not the parent strains, can synthesize β-galactosidase. The following cross provides such conditions:

$$\text{Hfr } z^+y^+i^+SmS \text{ x } F^-z^-y^+i^+SmR$$

SmS and *SmR* being the alleles of the *Sm* gene determining streptomycin sensitivity and resistance, respectively.

The cross is made in the presence of the inducer (isopropylthiogalactoside, IPTG) and streptomycin. The F^- strain cannot synthesize the enzyme since its structural gene is inactivated. The Hfr strain cannot synthesize it either, because its growth is inhibited by streptomycin; the zygotes are not inhibited by streptomycin because they inherit the F^- cytoplasm and chromosome and the chosen Hfr strain transfers the *Sm* gene at a rather low frequency. Under these conditions, the enzyme is formed by the population with a time and amount dependency which is compatible only with a synthesis by the zygotes having received the z^+ gene (Fig. 14.2).

The synthesis of enzyme begins a few minutes after the introduction of the z^+ gene (10 min on the chromosome). If one takes into account that the number of zygotes having received z^+ is 4 to 5 times greater than the measured number of

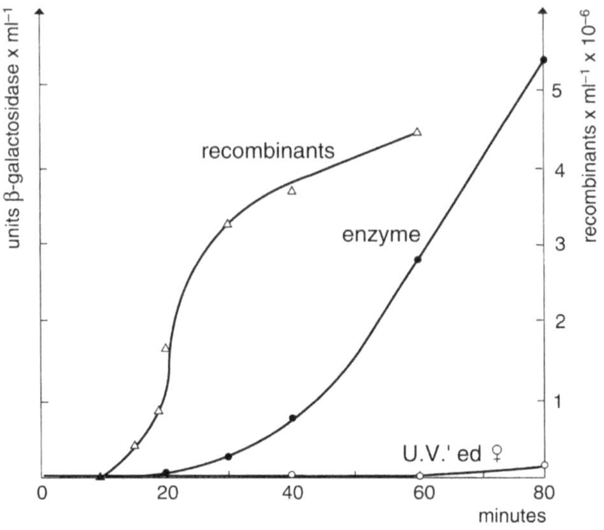

Fig. 14.2 Synthesis of enzyme and appearance of recombinants during the cross described above, performed in the presence of inducer and streptomycin. A control where the F^- strain has been inactivated by UV irradiation is included. The z^+SmR recombinants are selected on plates containing lactose and streptomycin, after interruption of the cross by violent shaking (From A.B. Pardee, F. Jacob and J. Monod, with permission of the Journal of Molecular Biology)

z^+SmR recombinants, and that each cell is on the average trinucleate (possesses three copies of z^+), one calculates that the rate of synthesis of the enzyme by the z^+ gene is normal.

Since all z^- mutants belong to the same cistron and since the z^+ allele is obviously dominant over z^-, one can then make the following cross:

$$Hfr\ z^+i^+\ x\ F^-\ z^-i^-$$

in the absence of inducer

Under those conditions, none of the parents can synthesize the enzyme. However, zygotes do synthesize β-galactosidase (Fig. 14.3): during the first hour following the mixture, one observes a rapid constitutive synthesis. The cross has thus allowed an immediate and complete interaction of the male i^- and female z^+ genes. The possibility that this interaction is due to z^+i^- recombinants is excluded: the synthesis starts immediately after the injection of z^+ whereas genetic recombination takes place only after 60–90 min after injection; on the other hand the *z* and *i* genes are so close that recombination between them is a rare event (such recombinants are 10000 times less frequent than zygotes); the kinetics of enzyme synthesis is such that all zygotes must be synthesizing β-galactosidase.

There is the possibility that the interaction leading to enzyme synthesis is a consequence of the pairing of the homologous chromosome segments of the two parents. This possibility has been excluded by performing the following cross in the absence of inducer:

$$Hfr\ z^-i^-\ x\ F^-\ z^+i^+$$

The zygotes obtained in this cross are identical to those of the preceding cross. However, no synthesis of enzyme is observed at any time, although adequate controls

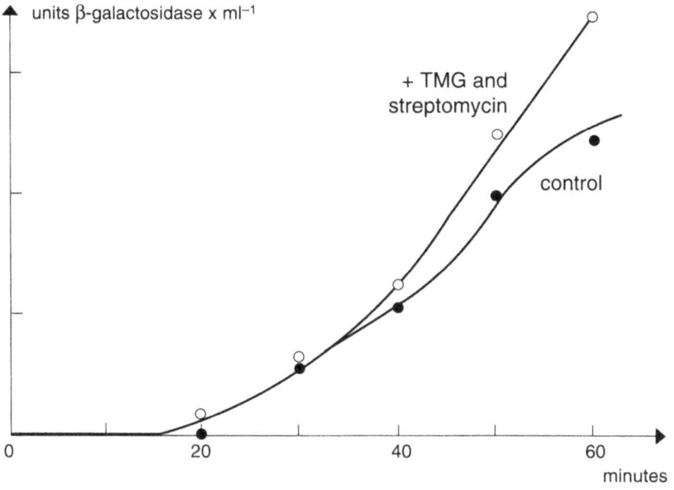

Fig. 14.3 Synthesis of enzyme by the zygotes from the cross: $Hfr\ z^{+}i^{+}\ x\ F^{-}\ z^{-}i^{-}$. At 20 and 25 mn respectively, streptomycin and TMG are added to an aliquot of the culture, in order to compare enzyme synthesis with and without inducer (From A.B. Pardee, F. Jacob and J. Monod, with permission of the Journal of Molecular Biology)

show that the conjugation and the injection of the male chromosome have occurred normally. The only difference between the two crosses is that in the second, the cytoplasm of the zygote is provided by the inducible partner. It can therefore be concluded that the interaction between i^{-} and z^{+} occurs through the cytoplasm. The simplest explanation is that the *i* gene is responsible for a diffusible cytoplasmic message which is received by the *z* gene or by a structure commanding the expression of the *z* gene. In addition, the hypothesis is made that one of the alleles of *i* causes the synthesis of the message whereas the other allele does not. If these hypotheses are exact, one of the alleles must be strictly dominant. The dominance must be expressed immediately when the cytoplasm is contributed by the dominant parent and be expressed only gradually when the cytoplasm is that of the recessive parent.

The fact that β-galactosidase is never synthesized in the cross:

$$Hfr\ z^{-}i^{-}\ x\ F^{-}\ z^{+}i^{+},$$

even after several hours means that the i^{-} constitutive allele is never expressed and that the dominant allele is the i^{+} one. If this is the case, the i^{+} allele should be expressed after a certain time in the cross:

$$Hfr\ z^{+}i^{+}\ x\ F^{-}\ i^{-}z^{-},$$

which produces the enzyme constitutively.

In order to verify this assumption, the cross is repeated with strains containing additional markers:

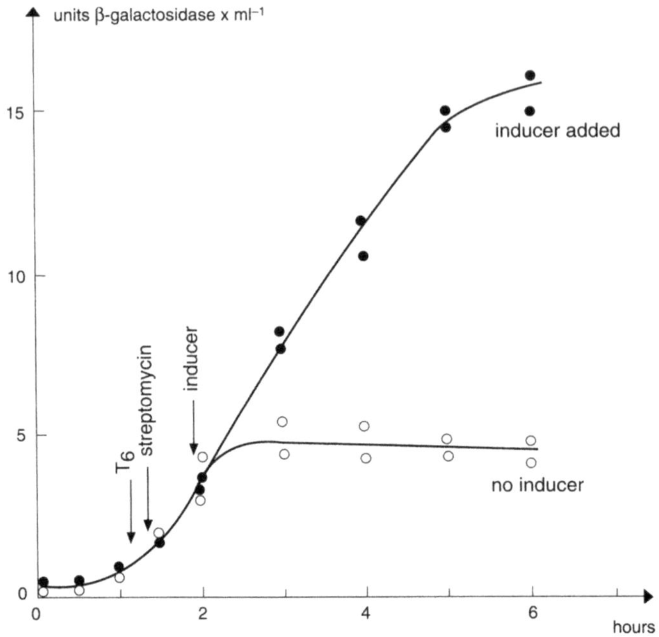

Fig. 14.4 Cross performed in quadruplicate in the absence of inducer. At the times indicated by the *arrows*, a suspension of T_6 phage and streptomycin is added. Inducer is added to two of the cultures (•), omitted in the two others (○). Same source as Fig. 14.2.

$$Hfr\ z^+i^+\ SmsT_6r\ x\ F^-z^-i^-SmrT_6\,r$$

and the synthesis of enzyme is followed, with or without inducer, in the presence of streptomycin and of bacteriophage T_6 (to prevent the induction of males and new crosses between recombinants). Figure 14.4 shows that, in the absence of inducer, the constitutive enzyme synthesis stops after 90 min after the entry of z^+i^+. If at this point, inducer is added, enzyme synthesis is resumed, showing that the zygotes z^+i^+/z^-i^- which were carrying a constitutive synthesis, have not been inactivated, but are now inducible. A necessary control shows that this transformation of constitutivity to inducibility is not due to the segregation of homozygous recombinants of the type $z^+i^+SmrT_6$, which happens much later.

From the above experiments, the following conclusions can be drawn:

- the *i* gene, under its active form (i^+) determines the synthesis of a diffusible cytoplasmic product which inhibits the expression of the structural genes *z*, *y* and *ac*, unless an external inducer is added. When the product of the *i* gene is absent or inactive, there is no need for the addition of an external inducer to express the structural genes. The product of the *i* gene is highly specific and exerts no effect on other known systems. The above considerations do not inform us on the nature and on the mode of action of the *i* gene. Two models are possible:

Table 14.2 Non-inducibility of i^s *mutants*

Genetic structure	Inducer	β-galactosidase units/mg dry weight
Wild: i^+z^+	0 IPTG 10^{-4} M	< 5 7000
Non inducible i^sz^+	0 IPTG 10^{-2} M	10–20
Diploid: i^s/i^+	0 IPTG 10^{-3} M	5 8
Diploid: i^s/i^-	0 IPTG 10^{-3} M	8 8

(a) the β-galactosidase forming system requires the presence of an internal inducer, synthesized by both the i^+ and the i^- strains. The i^+ allele is responsible for the synthesis of an enzyme which destroys or inactivates this internal inducer. The i^- mutants do not produce this enzyme and accumulate the endogenous inducer. This model accounts for both the dominance relationships observed and for the kinetics of transformation of the constitutive zygotes into inducible ones.
(b) the other model predicts that in the wild type (i^+), a specific repressor is synthesized. The inducer acts then as an antagonist of the repressor. The constitutive mutant i^- does not form an active repressor and therefore, an exogenous inducer is not necessary. This model equally accounts for the observed facts.

As induction is highly stereospecific, a recognition site for the inducer must be carried either by the hypothetical repressor, or by an element of the transcription or translation machinery recognized by the repressor. Whatever the case, one expects that mutations in the DNA segment coding for the repressor or for its target could abolish the recognition site for the inducer without affecting the interaction of the target with the repressor. A mutation of the repressor causing a loss of recognition for the inducer must be reflected in the zygotes in a permanent loss of the capacity for lactose utilization, and this incapacity should be dominant versus both the alleles i^+ and i^-.

Such mutants have been searched for in the following manner: in the immense majority of the *lac*⁻ mutants (798 out of 800 tested), the wild allele *lac*⁺ was dominant over the mutated alleles. In the two remaining cases, the mutated alleles were dominant on the wild type ($i^+z^+y^+$) and the fine localization of the mutations has assigned them to the *i* gene. The diploids are not inducible (Table 14.2). These mutations are called i^s.

Since an i^s mutation inactivating an (hypothetical) enzyme destroying an (hypothetical) endogenous inducer cannot be dominant over the wild allele i^+ allowing the synthesis of this enzyme, we are forced to accept that i^s determines the synthesis of a super-repressor which cannot be antagonized by inducer and as a consequence to adopt the model of cytoplasmic repressors antagonized by exogenous inducers.

Operators and Operons

That a given repressor specifically inhibits the synthesis of one or more proteins implies that at a given point of the expression of the gene into its encoded protein(s), this repressor must form a stereospecific combination with an element of the system. When this element is combined with the repressor, the expression is blocked. This hypothetical element which can switch on and off the expression of the system has been called *operator* by Jacob and Monod.

Whatever the nature of the operator and whatever the level at which it is involved in the regulation process, the specific configuration it must be endowed with to respond to the stereo-specificity of the repressor should be altered by mutation. In particular, there should exist mutations of the operator rendering it unable to recognize the repressor.

In a diploid cell with a chromosome carrying the wild operator O^+ and the other the mutated operator O^c (unable to bind the repressor), the phenotype should be that of a dominant constitutive because the repressor formed by i^+ will combine with the wild operator O^+ and the z^+ gene carried by chromosome A will not be expressed. But since O^c cannot combine with the diffusible repressor, the z^+ gene carried by chromosome B will be expressed freely.

i^+	O^+	z^+	A
i^-	O^c	z^+	B

The confirmation of the theory requires thus the isolation of dominant constitutive mutants, different from the i^- mutants, which are recessive.

An ingenious way to find such mutants is to select constitutive lactose fermenters among inducible heterogenotes homozygous for the i^+ gene (i^+z^+/i^+z^+); only the very improbable event of two independent i^- mutations in the two i^+ genes can lead to the constitutivity due to the existence of an inactive repressor.

Constitutives have actually been isolated from these heterogenotes. Their genetic analysis shows that they possess the expected characteristics:

(a) O^c is dominant over O^+ in O^+/O^c heterogenotes.
(b) O^c possesses the same degree of constitutivity for galactosidase and permease. Many O^c mutants are partially constitutive and produce 10 to 40 per cent of the inducible synthesis, which represents from 100 to 500 times more than the non-induced wild type.
(c) O^c is insensitive to the super-repressor: i^s O^+/i^+ O^c heterogenotes behave as constitutives.

In addition, O^c mutations display two additional properties which suggest the nature of the operator and allow to guess at which level it intervenes in the mechanism of expression. First, all the O^c mutations are localized in a small segment situated at the extremity of the *z* gene, between *i* and *z*:

i	*O*	*z*	*y*	*ac*

Second, when their behavior in heterogenotes is examined, one notes that only the genes situated on the same DNA segment than the O^c mutation are expressed constitutively.

Table 14.3 Synthetis of β-galactosidase, of CRM and of thiogalactoside transacetylase by haploid or diploid mutants of the operator gene

Genotype	Galactosidase		Transacetylase	
	Non induced	Induced	Non induced	Induced
$O^+z^+ac^+$	1	100	<1	100
$O^cz^+ac^+$	25	95	15	110
$O^+z^+ac^-/ F\ O^cz^+ax^+$	70	220	50	160
$O^+z^-ac^+/ F\ O^cz^+ac^-$	180	440	<1	220
$i^sO^+z^+ac^+/ Fi^+O^cz^+ac^+$	190	210	150	200
Genotype	Galactosidase		CRM	
	Non induced	Induced	Non induced	Induced
O^+z^+	<1	100	–	–
O^+z^+/ FO^+z^-	<1	105	<1	310
O^cz^+	15	90	–	–
O^+z^+/ FO^cz^-	<1	90	30	180
O^+z^-/ FO^cz^+	90	250	<1	85

Bacteria are grown in glycerol and induced, when necessary, by 10^{-4} M IPTG. The inactive protein (CRM) formed by certain mutants of the *z* structural gene, has the same antigenic properties as galactosidase, but is devoid of enzyme activity. Symnols *z* and *ac* refer to the genes coding respectively for β-galactosidase and thiogalactoside transacetylase. Symbols i^+ and i^- refer respectively to the inducible and constitutive alleles of the gene coding for the repressor. Symbols O^+ and O^c refer to the wild and constitutive dominant alleles of the operator gene
F refers to the sexual factor of *E. coli*; when this factor incorporates the *lac* region of the bacterial genome, it forms a new genetic unit, F-*lac*, an episome, which can be added to the bacterial genome to form diploids for the *lac* region. Several F-*lac* episomes can coexist within the same cell, which explains why the organisms bearing the wild allele of the gene on the episome synthesize more enzyme than the wild type
The values are given in per cent of the values obtained with the induced wild type, except for the inactive CRM protein, which is estimated in β-galactosidase antigenic equivalents. In O^c mutants, values obtained vary somewhat with the culture conditions, especially in non-induced cultures

In genetic language, one says that the O^c mutations act only in the *cis* position; for example, we know that an heterogenote z^+/z^- CRM which is homozygous for O^+i^+ synthesizes in the presence of inducer both active and inactive (CRM) galactosidase. If O^c is introduced on the chromosome or on an episome, one observes that the bacteria $O^c\ z^+/FO^+\ z^-$ CRM produce active enzyme constitutively whereas the CRM protein is synthesized only in the presence of inducer. On the other hand the bacteria $O^+z^+/F\ O^c\ z^-$ CRM produce the CRM protein constitutively while galactosidase is normally inducible. Table 14.3 shows examples of diverse possible combinations.

The β-galactosidase synthesized by the O^c mutants is identical to the inducible enzyme and to that produced by the constitutive i^- mutants, by all criteria available. The preceding experiments lead to reject a certain number of models to explain the repression and the derepression (constitutivity). For example, each gene of a pathway regulated in a coordinated manner cannot have its own operator. One would expect in this case that a $O^+ \rightarrow O^c$ mutation would cause the constitutivity of one enzyme only.

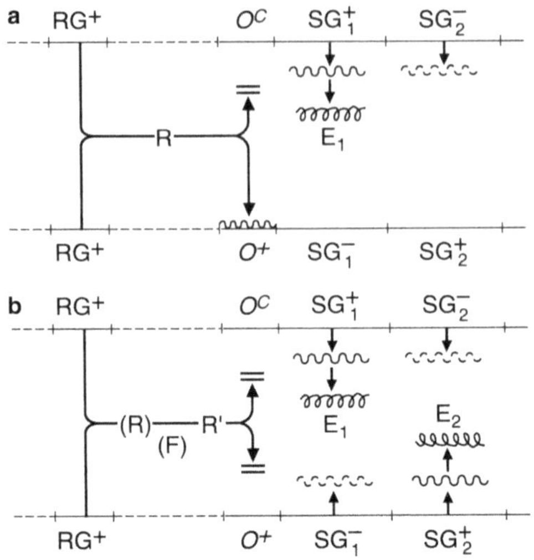

Fig. 14.5 Schematized behavior of two diploids heterozygous for the operator and for the structural genes. Heterozygote for a constitutive operator (O^c) mutation/wild type (O^+). (**a**). Non induced. The repressor made by the regulator genes (*RG*) blocks gene expression in the lower chromosome which carries a normal operator (O^+). It does not act on the upper chromosome which carries a mutated operator (O^c) with a low affinity for the repressor. Only enzyme E1 is produced constitutively since the upper chromosome carries a mutation in SG2 (SG2$^-$). (**b**). Induced. In the presence of inducer F, the repressor R is converted into the inactive form R′. Gene activity is allowed in both chromosomes. Both enzymes E1 and E2 are made (Adapted from F. Jacob and J. Monod)

The only model which has withstood analysis is that a single operator allows the expression of a sequence of genes and that this operator and the accompanying structural genes constitute an integrated topological and functional unit. We shall call such a unit an *operon*.

We know that the genetic information coded by DNA is transcribed as messenger RNA molecules (see next chapter); these molecules migrate toward the ribosomes where their information is translated into protein molecules. Messenger RNA is thus (in procaryotes) the primary product of the genes.

A remarkable and general property of the induction mechanisms is that all the genes whose expression depends on the same operator are contiguous on the chromosome; the operator itself being localized at the 5′ extremity of the DNA sequence coding for the different genes. The most satisfactory explanation of all these facts is that the operator is (or neighbors) the genetic locus where the transcription of the information of a group of genes coding for the structure of contiguous genes involved in a given metabolic sequence occurs. We can then define the operon as a genetic unit of coordinated transcription. The repressor blocks transcription by specifically binding the operator, in a way that no message is sent from the operon to the ribosomes, resulting in the arrest of the synthesis of the proteins coded by the structural genes. Figure 14.5 summarizes this theory.

Selected References

Specificity of induction. De novo synthesis

Monod J, Cohen-Bazire G, Cohn M (1951) Biochim Biophys Acta 7:585–599
Hogness DS, Cohn M, Monod J (1955) Biochim Biophys Acta 16:99–116

Pleiotropy of the i gene

Rickenberg HV, Cohen GN, Buttin G, Monod J (1956) Ann Inst Pasteur 91:829–857

Thiogalactoside transacetylase

Zabin I, Kepès A, Monod J (1962) J Biol Chem 237:253–257

Genetics

Jacob F, Wollman EL (1961) Sexuality and the genetics of bacteria. Academic, New York
Hayes W (1968) The genetics of bacteria and their viruses, 2nd edn. Blackwell Scientific, Oxford and Edinburgh

Repressors and operators

Pardee AB, Jacob F, Monod J, Mol J (1964) Biol 1:165–178
Willson C, Perrin D, Cohn M, Jacob F, Monod J, Mol J (1964) Biol 8:582–592
Jacob F, Monod J (1961) Cold Spring Harbor Symp Quant Biol 16:193–211

Chapter 15
Transcription: RNA Polymerase

At the end of the 1950s Hoagland and Zamecnik had succeeded in obtaining an in vitro protein synthesis from ribosomes and crude acellular extracts. Ribosomes cannot by themselves achieve a specific protein synthesis and require an information originating in last analysis in the genes.

A few minutes after the penetration of a virulent bacteriophage into *E. coli* cells, that is after the entry of the phage genes, the bacteria stop synthesizing their own proteins and produce only the proteins and enzymes necessary to manufacture a few hundred copies of the infecting phage with its nucleic acid and its protein capsid, the new phages causing bacterial lysis in a few minutes. It was then shown that phage infection caused the immediate arrest of the synthesis of bacterial RNA while a new RNA species subject to a rapid turnover appears. Volkin and Astrachan confirmed this observation in 1956 and showed that this new RNA species corresponds to only 3 % of the total RNA and that its base composition reflects that of the phage DNA and not that of the bacterial DNA. In 1961, Spiegelman showed that the RNA synthesized as the result of infection can form a complementary specific molecular hybrid with phage DNA, but not with the bacterial DNA.

Everything became clear with the concept of messenger RNA formulated the same year by Jacob and Monod. This concept originated from their work on the regulation of enzyme synthesis which is treated elsewhere in this volume: they postulated the existence of an unstable species, the messenger RNA, which transfers the genetic information from DNA to the site of protein synthesis, i.e. the cytoplasmic ribosomes.

Three independent groups showed that a shortlived variety of RNA with a base composition analogous to that of DNA and different from that of ribosomal RNA existed also in uninfected bacteria.

G.N. Cohen, *Microbial Biochemistry*,
DOI 10.1007/978-94-017-8908-0_15,

The Synthesis of Messenger RNA. The Bacterial RNA Polymerase

The synthesis of the messenger (mRNA) is catalyzed by an enzyme, DNA-dependent RNA polymerase, present in the cell-free extracts of bacteria, of animal and plant cells. In these preparations, it can be precipitated with DNA by addition of protamine or streptomycin. In eukaryotic cells, it is localized in the nuclei.

In addition to DNA, the polymerization reaction requires the presence of the four nucleoside triphosphates, that of a divalent metal (Mg^{++} or Mn^{++}). The matrix DNA can be single or double stranded, or can be present as short oligonucleotide fragments. Finally and above all, the reaction is characterized by the fact that the base composition of the single-stranded RNA product is defined by that of the template DNA. As in the case of DNA duplication, the new chain is complementary of that of the template DNA, which is of paramount importance for the transfer of genetic information. This process, catalyzed by RNA polymerase has been called transcription.

In vitro, any of the two DNA strands can serve as matrix. In vivo however, only one of the DNA strands can serve as model. This was shown by using a microorganism whose two strands differ by their density in a cesium chloride gradient. After separating the two strands by column chromatography, it is found that only one of the chains hybridizes with that of the mRNA from the same organism.

The basic unit of transcription by RNA polymerase is the holoenzyme which can initiate and elongate the RNA chain. The holoenzyme contains five major species of subunits, α, β, β', σ and ω present in the molar ratios 2:1:1:1:1. The preparations contain strongly bound zinc. When the enzyme is deprived of the σ subunit, the enzyme is called "core polymerase", which can be obtained by a single chromatography of the holoenzyme on a phosphocellulose column under non-denaturing conditions.

The widely conserved ω subunit encoded by *rpoZ* is the smallest subunit of *Escherichia coli* RNA polymerase but is dispensable for bacterial growth. Function of ω is known to be substituted by the chaperone GroEL in ω-null strain, which thus does not exhibit a discernable phenotype. Variants of ω whose expression in vivo leads to a dominant lethal phenotype have been isolated. Studies show that in contrast to ω, which is largely unstructured, ω mutants display substantial acquisition of secondary structure. By detailed study with one of the mutants, ω_6 bearing an asparagine to aspartic acid substitution, the mechanism of lethality has been deciphered. Biochemical analysis reveals that ω_6 binds to β' subunit in vitro with greater affinity than that of ω. The reconstituted RNAP holoenzyme in the presence of ω_6 in vitro is defective in transcription initiation. Formation of a faulty RNA polymerase in the presence of mutant ω results in death of the cell. Furthermore, lethality of ω_6 is relieved in cells expressing the *rpoC2112* allele encoding β'_{2112}, a variant β' bearing a tyrosine to sulfur substitution, immediately adjacent to the β' catalytic center. These results suggest that the enhanced ω_6-β' interaction may perturb the plasticity of the polymerase active center, implicating a role for ω and its flexible state.

The core polymerase is able to catalyze RNA synthesis, but the σ subunit is essential for transcription initiation from promoters. In addition, accessory factors, designated generally as "transcription factors" may be required for transcription initiation from some promoters. Some of these transcription factors associate with RNA polymerase in the absence of DNA. However, most factors interact only when both polymerase and factors are aligned along the same DNA strand, and thus the activation also requires a specific DNA-binding function.

The synthesis of RNA rests on the integrity of the β subunit, which is altered in rifampicin- and streptolydigin-resistant mutants. These antibiotics block RNA synthesis by interacting with the β-subunit of the wild-type polymerase. It has been determined that four contiguous amino acids of the β subunit define the target for streptolydigin resistance; their mutations can influence transcription elongation, but these amino acids are not directly involved in catalysis. The σ and β′ subunits are respectively essential for the selection of the promoter site and for the reconstitution of the holoenzyme from its isolated subunits. The genes corresponding to the β and β′ polypeptides are adjacent on the *E. coli* chromosome and are synthesized in a coordinate manner.

The polymerase, when engaged in a ternary complex with its DNA template and with a nascent RNA chain is devoid of σ subunit, which favors the idea that the "core polymerase" is responsible for the elongation. The fact that the σ subunit can be lost during the purification of the enzyme has suggested that other components might also have been lost. This approach has led to the isolation of ρ, a termination factor. Two pathways of transcription termination, factor-independent and ρ-dependent, exist in bacteria. The latter pathway operates on nascent transcripts that are not simultaneously translated and requires, in addition to ρ, two proteins NusG, and NusA, each of which is essential for viability of wild-type*Escherichia coli*. NusG and NusA are also involved in anti-termination of transcription at the ribosomal RNA operons, as well as in regulating the rates of transcription elongation of all genes. A bisulfite-sensitivity assay has demonstrated genome-wide increase in the occurrence of RNA-DNA hybrids (R-loops), including from antisense and read-through transcripts, in a *nusG* missense mutant defective for Rho-dependent termination. Lethality associated with complete deficiency of Rho and NusG (but not NusA) was rescued by ectopic expression of an R-loop-helicase especially so on defined growth media. (Helicases are motor proteins that move directionally along a nucleic acid, separating two annealed nucleic acid strands (i.e., DNA,RNA, or RNA-DNA hybrid) using energy derived from ATP hydrolysis).

RNA synthesis can be decomposed in several steps, each one complex. It is generally accepted that for the transcription units which need only the holoenzyme for a selective transcription, four steps are needed:

1. the attachment of the holoenzyme to the DNA matrix, during which the polymerase localizes the specific site where initiation can take place, and adopts an active conformation.
2. the initiation of the RNA chain, during which the enzyme catalyzes the coupling of ATP or GTP with a second ribonucleoside diphosphate to generate a

dinucleoside tetraphosphate pppPupX with elimination of pyrophosphate; the dinucleoside tetraphosphate remains strongly bound to the RNA polymerase-DNA complex.
3. the elongation of the RNA chain during which successive nucleoside monophosphate residues originating from the nucleoside triphosphates are added to the 3′-OH terminus of the growing chain.
4. the termination of the RNA chain, accompanied by the liberation of the free enzyme and of the newly synthesized RNA chain from the DNA matrix.

If one considers the selectivity of transcription, steps 1 and 4, the attachment to DNA and the termination, are the most important ones: it has been shown unequivocally that the selection of a specific site for initiation does not require the presence of nucleoside triphosphates; the correct termination of a growing RNA chain occurs on a specific site of the DNA matrix with or without the accessory factor ρ.

The DNA region which controls a correct initiation is called the *promoter region*. This region may contain a variety of individual functions, which may or may not be physically separated. One can note for example, the site(s) of recognition of the polymerase, the initiation sites proper, sites for the binding of positive regulators of the CAP type, and operator sites for the binding of negative regulators (repressors). Whereas the term promoter was initially meant, on the basis of genetic experiments, "to designate the site at which transcription is initiated", its meaning has been extended to designate the region of a transcription unit responsible both for the initiation of transcription and for its regulation.

An exhaustive study of the binding of the holoenzyme to bacteriophage T7 DNA has been carried out by Chamberlin in 1972. The productive binary complexes bind at the initiation sites of the promoter regions and are able to selectively initiate the transcription. The kinetics of the selection suggests the holoenzyme separates the two DNA strands at the initiation site and that these open complexes are ready to initiate. Other authors have shown that the binding of RNA polymerase to phage DNA is accompanied by the opening of about 24 base pairs (bp) per DNA molecule. Four regions appear to be recognized by the enzyme, meaning that an average 6 bp are opened per promoter site. At high ionic strength, the enzyme still binds, but there is no local opening and therefore no initiation. The corresponding non productive complexes are called closed complexes.

At least three processes are affected by the binding of the σ subunit to the polymerase: (a) the non specific binding to the DNA helix is strongly reduced, due to a 10^6-fold increase in the dissociation rate of the non specific complexes; (b) the σ subunit is required for the formation of the open complexes; (c) its binding to the core enzyme suppresses the spurious initiation at breaks occurring on a single strand and imposes a more strict initiation specificity.

The sequence of the promoter sites has been obtained by several methods, one of which is their isolation from the bulk of the DNA by nucleolytic digestion in the presence of the polymerase or of a repressor which are strongly bound to and protect the DNA region from the cleavage. The protected regions comprise the region recognizing the polymerase and the regions where the initiation takes place. By comparing a small number of promoter sequences, it was observed that each contained a sequence of seven base pairs, homologous to the TATAATG sequence, centered about 10 bp before the initiation site of messenger RNA. Later, Rosenberg and Court have compared 46 promoter sequences and observed that the 6th position of this heptamer was invariably a thymine residue (TATAATG). It appears that this base pair is essential to the promoter function and interacts in the same way with the polymerase in all promoters. The other positions of the heptamer are somewhat variable and one can extract from the comparison of all the promoters examined the consensus sequence

TAtaa**T**g (Pribnow box)

in which the base pairs represented by the capital bold letters are the less variable.

Upstream this "Pribnow box" centered at −10 bp from the initiation site, another region of homology has been detected and a consensus sequence for this region centered at – 35 bp, has been extracted

tg**TTG**aca

Initiation of transcription predominantly occurs 6–9 bp downstream the invariable T reesidue of the −10 region. The starting nucleotide is usually a purine (more often an adenine than a guanine). The mutations in the −10 sequence which affect most the promoter function are those of the most conserved base pairs. Physical studies show that polymerase unwinds the double helix of DNA at this level.

Certain purine bases are protected by RNA polymerase from chemical modification by dimethylsulfate, whereas other bases are normally modified or even are more susceptible to methylation. The specific interference by the polymerase has allowed to detect interactions between the enzyme and guanine residues in the large groove of DNA, the majority of contacts occurring in the −10 and −35 regions.

Sequence-specific DNA binding proteins must quickly bind target sequences, despite the enormously larger amount of non-target DNA present in cells. RNA polymerases (or associated general transcription factors) are hypothesized to reach promoter sequences by facilitated diffusion . In facilitated diffusion, a protein first binds to non-target DNA and then reaches the target by a one-dimensional sliding search. Experiments directly compared the rates of initial σ-containing polymerase binding to and dissociation from promoter and non-promoter DNAs measured in the same sample under identical conditions. Binding to a non-promoter DNA was much slower than binding to a promoter-containing DNA of the same length, indicating that the detected nonspecific binding events are not on the pathway to promoter binding. Truncating one of the DNA segments flanking the promoter to a length as short as 7 bp or lengthening it to around 3,000 bp did not alter the promoter-specific binding rate. These results exclude facilitated diffusion over distances corresponding to the length of the promoter or longer from playing any significant role in accelerating promoter search. Instead, the data support a direct binding mechanism, in which σ-RNA polymerase reaches the local vicinity of promoters by 3D diffusion through solution, and suggest that binding may be accelerated by atypical structural or dynamic features of promoter DNA.

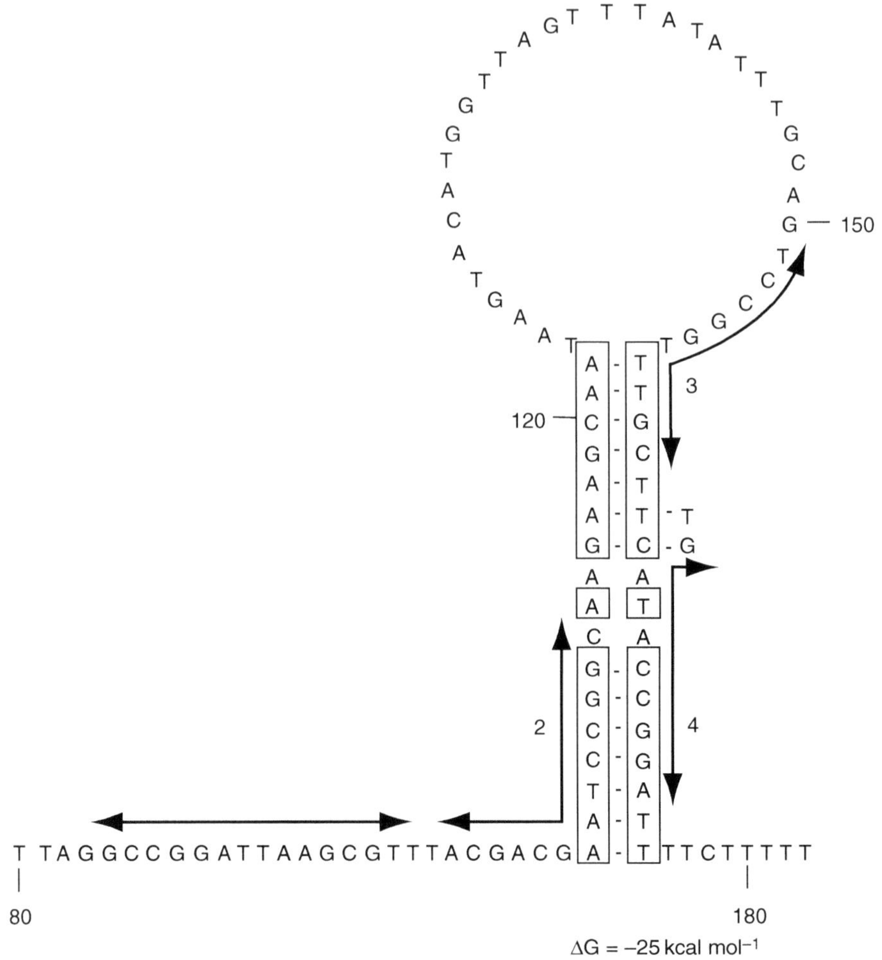

Fig. 15.1 The ρ-independent terminator region of the metBL operon of *E. coli*. The free energy of the stem and loop structure is shown (From N. Duchange, M. M. Zakin, P. Ferrara, I. Saint-Girons, I. Park, S. V. Tran, M-C Py and G. N. Cohen, with permission of the Journal of Biological Chemistry)

Termination of Transcription in Prokaryotes

The principal role of the termination of transcription is to maintain the different transcription units separate, so that there is no interference between the transcription of adjacent genes or operons. Comparing most prokaryotic termination sites has revealed three common characteristics: a region of binary symmetry (that is an inverted repeated sequence), G-C rich sequences of variable length (3–11 contiguous base pairs) followed by a poly U sequence which terminates the transcript (Fig. 15.1). In those cases, termination of transcription does not need additional factors. But there are other types of termination transcription sites which require the presence of the ρ factor. They are called ρ-dependent and do not present the structural characteristics of the above ρ-independent termination sites.

The active form of the ρ protein is a hexamer endowed with a nucleoside triphosphatase (NTPase) activity required for the liberation of the nascent transcript from the transcriptional complex. Covalent incorporation of a single molecule of 8-azido-ATP (an ATP photoaffinity analog of ATP) per ρ hexamer is sufficient to cause inactivation, indicating that the active sites of ρ interact in the RNA-dependent ATP hydrolysis. The function of ρ depends on interactions with nascent RNA molecules that contain unpaired cytidylate residues. Results of a study where cytidine and cytidine analogs are shown to inhibit the binding of ρ to cro mRNA and its ATPase activity suggests that ρ has a specificity pocket in its polynucleotide binding site that can make hydrogen bond interactions with the side of the unpaired cytidine ring that normally faces away from the sugar ring. A few common characteristics of the ρ-dependent terminators are known, among which the presence upstream of the termination site of an unstructured guanine-poor region. The establishment of a consensus sequence is rendered difficult by the scarcity of ρ-dependent termination systems.

Some activators of transcription do not act by helping RNA polymerase to start transcription; rather, they prevent termination and as a consequence allow transcription of the downstream genes. Termination-anti-termination as the means to switch genes on and off is employed in numerous systems in prokaryotes, and a few examples have been uncovered in eukaryotes as well. Two general mechanisms are responsible for anti-termination. One, attenuation, will be examined in Chap. 19. The other was discovered while studying anti-termination in phage λ: the N protein, coded by λ, interacts with RNA polymerase by masking its response to protein termination factors.

Yeast RNA Polymerases

The synthesis of cellular RNA which is accomplished in bacteria by a single polymerase is achieved in all eukaryotes by three distinct enzymes. RNA polymerase I (pol I) synthesizes the ribosomal RNAs, whereas pol II synthesizes the messenger RNAs and pol III is responsible for the synthesis of 5.6S ribosomal RNA, transfer RNAs and of some other small molecules. Like the bacterial enzymes, they are complex enzymes (MW ca. 500,000 Da). They differ however from their prokaryotic counterpart in their subunit number: they contain some ten different subunits, certain of which are common to pol I, II and III. In addition, the genomes from mitochondria are not transcribed by pol II.

As in the case of the bacterial polymerases, promoters and terminators limit the transcription units and regulate their activity. These signals are different for each of the three polymerases. The promoters of pol III have been well characterized and present the unexpected property to lie within the transcribed sequence.

Since pol II is responsible for the synthesis of the messengers, it has received the greater attention. The purified enzyme is not active with natural promoters: in order to achieve a specific transcription, it requires other "general transcription factors" (TFIIA, TFIIB, TFIID and TFIIE), the role of which is to recognize the TATA and the initiation sequences (see below) and to integrate the polymerase into the transcriptional complex. Five complexes have been identified by successive binding of TFIID, TFIIA, TFIIB, pol II and TFIIE. The addition of ribonucleoside triphosphates generates then new complexes which contain the correctly initiated complexes associated with the rest of the transcriptional machinery and with the DNA matrix.

The polypeptide that binds directly to the TATA box (TBP or TFIID is required for transcription by all three nuclear RNA polymerases. It thus resembles the essential subunits common to the three RNA polymerases in being a universal transcription factor. TBP is associated with other polypeptides, the TAFs to form the TFIID multiprotein complex.

Archaeal RNA Polymerases

The basal transcriptional components of Archaea share homology with the eukaryotic RNA polymerase II system at both subunit complexity and sequence level. For example, in the *Pyrococcus abyssi* genome, twelve RNA polymerase subunits are encoded in six loci. They are generally found in conserved clusters that also encode ribosomal proteins or lie in the immediate vicinity of ribosomal protein genes.

The initiation of transcription in Archaea appears to be a shortened version of the eukaryotic system required for RNA polymerase II initiation: the archaeal genome encodes two basal transcription factors, TBP, the TATA-binding protein, and TFB, a homolog of TFIIB. In addition, other proteins sharing similarity with subunits of TFIIE and TFIIH.

Despite this similarity of the eukaryotic and archaeal transcription machinery, transcription regulation in Archaea involves a plethora of bacterial regulators, like the Asn/Lrp family proteins.

The finding that a homologue of the eukaryotic transcription factor TFIIB is found in several archaea is of considerable interest. The degree of sequence similarity is very significant: this conservation implies a similar three-dimensional structure and is suggestive of a related function in a transcriptional initiation complex. It also is indicative that essential aspects of the eukaryotic transcriptional machinery were in place before the divergence of archaea and eukaryotes, that is much earlier in evolution than previously thought. The analysis of mutants show that the DNA region which affects transcription is much longer than the bacterial promoters. Examination of several hundred eukaryotic genes has shown certain homologies on lengths extending over several hundred nucleotides upstream the initiation site by pol II. One of the most frequent homologies is the TATA sequence (so-called TATA box) centered around position −25. It appears in many genes to be required for the precise positioning of the transcription start, in a way like the −10 Pribnow box in prokaryotes.

When yeast cells are starved for amino acids, the synthesis of a protein called GCN4 is induced; it is a leucine zipper protein which binds upstream many amino acid biosynthetic genes and activates their transcription. GCN4 interacts directly in vitro with RNA polymerase, which strongly suggests that this interaction might be important in the formation of the transcriptional complex.

Transcription Termination and PolyA Tails

At the 3′-extremity of most eukaryotic messengers, one finds long sequences of adenylic acid residues. These polyA sequences and their nuclear precursors are not coded by DNA, but are added by an enzyme called polyA polymerase, after the gene transcription to pre mRNA by polII.

PolyA polymerase catalyzes the reaction

$$\mathrm{RNA} + \mathrm{n\ ATP} \rightarrow \mathrm{RNA}(\mathrm{A})\mathrm{n} + \mathrm{n\ PPi}$$

Ten to thirty nucleotides upstream the resulting polyA sequence, in a section of RNA coded by DNA, a sequence AAUAAA (or a similar one) is always found. However, this sequence could not by itself be the signal for polyadenylation, since it is also found in other parts of mRNA or pre-mRNA not followed by "polyA tails". In fact, polII continues to elongate the RNA chain further from the polyadenylation site, sometimes by several thousand nucleotides. The nascent pre-mRNA is then cleaved by an endonuclease which creates a correct 3′-terminus which is then polyadenylated; an endonuclease degrades rapidly the downstream sequences. The upstream AAUAAA hexanucleotide is part of the cleavage signal. There is evidence that complexes between proteins and small nuclear RNAs are also involved in termination.

Substantial amounts of polyA sequences have been found in prokaryotic mRNAs from *Bacillus subtilis, B. brevis* and *E. coli*. Two specific mRNAs in *E. coli*, those of the outer membrane lipoprotein and tryptophan synthetase α subunit were found to be extensively polyadenylated. The gene coding for polyA polymerase in *E. coli* has been identified (*pcnB*) and overexpressed after insertion into an expression vector, opening the way for the study of biosynthesis and function of bacterial polyadenylated mRNA.

Selected References

RNA Polymerases and Sigma Factors

Buc H, Strick T (eds) (2009) RNA polymerases as molecular motors. The Royal Society of Chemistry, Cambridge, 331pp

Friedman LJ, Mumm JP, Gelles J (2013) Proc Natl Acad Sci USA 110:9740–9745

Murakami KS (2013) J Biol Chem 288:9126–9134

RNA Polymerase

Chamberlin M (1982) In Boyer P (ed) The enzymes, vol 15, part B. Academic Press, New York, p 61
Gross CA, Chan CL, Lonetto MA (1996) Philos Trans R Soc Lond B Biol Sci 351:475–482
Sarkar P, Sardesai AA, Murakami KS, Chatterji D (2013) J Biol Chem 288:25076–25087

Trans-acting Transcription Factors

Roeder RG (1991) Trends Biochem Sci 16:402–408

Eukaryotic RNA Polymerases

Cramer P, Bushnell DA, Fu J, Gnatt AL, Maier-Davis B, Thompson NE, Burgess RR, Edwards AM, David PR, Kornberg RD (2000) Science 288:640–649
Moqtaderi Z, Yale JD, Struhland K, Buratowski S (1996a) Proc Natl Acad Sci USA 93:14654–14658
Moqtaderi Z, Bai Y, Poon D, Weil PA, Struhl K (1996b) Nature 383:188–191
Sentenac A (1985) CRC Curr Rev Biochem 18:31–90

Termination of Transcription

Adhya S, Gottesman M (1978) Annu Rev Biochem 47:967–996
Friedman DI (1988) In The bacteriophage, vol. II. Plenum Publishing Corporation, New York
Leela JK, Syeda AH, Anupama K, Gowrishankar J (2013) Proc Natl Acad Sci USA 110:258–263
Platt T (1994) Mol Microbiol 11:983–990
Roberts JW (1968) Nature (London) 224:1168–1174

Chapter 16
Negative Regulation

Each of the hypotheses implicitly or explicitly put forward in Chap. 13 has been borne out by relevant experiments: enzyme induction is correlated with the synthesis of a specific messenger RNA; the cytoplasmic repressor coded by the *i* gene has been isolated and characterized; its affinities for the inducer and for the operator have been determined; the repressor extracted from an i^s strain has lost its affinity for IPTG; the operator is a DNA segment which specifically binds the repressor; finally the inducer displaces the repressor from its complex with the operator. We shall now describe briefly these experiments.

Induction Is Correlated with the Synthesis of a Specific Messenger

The study of the early induction kinetics and the effects of very short inducer pulses have led to the conclusion that an unstable polynucleotide precursor is synthesized before the appearance of the enzyme. In addition, when the inducer is removed, the half-life of the residual β-galactosidase synthesis is of the same order of magnitude as that inferred for this unstable RNA.

However, the hypothesis that induction (and repression) act at the genetic level can only be demonstrated by the identification of a messenger RNA species specific for a given inducible system and the establishment of a correlation between the rate of its synthesis and that of the corresponding induced enzymes.

The detection and the estimation of a specific mRNA corresponding to a gene or to a group of genes are based on the property of mRNAs to form specific molecular hybrids with the homologous single stranded DNA when a heated mixture of the two polymers is cooled slowly.

In order to apply this technique to a single or to a few mRNA species, it was necessary to bring the sensitivity of hybrid detection to 0.01–0.02 % of the *E. coli* genome, and to considerably reduce the noise due to the hybrids not pertaining to

G.N. Cohen, *Microbial Biochemistry*,
DOI 10.1007/978-94-017-8908-0_16,

the specific genes studied. This can be achieved by obtaining an mRNA of high specific radioactivity and by increasing the relative concentration of the DNA segment carrying the genes under study. Such an increase in concentration can be obtained by using as DNA source the DNA from the defective phage λdg or from the F-lac episome in which specific bacterial genes responsible for galactose or lactose have been specifically inserted by recombination. The non specific mRNAs can be eliminated by hybridization with the appropriate DNAs (that of phage λ or of *E. coli* F^+) devoid of these specific genes. We shall give here as an example the case of the galactose genes.

Four of the structural genes governing galactose degradation are closely linked on a small segment of the *E. coli* chromosome (the *gal* region), situated near the attachment site of prophage λ. Their coordinate expression is under the control of the same operator. The *gal* segment can be incorporated in the DNA of the transducing phage λdg. This DNA can then be used as acceptor in the hybridization experiments. As the DNA content of λdg is of the order of 1 % of that of *E. coli*, and since the two DNAs contain the same amount of *gal* genes, the latter are 100 times more concentrated in the λdg DNA. In addition, the noise is greatly reduced by the replacement of *E. coli* DNA by the heterologous λDNA.

The mRNA labeled with ^{32}P is purified from a non-lysogenic strain of *E. coli* whose galactose degrading enzymes are induced by fucose, a gratuitous inducer. The bacteria are grown for several generations in presence or absence of inducer and exposed to ^{32}P for a period not exceeding 0.7 % of their doubling time. Under these conditions, ribosomal RNAs are essentially unlabeled and most of the radioactivity sediments between 8 and 10S. The radioactive material is mixed with DNA (from λ or λdg) previously denatured by heating. The mixture is incubated for one hour at 50 °C, for 14 hours at 40 °C; and then slowly cooled. The mRNA-DNA hybrids are separated from unhybridized mRNA by centrifugation in a cesium chloride gradient and treated by ribonuclease; the amount of radioactivity associated to DNA in the ribonuclease-resistant species is then determined.

The radioactivity associated to DNA (measured by optical density) is not significantly different in the λdg-mRNA hybrids obtained from non-induced cells and in λ–mRNA hybrids obtained from induced cells. This shows that the noise is essentially due to interactions between *E. coli* mRNA and λ DNA. On the other hand, the amount of radioactivity associated to λdg is much higher when the mRNA has been obtained from induced than from non-induced bacteria. After subtraction of "non-specific" radioactivity, seven times more radioactivity is found associated to λdg DNA when cells are induced by fucose than when the cells were grown on glycerol alone. Note that the mRNA from glucose-grown cells is unlabeled (Table 16.1).

There is a good correlation between the relative amounts of mRNA specifically hybridizable with the genes of the galactose operon and the rate of synthesis of the corresponding enzymes. A constitutive dominant strain (O^c) synthesizes more specific mRNA (Table 16.2).

Table 16.1 Specific complexes between λdg or λ DNA and the ^{32}P labeled mRNAs originating from either non induced or fully induced bacteria

	O.D. (260 mm)	cpm	pm/O.D.	pm/O.D. corrected for λ DNA hybridization
λdg DNA + mRNA from induced cells (glycerol + fucose)	2.02	1222	602	353
λdg DNA + mRNA from non-induced cells (glycerol)	2.07	627	303	54
λDNA + mRNA from induced cells (glycerol + fucose)	1.44	359	249	–
λdg DNA + mRNA from non-induced cells (glucose)	2.16	490	227	0

Table 16.2 Synthesis of the enzymes of the galactose operon

	Galactokinase	G-1-P-uridyltransferase
Growth conditions	mmoles/h/mg bacteria	
Non induced wild type (glycerol)	0.87	0.83
Induced wild type (glycerol + fucose)	13.6	12.3
O^c strain (glycerol)	24.5	23

Isolation of the Lac Repressor

A central question of molecular biology concerns the nature of the regulatory substances. The model of negative regulation proposed by Jacob and Monod implies that certain genes, called regulatory genes, responsible for diffusible repressors which prevent the expression of other genes, called structural genes. The latter are organized as operons comprising *cis*-dominant operators, which are the targets of the repressors. Small appropriate molecules act either as inducers, by preventing repression (or as co-repressors leading to active repressors, as we shall see later). The simplest hypothesis, to explain the behavior of inducible systems, is that the repressor is the direct product of the regulatory gene which binds to the operator site of a DNA molecule in such a way as to prevent the transcription of the operon by RNA polymerase. This model implies that the inducer combines with the repressor to form a complex which cannot bind any more to the operator, thus allowing the free transcription of the messenger and the synthesis of the enzymes specific to the operon. However, other models could also account for the experimental results. The isolation of several repressors has led to the understanding of the regulation mechanisms at the molecular level. We describe here the isolation and some of the properties of the Lac repressor.

The detection method used involves the affinity of the hypothetical repressor with a radioactive inducer, isopropylthiogalactoside (IPTG) measured by equilibrium dialysis. In order to detect the binding, it is necessary to obtain a concentration of the receptor of the same order of magnitude as that of the dissociation constant of the repressor-inducer complex. In a *y*-strain, semi-maximal induction

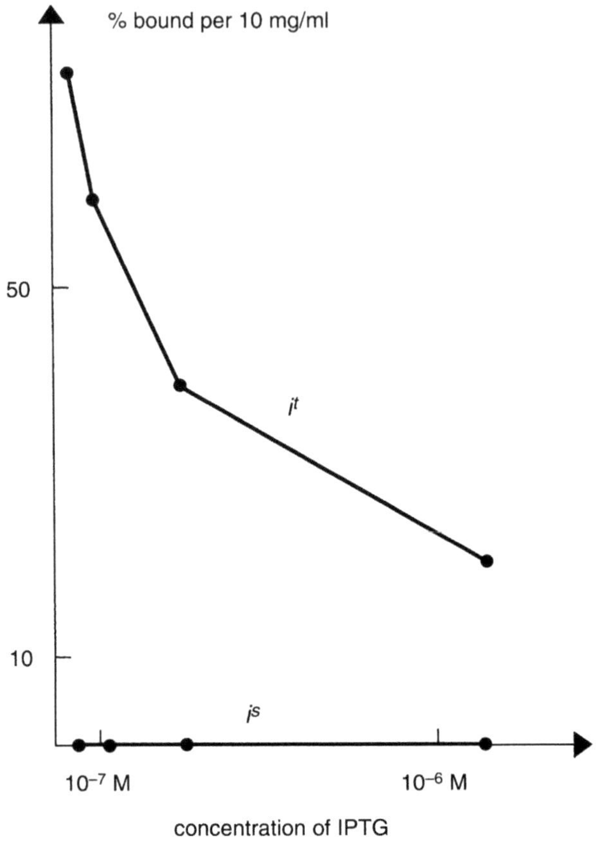

Fig. 16.1 The repressor fraction is isolated in parallel from 50 g batches of otherwise isogenic F′ i^s/i^s and F′ i^t/i^t strains. Aliquots of each extract are dialyzed against various concentrations of ^{14}C-IPTG during 8 h in an appropriate buffer. The excess radioactivity inside each dialysis bag is normalized for a protein concentration of 10 mg/ml (From W. Gilbert and B. Müller-Hill, with permission from Dr. W. Gilbert)

requires a concentration of IPTG of 2×10^{-4} M; however, this value does not lead directly to an estimation of the equilibrium constant: since the enzyme level increases by a factor 1,000 during induction and since the rate of enzyme synthesis is inversely proportional to the repressor concentration, it follows that the concentration of free repressor must have fallen of a factor 1,000 during induction. From the shape of the induction curve, one can deduce that the repressor extracted from the wild type must have an affinity of about 6×10^{-6} M for IPTG. If the binding is to be detected, one needs a repressor concentration of the same order. Since one expects very few repressor molecules per operator site, it appears probable that the repressor should be concentrated at least 100-fold before even a weak binding is detectable. Fortunately, a superinducible mutant of the *i* gene, i^t, has been isolated which synthesizes β-galactosidase with much lower concentrations of IPTG. With a diploid strain carrying two copies of i^t, the detection is still marginal (1,000 cpm/0.1 ml outside the dialysis bag, 1040 inside the bag!). Further purification by classical techniques leads rapidly to satisfactory results (50–100 % excess radioactivity inside the bag) for a protein concentration of 10 mg/ml.

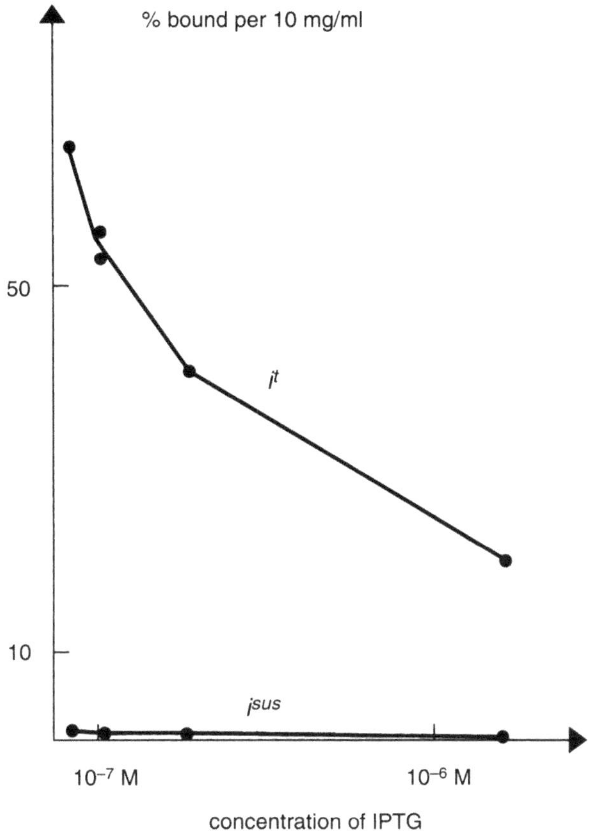

Fig. 16.2 A suppressible i- strain does not bind IPTG. The repressor fraction is isolated from two otherwise isogenic strains F' i^{sus}/i^{sus} and F' i^t/i^t; the fractions are tested as mentioned in the legend of Fig. 16.1. Same source as Fig. 16.1

Radioactive IPTG is reversibly bound and can dialyze in and out of the bag. The material which binds IPTG is indeed the product of the *i* gene: an i^s strain does not yield a material binding reversibly IPTG (Fig. 16.1); the same is true for a suppressible *i*- amber strain, which synthesizes only an inactive fragment of the *i* gene product (Fig. 16.2) or for a strain carrying a deletion of the *i* gene. All these negative controls are significant, since extracts from these strains do not inhibit the binding from the i^t extracts.

If one plots the amount of bound IPTG against the ratio free/bound IPTG, one obtains a straight line, indicating a single type of binding site. The intercept on the ordinate gives the molarity of the binding sites and the slope is the inverse of the association constant. Figure 16.3 gives such straight lines for the i^t strain and for the wild type. One notices that, as expected, the repressor from the i^t strain has a higher affinity.

The equilibrium dialysis method allows an estimation of the affinities of IPTG competitors for the repressor binding site for the repressor: TMG is a less good inducer than IPTG. Galactose is a weak inducer of the i^t strain *in vivo*. Glucose has practically no affinity for the site. The values of the association constants are consistent with the identification of the purified material with the repressor (Table 16.3).

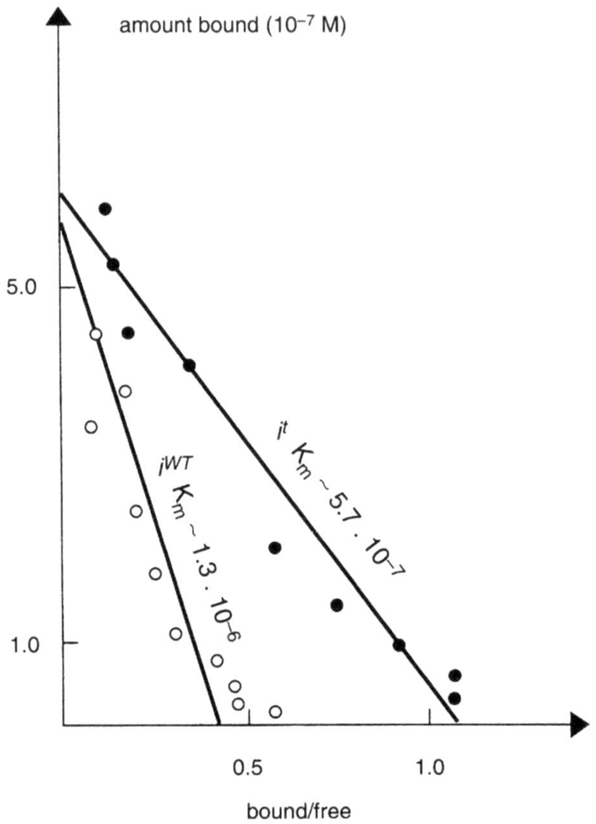

Fig. 16.3 The association constants of the repressors isolated from the wild type and from the it mutant strain. The repressor fractions from otherwise isogenic wild type or i^t strains are dialyzed against different concentrations of ^{14}C-IPTG, as in Fig. 16.1. Ordinates: amount of IPTG bound for each IPTG external concentration, normalized for a protein concentration of 10 mg/ml. Abscissas: Ratio bound IPTG/free IPTG. If there is only one type of binding site, Km is deduced from the relation $\text{IPTG bound} = \text{Number of sites} - \text{Km}\dfrac{\text{bound}}{\text{free}}$. Same source as Fig. 16.1.

Table 16.3 Dissociation constants of potential ligands of the Lac repressor

Competitors	K_i for the i^t gene product
Non radioactive IPTG	5×10^{-7} M
Thiomethylgalactoside	2×10^{-6} M
o-nitrophenylfucoside (ONPF)	2×10^{-5} M
Galactose	5×10^{-4} M
Glucose	3×10^{-2} M

The capacity to bind IPTG is insensitive to DNAse and to RNAse, but is destroyed by heating at 50 °C and by pronase. It sediments at 7.8 S on a glycerol gradient. It is a non inducible protein and calculations show that there are only 10–20 molecules of it per gene in the wild type, which amounts to about one ten thousandth of the total weight of the *E. coli* proteins.

The *lac* Operator Is a DNA Sequence

Radioactive repressor can be obtained by blindly following the techniques used for the purification of non radioactive repressor. It is not necessary that the repressor be pure, but that it represent a reasonable proportion of the labeled material and that other proteins bound to DNA be eliminated, so that the observed results are really specific. One starts from a strain triploid for the *lac* segment (on the chromosome, on a defective ϕ 80*dlac* and on an F*lac* episome), the proteins of which have been labeled by growth on ^{35}S. The proteins are fractionated by ammonium sulfate and on a DEAE-Sephadex column. After concentration of the material, it is deposited onto a glycerol gradient and the fraction sedimenting at 7.6 S is retained. This material is mixed with phage DNA containing the *lac* region, incubated and sedimented on a glycerol gradient: a small radioactive peak is displaced from the 8S region and migrates with DNA, at 35–40S. Figure 16.4a shows a centrifugation in the presence of 10^{-2} M Mg^{++}: a radioactive peak migrates with DNA, representing only 1 % of the total radioactivity present in the mixture, although DNA is in great excess. Figure 16.4b proves that the material bound to DNA is actually the Lac repressor: no radioactivity migrates with DNA if 10^{-4} M IPTG is present. This effect is specific: ONPF, which binds to the repressor, but does not induce the synthesis of β-galactosidase, is without effect on the binding (Fig. 16.4c).

Ribonuclease does not affect the association. The purified non radioactive Lac repressor competes with the binding. There is no or insignificant binding of the repressor with phage DNA not carrying the *lac* genes (Fig. 16.5).

A more significant test is provided by the DNA from phages carrying a mutated O^c operator gene in the *lac* segment: Fig. 16.6a–c show the compared binding on the ADN bearing the O^+ allele and the ADN from two different O^c alleles; one being a low-level constitutive (1 % of the maximum specific activity after induction), the other possessing a higher degree (20 %) of constitutivity. Whereas the binding of the repressor to the operator is still detectable in the first case, in the second, the radioactive characteristic peak is absent.

These experiments demonstrate that the repressor binds to a well-determined DNA sequence, the operator region. It binds only to double stranded DNA.

Specific protein-DNA interaction is essential for gene control in both prokaryotes and eukaryotes. Extensive biochemical and genetic studies with the Lac repressor indicate that the protein is a tetramer of identical subunits. Each subunit, 360 amino acid residues long (Fig. 16.7), is composed of three major domains: (1) a small N-terminal domain, the "headpiece" (59 amino acid residues) recognizes the operator; mutations in the corresponding part of the *i* gene diminish or abolish operator

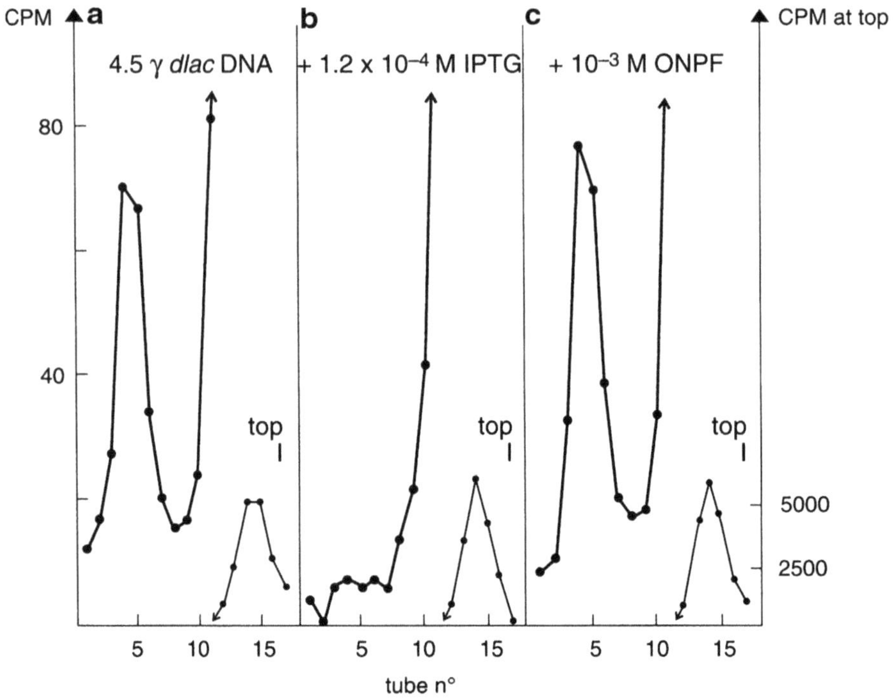

Fig. 16.4 The binding of Lac repressor to the DNA of phage dlac and its displacement by the inducer. Three identical mixtures of 4.5 μg of *dlac* DNA and of radioactive protein (7.8S) are centrifuged on a glycerol gradient, during 2 h at 35,000 rpm. DNA alone would have generated an optical density peak in tubes 4 and 5. (*a*) a distinct radioactivity peak sediments with DNA. The latter is in a tenfold excess and sediments with the maximum radioactivity it can bind. (*b*) the gradient solution contains IPTG. The binding of the repressor to DNA is abolished. (*c*) ONPF eight times more concentrated than IPTG has no effect on the binding (From W. Gilbert and B. Müller-Hill, with permission of Dr Walter Gilbert)

binding without affecting inducer binding. If the repressor is treated by trypsin or chymotrypsin, it loses residues 1–79 as well as about 20 amino acids at the C-terminal end (2) the core-protein, resistant to trypsin hydrolysis, (about 270 amino acids) binds the inducer and leads to the aggregation of dimers, and (3) the C-terminal 30 amino acids, a putative leucine mini zipper, mediate tetramer assembly.

How does Lac repressor, a tetrameric protein, recognize its target? It is necessary to know the operator structure to try to answer this question.

The *lac* operator has been isolated as a double-stranded DNA segment protected by the Lac repressor against pancreatic DNAase digestion. The starting material are bacteria lysogenic for an inducible defective and thermoinducible prophage carrying part of the *lac* genome including the operator region; the bacteria are induced at 42 °C and their DNA is fractionated by sonic disruption into fragments having an average length of 1000 nucleotides. The Lac repressor is added to these fragments under conditions optimal for its binding, and the mixture is digested by pancreatic DNAase, which hydrolyzes double stranded DNA, but is without action on the protected segment. The mixture is filtered on a cellulose nitrate membrane which

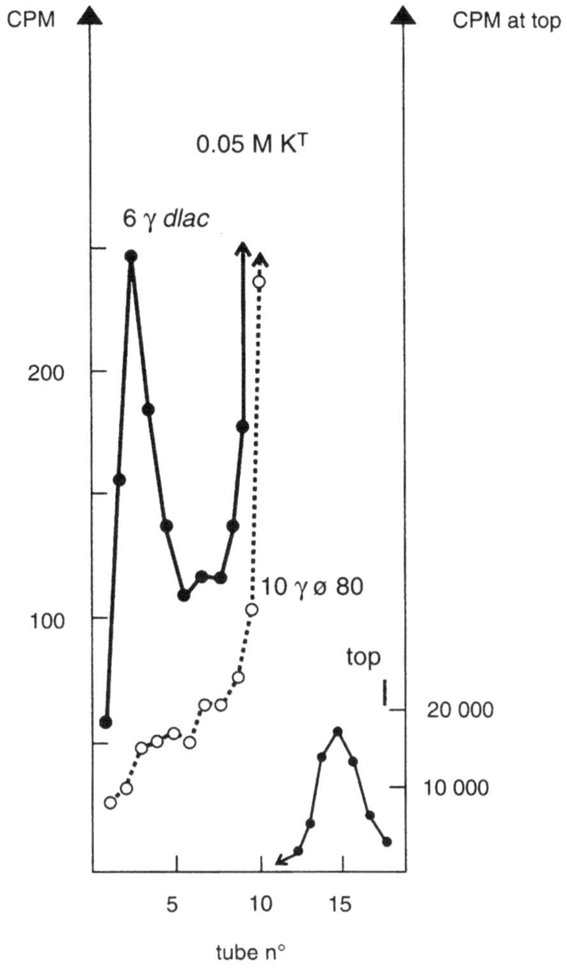

Fig. 16.5 Binding specificity. The same repressor preparation is centrifuged on two parallel gradients with two different DNAs. A mixture contains 6 μg of purified DNA from *dlac* phage: the other contains 10 μg of DNA of the parental phage (which does not carry the *lac* region). Same source as Fig. 16.4

retains only the repressor-operator complex. The operator is eluted from the complex by isopropylthiogalactoside. The protected fragment is able to combine again with repressor and is a 24 bp long double stranded segment. The DNA is denatured and each strand is used as a template for transcription by DNA-dependent RNA polymerase. The RNA copies of DNA are then sequenced (at the time the experiment was done, no DNA sequencing method was available). If they are transcribed on paper as DNA sequences, one obtains:

TGG AATTGT G A G C G G A T AAGAATT
ACC TTAACA C T C G C C T A TTGTTAA

This sequence show obvious symmetries. Sixteen bases are arranged on both sides of a binary symmetry axis which is not at the center of the total fragment: on the left, there are six bp symmetrically oriented to six other bp situated at the extreme right of the fragment; there are other base pairs symmetrically arranged along nine bases situated between the two principal regions.

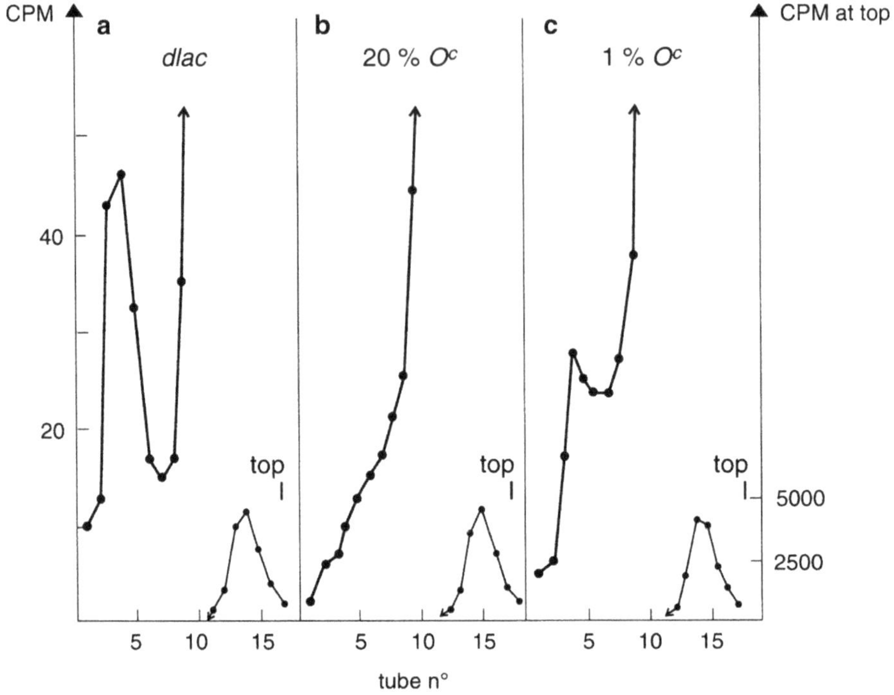

Fig. 16.6 The Lac repressor binds specifically the *lac* operator. The same preparation of Lac repressor is centrifuged in three parallel gradients with three purified DNAs originating from three different phages. (a) 4.5 μg of DNA from a phage *dlacO*⁺ (b) same amount of DNA carrying an O^c mutation leading to a 20 % constitutivity: the repressor is not bound (c) same amount of DNA carrying an O^c mutation leading only to a 1 % constitutivity (10 times the basal level of an uninduced wild type strain): one can still detect a weak binding. Same source as Fig. 16.4

This symmetry could allow the Lac repressor to interact with DNA along its binary symmetry axis. Note that the two longer sequences are separated by one turn of the double helix (B form).

The operator sequence does not immediately lead to an understanding of the recognition between the two molecules. Although crystals of the repressor have been obtained, they are too small to lend themselves to an analysis by X-ray diffraction. Nuclear magnetic resonance studies of the Lac repressor operator complex have been carried out and it was found that the sequence from amino acid 17 to 25 is a DNA-recognizing α-helix Genetic analyses of amino acid exchanges in this recognition helix and symmetric base-pair substitutions of the ideal *lac* operator revealed specific contacts between amino acids and base pairs and determined the orientation of the recognition helix within the major groove of the DNA which is opposite to that of λ and *cro* repressors (see Chap. 36 on protein-DNA interactions).

Recently, the core domain of Lac repressor (residues 63–323) has been aligned to other sugar-binding proteins of known structure, using primary sequence similarities and a structural model concordant with genetic and chemical data has been

Fig. 16.7 The sequence of the *E. coli* Lac repressor

Met Lys Pro Val Thr Leu Tyr Asp Val Ala Glu Tyr Ala Gly Val

Ser Tyr Gln Thr Val Ser Arg Val Val Asn Gln Ala Ser His Val
30
Ser Ala Lys Thr Arg Glu Lys Val Glu Ala Ala Met Ala Glu Leu

Asn Tyr Ile Pro Asn Arg Val Ala Gln Gln leu Ala Gly Lys Gln
60
Ser Leu Leu Ile Gly Val Ala Thr Ser Ser Leu Ala Leu His Ala

Pro Ser Gln Ile Val Ala Ala Ile Lys Ser Arg Ala Asp Gln Leu
90
Gly Ala Ser Val Val Val Ser Met Val Glu Arg Ser Gly Val Glu

Ala Cys Lys Ala Ala Val His Asn Leu Leu Ala Gln Arg Val Ser
120
Gly Leu Ile Ile Asn Tyr Pro Leu Asp Asp Gln Asp Ala Ile Ala

Val Glu Ala Ala Cys Thr Asn Val Pro Ala Leu Phe Leu Asp Val
150
Ser Asp Gln Thr Pro Ile Asn Ser Ile Ile Phe Ser His Glu Asp

Gly Thr Arg Leu Gly Val Glu His Leu Val Ala Leu Gly His Gln
180
Gln Ile Ala Leu Leu Ala Gly Pro Leu Ser Ser Val Ser Ala Arg

Leu Arg Leu Ala Gly Trp His Lys Tyr Leu Thr Arg Asn Gln Ile
210
Gln Pro Ile Ala Glu Arg Glu Gly Asp Trp Ser Ala Met Ser Gly

Phe Gln Gln Thr Met Gln Met Leu Asn Glu Gly Ile Val Pro Thr
240
Ala Met Leu Val Ala Asp Asn Gln Met Ala Leu Gly Ala Met Arg

Ala Ile Thr Glu Ser Gly Leu Arg Val Gly Ala Asp Ile Ser Val
270
Val Gly Tyr Asp Asp Thr Glu Asp Ser Ser Cys Tyr Ile Pro Pro

Leu Thr Thrl le Lys Gln Asp Phe Arg Leu Leu Gly Gln Thr Ser
300
Val Asp Arg Leu Leu Gln Leu Ser Gln Gly Gln Ala Val Lys Gly

Asn Gln Leu Leu Pro Val Ser Leu Val Lys Arg Lys Thr Thr Leu
330
Ala Pro Asn Thr Gln Thr Ala Ser Pro Arg Ala Leu Ala Asp Ser

Leu Met Gln Leu Ala Arg Gln Val Ser Arg Leu Glu Ser Gly Gln
360

proposed. Bound Lac repressor inhibits the unwinding reaction catalyzed by some *E. coli* DNA helicases. This inhibition is substantially weaker in the presence of IPTG. It is likely that replication forks encounter transcription complexes. The replication fork must also replicate regions of the chromosome that are complexed with other site-specific DNA-binding proteins like the various repressors of transcription. Such potential barriers to replication must be removed to allow DNA replication to proceed at a normal rate. Further experiments are needed to clarify this interesting problem.

Additional Lac repressor binding sites, the so-called pseudo-operators 02 and 03 have been discovered. 02 lies 92 bp from the main operator 01, wheras 03 is located within the coding sequence 401 bp downstream from 01. Contrary to 01, their affinities for Lac repressor are relatively low. Absence of either one of the pseudo-operators decreases repression by wild-type Lac repressor only 2–3 fold; however, absence of both decreases repression more than 50-fold: 01 represses only 20-fold instead of 1,000-fold. *In vitro* experiments indeed showed that tetrameric Lac repressor molecules bind simultaneously two lac operator sequences located on the same DNA fragment, thereby forcing the intervening DNA into a loop structure.

Direct Observation of Transcription Factor Dynamics in a Living Cell

Using fluorescence imaging, the kinetics of the binding of the lac repressor binding to its operator has been studied. Taking advantage of the fact that DNA bound proteins glow like bright dots, while the fluorescence of proteins diffusing in cytoplasm gets lost in the background, it was found that the repressor spends a few milliseconds weakly and nonspecifically bound to DNA, diffusing along the chromosome; then dissociates for a fraction of a millisecond. This cycle of unbinding and rebinding various DNA segments repeats for a few minutes until the repressor meets its specific target.

The lac operon induction depends on whether the repressor dissociates partially or completely from its operator, a single-molecule stochastic event.

Selected References

Messenger RNA

Attardi G, Naono S, Rouvière J, Jacob F, Gros F (1963) Cold Spring Harb Symp Quant Biol 28:363–372

Isolation of Lac Repressor and Lac Operator

Gilbert W, Müller-Hill B (1966) Proc Natl Acad Sci USA 56:1891–1898
Gilbert W, Müller-Hill B (1967) Proc Natl Acad Sci USA 58:2415–2421
Yancey-Wrona JE, Matson SW (1992) Nucleic Acids Res 20:6713–6721

Operators 02 and 03

Oehler S, Eismann ER, Krämer H, Müller-Hill B (1990) EMBO J 9:973–979

Formation of Loop Structures in DNA

Révet B, von Wilcken-Bergmann B, Bessert H, Barker A, Müller-Hill B (1999) Curr Biol 9:151–154

Quantifying Lac Repressor Kinetics

Choi PJ, Cai L, Frieda K, Xie XS (2008) Science 322:442–446

Elf J, Li GW, Xie XS (2007) Science 316:1191–1194

Chapter 17
Enzyme Repression in Anabolic Pathways

Description of the Phenomenon

Fifty years ago, the following observation was made at the Pasteur Institute: if the wild-type of *E. coli* was grown in the presence of an exogenous amino acid, the content of an enzyme involved in the biosynthesis of this amino acid was significantly lower in the cell suspensions or the extracts from these bacteria. Table 17.1 illustrates this observation in the case of methionine biosynthesis. The experiment has been carried out with a strain constitutive for β-galactosidase, in order to ascertain the specificity of the effect observed. All the amino acids reduce somehow the galactosidase specific activity; there is no correlation between this reduction and the effects on the levels of methionine synthase. Whereas the addition of methionine does not affect β-galactosidase, it has a spectacular effect on methionine synthase levels. This effect can be obtained with rather low methionine concentrations (5 mM of the L-isomer) (Table 17.1).

Since methionine synthase activity was in 1953 only measurable in intact cells, there was no possibility to ascertain that the observed inhibition was due to the absence of synthesis of a protein component of the system (since then, it has been found that this indeed the case). The following control was made however: cell suspensions do not lose their enzyme activity in presence of methionine, the effect being observed only after growth on this amino acid. In the case of tryptophan synthase (Table 17.2), the activity of which could be measured in cell extracts, there is no doubt possible: only growth in presence of tryptophan leads to a significant decrease in the level of tryptophan synthase. Mixtures of cell extracts from cultures grown in presence and in absence of tryptophan yield purely additive activities, eliminating the hypothesis that the observed effect could be due to an inhibitor. There is no indication of enzyme inactivation or of a direct inhibition: non-proliferating cell suspensions shaken in the presence of tryptophan retain a normal tryptophan synthase activity. The only alternative is that the presence of tryptophan inhibits the synthesis of the enzyme.

G.N. Cohen, *Microbial Biochemistry*,
DOI 10.1007/978-94-017-8908-0_17, © Springer Science+Business Media B.V. 2014

Table 17.1 Methionine synthase and β-galactosidase activities of cell suspensions of *E. coli* obtained after an overnight growth in presence of various amino acids

Amino acid during growth (6 mM)	% effect on β-galactosidase	% effect on methionine synthase
0	0	0
DL-isoleucine	−7	−36
DL-valine	−24	−18
DL-phenylalanine	−15	−27
DL-threonine	−6	−18
L-histidine	−29	+23
DL-serine	−30	+36
L-arginine	−7	+36
L-proline	−26	+36
L-tryptophan	−10	+41
DL-methionine (500 μM)	−4	−99
DL-methionine (5 mM)	0	−99

From Cohn et al. (1953)

Table 17.2 Tryptophan synthase activity of *Aerobacter aerogenes* having grown with various supplements

Supplement	Specific activity*	% effect on tryptophan synthase
None	1,300	0
L-histidine 2.5 mM	1,500	+15
L-arginine 5 mM	1,600	+20
DL-phenylalanine 5 mM	2,100	+60
DL-methionine 2 mM	2,300	+75
L-glutamate 2 mM	1,200	−8
L-proline 2 mM	1,100	−15
DL-serine 4 mM	1,500	+16
L-tryptophan 2 mM	360	−72

The reaction followed is indole + serine → tryptophan and is performed in presence of pyridoxal phosphate

*Activity expressed as nmoles of indole consumed/h/mg bacterial N (From Monod and Cohen-Bazire (1953)

The observations made at the Pasteur Institute were extended to acetylornithinase, whose synthesis is inhibited by arginine, to the pyrimidine, purine, histidine, valine, isoleucine, biosynthetic enzymes, etc… So the phenomenon of inhibition of enzyme synthesis is largely observed in biosynthetic systems. The word *repression* has been coined by Vogel to describe it. Repression is not limited to microorganisms.

A remarkable advance in the study of repression is due to an elegant experiment of Gorini and Maas on the regulation of the synthesis of ornithine transcarbamylase (OTCase), one of the enzymes involved in arginine biosynthesis. When arginine is added to an exponentially growing culture of *E. coli*, the synthesis of ornithine transcarbamylase stops immediately and the enzyme present in the cells is exponentially diluted during further growth. Thus, after about ten divisions, the intracellular concentration of the enzyme is about 100th that found in a control culture grown without added arginine. If the repressed bacteria are washed and resuspended in a medium without arginine, synthesis of OTCase resumes. This system provides an experimental study of the kinetics of enzyme formation, analogous to that consisting to add an inducer to non-induced cells in order to study the kinetics of synthesis in an inducible system.

In the case of β-galactosidase, the addition of inducer causes the quasi-immediate synthesis of the enzyme at a constant rate. The increase of enzyme is described by

$$\%E_{max} = 100\left(1 - e^{-\mu t}\right)$$

where Emax is the final enzyme concentration per cell and μ the rate constant of the equation for exponential growth, dN/dt = μN (N = number of bacteria per unit volume). This means that the maximal enzyme concentration is reached only after several generations. However, the kinetics of OTCase synthesis is paradoxical: the level of enzyme reaches a maximum very rapidly (after about half a division) and then decreases during the rest of the exponential phase. Gorini and Maas, in order to explain this unexpected result, made the hypothesis that the synthesis of the enzyme is repressed not only by exogenous arginine, but also by the arginine endogenouly synthesized. In the first part of the experiment, after growth on arginine and the washing, the cells contain very little enzyme and very little arginine; the enzyme is therefore synthesized very rapidly and arginine accumulates, repressing further enzyme synthesis (Fig. 17.1).

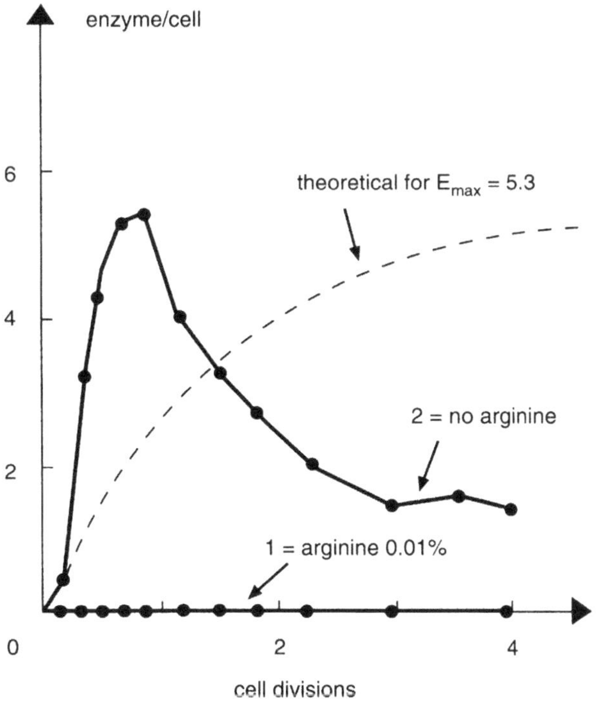

Fig. 17.1 Synthesis of ornithine transcarbamylase in discontinuous culture. A preculture is made in the presence of excess arginine, and then transferred in the same medium (*curve 1*) or in a medium devoid of arginine. The enzyme activity is measured (arbitrary units) at different times during growth (From L. Gorini and W.K. Maas, with permission of The Johns Hopkins Press)

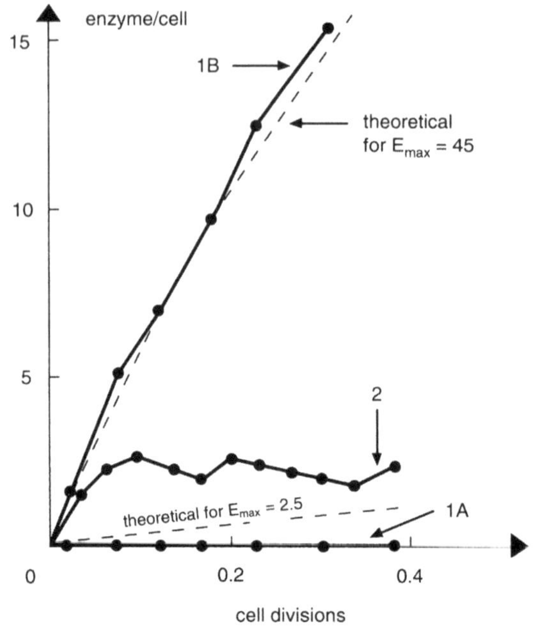

Fig. 17.2 Synthesis of ornithine transcarbamylase in continuous culture (chemostat), using an his-arg-double auxotroph. *Curve A*: histidine is the limiting factor. *Curve B*: Arginine is the limiting factor (From L. Gorini and W.K. Maas, with permission of The Johns Hopkins Press)

In order to verify this hypothesis, it was necessary to find conditions where a level of intracellular arginine lower than that of the wild type growing on a minimal medium could be achieved. This was done by using an auxotrophic mutant which required arginine and was blocked before OTCase

$$\text{N-acetylornithine} \rightarrow \text{Ornithine} \xrightarrow{\text{OTCase}} \text{Citrulline} \rightarrow \text{Arginine}$$

The growth of this mutant can be limited in a chemostat (Chap. 1) by the rate V_1 at which arginine is delivered. If the pool size of the intracellular arginine is due to the difference between this rate V_1 and the rate V_2 of incorporation of arginine into proteins ($V_1 > V_2$), conditions are realized in the chemostat where $V_1 = V_2$ and the pool of free arginine is equal to zero. The mutant utilized was also auxotrophic for histidine, which allowed to study the specificity of the observed effects. Figure 17.2 describes the results of the experiment. After a preculture with arginine and washing, the bacteria are transferred into the chemostat. When arginine limits growth, enzyme synthesis starts as the same rate as in the previous experiment, but the synthesis continues at the same rate since the intracellular conentration of arginine remains at best at a negligible level. If histidine limits growth and if arginine is delivered at a concentration exceeding by 1 μg/ml that necessary for keeping the growth rate at the imposed rate, the synthesis of OTCase is totally repressed.

One should note that, under these conditions of physiological derepression, the enzyme reaches an intracellular concentration 25 times higher than that obtained in the wild type growing in minimal medium. The important conclusion of this work is that the intracellular concentration of arginine (or of one of its derivatives) controls the rate of synthesis of ornithine transcarbamylase; this type of regulation had only been suggested by the experiments where it was shown that the final product of a biosynthetic pathway, provided exogenously, inhibited the synthesis of the enzymes involved in its own synthesis.

This inhibition of synthesis (repression) thus appears to be a general physiological mechanism of regulation of enzyme systems. One of the striking aspects of these studies is the high potential of the cells to synthesize a given enzyme, at a level by far higher of that required for growth. This high potential and repression allow the cell to adapt rapidly to a change in the environment, from a medium in which a given metabolite is scarce to one where it is in excess. The capacity of small molecular weight metabolites to control the cellular enzyme level, observed in the above experiments, is reminiscent of enzyme induction. We shall see that this analogy is more than formal.

Isolation of Derepressed (Constitutive) Mutants in Biosynthetic Pathways. The Use of Structural Analogues

Any attempt to formulate a unitary model to acount for the mechanisms of enzyme induction of catabolic enzymes and of enzyme repression of biosynthetic enzymes rests on the isolation in the latter systems of mutants of the type we have called *i*- and O^c.

In biosynthetic pathways, such mutants can be isolated with the aid of structural analogs.

An unexpected observation was made during the study of β–galactoside permease. One will recall that, in order to study its kinetics, chloramphenicol was added to stop the further synthesis of the permease during the experiment. One day, the chloramphenicol solution could not be readily found and *p*-fluorophenylalanine, which was reported to stop protein synthesis in yeast, was used instead. With a great surprise, it was found that the addition of the analogue to an exponential culture of *E. coli* did not stop the increase in mass, but rather turns the exponential growth into a linear one.

A detailed analysis of the phenomenon revealed that *p*-fluorophenylalanine is incorporated instead of phenylalanine in proteins synthesized after its addition (Fig. 17.3) and that the average degree of substitution in the overall proteins depends on the ratio endogenous phenylalanine / exogenous analogue. Similar results were obtained with norleucine, an analogue of methionine where the sulfur atom is replaced by an isosteric methylene group:

$$CH_3-S-(CH_2)_2-CH(NH_2)-COOH$$

Methionine

$$CH_3-CH_2-(CH_2)_2-CH(NH_2)-COOH$$

Norleucine

substitution of methionine by selenomethionine. It offers a general method (MAD) for introduction of anomalous scatters into cloned proteins. In 1957, Cohen and Cowie discovered that an *E. coli* strain auxotrophic for methionine, grows exponentially in a medium containin selenomethionine. The strain could grow, because selenomethionine becomes not only a component of proteins, but is also a substrate for S-adenosylmethionine synthetase, Se-adenosylmethionine being able to serve as a methyl donor and as a substrate of spermidine synthesis (see Chap. 25). W. Hendrickson and co-workers, 30 years later, showed that selenium is a useful anomalous scatterer. They solved the structure of selenobiotyl streptavidin and of selenomethionyl proteins using multiwavelength anomalous dispersion (MAD). Selenomethionine has since become a classic tool for protein crystallography.

Selected References

Repression of the Biosynthesis of Anabolic Enzymes

Cohn M, Cohen GN, Monod J (1953) Comptes Rendus 236:746–749
Gorini L, Maas WK (1957) Biochim Biophys Acta 25:208–209
Gorini L, Maas WK (1958). In: McElroy WD, Glass B (eds) The chemical basis of development. The Johns Hopkins Press, pp 469–478
Monod J, Cohen-Bazire G (1953) Comptes Rendus 236:530–532
Vogel HJ (1957). In: McElroy WD, Glass B (eds) The chemical basis of heredity. The Johns Hopkins Press, Baltimore, pp 267–289

Incorporation of Amino Acid Analogs into Proteins

Cohen GN, Munier R (1959) Biochim Biophys Acta 31:347–356
Cowie DB, Cohen GN (1957) Biochim Biophys Acta 26:252–261
Cowie DB, Cohen GN, Bolton ET, de Robichon-Szulmajster H (1959) Biochim Biophys Acta 34:39–46
Gilles AM, Marlière P, Rose T, Sarfati R, Longin R, Meier A, Fermandjian S, Monnot M, Cohen GN, Bârzu O (1988) J Biol Chem 263:8204–8209
Hendrickson WA, Pahler A, Smith JL, Satow Y, Merritt EA, Phizackerley RP (1989) Proc Natl Acad Sci USA 86:2190–2194
Hendrickson WA, Horton JR, LeMaster DM (1990) EMBO J 9:1665–1672
Hendrickson TL, de Crécy-Lagard V, Schimmel P (2004) Annu Rev Biochem 73:147–176
Munier R, Cohen GN (1959) Biochim Biophys Acta 31:378–391
Yang W, Hendrickson WA, Crouch RJ, Satow Y (1990) Science 272:1606–1614

Chapter 18
Positive Regulation

It has long been known that the presence of glucose (or of another easily degradable carbon source) inhibits the synthesis of certain bacterial enzymes such as β-galactosidase or tryptophanase from *E. coli*, or inositol dehydrogenase and histidase from *Bacillus subtilis*. This phenomenon was called glucose effect. In 1961, Magasanik gave it the more general name of catabolic repression. Many catabolic enzymes being inducible, the rate of their synthesis must then depend on the intracellular concentrations of the inducer and of the catabolite exerting its repressive effect.

The synthesis of catabolic enzymes is regulated in a manner such as their intracellular content be adequate for growth in a given environment. Inducers are signals indicating that the medium contains substances able to be degraded by the inducible enzymes to yield ATP, pyruvate, etc.... Catabolic repressors provide the information that these compounds exist in a sufficient amount. The synthesis of inducible enzymes will take rapidly place only when these repressors are at low level.

However, this is only a description of the physiological aspects of catabolic repression and says nothing on its genetic basis or on the molecular basis of this type of regulation. In addition to the nature and specificity of the catabolic repressors, it is important to define their targets and their relation with the specific repressors we have encountered in the preceding chapters and whose action is counteracted by specific inducers.

An observation made by Makman and Sutherland in 1965 helped to clarify the problem: the addition of glucose to a culture of *E. coli* growing on glycerol causes a rapid fall in the intracellular concentration of 3′-5′-cyclic adenosine monophosphate (cyclic AMP or cAMP).

It is now clear that the glucose effect is somehow linked to this depletion: Ullman and Monod, and Perlman and Pastan have independently observed in 1968 that the repressing effect of glucose on the synthesis of catabolic enzymes can be readily reverted by the addition of cyclic AMP to *E. coli* cultures (Fig. 18.1).

G.N. Cohen, *Microbial Biochemistry*,
DOI 10.1007/978-94-017-8908-0_18,

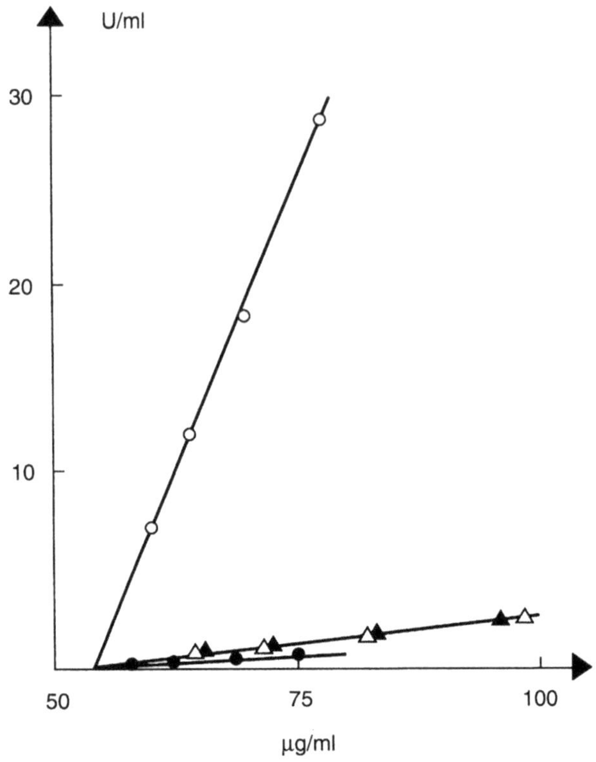

Fig. 18.1 Effects of glucose and cyclic AMP on the synthesis of β-galactosidase by a culture of ***E. coli***

How does cyclic AMP control enzyme synthesis?

Let us look again at the structure of the lactose operon, taking as a model the inhibition by glucose of β-galactosidase synthesis.

The Promoter Region

Constitutive strains respond normally to catabolic repression, whether they are recessive (i^-) or dominant (O^c). On the other hand, the initial constitutive synthesis of β–galactosidase by the zygotes from the cross Hfr (i^+O^+) and an F^- strain with a deletion encompassing the *i* gene is subjected also to catabolic repression. Since these zygotes do not contain any product of the *i* gene, a role of the Lac repressor in the mechanism of catabolic repression is excluded.

A study of mutants sensitive to repression by the product of *i*, but whose β-galactosidase synthesis is no more sensitive to the presence of glucose in the medium, has led to the localization of a new region, *lacP*, which has been called promoter region. The promoter is characterized as an essential site for the formation of the products of the structural genes *z*, *y* and *a*. Promoter deletions lead to a considerable reduction of the rate of synthesis of these proteins when they are coded by structural genes in *cis* position (on the same DNA strand as the deletion), but do not alter the response of the *lac* operon to the Lac repressor when they are in *trans* position.

The order of the genes in the *lac* region is

i	*p*	*o*	*z*	*y*	*a*

Strains of *p*-cells with a point mutation in the promoter can be induced by IPTG: they synthesize β-galactosidase at a rate of the order of 5–10 % of the rate of the parent p^+ strain. Such strains are sensitive to catabolic repression, but the effect of glucose is suppressed if cyclic AMP is added to the culture medium. Mutations of the promoter of the *lac* region are specific: the synthesis of tryptophanase by a strain which does not show a glucose effect on β-galactosidase synthesis is normally subjected to catabolic repression by glucose. Since the non sensitivity to glucose is *cis*-dominant, this excludes the intervention in the process of a diffusible product whose synthesis would be under the control of *p*.

Role of Cyclic AMP and of the CAP Protein in the Binding of RNA Polymerase to the Promoter Region

Experiments from numerous laboratories strongly suggested that catabolic repression affects gene expression at the transcription level (that of messenger RNA synthesis) rather than at the translation level (protein synthesis). Actually, the promoter region is the DNA site where RNA polymerase binds to initiate DNA transcription. In fact, it was during the study of catabolic repression that the promoter was first characterized, and the groups led by Beckwith, Miller and Reznikoff have established from 1967 to 1975 that the promoter is a region of *lac* DNA necessary for the initiation of transcription of the *lac* operon. A detailed analysis of the promoter region and of its role in transcription has been made in Chap. 13. The sequence of the *lac* promoter region contains the information for binding another protein in addition to RNA polymerase: In 1969, Chambers and Zubay observed that cyclic AMP is required for the DNA-dependent synthesis of the enzymes of the *lac* operon in the cell-free system they had developed (Fig. 18.2).

A year later, Zubay, Schwartz and Beckwith, theorizing that it was highly improbable that a mononucleotide intervenes directly and with the required specificity on the synthesis of catabolic enzymes, hypothesized that an intermediary protein existed which they called CAP (for catabolite gene activator protein). This hypothetical protein would have affinity for cyclic AMP, the complex CAP-cAMP once bound to DNA allowing the initiation of the messenger RNA at the promoter locus.

The detection of the CAP protein has been facilitated by the existence of a mutant which synthesizes a defective CAP: this mutant has been found during a systematic

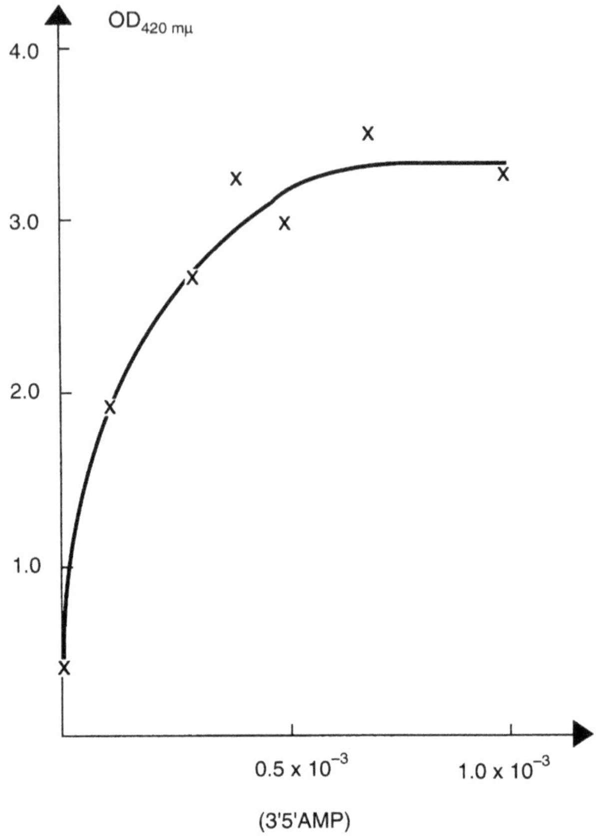

Fig. 18.2 Cyclic AMP requirement for the DNA-dependent synthesis of b-galactosidase. Ordinate: β-galactosidase synthesized in vitro (arbitrary units). Abscissa: cAMP concentration. The incubation mixture in Tris buffer pH 8.2 contains in addition to the energy source and the inorganic ions required for optimal synthesis, all the amino acids, the four ribonucleoside triphosphates, the tRNAs, DNA and proteins from an *E. coli* strain not producing the Lac repressor (From D.A. Chambers and G. Zubay, with permission of Dr G. Zubay)

Table 18.1 Requirements for the cell-free synthesis of β-galactosidase by the mutant unable to synthesize an active CAP protein cAMP: 5 mM. Partially purified extract containing CAP: 10 μg/ml

Cyclic AMP	CAP	β-galactosidase synthesized (arbitrary units)
−	−	1
+	−	1
−	+	1
+	+	5

screening of *E. coli* variants unable to grow on lactose and arabinose; they all possess very low levels of the enzymes known to be subjected to catabolic repression. Certain are devoid of adenylate cyclase, the enzyme responsible of the synthesis of cAMP from ATP: they revert (phenotypically) in the presence of cAMP. Other produce normal quantities of cAMP; one of those is devoid of the CAP protein necessary for the "activation" of the lactose operon. If the cell-free system from this mutant is used to synthesize β-galactosidase, it needs the addition of CAP and thus allows the purification of the protein. The stimulation by cAMP requires the presence of CAP (Table 18.1).

The synthesis of β-galactosidase is proportional to the amount of CAP (Fig. 18.3).

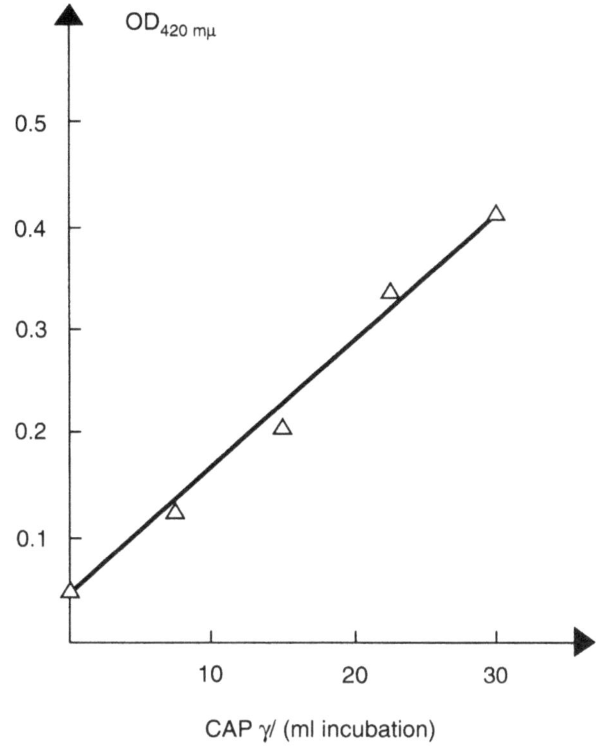

Fig. 18.3 Synthesis of β-galactosidase as a function of the concentration of partially purified CAP. The cell-free system is similar to that of Fig. 18.1, but the extract is from a strain which does not synthesize CAP. Ordinate: β-galactosidase synthesized, arbitrary units (From G. Zubay, D. Schwartz and J. Beckwith, with permission of Dr G. Zubay)

The same group of workers has repeated the experiment under conditions where the product detected during the cell-free synthesis is not β-galactosidase, but the messenger RNA corresponding to the transcription of the DNA segment coding for the lactose operon, by specific hybridization of mRNA to a defective bacteriophage carrying the operon. The results show that both cAMP and the CAP protein are required for the in vitro transcription of the *lac* operon. This transcription is inhibited by the Lac repressor.

The CAP protein has been purified and obtained in the homogeneous state by Pastan in 1971. It binds *lac* DNA, and the association constant is increased in the presence of cAMP. The binding to DNA is strongly inhibited by cyclic guanosine 3′-5′-mono-phosphate (cGMP), an effect reproducing an in vivo observation.

The CAP protein, coded by the *crp* gene, is a homodimer whose sequence has been established independently by Aiba and by Cossart and Gicquel-Sanzey in 1982. Each chain is made of 209 amino acid residues. Its mode of action will be examined in Chap. 36.

The Synthesis and Degradation of Cyclic AMP

Cyclic AMP is formed from ATP in a reaction catalyzed by adenylate cyclase (*cyaA*):

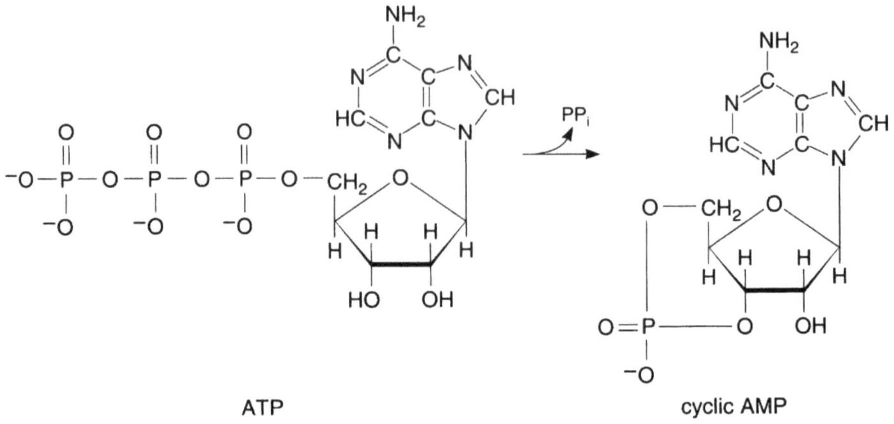

This reaction is driven toward the synthesis of cAMP by the subsequent action of a pyrophosphatase.

The cyclase is an integral membrane protein in eukaryotes; whereas early studies had reported this to be the case in *E. coli* this result has been challenged by workers who found the cyclase in the cytoplasmic fraction of disrupted cells. The partially purified enzyme is very labile. The *cya* gene has been cloned and sequenced; it codes for a polypeptide of 848 residues.

Cyclic AMP is a very stable molecule. It takes a specific phosphodiesterase to hydrolyze it to AMP:

$$\text{Cyclic AMP} + H_2O \xrightarrow{Mg^{2+}} \text{AMP} + H^+$$

How Does Glucose Exert Its Inhibitory Effect on *E. coli* β-galactosidase Synthesis?

We must go back to the description of diauxie and of the PTS system. It will be noted that sugars A, defined in Monod's monography all exert a "glucose effect", in fact a catabolic repression. They are also all substrates of permeation through the phosphoenolpyruvate phosphotransferase system (PTS).

The underlying mechanisms for PTS-mediated repression occur first at the metabolic level. PTS controls the concentrations of both cAMP (and therefore the cAMP/CAP adduct) and the inducer of the operon. A progress in the understanding of the molecular mechanisms underlying these controls resided in the isolation of mutations (*crr*), which relieved simultaneously four non-PTS operons from PTS-mediated regulation. The *crr* gene was found to code for IIIGlc. Both in vivo and in vitro experiments have shown that IIIGlc interacts with a variety of PTS and non-PTS proteins. For example, when IIIGlc is enclosed within natural membrane vesicles (see Chap. 4), it inhibits directly, but incompletely, the activity of β–galactoside permease while extravesicular III Glc is inactive. IIIGlc might thus participate in the glucose effect by excluding the inducer of β-galactosidase. On the other hand, a phosphorylated form of IIIGlc activates strongly adenylate cyclase and increases the level of cAMP (and of cAMP-CAP), thus increasing the expression of the *lac* genes. The ratio of phosphorylated to unphosphorylated PTS proteins appears to regulate the activities of non-PTS proteins. The situation is far from being clear and it will require more work to correlate catabolite repression and diauxic curves at the molecular level.

Coordination of Bacterial Proteome with Metabolism by Cyclic AMP Signalling

The cyclic AMP-dependent catabolite repression effect in Escherichia coli is among the most intensely studied regulatory processes in biology. However, the physiological function(s) of cAMP signaling and its molecular triggers remained elusive. A quantitative physiological approach shows that cAMP signaling tightly coordinates the expression of catabolic proteins with biosynthetic and ribosomal proteins, in accordance with the cellular metabolic needs during exponential growth. The expression of carbon catabolic genes increased linearly with decreasing growth rates upon limitation of carbon influx, but decreased linearly with decreasing growth rate upon limitation of nitrogen or sulfur influx. In contrast, the expression of biosynthetic genes showed the opposite linear growth-rate dependence as the catabolic genes. A scheme of integral feedback control featuring the inhibition of cAMP signaling by metabolic precursors is proposed and validated. These results reveal a key physiological role of cAMP-dependent catabolite repression: to ensure that proteomic resources are spent on distinct metabolic sectors as needed in different nutrient environments. These findings underscore the power of quantitative physiology in unravelling the underlying functions of complex molecular signaling networks.

Selected References

Catabolic Repression

Magasanik B (1961) Cold Spring Harb Symp Quant Biol 26:249–256

Effects of Cyclic AMP on the Glucose Effect

Makman RS, Sutherland EW (1965) J Biol Chem 240:1309–1313
Perlman R, Pastan I (1968) Biochem Biophys Res Comm 30:656–664; and 37:151–157 (1969)
Ullman A, Monod J (1968) FEBS Lett 2:57–60

The Promoter

Jacob F, Ullman A, Monod J (1964) Comptes Rendus 258:3125–3128

The CAP Protein

Chambers DA, Zubay G (1969) Proc Natl Acad Sci USA 63:118–122
Zubay G, Schwartz D, Beckwith J (1970) Proc Natl Acad Sci USA 66:104–110

Mode of Action of Cyclic AMP

Meadow ND, Fox DK, Roseman S (1990) Annu Rev Biochem 59:497–542
Ullmann A, Danchin A (1983) Adv Cyclic Nucleotide Res 15:1–53
You C, Okano H, Hui S, Zhang Z, Kim M, Gunderson CW, Wang Y-P, Lenz P, Yan D, Hwa T (2013) Nature 500:301–306

Chapter 19
The Ribosomes, Translation, Chaperones and Chaperonins

In the 1930s, cell biologists had surmised that the greatest part of RNA was localized in the cytoplasm and that there was a correlation between the amount of RNA and the protein synthesizing capacity of the cell. In the 1940s, Brachet and Caspersson established unequivocally this correlation with materials as diverse as the root tips of *Allium cepa*, imaginal disks of *Drosophila* larvae, secreting pancreatic cells and silk glands from the silkworm and with various methods such as specific dying techniques, specific nucleases and spectrophotometric methods on thin tissue slices.

By differential centrifugation, Claude identified a cellular fraction composed of small granules he called microsomes. Jeener and Brachet found that the microsomes contained RNA and made the hypothesis that they were involved in protein synthesis. The presence of Claude's microsomes in cytoplasm was confirmed by electron microscopy by Palade and Siekevitz who showed that these ribonucleoprotein particles, 200 Å in diameter, existed in all tissues. These particles, today called ribosomes, can be either attached to the endoplasmic reticulum in cells where this system is well developed, or exist as free particles.

From 1955 to 1957, studies in Zamecnik's laboratory showed with whole animals injected with radioactive amino acids and with cell-free preparations that the amino acids are first incorporated into the ribosomes from which they can be chased in the soluble fraction. It was shown that only 1 % of the amino acids from the ribosomal fraction turn rapidly over, which suggested a steady state between the rates of formation and release of the polypeptide chains synthesized.

A great advance in the understanding of these particles was due to studies on bacterial ribosomes. Luria, Delbrück and Anderson had observed as early as 1943, by electron microscopy of *E. coli* cells lyzed by bacteriophage, that in addition to the virus particles, large amounts of granules of homogeneous size. Schachman, Pardee and Stanier (1952), by ultracentrifugation of *E. coli* extracts, identified three well separable constituents of respective sedimentation coefficients (uncorrected) 40S, 29S and 5S. These constituents contain the essential of cellular RNA, the 40S constituent being composed of quasi-spherical particles. The sedimentation coefficients were corrected later by centrifuging in presence of Mg^{++} ions, which stabilize the ribosomes.

G.N. Cohen, *Microbial Biochemistry*,
DOI 10.1007/978-94-017-8908-0_19, © Springer Science+Business Media B.V. 2014

Thus, by the middle of the 1950s, it had been established that quasi-spherical particles of a diameter 200–300 Å, formed in approximately equal proportion of RNA and proteins, were the site of protein synthesis, in plants, animals, yeasts and bacteria. In 1956, Chao showed that the particles were composed of two unequal components, which could be separated in absence of magnesium, the metal being a bridge between the two parts. Tissières and Watson started to study the physico-chemical properties of ribosomes at Harvard; there were at that time very few laboratories working on the subject, the most active being the Biophysics group of the Department of Terrestrial Magnetism of the Carnegie Institution of Washington, headed by Richard Roberts. A special meeting of the recently created Biophysical Society was held in Boston in 1958. It is during this meeting that Roberts first introduced the term of ribosomes. In 1959, McQuillen, Britten and Roberts showed that ^{35}S-amino acids originating from radioactive sulfate were first incorporated in the ribosomes, before they could appear or be chased in other parts of the cell. Their elegant experiments were using ultracentrifugation in sucrose gradients. It became evident that with *E. coli*, experiments similar to the ones of Zamecnik in eukaryotes could be more easily amenable to understand the mechanism of protein synthesis. The names of Watson, Kurland, Tissières, Schlesinger, Gros and Zillig are associated to this early period (1959–1960).

The Components of *E. coli* Ribosomes

E. coli ribosomes have a sedimentation coefficient of 70S. They are composed of two subunits of 30S and 50S, containing respectively RNA molecules of 16S and 23S. They also contain a 5S RNA component, discovered by Monier and Rosset in 1963, and which was sequenced by Brownlee, Sanger and Barrell in 1968.

Post-transcriptional modification is a ubiquitous feature of ribosomal RNA in all kingdoms of life. Modified nucleotides are generally clustered in functionally important regions of the ribosome, but the functional contribution to protein synthesis is not well understood. High resolution crystal structures for the N^2-guanine methyltransferase RsmC that modifies residue G1207 in 16S rRNA near the decoding site of the 30S ribosomal subunit have been obtained. RsmC is a *S*-adenosyl-L-methionine-dependent methyltransferase composed of two methyltransferase domains.

The situation of ribosomal proteins is much more complex, as shown by Waller and Harris in 1961. Waller showed in 1964 by electrophoresis on starch gels and on carboxymethylcellulose, that the proteins of the 30S subunit were different from those of the 50S subunit. Two more years were necessary before the individual proteins could be identified with the techniques of protein chemistry.

We know today that the *E. coli* 50S subunit contains 31 polypeptides numbered from L1 to L34 (the discrepancy comes from errors in the early work) and the 30S subunit contains 21 proteins numbered from S1 to S21. All the proteins and the 16 and 23S RNAs have been sequenced through the efforts of two groups (Ebel in Strasbourg and Wittmann in Berlin).

Secondary structures for the RNAs have been proposed and the binding site of many of the ribosomal proteins on the RNA have been determined.

Nomura and his colleagues have devised techniques of partial or total disassembly of the two subunits and have succeeded in reconstituting them by adding to the RNA of each subunit the proteins that had been eliminated. The order in which the proteins are added is not indifferent and Nomura has established a very detailed reassembly chart. The information necessary for assembly is entirely contained in the structures of the proteins and the RNA and there is no need for auxiliary proteins or enzymes. The *in vitro* auto-assembly differs only slightly from the *in vivo* assembly process and reflects the physiological process.

The predictions of the ribosomal proteins secondary structure have been calculated from their primary structure (a hazardous method), but also by circular dichroism and small angle X-ray diffraction. Mutants of practically all the ribosomal proteins have been obtained and have greatly helped the functional studies.

Different approaches have been used to unravel the topology of the ribosomes which is of paramount importance for the understanding of structure-function relationships:

(a) antibodies acting on individual proteins form dimers between two subunits which can then be examined by electron microscopy; the antibody can be localized on the surface of the subunit and the position of the antigenic determinant of a given protein can be determined. This method rests on the divalent nature of the antibodies and on the facts that antibodies react with their antigen within the intact ribosome and that the two subunits possess distinguishable morphological features. This study has been principally carried in Wittmann's laboratory from 1973 to 1980 and the localization within the particle of the proteins belonging to one or the other subunit have been determined.
(b) the proximity relationships of the different proteins within a subunit have been also determined by cross-linking bifunctional reagents which create a covalent attachment between two neighboring proteins.
(c) finally, pairs of proteins have been labeled with fluorescent markers and the measurement of the energy transfer has provided a measure of the relative distance between the centers of the proteins tested.

It appears that the ribosome constitutes an extraordinarily complex machinery. A prominent common feature of the ribosomes is cavities in both the 50S subunit as well as in the 70S ribosome. The main cavity in the 70S ribosome comprises the space that is bordered by the 30S and the 50S subunits. This cavity is the functional center of the ribosome where mRNA decoding takes place. Using electron microscope tomography, it has been shown that starvation *in vivo* for aminoacyl-t-RNA (see Chap. 20) increases the separation between the two ribosomal subunits.

The Ribosomes of Eukaryotes and of Archaea

The eukaryotic ribosomes are composed of two subunits of 60S and 40S sedimentation coefficients (vs. 50S and 30S in prokaryotes). The association of the subunits form a ribosome of 80S (vs. 70S). The small subunit contains a molecule of 18S RNA and about 30 proteins, whereas the large subunit comprises three RNA molecules (5S, 5.8S and 28S) and 45–50 proteins. In spite of the difference in size, the electron micrographs of the eukaryotic ribosomes reveals a morphology very similar to that of the prokaryotic ones.

We have seen in a preceding chapter that the ribosomes from Archaea are chemically distinct from bacterial ribosomes. In fact they have similar drug sensitivities as 80S eukaryotic ribosomes.

Post-translational Modifications of Ribosomal Proteins

Post-translational modifications of ribosomal proteins are important for the accuracy of the decoding machinery. The *rimO* gene is involved in generation of the 3-methylthio derivative of residue Asp-89 in ribosomal protein S12 (Anton et al. 2008). This reaction is formally identical to that catalyzed by MiaB on the C2 of adenosine 37 near the anticodon of several tRNAs. Spectroscopic evidence shows that *Thermotoga maritima* RimO, like MiaB, contains two [4Fe-4S] centers, one presumably bound to three invariant cysteines in the central radical *S*-adenosylmethionine (AdoMet) domain and the other to three invariant cysteines in the N-terminal domain. Holo-RimO can specifically methylthiolate the aspartate residue of a 20-mer peptide derived from S12, yielding a mixture of mono- and bismethylthio derivatives.

Mechanistic Aspects of Translation of Messenger RNA to Protein by Ribosomes

The process of translation is divided into three main steps: initiation, elongation and termination. Initiation involves the binding of ribosomes to a specific start site on the messenger RNA sequence. During elongation, the ribosome moves along the mRNA, translating its sequence into a chain of amino acids that are supplied by transfer RNAs. Termination occurs when the ribosome and the nascent polypeptide both detach from the mRNA after reaching the end of the protein coding region on the mRNA. Specific initiation, elongation and termination protein factors aid each translation step.

Many of these aspects are being treated at length in many books, including the exhaustive one referenced to at the end of the chapter. We shall not deal with them in detail, but shall give a short summary of the events taking place on the ribosomes.

The components necessary for the initiation of translation include the interaction of the small and large ribosomal subunits, mRNA, an initiator aminoacyl-tRNA, GTP, and a large group of initiation factors. The initiation complex is formed by the small ribosomal subunit, the mRNA and Met-tRNA (methionine attached to a specialized initiator tRNA type). The Met-tRNA is hydrogen bonded to the AUG initiation codon on the mRNA. Formation of the ribosomal initiation complex is completed with the addition of the large ribosomal subunit. The AUG codon and the Met-tRNA are positioned in the P (peptidyl) site of the ribosome.

A new aminoacyl-tRNA is then positioned in the A (aminoacyl) site. The covalent bond between the amino acid and the tRNA in the P site is broken. A peptide bond is formed between the two amino acids. The empty tRNA then dissociates from the P site.

Translocation of the ribosome occurs such that the peptidyl-tRNA in the A site is translocated into the P site. The complex now looks very similar to that at the initiation of translation. The peptidyl-tRNA is in the P site and the A site is empty and ready to accept the next aminoacyl-tRNA.

The amino acid in the P site is separated from its tRNA and peptide bond formation takes place with the aminoacyl-tRNA in the A site. The tRNA is liberated from the P site and the ribosome translocates such that the new peptidyl-tRNA is in the P site; the A site is ready to accept the next aminoacyl-tRNA.

Initiation factor 2 (IF2) is a GTPase that functions within the 30S ribosomal initiation complex and promotes its joining with the 50S ribosomal subunit to form a 70S ribosome. An atomic resolution crystal structure of the full-length IF2, explains the mechanism by which IF2 guides ribosome assembly. This mechanism is unique among translational GTPases and acts more as a novel conformational switch.

Elongation factor P (EF-P) is an essential protein that stimulates the formation of the first peptide bond in protein synthesis. The crystal structure of EF-P bound to the *Thermus thermophilus* 70S ribosome along with the initiator transfer RNA N-formyl-methionyl-tRNA and a short piece of messenger RNA has been solved at a resolution of 3.5 Å. EF-P binds to a site located between the binding site for the peptidyl tRNA and the exiting tRNA site. It spans both ribosomal subunits with its amino-terminal domain positioned adjacent to the aminoacyl acceptor stem and its carboxyl-terminal domain positioned next to the anticodon stem-loop of the P site–bound initiator tRNA. Domain II of EF-P interacts with the ribosomal protein L1, which results in the largest movement of the L1 stalk that has been observed in the absence of ratcheting of the ribosomal subunits. EF-P facilitates the proper positioning of the fMet-tRNAMet for the formation of the first peptide bond during translation initiation.

The elongation cycle (the addition of amino acids one at a time to a growing polypeptide chain) continues for more peptide bond formation and translocation. After peptidyl transfer, the movement of messenger RNA) and transfer RNAs with respect to the ribosome places the next mRNA codon in the A site. This process of translocation proceeds via an intermediate state in which the acceptor ends of the tRNAs have moved with respect to the 50*S* subunit but not the 30*S* subunit, to result in A/P and P/E tRNA hybrid states. The guanosine triphosphatase elongation factor G (EF-G) catalyzes the subsequent movement of mRNA and tRNA with respect to the 30*S* subunit. How EF-G binds to the intermediate state of the ribosome and how this results in guanosine 5′-triphosphate (GTP) hydrolysis and translocation are questions that are greatly facilitated by a high-resolution structure of the *T. thermophilus* complex.

Distinct combinations of amino acids, such as polyproline stretches, cannot be efficiently polymerized by ribosomes, leading to translational stalling. The stalled ribosomes are rescued by the translational elongation factor EF-P) which by stimulating peptide-bond formation allows translation to resume. Using metabolic stable isotope labeling and mass spectrometry, it has been shown *in vivo* that EF-P is important for expression of not only polyproline-containing proteins, but also for specific subsets of proteins containing diprolyl motifs (XPP/PPX). Together with a systematic *in vitro* and *in vivo* analysis, a distinct hierarchy of stalling triplets has been established, ranging from strong stallers, such as PPP, DPP, and PPN to weak stallers, such as CPP, PPR, and PPH, all of which are substrates for EF-P. These findings provide mechanistic insight into how the characteristics of the specific amino acid substrates influence the fundamentals of peptide bond formation.

Elongation factor-G-catalyzed translocation of mRNA and tRNAs during protein synthesis involves large-scale conformational changes in the ribosome. Formation of hybrid-state intermediates is coupled to counterclockwise (forward) rotation of the body of the 30S subunit. Recent structural studies implicate intrasubunit rotation of the 30S head in translocation: Rotation of the head during translocation in real time using ensemble stopped-flow FRET with ribosomes containing fluorescent probes attached to specific positions in the head and body of the 30S subunit allow ordering of the rates of movement of the 30S subunit body and head during translocation: body forward > head forward > head reverse ≥ body reverse. The rate of quenching of pyrene-labeled mRNA is consistent with coupling of mRNA translocation to head rotation.

A stop codon (UAA, UAG, or UGA) positions in the A site; a stop codon has no anticodon (an aminoacyl tRNA that will hydrogen bond to this codon); for this reason, no other amino acid will be added. The stop codon is recognized and bound by a protein called the termination or release factor.

The release factor binds to the stop codon at the A site and begins a sequence of events that brings about the termination of translation. The termination sequence begins with the dissociation of the newly synthesized protein from the peptidyl-tRNA in the P site. Separation of the newly synthesized proteins from the ribosome is followed by the dissociation of all the remaining subunits.

The ribosomal subunits and the mRNA can then reassemble with Met-tRNA to form new initiation complexes and protein translation can begin again to produce additional copies of the protein.

When ribosomes stall on mRNA in eubacteria, a molecule called tmRNA and its protein partner SmpB come in and restart the ribosomes by a process called *trans*-translation. tmRNA and SmpB enter the stalled ribosome together, transfer the polypeptide to the tRNA-like domain of tmRNA, and get the ribosome going with translation. The highly conserved glycine 132 in SmpB worked as a flexible hinge so that the protein's C-terminal tail moved freely and allowed proper positioning and establishment of the tmRNA open reading frame as the substitute template. The C-terminal tail of SmpB is important in accurately identifying the resume codon for the ribosome. Taken together, these data suggest that the engagement of tmRNA and the selection of the correct translation resumption point are distinct activities that are influenced by independent tmRNA and SmpB determinants

Recruitment of translation factors is mediated by the ribosomal protein L12 stalk composed of protein L10 and several copies of protein L12, the only multi-copy of the ribosome. The majority of bacteria have L12 stalks allowing for binding of four or six copies of L12, largely independent of the taxonomic group or living conditions of the bacteria, whereas some cyanobacteria have eight copies. The last universal common ancestor probably had six molecules of L12 molecules bound to L10. Changes of the stalk composition provide a unique possibility to trace the evolution of protein components of the ribosome.

In 1980, Yonath managed to generate the first low-quality crystals of a ribosome. By 1990, she had upped the quality of her crystals, but she still struggled to a good structure. In 1995, Steitz in 1995, followed Yonath's recipe for making ribosomal crystals and by 1998, used additional data obtained from electron microscopy studies and obtained a low-resolution 9 Å structure of the ribosome. In 2000 Steitz's group published a higher 2.4 Å resolution structure of the 50S subunit. The following month, Yonath's and Ramakrishnan's groups published slightly lower resolution structures of the smaller subunit. Since then, the three groups, plus other teams, have used those structures to understand in atomic detail how ribosomes translate genetic information into proteins. This formidable accomplishment, result of 30 years of hard work has been recognized by the attribution of the Nobel Prize in chemistry in 2009 to Ada Yonath, Peter Steitz and Venkatraman Ramakrishnan. As making proteins is essential for bacteria survival, the ribosome is a practical target for drugs. Ramakrishnan, Steitz, and Yonath have provided vital information for the design of new antibiotics.

Chaperones and Chaperonins

Newly made proteins must fold from a linear chain of amino acids into a three-dimensional form. The understanding of protein folding was obtained from computer models (in silico) or from experiments in the laboratory (in vitro) in which an individual protein was denatured to observe it folding back into its original form. Although protein folding follows the same physical principles in vivo and in vitro, the crowded intracellular milieu poses additional challenges. Unstructured peptide backbones or hydrophobic amino acid residues exposed in nascent polypeptide chains, folding intermediates or partially unfolded proteins can engage in non-native intermolecular interactions. Such contacts compete with productive folding and promote protein aggregation. Chaperones prevent these off-pathway reactions by temporarily binding to partially folded intermediates and subsequently releasing them.

Although the fundamental energy rules also apply here, folding (at least of large proteins) rarely takes place spontaneously, Cells contain a vast number of proteins and other biomolecules at the extraordinarily high concentration of 340 g/l. Ordered protein folding in this cramped chaos is only possible under the supervision of specialized molecules, called chaperones, which accompany proteins and make sure that those that are being formed at the ribosomes do not clump together prematurely. Chaperones do not merely oversee the folding of the protein, they also protect its tertiary structure in situations in which the cell is under stress; for example, elevated body temperature, so these chaperones have also been classified as heat-shock proteins (HSPs). The HSP70s, so called because they have a molecular weight of 70 kDa, are the most important class of chaperones. They bind to the developing protein chain and protect those parts of the newly formed protein that are particularly sensitive to premature reaction with the environment and therefore to malformation. When they let go of the new protein chain it is ready to fold. It is now taken over by a chaperonin, a molecule shaped like a double ring, which fits round the protein chain like a cylinder so that it can fold undisturbed inside Although the cylindrical folding cage opens every ~10 s, the protein only leaves the chaperonin when it has achieved its required form. The energy to fold proteins is supplied by ATP.

Chaperonins are cage-like complexes in which polypeptides prone to aggregation reach their native state.

The GroEL/GroES complex in E. coli is the best characterized large (~1 MDa) Group I chaperonin complex.

- GroEL is a double-ring 14mer with a greasy hydrophobic patch at its opening and can accommodate the native folding of substrates 15–60 kDa in size.
- GroES is a single-ring heptamer that binds to GroEL in the presence of ATP or transition state analogues of ATP hydrolysis. It is like a lid that covers GroEL (box/bottle).

GroEL/GroES may not be able to undo protein aggregates, but kinetically it competes in the pathway of misfolding and aggregation, thereby preventing aggregate formation. Chaperonins undergo large conformational changes during a folding reaction as a function of the enzymatic hydrolysis of ATP as well as binding of substrate proteins and co-chaperonins, such as GroES. These conformational changes allow GroEL to bind an unfolded or misfolded protein, encapsulate that protein within one of the cavities formed by its two rings, and release the protein back into solution. Upon release, the substrate protein will either be folded or will require further rounds of folding, in which case it can again be bound by a chaperonin. A number of models of chaperonin action have been proposed, which generally focus on two (not mutually exclusive) roles of chaperonin interior: passive and active. Passive models treat the chaperonin cage as an inert form, exerting influence by reducing the conformational space accessible to a protein substrate or preventing intermolecular interactions e.g. by aggregation prevention. The active chaperonin role is in turn involved with specific chaperonin–substrate interactions that may be coupled to conformational rearrangements of the chaperonin. The overall action of GroEL involves binding of the non-native protein via a hydrophobic lining of the cavity of an open GroEL ring, preventing aggregation and potentially reversing misfolded states, followed by an ATP-GroES (lid) directed release of non-native polypeptide into a now GroES-encapsulated cavity where folding ensues, the longest step of the reaction cycle. After ATP hydrolysis, binding of ATP to the opposite ring leads to release of GroES and folded polypeptide.

Group II chaperonins, found in archaea are more poorly characterized. A chaperonin), found in the archaeon *Methanococcus maripaludis*, is composed of 16 identical subunits (eight per ring). Group II chaperonins are not thought to utilize a GroES-type cofactor to fold their substrates. They instead contain a "built-in" lid that closes in an ATP-dependent manner to encapsulate its substrates, a process that is required for optimal protein folding activity.

Extraribosomal Functions

Ribosomal proteins constitute a considerable part of the cell proteome. Although their primary role in the cell is to serve as integral components of protein synthesis machinery, many of them have functions beyond the ribosome (a phenomenon known as moonlighting), acting either as individual regulatory proteins or in complexes with other cell components. Extraribosomal activities of some ribosomal proteins were observed as early as in the 1970–1980s. In recent years, the list of moonlighting ribosomal proteins and their additional functions beyond the ribosome was greatly expanded, mainly owing to new techniques developed for dissecting RNA/DNA – protein or protein-protein interactions within functional complexes involved in various cell processes.

For example, L4 protein, a component of the 50S subunit of E. coli ribosome, regulates RNA degradation, interacting with a site outside of the catalytic domain of RNaseE. Analysis of mRNA stability shows that the half-lives of stress-responsive transcripts are increased by ectopic expression of L4, which normally increases along with other ribosomal proteins under stress conditions, and also by inactivation of RNaseE. The fact that L4 can inhibit RNaseE dependent decay may account at least in part for the elevated production of stress-induced proteins during bacterial adaptation to adverse environments.

S1 is the largest ribosomal protein and is vitally important for the cell. S1 is also a subunit of Qβ replicase, the RNA-directed RNA polymerase of bacteriophage Qβ. In both protein and RNA syntheses, S1 was believed to bind to a template RNA at the initiation step, and not to be involved in later events. However, it was recently shown that S1 functions at the termination step of Qβ replicase-mediated RNA synthesis by promoting release of the product strand in a single-stranded form.

Many more examples are described in the excellent review by Aseev and Boni.

There are a lot of grounds for believing that the list of extraribosomal functions of ribosomal proteins is far from complete and that new and even unexpected activities of structural components of the ribosome will be discovered in the future.

Selected References

Davydov II, Wohlgemuth I, Artamonova I, Urlaub H, Tonevisky AG, Rodnina MV (2013) Nat Commun, article number 1387. doi: 10.1038/ncomms2373

Demirci H, Gregory ST, Dahlberg AE, Jogl G (2008) J Biol Chem 283:26548–26556

Guo Z, Noller HF (2012) Proc Natl Acad Sci USA 109:20391–20394

Hershey JWB, Merrick WC (2000) Translational control of gene expression. Cold Spring Harbor Press, Cold Spring Harbor, pp 33–88

Hill WE, Dahlberg A et al (eds) (1990) The ribosome: structure, function and evolution. American Society for Microbiology, Washington, DC, 678 pp

Schachman HK, Pardee AB, Stanier RY (1952) Studies on the macro-molecular organization of microbial cells. Arch Biochem Biophys 38:245–260

Spirin AS (2009) J Biol Chem 284:21103–21119

Waller J-P (1964) J Mol Biol 10:319–336

Mechanistic Aspects of Translation

Peil L, Starosta AL, Lassak J, Atkinson GC, Virumäe K, Spitzer M, Tenson T, Jung K, Remme J, Wilson DN (2013) Proc Natl Acad Sci USA 110:15265–15270

Translocation

Tourigny DS, Fernandez IS, Kelley AC, Ramakrishnan V (2013) Science 340(6140):1235490

Zhou J, Lancaster L, Donohue JP, Noller HF (2013) Science 340(6140):1236086

Ribosome Stalling

Camenares D, Dulebohn DP, Svetlanov A, Wali Karzai A (2013) J Biol Chem 288:30527–30542

Post-translational Modifications of Ribosomal Proteins

Anton BP, Saleh L, Benner JS, Raleigh EA, Kasif S, Roberts RJ (2008) Proc Natl Acad Sci USA 105:1826–1831

Arragain S, Garcia-Serres R, Blondin G, Douki T, Clemancey M, Latour J-M, Forouhar F, Neely H, Montelione GT, Hunt JF, Mulliez E, Fontecave M, Atta M (2010) J Biol Chem 285:5792–5801

Eukaryotic Ribosomes

Dinman JD (2009) J Biol Chem 284:11761–11765
Wool IG (1979) Ann Rev Biochem 48:719–754

Crystallography

Bashan A, Yonath A (2008) Trends Microbiol 16:326–335
Blaha G, Stanley RE, Steitz TA (2009) Science 325:966–970
Moore PB, Steitz TA (2005) Trends Biochem Sci 30:281–283
Ogle JM, Murphy FV, Tarry MJ, Ramakrishnan V (2002) Cell 111:721–732
Schmeing TM, Ramakrishnan V (2009) Nature 461:1234–1242
Yonath A (2002a) Annu Rev Biophys Biomol Struct 31:257–273
Yonath A (2002b) Curr Protein Pept Sci 3:67–78

Protein Folding. Chaperones and Chaperonins

Anfinsen CB (1973) Science 181:223–230
Clare DK, Vasishtan D, Stagg S, Quispe J, Farr GW, Top M, Horwich AL, Saibil HR (2012) Cell 149:113–123
Saibil HR, Fenton WA, Clare DK, Horwich AL (2013) J Mol Biol 425:1476–1487
Wun C-C, Naveen V, Chien C-H, Chang Y-W, Hsiao C-D (2012) J Biol Chem 287:21461–21470

Extraribosomal Functions

Aseev LV, Boni IV (2011) Mol Biol 45:739–750
Singh D, Chang SJ, Lin PH, Averina OV, Kaberdi VR, Lin-Chao S (2009) Proc Natl Acad Sci USA 106:864–869
Vasilyev NN, Kutlubaeva ZS, Ugarov VI, Chetverina HV, Chetverin AB (2013) Nat Commun 4, article number 1781. doi: 10.1038/ncomms2807

Chapter 20
The Genetic Code, the Transfer RNAs and the Aminoacyl-tRNA-Synthetases

The Genetic Code

The first ideas on an intimate connection between RNA and protein synthesis go back to Brachet and Caspersson. We know now that their observations did not bear on the messenger RNA, but on ribosomal RNA. Nevertheless, it is thanks to them that this connection became familiar. Then Beadle formulated the hypothesis called the "one gene-one enzyme correlation", but his ideas did never reach the formulation of a linear code relating genes and proteins. In 1950, Caldwell and Hinshelwood published a theory according to which nucleic acid, by a process similar to crystallization, ensures the order of the amino acids in the protein. Their ideas were rather confusing, since they were considering 23 amino acids and 5 units in the nucleic acids, namely the four bases and ribose phosphate. The authors were not mentioning whether their doublet code was overlapping or not. This article, only occasionally quoted exerted no influence on the later theories. On the other hand, a paper published by Dounce in 1952 was well read: it treats mainly of the possible chemical mechanisms of protein synthesis, but it suggests a code where each nucleotide codes for an amino acid, which automatically leads to an overlapping code. He did not realize that his code could be rejected by the mere inspection of the few known polypeptide sequences then available.

The determination of the structure of the code was greatly helped by the discovery of the structure of DNA by Watson and Crick in 1953. It simplicity impressed many scientists, including the cosmologist George Gamow who put forward in 1954 a theory according to which protein synthesis occurred on the surface of the DNA double helix, the sequence of bases within the structure forming a series of cavities, each of which was specific of a given amino acid. The theory does not mention how the amino acids recognize the cavities, but suggests that their side chains adapt stereochemically to them, without the aid of any enzyme. Gamow, as Dounce was taking into account that the distance between residues of a stretched polypeptide chain is only 3.6–3.7 A and that the code had therefore to be overlapping. Gamow's abstract model, historically important, was not taking into account useless chemical

DOI 10.1007/978-94-017-8908-0_20,

details: his ideas that protein synthesis did happen on a double-stranded DNA matrix turned out to be entirely wrong. He had the merit to have clearly understood that an overlapping code imposed restrictions on the amino acid sequences and that it should be possible to confirm or reject the several possible overlapping codes by inspecting the known sequences.

Francis Crick wrote in 1955 an article which was never published, but circulated among specialists, called "On degenerate templates and the adaptor hypothesis". This paper lists the amino acids to be included in the code and demonstrates that Gamow's code is incompatible with the sequences of insulin and β-corticotropin and does not specify the direction of protein synthesis: he was already thinking that some mechanism should impose the direction in which DNA was read. This unpublished article contains for the first time the adaptor hypothesis, the adaptor being a small molecule able to form specific hydrogen bonds with a nucleic acid template, activation enzymes providing the specificity required for the combination between a given amino acid and its adaptor.

The next idea, formulated in 1957, was that of a code without punctuation (Crick, Griffith and Orgel). The same year, Sydney Brenner demonstrated that overlapping codes were impossible.

In 1959, the situation seemed hopeless. One expected that the sequences of a gene and of the protein it encoded should be co-linear, but this was shown independently by Yanofsky and by Brenner in 1964 only.

The missing link was that of messenger RNA. When the experimental evidence for its existence became available, the code problem was solved. In 1961, Nirenberg and Matthaei discovered that polyU could serve as a messenger and that it coded for polyphenylalanine. Crick and Brenner showed unequivocally that the code was a triplet, non overlapping and without punctuation. The groups of Nirenberg, of Ochoa, and Khorana prepared numerous artificial messengers, either by using the polynucleotide phosphorylase discovered by Grunberg-Manago and Ochoa in 1956 or by chemical synthesis. These polynucleotides had either random sequences or definite repeated sequences, and served for the preparation of polypeptides whose composition was analyzed. Finally, a binding method was devised by Leder and Nirenberg in 1964, which allowed to define a correspondence between each triplet (a codon) and an amino acid. This work led to the establishment of the genetic code: it is highly degenerated (leucine and arginine being coded by six different codons), non-ambiguous and it contains three nonsense codons, which do not code for an amino acid and cause a polypeptide chain termination when the message is translated by the ribosome (Table 20.1).

All we have learned since 1966 by the sequence of many proteins and nucleic acids has confirmed that the genetic code is practically universal. There are some deviations of the standard code in some systems, and in many cases, variations occur at well defined specific sites. However, these differences do not indicate a different origin of life. For example, in mitochondrial systems, the UGA termination codon codes for tryptophan and AUA codes for methionine instead of isoleucine. AGA is a termination codon in the vertebrate mitochondrial RNA, but codes for an amino acid in the *Drosophila* mitochondria. There are also some deviations which

Table 20.1 The genetic code

	Second base			
	U	C	A	G
	UUU Phe	UCU Ser	UAU Tyr	UGU Cys
U	UUC Phe	UCC Ser	UAC Tyr	UGC Cys
	UUA Leu	UCA Ser	UAA TER	UGA TER
	UUG Leu	UCG Ser	UAG TER	UGG Trp
	CUU Leu	CCU Pro	CAU His	CGU Arg
C	CUC Leu	CCC Pro	CAC His	CGC Arg
	CUA Leu	CCA Pro	CAA Gln	CGA Arg
	CUG Leu	CCG Pro	CAG Gln	CGG Arg
	AUU Ile	ACU Thr	AAU Asn	AGU Ser
A	AUC Ile	ACC Thr	AAC Asn	AGC Ser
	AUA Ile	ACA Thr	AAA Lys	AGA Arg
	AUG Met	ACG Thr	AAG Lys	AGG Arg
	GUU Val	GCU Ala	GAU Asp	GGU Gly
G	GUC Val	GCC Ala	GAC Asp	GGC Gly
	GUA Val	GCA Ala	GAA Glu	GGA Gly
	GUG Val	GCG Ala	GAG Glu	GGG Gly

All triplet codons have a meaning. All the amino acids except methionine and tryptophan are represented by more than one codon. Synonymous codons form groups in which the third base is the less significant. Three codons lead to termination of translation (TER). The order of bases is written in the direction $5' \rightarrow 3'$

have been found in the mitochondrial DNA of yeasts, of filamentous fungi, protozoa and plants. Some differences have also been observed in the genome of some mycoplasmas and in the nuclear genes of at least four species of ciliated protozoa where standard termination codons code for glutamine.

Most amino acids are encoded by multiple codons, and codon choice has strong effects on protein expression. Rare codons are enriched at the N terminus of genes in most organisms, although the causes and effects of this bias are unclear. Expression from >14,000 synthetic reporters in *Escherichia coli* show that using N-terminal rare codons instead of common ones increases expression by ~14-fold (median 4-fold). Individual N-terminal codons affect expression and show that these effects shape the sequence of natural genes. Reduced RNA structure and not codon rarity itself is responsible for expression increases. These observations resolve controversies over the roles of N-terminal codon bias and suggest a straightforward method for optimizing heterologous gene expression in bacteria.

Selenocysteine is recognized as the 21st amino acid in ribosome-mediated protein synthesis and its specific incorporation is directed by the UGA codon (normally a termination codon). Unique tRNAs that have complementary UGA anticodons are aminoacylated with serine, the seryl-tRNA is converted to selenocysteyl-tRNA and the latter binds specifically to a special elongation factor and is delivered to the ribosome. Recognition elements within the mRNAs are essential for translation of UGA as selenocysteine. A reactive oxygen-labile compound, selenophosphate, is the selenium donor required for synthesis of selenocysteyl-tRNA. Selenophosphate synthetase, which forms selenophosphate from selenide and ATP, is found in various prokaryotes, eukaryotes, and archaebacteria. The distribution and properties of some selenocysteine-containing enzymes and proteins that have been discovered to date will be described in Chap. 30. Artificial selenoenzymes such as selenosubtilisin have been produced by chemical modification. Genetic engineering techniques also have been used to replace cysteine residues in proteins with selenocysteine. In some cases a marked decrease in catalytic activity of an enzyme is observed when a selenocysteine residue is replaced with cysteine. This substitution caused complete loss of glycine reductase selenoprotein A activity.

The first genetically encoded amino acid was identified more than two centuries ago, but new ones are still being found. The twenty-second is pyrrolysine (Pyl) which was found in 2002 at the active sites of methyltransferase enzymes obtained from a methane-producing archaeon. Like all other amino acids, pyrrolysine is synthesized in the cytoplasm and incorporated at a specific position in a growing polypeptide chain during translation. A specific stop codon, UAG, in the sequence of methyltransferase enzymes from methanogenic archaea encodes pyrrolysine. The machinery associated with its synthesis and incorporation into proteins is encoded by the pylTSBCD cluster of genes . The decoding of UAG as pyrrolysine requires pyrT, which produces tRNA-pyl, and pyrS, which encodes a pyrrolysyl-tRNA synthetase. The pylB, pylC and pylD genes are each required for tRNA independent pyrrolysine synthesis. Genetic and mass spectrometric evidence shows that pyrrolysine arises from two lysines. Figure 20.1 describes the pathway by which two molecules of L-lysine are transformed into a molecule of L-pyrrolysine by the products of pylB, pylC and pylD.

Despite the fact that the genetic code is known to vary between organisms in rare cases, it is believed that during the lifetime of a single cell the code is stable. However, Söll and collaborators have found that although *Acetohalobium arabaticum* grown on pyruvate encode 20 amino acids, it expands its genetic code in the presence of trimethylamine to 21 amino acids including pyrrolysine., thus modulating the size of its genetic code in response to its environment and energy source. The gene cassette pylTSBCD is present in the genome of 24 anaerobic archaea and bacteria. Unlike archaeal Pyl-decoding organisms that constitutively encode pyrrolysine, *A. arabaticum* controls Pyl encoding by down-regulating transcription of the entire Pyl operon under growth conditions lacking trimethylamine to the point that no detectable Pyl-tRNApyl is made in vivo.

The natural four-letter genetic alphabet, comprised of just two base pairs (dA-dT and dG-dC), is conserved throughout all life, and its expansion by the development of a third, unnatural base pair has emerged as a central goal of chemical and synthetic biology. A class of candidate unnatural base pairs is exemplified by the pair formed between d5SICS and dNaM (Fig. 20.2). A PCR amplification of DNA containing one or more d5SICS-dNaM pairs in a wide variety of sequence contexts has been obtained by PCR amplification. Under standard conditions, this DNA is amplified with high efficiency and greater than 99.9 % fidelity. We the unnatural base pair is efficiently replicated with high fidelity in virtually all sequence contexts. The results show that, for PCR and PCR-based applications, d5SICS-dNaM is functionally equivalent to a natural base pair, and when combined with dA-dT and dG-dC, it provides a fully functional six-letter genetic alphabet.

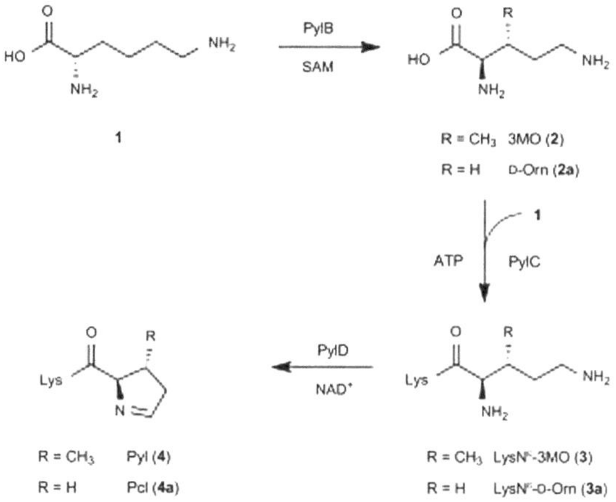

Fig. 20.1 Proposed pathway of pyrrolysine biosynthesis from two molecules of lysine by the products of pylB, pylC and pylD. The radical S-adenosylmethionine (*SAM*) protein PylB mediates a lysine mutase reaction that produces 3-methylornithine(3-month), which is then ligated to a second molecule of lysine by PylC before oxidation by PylD resulting in pyrrolysine

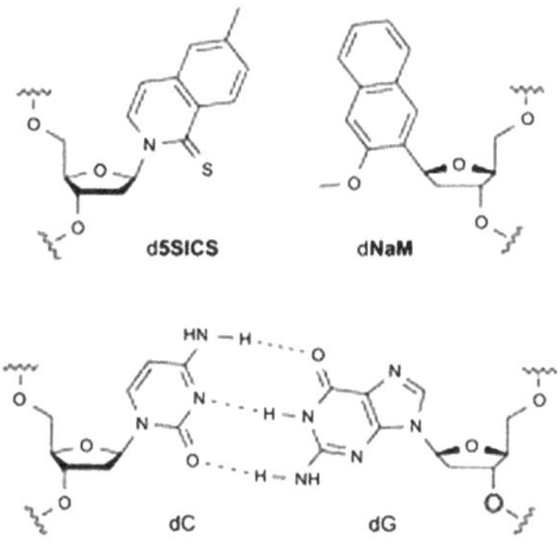

Fig. 20.2 Unnatural (**d5SICS**-d**NaM**) and natural Watson–Crick (dC-dG) base pairs

The Transfer RNAs

It appears useful to quote some lines from Crick's unpublished 1955 paper:

> … each amino acid would combine chemically, at a special enzyme, with a small molecule which, having a specific hydrogen-bonding surface, would combine specifically with the nucleic acid template. This combination would also supply the energy necessary for the polymerization. In its simplest form, there would be 20 different kinds of adaptor molecules, one for each amino acid and 20 different amino acids to their adaptors. Sidney Brenner, with whom I have discussed this idea, calls this the "adaptor hypothesis", since each amino acid is fitted to an adaptor to go on the template….

At about the same period, Zamecnik's group was persuaded that an additional reaction should exist between the "activated" amino acid and its appearance in the microsome as a nascent polypeptide chain. The biochemists approached the problem, not in terms of hydrogen bonds and of translation of one language to another, but in terms of enzymes and energy. Holley (1956) followed the activation reaction of alanine in liver extracts and obtained indications of a reaction occurring after the activation, which was sensitive to ribonuclease whereas the activation proper was not sensitive. The same year, Berg prepared for the first time an enzyme specific for the activation of a single amino acid, namely methionine, from a mixture obtained from yeast. There was a stoechiometric reaction between the enzyme and methionine, a result in contradiction with the catalytic nature of enzymes. The enzyme was not liberated and Berg thought that the reaction must be incomplete and started looking for the natural acceptor of the activated methionine, without having ever heard of Crick's adaptor hypothesis. He found it and it was a small-size RNA, unlinked to microsomes and specific both for the activating enzyme and for methionine.

In 1956 also, Zamecnik and Hoagland added radioactive leucine to their cell-free extract and isolated a radioactive fraction of soluble RNA. When this fraction was added to a new cell-free extract, radioactive leucine was liberated from the soluble RNA in a few minutes and was found incorporated in the protein growing chain on the microsome.

The discoveries of the biochemists were then converging to a definition of a molecule that had all the characteristics of Crick's hypothetical adaptor. Soon, soluble RNAs and enzymes specific for each amino acid were found by Berg, Lipmann, Holley and many others. Zamecnik found that all the 3′ ends of soluble RNAs presented the CCA sequence to which the amino acids were linked by an energy-rich bond. The soluble RNA was named transfer RNA (tRNA) and the activation enzymes were named aminoacyl-tRNA transferases.

Holley could separate the different species of tRNA and sequenced the yeast alanyl-tRNA in 1966. His long work was rendered somewhat easier by the fact that nine of the 77 bases of the molecule were not the classical RNA bases A, G, C, and U, but belonged to different nucleosides (inosine, pseudouridine) or to modified nucleosides such as dimethylguanosine. This was the first sequence of a nucleic acid ever obtained. There were several G and C residues which could be paired so as to give the molecule the appearance of a cloverleaf. At the extremity of the central lobe of the cloverleaf, three nucleotides (anti-codon) could pair with the triplet specific for the alanine codon of the template (by that time, messenger RNA and its properties were known). All tRNAs were subsequently found to have similar structures (Fig. 20.3).

In addition to the anticodon loop, there are two other loops, one always containing dihydrouracil, and one, called the pseudouridine loop, containing always the sequence GTΨC, which appears to play a role in the binding aminoacyl-tRNA to the 50S subunit of the ribosome. A fourth loop may be present, but is not a constant feature of all tRNAs.

Very soon after, attempts were made to crystallize the tRNAs in order to study their tridimensional structure. Rich and Klug determined the structure of phenylalanyl-tRNA from yeast in 1974–1975. Five years later, Rich and Moras determined respectively the tridimensional structure of N-formylmethionyl-tRNA and of aspartyl-tRNA which presented the same L structure as had been found for phenylalanyl-tRNA (Fig. 20.4).

The tRNAs are not transcribed as such from their structural genes, but derive from precursors longer both in the 3′ and 5′ sides. These precursors are shortlived

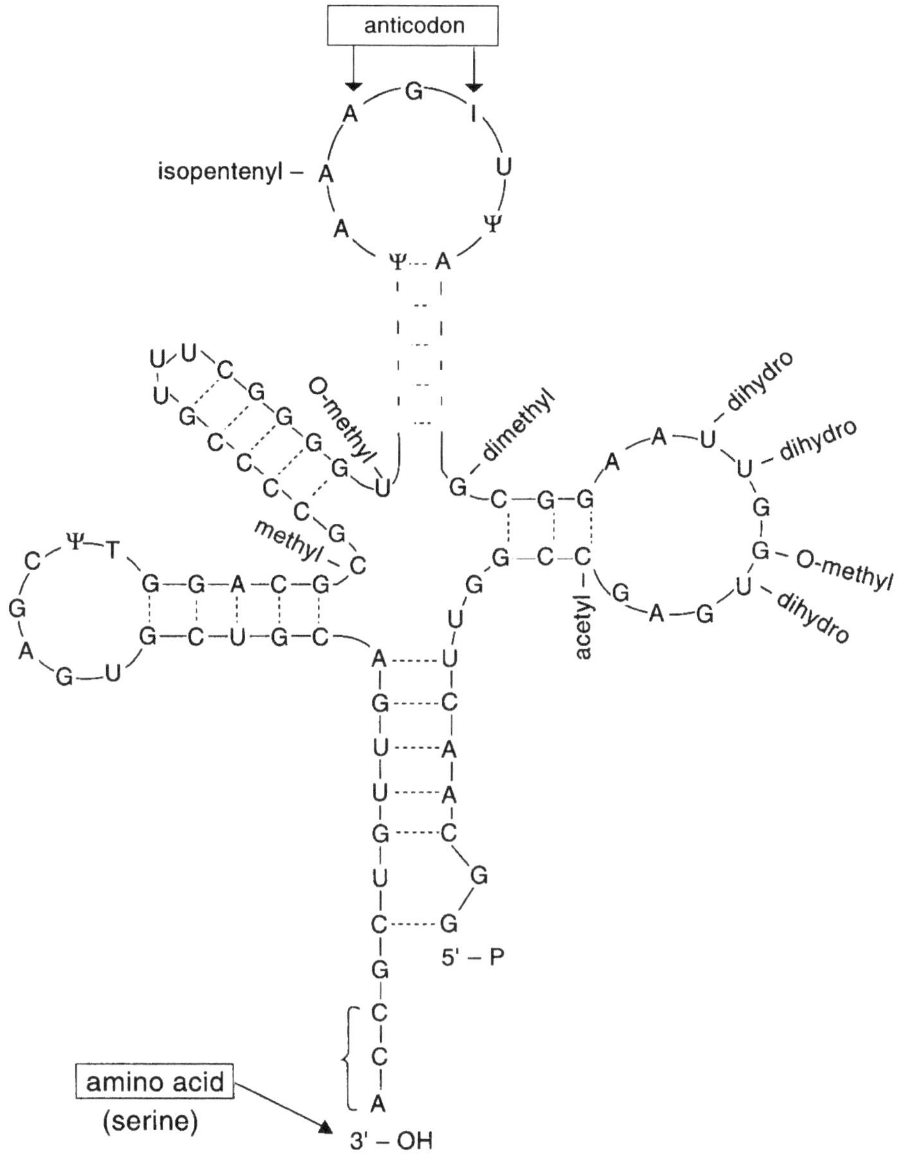

Fig. 20.3 Cloverleaf model of a seryl-tRNA from yeast. *A* = Adenine, *C* = Cytosine, *G* = Guanine, *I* = Inosine, *U* = Uracil, *T* = Thymine, ψ = Pseudouracil (From *Concise Encyclopedia of Biochemistry*, 2nd edn. Walter de Gruyter, Berlin/New York (1988) with permission of the publisher)

and are cleaved on the 5′ side by a specific nuclease, ribonuclease P, which is a non-covalent complex of a small RNA molecule of 337 residues and a small protein of 20,000 Da; the cleavage reaction is catalyzed at almost the same rate by the RNA component in the absence of the protein. Another enzyme, ribonuclease D cleaves

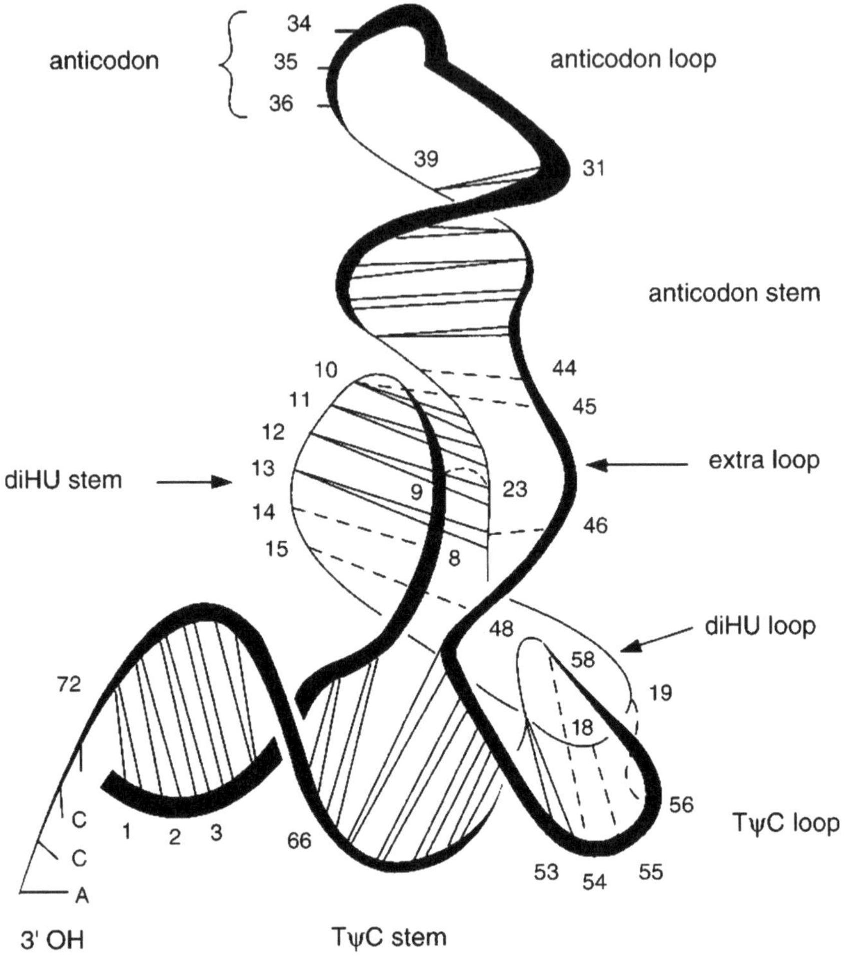

Fig. 20.4 Three dimensional structure of phenylalanyl-tRNA as determined by X-ray crystallography. Double solid lines represent hydrogen bonding between bases in double helical stems. Dotted lines represent hydrogen bonding between bases outside the helices (From *Concise Encyclopedia of Biochemistry*, 2nd edn. Walter de Gruyter, Berlin/New York (1988), with permission of the publisher)

the precursors on the 3′ side until it reaches the CCA extremity. Other specific enzymes catalyze a host of posttranscriptional modifications of the ribose residues and of the bases which are found in all the mature tRNAs.

These modifications include phosphorylation, mono- and dimethylation and isopentenylation. The anticodon stem-loop of transfer RNAs drives decoding by interacting directly with the mRNA through codon/anticodon pairing. Chemically complex nucleoside modifications found in the loop at positions 34 or 37 are known to be required for accurate decoding. Although over 100 distinct modifications have been structurally characterized in tRNAs, only a few are universally conserved, among them threonylcarbamoyl adenosine (t^6A), found at position 37 in the anticodon loop of a subset of tRNA. Structural studies predict an important role for t^6A in translational fidelity, and in vivo work supports this prediction. Although pioneering work in the 1970s identified the fundamental substrates for t^6A biosynthesis, the enzymes responsible for its biosynthesis have remained an enigma until the discovery that in bacteria four proteins are both necessary and sufficient for t^6A biosynthesis in vitro.

More than 70 different modifications have been described, and almost all tRNAs have multiple modifications. Several tRNA modifications affecting the anticodon region of tRNA are important for avoiding frameshift errors introduced by peptidyl-tRNA slippage. Specifically, there is extensive evidence that position 37 is important for the anticodon loop structure and to preposition the anticodon for interaction with the codon. Bulky hydrophobic modified nucleotides are often used at this position, presumably to facilitate stacking interactions. Almost all tRNAs that read codons starting with U have the modified nucleotide *N*6-isopentenyladenosine (or a further modified derivative) at position 37. The modification is formed by the transfer of an isopentenyl moiety from dimethylallyl pyrophosphate to A37 of tRNAs. This transfer is catalyzed by isopentenyl-tRNA transferases.

Queuine (Q) is a hypermodified base found in the first (or wobble) position of the anticodon of tRNAs specific for Asn, Asp, His, and Tyr, in most eukaryotes and prokaryotes. The nucleoside of queuine is queuosine. Queuine is not found in the tRNA of archaea; however, a related 7-deazaguanine derivative, the nucleoside of which is archaeosine, at the 15 position of most tRNAs. This modification is believed to help stabilize tRNA structure through ionic interactions, which is critical as many Archaea are thermophiles.

Due to the degeneracy of the genetic code, the third base is less discriminatory for the amino acid than the other two bases. This third position in the codon is referred to as the wobble position. If the tRNA contains an inosine at the wobble position in the anticodon, this tRNA may read codons having As, Us or Cs in the third position.

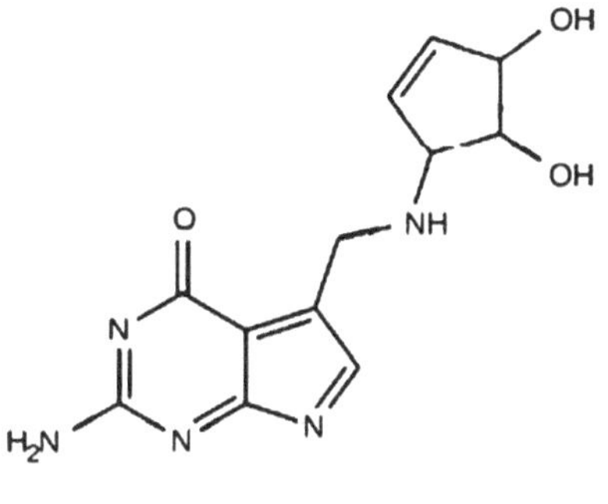

Queuine

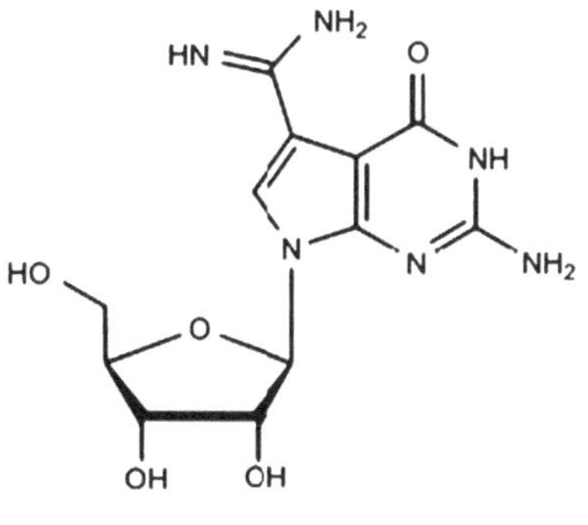

Archaeosine

Several methanogenic archaea lack cysteinyl–tRNA synthetase, the essential enzyme that provides cysteine for translation in most organisms. Partial purification of the corresponding activity from *Methanocaldococcus jannaschii* indicated that tRNACys becomes acylated with O-phosphoserine(Sep), but not with cysteine. Further analyses identified O-phosphoseryl-tRNA synthetase (SepRS) and Sep-tRNA:Cys-tRNA synthase (SepCysS). SepRS specifically forms Sep-tRNACys, which is then converted to Cys-tRNACys by SepCysS. Comparative genomic analyses suggest that this pathway, encoded in all organisms lacking CysRS, can also act as the sole route for cysteine biosynthesis. This was proven for *Methanococcus maripaludis*, where deletion of the SepRS-encoding gene resulted in cysteine auxotrophy. As the conversions of Sep-tRNA to Cys-tRNA or to selenocysteinyl-tRNA are chemically analogous, the catalytic activity of SepCysS provides a means by which both cysteine and selenocysteine may have originally been added to the genetic code.

The genes coding for the pre-tRNAs are found in two main regions of the *E. coli* chromosome, in which clustered genes are found, often organized in operons coding for up to seven different tRNAs. However, genes coding for tRNAs are also found in the coding space between the genes coding for the 16S and the 23S ribosomal RNAs and even 5S RNA. Their presence in the precursors of pre-ribosomal RNAs necessitates a coupling between the processing events of the two types of RNA. These tRNAs are obviously the most abundant ones since they are transcribed in a coordinate manner with the ribosomal RNAs.

Several iso-accepting tRNAs, the existence of which correlates with the degeneracy of the code, may have the same anticodon triplet; they are coded by different genes and are present at different concentrations; the codon usage to specify the same amino acid is not random: it is correlated with the occurrence of the corresponding codons in the mRNAs of the different organisms studied.

In contrast, there is no aminoacyl-tRNA synthetase specific to each tRNA corresponding to the same amino acid: the isoaccepting tRNAs are activated by the same synthetase.

The energy required for the synthesis of the aminoacyl bond originates from ATP. The enzyme first forms an amino acid adenylate:

$$\text{Amino acid} + \text{ATP} \underset{(\text{exchange reaction})}{\rightarrow} \text{Aminoacyladenylate} + \text{pyrophosphate}$$

The intermediate remains strongly bound to the enzyme and, in a second step, transfers its amino acid to the specific tRNA:

$$\text{Aminoacyladenylate} + \text{tRNA} \underset{(\text{charge reaction})}{\rightarrow} \text{Aminoacyl-tRNA} + \text{AMP}$$

The enzyme is thus able to recognize both a given amino acid and its cognate specific adaptor. It must have a binding site for each. Several of these transferases have been crystallized and their X-ray analysis has been carried out at high

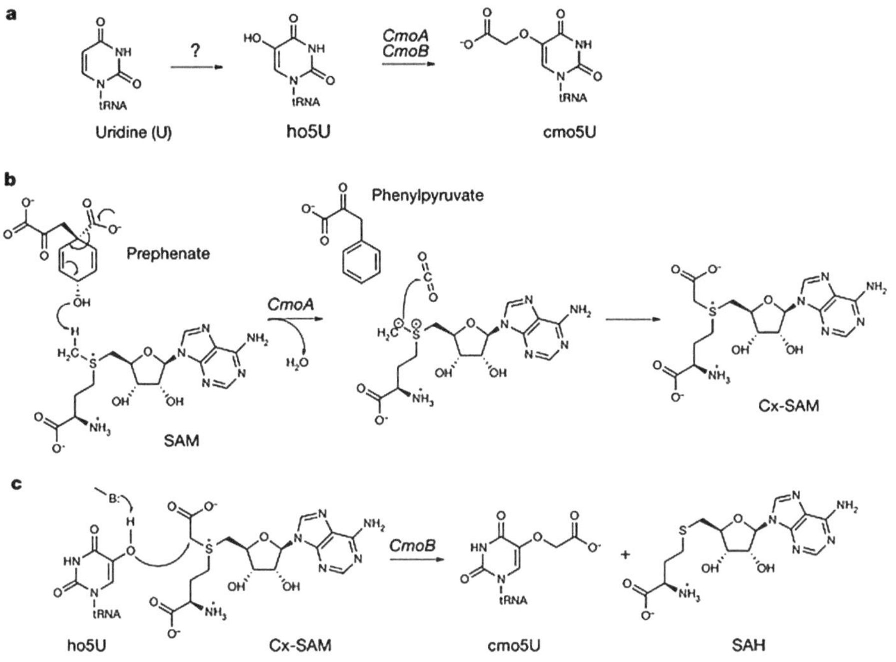

Fig. 20.5 (**a**) Previously identified biosynthetic pathway for cmo5U at wobble uridines. First, the wobble uridine is converted to 5-hyroxyuridine (ho5U) by an unknown mechanism, and this is followed by the action of CmoA and CmoB. (**b**) Mechanism for CmoA-catalysed Cx-SAM formation from SAM and prephenate. (**c**) Mechanism for CmoB-catalysed formation of cmo5U from ho5U and Cx-SAM

resolution. Co-crystals of the aspartyl-tRNA and its transferase have been obtained with yeast aspartyl-tRNA and their analysis should reveal how an aminoacyl-tRNA synthetase can achieve the discrimination between tRNAs corresponding to different amino acids and nonetheless bind the different isoaccepting tRNAs.

Modulation of tRNA Function

A new metabolite, carboxy-S-adenosylmethionine (Cx-SAM) has recently been discovered. Figure 20.5 describes its biosynthetic pathway.

This work shows an unprecedented role for prephenate as the carboxyl donor and the involvement of a unique ylide intermediate as the carboxyl acceptor in the CmoA-mediated conversion of SAM to Cx-SAM. A second member of the SAM-dependent methyltransferase superfamily, CmoB, recognizes Cx-SAM and acts as a carboxymethyltransferase to convert 5-hydroxyuridine into 5-oxyacetyl uridine at the wobble position of multiple tRNAs in Gram-negative bacteria, resulting in expanded codon-recognition properties. In most Gram-negative species, cmoA and cmoB are co-conserved and immediately adjacent to each other in the genome, supporting the demonstrated functional relationship.

Ribosomal protein synthesis is stereospecific, with only the L-stereoisomer being incorporated into the growing polypeptide chain. It was therefore surprising when Calendar and Berg demonstrated that tyrosyl-tRNA synthetases from *Escherichia coli* and *Bacillus subtilis* catalyze the activation and attachment of D-tyrosine to $tRNA^{Tyr}$ only one to two orders of magnitude less efficiently than for l-tyrosine. Tyrosyl-tRNA synthetase exhibits sigmoidal binding with respect to ATP in the D-tyrosine activation reaction.

Selected References

Colinearity of Genes and Proteins

Sarabhai AS, Stretton OW, Brenner S, Bolle A (1964) Nature 201:13–17
Yanofsky C, Carlton BC, Guest JR, Helinski DR, Henning U (1964) Proc Natl Acad Sci U S A 51:266–272

The Genetic Code

The Genetic Code (1966) Cold Spring Harb Symp Quant Biol 31:1–762. Cold Spring Harbor Laboratory Press (Woodbury). The whole book is devoted to the unraveling of the genetic code
Goodman DB, Church GM, Kosuri S (2013) Science 342:475–479
Malyshev DA, Dhami K, Quash HT, Lavergne T, Ordoukhanian P, Torkamani A, Romesberg FE (2012) Proc Natl Acad Sci USA 109:12005–12010

Selenocysteine and Pyrrolysine

Flohé L (2009) Biochim Biophys Acta 1790(11):1389–403
Gaston MA, Zhang L, Green-Church KB, Krzycki JA (2011) Nature 471:647–650
Kavran JM, Gundllapalli S, O'Donoghue P, Englert M, Söll D, Steitz TA (2007) Proc Natl Acad Sci USA 104:11268–11273
Prat L, Heinemann IU, Aerni HA, Rinehart J, O'Donoghue P, Söll D (2012) Proc Natl Acad Sci USA 109:21070–21075
Stadtman TC (1996) Ann Rev Biochem 65:83–100

Transfer RNAs

Cold Spring Harbor monograph series. Volume 09A: structure, properties and recognition and Volume 09B: biological aspects (1980) Cold Spring Harbor Laboratory Press (Woodbury)

Deutsch C, El Yacoubi B, de Crécy-Lagard V, Iwata-Reuyl D (2012) J Biol Chem 287:13666–13673
Kim S-H (1978) Adv Enzymology 46:279–315
Phillips G, Chikwana VM, Maxwell A, El-Yacoubi B, Swairjo MA, Iwata-Reuyl D, de Crécy-Lagard V (2010) J Biol Chem 285:12706–12713
Seif E, Hallberg BM (2009) J Biol Chem 284:6600–6604

Aminoacyl-tRNA Synthetases

Hoagland MB, Keller EB, Zamecnik PC (1956) J Biol Chem 218:345–358
Schimmel PR (1987) Ann Rev Biochem 56:125–158
Schimmel P (2008) J Biol Chem 283:28777–28782
Schimmel PR, Söll D (1979) Ann Rev Biochem 48:601–648
Sheoran A, Sharma G, First EA (2008) J Biol Chem 283:12960–12970

RNA-Dependent Cysteine Biosynthesis in Archaea

Sauerwald A, Zhu W, Major TA, Roy H, Palioura S, Jahn D, Whitman WB, Yates JR, Ibba M, Söll D (2005) Science 307:1969–1972

Modulation of tRNA Function

Kim J, Xiao H, Bonanno JB, Kalyanaraman C, Brown S, Tang X, Al-Obaldi NF, Patskovsky Y, Babbitt PC, Jacobson MP, Lee Y-S, Almo SC (2013) Nature 498:123–126

Chapter 21
Attenuation

Ames and his colleagues have shown in 1972 that the transcription of the histidine operon in *S. typhimurium* was not regulated by a repressor protein, but that the level of transcription depended on the intracellular level of tRNAhis charged by histidine. These observations have been confirmed and extended by Kasai in 1974. In the case of the *trp* operon, observations made from 1973 to 1976 indicated that repression could not be the only way by which *E. coli* regulates transcription. Mutants of tryptophanyl-t-RNA synthetase presented anomalies in the regulation of the operon: addition of tryptophan to the cultures of these mutants did not repress maximally the expression of the operon; not only transcription was not initiated, but its progress from the initial segment was also inhibited. Finally, *trpR* mutants devoid of a functional repressor did still respond to a tryptophan starvation by increasing the rate of synthesis of the *trp* messenger RNA.

When the transcription of the initial segment of the operon was analyzed in detail *in vitro* and *in vivo*, a ρ-independent transcription termination site was found upstream the structural genes of the operon. Tryptophan starvation reduces the termination at this site. This phenomenon has been called transcription attenuation by Yanofsky and describes the early termination of transcription. In order to understand the mechanism of this regulation process, it is necessary to examine the anatomy of the regulatory and structural gene regions of the *trp* operon (Fig. 21.1).

The results of the investigations carried out by Yanofsky on the tryptophan operon and by others on other amino acid biosynthetic operons can be summarized as follows:

Transcription is regulated by attenuation, the regulatory signal is charged tRNA, and the sensory event is the capacity to translate a short peptide coding region, rich in codons for the relevant amino acid and located in the initial segment of the transcript of the operon.

An RNA polymerase molecule which has escaped the regulation by the repressor initiates the transcription of the operon. Soon after the transcribed segment containing the ribosome binding site has been synthesized, a ribosome attaches to it.

DOI 10.1007/978-94-017-8908-0_21,

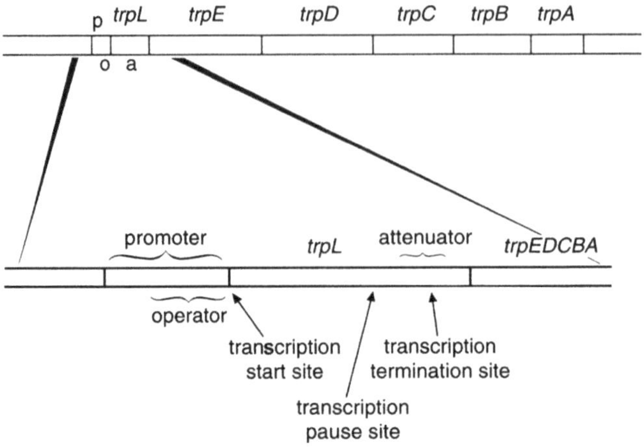

Fig. 21.1 The regulatory and structural gene regions of the trp operon of *E. coli*. Transcription initiation occurs at the promoter-operator region. Transcription termination is regulated at an attenuator, in a transcribed 162 base-pair region, *trpL*, which encompasses the segment coding for the putative leader peptide. All RNA polymerase molecules transcribing the operon pause at the transcription pause site before proceeding further (From C. Yanofsky, with permission of *Nature*)

The polymerase which has started transcribing at the promoter can synthesize a small transcript ("leader transcript") and either stop before reaching the first structural gene, or synthesize a much longer transcript corresponding to the structural genes of the operon, depending on the physiological conditions.

The short transcript exhibits two very special characteristics:

(a) it comprises a small open reading frame coding for a peptide ("leader peptide") 15–32 amino acid residues long, according to the cases which have been analyzed, in which is found a high proportion of the codon(s) corresponding to the amino acid(s) regulating the expression of the operon; these codons are called regulator codons. Depending on the levels of the appropriate tRNAs, translation of these regulator codons will or will not proceed, and this event determines whether transcription will terminate in the leader region. The actual synthesis of these peptides has been demonstrated.
(b) certain regions of the transcript can adopt secondary "stem and loop" structures, mutually exclusive, one of these structures being recognized by RNA polymerase as a signal of transcription termination.

The secondary structure adopted by the RNA is determined by the position of the ribosome while it is translating the leader peptide. The progression of the ribosome, and thence its position on messenger RNA depends on the intracellular concentration of the amino acid regulating the operon through the charged tRNAs corresponding to the regulator codons. On the scheme Fig. 21.2, the regions noted 1, 2, 3

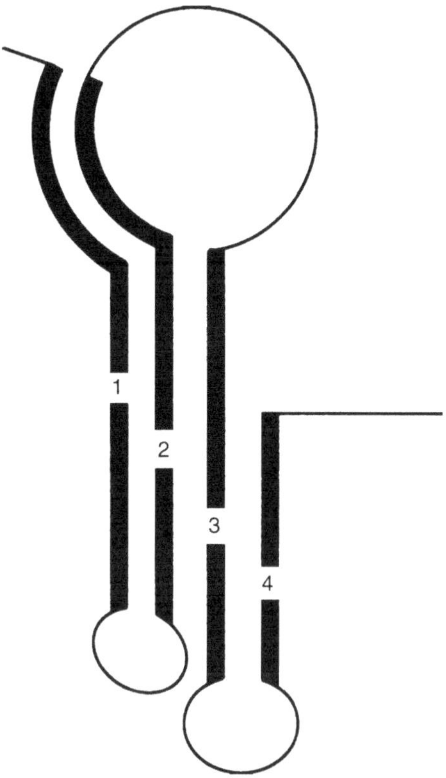

Fig. 21.2 Scheme representing the four regions of possible pairing of the messenger RNA upstram the translation start of the structural genes. Same source as Fig. 21.1

and 4 correspond to the regions of possible pairing of the messenger: the 3–4 pairing constitutes the terminator (signal of transcription termination); the 1–2 pairing is called the protector (structure which allows the formation of the 3–4 structure) and the 2–3 pairing, which prevents the formation of the terminator is called the preemptor.

When the amino acid regulating the expression of the operon becomes limiting, the progression of the ribosome on the messenger ceases at the level of the regulator codons (since the corresponding tRNAs are not charged); the ribosome covering region 1 inhibits the formation of the protector (1–2 pairing). When region 3 is transcribed, it can therefore pair with region 2 and the premptor is formed. As a consequence, region 4 cannot pair with region 3 and in the absence of the terminator structure, polymerase transcribes the structural genes (Fig. 21.3). It is seen that not only tryptophan starvation, but also arginine starvation, which causes also a stalling of the ribosome because of the strategic position of the arginine codon in the leader transcript results in the transcription of the tryptophan biosynthetic structural genes.

On the other hand, if the regulating amino acid is in excess or if the cell is starved for an amino acid whose codons are not topologically important in the leader

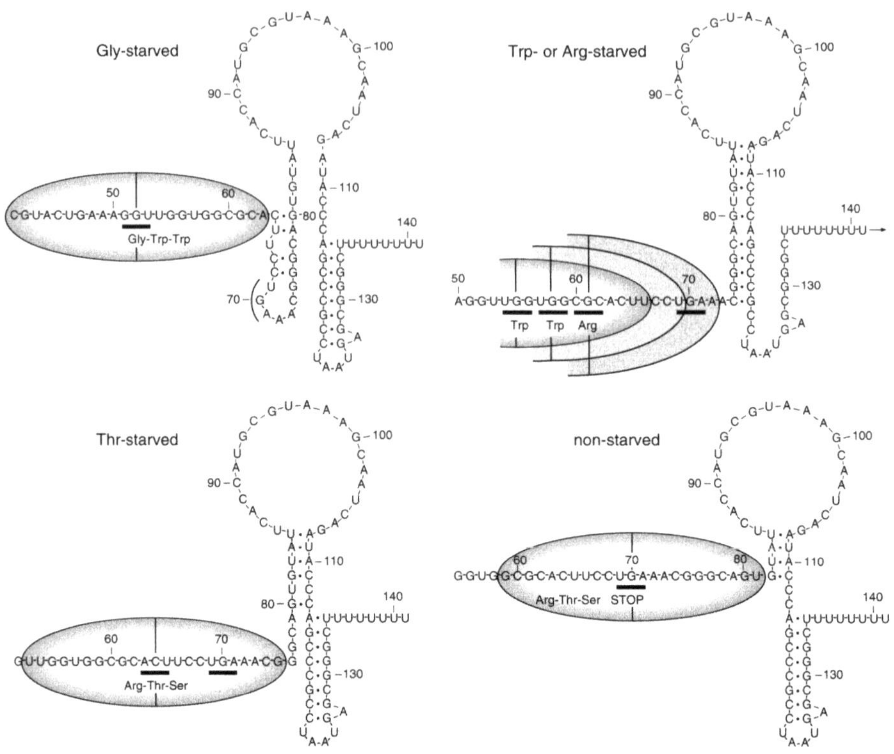

Fig. 21.3 Representation of the ribosome on the trp leader transcript in cells starved of Gly, Thr, Arg or Trp or not starved cells. A ribosome stalled over one of the codons of the transcript blocks about ten nucleotides in the downstream segment of the transcript, thereby determining which alternative secondary structure will form during transcription. The structure 2–3 will form only when cells are starved of Trp and Arg: the structure 3–4 (the ρ-independent terminator) will not form and transcription will proceed into the structural genes. Same source as Fig. 21.1

transcript, the ribosome progresses normally on the messenger and translates region 2 as soon as it is transcribed. When region 3 is transcribed, it cannot pair with region 2 which is covered by the ribosome and it is available for pairing with region 4 to form the terminator structure. The polymerase recognizes this termination signal and the structural genes are not transcribed (Fig. 21.3).

If the stability of the 3–4 structure is decreased by mutations that affect the base pairing, transcription termination is relieved, in keeping with the model (Fig. 21.4).

High field proton nuclear magnetic resonance studies of a synthetic 21-mer RNA fragment; corresponding to residues +114 to +134 within the *trp* leader mRNA transcript, have been carried out. Seven well resolved imino proton resonances corresponding to six C-G and one A-U hydrogen bonded base pairs, together with their characteristic NOE patterns can be identified in the NMR spectrum. This experimental result provides direct evidence for the postulated stem-loop secondary structure 3–4.

Such a mechanism implies a very precise coupling between transcription and translation. The sequences of the leader peptides of the *trp*, *pheA*, *his*, *leu*, *thr*,

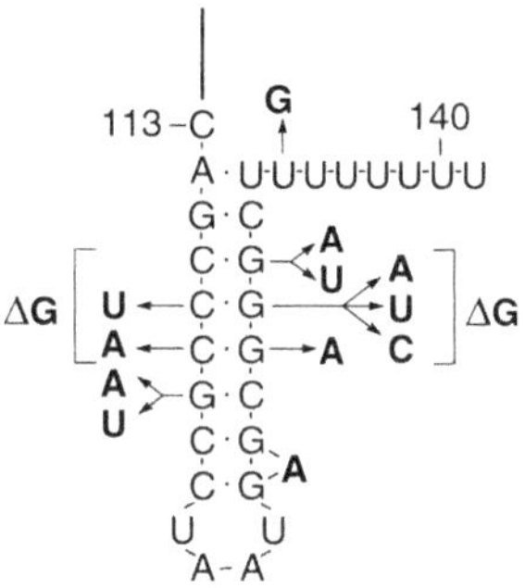

Fig. 21.4 Base pair changes in the trp leader region that relieve transcription termination at the attenuator. The changes in the transcript are shown: all are in the 3–4 structure and reduce its stability except the single U, G change at position 135. Same source as Fig. 21.1

ilvGMEDA operons have been determined and will be found in each corresponding chapter of our study of the amino acid biosynthetic pathways.

Repression rather than attenuation exerts the largest regulatory effect on the *in vivo* transcription of the *trp* operon. Repression can reduce transcription up to 70-fold, whereas attenuation may reduce only by a maximum of 10-fold. Since tryptophan starvation augments expression by using both mechanisms, transcription of the *trp* operon can vary by a factor 700.

The *Bacillus subtilis* tryptophan operon is also regulated by transcription termination. But the mechanism of selection between the two alternative RNA secondary structures (the transcription terminator and the antiterminator allowing transcription of the operon) is different. Whereas *E. coli* makes use of a ribosome to translate a leader peptide, *B. subtilis* uses the product of the *mtrB* gene, called TRAP (for trp attenuation protein) which, when bound to tryptophan, binds specifically the *trp* leader RNA and causes transcription of the operon to terminate in the leader region. In the absence of tryptophan, TRAP appears to be composed of 11 or 12 identical 8 kDa subunits.

Whereas tryptophan biosynthesis is regulated by both repression and attenuation, this is not general. No evidence for repression has been found for the threonine, histidine and leucine operons which appear to be regulated only by attenuation. In contrast, the methionine and arginine regulons respond to repression only: no structures corresponding to a leader transcript and to alternative pairing of the messenger RNA, nor evidence for an attenuation mechanism have been found.

If we compare the specific elements of repression to those of attenuation from the point of view of their information content, attenuation is more economical since its only specific element is a sequence of 50–100 base pairs of the leader sequence of each concerned operon.

Regulation of the trp Operon in Bacillus subtilis

In this organism, six of the seven trp genes exist as a trp suboperon within an aromatic(aro) supraoperon. This supraoperon has three genes preceding and the three genes following the trp suboperon. These downstream and upstream genes encode proteins involved in chorismate synthesis or in the synthesis of phenylalanine, tyrosine and histidine. Two promoters initiate transcription of the trp suboperon. Transcription from either promoter is regulated by transcription attenuation. Regulation is achieved by the action of the tryptophan-activated RNA-binding attenuation protein TRAP. The transcript segment immediately preceding the first gene of the suboperon can fold and form either of two hairpin structures, an antiterminator or a terminator. Choosing between these structures is the role to the TRAP protein. This protein, when activated by tryptophan, binds to the RNA segment that can form the RNA antiterminator structure and prevents its formation. The alternative termination structure then forms, which terminates transcription. TRAP is a circular doughnut-shaped protein consisting of 11 identical subunits. A tryptophan binding site is formed between each pair of TRAP subunits.

General Remarks on Regulatory Mechanisms

A simple analogy of the regulatory mechanisms controlling respectively enzyme regulation and enzyme synthesis can be made with the brick industry.

Allosteric inhibition or activation, as well as covalent enzyme modification, may be compared to a factory making less bricks or more bricks according to demand: if there are already too many bricks available, the factory will reduce its output or even close temporarily, but will be ready to function again as soon as there is a dearth of bricks.

Repression and attenuation correspond to a situation where less, or no more, brick factories will be built. As a result, when the bricks will be needed again, new factories will have to be built until the production corresponds to the demand.

It is impossible to conceive a nuclear plant or an oil refinery without controls, captors, relays and overflows. In the same vein, it is not surprising that an entangled network as complex as that of biosynthetic metabolism is endowed with microscopic feedback mechanisms: the contrary would be astonishing. The parallelism between a factory and metabolism is not a simple metaphor, and the underlying logic is similar. In most cases examined in this book, as in a thermostat or in Ktesibios clepsydra installed in Egypt three centuries before our era, or in the wine fountain of Heron of Alexandria: the logic is that of retroaction.

In the simplest case, that of feedback inhibition or of repression; the captor and the actuator are the same molecule, respectively the allosteric enzyme or the repressor, the final metabolite being an interdiction signal. We are in presence of a true analogic device. In the case of the covalent modification of *E. coli* glutamine synthetase, the metabolic signal is not perceived by the actuator, the modified enzyme, but by the modifying enzyme. In the case of attenuation, the signal, the aminoacyl-tRNA, is transmitted to the RNA polymerase actuator, through the relay of a site of transcription termination.

Another remarkable property of metabolic regulation concerns its hierarchy, its demultiplication. The double control of enzyme activity and of enzyme synthesis allows the establishment of a metabolic regime and a fine adjustment of the metabolic fluxes.

Finally, the economy of the setting up of the regulatory mechanisms during evolution should be emphasized. Despite their common logics, the great structural and functional diversity of allosteric enzymes suggests that in each case, the control site was created independently from the catalytic site. Along the same line, any protein able to bind DNA is a candidate for becoming a repressor, through the elaboration of a site having affinity for a metabolite. Experiments of reconstruction of regulatory proteins and earlier experiments of selection of such proteins leave little doubt that the miracles of microscopic clockwork we detect in cells have been the result of molecular tinkering, that is of expedient.

Selected References

Attenuation

Gollnick P (1994) Mol Microbiol 11:991–997
Kasai T (1974) Nature 249:523–527
Yanofsky C (1981) Nature 289:751–758
Yanofsky C (1988) J Biol Chem 263:609–612

TRAP Protein

Yanofsky C (2007) RNA 13:1141–1154

Chapter 22
Regulation by RNAs and Riboswitches

Gene control at the level of transcription initiation constitutes only a portion of the regulatory mechanisms used by bacteria. Many different posttranscriptional control mechanisms have been identified, regulating transcription initiation, elongation, and termination. A number of discoveries have revealed that regulatory RNA structures are often used for posttranscriptional control of essential genes.

The study of regulatory RNAs in the control of prokaryotic genomes has become a very active and rapidly growing field. New small (sRNA) and large noncoding RNA molecules continue to be discovered at a staggering rate in bacterial model organisms.

The sRNAs act by base pairing with target messenger RNAs (mRNAs) with which they share limited or extended complementarity, or by modulating protein activity, in some cases by mimicking other nucleic acids. Mechanistic insights into how sRNAs bind mRNAs and proteins, how they compete with each other, and how they interface with ribonucleases are active areas of discovery.

Examples of RNA regulatory elements responsive to proteins can be split into two categories: (1) RNAs that directly sense a particular protein through binding interactions and (2) proteins whose RNA-binding activity is modifiable by intracellular signals. An example of the former class is the feedback repression mechanisms observed for numerous ribosomal proteins. Ribosomal proteins (r-proteins) are usually organized into operons wherein most of the genes encode for ribosomal components. Their expression must be regulated to ensure that excess r-proteins, unassociated with ribosomes, are not wastefully synthesized. To achieve this control, a particular r-protein encoded in the operon interacts with an RNA structure in the 5′ untranslated region and negatively regulates associated genes.

Riboswitches, mainly studied by the laboratories of Ronald Breaker (Yale) and Evgeny Nudler (New York University) are metabolite binding domains within certain messenger RNAs that serve as precision sensors for their corresponding targets. Allosteric rearrangement of mRNA structure is mediated by ligand binding, and this results in modulation of gene expression. Thus, mRNAs sense metabolites and control gene expression without the need for protein factors.

G.N. Cohen, *Microbial Biochemistry*,
DOI 10.1007/978-94-017-8908-0_22, © Springer Science+Business Media B.V. 2014

These small RNA transcripts (a few hundred nucleotides long) are part of an mRNA molecule that can directly bind a small metabolite target molecule, and whose binding of the target affects the activity of the gene. Thus, an mRNA that contains a riboswitch is directly involved in regulating its own activity, depending on the presence or absence of its target molecule.

Although the metabolic pathways in which some riboswitches are involved have been studied for decades, the existence of riboswitches has only been relatively recently discovered, with the first experimental validations of riboswitches being published in 2002 independently by two groups led by Breaker and Nudler This oversight is probably due to an earlier assumption that genes are regulated only by proteins, not by the mRNA transcript itself.

Since 2002, several riboswitches regulating over 3 % of all bacterial genes have been described. Diverse groups of genes are regulated by riboswitches. They are involved in sulfur metabolism and in the biosynthesis and transport of vitamins, coenzymes, amino acids, and nucleotides. In all cases, one of the end products in a metabolic pathway serves as a small molecule ligand for a cognate riboswitch that triggers repression of corresponding genes and feedback regulation, or, in some rare instances, gene activation.

Most known riboswitches occur in bacteria, but functional riboswitches of one type (the TPP riboswitch) have been discovered in plants and certain fungi. TPP riboswitches have also been predicted in Archaea, but have not been experimentally tested yet.

The following riboswitches have been characterized:

(a) TPP riboswitch binds thiamin pyrophosphate and regulates thiamin biosynthesis and transport, as well as transport of similar metabolites. Depending on the configuration of the expression platform, the thiamine pyrophosphate (TPP) sensing riboswitch can terminate transcription of downstream genes in Gram-positive bacteria or suppress translation initiation in Gram-negative bacteria.

(b) FMN riboswitch binds flavin mononucleotide and regulates riboflavin biosynthesis and transport.

(c) Cobalamin riboswitch binds adenosylcobalamin and regulates cobalamin biosynthesis and transport. In *Listeria monocytogenes*, a vitamin B12-binding (B12) riboswitch was identified, not upstream of a gene but downstream, and antisense to the adjacent gene, *pocR*, suggesting it might regulate *pocR* in a nonclassical manner. In *Salmonella enterica*, PocR is a transcription factor that is activated by 1,2-propanediol, and subsequently activates expression of the *pdu* genes, which mediate propanediol catabolism and are implicated in pathogenesis. As enzymes involved in propanediol catabolism require B12 as a cofactor, it is hypothesized that the *Listeria* B12 riboswitch might be involved in *pocR* regulation. Actually, the B12 riboswitch is transcribed as part of a non-coding antisense RNA, herein named AspocR. In the presence of B12, the riboswitch induces transcriptional termination, causing *aspocR* to be transcribed as a short transcript. In contrast, in the absence of B12, *aspocR* is transcribed as a long antisense RNA, which inhibits *pocR* expression. Regulation by AspocR ensures that *pocR*, and consequently the *pdu* genes, are maximally expressed only when both propanediol and B12 are present. Strikingly, AspocR can inhibit *pocR* expression *in trans*, suggesting it acts through a direct interaction with *pocR* mRNA. Together, this study demonstrates how *pocR* and the *pdu* genes can be regulated by B12 in bacteria and extends the classical definition of riboswitches from elements governing solely the expression of mRNAs to a wider role in controlling transcription of noncoding RNAs.

(d) SAM riboswitches bind S-adenosyl methionine (SAM) to regulate methionine and SAM biosynthesis and transport. Three distinct SAM riboswitches are known: SAM-I (widespread in bacteria), SAM-II (found only in alpha-, beta- and a few gamma-proteobacteria) and SMK (found only in the order Lactobacillales). These three varieties of riboswitch have no obvious similarities in terms of sequence or structure.

(e) Two types of PreQ1 riboswitches bind pre-queuosine1, and regulate genes involved in the synthesis or transport of this precursor to queuosine. For reasons that are not presently understood, the classical pseudoknot fold of this system harbors an extra stem–loop structure within its 3′-terminal region immediately upstream of the Shine–Dalgarno sequence that contributes to formation of the ligand-bound state. Imaging ligand-dependent $preQ_1$ riboswitch folding from multiple structural perspectives, reveals that the extra stem–loop strongly influences pseudoknot dynamics in a manner that decreases its propensity to spontaneously fold and increases its responsiveness to ligand binding. It is concluded that the extra stem–loop sensitizes this RNA to broaden the dynamic range of the ON/OFF regulatory switch.

(f) SAH riboswitches bind S-adenosylhomocysteine and regulate genes involved in recycling this metabolite that is produced when S-adenosylmethionine is used in methylation reactions.

(g) Purine riboswitches bind purines and regulate purine metabolism and transport. Different forms of the purine riboswitch bind guanine or adenine. The specificity for either guanine or adenine depends completely upon Watson-Crick interactions with a single pyrimidine in the riboswitch at position 74. In the guanine riboswitch this residue is always cytosine C74, in the adenine riboswitch it is always uracil U74. Homologous types of purine riboswitches bind deoxyguanosine, but have more significant differences than a single nucleotide mutation.

(h) Lysine riboswitch binds lysine and regulate lysine biosynthesis, catabolism and transport. Upon lysine binding, the lysC riboswitch adopts a conformation that not only inhibits translation initiation but also exposes RNase E cleavage sites located in the riboswitch expression platform. However, in the absence of lysine, the riboswitch folds into an alternative conformation that simultaneously allows translation initiation and sequesters RNase E cleavage sites. Both regulatory activities can be individually inhibited, indicating that translation initiation and mRNA decay can be modulated independently using the same conformational switch. Because RNase E cleavage sites are located in the riboswitch sequence, this riboswitch provides a unique means for the riboswitch to modulate RNase E cleavage activity directly as a function of lysine. This dual inhibition is in contrast to other riboswitches, such as the thiamin pyrophosphate-sensing thiM riboswitch, which triggers mRNA decay only as a consequence of translation inhibition.

(i) glmS riboswitch, which is a ribozyme that cleaves itself when there is a sufficient intracellular concentration of glucosamine-6-phosphate.
(j) Glycine riboswitch binds glycine to regulate glycine metabolism genes, including the use of glycine as an energy source. As of 2008, this riboswitch is the only known natural RNA that exhibits cooperative binding, which is accomplished by two adjacent domains in the same mRNA.
(k) cyclic di-GMP riboswitches bind the signaling molecule cyclic di-GMP in order to regulate a variety of genes controlled by this second messenger.

In addition, a Moco riboswitch is presumed to bind molybdenum cofactor, to regulate genes involved in biosynthesis and transport of this coenzyme, as well as enzymes that use it or its derivatives as a cofactor. Now that riboswitches are a known mechanism of genetic control, it is reasonable to speculate that more riboswitches will be found.

Mechanisms of Riboswitches

Riboswitches reside in the leader sequences of bacterial operons and consist of an evolutionarily conserved metabolite-sensing domain (aptamer) coupled to a variable "expression platform."

The aptamer directly binds the small molecule. The expression platform undergoes structural changes in response to the changes in the aptamer, it adopts one of two alternative conformations in response to ligand-induced changes in the sensor domain, as shown in Fig. 22.1.

Expression platforms typically turn off gene expression in response to the small molecule, but some turn it on. Expression platforms include:

- The formation of rho-independent transcription termination hairpins
- Folding in such a way as to sequester the ribosome-binding site, thereby blocking translation
- Self-cleavage (i.e. the riboswitch contains a ribozyme that cleaves itself in the presence of sufficient concentrations of its metabolite)
- Folding in such a way as to affect the splicing of the pre-mRNA.

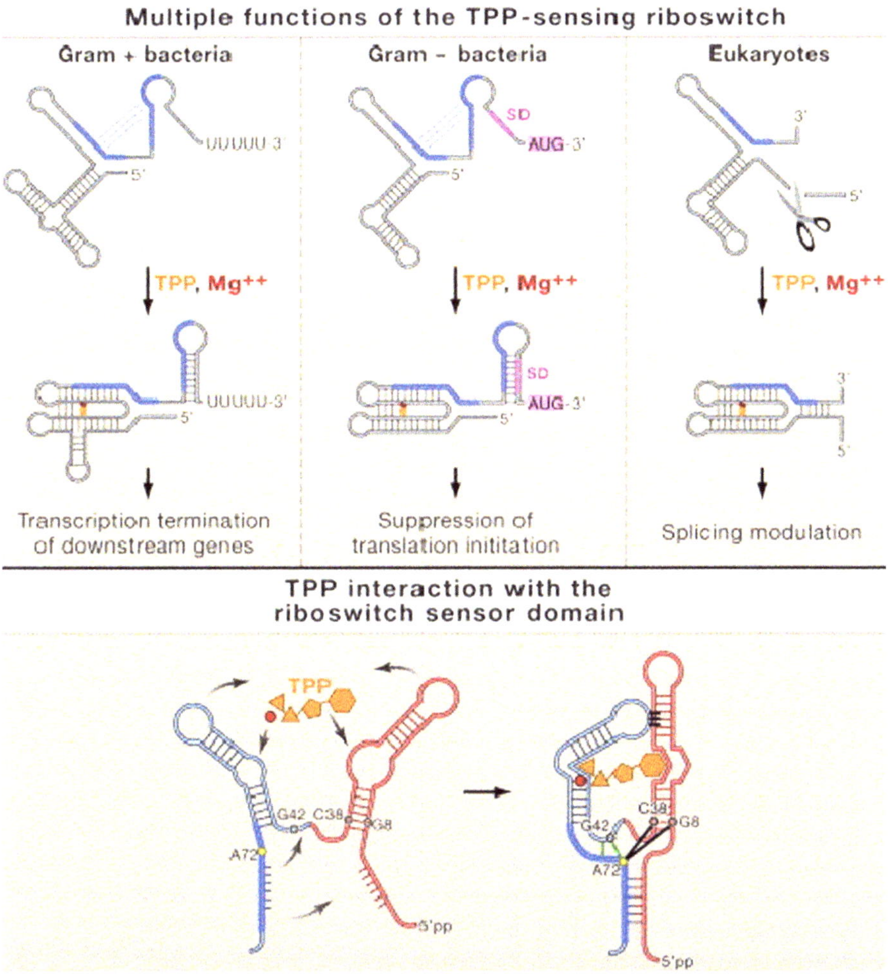

Fig. 22.1 The structural basis of riboswitch function. The TPP-sensing riboswitch, for which the sensor domain structure has been solved, is used here as a representative example. (*Top*) Depending on the expression platform, the TPP riboswitch acts as a transcription terminator (*left*), as a suppressor of translation initiation (*middle*; the ribosome binding site is indicated in *violet*), or as a modulator of splicing (*right*). (*Bottom*) The contacts between TPP and the riboswitch sensor domain are summarized. The TPP binding pocket is formed by two parallel helices: a pyrophosphate binding helix (*blue*) and a pyrimidine binding helix (*pink*). TPP is shown in *orange*, Mg2+ is in *red*, nonstandard base pairs are in *green*, and hydrogen bonds are shown by *black lines*. TPP binding stabilizes the three-way junction with A72, thus helping the sensor domain to sequester the segment (*blue*, highlighted) that would otherwise participate in the alternative structure of the expression platform (From Nudler E (2006) *Cell* 126:19–22, with permission from Elsevier and Dr. Evgeny Nudler)

Selected References

Caron MP, Bastet L, Lussier A, Simoneau-Roy M, Massé E, Lafontaine DA (2012) Proc Natl Acad Sci USA 109:3444–3453
Eiler D, Lin J, Simonetti A, Klaholz BP, Steitz TA (2013) Proc Natl Acad Sci USA 110: 15662–15667
Mandal M, Breaker RR (2004) Nat Rev Cell Biol 5:451–463
Mellin JR, Tiensuu T, Bécavin C, Gouin E, Johansson J, Cossart P (2013) Proc Natl Acad Sci USA 110:13132–13137
Mironov AS, Gusarov I, Rafikov R, Lopez LE, Shatalin K, Kreneva RA, Perumov DA, Nudler E (2002) Cell 111:747–756
Nahvi A, Sudarsan N, Ebert MS, Zou X, Brown KL, Breaker RR (2002) Chem Biol 9:1043
Serganov A, Polonskaia A, Phan AT, Breaker RR, Patel DJ (2006) Nature 441:1167–1171
Soulière MF, Altman RB, Schwarz V, Haller A, Blanchard SC, Micura R (2013) Proc Natl Acad Sci USA 110:3256–3264
Winkler W, Nahvi A, Breaker RR (2002) Nature 419:952–956

Chapter 23
The Biological Fixation of Nitrogen

Nitrogen fixation and photosynthesis are the two processes that ultimately feed the world.

Gaseous nitrogen (N_2) with a valence of 0 must be reduced to ammonia prior to incorporation into nitrogenous components of the cell. This process, called *nitrogen fixation*, is limited to certain procaryotic organisms, called diazotrophs able to synthesize the enzyme nitrogenase.

The contrast between the expensive chemical Haber-Bosch process for ammonia production from dinitrogen $N \equiv N$, which requires an elevated temperature and pressure, and the biological fixation of either the free-living nitrogen-fixing bacteria or the symbiotic *Rhizobia*, which occurs at ambient temperature and pressure, is striking. Each route accounts for approximately hundred million tons of N_2 converted per year to ammonia.

Nitrogenase, the enzyme responsible for the nitrogen fixation uses approximately 34 iron and 2 molybdenum atoms to catalyze the reaction:

$$N_2 + 16MgATP + 8e^- + 8H^+ \rightarrow$$
$$2NH_3 + 16MgADP + 16P_i + H_2$$

Biological reduction of dinitrogen (N_2) to ammonia (NH_3) occurs in a select group of bacteria that contain the metalloenzyme nitrogenase. To catalyze this difficult reaction, nitrogenase requires electrons, protons, and ATP. From the earliest studies of nitrogenase, it was realized that ATP hydrolysis is coupled to delivery of electrons and reduction of N_2, yet the order of ATP hydrolysis and electron transfer, which determines the nature of the coupling, was never established. Recently however, the order of all the key events during one catalytic cycle of electron delivery in nitrogenase was determined, showing that ATP hydrolysis follows electron transfer. These findings guide future studies aimed at understanding what roles ATP binding and hydrolysis play in the nitrogenase mechanism.

The substrate specificity of nitrogenase is relatively low; a number of other compounds, including N_3^-, N_2O, HCN, CH_3NC, CH_2CHCN and C_2H_2 are also reduced

G. N. Cohen, *Microbial Biochemistry*,
DOI 10.1007/978-94-017-8908-0_23,

by it. Some of these reactions involve the transfer of fewer electrons than the eight required to reduce gaseous nitrogen. The study of biological nitrogen fixation in whole cells and in extracts has been greatly aided by the introduction of an assay method using the substrate acetylene, which is reduced to ethylene

$$CH \equiv CH \xrightarrow[2H^+]{2e^-} CH_2 = CH_2$$

The product can be easily quantitated by gas chromatography, and the reaction is a highly specific one since no enzyme system other than nitrogenase can reduce acetylene to ethylene.

Nitrogenase is composed of two proteins, the Fe-protein (also called dinitrogenase reductase) and the MoFe-protein (also called dinitrogenase).

A feature common to both proteins is that they contain metalloclusters composed of Fe-S cores. The most familiar iron-sulfur clusters are the dinuclear Fe_2S_2 and the cubane Fe_4S_4 types found in a class of simple electron carrier proteins called ferredoxins (see Fig. 8.2). Advances in understanding the structure, organization and reactivity of certain biologically relevant iron-sulfur clusters have involved the characterization of clusters chemically extruded from their polypeptide matrices, preparation of synthetic iron-sulfur cluster analogs, and *in vitro* formation of protein-bound iron-sulfur clusters accomplished by incubation of the appropriate apoprotein in the presence of inorganic sulfide and iron.

These metalloclusters are necessary for nitrogenase activity and they appear to participate in various aspects of electron transfer or substrate reduction. The primary translation products of the genes that encode the nitrogenase structural components in *Azotobacter vinelandii* are not active. Immature forms of the nitrogenase component proteins are activated through the formation and insertion of their complementary metalloclusters in processes that require the activities of a consortium of auxiliary proteins. One such processing step is the mobilization of the inorganic sulfide required for formation of the Fe-S cores. A pyridoxal phosphate enzyme catalyzes the desulfurization of cysteine to yield alanine and elemental sulfur. The mechanism of the reaction has been analyzed and involves the intermediate formation of a cysteinyl persulfide formed through the nucleophilic attack of the cysteine thiolate of the enzyme on the cysteine-PLP ketimine adduct. The persulfide is thought to be the sulfur donor in Fe-S core formation. It appears that pyridoxal phosphate chemistry plays a general role in iron-sulfur cluster assembly.

The Fe-protein is a homodimer (total molecular weight: 60,000) containing a single Fe_4S_4 unit per dimer with the atoms at the corner of a cube and can be understood on the basis of the structure of ferredoxins, a class of low-potential Fe-S electron transfer proteins. The Fe-protein binds two MgATP per molecule which are hydrolyzed with the concomitant transfer of an electron to the MoFe-protein. The Fe-protein possesses all or at least part of the apparatus necessary to transduce the hydrolysis generated by ATP hydrolysis into a very negative electron potential which initiates the saturation of the triple bond in N_2. In the absence of the MoFe-protein and reductant, the Fe-protein exhibits no ATPase activity.

The amino acid sequence of more than 20 different Fe-proteins has been determined and indicate that this protein is highly conserved (the most divergent sequences are 69 % identical). The crystallographic structure of *Azotobacter vinelandii* Fe-protein has been resolved at 2.9 Å resolution. Many of the features uncovered are relevant to other systems in addition to nitrogenase: for example, the enzyme protobacteriochlorophyll reductase, which catalyzes the first committed step in photosynthetic growth, shows extensive similarity with the Fe-protein, especially for residues involved as cluster ligands, nucleotide binding and interface regions, suggesting that it performs a similar mechanistic function.

The *A. vinelandii* MoFe-protein, which possesses a dinitrogen binding site, consists of four subunits in an a_2b_2 arrangement (total molecular mass: 240 kD) the a and b-subunits being composed of 491 and 522 amino acids, respectively. Each molecule yields two equivalents of molybdenum as the MoFe cofactor (FeMoco),[1] which is a complex of approximate composition $MoFe_7S_8$ (homocitrate) and has been detected by its ability to activate a Mo-deficient protein found in the extracts of mutant strains of bacteria. Several physical measurements have shown beyond doubt that the cofactor in solution is essentially identical to the Mo-containing center in the native protein, both being sulfur-bridged polynuclear clusters: the Mo environment in the protein includes coordination to oxygen or nitrogen containing ligands as well as sulfur bridged to iron. The cofactor, present in two discrete centers, accounts for 14 of the 30 iron atoms in a molecule of the MoFe protein. The remaining 16 atoms are present in four Fe_4S_4 clusters which transfer one electron per cluster.

FeMoco has been synthesized *in vitro* in the presence of molybdate, Mg-ATP, homocitrate and the products of four genes involved in nitrogen fixation.

A crystallographic structure at 2.7 Å resolution has been obtained for the MoFe-protein from *Azotobacter vinelandii*. Each center consists of two bridged clusters; The MoFe cofactor (FeMoco) has 4Fe:3S and 1Mo:3Fe:3S clusters bridged by three non-protein ligands, and the 4 Fe4S4 clusters are bridged by two cysteine thiol ligands. Six of the seven Fe sites of the cofactor seem to have trigonal coordination symmetry while the seventh has tetrahedral geometry and is liganded to the side chain of a cysteine. The molybdenum atom is liganded by three sulfurs in the cofactor, two oxygens from the homocitrate, and the imidazole side chain of a histidine. Neither the cofactor nor the Fe_4S_4 clusters are exposed to the surface (they are buried at least 10 Å below it), suggesting that substrate entry, electron transfer, and product release must involve a carefully regulated sequence of interactions between the MoFe-protein and the Fe-protein in the native nitrogenase.

[1] With the exception of nitrogenase, the molybdenum atom in all molybdoenzymes from animals, plants and microorganisms is part of an organometallic structure called the molybdenum cofactor. Although the free cofactor is extremely labile, the structural analysis of inactive derivatives of the cofactor a dithiolene-containing pterin structure termed molybdopterin has been proposed to be the organic moiety of the factor from sulfite oxidase.

FeMoCo is probably the site of N_2 binding, although it is not clear whether the N_2 interacts primarily with the Mo, Fe or S^- components.

The overall mechanism is thought to be the following:

The Fe-protein pumps one electron per two molecules of ATP hydrolyzed into the MoFe-protein. Each stroke advances the redox status of the MoFe-protein one step around an eight-step circular pathway: actually hydrazine N_2H_4 has been detected as an intermediate between N_2 and NH_3 upon denaturation of MoFe-protein during reduction of N_2.

A doubly substituted form of the nitrogenase MoFe protein (α-$70^{Val \rightarrow Ala}$, α-$195^{His \rightarrow Gln}$) has the capacity to catalyze the reduction of carbon dioxide to yield methane. Under optimized conditions, 1 nmol of the substituted MoFe protein catalyzes the formation of 21 nmol of CH_4 within 20 min. The catalytic rate depends on the partial pressure of CO_2 (or concentration of HCO_3^-) and the electron flux through nitrogenase. The doubly substituted MoFe protein also has the capacity to catalyze the unprecedented formation of propylene ($H_2C{=}CH$-CH_3) through the reductive coupling of CO_2 and acetylene ($HC{\equiv}CH$). In light of these observations, it is suggested by the authors of this work that an emerging understanding of the mechanistic features of nitrogenase could be relevant to the design of synthetic catalysts for CO_2 sequestration and formation of olefins.

When the *A. vinelandii* nitrogenase MoFe protein α-Val^{70} residue is substituted by alanine or glycine, the resulting variant proteins will catalyze the reduction and coupling of CO to form methane (CH_4), ethane (C_2H_6), ethylene (C_2H_4), propene (C_3H_6), and propane (C_3H_8). The rates and ratios of hydrocarbon production from CO can be adjusted by changing the flux of electrons through nitrogenase, by substitution of other amino acids located near the FeMo-cofactor, or by changing the partial pressure of CO. Increasing the partial pressure of CO shifted the product ratio in favor of the longer chain alkanes and alkenes.

Control of Nitrogenase Synthesis and Activity

Nitrogen fixation is an intrinsically anaerobic process, due to the rapid irreversible inactivation of nitrogenase by even low partial pressures of oxygen. The mechanisms that protect nitrogenase from oxygen are varied. In many diazotrophs, more than one mechanism may be present. Obligate anaerobes such as *Clostridium pasteurianum* and *Desulfovibrio desulfuricans* are apparently devoid of any specific device to protect nitrogenase from oxygen inactivation. Facultative aerobes (*Klebsiella pneumoniae*, *Bacillus polymyxa*, *Rhodospirillum rubrum*) are able to grow on combined nitrogen both aerobically and anaerobically, but can fix nitrogen only in anaerobiosis. Microaerophilic bacteria (*Azospirillum*) show a preference for low partial oxygen pressures when fixing nitrogen. Finally, *Azotobacter* can fix nitrogen in the presence of air, the nitrogenase being protected by a 2Fe-2S protein.

Cyanobacteria are organisms able to perform oxygenic photosynthesis than can also fix nitrogen. They have the simplest nutritional requirements and can grow with light in a mineral medium, using CO_2 as a carbon source and N_2 as a nitrogen source. This presents an obvious paradox. In most cyanobacteria, like *Anabaena variabilis*, nitrogenase is localized in the heterocysts, which are thick-walled cells lacking photosynthetic oxygen evolution. Heterocysts rely on carbohydrates, imported from the photosynthetically active vegetative cells. High respiration rates and a diffusion barrier for gases enables heterocysts to fix nitrogen aerobically (in *Anabaena flos-aquae*, the glycolipid area of the heterocyst envelope thickens as exogenous oxygen partial pressure increases, showing an O_2-induced mechanism for providing a greater diffusion barrier against inhibition of nitrogenase). Heterocystous cyanobacteria do not form heterocysts when grown photosynthetically with a combined nitrogen source (ammonia or nitrate); and nitrogenase synthesis is also repressed under these conditions; both heterocyst formation and nitrogenase synthesis resume as soon as the combined nitrogen source is removed: this occurs even in the absence of N_2, in an atmosphere of argon.

Expression of nitrogenase in heterocyst-containing cyanobacteria depends on the availability of combined nitrogen, oxygen and upon cell differentiation. In addition to being under transcriptional control, enzyme activity is influenced by external factors which limit the carbon supply of heterocysts by vegetative cells and by excess ammonium chloride under certain pH conditions. These treatments cause inactivation of the enzyme, correlated with a decrease of electrophoretic mobility of the Fe-protein, which suggests a posttranslational enzyme modification. This modification has been shown to occur in many different nitrogen-fixing microorganisms and in the case of *Rhodospirillum rubrum*, it has been shown to be due to an ADP-ribose covalently attached to an arginine of the Fe-protein, which interrupts the catalytic function of nitrogenase. The modification, catalyzed by dinitrogenase reductase. ADP-ribosyltransferase, requires NAD^+ as a substrate (like in the case of the ADP-ribosylation of the elongation factor catalyzed by diphtheria toxin) and ADP plus a divalent cation as cosubstrates. The modification is reversed under the action of dinitrogenase ADP-ribosyl glycohydrolase in the presence of MgATP and a divalent cation. The glycohydrolase is associated with the membrane and is extremely oxygen-labile. The structure of the ADP-ribosylhydrolase that acts on *Azospirillum brasilense* nitrogenase (dinitrogenase reductase-activating glycohydrolase, DraG) has been solved at a resolution of 2.5 A. This bacterial member of the ADP-ribosylhydrolase family acts specifically towards a mono-ADP-ribosylated substrate. The protein shows an all-alpha-helix structure with two magnesium ions located in the active site. Comparison of the DraG structure with orthologues deposited in the Protein Data Bank from Archaea and mammals indicates that the ADP-ribosylhydrolase fold is conserved in all domains of life. Modeling of the binding of the substrate ADP-ribosyl moiety to DraG is in excellent agreement with biochemical data. This reversible inactivation of nitrogenase upon a nutritional switch-off is reminiscent of the *E. coli* glutamine synthetase case which will be studied in detail in a later section.

Studies with nonheterocystous cyanobacteria have revealed no universal strategy that would allow nitrogenase to function in these organisms under ambient atmospheric conditions and concomitant with oxygen photosynthesis. One case is worth mentioning, that of *Frankia* strains. Oxygen exclusion in *Frankia* species, members of an actinomycetal genus that forms root-nodule symbioses in a wide range of woody Angiosperms, is accomplished within specialized vesicles, where nitrogen fixation is localized. The lipidic vesicle envelope is apparently a functional analogue of the cyanobacterial heterocyst envelope, forming a multilamellar external gas-diffusion barrier around the nitrogen-fixing cells. The number of envelope lamellae, each 3–4.5 nm in thickness is regulated by external O_2 levels. The purified vesicle envelopes consist primarily of two hopanoid[2] lipids, rather than of glycolipids, as in the case of cyanobacteria. One envelope hopanoid, bacteriohopanetetrol phenylacetate monoester, is vesicle-specific. The *Frankia* vesicle envelope thus represents a layer specific to the locus of nitrogen fixation that is biosynthetically uniquely derived.

One of the most exciting discoveries in nitrogen fixation research in recent years was the finding that in addition to the conventional molybdenum nitrogenase, *A. vinelandii* has two nitrogenases that do not contain molybdenum. The three nitrogenases are similar to each other in structure and physico-chemical properties and all include iron in three different types of cluster. One of the two "new" nitrogenases contains a vanadium analogue of FeMoco. No metals other than iron appear to be present in significant amounts in the third enzyme. Under normal conditions, synthesis of the two non-molybdenum nitrogenases is repressed in presence of nanomolar amounts of molybdenum. In addition, the synthesis of the third nitrogenase is also repressed by vanadium. The subunits of the three enzymes are encoded by three different genes. The non-molybdenum nitrogenases contain an additional subunit that has no equivalent in the molybdenum nitrogenase enzyme.

In the absence of fixed nitrogen and aerobically at 30 °C, *A. vinelandii* synthesizes one of the three nitrogenases according to the metal content of the medium. Molybdenum prevents expression of the alternative nitrogenases and is required for expression of the genes (*nifHDK* operon) encoding molybdenum nitrogenase; vanadium prevents expression of a third nitrogenase and is required for expression of the *vnfDGK* operon. Although the detailed mechanism by which metals regulate nitrogenase synthesis is not known, some of the genes that affect this regulation have recently been identified.

---

[2] Hopanoids, a family of polycyclic triterpenes, occur in cell membranes in a wide range of microorganisms, where these lipids contribute to membrane stability and alter phase-transition properties. They are functional analogues of cholesterol found in eucaryotic membranes. They have the dimensions and the amphiphilic character of cholesterol. They are important for the survival of organisms which thrive in extreme thermal environments and have been identified as a major component of oil shales. Although the biological role of these molecules as membrane stabilizers is well established, their assembly into an extracellular layer represents an unusual adaptation.

Selected References

Burris RH (1991) J Biol Chem 266:9339–9342
Duval S, Danyal K, Shaw S, Lytle AK, Dean DR, Hoffman BM, Antony E, Seefeldt LC (2013) Proc Natl Acad Sci USA 110:16414–16419
Elmerich C (1993) Ann Inst Pasteur/Actualités 4:133–153

Fe-Protein

Kim J, Rees DC (1992a) Science 257:1677–1682
Li XD, Huergo LF, Gasperina A, Pedrosa FO, Merrick M, Winkler FK (2009) J Mol Biol 390:737–746

Mo-Fe Protein

Kim J, Rees DC (1992b) Nature 360:553–560

Fe-MoCo

Ugalde RA, Imperial J, Shah VK, Brill WJ (1984) J Bacteriol 158:888–893
Yang Z-Y, Dean DR, Seefeldt LC (2011) J Biol Chem 286:19417–19421
Yang Z-Y, Moure VR, Dean DR, Seefeldt LC (2012) Proc Natl Acad Sci USA 109:19644–19648

Oxygen Relations of Nitrogen Fixation in Cyanobacteria

Fay P (1992) Microbiol Rev 56:340–373

Chapter 24
How Biosynthetic Pathways Have Been Established

In order to determine the pathways of biosynthesis, it is preferable to use organisms which are growing actively. Newly synthesized material can be recognized by the increase in cell numbers in a bacterial culture or the increase in weight in a growing animal such as a rat. Animal and plant cells grow slowly in general: the cells of the mammalian central nervous system grow only during the early life of the animal and is believed to divide no further; muscle cells divide and grow slowly; liver cells divide about every three months; those in the intestinal mucosa have a division time easily measured in days. In contrast, bacterial cells can double their number every twenty minutes. Newly synthesized material, in a bacterium such as *E. coli*, arises solely from a single carbon source such as glucose or acetate, ammonia, sulfate and inorganic phosphate. Such a bacterium is capable of intense chemical activity.

Nearly all the metabolic energy in the bacterium is used for biosyntheses. It is obvious that bacteria are the material of choice for the study of anabolic phenomena.

Several methods have been used to elucidate the pathways of biosynthesis for small molecules, amino acids, purines, pyrimidines, which are the materials from which proteins and nucleic acids are synthesized, and vitamins which are the cofactors or the substrates of important enzyme activities. The most important, which we are going to survey, and of which we shall see examples throughout this volume, are the use of isotopes, the use of auxotrophic mutants (unable to synthesize a given essential metabolite) and finally the examination of individual enzyme reactions.

Use of Isotopes

Organisms can be grown on specific carbon sources, such as glucose uniformly labeled with ^{14}C or labeled in the C1 or C6 position, acetate labeled with radiocarbon or acetate doubly labeled in the methyl group with ^{14}C and in the carboxyl group with the non-radioactive heavy isotope ^{13}C. After growth, the organisms are collected, disrupted and the protein fraction (for example) is isolated and

G.N. Cohen, *Microbial Biochemistry*,
DOI 10.1007/978-94-017-8908-0_24,

subjected to acid or alkaline hydrolysis; the individual amino acids are isolated and the distribution of the individual carbon atoms is determined by laborious degradation methods. Such methods are generally very difficult, and seldom provide unambiguous answers to the question of the detailed chemical reactions by which the carbon atoms of the nutrient are incorporated into an amino acid. More often, they at best allow to exclude a pathway which might have seemed possible.

A much more useful and practical method is isotope competition, which is described below.

The principle is as follows: if uniformly labeled glucose is provided as the sole carbon source for the bacterium, all the essential metabolites will be uniformly labeled. If a suspected intermediate is added to the growing culture as an unlabeled compound, it will dilute enormously the radioactivity of the labeled intermediate. As a result, the final metabolite will be non-radioactive, or at least its specific radioactivity will be greatly reduced. In addition to its dilution effect, the exogenous addition often causes an allosteric inhibitory effect or a repression effect on the activity or synthesis of one or more enzymes of the pathway which result in a reduced or suppressed synthesis of the final metabolite. Let us look as some applications of this method.

(a) a culture of *E. coli* is grown in a synthetic medium containing uniformly labeled glucose and non-radioactive homoserine. After growth, the culture is centrifuged, washed, the cells are disrupted by sonication and the protein fraction isolated and hydrolyzed with 6N HCl. The resulting amino acids are separated by a bidimensional paper chromatography with different appropriate solvents so that all amino acids can be identified. The chromatograms are placed on a film sensitive to radiation; the spots corresponding to radioactive amino acids will blacken the film. In addition, non-radioactive amino acids, if any, can be identified on the chromatogram by ninhydrin. The amino acids threonine, methionine and isoleucine are found to have little or no radioactivity, while the specific activity of all the other amino acids is identical to that of the original glucose. The conclusion is reached that homoserine is probably an intermediate in the biosynthesis of threonine, isoleucine and methionine.
(b) if the same type of experiment is carried out, but this time adding non-radioactive aspartate, the radioactivity of the protein aspartate is greatly reduced, and so also is that of diaminopimelic acid (a cell-wall constituent in *E. coli*), lysine and the same three amino acids (threonine, methionine and isoleucine) of the previous experiment.
(c) upon addition of non-radioactive threonine as a competitor, the radioautograms of the protein hydrolysate show that only threonine and isoleucine from protein are non-radioactive (Figs. 24.1 and 24.2).
(d) finally, when non-radioactive methionine and isoleucine are used as competitors, they affect only their own specific radioactivities.

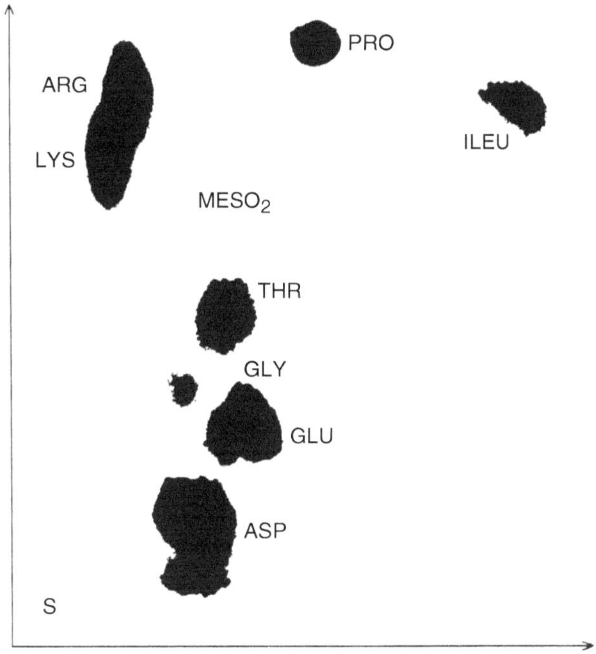

Fig. 24.1 Autoradiogram of a bidimensional chromatogram of a protein hydrolyzate from *E. coli* cells grown in the presence of uniformly labeled glucose (From Roberts et al. (1955), with permission of the Carnegie Institution of Washington)

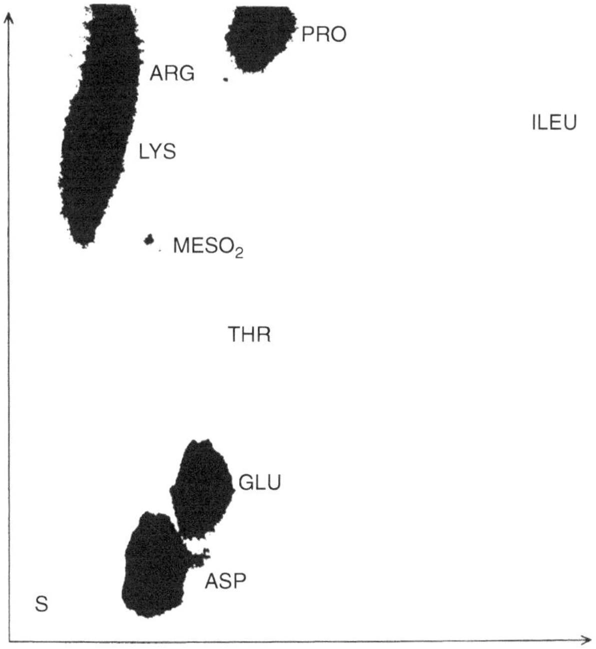

Fig. 24.2 Same experiment as in Fig. 24.1, but *E. coli* was grown in the presence of non-radioactive threonine. Note that the radioactivity of threonine and isoleucine have disappeared. Same source as Fig. 24.1.

Taken collectively, the results of these experiments suggest the following biosynthetic pathway:

Aspartate	→	Homoserine	→	Threonine	→	Isoleucine
↓		↓				
Diaminopimelate and Lysine		Methionine				

Many similar experiments of "isotopic competition" have made it possible to establish clearly that the majority of the amino acids in the proteins of *E. coli* are derived from a small number of intermediates from the glycolytic or the tricarboxylic cycle: phosphoglycerate (serine, glycine, cysteine), pyruvate (alanine, valine, leucine), oxaloacetate (aspartate, asparagine, lysine, diaminopimelate, threonine, methionine, isoleucine), α-ketoglutarate (glutamate, glutamine, proline and arginine).

We shall examine these families in turn in later chapters. The aromatic aminoacids, phenylalanine, tyrosine and tryptophan, and the heterocyclic amino acid histidine are synthesized by a series of reactions which will be examined separately.

Use of Auxotrophic Mutants

In every wild type bacterial population, there exist spontaneous mutants unable to synthesize such or such essential metabolite. They occur at a very low frequency, of the order of 10^{-8}. These mutants, whose growth depends on the exogenous addition of the metabolite they cannot synthesize, are called auxotrophic mutants, as against the wild type prototroph from which they derive. The frequency of mutation can be increased by ultraviolet radiation or by treatment with chemical mutagens, some of which are highly effective. The problem which arises, after mutagenesis, is to select the desired mutant from amidst the whole population of prototrophs and mutants in which one has no interest.

One of the most frequently employed methods is to use selection by penicillin, developed independently by Davis, and by Lederberg and Zinder, and ameliorated by Adelberg and by Gorini.

The basis of the method is the property of penicillin to kill only actively dividing cells (actually, the antibiotic acts by inhibiting the synthesis of the cell wall; as a result, the bacteria who do not grow are protected). After mutagen treatment, the bacteria are allowed to divide a number of times in the presence of the growth factor for which auxotrophs are desired. The reasoning is as follows: bacteria have several nuclei, and therefore several identical chromosomes; the mutagen has acted on only one of the chromosomes at the specific locus coding for the synthesis of the required growth factor, for which the unaffected chromosomes still possess all the information. After a few divisions, the nuclei have become segregated among the different daughter cells, and the bacteria have become arranged into groups with a homogeneous make-up. Excess growth factor is eliminated by washing several times, and the culture is resuspended in a medium containing every growth factor (amino acids, purines, pyrimidines, vitamins) except the one for which auxotrophic mutants are being sought. The new medium also contains penicillin. Bacteria which have not been killed by the mutagen make up a population comprising prototrophic bacteria, mutants which are of no interest, and the mutants desired. The first two classes can divide and therefore will be killed by penicillin. The third class cannot grow and so will escape death. After removal of penicillin, plating on a suitable medium will in principle give colonies which are largely those of the mutants desired.

Let us consider a group of mutants obtained independently by this method, all of which are unable to synthesize a given amino acid, for example tryptophan. None of the mutants will grow in a synthetic medium which contains glucose as the only carbon source, but all will be able to grow if tryptophan is added. A class of mutants may be found, which under certain conditions secretes into the growth medium a substance not found in the growth medium of the parental type. This substance has been identified as indole; it allows the growth of a second class of tryptophan-requiring mutants. This class, like the first, comprises individuals unable to synthesize tryptophan, but unlike the first class, also unable to synthesize indole. Mutants of this second kind can also accumulate a substance, anthranilic acid, which allows a third class to grow. Thus the third class is unable to carry out one or several reactions in the synthesis of tryptophan earlier than anthranilate production, and can only grow in the presence of anthranilate, indole or tryptophan. The biosynthesis of tryptophan can therefore be written schematically in the following provisional way:

$$\text{glucose} + NH_4^+ \underset{(3)}{\rightarrow} \text{anthranilate} \underset{(2)}{\rightarrow} \text{indole} \underset{(1)}{\rightarrow} \text{tryptophan}$$

The numbers in parentheses indicate "genetic blocks" of different kinds, that is reactions that cannot be carried out by the corresponding mutants.

Likewise, a mutant requiring both threonine and methionine for growth can see this double requirement satisfied by homoserine, in keeping with the results of isotopic competition described above.

Auxotrophic mutants have been obtained for practically every pathway, in particular for amino acid, purine, pyrimidine and vitamin biosyntheses, and have proved to be of immense use for the elucidation of the pathways. However, it remained to be shown, in each case, that the accumulated products are the truly direct intermediates and not metabolic products of the actual intermediates.

Enzymatic Analysis

The interpretation of the results obtained by the above methods is not always as easy and unequivocal as the concise summary just given would lead us to believe. The results must be considered as a point of departure for more detailed studies, and not as a definite proof of the existence of the reactions postulated. As a consequence, it is necessary to try to confirm them independently by using cell suspensions or better cell-free extracts to show that every postulated reaction does indeed take place and is missing in the mutants which are suspected to be unable to carry it out. For example, mutants requiring homoserine (or threonine plus methionine) for growth have been obtained in several bacterial and fungal species and have been shown to lack the enzyme homoserine dehydrogenase which catalyzes the reduction of aspartic semialdehyde to homoserine. This result confirms the evidence from the methods of isotopic competition and nutrition, and makes it certain that homoserine is an intermediate in the synthesis of homoserine and methionine.

Selected References

Isotopic Competition

Roberts RB, Abelson PH, Cowie DB, Bolton ET, Britten RJ (1955) Studies in biosyntheses in Escherichia coli. Carnegie Inst. Pub. No. 607 Kirby Lithographic Co. (Washington)

Isolation of Auxotrophic Mutants

Davis BD (1948) J Am Chem Soc 70:4267
Gorini L, Kaufman H (1960) Science 131:604–605
Lederberg J (1950) In: Comrie JH Jr (ed) Methods in medical research, 3rd edn. Year Book Publishers, Chicago, pp 5–21
Lederberg J, Zinder N (1948) J Am Chem Soc 70:4267–4268

Chapter 25
The Aspartic Acid Family of Amino Acids: Biosynthesis

The Biosynthesis of Aspartic Acid and Asparagine

Aspartic acid can be synthesized by two different reactions, both of which involve a substrate that is a participant in the tricarboxylic acid cycle. It can be obtained by transamination of oxaloacetic acid with glutamic acid, whose biosynthesis will be examined later:

$$\mathrm{COOH-CH_2-CO-COOH+COOH-(CH_2)_2-CH(NH_2)-COOH}$$
$$\downarrow\uparrow$$
$$\mathrm{COOH-CH_2-CH(NH_2)-COOH+COOH-CH_2-CH_2-CO-COOH}$$

Of the different transaminases found in *E. coli* by Rudman and Meister, only two can synthesize aspartate. The major one has a high affinity for oxaloacetate and a low affinity for the keto analogs of phenylalanine and tyrosine. It is specified by the *aspC* gene, is a homodimer whose subunit has a molecular weight of 43,573. The other one, coded by *tyrB* has a low affinity for oxaloacetate; furthermore, its synthesis is repressed by tyrosine. A double mutation in *aspC* and *tyrB* causes a double requirement for aspartate and tyrosine (the theoretical phenylalanine requirement is met by another transaminase).

One would have thought that aspartate could also have been synthesized by aspartase, which catalyzes a direct amination of fumarate by ammonia:

$$\mathrm{COOH-CH=CH-COOH+NH_3\rightarrow}$$
$$\mathrm{COOH-CH_2-CH(NH_2)-COOH}$$

However, an aspartate auxotroph lacking the *aspC* and *tyrB* gene products is not deficient in aspartase. It may be that the intracellular concentration of fumarate is too low for aspartate synthesis in a synthetic growth medium.

G.N. Cohen, *Microbial Biochemistry*,
DOI 10.1007/978-94-017-8908-0_25,

In *E. coli*, asparagine auxotrophy results from two mutations in two unlinked genes, *asnA* which encodes an ammonia-dependent asparagine synthetase A, and *asnB* which specifies an enzyme (asparagine synthetase B) in which the obligatory amino donor is glutamine. Whereas the presence of the ammonia-dependent amidation is restricted to procaryotes, the glutamine-dependent one is probably universally distributed.

Both reactions require ATP, the products being asparagine, AMP and pyrophosphate (and glutamate, in the case glutamine is the amino donor):

$$\text{Aspartate} + \text{ATP} + \text{NH}_3 \rightarrow \underbrace{\text{H}_2\text{N}-\text{CO}-\text{CH}_2-\text{CH}(\text{NH}_2)-\text{COOH}}_{\text{L - asparagine}} + \text{AMP} + \text{PPi}$$

Whereas the ammonia-dependent enzyme is a homodimer of subunit molecular weight 36,688 (330 amino acid residues), the glutamine-dependent one is a homotetramer of subunit molecular weight 57,000 (554 residues). The latter enzyme exhibits aspartate-independent glutaminase activity, and in that respect, resembles many glutamine-dependent amidotransferases.

Both enzymes are inhibited by asparagine; the synthesis of both enzymes is repressed by asparagine, due to the presence of the product of the *asnC* gene, adjacent to *asnA* but transcribed in the opposite direction and coding for a 17,000 molecular weight modulator polypeptide. The activation of *asnA* transcription by this product is inhibited in the presence of asparagine.

A highly conserved protein motif, characteristic of a class of aminoacyl tRNA synthetases and containing an invariant arginine residue was found to align with a region of asparagine synthetase A. Every substitution of this arginine by site-directed mutagenesis, including its replacement by a lysine, results in a loss of asparagine synthetase activity. The results are consistent with the possibility that asparagine synthetase A evolved from an ancestral aminoacyl tRNA synthetase.

In keeping with this hypothesis, it is worth noting that biochemical experiments and genomic sequence analysis showed that *Deinococcus radiodurans and Thermus thermophilus* do not possess asparagine synthetase (encoded by *asnA* or *asnB*). Instead these organisms derive asparagine from asparaginyl-tRNA, which is made from aspartate by a tRNA-dependent transamidation pathway. A genetic knockout disrupting this pathway deprives *D. radiodurans* of the ability to synthesize asparagine and confers asparagine auxotrophy. The capacity to make asparagine could be restored by transformation with *E. coli asnB*. This result demonstrates that in *Deinococcus*, the only route to asparagine is via asparaginyl-tRNA. Analysis of the completed genomes of many bacteria reveal that, barring the existence of an unknown pathway of asparagine biosynthesis, a wide spectrum of bacteria rely on the tRNA-dependent transamidation pathway as the sole route to asparagine.

The biosynthesis of aspartate semialdehyde, the common intermediate in the biosynthesis of lysine, methionine, threonine and isoleucine.

A specific β-aspartokinase catalyzes the phosphorylation of aspartic acid in the β-position:

$$\mathrm{COOH-CH_2-CH(NH_2)-COOH+ATP} \rightarrow \mathrm{H_2O_3POOC-CH_2-CH(NH_2)-COOH} + \mathrm{ADP}$$

aspartyl phosphate

The β-aspartyl phosphate thus formed is an unstable compound, and when aspartokinase is to be studied, the reaction is performed in the presence of hydroxylamine; the hydroxamate produced reacts with ferric ions to produce a colored compound for which a rather insensitive colorimetric assay exists. When purified enzyme is used, and ATPase activity has been eliminated, the reaction can be coupled with phosphoenolpyruvate and pyruvate kinase, the ATP being regenerated. The pyruvate formed is then reacted with lactate dehydrogenase and the oxidation of NADPH followed spectrophotometrically.

The substrate specificity of aspartokinase I, one of the three isofunctional aspartokinases has been investigated. Whereas the integrity of the α-amino group is required, the amide or esters of the α-carboxyl group are competent alternative substrates.

The following step is the reduction of β-aspartyl phosphate to aspartate semialdehyde, catalyzed by aspartate semialdehyde dehydrogenase (coded for by the *asd* gene):

$$\mathrm{H_2O_3\,POOC-CH_2-CH(NH_2)-COOH+NADPH+H^+} \rightarrow \mathrm{OHC-CH_2-CH(NH_2)-COOH+Pi+NADP^+}$$

Aspartate semialdehyde

This reaction is reminiscent of that catalyzed by triose phosphate dehydrogenase, which similarly involves the reduction of an acyl phosphate group to an aldehyde. In both cases, the suggested mechanism involves the formation of a thioester intermediate resulting from the attack of a cysteine thiolate on the substrate carbonyl group, followed by hydride transfer to $NADP^+$. Subsequent attack on the thioester intermediate by an oxygen anion of bound inorganic phosphate leads to an expulsion of the cysteine thiolate and formation of the phosphorylated product.

Aspartate semialdehyde dehydrogenase from *E. coli* has been purified: it is a dimer of 77,000 Da composed of two identical subunits. Alkylation of a cysteine at the active site by 2-amino-4-oxo-5-chloropentanoate, a substrate analogue, inactivates irreversibly the enzyme; aspartate semialdehyde protects against the inactivation. Substitution of this cysteine with an alanine by site-directed mutagenesis results in complete loss of enzyme activity. However, changing this cysteine to a serine yields a mutant enzyme with a maximum velocity that is only 0.3 % that of the native enzyme, but which has retained the same affinity for substrates as the native enzyme, and the same overall conformation. The dehydrogenase is also inactivated by the coenzyme analogue chloroacetylpyridine-ADP. $NADP^+$ and NADPH, but not the substrate protect against the alkylation.

Mutants with an inactivated *asd* gene require diaminopimelate, lysine, methionine and threonine for growth. This requirement can be met if only diaminopimelate and homoserine are provided, since homoserine is the common precursor of methionine and threonine (see below).

Biosynthesis of Lysine from Aspartate Semialdehyde in Bacteria

Aspartate semialdehyde (ASA) condenses with one molecule of pyruvate to eliminate two molecules of water. The reaction product is dihydrodipicolinic acid:

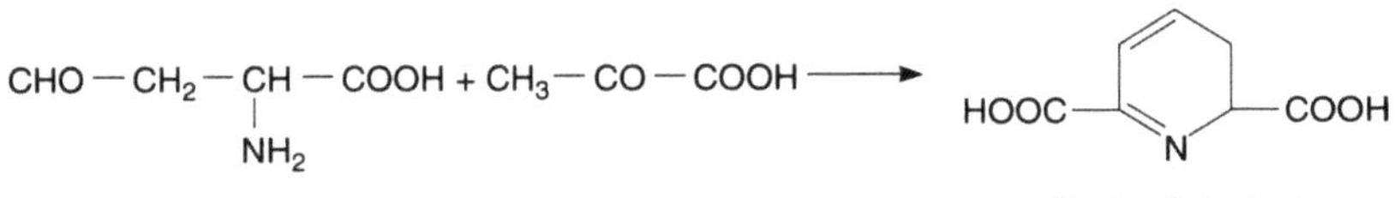

The enzyme catalyzing this reaction, dihydropicolinate synthase is coded by the *dapA* gene The *E. coli* protein has been purified to homogeneity. Pyruvate binds first to the enzyme, forming a Schiff base with the ∈-amino group of Lys-161, followed by binding of ASA. The native protein has a molecular weight of 134,000.

The enzyme is inhibited by lysine; the lysine inhibition is cooperative and non competitive toward both substrates.

Dihydrodipicolinate reductase (coded by *dapB*) has a native apparent molecular weight of 115,000. The *dapB* gene has been sequenced. It codes for 273 residues. The enzyme is a tetramer, which has been purified to homogeneity. Its tridimensional structure has been determined at 2.2 Å resolution. The product of the enzyme is tetrahydrodipicolinate. The reducing agent is NADH, twice as effective as NADPH. A chemical mechanism has been proposed.

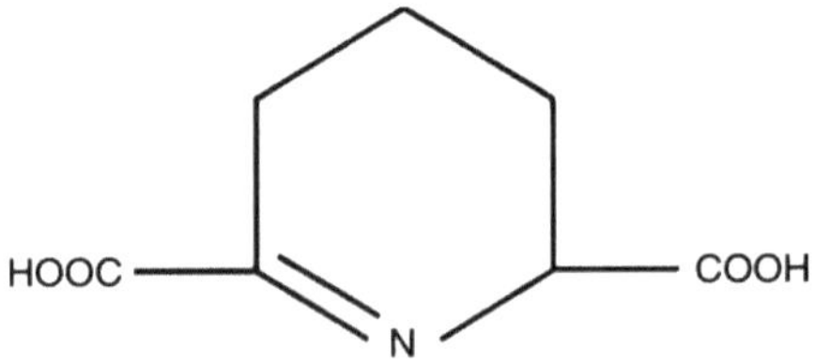

tetrahydrodipicolinate

The heterocyclic ring is then opened with a concomitant succinylation by tetrahydrodipicolinate succinylase (*dapD*) in the presence of succinyl CoA to form N-succinyl ε–keto-L-α-aminopimelic acid

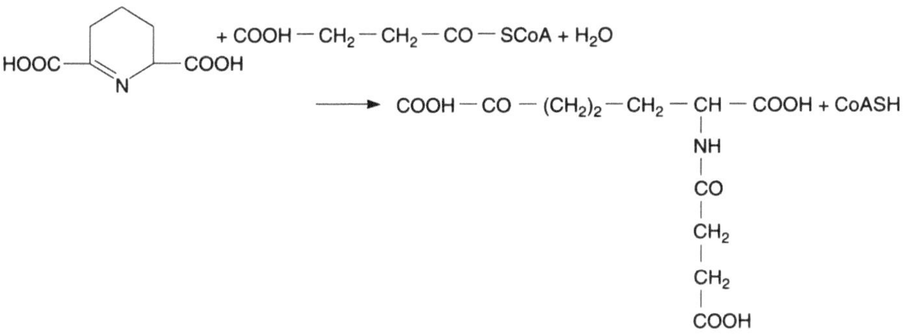

N-succinyl-ε-keto-L-α-aminopimelate

The nucleotide sequence of the gene has been determined: it encodes a polypeptide of 274 residues. The X-ray crystal structures of DapD in ternary complexes with pimelate/succinyl-CoA, and L-2-aminopimelate with the nonreactive cofactor analog, succinamide-CoA have been solved. The binding conformation of the cofactor succinyl group is defined: its interactions with the enzyme place its thioester carbonyl carbon in close proximity to the nucleophilic 2-amino group of the acceptor, in support of a direct attack ternary complex mechanism.

The *E. coli* aminotransferase leading to succinyl-diaminopimelate has been partially purified and characterized. It was assumed to be the product of the *dapC* gene which has not been clearly identified. Recently, the *dapC* gene has been characterized in *Bordetella pertussis*: apart from the pyridoxal-5′-phosphate binding motif, it does not show further aminoacid sequence similarities with the only other known enzyme with N-succinyl-L,L-DAP aminotransferase activity, ArgD of *E. coli*.

Some *E. coli* auxotrophs simultaneously requiring lysine *plus* methionine can grow on succinate in the absence of the amino acids. They have been identified as *suc* mutants with impaired synthesis of succinyl coenzyme A, which participates in lysine and methionine biosyntheses through succinylation reactions.

N-succinyldiaminopimelate desuccinylase has been partially purified. It requires Co+ + for activity. The enzyme is encoded by *dapE*., and its product is LL-diaminopimelate.

$$\underset{\text{Diaminopimelate (DAP)}}{COOH-CH(NH_2)-(CH_2)_3-CH(NH_2)-COOH}$$

Diaminopimelate epimerase (*dapF*) catalyzes the interconversion of LL-and *meso*-diaminopimelate. Mutants of this gene can grow on *meso*-, but not on LL-diaminopimelate.

Previous studies report that DAP epimerase exists as a monomeric enzyme. However, using analytical ultracentrifugation, X-ray crystallography, and enzyme kinetic analyses, it was shown that DAP epimerase from Escherichia coli exists as a functional dimer in solution and the crystal state. Furthermore, the 2.0-Å X-ray crystal structure of the *E. coli* DAP epimerase dimer shows that the enzyme exists in an open, active conformation. The importance of dimerization was subsequently probed by using site-directed mutagenesis to generate a monomeric mutant catalytically inactive, thus demonstrating that dimerization of DAP epimerase is essential for catalysis.

Finally, diaminopimelate decarboxylase (*lysA*) catalyzes the conversion of DAP to lysine.

$$\underset{meso-\text{DAP}}{\text{COOH}-\text{CH}(\text{NH}_2)-(\text{CH}_2)_3-\text{CH}(\text{NH}_2)-\text{COOH}}$$
$$\rightarrow \underset{\text{L-lysine}}{\text{NH}_2-(\text{CH}_2)_4-\text{CH}(\text{NH}_2)-\text{COOH}} + \text{CO}_2$$

The enzyme is specific for *meso*-DAP; the LL-isomer is not a substrate. The purified protein has a molecular weight of 200,000 as calculated from its sedimentation coefficient. Mutants devoid of the decarboxylase accumulate large amounts of DAP. In some *lysA* mutants the affinity of the decarboxylase for its coenzyme, pyridoxal phosphate, is altered: as a result, they require lysine or pyridoxine for growth. The nucleotide sequence of *E. coli lysA* indicates that it encodes a polypeptide of 420 residues (MW=46,099) and suggests that the native decarboxylase is a tetramer.

Cocrystal structures of *Methanococcus jannaschii* diaminopimelate decarboxylase (DAPDC) bound to a substrate analog, azelaic acid, and its L-lysine product have been determined at 2.6 Å and 2.0 Å, respectively.

In *Corynebacterium glutamicum*, L-lysine is synthesized simultaneously via the above described pathway (succinylase pathway) and a dehydrogenase variant of the diaminopimelate pathway, where tetrahydrodipicolinate is the substrate of a diaminopimelate dehydrogenase converting it directly to *meso*-DAP in the presence of ammonia and NADPH. The gene from *C. glutamicum* can complement defective *E. coli* lysine auxotrophs affected in the succinylase pathway. Another pathway of lysine biosynthesis, operative in yeasts and higher fungi, will be examined in Chap. 25.

The Synthesis of Dipicolinic Acid, a Substance Present in the Spores of Gram-Positive Bacilli

The structural events associated with spore formation have been elucidated by a combination of light and electron microscopy observations.

At the shift from exponential vegetative growth to linear growth that just precedes the onset of sporulation, each cell contains two nuclear bodies which coalesce to form an axial chromatin thread. Then, a transverse septum separates the cytoplasm and the DNA of the smaller cell (the future spore) from the rest of the cell.

In sporulating bacteria, a smaller cell becomes engulfed within the cytoplasm of the larger cell to produce a forespore. The cell is then committed to undergo sporulation and there is a rapid synthesis and deposition of new structures, the cortex and the spore coat that enclose the forespore. The maturing spore becomes refractile, but is still not heat-resistant. The development of heat-resistance closely follows the massive uptake of Ca^{++} ions by the sporulating cell and the synthesis in large amounts of dipicolinic acid, a compound absent in vegetative cells.

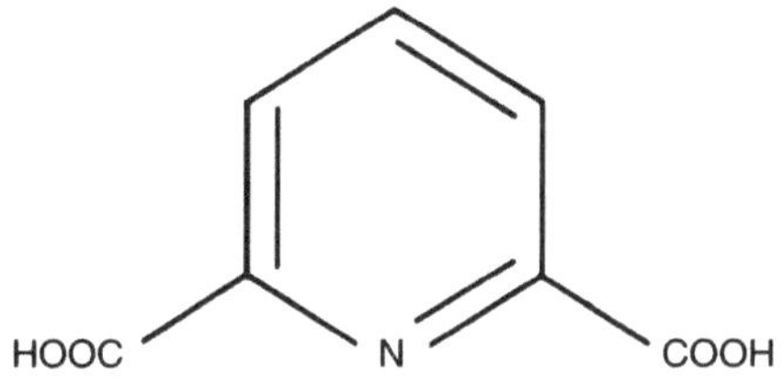

dipicolinic acid

In mature spores, the molar ratio dipicolinate/Ca++ is close to unity and actually dipicolinate occurs as a calcium chelate. This compound represents 10–15 % of the spore dry weight and is located within the spore. The synthesis of dipicolinic acid represents an offshoot of the lysine branch of the aspartate family pathway, at the dihydrodipicolinate level:

$$\begin{array}{l}\text{Dihydrodipicolinic acid} \rightarrow \text{Dipicolinic acid}\\ \qquad\downarrow\\ \qquad\text{Lysine}\end{array}$$

The Reduction of Aspartate Semialdehyde to Homoserine, the Common Precursor of Methionine and Threonine

An enzyme called homoserine dehydrogenase catalyzes the reduction of the aldehyde group of aspartate semialdehyde to a primary alcohol. Whereas all the aspartic semialdehyde dehydrogenases are specific for NADPH, the homoserine dehydrogenases from different bacteria can use either NADH or NADPH. Depending on the source, either one or the other of the pyridine nucleotides is the more active

$$\begin{array}{l}\mathrm{COOH-CH(NH_2)-CH_2-CHO+NADPH+H^+}\\ \quad\rightarrow \mathrm{COOH-CH(NH_2)-CH_2-CH_2OH+NADP^+}\\ \qquad\qquad\qquad\text{Homoserine}\end{array}$$

The affinity label 2-amino-4-oxo-5-chloropentanoate inactivates homoserine dehydrogenase. Homoserine and aspartate semialdehyde protect against this inactivation caused by the covalent binding of one equivalent of the analogue per subunit.

For reasons that will be developed in the next chapter and that pertain to the allosteric inhibition of aspartokinase (the first enzyme of the pathway) and homoserine dehydrogenase (the third enzyme of the pathway) attempts have been made to find a physical association of aspartate semialdehyde dehydrogenase with these two enzymes. These attempts have not met with success.

Biosynthesis of Methionine from Homoserine

Apart from its role as protein constituant, methionine participates in the initiation of protein synthesis (as N-formylmethionine), as a universal methylation agent (as S-adenosylmethionine, SAM) and as a precursor of spermidine through SAM.

The methionine biosynthetic pathway is complex. The carbon skeleton derives from aspartate, the sulfur atom from cysteine and the methyl group from the β-carbon of serine (Fig. 25.1).

The first step of methionine biosynthesis is catalyzed by homoserine succinyltransferase, product of the *metA* gene which transforms in bacteria homoserine to O-succinylhomoserine in the presence of succinyl coenzyme A (in yeast and *Neurospora crassa*, the acylation reaction is an acetylation, catalyzed by homoserine acetyltransferase).

$$\underset{\text{Homoserine}}{COOH-CH(NH_2)-CH_2-CH_2OH} + \text{Succinyl CoA} \rightarrow \text{O-succinylhomoserine} + \text{CoA}$$

The enzyme has been partially purified in *E. coli*. The *metA* gene has been cloned and sequenced; it codes for a polypeptide of molecular weight 35,673. The enzyme appears to be a homodimer, showing unusual temperature sensitivity in many *Enterobacteriaceae*, which appears to limit growth at elevated temperatures. Controlling growth rate by regulating the availability of methionine is a very effective strategy, as methionine is required for initiation and elongation of proteins, biosynthesis of purines and pyrimidines, various methylation reactions and biosynthesis of polyamines. There is evidence that homoserine transsuccinylase is a heat shock protein and is involved in other types of stress than elevation of temperature. In this respect, it is to be noted that in *S. cerevisiae*, one of the genes (of unknown biochemical function) involved in methionine biosynthesis is identical to a gene which is responsible for high salt tolerance. Actually, methionine supplementation increases the tolerance of yeast to NaCl.

O-succinylhomoserine is transformed to cystathionine, a seven-carbon diaminoacid (presenting a second asymmetric center), in the presence of cysteine:

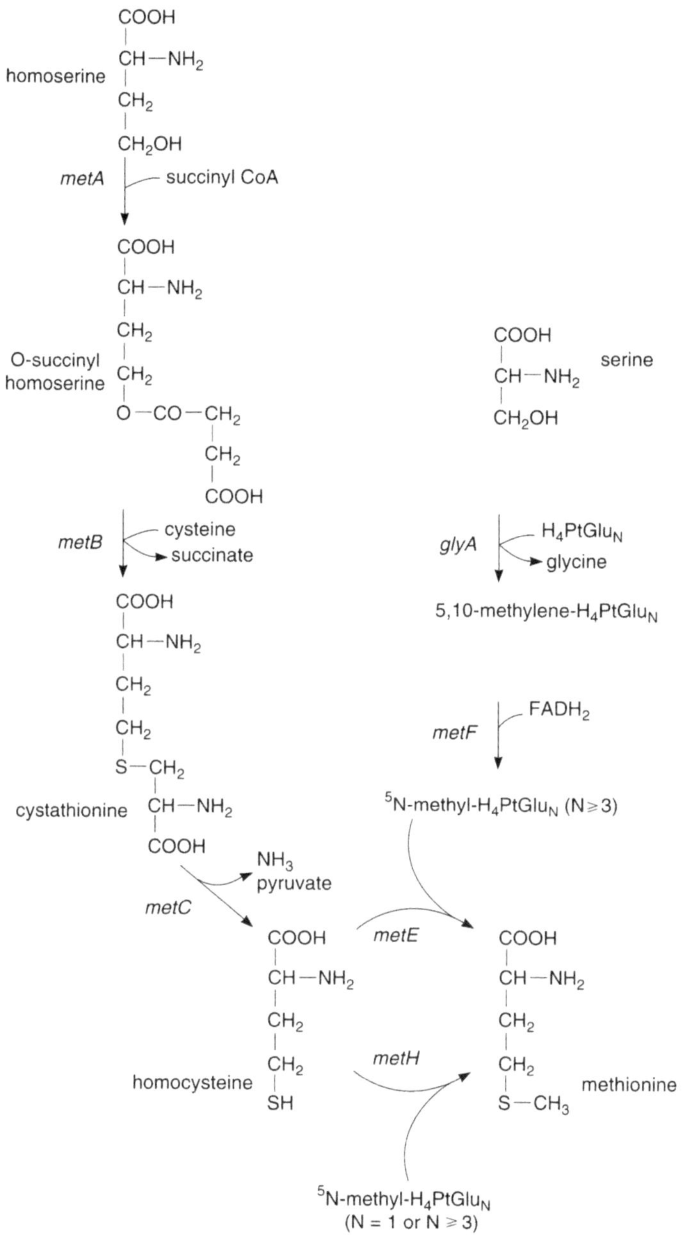

Fig. 25.1 Biosynthesis of methionine from homoserine and serine in *Enterobacteriaceae*. Relevant enzymes of the *E. coli* genes are discussed in the text. ^{5}N-H_4PtGlu stands for tetrahydropteroylglutamate. The product of the *metH* gene can use the mono- or polyglutamate forms, whereas the product of the *metE* gene can only use the polyglutamate form of ^{5}N-methyl-H_4PtGlu as substrate

$$\underset{\text{O - succinylhomoserine}}{COOH-CH(NH_2)-CH_2-CH_2O-CO-(CH_2)_2-COOH} + \underset{\text{Cysteine}}{COOH-CH(NH_2)-CH_2SH}$$
$$\downarrow$$
$$\underset{\text{Cystathionine}}{COOH-CH(NH_2)-CH_2-CH_2-S-CH_2-CH(NH_2)-COOH} + \text{succinic acid}$$

This reaction is catalyzed by cystathionine-γ-synthase (CGS), encoded by the *metB* gene. The reaction mechanism of the enzyme is thought to proceed via a pyridoxamine derivative of vinylglyoxylate. The enzyme has been obtained in the pure state in *S. typhimurium* and in *E. coli*. In both cases, it is a homotetramer whose subunit has a molecular weight of 40,000; in *S. typhimurium*, it has been shown to contain 4 pyridoxal phosphate molecules. The *metB* gene of *E. coli* has been cloned and sequenced; it corresponds to a polypeptide of 386 amino acid residues. The deduced N-terminal sequence agrees perfectly with the experimentally determined N-terminal protein sequence.

Cystathionine is cleaved to give homocysteine, ammonia and pyruvate in a reaction catalyzed by cystathionase (cystathionine-β-lyase, CBL), product of the *metC* gene.

$$COOH-CH(NH_2)-CH_2-CH_2-S-CH_2-CH(NH_2)-COOH$$
$$\downarrow$$
$$\underset{\text{Homocysteine}}{COOH-CH(NH_2)-(CH_2)_2-SH} + CH_3-CO-COOH + NH_3$$

This is also a pyridoxal phosphate enzyme which has been purified in *E. coli* and *S. typhimurium*. The *E. coli* enzyme has been purified from a strain harboring a multicopy plasmid carrying the *metC* gene. Its pH optimum, substrate specificity and kinetic parameters have been studied. The protein is a tetramer. The *E. coli metC* gene was cloned and its nucleotide sequence determined; the deduced sequence of the protein (395 residues) supported by the experimentally determination of the first ten residues, shows strong homology with that of cystathionine-γ-synthase and thus points to a common ancestor of the two proteins (see section on protein evolution in Chap. 40).

The crystal structure of the *E. coli* cystathionine-γ synthase (CGS) has been solved by molecular replacement with the known structure of cystathionine beta-lyase (CBL) at 1.5 Å resolution. The enzyme crystallizes as an α4 tetramer. The spatial fold of the subunits, with three functionally distinct domains and their quaternary arrangement, is similar to that of CBL (see below). Previously proposed reaction mechanisms for CGS can be checked against the structural model, allowing interpretation of the catalytic and substrate-binding functions of individual active site residues. Enzyme-substrate models pinpoint specific residues responsible for the substrate specificity, in agreement with structural comparisons with CBL. Both steric and electrostatic designs of the active site seem to achieve proper substrate selection and productive orientation. Amino acid sequence and structural alignments of CGS and CBL suggest that differences in the substrate-binding characteristics are responsible for the different reaction chemistries.

The crystal structure of the MetC protein (CBL) has been determined at 1.83 Å resolution. Each monomer can be described in terms of three spatially and functionally different domains. The N-terminal domain (residues 1–60) consists of three α-helices and one β-strand. It contributes to tetramer formation and is part of the active site of the adjacent subunit. The second domain (residues 61–256) harbors pyridoxal-phosphate (PLP) and has an α/β-structure with a seven-stranded β-sheet as the central part. The remaining C-terminal domain (residues 257–395), connected by a long α-helix to the PLP-binding domain, consists of four helices packed on the solvent-accessible side of an antiparallel four-stranded β-sheet. The fold of the C-terminal and of the PLP-binding domains as well as the location of the active site are similar to aminotransferases. Most of the residues in the active site are strongly conserved among the enzymes of the transsulfuration pathway.

3,3,3-trifluoroalanine binds covalently to the enzyme and inhibits it irreversibly. The structure of the inactivated MetC has been refined at 2.3 Å resolution.

A mutation, *metQ*, allows *metC* mutants to directly catalyze the formation of homocysteine from homoserine, thus by-passing the cystathionine normal intermediate.

The last step in methionine biosynthesis involves the methylation of homocysteine.

$$COOH-CH(NH_2)-(CH_2)2-SH+R-CH_3$$

$$\downarrow$$

$$\underset{\text{Methionine}}{COOH-CH(NH_2)-(CH_2)2-CH-S-CH3}+R$$

The methyl donor R is ^{5}N-methyl tetrahydropteroylglutamate (N > 1) a member of the folate family, which is derived from ^{5}N, ^{10}N-methylene tetrahydrofolate (generated by the conversion of serine to glycine) by a reductase specified by the *metF* gene. This reaction gives the methyl group specific for methionine biosynthesis. Two enzymes can catalyze this reaction: a transmethylase with an activity depending on the presence of vitamin B_{12} (product of the *metH* gene) which can use the mono- or polyglutamate form of ^{5}N-methyl tetrahydropteroylglutamate as methyl donor; and a vitamin B_{12}-independent transmethylase (product of *metE*) which can utilize as a substrate only the polyglutamate forms of ^{5}N-methyl tetrahydropteroylglutamate (N > 3). *E. coli* possesses both enzymes whereas most organisms possess only one. It does not synthesize vitamin B_{12} and uses one or the other transmethylase according to the availability of vitamin B_{12} in the medium. Strains with a mutation in *metE* require either methionine or vitamin B_{12} for their growth. A word of caution is necessary: vitamin B_{12} synthesis under anaerobiosis has been shown to occur. There is evidence that the vitamin B_{12}-independent enzyme, coded by *metE*, is a rather inefficient enzyme.

The mechanism of the vitamin B_{12} dependent enzyme involves the methylation of the cobalt atom of the vitamin. Catalytic amounts of SAM are required for this priming step. After the methyl group is transferred to homocysteine, subsequent methyl groups come from ⁵N-methyltetrahydrofolate. In addition to SAM, methyl B_{12} and the folate derivative, the enzyme requires $FADH_2$ for activity.

The B_{12} independent transmethylase does involve neither a methylated coenzyme nor the other requirements of the vitamin-dependent enzyme such as SAM or a reducing agent. Alkylation of Cys 726 results in complete loss of activity. The possibility has been suggested that this thiol functions as an intermediate methyl acceptor in catalysis, analogous to the role of cobalamin in the reaction catalyzed by the vitamin B12 dependent enzyme.

The sequence of *metE* has actually been determined, it has been cloned and expressed as a polypeptide of 84,654 Da. The sequence of the *metH* gene is also known, it codes for an exceptionally long polypeptide of 132,628 Da showing no similarity with the *metE* product.

The methyl group which is transferred to homocysteine derives from the β-carbon of serine: the interconversion of serine and glycine involves a single enzyme, serine hydroxymethyltransferase whose other substrate is tetrahydrofolate; this reaction also produces ^{5}N-^{10}N-methylenetetrahydrofolate, the substrate of the reductase coded by *metF*. which reduces it to ^{5}N-methyltetrahydrofolate which in turns transfers its methyl group to homocysteine by any one of the two transmethylases.

In *Corynebacterium glutamicum*, two pathways of homocysteine synthesis coexist: the one described above for *E. coli*, and a direct sulfhydrylation pathway catalyzed by the *metY* gene, encoding O-acetylhomoserine sulfhydrylase.

$$\text{O}-\text{acetylhomoserine} + \text{H}_2\text{S} \rightarrow \text{Homocysteine}$$

An *E. coli metB* mutant could be complemented by transformation of the strain with a DNA fragment carrying corynebacterial *metY* and *metA* genes.

The same enzyme has also been found in the archaeon *Thermus thermophilus*: this enzyme might be inactive under ordinary conditions but might become active as an alternative homocysteine synthase, only under such conditions as deficiency in transsulfuration, bringing about a sufficient amount of sulfide available in the cell.

In *Leptospira meyeri*, the transsulfuration pathway via cystathionine does not exist and this organism is able to perform direct sulfhydrylation for methionine biosynthesis by using O-acetylhomoserine as a substrate.

S-Adenosylmethionine (SAM) Biosynthesis

S-adenosylmethionine (SAM or AdoMet) is a biological sulfonium compound known as the major biological methyl donor in reactions catalyzed by methyltransferases. SAM is also used as a source of methylene groups (in the synthesis of cyclopropyl fatty acids), amino groups (in the synthesis of 7,8-diaminoperlagonic acid, a precursor of biotin), ribosyl groups (in the synthesis of epoxyqueuosine (a modified nucleoside in tRNAs)) and aminopropyl groups (in the synthesis of ethylene and polyamines). Even though the mechanism of most of these reactions has not been extensively characterized, it is likely that is mainly driven by the electrophilic character of the carbon centers that are adjacent to the positively charged sulfur atom of SAM. In addition, SAM, upon one-electron reduction, is a source of 5′-deoxyadenosyl radicals, which initiate many metabolic reactions and biosynthetic pathways by hydrogen-atom abstraction. SAM presents a unique situation in which all constituent parts have a chemical use.

The radical *S*-adenosylmethionine (AdoMet) superfamily currently comprises thousands of proteins that participate in numerous biochemical processes across all kingdoms of life. These proteins share a common mechanism to generate a powerful 5′-deoxyadenosyl radical, which initiates a highly diverse array of biotransformations. Recent studies are beginning to reveal the role of radical AdoMet proteins in the catalysis of highly complex and chemically unusual transformations, *e.g.* the anaerobic ribonucleoside triphosphate reductase reaction, Chap. 34. The unique features and intriguing chemistries of these proteins thus demonstrate the remarkable versatility and sophistication of radical enzymology.

S-adenosylmethionine synthetase catalyzes the only known route of biosynthesis of the intracellular alkylating agent S-adenosylmethionine and ATP:

$$\text{Methionine} + \text{Adenosine} - \underset{\alpha}{\text{P}} - \underset{\beta}{\text{P}} - \underset{\gamma}{\text{P}} \rightarrow \text{S} - \text{adenosylmethionine} + \text{PPi} + \text{Pi}$$

This reaction is interesting because it is the phosphorus moieties in position α and β which are the source of the pyrophosphate produced. Until this discovery, it was believed that ATP reacted in three ways: (1) as phosphate donor to form ADP; (2) as pyrophosphate donor to form AMP; (3) as adenylate donor to form pyrophosphate. SAM synthesis showed for the first time that ATP can also react as an adenosine donor and that the ribose-phosphate bond is not as inert as had been thought.

The enzyme has been purified to homogeneity. It was found to have a molecular weight of 140 kDa and to be composed of four identical subunits. It is coded by the *metK* gene, encoding a 384 residues polypeptide. In addition to the transferase reaction, the pure enzyme catalyzes a tripolyphosphatase reaction stimulated by SAM. The mechanism of the two reactions has been studied extensively. The three dimensional structure of the tetrameric enzyme encoded by the *E. coli metK* gene has been determined by X-ray crystallography.

These studies of *E. coli* SAM synthetase have defined a flexible polypeptide loop that can gain access to the active site without contacting the substrates. The influence of the length and sequence of this active site loop on catalytic efficiency has been characterized in a mutant in which the *E. coli* sequence (DRADPLEQ) has been replaced with the distinct sequence of the corresponding region of the otherwise highly homologous rat liver enzyme (HDLRN EEDV). Four additional mutants in which the entire DRADPLEQ sequence was replaced by five, six, seven, or eight glycines have been studied to unveil the effects of loop length and the influence of side chains. In all of the mutants, the maximal rate of S-adenosylmethionine formation is diminished by more than 200-fold whereas the rate of hydrolysis of the tripolyphosphate intermediate is decreased less than 3-fold. Thus, the function of the loop is localized to the first step in the overall reaction.

Several *E. coli metK* mutants have lower than normal levels of synthetase and an increased level of methionine biosynthetic enzymes. Since a total inactivation of the *metK* gene results in residual enzyme activity, it seemed likely that *E. coli* possesses two different S-adenosylmethionine synthetases, as has been reported for *S. cerevisiae*. Actually, a second enzyme has been found in *E. coli*, coded by the *metX* gene. Although antigenically related, the two enzymes can be distinguished. The *metX* gene is only expressed in rich media while *metK* is only expressed in minimal medium.

The S-adenosylhomocysteine (SAH) formed as a by-product of SAM-dependent methylation reactions is salvaged as the adenine base and homocysteine by a cyclical regenerative branch of the pathway, which is in the process of being characterized genetically.

Biosynthesis of Threonine from Homoserine

Threonine is an isomer of homoserine in which the alcohol group is a secondary alcohol. The *thr B* gene encodes homoserine kinase which catalyzes the phosphorylation of homoserine

$$R-CH_2OH+ATP \rightarrow \underset{\text{Homoserine phosphate}}{R-CH_2O-PO_3H_2}+ADP+Pi$$

$$\text{where } R = COOH-CH(NH_2)-CH_2-.$$

The phosphate bond so formed in homoserine phosphate is not energy-rich unlike in aspartyl phosphate, where the phosphorylated group is a carboxyl. The native protein consists of two identical subunits of 29,000 daltons. The protein has been purified to homogeneity and requires Mg^{++} for activity, which is stimulated by K^+. The nucleotide sequence of *thrB* has been determined.

The transformation of homoserine phosphate to L-threonine is catalyzed by threonine synthase, a pyridoxal phosphate enzyme:

$$\underset{\text{L - threonine (Note the two asymmetric centers)}}{CH_3-CHOH-CH(NH_2)-COOH}$$

The reaction mechanism of threonine synthase has been studied extensively with the enzyme from *Neurospora crassa*. It has been proposed that a water molecule is removed from homoserine, leading to vinylglycine, which is then rehydrated to threonine.

$$\underset{\text{Vinylglycine}}{CH_2=CH-CH(NH_2)-COOH}$$

The positions where solvent hydrogen from water is introduced during the reaction have been determined. In *E. coli*, threonine synthase is encoded by the *thrC* gene which has been cloned and sequenced. The molecular weight of 47,000 of the polypeptide expressed in minicells, as well as the apparent molecular weight of the pure protein, both determined by polyacrylamide gel electrophoresis under denaturing conditions agree with the molecular weight deduced from the DNA sequence.

The crystal structure of threonine synthase from *S. cerevisiae* has been solved at 2.7 Å resolution using multiwavelength anomalous diffraction. The structure reveals a monomer as active unit, which is subdivided into three distinct domains: a small N-terminal domain, a PLP-binding domain that covalently anchors the cofactor and a so-called large domain, which contains the main of the protein body. All three domains show the typical open α/β architecture. The cofactor is bound at the interface of all three domains, buried deeply within a wide canyon that penetrates the whole molecule. Based on structural alignments with related enzymes, an enzyme-substrate complex was modeled into the active site of yeast threonine synthase, which revealed essentials for substrate binding and catalysis. Furthermore, the comparison with related enzymes of the family of PLP-dependent enzymes indicated structural determinants of the oligomeric state and thus rationalized for the first time how a PLP enzyme acts in monomeric form.

Threonine synthase catalyzes the elimination of the γ-phosphate group from *O*-phospho-L-homoserine (OPHS) and the subsequent addition of water at Cβ to form L-threonine. The catalytic course of threonine synthase is the most complex among the PLP enzymes, and it is an intriguing problem how the elementary steps are controlled in threonine synthase to carry out selective reactions. When L-vinylglycine was added to the *Thermus thermophilus* enzyme in the presence of phosphate, L-threonine was formed with k_{cat} and reaction specificity comparable with those when OPHS was used as the substrate. However, in the absence of phosphate or when sulfate was used in place of phosphate, only the side reaction product, α-ketobutyrate, was formed. Global analysis of the spectral changes in the reaction of the synthase with L-threonine showed that compared with the more acidic sulfate ion, the phosphate ion decreased the energy levels of the transition states of the addition of water at the Cβ of the PLP-α-aminocrotonate aldimine (AC) and the transaldimination to form L-threonine. The x-ray crystallographic analysis of the synthase complexed with an analog for AC gave a distinct electron density assigned to the phosphate ion derived from the solvent near the Cβ of the analog. These results indicated that the phosphate ion released from OPHS by γ-elimination acts as the base catalyst for the addition of water at Cβ of AC, thereby providing the basis of the reaction specificity. The phosphate ion is also considered to accelerate the protonation/deprotonation at Cγ.

Biosynthetic Threonine Dehydratase

Threonine is deaminated by a pyridoxal phosphate enzyme and yields α-ketobutyrate:

$$CH_3-CHOH-CH(NH_2)-COOH \rightarrow$$
$$CH_3-CH_2-CO-COOH+NH_3$$

This enzyme (*ilvA*) is distinct from another enzyme called degradative threonine dehydratase which performs the same catalytic function, but is not involved in isoleucine biosynthesis.

The *S. typhimurium* enzyme has been purified and consists of four apparently identical subunits, of a molecular weight of about 48,500. Only two of the subunits can bind pyridoxal phosphate. This half-of-the-sites reactivity has been encountered in several other cases.

Isoleucine Biosynthesis

α-ketobutyrate condenses with a molecule of acetaldehyde derived from pyruvate to give α-aceto-α-hydroxybutyric acid, a reaction catalyzed by α-aceto-α-hydroxy acid synthase (*ilvB*):

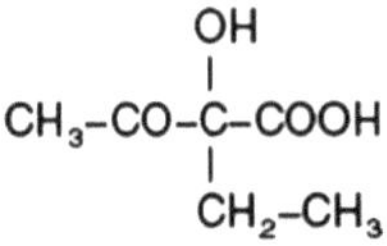

α-aceto-α hydroxybutyric acid

The enzyme contains flavin adenine dinucleotide (FAD), essential for activity. The function of FAD is unclear since the enzyme does not catalyze an oxidation-reduction reaction. The hypothesis has been proposed that the requirement for FAD is a vestige of evolution and that the acetohydroxyacid synthase evolved from pyruvate oxidase whose FAD binding site may have been retained to maintain the enzyme structure. The enzyme contains a small subunit (ca. 10 kDa) that is required for feedback inhibition by the branched-chain amino acids.

α-aceto-α-hydroxybutyric acid has the same number of carbon atoms as isoleucine but the carbon skeleton must be rearranged. This is brought about by an acetohydroxyacid isomeroreductase coded by *ilvC*, which catalyzes a pinacol rearrangement accompanied by a reduction of the carbonyl group, leading to α, β-dihydroxymethylvaleric acid (trivial name, dihydroxyisoleucine)

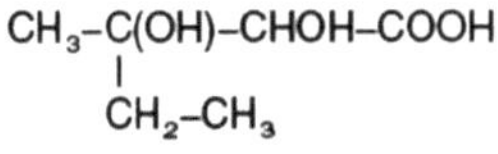

α, β-dihydroxy-β-methylvaleric acid

The *ilvD* gene specifies a dihydroxyacid dehydratase which catalyzes the loss of a molecule of water and the α-keto acid corresponding to isoleucine is produced:

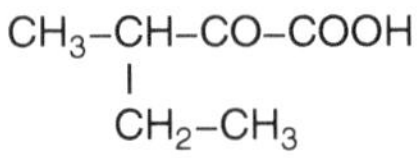

α-keto-β-methylvaleric acid (α-ketoisoleucine)

This dehydratase contains a Fe-S cluster (as several hydrolyases, such as aconitase and fumarase where they participate as Lewis acids in the dehydration reaction catalyzed by these enzymes). *E. coli* cells treated with hyperbaric oxygen lose their dehydratase activity, which is reactivated by return to ambient oxygen. This reversible inactivation is due to the destruction and reactivation of the Fe-S cluster.

The last step of isoleucine synthesis (Fig. 25.2) is due to a transamination between α-ketoisoleucine and glutamic acid, catalyzed by a transaminase coded by the *ilvE* gene.

COOH
I
$CH{-}NH_2$
I
$CH{-}NH_3$
I
$CH_2{-}NH_3$
L-isoleucine

(Note that L-isoleucine possesses a second asymmetric carbon atom).

Ectoine Biosynthesis

Ectoine (1,4,5,6-tetrahydro-2-methyl-4-pyrimidine carboxylic acid) is an osmoprotectant present in *Halomonas elongata* as well as other halophilic bacteria. The pathway begins with the phosphorylation of aspartate and the first two steps are shared with the biosynthesis of amino acids in the aspartate family.

A specific transaminase, coded by the *ectB* gene, catalyzes the following reaction

$$\text{Aspartatesemialdehyde} + \text{glutamate} \rightarrow \alpha - \text{ketoglutarate} + \underset{\text{L}-2,4\,\text{diaminobutyrate}}{NH_2 - CH_2 - CH\left(NH_2\right) - COOH}$$

Diaminobutyrate N-acetyltransferase (*ectA*) leads to N-acetyl-L-2,4 diaminobutyrate

$$\text{L - 2, 4 diaminobutyrate} + \text{acetyl CoA} \rightarrow \text{N - acetyl - L - 2, 4 diaminobutyrate} + \text{CoA}$$

This compound loses a molecule of water and is cyclized to ectoine by ectoine synthase (*ectC*)

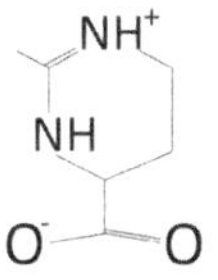

Ectoine

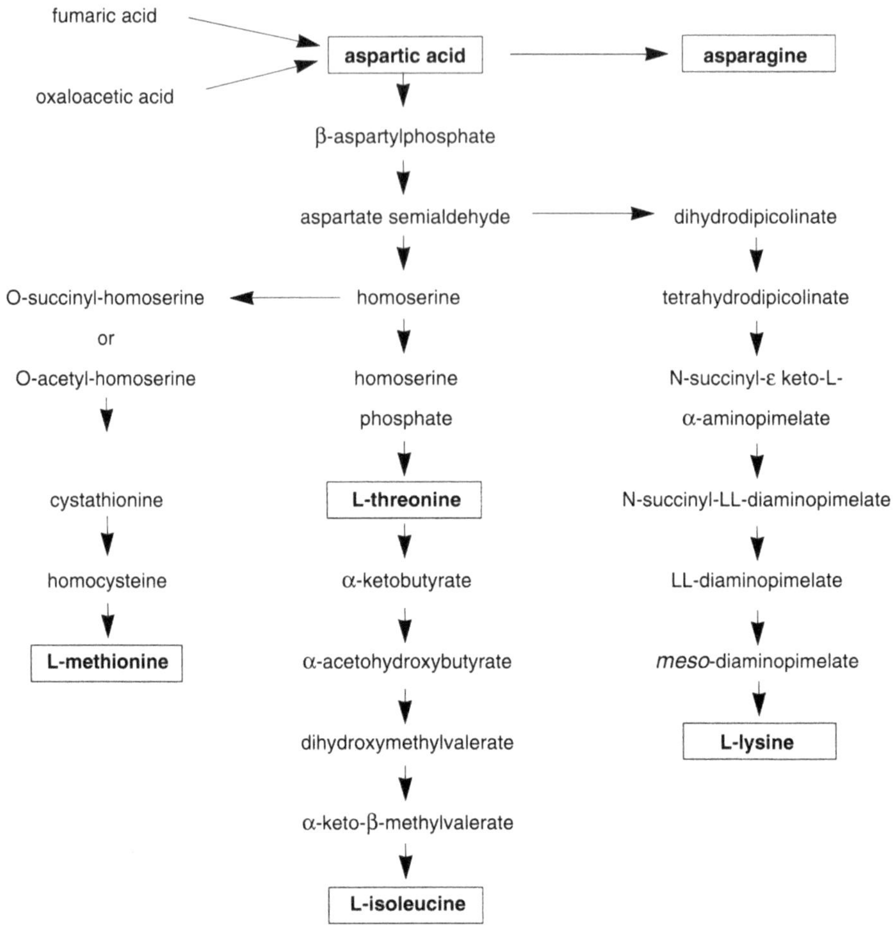

Fig. 25.2 Summary of the biosynthetic pathway of the aspartate family of amino acids

Bacterial osmoadaptation involves the cytoplasmic accumulation of compatible solutes to counteract extracellular osmolarity. The halophilic and highly halotolerant bacterium *Chromohalobacter salexigens* is able to grow up to 3 M NaCl in a minimal medium due to the *de novo* synthesis of ectoine. This is an osmoregulated pathway that burdens central metabolic routes by quantitatively drawing off TCA cycle intermediaries. Consequently, metabolism in *C. salexigens* has adapted to support this biosynthetic route. Metabolism of *C. salexigens* is more efficient at high salinity than at low salinity, as reflected by lower glucose consumption, lower metabolite overflow, and higher biomass yield. At low salinity, by-products (mainly gluconate, pyruvate, and acetate) accumulate extracellularly. Using [1-13C]-, [2-13C]-, [6-13C]-, and [U-13C6]glucose as carbon sources, the main central metabolic pathways involved in ectoine biosynthesis from glucose have been determined. *C. salexigens* uses the Entner-Doudoroff pathway rather than the standard glycolytic pathway for glucose catabolism, and anaplerotic activity is high to replenish the TCA cycle with the intermediaries withdrawn for ectoine biosynthesis. Metabolic flux ratios at low and high salinity were similar, revealing a certain metabolic rigidity, probably due to its specialization to support high biosynthetic fluxes and partially explaining why metabolic yields are so highly affected by salinity.

Selected References

To be consulted for further information on the topics treated in this chapter and the following ones up to Chap. 35 included:

Herrmann KM, Somerville R (eds) (1983) Amino acids: biosynthesis and genetic regulation. Addison-Wesley Publishing Company, Reading, p 453 pp

Meister A (1965) Biochemistry of the amino acids, vol 1 and 2, 2nd edn. Academic, New York

Neidhardt F, Curtiss R III, Ingraham JL, Lin ECC, Low KB, Magasanik B, Reznikoff WS, Riley M, Schaechter M, Umbarger HE (eds) (1996) *Escherichia coli* and *Salmonella typhimurium*. Cellular and molecular biology, 2nd edn. American Society for Microbiology, Washington, DC

Asparagine

Curnow AW, Tumbula DL, Pelaschier JT, Min B, Soll D (1998) Proc Natl Acad Sci USA 95:12838–12843

Min B, Pelaschier JT, Graham DE, Tumbula-Hansen D, Soll D (2002) Proc Natl Acad Sci USA 99:2678–83

Threonine Synthase

Garrido-Franco M, Ehlert S, Messerschmidt A, Marinkovic S, Huber R, Laber B, Bourenkov GP, Clausen T (2002) J Biol Chem 277:12396–12405

Murakawa T, Machida Y, Hayashi H (2011) J Biol Chem 286:2774–2784

Diaminopimelate Epimerase

Hor L, Dobson RC, Downton MT, Wagner J, Hutton CA, Perugini MA (2013) J Biol Chem 288:9238–9248

Diaminopimelate Decarboxylase

Ray SS, Bonanno JB, Rajashankar KR, Pinho MG, He G, De Lencastre H, Tomasz A, Burley SK (2002) Structure (Camb) 10:1499–1508

Methionine Biosynthesis; Direct Sulfhydrylation Pathway

Hwang BJ, Yeom HJ, Kim Y, Lee HS (2002) J Bacteriol 184:1277–86

Ectoine Biosynthesis and Osmoregulation

Pastor JM, Bernal V, Salvador M, Argandona M, Vargas C, sonka LC, Sevilla A, Iborra JL, Nieto JJ, Canovas M (2013) J Biol Chem 288:17769–17781

Ectoine Biosynthesis. A Specialized Aspartokinase

Stöveken N, Pittelkow M, Sinner T, Jensen RA, Heider J, Bremer E (2011) J Bacteriol 193:4456–4468

S-Adenosylmethionine

Markham GD, Hafner EW, Tabor CW, Tabor H (1980) J Biol Chem 255:9082–9092

Zhang Q, Liu W (2011) J Biol Chem 286:30245–30252

Chapter 26
Regulation of the Biosynthesis of the Amino Acids of the Aspartic Acid Family in *Enterobacteriaceae*

Cells have two main means at their disposal for regulating their metabolic reaction rates:

1. they can use small molecules for certain target enzymes, which as a rule belong to a certain class of proteins, the allosteric enzymes. These ligands bring about conformational changes which result sometimes in a lower enzyme activity (allosteric inhibition) and sometimes in a higher (allosteric activation).
2. the presence of essential metabolites in the cytoplasm at a certain concentration causes the repression, at the level of transcription, of the biosynthesis of enzymes leading to the biosynthesis of such metabolite.

A Paradigm of Isofunctional and Multifunctional Enzymes and of the Allosteric Equilibrium

We have seen in the preceding chapter that the biosynthetic pathway leading from aspartic acid to diaminopimelic acid and lysine on the one hand, and to methionine, threonine and isoleucine on the other is not a linear sequence of reactions, but a highly branched chain. Let us look at the implications of this fact for growing bacterial cells.

(a) Consider a linear sequence of reactions:

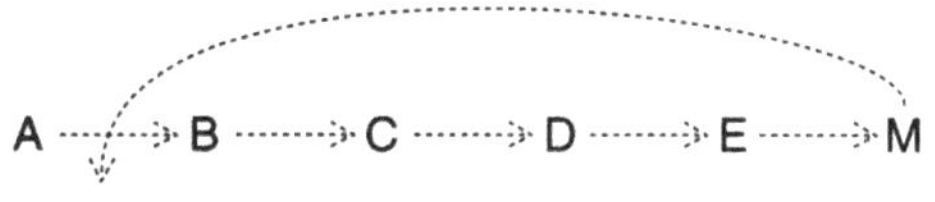

Leading to the synthesis of the essential metabolite M. In the presence of excess M in the intracellular pool, the enzyme carrying out the conversion A → B will be inhibited by M if, as is generally the case, this enzyme is an allosteric protein. The presence of M may also repress the synthesis of the enzymes of the

G.N. Cohen, *Microbial Biochemistry*,
DOI 10.1007/978-94-017-8908-0_26, © Springer Science+Business Media B.V. 2014

whole chain. These control phenomena, which are economical, do not cause any detrimental side effects, because the final product M is in excess; when production of M falls sufficiently, the activity of the first enzyme will increase and the synthesis of the enzymes of the chain will take place at a rate which is greater than under the conditions of repression. This new state will hold until M again reaches a concentration high enough to cause inhibition and repression. Oscillations in the size of the pool are therefore expected and have been observed experimentally in several instances.

(b) Now consider a branched reaction sequence such as that of the aspartate family of amino acids:

aspartate
↓
aspartyl phosphate
↓
aspartate semialdehyde → lysine
↓
homoserine → threonine → isoleucine
↓
↓
↓
methionine

Examination of the diagram of the reaction sequence shows that accumulation of any one of the end products and control by either of the above regulatory processes would reduce the production of aspartyl phosphate consecutive to the decreased aspartokinase activity and consequently to a deficiency in the production of the other end products.

The solution of this potential dilemma differs according to the organism studied. Let us first consider the case of the Enterobacteriaceae, in particular that of *E. coli*.

Two Aspartokinases in *E. Coli*

Of all the natural amino acids known to occur naturally as protein constituents, only two, L-lysine and L-threonine, were found to influence the aspartokinase activity of crude extracts. Neither of the two amino acids inhibit totally the activity, but as shown in Fig. 26.1, when they are added simultaneously, the total inhibition is the sum of that observed for each independently. This suggests that crude extracts contain at least two aspartokinases and this interpretation is borne out by the physical separation of the two enzyme activities by ammonium sulfate fractionation.

A kinetic analysis of the inhibitions vs. aspartate concentrations shows that the threonine sensitive enzyme is competitively inhibited whereas the inhibition by lysine of the other kinase is non-competitive (Fig. 26.2).

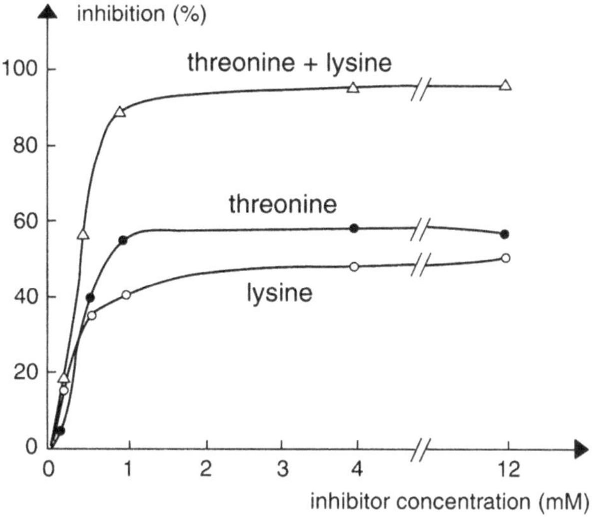

Fig. 26.1 The inhibitory effects of L-lysine and L-threonine or both on aspartokinase activity (From Stadtman et al. (1961), with permission from the *Journal of Biological Chemistry*)

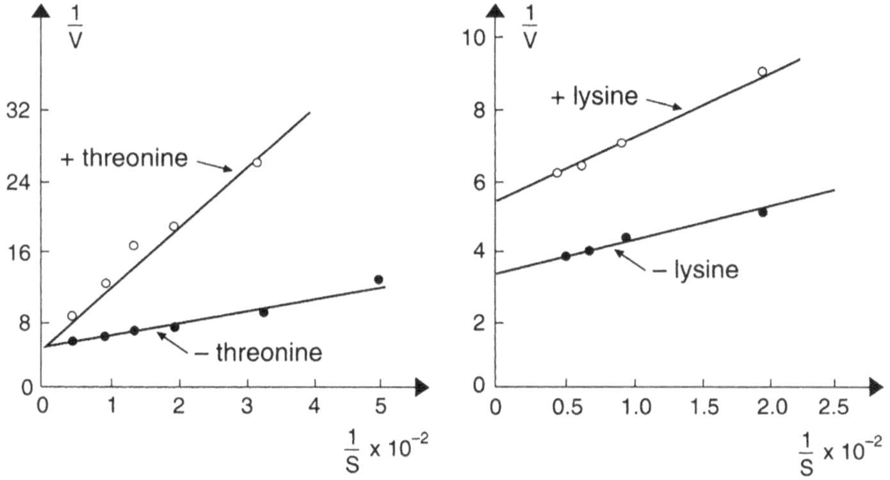

Fig. 26.2 Inhibition analysis of the lysine- and threonine-sensitive aspartokinases (From Stadtman et al. (1961), with permission from the *Journal of Biological Chemistry*)

The synthesis of the lysine-sensitive enzyme is repressed when the organism is grown in the presence of lysine: only the threonine-sensitive enzyme is present in the extracts. On the other hand, the synthesis of the threonine-sensitive enzyme is subject to a multivalent repression by threonine and isoleucine.

The finding that a single organism may contain more than one enzyme catalyzing the same biochemical reaction was not uncommon at the time of the above findings (for example the biosynthetic and degradative threonine dehydratases). The discovery, however, of the existence of two aspartokinases in *E. coli*, was of special significance because it offered a reasonable explanation for the multiplicity of enzymes. The rational solution to the dilemma posed by the regulation of branched biosynthetic pathways was the synthesis of multiple, isofunctional enzymes which catalyze the synthesis of the common precursor, aspartyl phosphate, each of which is independently subject to feedback inhibition and to repression by different end-product metabolites.

The Threonine-Sensitive Homoserine Dehydrogenase of E. Coli

This enzyme, which catalyzes the third reaction between aspartate and homoserine, is inhibited very specifically by L-threonine. The inhibition is non-competitive versus both substrates, homoserine and NADPH. L-threonine is the best stabilizing agent, along with KCl. By using an inhibitor analogue, α-amino-β-hydroxyvaleric acid, mutants resistant to the inhibition by threonine have been selected which have been useful in mapping the structural gene for this homoserine dehydrogenase. The synthesis of homoserine dehydrogenase, as in the case of the threonine-sensitive aspartokinase, is repressed when threonine and isoleucine are present in excess and is derepressed upon limitation of either of the two amino acids.

Isolation of a Mutant Lacking the Lysine-Sensitive Aspartokinase and of Revertants Thereof

The presence of several aspartokinases makes it awkward to apply the penicillin selection method to isolate directly mutants having lost one or the other aspartokinase. The principle of the selection can be understood easily: in the presence of threonine and isoleucine, the synthesis of homoserine is repressed and the activity of one aspartokinase is inhibited by threonine; as a result, growth is severely inhibited due to the artificial methionine limitation thus created. Now if, among the mutagenized population, there are some individuals who have lost the lysine-sensitive aspartokinase activity, since the remaining kinase has its synthesis repressed and its activity inhibited, these individuals will have no means to synthesize their diaminopimelate and lysine: they will not grow and will not be killed by penicillin. Such an individual was isolated; its phenotype is that of a prototroph growing perfectly on minimal medium, but exhibiting a conditional requirement for diaminopimelate in the presence of threonine, isoleucine and methionine. The defect causing this phenotype was localized on the *E. coli* chromosome.

Spontaneous revertants of this mutant were selected on the basis of their capacity to grow on threonine plus isoleucine plus methionine. Some of them were heavy threonine excretors. They had not reacquired the lysine-sensitive aspartokinase, but the other kinase was resistant to end-production inhibition by threonine, thus permitting a more rapid flow of aspartyl phosphate and aspartic semialdehyde and a normal rate of synthesis of DAP and lysine. Because they overproduce threonine, the levels of aspartokinase and of homoserine dehydrogenase are at

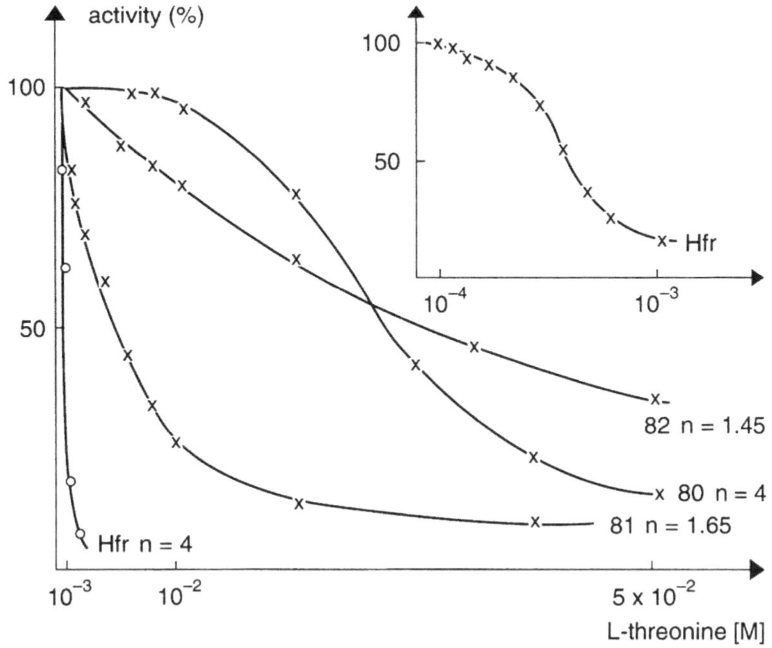

Fig. 26.3 Inhibition by threonine of threonine-sensitive aspartokinases. The *inset* is an enlargement of the wild type (Hfr) curve (From J.-C. Patte, P. Truffa-Bachi and G.N. Cohen, *Biophysica et Biochimica Acta*, with permission of Elsevier Science Publishers)

the repressed value, precluding the easy study of these enzymes. Therefore, the part of their genome responsible for the excretion character was transduced into an Ile- strain, in order to enable derepressed synthesis in a chemostat limited in isoleucine.

It was immediately apparent (Fig. 26.3) that the revertants possess a modified threonine-sensitive aspartokinase: a concentration of threonine which causes a complete inhibition in the wild-type extracts causes either no inhibition or an incomplete inhibition, according to the mutant extract tested.

A surprising fact was then uncovered: the homoserine dehydrogenase of the revertants was *also* modified with respect to their capacity to be inhibited by threonine (Fig. 26.4). The apparent substrate affinity constants of the two enzymes obtained from the revertants do not differ significantly from those obtained in the wild-type. The fact that the two activities are modified as a result of a single mutation prompted the reexamination of a mutant where feedback-resistance of the threonine-sensitive aspartokinase had been obtained by resistance to α-amino-β-hydroxyvalerate: its homoserine dehydrogenase was also resistant to inhibition by threonine.

The fact that the two enzymes are affected in the same way by a single mutational event led to the inescapable conclusion that they are carried by the same protein.

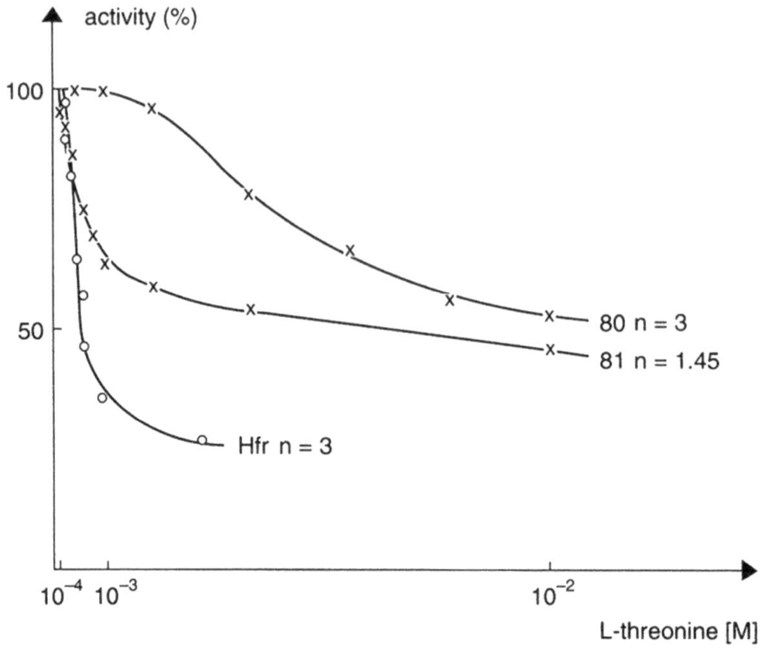

Fig. 26.4 Inhibition by L-threonine of homoserine dehydrogenases from the wild type and from mutants (From J.-C. Patte, P. Truffa-Bachi and G.N. Cohen, *Biophysica et Biochimica Acta*, with permission of Elsevier Science Publishers)

Evidence That the Threonine-Sensitive Aspartokinase and Homoserine Dehydrogenase of E. Coli *Are Carried by the Same Bifunctional Protein*

Indirect evidence was obtained by the observation that the kinase activity is weakly inhibited by homoserine and $NADP^+$, and that the dehydrogenase activity is notably inhibited by aspartate and ATP. A much stronger argument is provided by studies on heat stability: whereas NADPH is a very efficient protector against thermal inactivation of the dehydrogenase, it also protects the kinase for which it is not a substrate. As a control, it has been established that NADPH does not protect the lysine-sensitive aspartokinase (efficiently protected by L-lysine) against thermal inactivation, although it catalyzes exactly the same reaction as the threonine-sensitive aspartokinase.

The single protein carrying the two activities was purified to homogeneity by standard procedures. It has a molecular weight of 360,000 as determined by ultracentrifugation and light scattering studies. It is composed of four subunits as determined by cross linking with dimethylsuberimidate. Gel filtration and isoelectric focusing showed that the subunits have an identical molecular

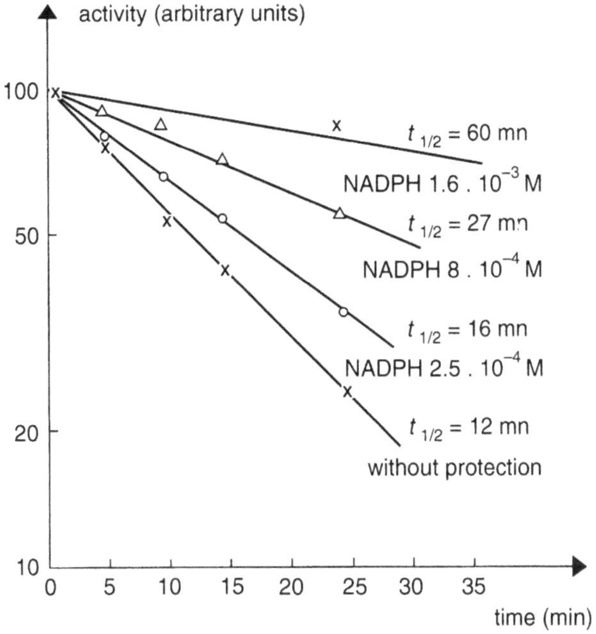

Fig. 26.5 Protection against thermal inactivation of the threonine-sensitive aspartokinase activity by NADPH. Temperature: 44.8 °C (From J.-C. Patte, P. Truffa-Bachi and G.N. Cohen, *Biophysica et Biochimica Acta*, with permission of Elsevier Science Publishers)

weight and are identically charged. Finally, sequence studies showed that the subunits are identical. The sequence of the polypeptide chain was determined by the methods of protein sequencing and by deducing it by the sequence of the cloned gene *thrA*. For reasons that will become evident later in this chapter, the bifunctional protein was called aspartokinase I-homoserine dehydrogenase I (AKI-HDHI) (Fig. 26.5).

When AKI-HDHI is subjected to mild proteolysis, the kinase is destroyed. A fragment carrying only the homoserine dehydrogenase activity resists to proteolysis and was purified. It has the same carboxy-terminal sequence as the native polypeptide. This fragment is a homodimer whose subunit has a molecular weight of ca. 58,000 as opposed to 89,000 for the subunit of the native protein. The dehydrogenase activity of the dimer is totally insensitive to threonine. On the other hand, a tetrameric threonine-sensitive aspartokinase fragment was purified from a nonsense (ochre) mutant. As expected, it has the same N-terminal sequence as the native polypeptide. Since the molecular weight of the fragment from the mutant is of ca. 48,000, the two fragments have a common overlapping sequence.

A more refined study of proteolysis has shown that the production of the dimeric dehydrogenase is preceded by the formation of a hybrid fragment H composed of a native chain and of a proteolyzed chain. A careful study led to the elucidation of the sequence of events summarized in Fig. 26.6 and to the triglobular model of quaternary structure depicted in Fig. 26.7.

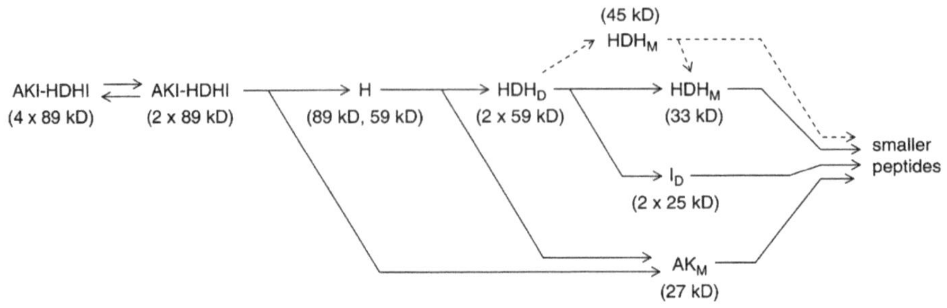

Fig. 26.6 Pathway of the proteolysis of AKI-HDHI. The native tetrameric AKI-HDHI is dissociated into a dimer. The dimer is cleaved into the hybrid (*H*) and the monomeric aspartokinase fragment (*AKM*). The intact chain of the dimer is then proteolyzed to a second AKM fragment and the homodimeric fragment HDHD is obtained. The next step in the proteolysis is the cleavage of HDHD to produce monomeric homoserine dehydrogenase (*HDHM*) and an inactive dimeric fragment (*ID*). AKM, HDHM and ID are further cleaved to give smaller fragments of about 10–12 kDa (From A. Fazel, K. Müller, G. LeBras, J-R. Garel, M. Véron and G.N. Cohen, with permission of *Biochemistry*)

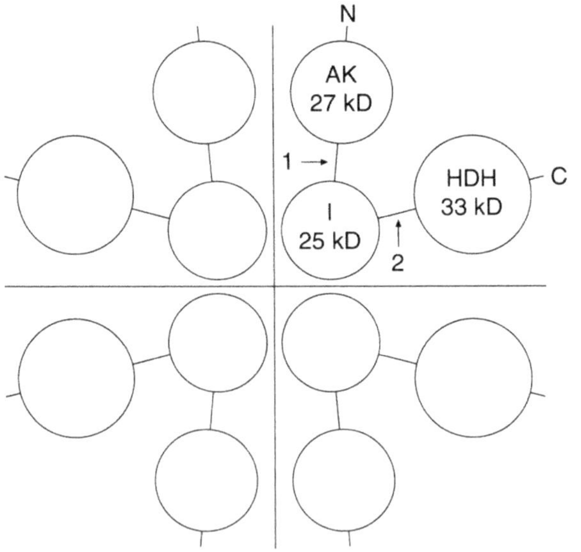

Fig. 26.7 Domain I is depicted as responsible for the subunit contacts which generate the dimeric structure of H and HDHD and possibly the tetrameric structure of the native enzyme. The actual polymeric fragments obtained by limited proteolysis or by extraction from the ochre mutant (see text) can be visualized as the sum of I and of either the two catalytically active fragments, yielding respectively the corresponding dimmer (HDHD) or tetramer (AK from the ochre mutant). The *arrows* 1 and 2 indicate the sites of the corresponding proteolytic or mutational cleavages (From A. Fazel, K. Müller, G. LeBras, J-R. Garel, M. Véron and G.N. Cohen, with permission of *Biochemistry*)

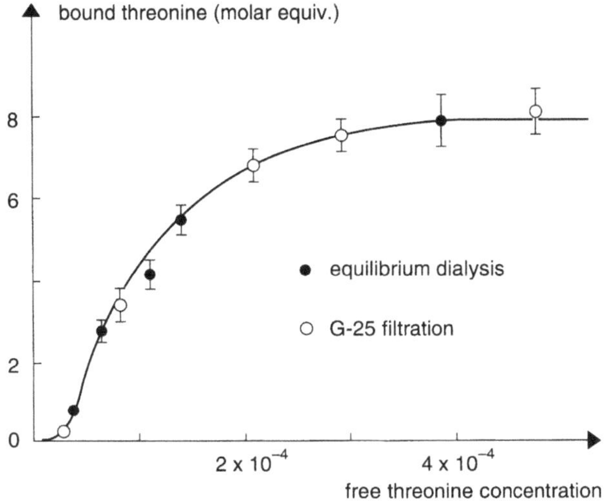

Fig. 26.8 Binding of L-threonine to AKI-HDHI (From P. Truffa-Bachi, M. Véron and G.N. Cohen, with permission of *CRC Critical Reviews in Biochemistry*)

The Binding of Threonine to Aspartokinase I-Homoserine Dehydrogenase I

The threonine binding experiments were performed using ^{14}C-L-threonine, either by equilibrium dialysis or by gel filtration on Sephadex G-25 preequilibrated with the ligand. Fig. 26.8 shows that saturation of the binding sites is reached at free ligand concentrations in the 0.3–0.5 mM range, where 8 moles of threonine are bound to 1 mole of enzyme. The same experiment performed after treatment with -SH reagents which desensitize the homoserine dehydrogenase activity (and destroy the aspartokinase activity) show the loss of binding capacity toward threonine. In the same vein, the homodimeric threonine-insensitive homoserine dehydrogenase proteolytic fragment does not bind threonine either.

We have seen that the inhibition of the two activities of the protein by threonine is strongly cooperative (see Figs. 26.3 and 26.4). It is interesting to find that the binding of threonine also shows cooperativity. At non-saturating concentrations of threonine, the binding depends upon a number of factors, especially the concentration of KCl, which is found to displace in a very similar manner the inhibition curve of homoserine dehydrogenase and aspartokinase activities. In the presence of aspartate, the binding of threonine is also found to decrease, a result that could be expected since the inhibition of aspartokinase is competitive toward aspartate. The large cooperativity of inhibition by threonine (Hill number equal to four, near complete inhibition of the aspartokinase activity) implies several allosteric sites for the protein. In this view, it is satisfactory to find that, at saturation, the protein composed of four identical subunits binds eight equivalents of threonine and it can be inferred that each native subunit carries two sites for binding the allosteric inhibitor (Fig. 26.8).

Fig. 26.9 Titration of -SH groups of AKI-HDHI with dithio-bis-nitrobenzoic acid (*DTNB*). The reaction was performed at 20°, with .8 μM protein and 25 mM DTNB, with or without allosteric effector added (From P. Truffa-Bachi, R. van Rapenbusch, J. Janin, C. Gros and G.N. Cohen, with permission of the *European Journal of Biochemistry*)

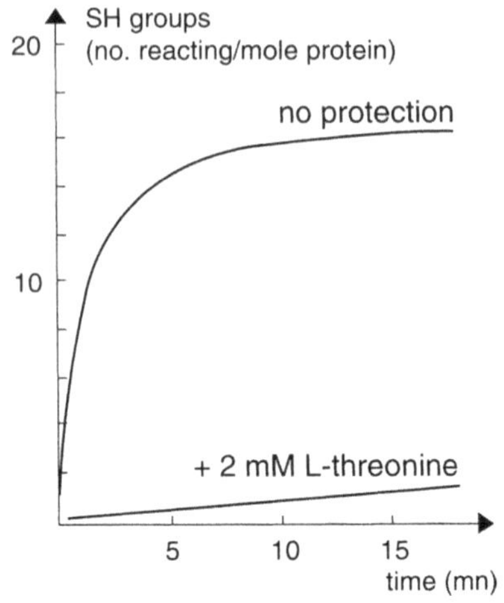

The Binding of Pyridine Nucleotides to Aspartokinase I-Homoserine Dehydrogenase I

The binding of $NADP^+$ has been studied by the gel filtration technique using the ^{14}C-labeled coenzyme. The linear Scatchard plot indicates a simple Michaelian type of variation with ligand concentration. Extrapolation leads to a value of 4 mol of $NADP^+$ bound per mole of enzyme.

The Effects of Threonine on Aspartokinase I-Homoserine Dehydrogenase I Are Not Only due to Direct Interactions

Quantitative measures on the protein showed that whereas many sulfhydryl groups could be titrated by specific reagents, essentially no -SH groups could be titrated in the presence of threonine (Fig. 26.9). The magnitude of this effect renders it unlikely that it is due to a direct steric effect of the threonine molecules "covering" 18 cysteine residues. More probably, in the presence of the allosteric effector, the protein undergoes a conformation change such that as 18 cysteine residues become buried.

Aspartokinase inactivation and desensitization of the homoserine dehydrogenase activity were concomitant with the titration. At any given moment, the remaining aspartokinase activity is fully sensitive to threonine inhibition (Fig. 26.10). When the reaction is performed in the presence of threonine, no effect on either enzymatic activity is found.

When a solution of the enzyme is prepared in a buffer containing threonine, slight differences in the absorption spectrum appear, which are evidenced by the difference spectrum of Fig. 26.11. The salient features are the peaks at 289 and

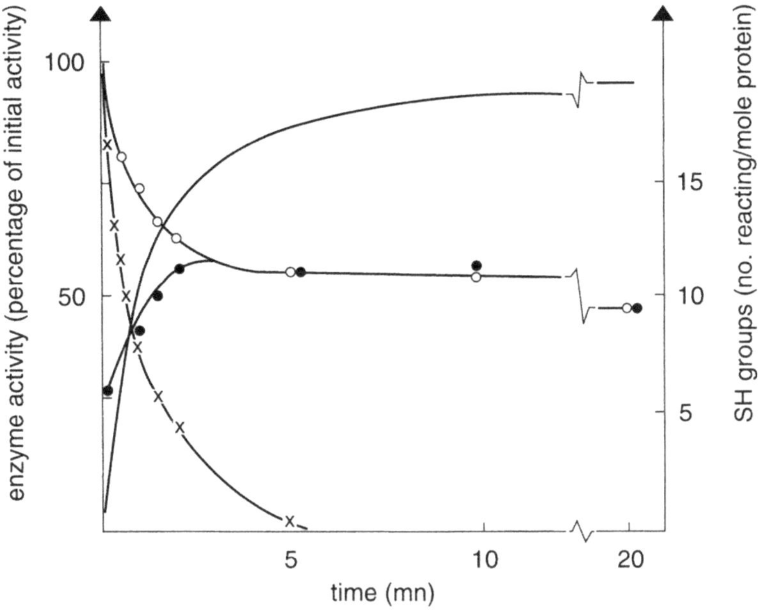

Fig. 26.10 Effect of DTNB on the enzymatic activities of aspartokinaseI-homoserine dehydrogenaseI. –: titration of -SH groups; ○, homoserine dehydrogenase activity; ●, homoserine dehydrogenase activity measured in the presence of 20 mM L-threonine; ×, aspartokinase activity. No threonine was present in the mixture. Conditions essentially as in Fig. 26.9. Samples were diluted 50 times at various times; the activities and their inhibition by L-threonine were immediately tested (From P. Truffa-Bachi, R. van Rapenbusch, J. Janin, C. Gros and G.N. Cohen, with permission of the *European Journal of Biochemistry*)

281 nm, and the complex trough around 269 nm. The analogy with the perturbation spectra of tyrosine and tryptophan residues dissolved in non-polar solvents suggests that both types of residues are involved and move to an environment less polar than water when threonine is added.

The fluorescence of most proteins is associated with their ultraviolet absorption with a major contribution of their tryptophan residues. When excited at 285 nm, the aspartokinase I-homoserine dehydrogenase I has a single emission band around 340 nm, entirely attributable to its aromatic residues (Fig. 26.12). The same figure shows that the effect of 1 mM L-threonine on the excitation spectrum is comparable to the difference spectrum observed above: the emission maximum is shifted toward the longer wavelengths, and its intensity decreases by 12–15 %. Some quenching, but no shift of the peaks, is also noted upon addition of other ligands.

As was found with other dehydrogenases, the complex formed with the reduced coenzyme has a characteristic fluorescence. When an excess of protein is added to a solution of NAPDH (Fig. 26.13), the blue fluorescence of the coenzyme excited in the 340 nm absorption band is considerably increased; moreover, a new excitation band appears which corresponds to the absorption of the protein, showing

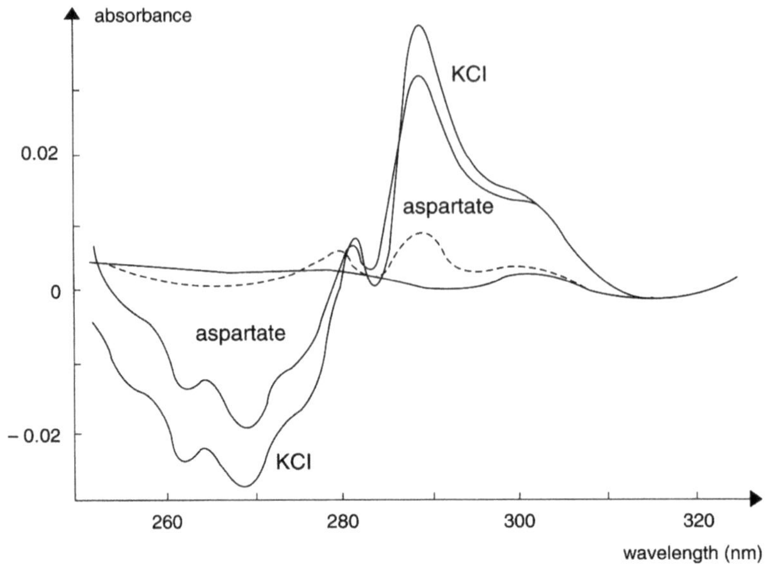

Fig. 26.11 Difference spectra in the presence of threonine. *Baseline*: reference and sample, no addition. *Dashed line*: reference, no addition; sample: 5 mM L-threonine. Aspartate spectrum: reference 15 mM L-aspartate; sample: 15 mM L-aspartate+.5 mM L-threonine. K spectrum: reference, 15 M KCl; sample, 15 M KCl+5 mM L-threonine (From J. Janin and G.N. Cohen, with permission of the *European Journal of Biochemistry*)

Fig. 26.12 Fluorescence of aspartokinase I-homoserine dehydrogenase I. The excitation emission observed at 345 nm and emission (excitation at 282 nm) spectra were recorded (*continuous line*). Then 1 mM L-threonine was added and the same measurements repeated (*broken line*) (From J. Janin, R. van Rapenbusch, P. Truffa-Bachi and G.N. Cohen, with permission of *European Journal of Biochemistry*)

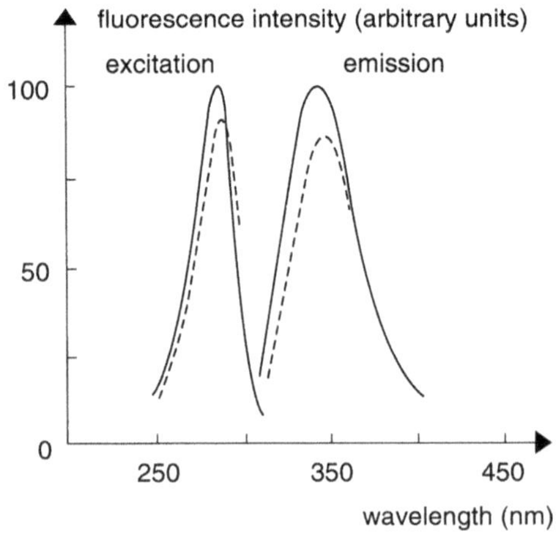

that excitation is transferred from an aromatic residue to the coenzyme. Finally, Fig. 26.13 also shows that the fluorescence of the protein-NADPH complex is strongly quenched upon addition of threonine. This considerable effect may be partly attributed to a perturbation of the same aromatic residues which leads to the difference spectrum, but the environment of the bound coenzyme is also modified,

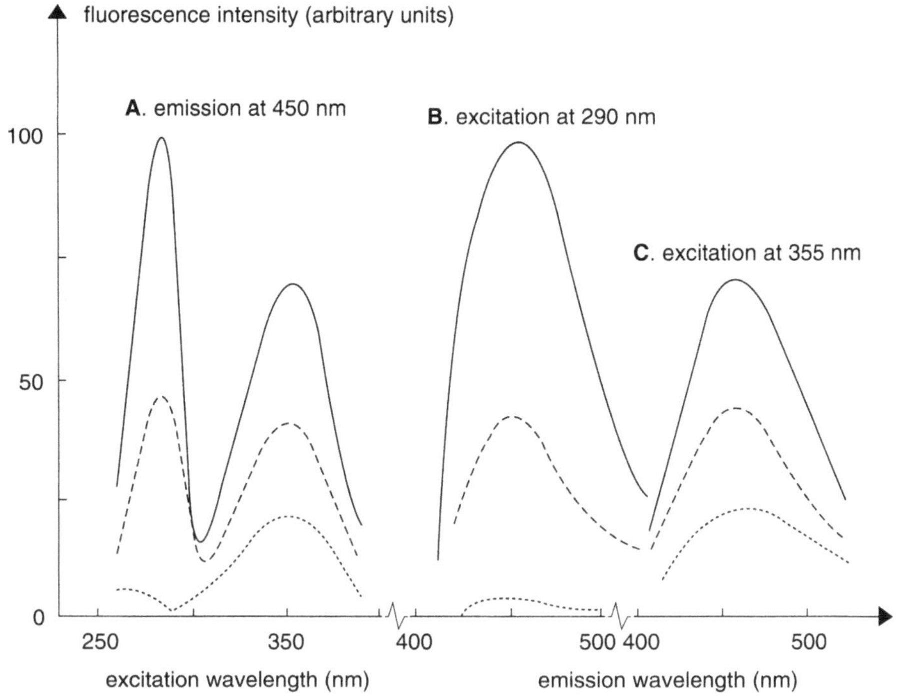

Fig. 26.13 Fluorescence of the aspartokinase I-homoserine dehydrogenase I-NADPH complex. The fluorescence spectra of a solution of 1.5 μM protein and 3 μM NADPH were recorded before (*continuous line*) and after (*broken line*) addition of 1 mM L-threonine. Corrections were made for absorption. Emission was observed at 450 nm for the excitation spectrum (**a**); the excitation wavelength was 290 nm for the transfer emission spectrum (**b**) and 355 nm for the emission spectrum of NADPH (**c**). The *dotted line* represents the fluorescence of 3 μM NADPH under the same conditions (From J. Janin, R. van Rapenbusch, P. Truffa-Bachi and G.N. Cohen, with permission of *European Journal of Biochemistry*)

since both the fluorescence upon excitation in the absorption band of the coenzyme and the transfer fluorescence are quenched.

Taken collectively, these results on the effect of threonine on the microenvironments of cysteine and aromatic residues can hardly be attributed to direct interactions of eight molecules of inhibitor with the groups involved in each case and therefore indicate a conformational change of the protein from a "relaxed" form in absence of threonine to a "tight" form in its presence.

The Allosteric Transition of Aspartokinase I-Dehydrogenase I

The difference spectra and the effects on the fluorescence of the protein can be used to define two conformations of the protein. Let T be the form with buried chromophores (low absorption at 269 nm and high absorption at 289 nm) and R the form with exposed chromophores. A straightforward interpretation of the data is that T is stable in the absence of added ligand or in the presence of threonine, and R is stable in the presence

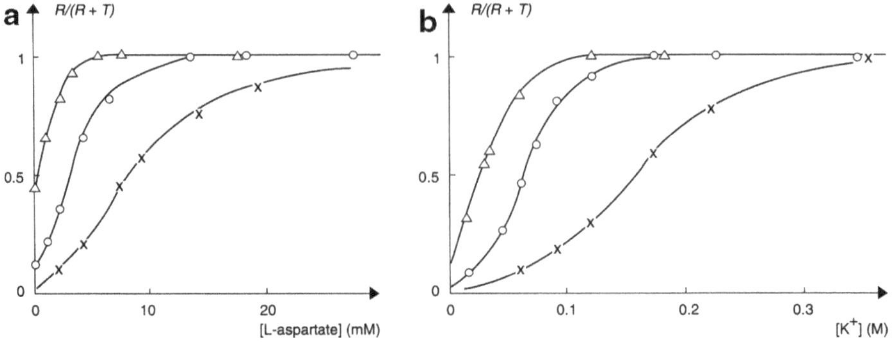

Fig. 26.14 Effects of aspartate and K+ on the allosteric equilibrium. The values of R/R+T in the presence of various concentrations of ligands are derived from fluorescence intensity of the protein. (**a**) Effects of aspartate. Supplements: O, none; Δ: 30 mM KCl; ×: 40 μM L-threonine (**b**) effects of K^+ Supplements: O: none; Δ: 2 mM L-aspartate; ×: 40 μM L-threonine (From J. Janin and G.N. Cohen, with permission of the *European Journal of Biochemistry*)

of potassium ions or of aspartate. As a consequence of this model, difference spectra similar to these should be obtained in the absence of added threonine when samples of protein in buffer containing KCl or aspartate are compared with similar solutions with no added ligands. This is found to be the case. Aspartate and K^+, at least, are not structurally related, and it is a priori likely that the three ligands under study have different binding sites. Short range perturbations of the aromatic residues of the protein would then be expected to have very different effects on the ultraviolet absorption when aspartate or K^+ are bound. The similarity of the actual spectra implies that the ligands act by shifting an equilibrium between conformations of the protein rather than merely modifying the environment of their respective binding sites.

Advantage has been taken of the easiness and sensitivity of the fluorimetric measurements for studying the effects of intermediate concentrations of the ligands and describing more quantitatively the changes of the protein conformation. Upon adding threonine to enzyme solutions containing KCl, the variations in absorbance and fluorescence are coordinated. It can be assumed that both techniques detect the same phenomenon. Figs. 26.14 and 26.15 describe the effects of the three ligands on the equilibrium. The ordinates R/R+T or T/R+T represent the fraction of the protein in the R or T states. It can be seen in Fig. 26.14a, b that aspartate and K^+ give positive homotropic effects (the curves are sigmoidal) and positive heterotropic effects (they help each other), but they give negative heterotropic effects with threonine. When threonine is added to protein solutions containing various amounts of the other ligands (Fig. 26.15a, b) similar features are observed.

As a first approximation, the above experimental facts may be visualised by a two-state model.

$$\overset{\text{Threonine}}{\underset{\text{Aspartate, K}^+}{\text{R} \Leftrightarrow \text{T}}}$$

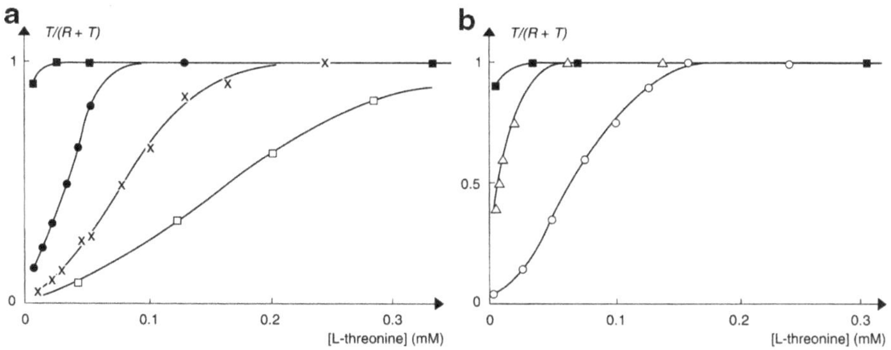

Fig. 26.15 Effects of L-threonine on the allosteric equilibrium L-threonine is added to a solution of protein containing the following supplements: (**a**) ■, none; ●, 60 mM KCl; ×: 0.15 M KCl; □ 0.45 M KCl. (**b**) ■, none; Δ, 5 mM L-aspartate; ○, 020 mM L-aspartate (From J. Janin and G.N. Cohen, with permission of the *European Journal of Biochemistry*)

Table 26.1 General properties of the R and T confirmations

Property	T form	R form	Remarks
Binding of ligands (dissociation constants)			Shifts the equilibrium towards
Threonine	K_T=30 μM	K_R>1 mM[a]	T form
Aspartate	K_T>20 mM	K_R=3 mM	R form
K^+ ions	low affinity	high affinity	R form
NADPH	K_T=0.3 μM	K_R=0.3 μM	No detectable effect
Spectophotometric properties			
Relative fluorescence intensity	100	116	Excitation at 280 nm, emission at 335 nm
Tryptophan and tyrosine residues	Buried	Exposed	
Enzymatic activities aspartokinase	None	Active	T does not bind the substrate
Homoserine dehydrogenase	15–20 %	100 %	Different configurations of the active sites

[a]Binding of threonine at saturation of KCI or aspartate

A strong correlation may be established between the measurements of absorption, of fluorescence, and of threonine binding, all of which are compatible for the existence of two states. Table 26.1 shows a similar correlation between the catalytic and regulatory properties of the protein. The competitive inhibition of aspartokinase by threonine (K type) can be accounted for by the preferential binding of aspartate to the R form. In contrast, the inhibition of the dehydrogenase activity and the analogous activation by potassium do not seem to be an effect of the R ⇔ T transition on the binding of the substrates, for a noncompetitive and incomplete inhibition are observed (Fig. 26.4). They may be due to an actual requirement for K^+, or more likely, to an improper configuration of the dehydrogenase active sites. Although the binding constants of the R and T forms for NADPH are similar, the fluorescence of the T form is much smaller, indicating a different environment of the bound nucleotide.

Fast kinetic experiments were carried by stopped-flow and temperature jump in order to study the presteady state kinetics. When threonine was rapidly mixed with AK I-HDH I, an initial lag in the onset of threonine inhibition was observed. The rate constants for the lag times were dependent on threonine concentration but independent of protein concentration, thus reflecting an isomerization between active and inactive protein. The same result was essentially found in the temperature jump experiments, in the absence of any bound ligands by raising the temperature from 20 to 28 °C. Thus, the binding of the ligands is irrelevant to the equilibrium which preexists in the absence of binding substrates or effectors. All the kinetic predictions of the allosteric model are fulfilled in the particular case of this protein.

Aspartokinase II-Homoserine Dehydrogenase II

If the homoserine dehydrogenase is carried by a single species, AK I-HDH I, it should be possible to obtain mutants that, following the loss of this activity, would require either homoserine or threonine plus methionine. Such mutants have been described in *Bacillus subtilis*, *Neurospora crassa* and *Saccharomyces cerevisiae*. Two such mutants have also been obtained in *E. coli*, strain B; despite numerous efforts, it has for a long time been impossible to obtain a similar phenotype in *E. coli*, strain K12, the favorite and most studied strain by microbial geneticists. As a consequence, a transduction has been carried out introducing into K12 the mutated chromosomal piece corresponding to the inactivated homoserine dehydrogenase from *E. coli* B. The transductants can grow with threonine alone; in minimal medium, they eventually grow in 48 h. They have no AK I-HDH I activity, but weak threonine-insensitive aspartokinase and dehydrogenase activities. These activities are greatly lowered by growth in the presence of methionine. Finally, methionine requiring mutants of the transductants have been prepared which, when grown in a chemostat under methionine limitation, produce a high level of both threonine-insensitive activities. These two activities, whose synthesis is regulated by the level of the intracellular methionine pool have been called aspartokinase II (AKII) and homoserine dehydrogenase II (HDH II), the name of aspartokinase III (AK III) being attributed to the lysine-sensitive enzyme.

During a systematic survey of spontaneous revertants of the transductants, that is of mutants (devoid of AK I- HDH I activities) which can grow rapidly without added threonine, a mutant was isolated that synthesized constitutively AKII and HDHII. From this mutant, the purification of the two activities was achieved. The two activities could not be separated by the standard procedures of protein chemistry and an homogeneous protein, aspartokinase II-homoserine dehydrogenase II was isolated in the pure state.

It is amusing to notice that in *S. typhimurium* and in other strains of *E. coli*, AK II-HDH II exists in such amounts that all the sophistication which was applied to its detection in the K12 strain of *E. coli* would have been unnecessary with these organisms.

The molecular weight of AK II-HDH II, determined by equilibrium sedimentation is 185,000 ± 10,000. It is a dimer of identical subunits. In contrast to AK I-HDH I, it is not subject to regulation by allosteric control: neither of the two activities is inhibited by L-methionine, S-adenosylmethionine, L-threonine, or combinations, two by two, of methionine, threonine, isoleucine and lysine. There is no cooperative binding of the substrates for either of the activities. ATP and aspartate do not inhibit homoserine dehydrogenase II; homoserine and NADPH do not inhibit aspartokinase II. NADPH, a good protector against thermal inactivation of homoserine dehydrogenase II, does not protect aspartokinase II against such a treatment.

Upon limited proteolysis, as in the case of AKI-HDHI, aspartokinase activity gradually disappears and is finally completely destroyed, whereas homoserine dehydrogenase is essentially conserved. Two dimeric fragments are recovered, one being a dimer of 2×37,000 mol. wt., endowed with dehydrogenase activity, and one of 2×24,000 mol. wt., inactive. A careful study shows that, as in the case of the threonine-sensitive protein (a) the enzyme is triglobular; (b) the kinase resides on the N-terminal part and the dehydrogenase in the C-terminal part, the inactive fragment being central.

The gene coding for AK II-HDH II, *metL*, has been cloned and sequenced. The sequence homology between the two bifunctional enzymes will be discussed in Chap. 37.

Aspartokinase III

The enzyme is coded by *lysC*. The gene has been cloned and sequenced. The enzyme has been purified to homogeneity. It is a dimeric protein of 2×50,000, which is in equilibrium with both a tetrameric species, favored by high ionic strength in the presence of lysine, and a monomeric species, favored by low ionic strength in the absence of lysine. Only the dimeric form is enzymatically active. However, lysine causes a cooperative allosteric transition leading to an inactive dimeric species, at protein concentrations below those required for tetramerization.

Certain so-called nonspecific hydrophobic amino acids such as leucine, isoleucine, phenylalanine, and others at much higher concentrations also inhibit aspartokinase III substantially. Synergism of inhibition by lysine and low concentrations of the nonspecific amino acids appears to be physiologically significant. The enzyme is also inhibited in an allosteric fashion by the lysine analogs β-aminoethylcysteine (thialysine) and selenalysine. Thialysine is a growth inhibitor to which resistant mutants have been isolated. Most excrete lysine in the growth medium. All the mutations analyzed lie within *lysC* and possess a desensitized aspartokinase III.

Crystal structures of AKIII in the inactive T-state with bound feedback allosteric inhibitor lysine and in the R-state with aspartate and ADP have been determined. Comparison of R- and T-state AKIII indicates that binding of lysine to the regulatory domain in R-state AKIII instigates a series of changes that release a "latch", from the catalytic domain, which in turn undergoes large rotational rearrangements, promoting tetramer formation and completion of the transition to the T-state.

Lysine-induced allosteric transition in AKIII involves both destabilizing the R-state and stabilizing the T-state tetramer. Rearrangement of the catalytic domain blocks the ATP-binding site, which is therefore the structural basis for allosteric inhibition of AKIII by lysine.

From Homoserine to Methionine

O-succinylhomoserine synthetase, the first committed step of methionine biosynthesis is synergistically inhibited by methionine and S-adenosylmethionine (SAM). This is the only allosteric control in the specific methionine branch. The enzyme is inhibited by α-methylmethionine and resistance to this analogue has been localized in the *metA* gene. Such mutants overproduce methionine and their homoserine succinyltransferase is now insensitive to methionine and SAM. The synthesis of the enzyme, as well as the synthesis of the rest of the enzymes leading to methionine and that of aspartokinase II-homoserine dehydrogenase II is repressed by the presence of methionine in the culture medium.

From Threonine to Isoleucine

The biosynthetic threonine dehydratase is inhibited by isoleucine in a cooperative manner. Actually, it was during his studies on this enzyme that Umbarger described the first case of feedback inhibition in a microorganism and drew attention to the fact that its kinetics as a function of substrate concentration did not obey the Henri-Michaelis law. Umbarger in his first paper on the subject described the inhibition by isoleucine as competitive. To quote his own words: "when the double reciprocal plot of Lineweaver and Burk is employed, it is necessary to square the substrate (or inhibitor) concentration. This property of the data would be expected if the enzyme combined with two molecules of substrate or inhibitor" (Figs. 5.4 and 5.5). Later it was confirmed that this enzyme does indeed not follow the simple Henri-Michaelis law, but neither does it obey a purely bimolecular law with respect to substrate or inhibitor concentration. The results obtained have been interpreted as showing that several substrate or inhibitor molecules can react with a molecule of enzyme and that under these conditions, cooperative interactions exist between several molecules of ligand or several of the specific receptors on the enzyme. This research, together with the work on aspartate transcarbamylase, is one of the foundations on which the theory of allosteric transitions has been built.

This enzyme is thus a useful model system for studying the structural basis of allosteric control mechanisms. Crystals have been obtained and studied at 2.8 Å resolution. The C-terminal regulatory domains project out from a core of catalytic PLP-containing N-terminal domains. The subunits, and especially the regulatory domains, associate extensively to form dimers, which associate less extensively

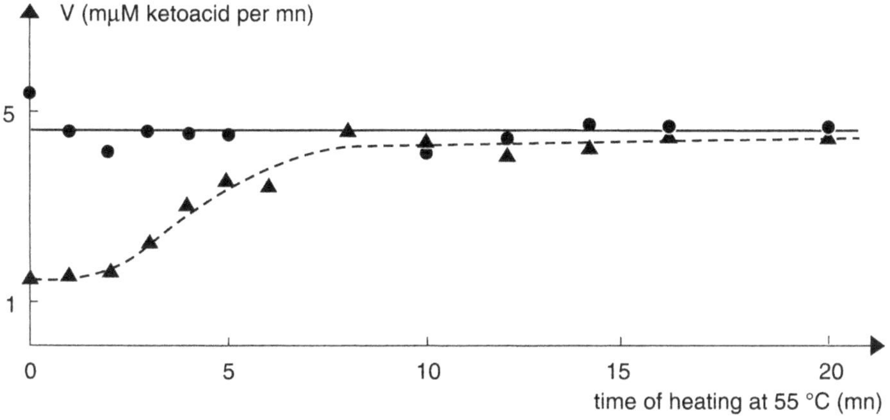

Fig. 26.16 Heat inactivation of biosynthetic L-threonine deaminase. The solid line represents the enzyme activity in the absence of inhibitor; the dotted line the activity measured with 10-2 M L-isoleucine (From J.-P. Changeux, with permission of the *Société Française de Biochimie et Biologie Moléculaire*)

to form the tetramer. Within the dimer, each monomer twists approximately 150° around a thin neck between the domains to place its catalytic domain adjacent to the regulatory domain of the other subunit.

L-isoleucine, the allosteric inhibitor, confers considerable stability to threonine dehydratase. During inactivation (spontaneous or thermal), the enzyme becomes desensitized, i.e., it is converted to a protein species insensitive to isoleucine inhibition (Fig. 26.16).

This desensitization can be obtained also by treatment by mercurials, by incorporation of amino acid analogues into the enzyme molecule, or as the result of mutations. This was the first case of desensitization demonstrated. An important property of the desensitized preparations is that they no longer exhibit cooperative interactions between substrate molecules, the kinetics adhering closely to the Michaelis law. Whereas, in several cases, desensitization and dissociation of an oligomeric protein occur together, desensitization does not seem to be accompanied by a change in the apparent molecular weight of the enzyme, in this particular case.

Of all the other enzymes of the pathway, only acetohydroxyacid synthase is subject to feedback inhibition by valine in the K12 strain of *E. coli*. As a result, the addition of valine inhibits the growth of this organism by creating a phenotypic isoleucine requirement

Multifunctional Proteins

Only proteins that combine several autonomous functions on one polypeptide chain should be accepted as multifunctional proteins. Autonomy in this sense implies that each function is assigned to a different region, that is a domain, on the polypeptide. The definition excludes enzymes which can catalyze different reactions using the same active site, such as asparaginase working as glutaminase, phosphoglyceromutase which can carry three reactions using the same reaction center, or cystathionine-γ-synthase which catalyze a whole series of analogous reactions due to its relative lack of substrate specificity. As constituents of multifunctional proteins, we find mainly catalytic functions (AK-HDH I and II, yeast fatty acid synthetase, DNA polymerase I of *E. coli*, phosphoribosylanthranilate isomerase-anthranilate synthase of *E. coli*, tryptophan synthase of *Neurospora crassa*, flavocytochrome *b2-lactate* dehydrogenase of yeast, DNA polymerase-5′ exonuclease of bacteriophage T4, to cite only a few). Certain polypeptide chains carry more than two catalytic functions. In addition to the already cited fatty acid synthetase, we can mention DNA polymerase I of *E. coli* and a pentafunctional protein of the aromatic amino acid biosynthetic pathway in *N. crassa*. In certain cases, catalytic and non-catalytic domains are fused in a single polypeptide, as the microsomal cytochrome *b5* reductase, the mammalian myosin, two enzymes which have structural roles in addition to their enzyme activity; the diphtheria toxin which has a domain responsible for the binding to the cell-membrane and a domain responsible for the ADP-ribosylation of eukaryotic elongation factor EF2. Finally, certain multifunctional proteins not endowed with a catalytic activity comprise independent binding sites for several ligands: repressors, some ribosomal proteins, immunoglobulins.

The existence of two different bifunctional enzymes carrying the same two activities in the same organism raises several questions; what is the selective advantage of bifunctionality for the aspartokinases-homoserine dehydrogenases? Do they derive from a common ancestor? And if so, was the common ancestor bifunctional, or did the corresponding gene fusion occur later?

We shall try to answer some of these questions when we shall review the evolution of proteins. Suffice it to say at this point that AKII-HDHII appears to be simpler: it is not inhibited by its end product, and it is a dimer, not a tetramer. Further comparative studies of these two proteins may lead to the explanation of some of their functional differences, especially as regards allosteric regulation, in terms of structural differences.

In the case of AKI-HDHI, one could argue that the end product, threonine, when in excess in the cell, inhibits the two activities simultaneously and that is a selective advantage that maintains a bifunctional enzyme. In the case of AKII-HDHII, no such argument holds. Since the reactions catalyzed by the two proteins are not sequential, but are the first and the third of the pathway, one could also think that one or two of the bifunctional enzymes form a multienzyme complex with aspartate semialdehyde dehydrogenase and maybe other enzymes of the threonine and methionine biosynthetic pathways. Despite many attempts, it has been impossible to demonstrate directly such protein associations. Finally, as we shall see in the next chapter, in many bacterial genera, the aspartokinase and dehydrogenase activities are carried by different proteins which are regulated in a variety of ways.

Thus, although AK I-HDH I was one of the first well demonstrated multifunctional enzymes, it is still not clear why it is bifunctional.

Regulations at the Genetic Level

The Threonine Operon

We have seen that threonine biosynthesis proceeds in bacteria through five consecutive steps. The first three are common to the biosynthesis of threonine and methionine; the first two are also the initial steps in lysine biosynthesis. The corresponding enzymes have been described above. Whereas the enzyme catalyzing the second step, aspartate semialdehyde dehydrogenase, is coded by the *asd* gene, localized at 75 min, the bifunctional enzyme AK I-HDH I, homoserine kinase and threonine synthase, catalyzing respectively the first and third, fourth and fifth step of threonine biosynthesis, are encoded by the genes *thrA*, *thrB* and *thrC*, which form a single transcription unit localized at 0 min on the *E. coli* chromosome, the threonine operon.

The whole operon has been sequenced, including the 5′ regulatory region and the termination region. The order of the genes, *thrA thrB thrC*, with transcription starting from *thrA* to *thrC*, has been established by analysis of phage Mu insertions, nonsense mutations in the *thr* cluster and by analysis of phage λ insertion into the *thr* regulatory region, upstream of and adjacent to the first structural gene of the operon, *thrA*. The λ insertion affects in a pleiotropic manner the expression of the three structural genes of the operon. Complementation analysis reveals that *thrB* and *thrC* consist of single cistrons, in keeping with the homopolymeric nature of their protein products, whereas *thrA* is composed of two cistrons, in accord with the bifunctional nature of aspartokinase I-homoserine dehydrogenase I. The intracistronic region between *thrA* and *thrB*, TGACATG, contains only one base pair while the boundary between *thrB* and *thr C* presents no gap. The ribosome binding sites which mark the start of translation of *thrB* and *thrC* are thus within the structural genes of *thrA* and *thrB*, respectively.

Coordinate regulation of the threonine biosynthetic enzymes was demonstrated in both *E. coli* and *S. typhimurium* and it was observed that both threonine and isoleucine were required for repression, limitation in any one of the two causing an increased expression of the enzymes. This form of regulation was called "multivalent repression", which is a misnomer, since the regulation occurs in this case by attenuation. In fact, studies with a variety of regulatory mutations leading to constitutive pleiotropic expression of the structural genes causing a threonine excretion have shown that the mutations are localized upstream of *thrA*. They all lie in the threonine attenuator.

The sequence of the leader peptide involved in the attenuation mechanism has been established by Gardner:

Met Lys ArgIleSerThr Thr Ile Thr Thr Thr Ile Thr Ile Thr Thr

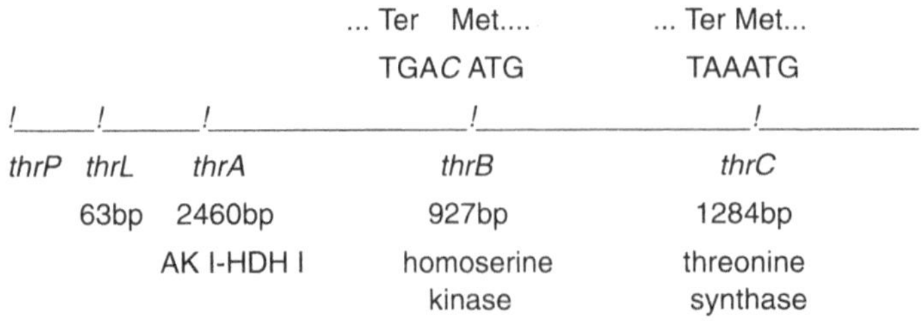

Fig. 26.17 The threonine operon

and is seen to contain several threonine and isoleucine residues at which the uncharged specific t-RNAs can cause stalling of the ribosomes translating the RNA, favoring the transcription of the structural genes (See Chap. 19).

Figure 26.17 describes the structure of the threonine operon of *Escherichia coli*. The abbreviations and symbols used are: Ter, termination codon; Met, initiation codon; thrP, promoter of the operon; thrL, sequence coding for the leader peptide.

Regulation of the Lysine Regulon at the Genetic Level

One of the most striking features of the diaminopimelate-lysine pathway is that the structural genes are scattered over the *E. coli* chromosome. There are no multigenic operons.

The concept of regulon arose from studies showing that a single transcriptional regulator can control several distinct operons. The first regulons described were similar to operons in that the member functions were involved in a specific physiological role, such as the uptake and utilization of maltose, or arginine biosynthesis.

Although the regulation of the synthesis of some of the lysine enzymes has not been demonstrated and no common regulatory protein has been implicated in the repression of the *lys* and *dap* genes, the term regulon is used here to convey the notion of a physiological unit.

The specific activity of aspartokinase III, the lysine-sensitive species, varies in response to the intracellular lysine concentration. Approximately a tenfold repression is observed when lysine is added to the minimal growth medium, whereas a 15-fold derepression is seen during lysine-limited growth in a chemostat. The number of active enzyme molecules may thus vary about 150-fold. So far, the molecular mechanism of this regulation remains obscure. The promoter and the start of transcription of *lysC* have been identified, and no signals similar to those described in attenuation have been found in the long sequence (308 bp) existing between the transcription and the translation starts. No repressor protein has been identified.

Aspartate semialdehyde dehydrogenase, which is not inhibited by lysine, is repressed about 50 % by it and lysine limitation causes a 20–30-fold derepression.

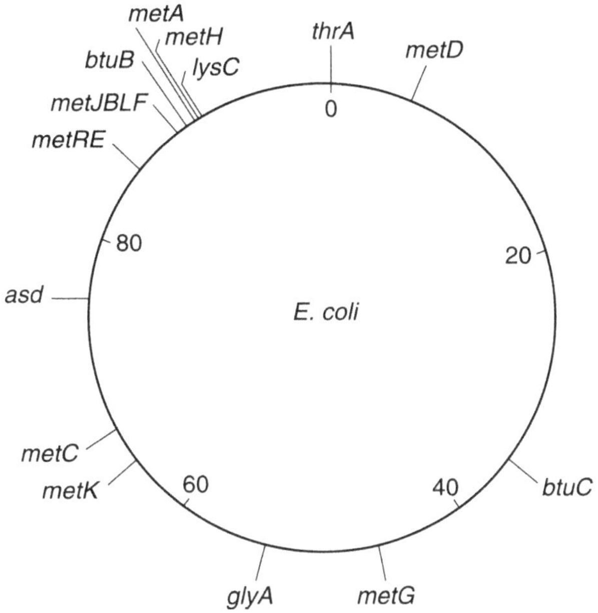

Fig. 26.18 E. coli chromosomal map showing the *met* genes. The *metR* gene was identified in *S. typhimurium. metG* is methionyl-t-RNA synthetase, met D a gene specifying the resistance to a methionine analogue. *btuC* is a gene coding for the transport of vitamin B_{12}. All the other genes are defined in the text

Variations of the lysine pool have no effect on dihydrodipicolinate synthetase, while the level of dihydrodipicolinate reductase may vary 15–20-fold from the repressed to the derepressed condition. The variation is minimal for the desuccinylase. The results reported for DAP-decarboxylase are the following: lysine represses the formation of the enzyme, which is induced by diaminopimelate.

In conclusion, the regulation of lysine biosynthesis at the genetic level is far from being completely understood.

Regulation of Methionine Biosynthesis at the Genetic Level

The *metA* gene has two transcriptional start sites located 74 nucleotides apart. The two promoters are expressed *in vivo*, but only one is regulated by intracellular methionine levels. The *metB*, *metC* and *metF* genes, coding respectively for cystathionine synthase, cystathionase and 5, 10-methylene tetrahydrofolate reductase, have each a single promoter. The *metA*, *metB*, *metC*, *metE* and *metF* genes are all repressed when excess methionine is added to the minimal growth medium, albeit to a different degree: the level of the enzymes varies from 300-fold for homoserine transsuccinylase, to 40-fold for cystathionine synthetase, 6–12-fold for cystathionase, 20-fold for the methylene tetrahydrofolate reductase and 60-fold for the vitamin B12-independent transmethylase. There is thus no coordinate repression, reflecting the fact that the *met* genes do not form a single operon, but are scattered through the bacterial chromosome (Fig. 26.18).

There is a notable exception: *metB* and *metL*, coding for cystathionine synthase and AKII-HDH II, constitute an operon, i.e., a single unit of transcription. These two genes belong to a cluster of genes, *metJ-BLF*, located at 89 min on the chromosome.

Methionine regulates also the synthesis of aspartokinase II- homoserine dehydrogenase II (*metL*) and also that of S-adenosylmethionine synthetase (*metK*). Although methionine affects the level of the vitamin B_{12}-dependent transmethylase, coded by *metH*, this effect appears to be the indirect result of repression of the synthesis of an activator protein (see below, Met R).

The expression of the *metE* gene is repressed by vitamin B_{12}; this is primarily due to a loss of MetR-mediated activation due to depletion of the coactivator homocysteine, rather than a direct repression by the MetH-B12 holoenzyme, as it had been suggested (see below, MetR).

Mutants which were selected on the basis of their resistance to some methionine analogues were found to overproduce methionine, to be derepressed for the methionine biosynthetic enzymes and to be resistant to repression by exogenous methionine. Many of them mapped to a chromosomal locus which was called *metJ*. The *metJ* gene product was thought to be a protein on the basis that some amber-suppressible *metJ* mutants could be isolated. Since *metJ*+ was found to be dominant to *metJ* in partial diploids, it seemed likely that *metJ* was coding for a repressor molecule.

Not all the analogue-resistant mutants map to *metJ*. Some affect the *metK* gene, coding for S-adenosylmethionine synthetase (we have seen in the preceding chapter that this is an essential enzyme, and since S-adenosylmethionine cannot be transported by *E. coli*, the mutants must possess some residual enzyme activity). A class of *metK* mutants did excrete methionine, and were derepressed for the methionine biosynthetic enzymes which were not repressed by the presence of methionine in the growth medium. This suggested that S-adenosylmethionine, not methionine itself, was the corepressor of the *metJ* product.

The Methionine Repressor

The two members of the 89 min *metJBLF* cluster, *metJ* and *met F*, constitute separate units of transcription, *metJ* being transcribed divergently from *metB* and, consequently a complex 276 base pair regulatory region is found between *metB* and *metJ*. There is a single promoter for *metB*, whereas *metJ* is transcribed from three separate points. Only two of the three *metJ* promoters are regulated by methionine. The repressor regulates its own synthesis.

The gene coding for *metJ* has been cloned and sequenced, and put under the control of a strong promoter in a multicopy plasmid. The amount of protein thus overexpressed represents 2 % of the total soluble proteins, a 200-fold increase over the wild type.

The *E. coli metJ* gene is 312 bp long and codes for the Met repressor, a molecule of 104 amino acid residues (Fig. 26.19). The repressor was found by sedimentation equilibrium to be a dimer.

```
AEWSGEYISPYAEHGKKSEQVKKITVSIPLKVLKILTDERTRRQVNNLRH
                               * *
                              ________ _______________
                                 β1          α1

ATNSELLCEAFLHAFTGNPLPDDADLRKERSDEIPEAAKEIMREMGINPE
   •
 ______________
      α2

TWEY
```

Fig. 26.19 Sequence of the E. coli methionine repressor. The asterisks indicate the amino acid residues making contacts with DNA inferred from the tridimensional structure of the repressor DNA complex. The dot under Ser 54 indicates that when this residue is replaced by an Asn residue by site-directed mutagenesis, the methionine biosynthetic enzymes are derepressed as a result of an inactive repressor, and methionine is overproduced. The underlined sections indicate the β-pleated strand and the α-helices of the protein respectively (see p. 576–578)

The effect of the repressor on methionine biosynthetic enzymes has been assessed by measuring the β-galactosidase synthesis in a cell-free system derived from bacteria where the synthesis of this enzyme is under the control of *metF* (*metF-lacZ* fusion). Fig. 26.20a shows that at a constant concentration of DNA, the expression of the galactosidase is progressively repressed by increasing concentrations of the aporepressor, complete repression being attained at 600 nM of the protein. SAM lowers the effective concentration of the MetJ protein as a repressor. On the other hand, Fig. 26.20b shows that at a constant concentration of aporepressor, SAM enhances the repression in a concentration-dependent manner, half-maximum inhibition being reached at 10 μM and complete repression at 100 μM. Methionine had to be present in this experiment (0.16 mM in Fig. 26.20a and 0.04 mM in Fig. 26.20b) as a building block necessary for protein synthesis. Other assay systems have demonstrated *in vitro* that the aporepressor plus SAM have the same effect on the expression of the *metB* and *metL* genes, and on that of the *metJ* gene itself.

By equilibrium dialysis, it has been found that the methionine repressor binds two molecules of S-adenosylmethionine per dimer. The corresponding Scatchard plot is linear, showing that there is no cooperativity in the corepressor binding to the aporepressor. Also, equilibrium dialysis confirms that methionine does not bind at all to the aporepressor (Fig. 26.21).

Since it was reasonable to assume that the repressor binding sites should be similar for all of the genes regulated by methionine, an extensive survey of the regions upstream each structural gene was carried out. The comparison between the four 5′ regions of *metC*, *metB*, *metA* and *metF* revealed the existence of a repetitive unit (R) eight nucleotides long which was named "Met box". As a consequence,

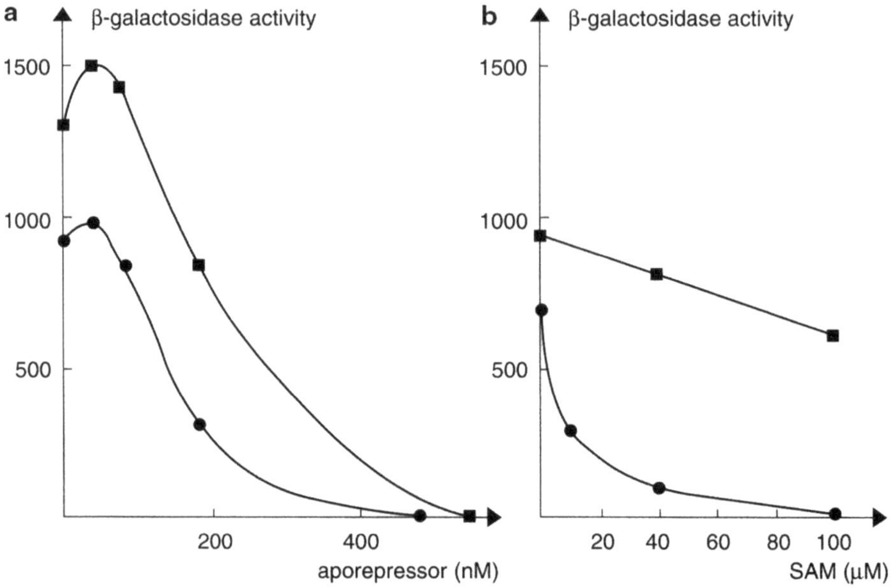

Fig. 26.20 Inhibition of the synthesis of the galactosidase encoded by the metF-lacZ gene in a cell-free system by different concentrations of aporepressor (**a**) and S-adenosylmethionine (**b**). Template DNA was a plasmid carrying the gene fusion. (**a**) ■ without or ● with 50 μM SAM. (**b**) ■ without or ● with 400 nM repressor (From I. Saint Girons, J. Belfaiza, Y. Guillou, D. Perrin, N. Guiso, O. Bârzu, and G. N. Cohen, with permission of the *Journal of Biological Chemistry*)

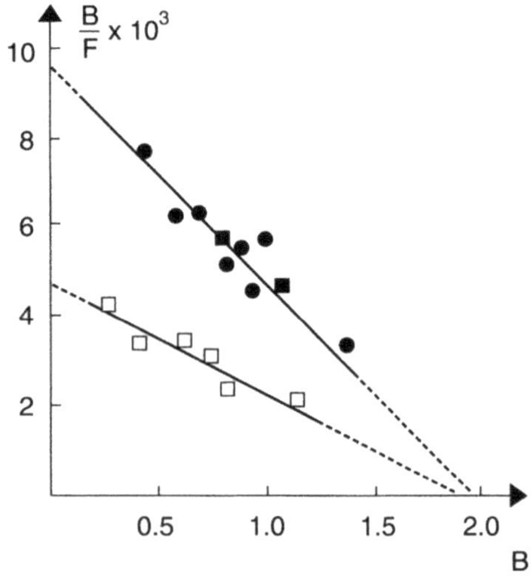

Fig. 26.21 Binding of tritiated S-adenosylmethionine to the methionine repressor in the absence (□) or in the presence (●) of 80 mM KCl, as determined by equilibrium dialysis. Two experimental points (■) correspond to samples incubated with 2 mM L-methionine, in the absence of KCl (From the same source as Fig. 26.20)

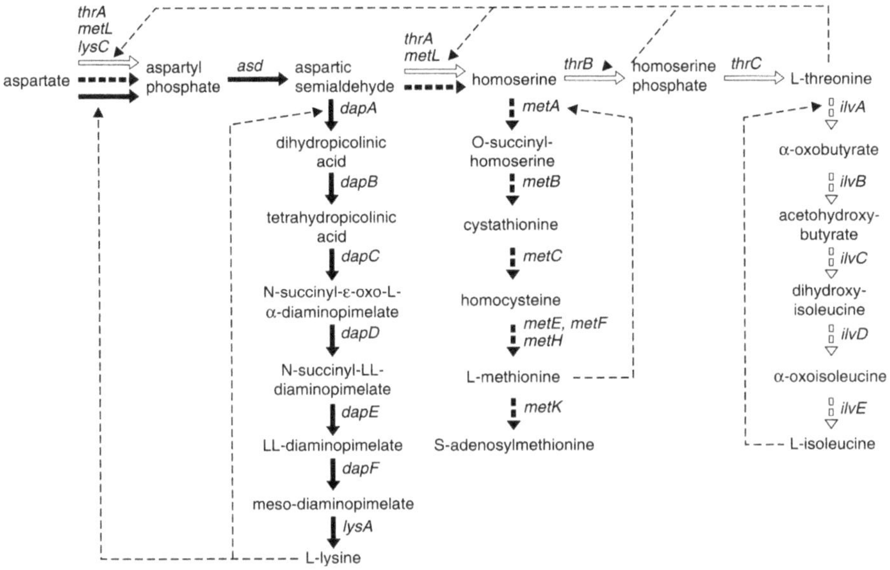

Fig. 26.22 The pathway leading to lysine, threonine, isoleucine, methionine and S-adenosylmethionine in *E. coli* and *S. typhimurium*. Details of the specific methionine pathway are given in Fig. 22.1. The names of the relevant enzymes corresponding to the mentioned genes are given in the text. Long dashed arrows represent regulation at the level of enzyme activity. Expression of the genes is regulated specifically by the different end products of the pathway (*white arrows, dotted arrows,* and *black arrows*: regulation by threonine and isoleucine, methionine, lysine and isoleucine, respectively)

multiple alignments could be drawn by sliding each sequence with an eight-nucleotide periodicity. In the alignment presented in Fig. 26.22, out of the 128 positions considered, 89 matches and 21 transitions are found when the repetitive units are compared to their consensus sequence R. This consensus sequence is a perfect palindrome, AGACGTCT, which is present under an altered form, two to five times in the Met boxes. The *metB* and *metJ* genes, being transcribed divergently, share the same Met box.

It was found later that the *metE* and *metR* genes also possess those repetitive units (3 for *metE*, and 4 for *metR*). The two exceptions are *metK* and *metH* which, although regulated by methionine, do not seem to possess the consensus sequence.

It is possible that the difference between the number of boxes and/or their sequences are related to the different extents of repression elicited by the Met repressor, although other explanations are of course possible.

In order to analyze the interaction between the Met repressor and its target, a nitrocellulose filter-binding assay using radioactive oligodeoxynucleotides has been

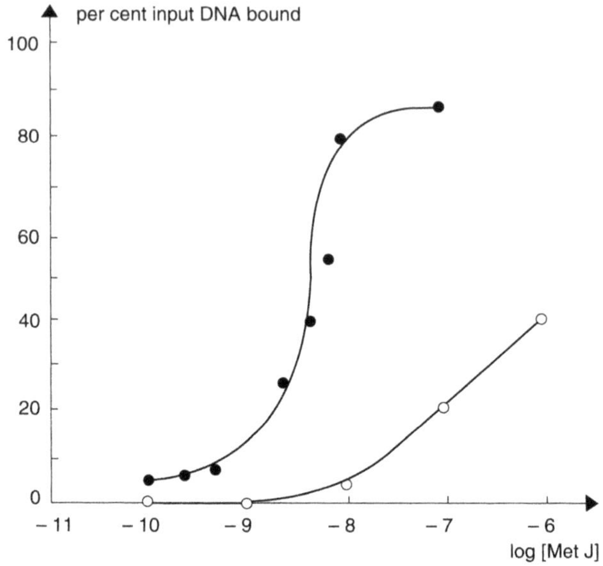

Fig. 26.23 Filter binding of a fragment comprising the perfect consensus sequence 5′-dAGACGTCTAGAC-GTCT-3′ 16-mer in the presence (●) or absence (○) of 100 mM SAM, and in presence of various concentrations of methionine repressor (Met J) (From S. E. V. Phillips, I. Mainfield, I.Parsons, B. E. Davidson, J. B. Rafferty, W. S. Somers, D. Margarita, G. N. Cohen, I. Saint Girons and P. G. Stockley, with the permission of *Nature*)

developed. These experiments have shown that DNA fragments containing two consecutive consensus Met boxes are tightly bound by repressor in the presence of saturating SAM. Nucleotide sequences containing only one Met box are not bound. Scatchard analysis of the binding to the synthetic operator shows that binding is cooperative with respect to repressor concentration (Fig. 26.23).

The repressor protects against the operator against nuclease digestion and the results strongly indicate the binding of an array of repressor dimers, centered on the 16 bp operator site, but extending in the neighboring non-operator DNA. The same is true for a fragment containing the *metF* regulatory region where the five boxes are protected, as shown by a footprint experiment. These observations are consistent with the following *in vivo* data: the level of repression of defined mutants of the *metC* operator, the smallest one known (two Met-boxes) is increased when the sequence becomes closer to the consensus sequence; the same type of experiments was carried with *metF* and a similar series of point mutations and deletions in any of the five Met boxes was analyzed with the same result; furthermore, the arrangement of all five boxes in tandem repeats is important for effective repression, since there were some point insertions within the operators which lowered repression ratios 100-fold.

In parallel with the experiments where the sequences of the operator Met boxes were altered by spontaneous or site-directed mutagenesis, a number of mutant repressor proteins have been prepared, also by site-directed mutagenesis, to probe the roles of various amino acids in operator binding. Advantage has obviously been taken from the knowledge of the tridimensional structure of the repressor-operator complex by making the mutations at residues contacting the DNA directly, and at those involved in the protein-protein contacts between adjacent repressor molecules. Effective repression, as expected, depends on cooperativity arising from these contacts.

The metR *Gene and Its Product*

Some mutants with normal *metF*, *metE* and *metH* genes were found to express the two transmethylases at a very low level, which resulted in methionine auxotrophy. This auxotrophy was overcome in strains carrying multicopy number plasmids with either the *metE* or the *metH* gene. These results implied that the methionine auxotrophy was due to an independent mutation resulting in an inability to synthesize enough homocysteine transmethylase enzymes to allow growth. The mutations are linked to *metE*, but lie outside the *metE* structural gene. Their locus was called *metR*. The *metR* genes from both *E. coli* and *S. typhimurium* have been cloned and sequenced. The *E. coli* polypeptide is 317 amino acid residues long. The native protein is a homodimer and is proposed to contain a leucine zipper, a motif characteristic of many eukaryotic DNA binding proteins.

The expression of *metR* is repressed by the Met repressor, product of *metJ* (the 5′ flanking region of *metR* contains four Met boxes) and by the product of *metR* itself. Conversely, the MetR protein plays no role in the regulation of the *metJ* gene.

MetR activates both *metE* and *metH* expression and homocysteine, the substrate of both transmethylases is the coactivator in the case of *metE* but not of *metH*. Many transcription factors activate by directly interacting with RNA polymerase The C terminus of the polymerase α subunit is a common target of activators. Residues that are crucial for MetR-dependent activation of *metE* and *metH*. localize to two distinct faces of this domain. The first is a complex surface consisting of residues important for α-DNA interactions, activation of both genes and activation of either *metE* or *metH*. The second is a distinct cluster of residues important for *metE* activation only. It has been suggested that this positive regulatory mechanism may apply when vitamin B_{12} is absent, i.e., when there is no B_{12}-dependent transmethylase activity (catalyzed by the *metH* product). The resulting methionine starvation would then cause the derepression of the biosynthetic enzymes and the accumulation of homocysteine. The latter then could act as a signal to activate the synthesis of the vitamin B_{12}-independent transmethylase, coded by *metE*.

In *S. typhimurium*, the MetR protein activates *metA*, the gene for homoserine succinyltransferase, homocysteine playing an inhibitory role in this MetR-mediated activation. Physiologically, this result makes sense.

Recently, it has been shown that MetR binds to the *lux* promoters of the luminescent bacterium *Vibrio harveyi* and represses its luminescence. Understanding how MetR regulates luminescence could connect amino acid regulation to luminescence and possibly even to the synthesis of the autoinducers A-1 and A-2 of luminescence, both of which probably originate from SAM in the same manner of other homoserine lactone-based quorum sensing autoinducers.

The Regulation of Isoleucine Synthesis at the Genetic Level

The regulation of carbon flow into and through the pathways to isoleucine (and valine) appears to rest upon end product inhibition of the two initial enzymes, threonine dehydratase (see section on allosteric enzymes in Chap. 5) and acetohydroxy acid synthase.

The regulation of the enzymes leading from α-ketobutyrate, the product of threonine dehydratase, to isoleucine occurs solely at the level of transcription, the enzymes being subject to multivalent repression (see valine biosynthesis) except in the case of the acetohydroxy isomeroreductase, whose synthesis is induced by either of its substrates (α-aceto-α-hydroxybutyrate or α-acetolactate). A detailed account will be given when we shall study the regulation of valine biosynthesis.

Appendix: More on Regulons

The concept of regulons was generalized with the recognition of global responses, in which a number of functions are coordinately controlled in response to an environmental signal (temperature, pH, starvation, DNA damage, cell density etc....). The factor mediating transcriptional regulation in a global response can be a repressor, an activator, an alternative sigma factor or an effector of RNA polymerase (Fig. 26.24).

The leucine-Lrp regulon (Lrp standing for leucine-responsive regulatory protein) was identified as having an activating or repressing effect on the transcription of many unrelated genes or operons. Some of these operons were found by chance, others were found by two-dimensional electrophoretic analysis of the proteins synthesized under different conditions, and others were found by analysis of mutants created by transpositions.

To cite only a few of the some 75 genes estimated to be regulated by Lrp:

(a) *ilvIH*, *serA*, *gltD*, *leu*: activating effect of Lrp antagonized by leucine
(b) *sdaA* (serine deaminase), *tdh* (threonine dehydrogenase), *opp* (oligopeptide permease): repressing effect of Lrp antagonized by leucine
(c) *livJK* (uptake of isoleucine and valine): repressing effect of Lrp, leucine required
(d) *gcv*, *lrp* (glycine cleavage and synthesis of Lrp): respectively activating and repressing effect of Lrp, with no effect of leucine on either process.

One of the most striking aspects of the Lrp regulon is the number of different patterns of regulation that are caused by the interaction of Lrp and leucine. As in the case of some other important global regulators, like Crp, Lrp activates the expression of some operons and represses the expression of others. In some cases, activation requires leucine, is counteracted by leucine, or is independent of leucine. We do not yet know what aspect of the environment leucine represents.

Lrp is a 19 kDa DNA-binding protein, existing in solution as a dimer, an octamer or a decahexamer. In one case where an *in vitro* study has been conducted, that of the *ilvIH* gene, it has been shown that the purified Lrp protein stimulates transcription from the promoter at concentrations similar to those required for Lrp binding to *ilvIH* DNA.

In the case of the *ilvIH* operon, six Lrp binding sites are present within a several hundred base pair region upstream of the promoter region. Binding of Lrp

```
        -50
metC :  ATATTCATGCTAGTTT A G A C A T C C
                                                  -3
                         A G A C G T A T AAAAACAGGAATCCCG
        -119
metB :  GGGATTTGCTCAATCT A T A C G C A A
                         A G A A G T T T
                         A G A T G T C C
                         A G A T G T A T
                                                  -48
                         T G A C G T C C ATTAACACAATGTTTA
        -97
metF :  CGCCCTTCGGCTTTTC C T T C A T C T
                         T T A C A T C T
                         G G A C G T C T
                         A A A C G G A T
                                                  -26
                         A G A T G T G C ACAACACAACATATAA
        -67
metA :  TTTTCTGGTTATCTTC A G C T A T C T
                         G G A T G T C T
                         A A A C G T A T
                                                  -4
                         A A G C G T A T GTAGTGAGGTAATCAG

R    :                   A G A C G T C T
```

Fig. 26.24 Comparison of the upstream regions of the metC, metB, metF and metA genes. The sequences 5′ to the structural genes are presented discontinuously and have been aligned in order to focus on the presence of the underlying palindromic repetitive unit. Numbers indicate the positions relative to the adenine of the respective start codon taken as +1. The −10 promoter sequences are overlined and arrowheads indicate the transcription start signals. In the case of *metB*, the overlined hexamer is the −35 box. The two underlined promoter sequences represent the −35 and −10 boxes of the first promoter of *metJ* (From J. Belfaiza, C. Parsot, A. Martel, C. Bouthier de la Tour, D. Margarita, G. N. Cohen and I. Saint Girons)

induces an important DNA bending that can organize the assembly of a higher order nucleoprotein structure. Equilibrium dialysis experiments show the formation of an Lrp-leucine complex *in vitro*. It is suggested that leucine negatively affects ilvH transcription because its interaction with Lrp reduces the efficiency of binding of the regulatory protein to the promoter region.

The crystal structure of the Lrp protein from the archaeon *Pyrococcus furiosus* has been determined.

Selected References

Isofunctional Aspartokinases

Stadtman ER, Cohen GN, Le Bras G, de Robichon-Szulmajster H (1961) J Biol Chem 236:2033–2038

Aspartokinases-Homoserine Dehydrogenases I and II. Structure and Regulation of Activity

Cohen GN, Dautry-Varsat A (1980) In: Bisswanger H, Schmincke-Ott E (eds) Multifunctional proteins. Wiley, New York, pp 49–121, 333 pp
Fazel A, Müller K, le Bras G, Garel J-R, Véron M, Cohen GN (1983) Biochemistry 22:158–165

The Threonine Operon and Its Regulation

Gardner JF (1979) Proc Natl Acad Sci U S A 76:1706–1710
Katinka M, Cossart P, Sibilli L, Saint Girons I, Chalvignac MA, Le Bras G, Cohen GN, Yaniv M (1980) Proc Natl Acad Sci USA 77:5730–5733
Parsot C, Cossart P, Saint Girons I, Cohen GN (1983) Nucl Acid Res 11:7331–7335

Aspartokinase III. Crystallography

Kotaka M, Ren J, Lockyer M, Hawkins AR, Stammers DK (2006) J Biol Chem 281:31544–31552

Regulation of Methionine Biosynthesis. The Methionine Repressor

Saint Girons I, Parsot C, Zakin MM, Barzu O, Cohen GN (1988) CRC Crit Rev Biochem 23(suppl 1):1–42
Phillips SEV, Mainfield I, Parsons I, Davidson BE, Rafferty JB, Somers WS, Margarita D, Cohen GN, Saint-Girons I, Stockley PG (1989) Nature 341:711–715

Regulation of the Synthesis of the Branched-Chain Amino Acids

Eisenstein E (1994) J Biol Chem 269:29416–29422, and 29423–29429
Umbarger HE (1996) Escherichia coli and Salmonella typhimurium. In: Ingraham JL, Low KB, Magasanik B, Schachter M, Umbarger HE (eds) Cellular and molecular biology, 2nd edn, vol I. American Society for Microbiology, Washington, DC, pp 442–457

The Leucine-Lrp Regulon

Brinkman AB, Dahlke I, Tuininga JE, Lammers T, Dumay V, de Heus E, Lebbink JH, Thomm M, de Vos WM, van der Oost J (2000) J Biol Chem 275:31624–31629
Hung SP, Baldi P, Hatfield GW (2002) J Biol Chem 277:40309–40323
Newman EB, Lin Annu R (1995) Rev Microbiol 49:747–775
Newman EB, d'Ari R, Lin RT (1992) Cell 68:617–619
Wang Q, Calvo JM (1993) EMBO J 12:2495–2501

Chapter 27
Other Patterns of Regulation of the Synthesis of Amino Acids of the Aspartate Family

It appears that all *Enterobacteriaceae* make use of the three isofunctional enzymes to accomplish a differential control of the aspartokinase activity. The inhibition and repression patterns found in *Escherichia coli* and *Salmonella typhimurium* are also found in other coliform bacteria such as *Edwardsiella tarda*, *Aerobacter aerogenes* and *A. cloacae*, *Serratia marcescens*, *Erwinia carotovora* and *E. aroideae*, *Proteus vulgaris* and *Providencia* sp. The relative activity of the three aspartokinases (measured in crude extracts) differ greatly from strain to strain, even within a single species and probably reflect a different degree of repression due to the different size of the individual free amino acids in the cytoplasm. Like *E. coli* and *S. typhimurium*, all strains examined contain a homoserine dehydrogenase partly inhibitable by threonine. The possible physical association between aspartokinase I and II with homoserine dehydrogenase in the other coliforms has not been examined, since this is a property determinable only after some degree of enzyme purification.

Concerted Feedback Inhibition of Aspartokinase Activity in *Rhodobacter capsulatus* (Formerly *Rhodopseudomonas capsulata*)

This species possesses a single aspartokinase which is insensitive to feedback inhibition by any one of the essential metabolites, lysine, threonine or isoleucine. However, if lysine and threonine are simultaneously present, a considerable inhibition of the enzymatic activity is found. This absolute requirement for two or more of the end products to accomplish inhibition has been called by Howard Gest, who discovered it, concerted feedback inhibition or multivalent feedback inhibition. This mechanism seems to be a less delicate control than the one using the isofunctional enzymes, since it does not allow the independent regulation of the first reaction of the branched pathway. Rather, it presents an alternative solution to the difficulty raised by the existence of such pathways. Figure 27.1 shows a typical experiment

G.N. Cohen, *Microbial Biochemistry*,
DOI 10.1007/978-94-017-8908-0_27, © Springer Science+Business Media B.V. 2014

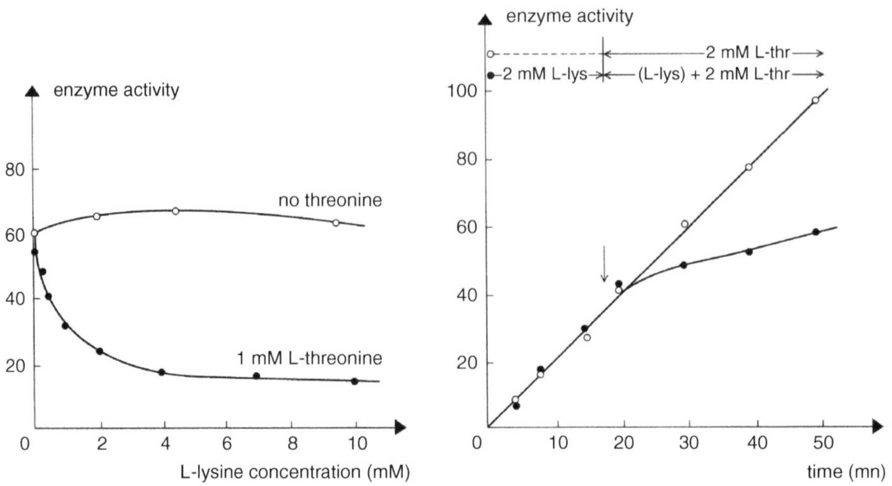

Fig. 27.1 Two experiments showing the concerted feedback inhibition of the single aspartokinase from *Rh. capsulatus*. The experiment on the left shows the absence of inhibition by lysine alone and the progressive inhibition on increasing the concentration in the presence of threonine. In the experiment on the right, white circles represent the enzyme activity measured in the absence of allosteric effector, black circles the same in the presence of lysine. At time $t = 18$ min, threonine is added to the two incubation mixtures. Inhibition occurs only when lysine is present (From P. Datta and H. Gest, with permission of Dr Howard Gest)

carried out with an extract of *Rh. capsulatus*. The concerted inhibition is never 100 %, a fact which would seem to be important to enable the synthesis of methionine to proceed even in the presence of both allosteric effectors.

The synthesis of the single aspartokinase of *Rh. capsulatus is* repressed by the presence of methionine in the culture medium. As a result, the growth of this organism, a prototroph in the wild type, is partially inhibited by excess methionine, or by an excess of the combination of threonine plus lysine, growth being re-established in the presence of the three amino acids (Fig. 27.2).

Pseudomonads

The fluorescent pseudomonads constitute the central taxonomic cluster of the genus *Pseudomonas*. Although they share many common phenotypic properties, a thorough analysis has shown that they can be subdivided in three main groups: *P. aeruginosa*, *P. putida* and *P. fluorescens*, of which the last two are further sub divisible on minor phenotypic traits into a number of different biotypes. It should be expressed that this type of classification is an extremely conservative one, relative to the current formal classification of coliform bacteria; the phenotypic differences between the three species of fluorescent pseudomonads are no smaller than those which

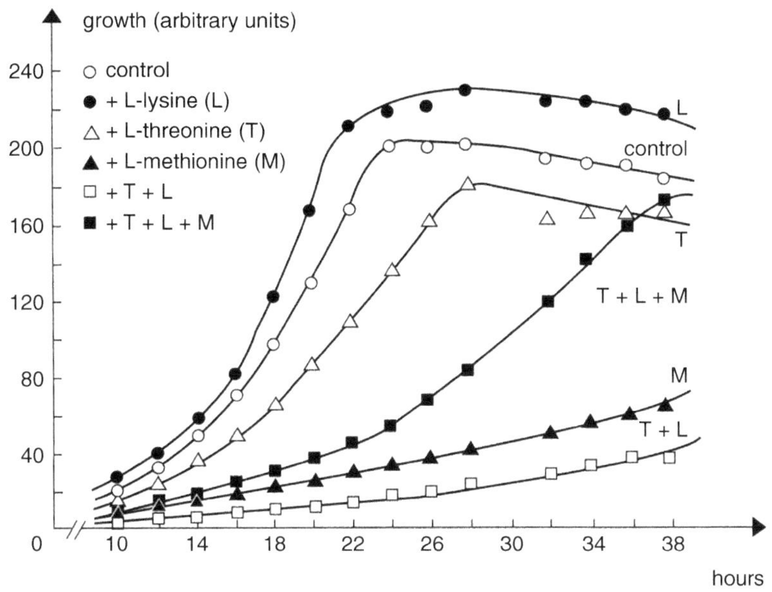

Fig. 27.2 Effects of adding amino acids on the growth of *Rh. capsulatus*. L-lysine is added at a concentration of 10^{-3} M,L-threonine and L-methionine at 5×10^{-4} M. Only methionine and the combination of threonine+lysine have an appreciable inhibitory effect on growth. The mixture of the three amino acids restores growth to near normal (From L. Burlant, P. Datta and H. Gest, with permission of Science)

differentiate between such genera as *Escherichia*, *Citrobacter*, *Salmonella* and *Shigella*.

Markedly different from the fluorescent pseudomonads in many phenotypic respects are the two *Pseudomonas* species of the *acidovorans* group, *P. acidovorans* and *P. testosteroni*, which likewise share many common properties. In the fluorescent pseudomonads, there is a common pattern of inhibition of a single aspartokinase by threonine and lysine, entirely different from that characteristic of coliform bacteria. The aspartokinase activity of these pseudomonads is subject to concerted inhibition by the two amino acids; each amino acid is, however, inhibitory when furnished at a high concentration. Consequently, the concerted nature of the inhibition can be clearly shown only when relatively low levels of the effectors are used (Table 27.1).

Methionine is without effect on the aspartokinase activity. This pattern has been confirmed for many strains belonging to all three species. It is different in several respects from the mode of concerted feedback inhibition originally discovered in *Rh. capsulatus* in which the single amino acids have little effect on aspartokinase activity, even at high concentrations.

In the strains representative of the *acidovorans-testosteroni* group, feedback inhibition is again concerted and resembles that characteristic of *Rh. capsulatus*

Table 27.1 Concerted feedback inhibition of the aspartokinase activity in *Pseudomonas aeruginosa*

Additions	Activity (nmole/min/mg)
None	26.8
L-lysine, 10^{-2} M	14.2 (47)
L-threonine, 10^{-2} M	6.5 (75)
Both effectors, 10^{-2} M	0 (100)
L-lysine, 10^{-3} M	18.8 (29)
L-threonine, 10^{-3} M	9.5 (65)
Both effectors, 10^{-3} M	2.6 (90)
L-lysine, 3×10^{-4} M	25.9 (3)
L-threonine, 3×10^{-4} M	28.7 (0)
Both effectors, 3×10^{-4} M	9 (66)

The values in parentheses represent the percent inhibition of the total activity

rather than that of fluorescent pseudomonads: the single amino acids do not inhibit appreciably even at 10^{-2} M.

A further difference between the fluorescent pseudomonads and those of the *acidovorans-testosteroni* group appears when their homoserine dehydrogenase activity is considered: whereas the enzyme is active with both pyridine nucleotides in the fluorescent group, the enzyme of the other group shows an absolute specificity for NADH.

The aeromonads have an uncertain taxonomic position, since they have properties that could permit assignment either to the pseudomonads or to the coliform group. The great weight traditionally assigned to the mode of flagellar insertion in bacterial taxonomy had led to their exclusion from the coliform group. In three species of aeromonads, *A. hydrophila*, *A. shigelloides* and *A. formicans*, the three aspartokinases were found however with the inhibition and repression pattern characteristic of coliforms.

The conclusion, also reached formerly by studies on the regulation of the aromatic biosynthetic pathway by Jensen, is confirmed by these studies: specific mechanisms of regulation are stable group characters, shared by the members of large taxa for which a common evolutionary origin can be postulated on other independent grounds.

Specific Reversal of a Particular Feedback Inhibition by Other Essential Metabolites. The Case of *Rhodospirillum Rubrum*

This organism, like *R. capsulatus*, possesses only one aspartokinase and apparently a single homoserine dehydrogenase. The feedback inhibition by L-threonine has been examined in some detail. Both enzymes are inhibited by threonine, the inhibition being reversed by the addition to the reaction mixture of isoleucine in the case of aspartokinase, or of methionine or isoleucine in the case of the dehydrogenase.

Table 27.2 Effect of certain amino acids on the aspartokinase activity of *Rhodospirillum rubrum*

Additions (10^{-4} M)	Enzyme activity (arbitrary units)
None	41
L-isoleucine	74
L-methionine	70
L-threonine 0	
L-threonine + L-isoleucine	47
L-threonine + L-methionine	4
L-isoleucine + L-methionine	69

Inspection of Table 27.2 shows that in the absence of the inhibitor, isoleucine and methionine have an activating effect on aspartokinase.

The effect of different ligands on homoserine dehydrogenase has been studied with regard to the state of aggregation of the enzyme. Centrifugation experiments in sucrose gradients show that L-threonine stimulates an aggregation of the enzyme to an inactive form. This aggregation is reversed by the allosteric modifiers, methionine or isoleucine. As can be seen in Fig. 27.3, both forms of the enzyme can also be distinguished by gel filtration in a buffer containing the different effectors. These observations were interpreted as reflections of the importance in this organism of monomer → polymer interconversions in controlling the activity of homoserine dehydrogenase.

In this organism, which possesses only one aspartokinase, an increase in the size of the intracellular pool of threonine above a critical level must produce a decrease in the concentrations of the common intermediates, due to allosteric inhibition of the kinase and the dehydrogenase. Now, it is found that the threonine dehydratase of *R. rubrum*, unlike that of *E. coli*, is practically insensitive to isoleucine inhibition; it is therefore probable that the synthesis of isoleucine proceeds freely from threonine under these circumstances, but the problem of achieving a normal synthesis of DAP, lysine and methionine remains. The specific reversal by isoleucine of the allosteric inhibition (caused by threonine) of aspartokinase and homoserine dehydrogenase provides the answer: it appears that it is the increase in the isoleucine/threonine ratio which is the signal for an accelerated production of common intermediates required for the synthesis of DAP, lysine and methionine.

Archaea

The archeal species *Methanococcus jannaschii* has only a single, monofunctional form of aspartokinase. The substrate L-aspartate binds to this enzyme in two different orientations, providing the first structural evidence supporting the relaxed regiospecificity previously observed with several alternative substrates of *Escherichia coli* aspartokinases. Binding of the nucleotide substrate triggers significant domain movements that result in a more compact quaternary structure. In contrast, the highly cooperative binding of the allosteric regulator L-threonine to multiple sites

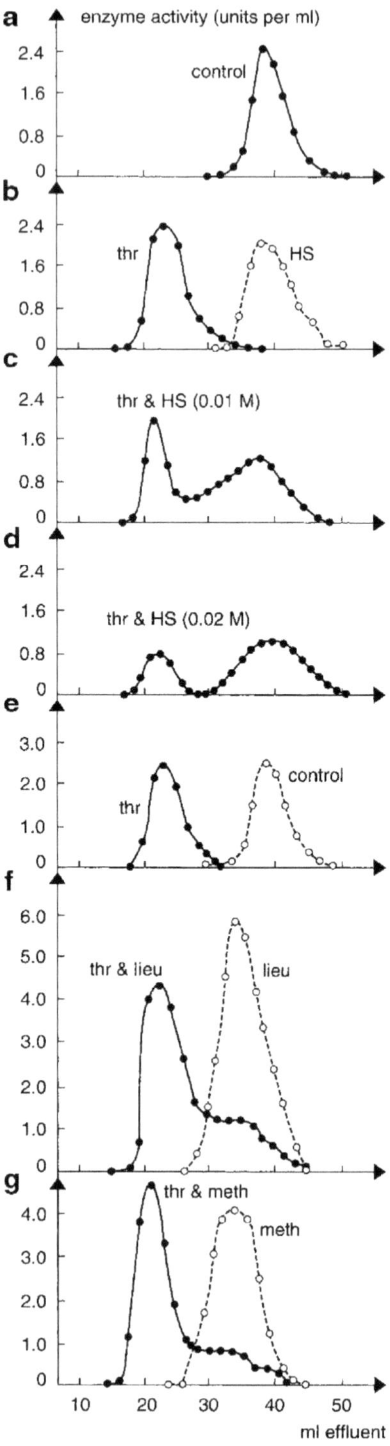

Fig. 27.3 Elution pattern of *Rhodospirillum rubrum* homoserine dehydrogenase on Sephadex G200 in the presence and absence of substrate and/or allosteric modifiers. (**a**) Enzyme filtered in buffer with no additions. (**b**) Enzyme filtered in the presence of threonine. (**c, d**) Enzyme filtered in the presence of various concentrations of threonine and homoserine. (**e**) Enzyme filtered in the presence of threonine. (**f, g**) Enzyme filtered in the presence of threonine and modifiers. The dotted curves come from different experiments and are included for comparison. Threonine evidently causes the enzyme to aggregate, the effect being reversed in the presence of sufficient substrate (**c, d**). In F and G, the high molecular weight characteristic of the threonine-treated enzyme is always present, but a considerable trailing is observed which extends into the region where species of lower molecular weight are expected (From P. Datta, H. Gest and H. J. Segal, with permission of Dr Howard Gest)

on this dimer of dimers leads to an open enzyme structure. A comparison of these structures supports a mechanism for allosteric regulation in which the domain movements induced by threonine binding causes displacement of the substrates from the enzyme, resulting in a relaxed, inactive conformation.

The Particular Case of Spore-Forming Bacilli

In the bacteria we have examined so far, the aspartate pathway has the relatively simple function of providing amino acids needed for growth; in non-growing cells, the pathway can be essentially quiescent. A more complex situation occurs in the genus *Bacillus* as a result of its characteristic life cycle. Under adverse nutrient conditions, members of this genus enter a resting state, the bacterial spore, which can effectively survive prolonged heat, dehydration and mechanical stress. We have seen (Chap. 22) that the development of heat resistance is accompanied, among others by the synthesis of large amounts of dipicolinic acid, a compound derived from dihydrodipicolinic acid, itself a precursor of diaminopimelic acid and lysine.

In addition, the process of sporulation involves the synthesis of large amounts of diaminopimelate for the synthesis of new structures such as the spore cortex mucopeptide. Aspartokinase has thus a dual function in *Bacillus* species: the synthesis of amino acids during growth and the production of dipicolinate and diaminopimelate during sporulation.

Is this dual function reflected in the regulation of aspartokinase? The predominant pattern that emerges from a survey of several members of the genus is that of two distinct aspartokinases whose relative amounts vary during vegetative growth and sporulation. The major enzyme in growing cells (vegetative type or AKII) is usually subject to feedback inhibition by lysine (e.g., *B. subtilis*) or by combination of lysine and threonine (e.g., *B. polymyxa*), consistent with the idea that its primary function is to provide amino acids for protein synthesis. The predominant aspartokinase activity during sporulation (AK I) is inhibited by the cell-wall constituent *meso*-diaminopimelate, allowing it to function even in the presence of high levels of protein amino acids to produce the structural components of the spore. To assure the efficient production of dipicolinate and diaminopimelate under these conditions, the lysine branch of the aspartate pathway in *Bacillus* species has two characteristic features. One is the absence of inhibition by lysine of dihydrodipicolinate synthase, the first committed step in the lysine branch, contrasting with the situation in *E. coli*, thus allowing the production of dipicolinate and diaminopimelate even in the presence of lysine. Another is the control of the conversion of diaminopimelate to lysine either by feedback inhibition of diaminopimelate decarboxylase by lysine or by inactivation of the enzyme during sporulation, thereby preventing the needless diversion of diaminopimelate to lysine.

Whereas the structure of aspartokinase I has not been examined in detail, this is not the case of aspartokinase II which has been thoroughly studied by the group of Henry Paulus in *B. polymyxa* and *B. subtilis*.

The enzyme is composed of two non-identical subunits of respective molecular weight 43,000 and 17,000, which are arranged in an α2β2 structure. The immunological and chemical comparisons of the subunits of *B. subtilis* aspartokinase II revealed that the smaller β-subunit is identical with the COOH-terminal portion of the α-subunit. It was first believed that the two subunits were related by a proteolysis mechanism, β being a proteolytic fragment of α. This is not so and unexpectedly, when the gene for aspartokinase II was cloned and sequenced, it became apparent that the two subunits are encoded by in-phase overlapping sequences and are the products of independent and alternative initiation of translation, the ribosome binding site for the β-subunit being part of the 3' sequence of the cistron coding for the α-subunit.

In *B. megaterium*, the existence of two aspartokinases are reported on the basis of their different pattern of repression and inhibition: the first, repressed by methionine is inhibited in a concerted fashion by lysine and methionine, while the second is repressed and inhibited by threonine. Unfortunately, the two forms of aspartokinase have not been physically separated.

Some Other Cases

The presence of a single aspartokinase was demonstrated in *Rhodospirillum tenue*. The enzyme was partially purified. Inhibition by lysine, by threonine, and concerted inhibition by these two end products are regulatory characters which have also been found in many other species. In *R. tenue*, aspartokinase is also subject to a hitherto not encountered type of concerted feedback inhibition, by L-threonine plus L-methionine. Methionine alone, even at high concentrations, is without effect.

The concerted feedback inhibition by threonine and methionine has been found in *R. tenue* only; the other photosynthetic bacteria examined, *Rhodospirillum molischianum*, *R. fulvum*, *R. rubrum* (see above), *Rhodobacter palustris*, *R. gelatinosa* display only the concerted inhibition by lysine and threonine, the pattern first discovered with *R. capsulatus*.

In *Rhodobacter spheroides*, there is no feedback inhibition by threonine, lysine, or methionine, single or in combination. The effector here is aspartate semialdehyde.

Although methionine does not participate to a concerted effect on the activity of the aspartokinase from *R. capsulata*, we have seen that the synthesis of the enzyme is repressed by the presence of methionine in the culture medium. As a result, the growth of this organism is partially inhibited by excess methionine. In contrast with this behavior, the synthesis of *R. tenue* aspartokinase is not repressed by methionine, nor does this amino acid have an effect of the growth rate of the organism. In both organisms, evolution has produced a different pattern leading to a similar result, whereby methionine can exert a regulatory function.

In *Micrococcus glutamicus* and in the yeast *S. cerevisiae* (from which the single aspartokinase has been cloned and sequenced), the synthesis of the single homoserine dehydrogenase is specifically repressed by methionine. As in *E. coli*, it is threonine which is the allosteric inhibitor in *M. glutamicus*, while in *S. cerevisiae*, the activity is inhibited by threonine and methionine, the latter being by far the more effective.

Conclusion

The above examples have been chosen to illustrate the very marked variations which are met with in different organisms in regard to the means of reaching the same goal, i.e., the harmonious integration of the rates of synthesis of the different cellular constituents.

Selected References

Concerted Feedback Inhibition

Datta P, Gest H (1964) Proc Natl Acad Sci USA, 52:1004–1009 & Nature 203:1259–1261
Paulus H, Gray E (1964) J Biol Chem 239:4008–4009

Pseudomonads

Cohen GN, Stanier RY, Le Bras G (1969) J Bacteriol 99:791–801

Rhodospirillum Rubrum

Datta P, Gest H, Segal HJ (1964) Proc Natl Acad Sci USA 51:125–130

Archaea

Liu X, Pavlovsky AG, Viola RE (2008) J Biol Chem 283:16216–16225

Spore-Forming Bacilli

Paulus H (1984) J Biosci 6:403–418

Chapter 28
Biosynthesis of the Amino Acids of the Glutamic Acid Family and Its Regulation

The Biosynthesis of Glutamine

Biosynthesis of Glutamine. Cumulative Feedback Inhibition

This reaction is carried out by an enzyme, glutamine synthetase, whose occurrence is not restricted to microorganisms. It is essential not only for the synthesis of glutamine, a building block of proteins. Glutamine contributes also through its amide group to the synthesis of many nitrogen containing substances; in addition, it is an obligatory intermediate in ammonia assimilation in ammonia-limited media.

Glutamine synthetase has been obtained in the pure state from different sources, in particular in *E. coli*, where it is encoded by the *glnA* gene. It catalyzes the following reactions:

$$\text{L-glutamate} + NH_3 + \text{ATP} \rightarrow \text{L-glutamine} + \text{ADP} + \text{Pi}$$
$$\text{L-glutamate} + NH_2OH + \text{ATP} \rightarrow \text{L-glutamohydroxamate} + \text{ADP} + \text{Pi}$$

The second reaction (glutamyltransferase) is obviously non-physiological, but provides an easy colorimetric assay by reacting the hydroxamate with ferric ions.

In the absence of ammonia, the enzyme can also catalyze the following reactions, which are not believed to be physiologically significant:

$$\text{L-glutamate} + \text{ATP} \rightarrow \text{pyrrolidone carboxylate} + \text{ADP} + \text{Pi}$$
$$\text{L-glutamine} + H_2O \rightarrow \text{L-glutamate} + NH_3$$
$$N_1TP + N_2DP \rightarrow N_2TP + N_1DP$$

In this last reaction N_1 and N_2 stand for different nucleosides.

G.N. Cohen, *Microbial Biochemistry*,

DOI 10.1007/978-94-017-8908-0_28,

In the presence of Mn^{++} and AMP, the unadenylylated enzyme (see below) can also catalyze the formation of glutamate and pyrophosphate from glutamine and inorganic phosphate:

$$\text{L-glutamine} + 2\,\text{Pi} \rightarrow \text{L-glutamate} + \text{PPi} + NH_3$$

The allosteric control of glutamine synthetase deserves a detailed examination, for it differs from the schemes we have encountered so far for the branched biosynthetic pathways. First, it is important to recall that the glutamine amide group supplies the nitrogen atoms of the following molecules by means of reactions which we shall study in due course: tryptophan, adenylic acid, cytidylic acid, glucosamine-6-phosphate, histidine, and carbamyl phosphate. In addition, it has been shown in *E. coli* that glutamine is used as a substrate for transaminases which convert respectively pyruvate to alanine and glyoxylate to glycine.

It is very interesting to note that the eight compounds listed, plus serine, are inhibitors of the glutamine synthetase of *E. coli*, and also of organisms as different *as S. typhimurium, P. fluorescens, Neurospora crassa, Bacillus licheniformis, Chlorella pyrenoidosa* and *Anabaena* sp. Many other nitrogen-containing compounds not derived from glutamine are not inhibitors. The majority of the organisms studied do not possess isofunctional glutamine synthetases (as found for *E. coli* aspartokinase and homoserine dehydrogenase activities) and no glutamine synthetase shows the phenomenon of concerted feedback inhibition (as found, for example, for *Rh. capsulatus* aspartokinase).

Each of the inhibitors causes only partial inhibition, even at near saturation, when investigated individually; however, in combination, their effects are cumulative. Each inhibitor acts independently and as a result, when two or more end-products are present together at saturating concentrations, the residual activity is equal to the product of the residual activities observed with each of the inhibitors at saturating concentration. The enzyme activity can be completely inhibited by the simultaneous presence of the eight inhibitors. Cumulative feedback inhibition provides an elegant regulatory mechanism for branched pathways which lead to the formation of numerous end products. Ideally, the fraction reduced by any given inhibitor is sufficient to curtail significantly the synthetic pathway to produce that end product while the other branches can still function more or less normally through the differential in binding constants. However, the inhibition of glutamine synthetase might prove disastrous for the economy of the cell, if there did not exist individual control mechanisms for each branch.

The fundamental mechanism underlying the cumulative feedback inhibition is not yet clear; in fact, it is difficult to visualize a simple model which accounts for the experimental observations; each model must explain the restricted inhibition by each of the effectors, together with the absence of synergistic effects or antagonism at saturating concentrations. This last phenomenon implies that the enzyme must possess distinct allosteric binding sites for each of the inhibitors. Indeed, early studies indicated that this is the case: inhibition by glycine, cytidylic acid or tryptophan is competitive with respect to glutamate, whereas inhibition by histidine or glucosamine-6-phosphate is competitive with respect to ammonia, and inhibition by alanine, adenylic acid or carbamyl phosphate is not competitive with respect to either of the two substrates. This result implies at least three binding sites. Furthermore, 1 M urea (in the absence of chelating agents) desensitizes the enzyme to inhibition by tryptophan, histidine and glucosamine-6-phosphate; there must be therefore at least one additional site. Treatment of the enzyme with 30 % acetone does not cause any loss in activity, but produces an increased inhibition by alanine and glycine: we have now six sites. Another difference arises from the observation that adenylic acid and histidine protect the enzyme against mercaptide formation with p-chloromethylsulfonate, while carbamyl phosphate and cytidylic acid accelerate the rate of inactivation by this reagent.

Taking all these results together, we reach the conclusion that each inhibitor must have its own binding site.

These observations were made on a mixture of adenylylated and unadenylylated glutamine synthetase (see below), which complicates the understanding of the mechanism of this cumulative feedback inhibition, since the adenylylated enzyme is highly sensitive to the above effectors, while the unadenylylated enzyme is much less susceptible to them.

As for all allosteric enzymes, the determination of the number of binding sites on a homogeneous preparation is absolutely necessary for constructing a model which would account for the observed facts.

It has been unequivocally shown that fully unadenylylated glutamine synthetase binds glycine, alanine and serine at the glutamate binding site. This result, obtained by X-ray crystallography, contradicts some of the above conclusions Further crystallographic studies with fully unadenylylated enzyme suggest that adenine, adenosine, AMP, ADP, GDP and the ATP analog AMP-PNP all bind to the same general site, the ATP substrate binding site, with a maximum of one nucleotide per subunit. Furthermore, kinetic measurements show clearly that AMP is competitive with respect to ATP. The contradictions are probably due to the fact that the earlier studies were made with a mixture of unadenylylated and partially or totally adenylylated enzyme, the covalent modification introducing complications of the glutamine synthetase properties. A full exploration of the effects of enzyme adenylylation requires the crystal structures of the adenylylated enzyme complexed with nucleotides.

Biosynthesis of Glutamine. The Covalent Modification of Glutamine Synthetase

E. coli cells grown under nitrogen limitation (glutamate as the nitrogen source) contain active glutamine synthetase. In the presence of excess ammonia, the synthesis of the enzyme is repressed. Addition of NH_4^+ ions to derepressed cells not only causes a repression, but also an "inactivation" of the enzyme already present, and this in a few seconds. If cells containing the inactivated enzyme are transferred in a medium where ammonia is limiting, the enzyme is reactivated. The

SAEHVLTMLNEHEVKFVDLRFTDTKGKEQHVTIPAHQVNA
EFFEEG**K**MFDGSSIGGWKGINESDMVLMPDASTAVIDPFF 80

ADSTLIIR**C**DILEPGTLQGTDRDPRSMSKRADEYLRSTGI
ADTVLFGPEPEFFLFDDIRFGSSISGSHVAIDDIEGAWNS 160

STQYEGGNKGHRPAVKGGYFPVPPVDSAQDIRSEM**C**LVME
QMGLVVEAHHHEVATAGQNEVATRFNTMTKKADEIDIYKY 240

VVHNVAHRFGKTATFMPKPMFGDNGSGM**HC**HMSLSKNGVN
LFAGDKYAGLSEQALYYIGGVIKHAKAINALANPTTNSYK 320

RLVPGWEAPVMLAYSARNRSASIRIPVVSSPKARRIEVRF
PDPAANPYL**C**FAALLMAGLDGIKNKIHPGEAMDKNL**Y**DLP 400

PEEAKEIPQVAGSLEEALNELDLDREFLKAGGVFTDEAID
AYIALRREEDDRVRMTPHPVEFELYYSV 468

Fig. 28.1 The amino acid sequence of E. coli glutamine synthetase

inactivation can be obtained *in vitro* in the presence of glutamine, ATP and Mg^{++} ions, and of a specific inactivating enzyme. Only the glutamine synthetase activity proper is inactivated, whereas most of the glutamyl transferase activity (measured by glutamyl hydroxamate synthesis) remains unaffected. The conclusion is drawn that only part of the catalytic site of glutamine synthetase is altered by the inactivating enzyme.

The main difference between the two forms of the enzyme is in their ultraviolet spectrum: the difference between the two spectra has a maximum at 261 nm, which suggests that the ammonium chloride grown cells contain a bound nucleotide absent from the cells grown in the presence of glutamate. Since the inactive enzyme is reactivated by the action of snake venom phosphodiesterase, it is concluded that the modification caused by the inactivating enzyme is an incorporation of a bound nucleotide into glutamine synthetase. This conclusion is confirmed by the incorporation of radioactivity into the inactivated enzyme when the inactivation is performed in the presence of radioactive ATP labeled either in the adenine or with ^{32}P. The bound nucleotide has been unequivocally identified as covalently attached 5′-AMP. It is presumed that the effect observed *in vivo* is due to a synthesis of glutamine from ammonia followed by an inactivation of glutamine synthetase by the inactivating enzyme.

Glutamine Synthetase Structure

The gene coding for the *E. coli* glutamine synthetase (*glnA*) has been cloned and sequenced. The deduced amino acid sequence is given in Fig. 28.1

Each subunit (MW = 51,814 Da) has 4 cysteines, not accessible to SH reagents when the enzyme is in its active conformation, 2 tryptophans, whose fluorescent quantum yield is very sensitive to the binding of substrates and effectors; 17 tyrosines, including Tyr397 which is the adenylylation site; 16 histidines, one of which (His 269) whose oxidation by mixed-function oxidation reagents causes the irreversible inactivation of the enzyme. In addition, Lys 47 is covalently modified by an ATP analogue, which inactivates the enzyme in a competitive manner to ATP binding, suggesting that it is part of the ATP binding site.

The sequence of the *Salmonella* enzyme is very similar, that of *Anabaena* is somewhat different and the mammalian enzyme is distinctly different.

The enzyme is dissociated by guanidine hydrochloride into 12 identical subunits. Electron micrographs of the native enzyme show that it is made up of 12 subunits arranged in two layers of six. This double-decked hexagon has a sixfold dihedral symmetry. The outer diameter of the hexagon is 140 Å and the subunit molecular dimensions are 45 × 45 × 53 Å (Fig. 28.2).

The tridimensional structure of the completely unadenylylated enzyme from *S. typhimurium*, which has a very similar primary sequence and which crystallizes isomorphously with the *E. coli* enzyme has been solved at 3.5 Å resolution. The 12 identical subunits are present in the asymmetric unit of the crystal. The model obtained is in general agreement with the electron microscopy. The outer diameter of the hexameric ring is 143 Å.

In each subunit, the small N-terminal domain (residues 1–103) is formed of six strands of antiparallel β-sheets, one large and one small α-helix. It is connected to a large C-terminal domain just before the long helix starting at residue 104. This domain contains seven long and six short α-helices and an extended β-sheet (six long and three short strands). At the top six prominent strands of β-structure from the C-domain form a partial barrel around two metal ions. In addition, there are 25 turns and loops (Fig. 28.3).

The view of glutamine synthetase down the sixfold axis (Fig. 28.4) reveals a petal-shaped molecule with an outer diameter of 143 Å. As anticipated from the electron micrographs taken some 20 years before the structure was elucidated, there is a central channel about 40 Å in diameter, but it is not completely open. The central loop referred to in the legend of Fig. 28.3 protrudes into the channel, so that the open inner diameter is only 24 Å. This loop is sensitive to various proteases. Six active sites, marked each by a pair of manganese ions are roughly midway between the periphery of the glutamine synthetase molecule and the central channel. Each active site is at the interface between two subunits. Each of the two metal atoms necessary for catalytic activity is connected to the protein backbone by four ligands. The formation of an active site from the association of two subunits had already been found in several enzymes among which aspartate aminotransferase and aspartate transcarbamylase.

This model, proposed by David Eisenberg and his colleagues, suggests a mechanism for the main functional role of glutamine synthetase: how the enzyme regulates the rate of synthesis of glutamine in response to covalent modification and feedback inhibition.

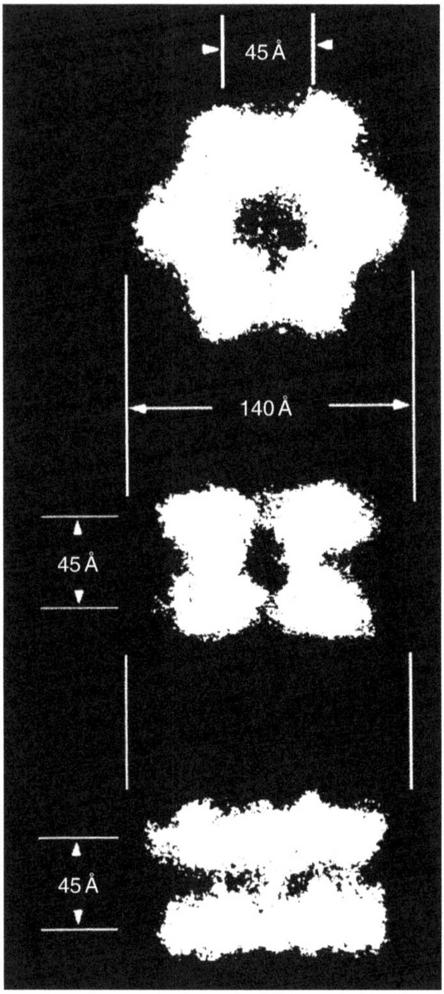

Fig. 28.2 A high-magnification picture of five superimposed images of unfixed molecules in the three characteristic orientations. The mean dimensions are indicated. When the molecule rests on a face, the subunits appear as a hexagonal ring (*top*). Molecules seen on edge show two layers of subunits, either as four spots (*center*) when viewed exactly down a diameter between subunits, or in general as two lines (*bottom*). The molecular structure is thus based on 12 identical subunits in two hexagonal layers, the units in one layer being directly above those in the other. Magnification: 3,160,000 (From R.C. Valentine, B.M. Shapiro and E.R. Stadtman, with permission of *Biochemistry*)

Reversible Adenylylation of the Glutamine Synthetase

Glutamine synthetase from *E. coli* and other enteric bacteria was found to exist into two interconvertible forms: an adenylylated form, which is highly sensitive to allosteric inhibitors (the activity measured in this case is the glutamyltransferase), and an unadenylylated form, relatively insensitive to feedback inhibition. Adenylylation implies the specific transfer of the adenylyl moiety of ATP to Tyr 397 of each subunit of the dodecamer and is catalyzed by an enzyme which has been called glutamine synthetase adenylyl transferase. This tyrosine is known by the crystallographic model to be located near the interface between two subunits and to be close to the outer periphery of the hexameric ring. The linkage between the adenylyl group and the tyrosine hydroxyl is a phosphodiester bond.

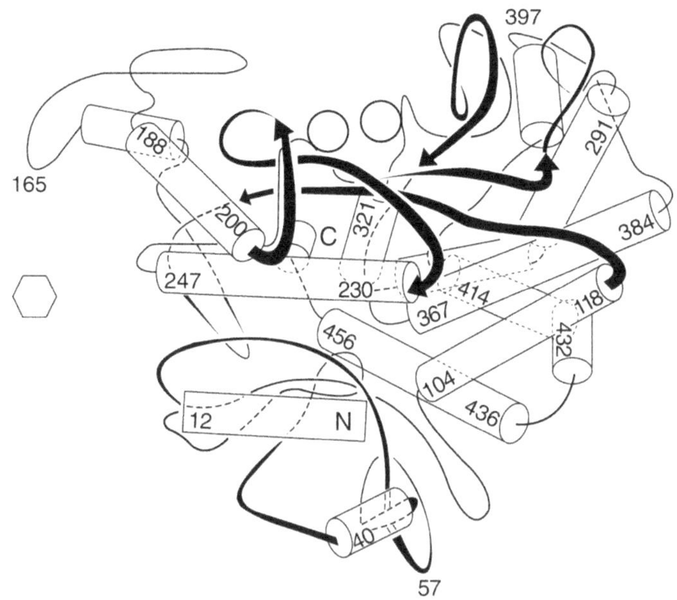

Fig. 28.3 A schematic drawing of the glutamine synthetase polypeptide chain. The sixfold axis is indicated by the hexagon at the left, and the metal ions by two *circles* in the upper center. Cylinders: α-helices; *heavy arrows*, the six strands surrounding the metals. *Bottom left*: the N-terminal domain (1–103). The C-terminal domain starts at residue 104. Note the central loop at the upper left (From R.J. Almassy, C.A. Janson, R. Hamlin, N-H. Xuong and D. Eisenberg, with permission of *Nature*)

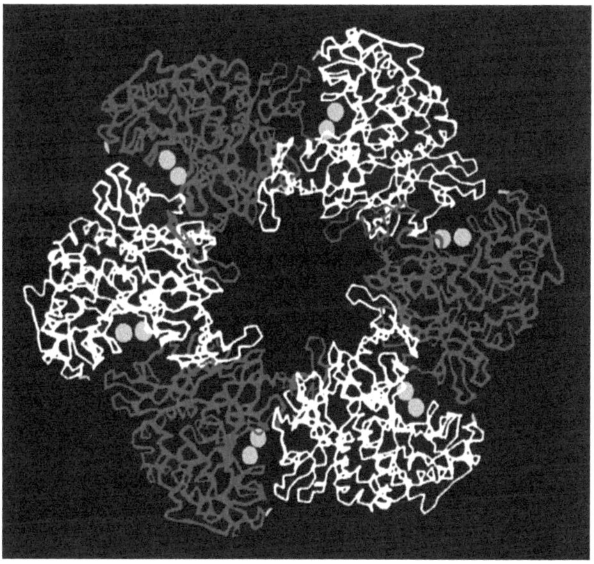

Fig. 28.4 The upper half of the glutamine synthetase molecule, projected down the sixfold axis of symmetry. The six subunits are shown as line segments connecting sequential α-carbon atoms. Six active sites are marked by pairs of Mn^{++} ions. The side-chain of Tyr 397, target of adenylylation, is between subunits at larger radius than the active site left (From R.J. Almassy, C.A. Janson, R. Hamlin, N-H. Xuong and D. Eisenberg, with permission of *Nature*)

The enzyme has been adenylylated by a spin-labeled analogue of ATP. The modified enzyme exhibits similar catalytic properties, pH profile and inhibitor susceptibility as that of the glutamine synthetase adenylylated with normal ATP. The nitroxyl spin label allows the determination of the distance between the paramagnetic metal (Mn^{++}) and the nitroxyl moiety of the probe. This permits a topographical analysis of the correlation between the structural metal binding site, the catalytic site and the regulatory adenylylation site. The data show that the two latter sites are close to each other and that binding of substrates changes their distances. In addition, the data on the rotation correlation time confirm that the adenylyl site is located on the surface of the enzyme.

Each subunit can be adenylylated, so that a molecule of glutamine synthetase can have a maximum of 12 adenylyl groups covalently attached:

$$\text{Glutamine synthetase} + \text{nATP} \xrightarrow{\text{Mg}^{++}} \text{glutamine synthetase}(\text{AMP})_n + \text{nPPi}$$

Adenylyltransferase can also catalyze the phosphate dependent removal of AMP from each subunit of glutamine synthetase.

The deadenylylation is not the reverse reaction of adenylylation

$$\text{Glutamine synthetase}(\text{AMP})_n + \text{n Pi} \xrightarrow{\text{Mg}^{++},\ \text{ATP}} \text{glutamine synthetase} + \text{nADP}$$

Adenylyltransferase (coded by *glnE*) has been purified and is a monomer (MW = 115,000). It can catalyze the two reactions *in vitro* in the absence of the two regulatory proteins (see below), PII and uridylyltransferase/uridylyl removing enzyme. Biochemical and genetic evidence indicate that the two activities are carried on separate sites on the same polypeptide.

Regulation of Glutamine Synthetase Activity by Covalent Adenylylation

The state of the cell with regard to nitrogen availability is defined by the intracellular ratio of glutamine to α-ketoglutarate. When ammonia is in excess, the rate of glutamine synthesis by glutamine synthetase increases as does the conversion of α-ketoglutarate to glutamate by glutamate synthase and glutamate dehydrogenase. If ammonia becomes limiting, the rate of glutamine synthesis decreases. As a result, excess ammonia in the medium causes a high glutamine/α-ketoglutarate ratio and a low intracellular concentration of partially unadenylylated glutamine synthetase. When ammonia is limiting, this ratio becomes low and the intracellular concentration of active, unmodified glutamine synthetase becomes high.

The regulation of glutamine synthetase activity requires the cooperation of three proteins: (1) the glutamine synthetase adenylyltransferase (ATase), (2) a protein called PII, which exists in two forms, a uridylylated form and a non-uridylylated form, (3) a bifunctional uridylyltransferase/uridylyl-removing enzyme (UTase/UR) A high concentration of glutamine activates the uridylyl removing enzyme which in turn causes the deuridylylation of protein PII which interacts with the adenylyl transferase which causes the adenylylation of glutamine synthetase. The deadenylylation is carried out as follows: a high concentration of α-ketoglutarate activates the uridylyltransferase activity of the bifunctional protein and PII is uridylylated; the uridylylated form of PII interacts with the adenylyltransferase which causes the deadenylylation of glutamine synthetase.

Here again, the deuridylylation reaction is not he reversal of the uridylylation one. The two reactions can be written as follows:

$$\mathrm{PII + nUTP} \underset{\mathrm{Mg^{++}}\alpha\ \text{ketoglutarate}}{\overset{\text{uridylyltransferase}}{\longrightarrow}} \mathrm{PII\ (UMPn) + n\,PPi}$$

$$\mathrm{PII - UMPn + n\,H_2O} \underset{\text{glutamine}}{\overset{\text{uridylyl removing enzyme}}{\longrightarrow}} \mathrm{PII + n\,UMP}$$

PII (coded by *glnK*) is a tetramer consisting of identical subunits, each of a molecular weight 44,000. The uridylylation of PII occurs at one specific tyrosine residue in one or more subunits. The UTase/UR enzyme (coded by *glnD*) is a monomer of MW = 95,000.

Figure 28.5 summarizes the sequence of events:

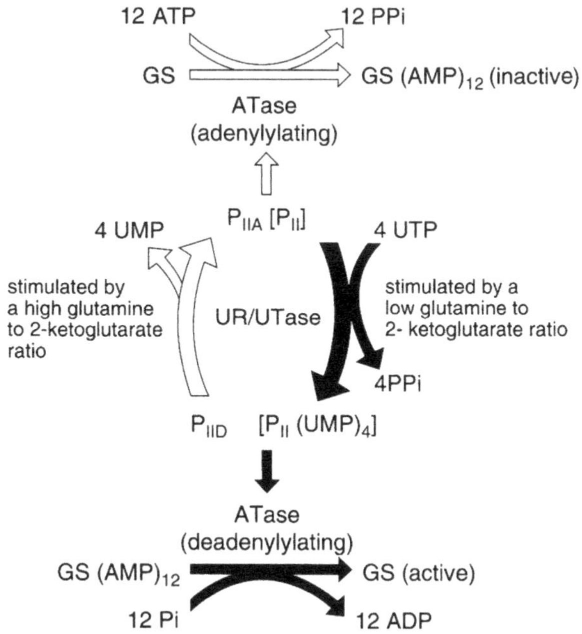

Fig. 28.5 Covalent modification of glutamine synthetase (*GS*). *UR/UTase* uridylyl removing enzyme/ uridylyl transferase

The Regulation of the Synthesis of Glutamine Synthetase Also Involves the Two Forms of PII and UTase/UR

The *glnA* gene, coding for glutamine synthetase, belongs to an operon which comprises three structural genes, *glnA* itself and two genes, *glnG* and *glnL* coding respectively for two regulator proteins, NRI and NRII. These proteins are homodimers whose subunit have a respective molecular weight of 110,000 and 68,000. NRII has been shown to be a kinase which phosphorylates NRI in the presence of ATP. The phosphorylated NRI is responsible for the activation of transcription of the *glnA* gene at one of the promoters of the operon (*glnA p2*). The phosphorylation is reversible, NRII being a bifunctional enzyme endowed with a phosphatase activity which regenerates un-phosphorylated NRI, which cannot activate the transcription of *glnA*.

The role of UR/UTase and of PII in the regulation of transcription of *glnA* is best described by Fig. 28.6, which outlines that the variations of the glutamine and α-ketoglutarate concentrations, which modulate the level of uridylylation of PII, affect not only the activity of glutamine synthetase by controlling its adenylylation but also its synthesis.

The reciprocal effects glutamine and α-ketoglutarate can be more fully appreciated if one considers that α-ketoglutarate is a precursor of glutamine and therefore the ratio glutamine/α-ketoglutarate will vary in response to the availability of ammonia as demonstrated by various *in vivo* experiments.

PII proteins are one of the most widespread families of signal transduction proteins in nature, being ubiquitous throughout bacteria, archaea, and plants. In all these organisms, PII proteins coordinate many facets of nitrogen metabolism by interacting with and regulating the activities of enzymes, transcription factors, and membrane transport proteins. The primary mode of signal perception by PII

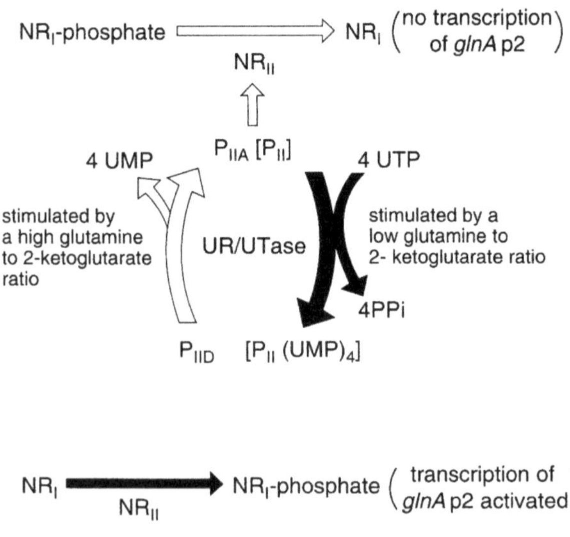

Fig. 28.6 Covalent modification of NRI, the regulator of glnA transcription (From L. J. Reitzer and B. Magasanik, with permission of the *American Society for Microbiology*)

proteins derives from their ability to bind the effector molecules 2-ketoglutarate and ATP or ADP. The role of 2- ketoglutarate as an indicator of cellular nitrogen status is well understood, but the function of ATP/ADP binding was unresolved. Until it was shown that the *Escherichia coli* PII protein, GlnK, has an ATPase activity that is inhibited by 2-ketoglutarate. Hence, when a drop in the cellular 2-ketoglutarate pool signals nitrogen sufficiency, 2-ketoglutarate depletion of GlnK causes bound ATP to be hydrolyzed to ADP, leading to a conformational change in the protein. The role of ATP/ADP binding in *E. coli* GlnK is to effect a 2-ketoglutarate-dependent molecular switch that drives a conformational change in the PII protein. Another PII protein, that of *Azospirillum brasilense*, has a similar ATPase activity, and it is suggested that this switch mechanism is likely to be a general property of most members of the PII protein family.

The levels of glutamine synthetase are also regulated by oxidation followed by proteolytic degradation.

Levels of intracellular proteins, reflecting a balance between the rate of synthesis and the rate of degradation, are regulated by the nutritional state of organisms. A mechanism for regulation of these levels has been proposed by Stadtman and his colleagues. First proteins undergo an oxidative modification which abolishes activity and renders them susceptible to proteolytic attack. The latter occurs through strictly specific proteases which degrade the oxidized, but not the native proteins. When degradation fails, as in aging and some pathological processes, oxidized proteins accumulate.

This hypothesis has been borne out by the following facts using *E. coli* and *S. typhimurium* glutamine synthetase: At least, two residues, His269 and Arg344 are oxidized in the presence of both oxygen and ferrous ions to Asn and γ-glutamyl semialdehyde, respectively. This destroys catalytic activity. If the exposure is prolonged, other residues are oxidized and increase the susceptibility of the enzyme to specific proteases Study of the tridimensional structure of engineered mutants where His269 and Arg344 have been replaced by non-oxidizable residues, and of oxidized native enzyme have allowed Eisenberg and his associates to propose a more detailed model.

Regulation of glutamine synthetase by metal-catalyzed oxidative modification has been shown to occur in the marine oxyphotobacterium *Prochlorococcus*.

Glutamine Synthetase in Other Microorganisms

Bacillus species do not have a global sophisticated system of regulation of nitrogen metabolism as demonstrated in *E. coli*. It is clear for example that *B. subtilis* lacks the wide range in adaptability with respect to the nitrogen source that is typical of *Enterobacteriaceae*. On the other hand, *B. subtilis* responds by sporulating when deprived of a rapidly metabolizable nitrogen source. In the same species, selective proteolysis may serve as an important regulatory mechanism: *in vitro* proteolysis of the enzyme is inhibited by the substrates metal-ATP and glutamate and accelerated by the product, glutamine. Thus, the activity of glutamine synthetase might be controlled directly through the ratio of the concentrations of the substrates and/or the products in the cell. In *B. subtilis*, glutamine synthetase is also a dodecamer consisting in a two-layered structure of hexagonal rings.

The extreme thermophile, *Bacillus caldolyticus*, expresses two glutamine synthetases as separated gene products, regulated in a complementary fashion by the end-product metabolites derived from L-glutamine.

In the photosynthetic bacteria, *Rhodobacter capsulatus*, *Rhodobacter spheroides* and *Rhodopseudomonas palustris*, and in an *Acinetobacter* sp., glutamine synthetase is regulated by an adenylylation-deadenylylation system, similar to that of *E. coli*. In contrast, in *Synechocystis*, a cyanobacterium which does not possess this system, glutamine synthetase is however inactivated by addition of ammonium ions to cultures growing on nitrate as the sole nitrogen source. This appears to be due to the non-covalent tight binding of a phosphorylated compound to the enzyme. Alkaline phosphatase, but not phosphodiesterase, and various physical or chemical treatments, reactivate the enzyme *in vitro*. The nature of the actual inhibitor is unknown at present.

The nitrogen-fixing actinomycete *Frankia* can fix nitrogen for its own growth in pure culture and can also be a root-nodule symbiont. In free culture, the main nitrogen assimilation occurs through glutamine synthetase. Two glutamine synthetases, GS1 and GS2 are synthesized under nitrogen-fixing conditions, but only GS1 is synthesized during growth on ammonia. Under N2-fixing symbiotic conditions, the synthesis of both GS1 and GS2 is repressed. Whereas GS1 consists of 12 subunits of about 59 kDa each and its synthesis is repressed by nitrogen starvation, the octameric GS2 gene has high sequence similarities to the genes of eukaryotic glutamine synthetases.

The existence of two glutamine synthetases in the bacterium *Bradyrhizobium japonicum* has been the subject of a very interesting debate. Whereas GSI from this organism is a dodecamer as found in eubacteria and archaebacteria, GSII was found to have an octameric arrangement and an amino acid sequence related to that eukaryotes. The obvious opportunity afforded by its symbiotic association with plants made the possibility of a horizontal gene transfer from the plant to the bacterium quite plausible.

Glutamine synthetase I (GSI) enzyme activity in *Streptomyces coelicolor* is controlled post-translationally by the adenylyltransferase (GlnE) as in enteric bacteria. Although other homologues of the *E. coli* Ntr system (*glnK*, coding for a PII family protein; and *glnD*, coding for an uridylyltransferase) are found in the *S. coelicolor* genome, the regulation of the GSI activity was found to be different. The functions of *glnK* and *glnD* were analysed by specific mutants. Surprisingly, whereas modification of GSI by GlnE occurs normally in all mutant strains, neither GlnK nor GlnD are required for the regulation of GlnE in response to nitrogen stimuli. Analysis of the post-translational regulation of GlnK *in vivo* by mass spectrometry indicated that it is subject to both a reversible and a non-reversible modification in a direct response to nitrogen availability. The irreversible modification was identified as removal of the first three N-terminal amino acid residues of the protein, and the reversible modification as adenylylation of the conserved tyrosine 51 residue that is known to be uridylylated in *E. coli*. The *glnD* insertion mutant expressing only the N-terminal half of GlnD was capable of adenylylating GlnK, but was unable to perform the reverse deadenylylation reaction in response to excess ammonium. The *glnD* null mutant completely lacked the ability to adenylylate GlnK. This work provides the first example of a PII protein that is modified by adenylylation, and demonstrates that this reaction is performed by a homologue of GlnD, previously described only as a uridylyltransferase enzyme.

The Biosynthesis of Glutamate

In principle, glutamate synthase and glutamate dehydrogenase are both able to synthesize glutamate. Mutants lacking both enzymes have an absolute requirement for glutamate. But glutamate dehydrogenase is a completely dispensable enzyme, although the cell can use it if provided with a high enough concentration of ammonia.

Glutamate Dehydrogenase

Glutamate dehydrogenase has been purified from *E. coli* and *S. typhimurium*. Its molecular weight is 300,000, with a relatively low apparent affinity for ammonia. It is a hexamer of identical subunits.

The reaction catalyzed,

$$\mathrm{COOH-(CH_2)_2-CO-COOH+NH_3+NADPH+H^+}$$
$$\downarrow$$
$$\mathrm{COOH-(CH_2)_2-CH(NH_2)-COOH+H_2O+NADP+Glutamate}$$

uses NADPH for the reductive amination, but not NADH.

Although there is no control of glutamate dehydrogenase synthesis by ammonia, repression is observed when glutamate is present in the growth medium of *E. coli*, but curiously this effect is not seen in *S. typhimurium*.

Glutamate Synthase

When the ammonium ions concentration is sufficiently high, ammonia is incorporated directly into glutamine, glutamate and asparagine. When ammonia is limiting, it is incorporated in glutamine only. Since glutamate is about eight times more abundant than glutamine in *E. coli* proteins, glutamine has to be somehow recycled in glutamate under these conditions. The enzyme responsible for this recycling is glutamate synthase, which catalyzes the following reaction:

$$\text{Glutamine} + \alpha\text{-ketoglutarate} + \text{NADPH} + \text{H}^+ \rightarrow 2\,\text{glutamate} + \text{NADP}^+$$

The enzyme has been purified from *E. coli* and *Klebsiella aerogenes*. It is a heterodimer whose two subunits which have molecular weights of 53,000 and 135,000 in *E. coli*. It contains both flavin adenine dinucleotide (FAD) and flavin mononucleotide (FMN), iron and labile sulfide. As many other amidotransferases, it can utilize a high level of ammonia instead of glutamine as the nitrogen donor. The ammonia dependent activity is catalyzed by the isolated small subunit, which then performs essentially the same reaction as glutamate dehydrogenase.

A mechanism for glutamate synthase has been proposed, where NADPH binds the small subunit and transfers its electrons to the large subunit, which reduces the flavin. Alpha-ketoglutarate binds the small subunit and glutamine binds the large one. The amide group of glutamine is then transferred to ketoglutarate and the reduced flavin reduces a postulated iminoglutarate intermediate to glutamate. The loss of either subunit results in the loss of capacity to carry the overall reaction. This mechanism is supported by chemical and kinetic data.

The genes coding for the two subunits form an operon *gltBD*, *gltB* and *gltD* coding respectively for the large and the small subunit.

Glutamate represses the synthesis of glutamate synthase, which is activated upon glutamate deprivation. Lrp regulates the transcription of *gltBD*: relative to expression in glucose minimal medium, *gltBD* expression in an *lrp*+strain is repressed 2.2-fold in the presence of 10 mM exogenous leucine and 16-fold in rich medium. Repression of *gltBD* expression by leucine or rich medium is not seen for an isogenic strain containing a Tn10 insertion in *lrp*. Lrp binds specifically to DNA fragments containing the *gltBD* promoter region.

Glutamate synthases from other bacteria have been studied: they are all iron-sulfur flavoproteins, with the exception of *Clostridium pasteurianum*, which possesses a different quaternary structure and lacks iron and flavin as prosthetic groups.

Glutamate Decarboxylase

Glutamate decarboxylase (*gadB*) from *E. coli* is a hexameric, pyridoxal 5′-phosphate-dependent enzyme catalyzing CO_2 release from the α-carboxyl group of L-glutamate to yield γ-aminobutyrate. GadB exhibits an acidic pH optimum and undergoes a spectroscopically detectable and strongly cooperative pH-dependent conformational change involving at least six protons. Crystallographic studies showed that at mildly alkaline pH GadB is inactive because all active sites are locked by the C termini and that the 340 nm absorbance is an aldamine formed by the pyridoxal 5′-phosphate-Lys^{276} Schiff base with the distal nitrogen of His^{465}, the penultimate residue in the GadB sequence.

Biosynthesis of Proline

In Enterobacteriaceae

The radioactivity from ^{14}C-glutamate is incorporated into proline and isotope competition experiments show that glutamate is a precursor of proline.

It was established during the 1950s that an intermediate of proline biosynthesis was glutamate γ-semialdehyde, arising from the reduction of the γ–carboxyl of glutamate.

$$\begin{gathered}\mathrm{OHC-CH_2-CH_2-CH(NH_2)-COOH}\\ \text{Glutamate } \gamma-\text{semialdehyde}\end{gathered}$$

This compound is the higher homologue of aspartate β-semialdehyde, the common precursor of lysine, methionine and threonine, which we have encountered in a preceding chapter. Unfortunately, this compound is extremely labile and cyclizes spontaneously to 5-oxopyrrolidine-2-carboxylate. It is expected that between glutamate and the semialdehyde, there should exist some activated intermediate. It was proposed in 1969 that such an intermediate could be γ-glutamyl phosphate. If the postulated corresponding kinase reaction is carried in the presence of hydroxylamine, a hydroxamate is obtained which is to be expected if an energy-rich phosphate intermediate is formed. On the other hand, if the second step in the pathway, glutamate semialdehyde dehydrogenase is studied in the reverse direction, the end product is 5-oxopyrrolidine-2-carboxylic acid, the cyclization product of glutamyl phosphate. It was concluded that the latter existed *in vivo* as en enzyme-bound intermediate.

In 1984, it was shown that the glutamyl kinase is active only in the presence of the enzyme catalyzing the second step, glutamate semialdehyde dehydrogenase. The complex is very labile in vitro and its dissociation leads to an inactive kinase.

The glutamate kinase has been finally purified from a strain containing a multi copy plasmid carrying the genes for the two enzymes, which are organized into a single operon, *proBA*. It is assayed in the presence of an excess of the dehydrogenase. It is a homopolymer of six subunits of molecular weight of 38,952, as deduced from the DNA sequence of its coding gene (*proB*). The reaction catalyzed

$$\text{Glutamate} + \text{ATP} \rightarrow \gamma\text{-glutamylphosphate} + \text{ADP}$$

is inhibited by proline at a very low concentration.

Glutamic semialdehyde dehydrogenase catalyzes the second step of proline biosynthesis, the reduction of glutamylphosphate to glutamate semialdehyde. Whereas the subunit molecular weight is 43,503 as determined by the nucleotide sequence of the *proA* gene, there is a controversy concerning the molecular weight of the native enzyme, which could be either a tetramer or a hexamer.

Glutamic semialdehyde exists in equilibrium with Δ^1-pyrroline-carboxylic acid; which is the substrate of a specific reductase which reduces it to L-proline in the presence of NADPH:

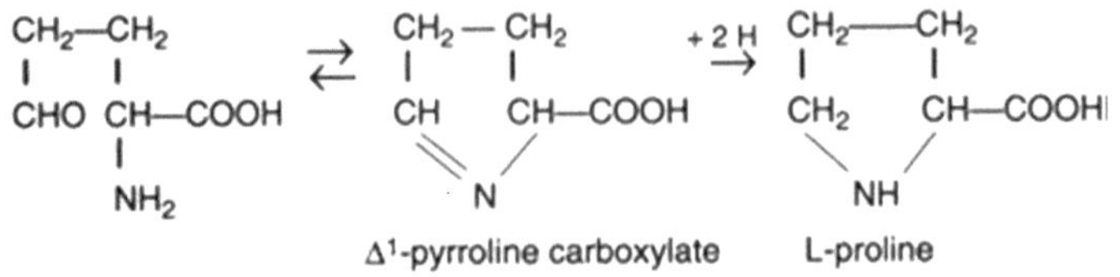

From a strain harboring a plasmid carrying *proC*, the gene coding for the reductase, a homogeneous preparation was obtained with a MW = ca. 300,000. The subunit, whose sequence was derived from the DNA sequence, has a molecular weight of 28,112.

The inhibition of glutamate kinase by proline is the major control mechanism of proline biosynthesis. Although it has been sought for in several laboratories, a negative control by end product repression has not been put into evidence. It has even been excluded in a study performed with *Serratia marcescens*, by analyzing the expression of the *lacZ* structural gene fused with the *proBA* promoter. Other control mechanisms, such as a translational control at the level of the mRNA specified by the *proBA* operon (transcribed from *proB* to *proA*), have not been excluded.

In Other Genera

The complete *B. subtilis* genome contains four genes (*proG*, *proH*, *proI*, and *comER*) with the potential to encode Δ^1-pyrroline-5-carboxylate reductase, Simultaneous defects in three of these genes (*proG*, *proH*, and *proI*) were required to confer proline auxotrophy, indicating that the products of these genes are mostly interchangeable with respect to the last step in proline biosynthesis.

Bacteria have developed other ways to synthesize proline. Proline can be derived from exogenous ornithine, which is an intermediate in arginine biosynthesis and degradation. The conversion of ornithine to proline can occur by two metabolic routes: in one, the δ-amino group of ornithine is lost, and in the other, the α-amino group is lost. Some like the strict anaerobes *Clostridium sporogenes* or *Cl. botulinum* use iornithine cyclase which transforms ornithine to proline in a single step, while others like *Aspergillus nidulans* or *Cl. sticklandii* rely on ornithine-δ-aminotransferase.

$$\text{Ornithine} + \alpha\ \text{ketoglutaric acid} \rightarrow \text{glutamate semialdehyde}$$

In a wide variety of organisms, from bacteria to plants, proline is accumulated in response to osmotic stress. From *Vigna aconitifolia*, a cDNA clone has been isolated which encodes a bifunctional enzyme, D[1]-pyrroline-5-carboxylate synthetase which corresponds to the two proteins encoded by *proA* and *proB* in *E. coli*.

Utilization of Proline

Proline utilization by *E. coli* and *S. typhimurium* requires the expression of genes *putP* (encoding a proline transporter) and *putA*. Genetic evidence indicates that the PutA protein is both Put repressor and a respiratory chain-linked dehydrogenase. The *putA* gene from *E. coli* has been cloned and sequenced and the corresponding protein purified to homogeneity; it is a dimer and binds the *put* DNA *in vitro*.

Degradation of Hydroxyproline

L-Hydroxyproline (4-hydroxyproline) mainly exists in collagen, and most bacteria cannot metabolize this hydroxyamino acid. *Pseudomonas putida* and *Pseudomonas aeruginosa* convert L-hydroxyproline to α-ketoglutarate via four hypothetical enzymatic steps different from known mammalian pathways, two novel enzymes, D-hydroxyproline dehydrogenase and Δ1-pyrroline-4-hydroxy-2-carboxylate (Pyr4H2C) deaminase, involved in this hypothetical pathway were identified and characterized. D-Hydroxyproline dehydrogenases from *P. putida* and *P. aeruginosa* were completely different from known bacterial proline dehydrogenases and showed similar high specificity for substrate (D-hydroxyproline) and some artificial electron acceptor(s). The former contains FAD as a prosthetic group, whereas the latter is a novel heterododecameric structure consisting of three different subunits ($\alpha_4\beta_4\gamma_4$), with two FADs, FMN, and [2Fe-2S] iron-sulfur cluster contained in αβγ of the heterotrimeric unit. Disruption of Pyr4H2C deaminase genes led to loss of growth on L-hydroxyproline (as well as D-hydroxyproline) but not on L- and D-proline, indicating that this pathway is related only to L-hydroxyproline degradation, which is not linked to proline metabolism.

The Biosynthesis of Arginine and Polyamines

Biosynthesis of Arginine

In the course of evolution, microorganisms have developed a mechanism analogous to the one we have already seen during the study of lysine biosynthesis to protect an amino group from involvement in spontaneous cyclization. But in this case, it is acetylation rather than succinylation that we find: a specific N-acetylglutamate synthase (NAGS) is both repressed and inhibited by arginine. In *E. coli*, where its structure is specified by *argA*, it consists of a single type of subunit of molecular weight of ca. 50,000. Arginine is a potent feedback allosteric inhibitor of the enzyme. Desensitized mutants have been obtained.

$$\text{Glutamate} + \text{Acetyl CoA} \rightarrow \underset{\substack{|\\ \text{NH-CO-CH}_3\\ \text{N-acetylglutamate}}}{\text{COOH–CH}}\text{–(CH}_2)_2\text{–COOH} + \text{CoASH}$$

Analysis of the crystal structures of NAGS from *Neisseria gonorrhoeae* in the inactive T-state with L-arginine bound and in the active R-state complexed with CoA and l-glutamate has uncovered the mechanism of allosteric inhibition of the enzyme by arginine.

The *argB* gene codes for the next enzyme, N-acetylglutamokinase (NAGK, MW of the primary polypeptide product, ca. 29,000) which phosphorylates the γ-carboxyl group of N-acetylglutamate, yielding an energy-rich compound, N-acetyl-γ-glutamylphosphate, which is labile and has not been isolated in the free state,

$$\text{N - Acetylglutamate} + \text{ATP} \rightarrow \text{N - Acetyl - }\gamma\text{ - glutamylphosphate} + \text{ADP}$$

and is hypothesized to be transferred onto the next enzyme without being liberated.

In *S. cerevisiae*, NAGS and NAGK associate in a complex, essential to synthase activity.

The next enzyme, N-acetylglutamylphosphate dehydrogenase or reductase (*argC*) catalyzes the reduction of the phosphoryl ester to N-acetylglutamate semialdehyde. It has been somewhat purified and the molecular weight of the *argC* encoded peptide has been reported to be 47,000, although the gene coding for it encodes only 334 residues. Comparison of the sequences of the enzyme from *E. coli*, *B. subtilis*, *Streptomyces clavuligerus*, *S. cerevisiae* and from the cyanobacterium *Anabaena* highlight a cysteine residue present in a highly conserved domain in the five proteins. This cysteine is probably located in the active site of the enzyme. The electron donor is NADPH.

The semialdehyde is the acetylated form of the intermediate occurring in proline biosynthesis, and were it not for the acetylation, all the semialdehyde would be converted to proline.

N-acetylglutamate semialdehyde undergoes a transamination reaction, catalyzed by N-acetylornithine aminopherase (*argD*) which has been obtained in the crystalline state (MW of the native enzyme = 119,000, of the subunit = 31,000)

$$\text{CHO–(CH}_2)_2\text{–}\underset{\substack{|\\ \text{NH–CO–CH}_3\\ \text{N-acetylglutamate semialdehyde}}}{\text{CH}}\text{–COOH} + \text{glutamate} \rightarrow \text{NH}_2\text{–(CH}_2)_3\text{–}\underset{\substack{|\\ \text{NH–CO–CH}_3\\ \text{N-acetylornithine}}}{\text{CH}}\text{–COOH} + \alpha\text{-ketoglutarate}$$

The acetyl group can now be removed. This reaction is catalyzed by N-acetylornithinase (also called ornithine acetyltransferase (*argE*)), an enzyme depending for its activity on the presence of divalent cobalt ions and a thiol compound. The enzyme has been obtained in the homogeneous state and is a monomer of MW 62,000; it is rather unspecific and deacylates other substrates than N-acetylornithine.

In *B. stearothermophilus*, N-acetylglutamate synthase (*argA*) and ornithine acetyltransferase (*argE*) activities co-elute in a single peak corresponding to M(r) 110,000. Both activities are heat-inactivated at the same temperature and strongly inhibited by ornithine. These results suggest that both activities can be ascribed to a single protein.

Ornithine carbamoyltransferase also known as ornithine transcarbamylase; (*argI*) catalyzes the sixth step of arginine biosynthesis, the coupling of ornithine and carbamyl phosphate to yield citrulline:

$$NH_2{-}(CH_2)_3{-}\underset{\substack{|\\ NH_2}}{CH}{-}COOH + NH_2{-}CO{-}O{-}PO_3H_2 \rightarrow \underset{\substack{|\\ CONH_2}}{NH}{-}(CH_2)_3{-}\underset{\substack{|\\ NH_2}}{CH}{-}COOH + Pi$$

Ornithine Carbamylphosphate Citrulline

Ornithine transcarbamylase and carbamyl phosphate synthetase are by far the best known enzymes of the arginine regulon. Carbamyl phosphate synthetase (coded by the *carAB* genes), which provides the carbamyl phosphate necessary not only for arginine, but also for pyrimidine synthesis will be examined later in this book. Ornithine transcarbamylase is a homotrimer. In the K12 strain of *E. coli*, there is a second gene, *argF*, located at another place than *argI* on the circular chromosome, which codes for another ornithine transcarbamylase, also a homotrimer, which displays 86 % amino acid homologies with the enzyme coded by *argI*. Other strains of *E. coli* and other *Enterobacteriaceae* do not possess this second gene. In K12, the products of each gene interact to form four trimeric isoenzymes. The same pattern of homotrimeric quaternary structure is found in the ornithine transcarbamylases of *S. typhimurium*, *Pseudomonas putida*, *P. aeruginosa*, *Aeromonas formicans* and *S. cerevisiae*. The molecular weight of their identical subunits varies between 35,000 and 40,000. A notable exception is the enzyme of *B. subtilis*, which exists as an equilibrium mixture of dimeric, tetrameric and hexameric forms of a 44,000-MW subunit. Interestingly, the pattern of regulation is also different in *B. subtilis*, the enzyme being repressed by arginine in exponentially growing cells, induced by arginine at the end of the exponential phase and inactivated during sporulation. The enzyme is the same under all conditions.

The kinetic properties of the *E. coli argI* enzyme have been thoroughly investigated: carbamylphosphate binds first, and Pi is released last.

The seventh step of this particular pathway is the addition to citrulline of an aspartic acid residue in the presence of ATP: ATP activates the ureido group of citrulline, a citrulline-adenylate intermediate being likely, and argininosuccinate, AMP and pyrophosphate are released:

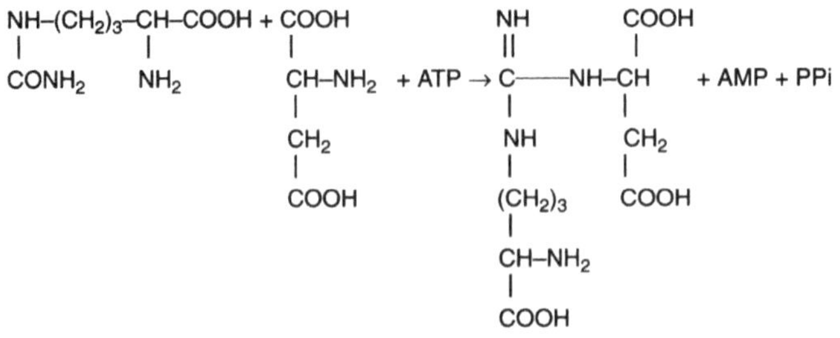

citrulline aspartate argininosuccinate

The crystal structure of *E. coli* argininosuccinate synthetase (*argG*) has been determined by the use of selenomethionine incorporation and MAD phasing. The structure has been refined at 1.6 Å resolution in the absence of its substrates and at 2.0 Å in the presence of aspartate and citrulline. Each monomer of this tetrameric protein has two structural domains: a nucleotide binding domain similar to that of the "N-type" ATP pyrophosphatase class of enzymes, and a novel catalytic/multimerization domain. The structure clearly describes the binding of citrulline at the cleft between the two domains and of aspartate to a loop of the nucleotide binding domain, whereas homology modeling with the N-type ATP pyrophosphatases has provided the location of ATP binding. The structures identify catalytically important residues and suggest the requirement for a conformational change during the catalytic cycle.

The structure of the enzyme from the archaeon *Thermus thermophilus* has also been obtained. It is quite similar to that of the *E. coli* enzyme, but the structure of the complex form is quite similar to that of the native one, indicating that no conformational change occurs upon the binding of ATP and the substrate analogues.

The yeast enzyme is also a homotetramer.

The last step is catalyzed by argininosuccinate lyase (*argH*) which promotes a trans-elimination and hydrolyzes argininosuccinate to fumaric acid and arginine:

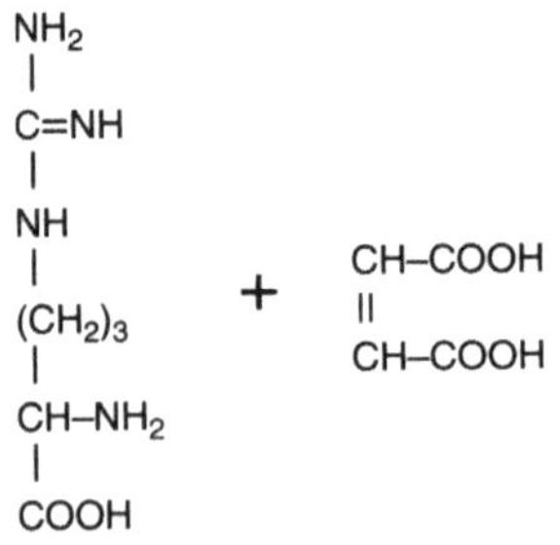

L-arginine

The information on the bacterial enzyme is scanty. The purified mammalian counterpart is a tetramer of 50,000 Da subunits. Indirect evidence is available that the *E. coli* polypeptide coded by *argH* is of a similar size.

Regulation of Arginine Biosynthesis at the Transcriptional Level

If we examine the circular chromosome of *E. coli*, we notice that the structural genes of arginine biosynthetic enzymes are scattered all over the chromosome. They must therefore belong to several operons. The only genes to be clustered are *argECBH*. A fine analysis has revealed that the *argECBH* cluster is actually a divergent operon consisting of two arms, *argE* and *argCBH*, transcribed in opposite directions with an internal operator common region flanked by two convergent promoters.

The study of the 5′-flanking region of *argI*, *argF* and of the carAB operon (specifying the structure of the two subunits of carbamylphosphate synthetase) has revealed the existence of operator regions which show a great similarity to that of the *argECBH* common operator and has allowed to derive a consensus sequence from the various operators, the Arg boxes, as we have seen in the case of the Met box operators of the methionine biosynthetic genes. Here like in the methionine case, the variation between the normally found level and maximally derepressed levels of the various enzymes of the pathway greatly differs: ornithine transcarbamylase levels can vary 400-fold, whereas N-acetylglutamate reductase, N-acetylglutamate kinase and argininosuccinase, which belong to the same transcription unit (*argCBH*) vary coordinately only 50- to 70-fold, and the other individual gene products are also repressed in a parallel but uncoordinated fashion, as well as the products of the *carAB* genes and that of *argR* coding for an arginine repressor.

The Arginine Repressor

In *E. coli*, the arginine repressor, in conjunction with arginine as the corepressor, controls the genes of the enzymes of arginine biosynthesis. The nucleotide sequence of the *argR* gene has been described. It corresponds to a monomeric subunit of a molecular weight 16,500. The repressor is precipitable with arginine, which facilitates its purification. The native repressor is a hexamer (MW=98,000). It binds to the Arg box of *argF* and to its own Arg box (auto regulation). In the presence of arginine, it inhibits the transcription of *argF* in an *in vitro* assay. In equilibrium dialysis experiments with arginine, the Arg repressor shows positive binding cooperativity, in contrast to the Met and Trp repressors.

An unexpected finding has been the implication of the arginine repressor as an auxiliary protein in site-specific recombination.

Using foot printing and interference with ethylation of the phosphate groups of the operator DNA, it has been shown that the arginine repressor makes direct contacts to minor and major groove determinants of the operators.

A similar arginine repressor, also a hexamer, plays a dual role in *B. subtilis*. It represses the arginine biosynthetic genes and activates the genes responsible for the enzymes of arginine catabolism in this organism. Although *E. coli* and *B. subtilis* are taxonomically distant, the *B. subtilis* repressor can functionally complement *argR* for its two distinct functions in repression and recombination, probably by binding to similar regulatory sequences. Both repressors having been crystallized, this should eventually allow detailed comparison of the two systems.

Polyamine Biosynthesis

Polyamines are small flexible organic polycations found in almost all cells. They likely existed in the last universal common ancestor of all extant life, and yet relatively little is understood about their biological function, especially in bacteria and archaea.

We have seen how the acetylation of glutamate, the first reaction specific to the synthesis of arginine, is allosterically inhibited by this amino acid. In addition, in *E. coli*, the presence of excess arginine represses in a parallel, but not coordinate fashion, the synthesis of all the enzymes involved in its biosynthesis.

However, the synthesis of arginine cannot be considered as a perfectly linear sequence of reactions. In fact, one of the intermediates, ornithine, is the precursor of a diamine, putrescine:

$$\underset{\text{Ornithine}}{H_2N-(CH_2)_3-CH(NH_2)-COOH} \rightarrow \underset{\text{Putrescine}}{H_2N-(CH_2)_4-NH_2} + CO_2$$

This reaction is catalyzed by ornithine decarboxylase (*speC*). This pyridoxal phosphate dependent enzyme enzyme has been isolated as a dimer of MW = 160,000. In a species of *Lactobacillus*, the enzyme has been crystallized: it is a dodecamer composed of six homodimers and a single point mutation (Gly121Tyr) prevents association of dimers into dodecamers. The activity does not depend on the dodecameric state. In the archaeon *Thermus thermophilus*, ornithine decarboxylase is a phosphorylated protein which loses all activity upon treatment with alkaline phosphatase.

The physiological role of putrescine and of the two other polyamines known to occur in *E. coli*, spermidine and cadaverine, is still under discussion, although it is clear that they are required for optimal growth and that they must act through their polybasic nature.

The level of polyamines in *E. coli* is high, of the order of 40 mmol/g dried cells, whereas the level of arginine is of 280 mmol/g of dried cells. Therefore, in the presence of excess arginine, the synthesis of ornithine would be reduced to a rate unsuitable for the synthesis of polyamines. Under these conditions, putrescine is synthesized via an alternative route: an arginine decarboxylase (*speA*) decarboxylates arginine to agmatine:

$$\underset{\text{Arginine}}{H_2N{-}\underset{\underset{NH}{\|}}{C}{-}NH{-}(CH_2)_3{-}\underset{\underset{NH_2}{|}}{CH}{-}COOH} \rightarrow \underset{\text{Agmatine}}{H_2N{-}\underset{\underset{NH}{\|}}{C}{-}NH{-}(CH_2)_4{-}NH_2} + CO_2$$

Arginine decarboxylase is a tetramer of 296,000 Da, whose activity is inhibited by putrescine and spermidine. Agmatine is then hydrolyzed by the dimeric agmatine amidinohydrolase ($Mr = 2 \times 40{,}000$); also called agmatine ureohydrolase, specified by *speB*:

$$\underset{\text{Agmatine}}{H_2N{-}\underset{\underset{NH}{\|}}{C}{-}NH{-}(CH_2)_4{-}NH_2} + H_2O \rightarrow \underset{\text{Putrescine}}{NH_2{-}(CH_2)_4{-}NH_2} + \underset{\text{Urea}}{H_2N{-}CO{-}NH_2}$$

Urea is not degraded by most strains of *E. coli* and its concentration is actually a measure of the agmatine pathway. This alternative pathway is a new example of the diversity of the solutions arising in the course of evolution to the problem posed by the existence of branched pathways. As can be seen in the scheme below the synthesis of arginine is effectively part of a branched system, the branch point being situated at ornithine. The most economical way to produce putrescine is obviously to decarboxylate ornithine. However, when repressible strains of *E. coli* are in the presence of excess arginine, the activity of the first enzyme, N-acetylglutamate synthase and the synthesis of the biosynthetic enzymes leading to arginine are inhibited and the conversion of arginine to putrescine is the only alternative for the synthesis of polyamines. The other important polyamine, spermidine is a triamine and has the following structure:

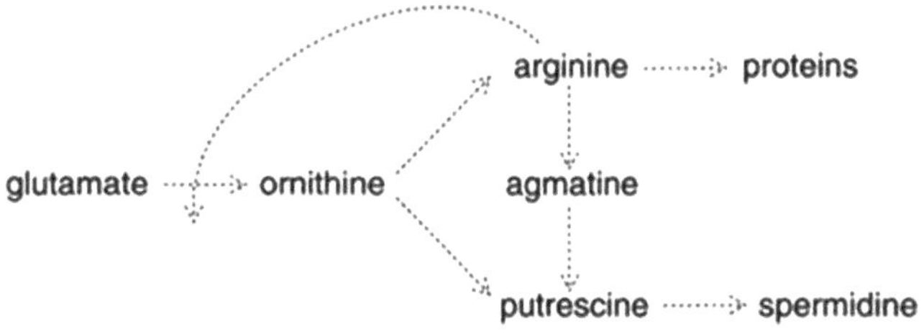

$$\underset{\text{Spermidine}}{H_2N{-}(CH_2)_3{-}NH{-}(CH_2)_4{-}NH_2}$$

which derives from putrescine by the addition of an aminopropyl group.

We shall see how this addition occurs.

An enzyme, S-adenosylmethionine (SAM)- decarboxylase (*speD*), transforms SAM to S-adenosyl-5′-δ-methylmercaptopropylamine or decarboxylated S-adenosylmethionine:

$$SAM \rightarrow H_2N{-}CH_2{-}CH_2{-}CH_2{-}\overset{+}{\underset{\underset{CH_3}{|}}{S}}{-}Ado + CO_2$$

Decarboxylated SAM

S-adenosylmethionine decarboxylase has been purified from *E. coli*. It has a molecular weight of 108,000 and is made of six identical subunits; as some other bacterial decarboxylases, it does not require pyridoxal phosphate, but covalently bound pyruvate. S-adenosylmethionine decarboxylase is first synthesized as a proenzyme, which is cleaved post translationally to form α and β subunits. The α subunit contains a covalently bound pyruvoyl group derived from serine that is essential for activity.

The crenarchaeon *Sulfolobus solfataricus* uses arginine to produce putrescine for polyamine biosynthesis. However, genome sequences from *S. solfataricus* and most crenarchaea have no known homologs of the previously characterized pyridoxal 5′-phosphate or pyruvoyl-dependent arginine decarboxylases that catalyze the first step in this pathway. Instead they have two paralogs of the *S*-adenosylmethionine decarboxylase . Whereas one paralog produces an adenosylmethionine decarboxylase, the other produces a novel arginine decarboxylase . Both thermostable enzymes self-cleave at conserved serine residues to form amino-terminal β-domains and carboxyl-terminal α-domains with reactive pyruvoyl cofactors. The arginine decarboxylase specifically catalyzed arginine decarboxylation more efficiently than previously studied pyruvoyl enzymes.

Despite spermidine concentrations in the Methanococcales that are several times higher than in *E. coli*, no SAM decarboxylase gene was recognized in the complete genome sequence of *Methanococcus jannaschii*. The gene encoding SAM decarboxylase in this archaeon has been identified as a highly diverged homolog of the *E. coli speD* gene (less than 11 % identity). The *M. jannaschii* enzyme has been expressed in *E. coli* and purified to homogeneity. Mass spectrometry showed that the enzyme is composed of two subunits of 61 and 63 residues that are also derived from a common proenzyme; these proteins associate in an $\alpha_2\beta_2$ complex. The pyruvoyl-containing subunit is less than one-half the size of that in previously reported SAM decarboxylases, but the holoenzyme has enzymatic activity comparable to that of other SAM decarboxylases. The sequence of the *M. jannaschii* enzyme is a prototype of a class of SAM decarboxylases that includes homologs in other archaea and diverse bacteria. The broad phylogenetic distribution of this group suggests that the canonical SpeD-type decarboxylase was derived from an archaeal enzyme. Both SpeD-type and archaeal-type enzymes have diverged widely in sequence and size from analogous eucaryal enzymes.

The mammalian enzyme is produced as a proenzyme which is cleaved to form an α and a β subunit in an apparently autocatalytic reaction that forms the pyruvate prosthetic group at the amino terminus of the α subunit. Regulation of the mammalian enzyme occurs at several levels: the synthesis of the proenzyme is inhibited by spermidine and spermine, whereas the processing of the proenzyme and the catalytic activity of the mature enzyme are stimulated by putrescine.

Spermidine synthetase catalyzes the reaction between decarboxylated SAM and putrescine to yield spermidine and 5′-thiomethyladenosine:

$$\text{Decarboxylated SAM + putrescine} \rightarrow \text{spermidine} + \underset{\substack{|\\ \mathrm{CH_3}}}{\overset{+}{\mathrm{S}}}\text{-Ado}$$

5′-methylthioadenosine

The availability of fully sequenced bacterial genomes has revealed that many species known to synthesize the polyamine spermidine lack the spermidine biosynthetic enzymes *S*-adenosylmethionine decarboxylase and spermidine synthase. Such species possess orthologues of the *sym*-norspermidine biosynthetic enzymes carboxynorspermidine dehydrogenase and carboxynorspermidine decarboxylase. By deleting these genes in the food-borne pathogen *Campylobacter jejuni*, it has been found that the carboxynorspermidine decarboxylase orthologue is responsible for synthesizing spermidine and not *sym*-norspermidine *in vivo*. In polyamine auxotrophic gene deletion strains of *C. jejuni*, growth is highly compromised but can be restored by exogenous *sym*-homospermidine and to a lesser extent by *sym*-norspermidine. The alternative spermidine biosynthetic pathway is present in many bacterial phyla and is the dominant spermidine route in the human gut, stomach, and oral microbiomes, and it appears to have supplanted the *S*-adenosylmethionine decarboxylase/spermidine synthase pathway in the gut microbiota.

Some *speD* mutants require the addition of exogenous spermidine for optimal growth. Spermidine synthetase is a dimer of molecular weight 72,000. It can transfer the aminopropyl group not only to putrescine, but also to the higher homologue cadaverine (the product of decarboxylation of lysine), but also to spermidine itself to form spermine, not normally found in *E. coli*, but among others in sperm cells where it stabilizes the tightly packed DNA.

$$\underset{\text{Spermine}}{H_2N-(CH_2)_3-NH-(CH_2)_4-NH-(CH_2)_3-NH_2}$$

The 5′-methylthioadenosine (MTA) formed during spermidine synthesis is not lost and serves to reconstruct methionine in the so-called methionine salvage pathway which has been fully analyzed in *Klebsiella pneumoniae*: methylthioadenosine phosphorylase cleaves it to adenine, itself salvaged, and 5′-methylthioribose-1-phosphate, further converted to α-ketomethylthiobutyrate which is converted to methionine by a specific transaminase.

Using in silico genome analysis and transposon mutagenesis in *B. subtilis*, the major steps of the dioxygen-dependent methionine salvage pathway have been established.

Although similar to that found in *Klebsiella pneumoniae*, its implementation depends on some entirely different proteins. The promoters of the genes have been identified by primer extension, and gene expression was analyzed. Among the most remarkable discoveries in this pathway is the role of an analog of ribulose diphosphate carboxylase (Rubisco, the plant enzyme used in the Calvin cycle which recovers carbon dioxide from the atmosphere) as a major step. In particular, a paralogue or Rubisco, MtnW, is used at one of the steps in the pathway. In the absence of MtnW, methylthioribose becomes extremely toxic to the cell, opening an unexpected target for new antimicrobial drugs. In addition to methionine salvage, this pathway protects *B. subtilis* against dioxygen produced by its natural biotope, the surface of leaves.

Utilization of Arginine as Sole Nitrogen Source by B. Subtilis

In *B. subtilis*, arginine is converted to glutamate by a series of three enzymatically catalyzed steps forming the so-called arginase pathway. Arginine is cleaved by arginase to give ornithine, the latter being converted to glutamate semialdehyde by ornithine aminotransferase. The conversion of the latter to glutamate is accomplished by a pyrroline-5-carboxylate dehydrogenase, this reaction being shared with the proline catabolic pathway. In this organism, arginine represses the synthesis of the arginine biosynthetic enzymes and activates the catabolic pathway through a regulatory gene (*ahrC*), encoding a protein similar in sequence and polymeric state to the arginine repressor ArgR.

Nitric Oxide Synthase in Bacteria

In mammals, nitric oxide (NO) is involved in many biological processes that range from regulation of blood pressure to protection against pathogens . Mammalian NO synthases (mNOSs) are highly regulated, complex enzymes that catalyze the oxidation of L-arginine to NO and citrulline (Fig. 28.7). Nitrous oxide (N2O) is a powerful greenhouse gas implicated in climate change. The dominant source of atmospheric N2O is incomplete biological denitrification, and the enzymes responsible for the release of N2O are NO reductases. It was recently reported that ambient emissions of

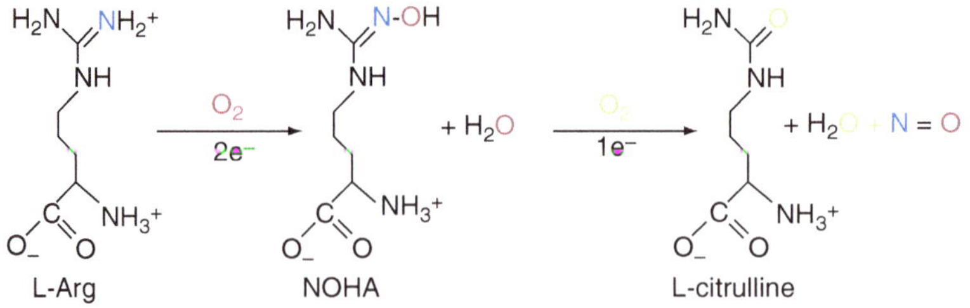

Fig. 28.7 Two-step oxidation of L-arginine (L-Arg) to L-citrulline and NO. Abbreviation: *NOHA* Nω-hydroxy-L-Arg

N2O from the Great Boiling Spring in the United States Great Basin are high, and attributed to incomplete denitrification by *Thermus thermophilus* and related bacterial species The NO reductase (NOR) from *T. thermophilus* has been isolated and characterized. The enzyme is a member of the cNOR family of enzymes and belongs to a phylogenetic clade that is distinct from previously examined cNORs. Like other characterized cNORs, the *T. thermophilus* cNOR consists of two subunits, NorB and NorC, and contains a one heme *c*, one Ca2+, a low-spin heme *b*, and an active site consisting of a high-spin heme *b* and FeB. The roles of conserved residues within the cNOR family were investigated by site-directed mutagenesis. The most important and unexpected result is that the glutamic acid ligand to FeB is not essential for function.

NO production by NOS from gram-positive bacteria has so far been characterized to be similar to that of mNOS, with the exception that the reducing equivalents cannot be supplied from a covalently attached reductase domain. A flavodoxin was demonstrated to be an efficient electron donor for NOS that could support NO synthesis in vitro. However, when the flavodoxin gene is deleted, B. subtilis still presents phenotypes indicative of NOS activity. Furthermore, bacterial NOSs expressed in Escherichia coli produce NO in vivo, presumably utilizing reductases from the host . Thus, unlike the mammalian proteins, which contain a dedicated reductase module, the bacterial proteins appear to accept electrons from several different reductases. Nonetheless, it may still be true that particular functions require specific, yet to be determined, reductases. The identification of NO synthase homologs encoded in bacterial genomes was received with excitement. Given the importance of mNOSs in multicellular signaling and immune response, the question of what the role of NOSs could be in bacteria immediately provoked interest. Subsequent investigations have progressed from assaying NOS activity in bacterial cell lysates to identifying the enzymes responsible for this activity. Bacterial NOSs have now been implicated in diverse biological functions that are decidedly different to those found in animals, including pathogenicity. Several X-ray crystal structures of bacterial NOSs have been determined, and similarities, as well as differences, to those of animal NOSs have been noted. Because of their stripped-down domain structure, ease of purification and in some cases, thermostability, these proteins have proved useful for investigations into the mechanism of NO synthesis.

The Biosynthesis of Lysine in Yeasts and Molds

It has long been known that lysine is synthesized via two mutually exclusive pathways. Aspartate is a precursor of the diaminopimelate pathway, found in bacteria, in a small number of Phycomycetes and in green plants. Glutamate, or rather α-ketoglutarate is the lysine precursor in Euglenids, some Phycomycetes, yeast and higher fungi, which use the so called aminoadipate pathway.

So far, in spite of urgent research, no organism has been found in which the two pathways exist together. It is tempting to think that the aminoadipic pathway may be more economical than the diaminopimelic acid pathway and has therefore been selected during evolution in yeasts and molds. Bacteria, in which diaminopimelic acid is a necessary compound for the cell wall construction, have only to acquire an additional decarboxylase to obtain their lysine, thus saving the necessity to synthesize the enzymes of the aminoadipic pathway.

It should be noted that the aminoadipic pathway has been found in *Thermus thermophilus* and other Archaea, which do not have a murein cell wall.

The Aminoadipic Acid Pathway

The methyl group of acetate reacts with the carbonyl group of α-ketoglutarate to yield homocitric acid. This reaction is similar to the one leading to the synthesis of citrate from acetyl CoA and oxaloacetate or to the formation of isopropylmalate (β-carboxy-β-hydroxyisocaproate), which we shall meet in the study of leucine biosynthesis in a forthcoming chapter:

$$\text{COOH–CO–(CH}_2)_2\text{–COOH} + \text{CH}_3\text{–CO–SCoA} \rightarrow \begin{array}{c}\text{COOH}\\ |\\ \text{HO–C–COOH}\\ |\\ (\text{CH}_2)_2\text{–COOH}\end{array}$$

α-ketoglutarate acetyl CoA Homocitrate

Whereas only one homocitrate synthase is found in *N. crassa* and *Saccharomyces lipolytica*, two isoenzymes are found in *S. cerevisiae*, each isoform being inhibited by lysine with a different sensitivity. This feedback inhibition of homocitrate synthase isoenzymes by lysine modulates the activation of the lysine biosynthetic genes gene expression by a transcriptional activator.

Homocitrate dehydrase and homoaconitate hydratase catalyze two successive steps leading to the formation of homoisocitrate with the intermediate formation of homoaconitate, calling to mind the well-known catalysis by aconitase during the tricarboxylic acid cycle.

$$\begin{array}{l}\text{CH–COOH}\\ \|\\ \text{CMCOOH}\\ |\\ \text{C–COOH}\\ |\\ (\text{CH}_2)_2\text{–COOH}\end{array} \qquad \begin{array}{c}\text{HO–CH–COOH}\\ |\\ \text{CH–COOH}\\ |\\ \text{CH–COOH}\\ |\\ (\text{CH}_2)_2\text{–COOH}\end{array}$$

Homoaconitate Homoisocitrate

Homoisocitrate dehydrogenase reduces homoisocitrate in an NAD^+-dependent reaction with the presumed formation of enzyme-bound oxaloglutarate, which is then decarboxylated to α-ketoadipate, the higher homologue of α-ketoglutarate, formed in a similar fashion from isocitrate in the tricarboxylic acid cycle.

$$\begin{array}{l} CO{-}COOH \\ | \\ CH{-}COOH \\ | \\ (CH_2)_2{-}COOH \end{array} \qquad \begin{array}{l} CO{-}COOH \\ | \\ CH_2 \\ | \\ (CH_2)_2{-}COOH \end{array}$$

oxaloglutarate α-ketoadipate

A specific aminoadipate aminotransferase catalyzes a transamination with glutamate to yield α-aminoadipate and α-ketoglutarate:

$$COOH{-}(CH_2)_3{-}CH(NH_2){-}COOH$$

α–aminoadipate

This diacid is reduced to the corresponding semialdehyde by a type of reaction we have already met in connection with aspartic and glutamic acids. It has been reported however that it is catalyzed by a single enzyme depending on ATP, Mg^{++} and NADPH. The intermediate energy-rich phosphate has not been isolated.

$$\begin{array}{l} O{=}C{-}(CH_2)_3{-}CH(NH_2){-}COOH \\ \;\;\;| \\ \;\;\;H \end{array}$$

α-aminoadipate semialdehyde

Saccharopine dehydrogenase condenses the semialdehyde with a molecule of glutamic acid with a concomitant reduction of the adduct by NADPH.

$$\begin{array}{ll} CH_2{-}NH{-} & CH{-}COOH \\ | & | \\ (CH_2)_3 & (CH_2)_2 \\ | & | \\ CH{-}NH_2 & COOH \\ | & \\ COOH & \end{array}$$

Saccharopine

This saccharopine dehydrogenase is called glutamate-forming, because glutamate is formed when it is studied in the reverse direction. The structure of the enzyme from the plant pathogen *Magnaporthe grisea* has been determined at 2 Å resolution. It is a homodimer, and each subunit consists of three domains, which are not consecutive in amino acid sequence. Domain I contains a variant of the Rossmann fold that binds NADPH. Domain II folds into a mixed seven-stranded beta sheet flanked by alpha helices and is involved in substrate binding and dimer formation. Domain III is all-helical. The structure analysis of the ternary complex reveals a large movement of domain III upon ligand binding. The active site is positioned in a cleft between the NADPH-binding domain and the second alpha/beta domain. Saccharopine is tightly bound to the enzyme via a number of hydrogen bonds to invariant amino acid residues.

The last reaction of this series is carried by a distinct saccharopine dehydrogenase, called lysine-forming, which in the presence of NAD^+ produces α-ketoglutarate and lysine.

Kinetic studies on the lysine-forming enzymes have indicated an ordered mechanism where NAD^+ and saccharopine bind, and lysine, α-ketoglutarate and NADH are released in this order.

$$\underset{\text{L-lysine}}{H_2N-(CH_2)_4-CH(NH_2)-COOH}$$

In *S. cerevisiae*, the enzymes of α-aminoadipate synthesis are located in the mitochondria, whereas a second aminoadipate aminotransferase and the last three enzymes of lysine biosynthesis are cytoplasmic. The first enzyme in the α-aminoadipate pathway, homocitrate synthase (HCS), is the target of the feedback regulation and is strongly inhibited by l-lysine. The structure of Schizosaccharomyces pombe HCS) in complex with l-lysine has been established. The structure illustrates that lysine directly competes with ketoglutarate for binding within the active site of HCS. Differential recognition of the substrate and inhibitor is achieved via a switch position within the (α/β)8 TIM barrel of the enzyme that can distinguish between the C5-carboxylate group of ketoglutarate and the ε-ammonium group of l-lysine. In vitro and in vivo assays demonstrate that mutations of the switch residues, which interact with the l-lysine ε-ammonium group, abrogate feedback inhibition, as do substitutions of residues within the C-terminal domain that were identified in a previous study of l-lysine-insensitive HCS mutants in Saccharomyces cerevisiae.

The following scheme summarizes the biosynthesis of the amino acids of the glutamic acid family in bacteria:

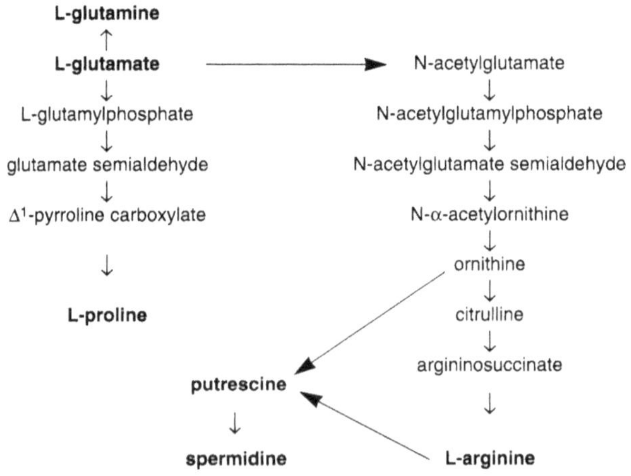

Selected References

Glutamine Synthetase Activity and Its Regulation by Covalent Modification: Structure

Almassy RJ, Johnson CA, Hanlyn R, Xuong NH, Eisenberg D (1986) Nature 323:3304–3309
Hesketh A, Fink D, Gust B, Rexer HU, Scheel B, Chater K, Wohlleben W, Engels A (2002) Mol Microbiol 46:319–330
Liaw SH, Pan C, Eisenberg D (1993a) Proc Natl Acad Sci USA 90:4996–5000
Rhee SG, Chock PB, Stadtman ER (1989) Adv Enzymol 62:36–92
Valentine RC, Shapiro BM, Stadtman ER (1968) Biochemistry 7:2143–2152

Glutamine Synthetase Regulation of Gene Expression

Keener J, Kustu S (1988) Proc Natl Acad Sci USA 85:4976–4980
Kustu S, Sei K, Keener J (1986) In: Booth IR, Higgins CF (eds) Regulation of gene expression-25 years on. Cambridge University Press, Cambridge, pp 139–154
Ninfa AJ, Magasanik B (1986) Proc Natl Acad Sci USA 83:5909–5913

The Levels of Glutamine Synthetase Are Also Regulated by Oxidation Followed by Proteolytic Degradation

Liaw S-H, Villafranca JJ, Eisenberg D (1993b) Biochemistry 32:7999–8003
Stadtman ER (1992) Science 257:1220–1224

Glutamate Synthase

Miller RE, Stadtman ER (1972) J Biol Chem 247:7407–7419
Rendina AR, Orme-Johnson WH (1978) Biochemistry 17:5388–5393
Radchenko MV, Thornton J, Merrick M (2013) Proc Natl Acad Sci USA 110:12948–12953

Glutamate Decarboxylase

Pennacchietti E, Lammens TM, Capitani G, Franssen MCR, John RA, Bossa F, De Biase D (2009) J Biol Chem 284:31587–31596

Proline Biosynthesis

Csonka LN, Baich A (1983) In: Herrmann KM, Somerville RL (eds) Amino acids: biosynthesis and genetic regulation. Addison-Wesley Publishing Company (Reading, Massachusetts), pp 35–51
Kenklies J, Ziehn R, Fritsche K, Pich A, Andreesen JR (1999) Microbiology 145:819–826

Degradation of Hydroxyproline

Watanabe S, Morimoto D, Fukumori F, Shinomiya H, Nishiwaki H, Kawano-Kawada M, Sasai Y, Tozawa Y, Watanabe Y (2012) J Biol Chem 287:32674–32688

Arginine Biosynthesis and Regulation

Abadjieva A, Pauwels K, Hilven P, Crabeel M (2001) J Biol Chem 276:42869–42880
Cunin R, Glansdorff N, Pièrard A, Stalon V (1986) Microbiol Revs 50:314–352
Lim D, Oppenheim J, Eckhardt T, Maas WK (1988) In: Gene expression and regulation: the legacy of Luigi Gorini. Excerpta Medica, Amsterdam, pp 55–63
Min L, Jin Z, Caldovic L, Morizono H, Allewell NM, Tuchman M, Shi D (2009) J Biol Chem 284:4873–4880

Polyamine Biosynthesis

Giles TN, Graham DE (2008) J Biol Chem 283:25829–25838
Kashiwagi JK, Taneja SK, Xie Q-W, Tabor CW, Tabor H (1988) Adv Exp Med Biol 250:73–79

The Arginine Repressor

Maas WK (1994) Bacteriol Revs 58:641–685
Wang H, Glansdorff N, Charlier D (1998) J Mol Biol 277:805–824

Spermidine Biosynthesis

Hanfrey CC, Pearson BM, Hazeldine S, Lee J, Gaskin DJ, Woster PM, Phillips MA, Michael AJ (2011) J Biol Chem 286:43301–43312
Tabor CW, Tabor H (1976) Annu Rev Biochem 45:285–306

The Methionine Salvage Pathway

Myers RW, Wray JW, Fish S, Abeles RH (1993) J Biol Chem 268:24785–24791
Sekowska A, Danchin A (2002) BMC Microbiol 25:8

Nitric Oxide Synthase

Sudhamsu J, Crane BR (2009) Trends Microbiol 17:212–218

Aminoadipic Acid Pathway

Bhattacharjee JK (1985) Crit Rev Microbiol 12:131–151
Bulfer SL, Scott EM, Pillus L, Trievel RC (2010) J Biol Chem 285:10446–10453
Feller A, Ramos F, Piérard A, Dubois E (1999) Eur J Biochem 261:163–170
Johansson E, Steffens JJ, Lindqvist Y, Schneider G (2000) Structure Fold Des 1037–1047

Chapter 29
Biosynthesis of Amino Acids Derived from Phosphoglyceric Acid and Pyruvic Acid

Biosynthesis of Glycine and Serine

The use of appropriate mutants and an adequate enzyme analysis has enabled to decide which of these amino acids is synthesized first, since the final step in the pathway, the interconversion of serine and glycine, is catalyzed by a single enzyme, serine hydroxymethyltransferase, and since mutations affecting the biosynthesis of serine result in a nutritional requirement which can be satisfied by one or the other of these amino acids. However mutations deficient in hydroxymethyltransferase require specifically glycine and cannot grow with serine.

The first enzyme of the pathway, phosphoglycerate dehydrogenase (*serA*) catalyzes a transformation of D-3-phosphoglycerate, an intermediate of the glycolytic cycle, into 3-phosphohydroxypyruvate

$$\underset{\text{3-phosphoglycerate}}{H_2O_3P-CH_2-CHOH-COOH} + NAD^+$$

$$\rightarrow \underset{\text{3-phosphohydroxypyruvate}}{H_2O_3P-O-CH_2-CO-COOH} + NADH + H^+$$

The enzyme is a homotetramer whose subunit has a molecular weight of ca. 41,000. In both *E. coli* and *S. typhimurium*, phosphoglycerate dehydrogenase is the target of an allosteric inhibition by serine which inhibits it non competitively. In these organisms, this inhibition constitutes the main mode of regulation of serine biosynthesis, since the levels of the dehydrogenase and of the other enzymes of serine biosynthesis are not increased by serine starvation or decreased by excess serine. Glycine is also an allosteric inhibitor, but it is necessary to reach a concentration more than a 100 times higher to reach the same degree of inhibition. It is not known whether glycine and serine bind at the same site or not. The conformational change resulting from serine binding is cooperative. The quaternary arrangement of the subunits has revealed an elongated ellipsoid. Each subunit consists of three domains: nucleotide, substrate and regulatory.

G.N. Cohen, *Microbial Biochemistry*,
DOI 10.1007/978-94-017-8908-0_29,

Extensive interactions are formed between nucleotide binding domains. A second subunit-subunit interaction occurs between regulatory domains creating an extended beta sheet. The serine-binding sites overlap this interface.

The 3-phosphohydroxypyruvate is the substrate of a transamination reaction, catalyzed by phosphoserine aminotransferase (*serC*), yielding phosphoserine:

$$H_2O_3P-O-CH2-CO-COOH+Glu \rightarrow$$
$$H_2O_3P-O-\underset{\text{Phosphoserine}}{CH_2-CH(NH_2)-COOH}+\ \alpha-Kg$$

The crystal structure of phosphoserine aminotransferase from *E. coli* has been determined at 2.3 Å resolution

A specific serine phosphate phosphatase hydrolyzes this compound to serine:

$$\underset{\text{L-serine}}{CH_2OH-CH(NH_2)-COOH}$$

Phosphoserine phosphatase (PSP), which has been found in all three domains of life, belong to the haloacid dehalogenase-like hydrolase superfamily. However, certain bacteria, lack a classical phosphoserine phosphatase gene, although they appear to possess a functional phosphoserine synthetic pathway. The apparent lack of a PSP ortholog in *Hydrogenobacter thermophilus*, an obligatory chemolithoautotrophic and thermophilic bacterium, represented a missing link in serine synthesis, since previous studies suggested that serine should be synthesized from phosphoserine. Phosphoserine phosphatase activity was found in cell-free extracts of *H. thermophilus* and two proteins two with PSP activity. Surprisingly, these proteins belonged to the histidine phosphatase superfamily and had been annotated as cofactor-dependent phosphoglycerate mutase (dPGM). However, because they possessed neither mutase activity nor the residues important for the activity, these proteins were defined as novel-type PSPs. Considering the strict substrate specificity toward phosphoserine, kinetic parameters, and PSP activity levels in cell-free extracts, these proteins were strongly suggested to function as PSPs *in vivo*. A number of bacterial genera, including Firmicutes and Cyanobacteria, were proposed to be strong candidates for possessing this novel type of PSP.

As stated above, the interconversion of serine and glycine is carried by serine hydroxymethyltransferase. The enzyme from *E. coli* has an apparent molecular weight estimated by molecular sieving to be 170,000. The relevant gene (*glyA*) has been cloned and sequenced, and corresponds to a polypeptide of 417 residues. The native enzyme is a homotetramer.

The enzyme is not subject to feedback inhibition, at least in Enterobacteriaceae, by any of the numerous compounds that have been tested.

The interconversion can essentially be written on paper as follows:

$$\text{Serine} \rightarrow \text{Glycine} + \text{Formaldehyde}$$

Actually, the C1 fragment obtained from serine is a tetrahydrofolate derivative of formaldehyde, 5,10-methylene tetrahydrofolic acid and the reaction is in fact:

$$\text{serine} + \text{FH4} \rightarrow \text{glycine} + \text{CH2OH} - \text{FH4}$$

$$\text{Sum}: \frac{\text{CH2OH} - \text{FH4} \rightarrow \text{HCHO} + \text{FH4}}{\text{Serine} \rightarrow \text{glycine} + \text{HCHO}}$$

Serine hydroxymethyltransferase catalyzes many other reactions involving other aldehydes than formaldehyde, but only the serine-glycine interconversion requires tetrahydrofolate. Compelling evidence shows that formaldehyde introduces constraints not encountered with other aldehydes and that the primary role of H4-folate is to add and remove formaldehyde from the active site. In addition to tetrahydrofolate, the reaction requires pyridoxal phosphate. Extensive spectroscopic studies suggest that it participates in the reaction through the formation of a Schiff base between the amino group of serine and the formyl group of pyridoxal phosphate (external aldimine).

The crystal structure of *E. coli* serine hydroxymethyltransferase has been elucidated at 2.4 Å resolution in complex with glycine substrate and 5-formyl tetrahydrofolate. This structure reveals the interactions of both cofactors and glycine substrate with the enzyme. Comparison with the aspartate aminotransferase (another pyridoxal phosphate enzyme) structure shows the distinctions in sequence and structure which define the folate cofactor binding site in serine hydroxymethyltransferase and the differences in orientation of the amino terminal arm, the evolution of which was necessary for elaboration of the folate binding site.

All forms of the enzyme for which a primary structure is known have five threonine residues near the active site lysyl residue that forms the internal aldimine with pyridoxal phosphate. The nine-residue conserved sequence is VVTT**T**THK(pyr)-T. In *E. coli*, each of the five threonines has been changed to an alanine by site directed mutagenesis. The five resulting mutant proteins have been purified and characterized with respect to affinity of coenzymes and substrates. Only the enzyme with the threonine in bold face in the sequence plays a critical role: the protein with an alanine replacing that threonine has normal affinities for the substrates and coenzymes but displays only 3 % of the catalytic activity of the wild type enzyme.

Under certain conditions of growth, it has been reported that *E. coli* and *Clostridium pasteurianum* are able to obtain glycine from threonine by the successive action of two enzymes in the presence of CoA with the intermediate formation of 3-keto-2-aminobutyrate (the genes coding for these two enzymes form an operon regulated by Lrp).

$$\underset{\text{Threonine}}{CH_3 - CHOH - CH(NH_2) - COOH} \rightarrow \underset{\text{Glycine}}{H_2N - CH_2 - COOH} + \text{Acetyl CoA}$$

Finally, let us recall that the methylenetetrahydrofolate produced by serine hydroxymethyltransferase is the substrate of the reductase yielding the methyltetrahydrofolate which is the source of the methyl group of methionine. This is how the β-carbon of serine is the source of that methyl group.

Regulation of Serine Hydroxymethyltransferase at the Transcriptional Level

The activity of serine hydroxymethyltransferase produces one-carbon units which are utilized in the synthesis of methionine, purines and thymine. The control of serine hydroxymethyltransferase synthesis is complex. Several compounds related to one-carbon metabolism (serine, glycine, methionine, purines, thymine) are apparently all involved in a kind of cumulative repression. In spite of the partial control by methionine, the regulation does not involve the methionine repressor, coded by *MetJ*: the controlling element is *MetR* which activates slightly the transcription of *glyA* in the presence of homocysteine. The controlling element in the case of purines is *PurR*. Much remains to be investigated in this very complex regulation.

Biosynthesis of Cysteine

The biosynthesis of cysteine is important not only because cysteine is a constituent of proteins, but also because the sulfur atom is involved in methionine, thiamine, biotin, pantetheine, lipoic acid, glutathione, and other cellular compounds.

Auxotrophic mutants of *E. coli* and *S. typhimurium* are known, which cannot utilize sulfate as a sulfur source, but can grow on sulfate. Other mutants cannot utilize either sulfate or sulfite but can grow on thiosulfate. These mutants were characterized by nutritional studies long before genetic mapping.

Cysteine biosynthesis is a branched convergent pathway where one branch leads from sulfate to sulfide, the other branch being the conversion of serine to O-acetylserine. The convergence point is the reaction of O-acetylserine with sulfide to yield cysteine.

The sole pathway of sulfate assimilation in bacteria and fungi involves an activation of sulfate with ATP, catalyzed by sulfate adenylyltransferase (or ATP sulfurylase) coded by the *cysD* gene:

$$\text{ATP} + \text{sulfate} \rightarrow \underset{(\text{APS})}{\text{adenosine} - 5' - \text{phosphosulfate}} + \text{PPi}$$

The equilibrium constant of this reaction does not favor the synthesis of APS and it follows that either pyrophosphatase, or the next reaction, or both are necessary to drive the sulfur further toward cysteine synthesis. Actually, ATP sulfurylase from *E. coli* is composed of two types of subunits, CysN (53 kDa) and CysD (35 kDa). The native (390 kDa) and subunit molecular weights suggest that the enzyme is a tetramer of CysN-CysD heterodimers. The primary sequence of CysN contains a GTP-binding motif common to the GTPases superfamily. The protein is actually a GTPase and the rate of APS synthesis is stimulated more than 100-fold at saturating concentrations of GTP.

In *Enterobacteriaceae*, APS cannot be reduced directly to sulfide and the next reaction catalyzed by adenylylsulfate kinase (or APS kinase, *cysC*, corresponding to a protein of 22,321 Da) produces a derivative phosphorylated at the 3′-position of ribose:

$$\text{ATP} + \text{APS} \rightarrow \underset{\text{(PAPS)}}{3'-\text{phosphoadenosine}-5'-\text{phosphosulfate}} + \text{ADP}$$

In the absence of APS, ATP phosphorylates the enzyme with a rate constant which is equivalent to the Vmax for the overall APS kinase reaction, suggesting that the phosphorylated enzyme is on the reaction path.

PAPS can now be reduced. This is achieved by a two-step process catalyzed by a specific PAPS sulfotransferase (*cysH*). This enzyme transfers the sulfonyl moiety of PAPS to thioredoxin to form an acceptor -S-SO_3- derivative. The intervention of a thiol such as glutathione or the second sulfhydryl group of the thioredoxin adduct eventually liberates free sulfite (for a discussion on thioredoxin, see Chap. 34). The *B. subtilis* gene was sequenced and found to encode a phosphoadenylylsulfate sulfotransferase with a molecular mass of 27 kDa. Expression of *lacZ* fused to the *cysH* promoter was repressed by cysteine and sulfide and induced by sulfur limitation, indicating that *cysH* is controlled at the level of transcription.

The sulfite reductase of *E. coli* is a very complex enzyme, which is able to achieve the 6-electron reduction of sulfite to sulfide without release of intermediates. It is an $\alpha_8\beta_4$ protein. The two component polypeptides, α and β (specified by *cysJ* and *cysI*) have respectively a molecular weight of about 60,000 and 55,000. By treating the holoenzyme with urea, one obtains a flavoprotein with an α8 structure and a monomeric hemoprotein. The flavoprotein contains 4 FAD and 4 FMN and is endowed with NADPH-cytochrome *c* reductase activity. The hemoprotein contains a Fe4S4 cluster and a heme group called siroheme, a product of uroporphyrinogen metabolism. The hemoprotein is able to reduce sulfite with reduced methylviologen as the electron donor.

The physiological electron flow is believed to be as follows:

$$\text{NADPH} \rightarrow \text{FAD} \rightarrow \text{FMN} \rightarrow \text{Fe4S4} \rightarrow \text{siroheme} \rightarrow \text{sulfide}$$

One cannot detect free flavoprotein of free hemoprotein in wild type extracts, but they can be demonstrated in extracts of *cysJ* or *cysI* mutants unable to synthesize one or the other protein. The two genes have been sequenced.

In *B. subtilis*, CysL, a transcriptional regulator, activates the transcription of the *cysJI* operon, which encodes sulfite reductase. Gel mobility shift assays and DNase I footprint experiments showed that the CysL protein specifically binds to *cysJ* and *cysL* promoter regions. This establishes also that CysL also negatively regulates its own transcription, a common characteristic of this type of regulators.

Another mutation leads to the incapacity of reducing sulfite. It occurs at another locus, called *cysG*. These mutations are pleiotropic and lead also to a deficiency in nitrite reductase (a siroheme containing enzyme) and cobalamin synthesis. It is probable that *cysG* (a misnomer) specifies a methylase necessary for siroheme and cobalamin synthesis.

Yeast sulfite reductase seems to have a different quaternary organization, with two α subunits of 116 kDa and two β subunits of 169 kDa. The enzyme has been reported to contain 2 FAD, 2 FMN and two siroheme chromophores per molecule. The gene coding for the α subunit of *S. cerevisiae* has been cloned and sequenced. Distinct homology has been found with the flavin-interacting parts of the bacterial flavoprotein subunit encoded by *cysJ*.

O-acetylation of Serine

In most plants and microorganisms that have been examined, O-acetylserine rather than serine is the precursor of the carbon moiety of cysteine. Serine transacetylase (*cysE*) catalyzes its synthesis from serine and acetyl CoA:

$$\text{Serine} + \text{acetyl CoA} \rightarrow \text{O}-\text{acetylserine} + \text{CoA}$$

However, in *Mycobacterium tuberculosis*, the cysteine synthase CysM is, in contrast to previous annotations, an *O*-phosphoserine-specific cysteine synthase. CysM belongs to the fold type II pyridoxal 5′-phosphate-dependent enzymes, as revealed by the crystal structure determined at 2.1-Å resolution. A model of *O*-phosphoserine bound to the enzyme suggests a hydrogen bonding interaction of the side chain of Arg^{220} with the phosphate group as a key feature in substrate selectivity. Replacement of this residue results in a significant loss of specificity for *O*-phosphoserine. Notably, reactions with sulfur donors are not affected by the amino acid replacement. The specificity of CysM toward *O*-phosphoserine) provides strong evidence for an *O*-phosphoserine-based cysteine biosynthesis pathway in *M. tuberculosis* that is independent of both *O*-acetylserine and the sulfate reduction pathway.

Cysteine Synthesis in Methanogens

The translation of cysteine codons in mRNA during protein synthesis requires cysteinyl-tRNA (Cys-tRNACys). Cys-tRNACys is normally synthesized from the amino acid cysteine and the corresponding tRNA isoacceptors (tRNACys) in an adenosine triphosphate (ATP)–dependent reaction catalyzed by cysteinyl-tRNA synthetase (CysRS). Genes encoding CysRS, *CysS*, have been detected in hundreds of organisms encompassing all three living domains . The only exceptions are certain methanogenic archaea, the completed genome sequences of which encode no open reading frames with obvious homology to known *CysS* sequences

Methanococcus maripaludis and *Methanocaldococcus jannaschii* produce cysteine for protein synthesis using a tRNA-dependent pathway. These methanogens charge tRNA(Cys) with phosphoserine, an intermediate in the predicted pathways for serine biosynthesis. To establish the mode of phosphoserine production in *Methanococcales*, cell extracts of *M. maripaludis* were shown to have phosphoglycerate dehydrogenase and phosphoserine aminotransferase activities. The heterologously expressed and purified phosphoglycerate dehydrogenase from *M. maripaludis* had enzymological properties similar to those of its bacterial homologs but was poorly inhibited by serine. While bacterial enzymes are inhibited by micromolar concentrations of serine bound to an allosteric site, the low sensitivity of the archaeal protein to serine is consistent with phosphoserine's position as a branch point in several pathways. A broad-specificity aspartate aminotransferase from *M. jannaschii* converted the phosphohydroxypyruvate product to phosphoserine. This enzyme catalyzed the transamination of aspartate, glutamate, phosphoserine, alanine, and cysteate. The *M. maripaludis* homolog complemented a *SerC* mutation in the *E. coli* phosphoserine aminotransferase. All methanogenic archaea apparently share this pathway, providing sufficient phosphoserine for the tRNA-dependent cysteine biosynthetic pathway. O-phosphoserine is converted to cysteine in a second step, catalyzed by O-phosphoserine sulfhydrylase.

Equilibrium sedimentation studies show that the serine acetyltransferase of *E. coli* is a hexamer. The results of velocity sedimentation and quasi-elastic light scattering experiments suggest that the identical subunits are loosely packed and/or arranged in an ellipsoidal fashion. Chemical cross-linking studies indicate that the fundamental unit of quaternary structure is a trimer. The likelihood, therefore, is that in solution the enzyme exists as an open arrangement of paired trimers.

Kinetic experiments with serine transacetylase are indicative of a ping pong mechanism, acetyl CoA being bound first, an acetylenzyme intermediate being formed and CoA liberated, serine being then bound and O-acetylserine being liberated. A similar mechanism has also been reported for the homoserine transacetylase of various microorganisms.

In crude extracts, the transacetylase is associated as a complex with the enzyme of the pathway which condenses O-acetylserine and sulfide to cysteine. This complex (MW ca. 300,000) dissociates in presence of O-acetylserine to free serine transacetylase and O-acetylserine sulfhydrylase (*cysK*). The dissociation is prevented by sulfide. The multifunctional complex is formed by serine transacetylase (MW = 160,000) and two molecules of a dimeric pyridoxal phosphate dependent O-acetylserine sulfhydrylase (OASS) composed of two identical subunits of 36,000 Da each, which makes acetate and cysteine from O-acetylserine and sulfide. The complex has been called cysteine synthase.

$$\underset{\text{Cysteine}}{H_2S-CH_2-CH(NH_2)-COOH}$$

OASS has been crystallized and its three-dimensional structure determined.

Upon substrate or product binding to OASS, evidence has been obtained for an energy transfer from one or the two tryptophan residues of the apoenzyme to the PLP internal aldimine accompanied by conformation change of the PLP Schiff base when either product, acetate or cysteine, is bound.

Because mutants devoid of serine transacetylase cannot be derepressed for the other cysteine biosynthetic enzymes, it is assumed that O-acetylserine plays the role of an internal inducer.

Allosteric Regulation of Cysteine Synthesis

As shown by isotope competition experiments, cysteine biosynthesis ceases if cells are grown in excess cysteine. Serine transacetylase is allosterically inhibited by cysteine at the exceedingly low concentration of 10^{-6} M.

There is no physiologically significant regulation by cysteine of any other enzyme of the biosynthetic pathway or of the sulfate transport.

Regulation of Cysteine Synthesis at the Genetic Level

Some words should be given about the organization of the various *cys* genes on the chromosome of *E. coli*.

The gene specifying serine transacetylase, *cysE*, is not linked to any other *cys* gene. The genes for sulfate permease, *cysA*, and for O-acetylsulfhydrylase, *cysK*, although both located at 49 min, are separated by 9 kb, and are not part of a single transcription unit.

In contrast, *cysC* and *cysD*, coding respectively for APS kinase and ATP sulfurylase are adjacent in *S. typhimurium* and the fact that they catalyze two consecutive reactions suggest that they may constitute a single transcription unit (an operon). There is a well-documented operon, the *cysJIH* transcriptional unit (coding for PAPS sulfotransferase, the apoflavoprotein and the apohemoprotein), defined by polarity effects and a promoter mutation.

Maximal expression of the cysteine regulon, encompassing all these genes, requires three factors: sulfur limitation, O-acetylserine availability, and a protein coded by a positive regulatory gene, *cysB* (28 min in *E. coli*).

Most *cysB* strains require cysteine for growth and like *cysE* mutants cannot be derepressed for the biosynthetic pathway. Merodiploids *cysB+/cysB* grow without added cysteine. The protein coded by *cysB* is a trans-acting element of positive control. The protein of *S. typhimurium* has been purified to homogeneity: it is a homotetramer of 4×36,000 Da, devoid of any known enzymatic activity. Its hypothetical mode of action is that a complex of the *cysB* protein and O-acetylserine facilitates initiation of transcription at the various *cys* promoters, in the way it has been described for other positive regulators. It has actually been shown that the

CysK protein binds to the *S. typhimurium cysJIH*, *cysK* promoters just upstream of their −35 regions.

Biosynthesis of Alanine

No alanine auxotroph has ever been isolated. This already indicates that there must be several ways to synthesize this amino acid.

Rudman and Meister have made a survey of transaminases in *E. coli*. Their work has been followed by other workers. At least two enzymes can synthesize alanine, but there may be as many as four. A glutamate-pyruvate transaminase

$$\begin{array}{c} HOOC{-}(CH_2)_2{-}CH(NH_2){-}COOH + CH_3{-}CO{-}COOH \\ \Downarrow\Uparrow \\ \underset{\alpha\text{-ketoglutarate}}{COOH{-}(CH_2)_3{-}CO{-}COOH} + \underset{\text{Alanine}}{CH_3{-}CH(NH_2){-}COOH} \end{array}$$

is induced by growth with glycerol or with alanine in the growth medium. This suggests a degradative function rather than a biosynthetic one.

Rudman and Meister discovered another transaminase, transaminase C which catalyzes the following reaction:

$$\text{Valine} + \text{Pyruvate} \Leftrightarrow \alpha - \text{ketoisovalerate} + \text{Alanine}$$

Since another transaminase, transaminase B catalyzes the reaction

$$\text{Glutamate} + \alpha - \text{ketoisovalerate} \Leftrightarrow \text{Valine} + \alpha - \text{ketoglutarate}$$

the sum of the two reactions results in the conversion of pyruvate to alanine, with glutamate as the amino donor, and serves this time a biosynthetic function.

In several microorganisms other than Enterobacteriaceae, pyruvate can undergo a reductive amination to form alanine:

$$\begin{array}{l} CH_3 - CO - COOH + NH_3 + NADH + H^+ \rightarrow \\ CH_3 - CH(NH_2) - COOH + NAD^+ + H_2O \end{array}$$

In addition, an aspartate decarboxylase has been found in a number of microorganisms (among others, *Alcaligenes faecalis*, *Desulfovibrio desulfuricans*, some species of *Achromobacter* and *Pseudomonas*) and its occurrence has been reported in the silkworm, the lobster and the crayfish

It catalyzes the following reaction:

$$HOOC-CH_2-CH(NH_2)-COOH \rightarrow CH_3-CH(NH_2)-COOH + CO_2$$

The enzyme from *Alcaligenes faecalis* has been obtained in the pure state. It is made of 12 subunits of 57 kDa each, organized as six dimers of molecular weight 114,000. Each monomer contains one mole of pyridoxal phosphate. The enzyme is allosterically inhibited by α-keto acids. Its catalytic mechanism and the mechanism of inhibition have been exhaustively analyzed by Meister.

Biosynthesis of Valine

This biosynthetic route is exactly parallel to that of isoleucine and uses the same enzymes. α-aceto-α-hydroxyacid synthetase which, as we have seen, catalyzes the condensation of one molecule of acetaldehyde (arising from pyruvate) with a molecule of α-ketobutyrate (deriving from threonine), is also able to carry out the condensation of a molecule of acetaldehyde with a second molecule of pyruvate. The enzyme requires the presence of Mg^{++} and thiamine pyrophosphate for its activity. The reaction takes place in two steps: a molecule of hydroxyethylthiamine pyrophosphate or active acetaldehyde is first formed and condenses with a second molecule of pyruvate to produce α–acetolactate (or with a molecule of α-ketobutyrate to produce α-aceto-α-hydroxybutyrate, in the isoleucine pathway).

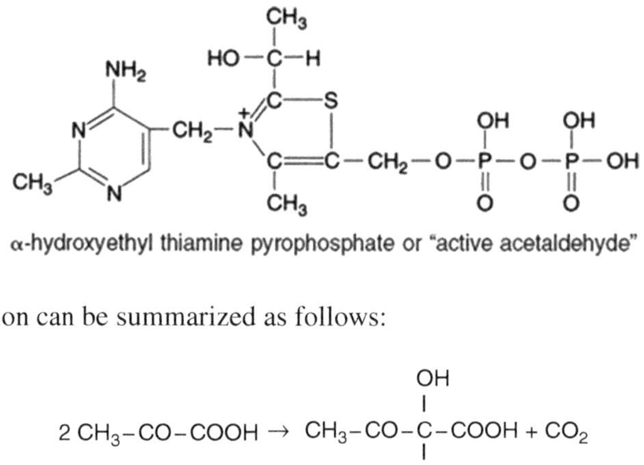

α-hydroxyethyl thiamine pyrophosphate or "active acetaldehyde"

The reaction can be summarized as follows:

$$2\,CH_3-CO-COOH \rightarrow CH_3-CO-\overset{\overset{\large OH}{|}}{\underset{\underset{\large CH_3}{|}}{C}}-COOH + CO_2$$

A rearrangement similar to the one already seen in isoleucine biosynthesis occurs under the influence of the reductoisomerase. It involves a reduction of the carbonyl group of α-acetolactate and the production of α,β-dihydroxy-β-methylbutyrate:

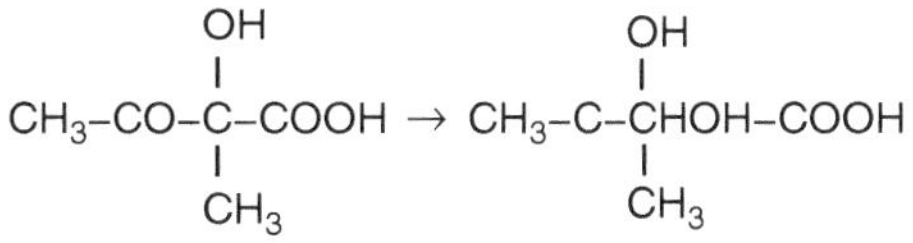

α, β–dihydroxy-β-methylbutyrate

The dehydratase specific for α,β-dihydroxyacids catalyzes the loss of one molecule of water and α-keto-β-methylbutyrate is obtained:

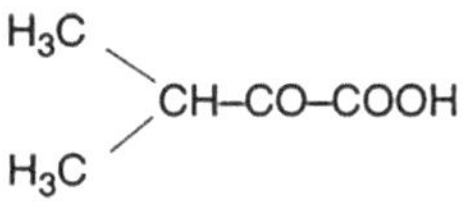

The same transaminase as was involved with isoleucine (*ilvE*), or alternatively with alanine as the amino donor, catalyzed by *avt* acts on this α-keto acid and L-valine is produced:

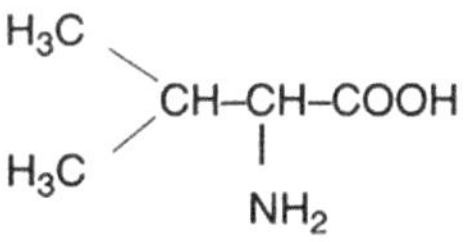

The implications of the existence of a series of enzymes common to the biosynthesis of valine and isoleucine in microorganisms become obvious when we examine the following scheme, which applies to *Neurospora* as well as to *E. coli*:

L-threonine		
↓	(threonine dehydratase)	
α-ketobutyrate		pyruvate
↓	A	↓
α-aceto-α-hydroxybutyrate		α-acetolactate
↓	B	↓
α,β-dihydroxy-β-methylvalerate		α,β-dihydroxy-β-methylbutyrate
↓	C	↓
α-keto-β-methylvalerate		α-keto-β-methylbutyrate
↓	D	↓
L-isoleucine		L-valine

Every mutant deficient in threonine dehydratase will be isolated as an isoleucine auxotroph; every mutant deficient in α-aceto-α-hydroxyacid synthetase, reductoisomerase, dehydratase or transaminase (reactions A, B, C, and D) will show a double requirement (valine plus isoleucine) for growth. We have seen that isoleucine is an allosteric feedback inhibitor of threonine dehydratase; this creates no problem for growth of the mutants since the end product, isoleucine has to be added to the medium. But, in contrast, an effective feedback inhibition of α-aceto-α-hydroxyacid synthetase by valine will present difficulties, since it automatically creates a deficiency in isoleucine. This is in fact the case in many organisms, including the K12 wild type strain of *E. coli*, whose growth is inhibited by valine, the inhibition being removed in the presence of isoleucine.

Biosynthesis of Leucine

α-keto-β-methylbutyrate, the immediate precursor of valine, is also the direct precursor of leucine. It is acetylated by α-isopropylmalate synthetase (*leuA*) which catalyzes the reaction of the keto acid with a molecule of acetyl CoA to yield α-isopropylmalate:

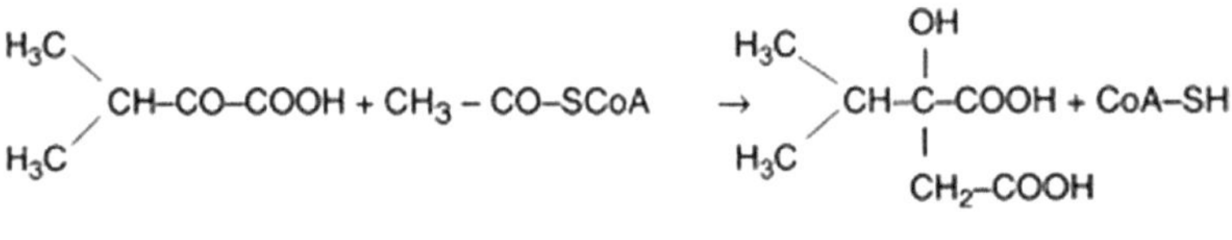

α-isopropylmalate

The α–isopropylmalate synthase from *S. typhimurium* has been purified and contains identical 50 kDa subunits. In the presence of both substrates, it exists as a tetramer, but leucine, its specific inhibitor (see below) dissociates it into monomers and dimers.

α-isopropylmalate isomerase dehydrates α-isopropylmalate and then rehydrates it to β-isopropylmalate:

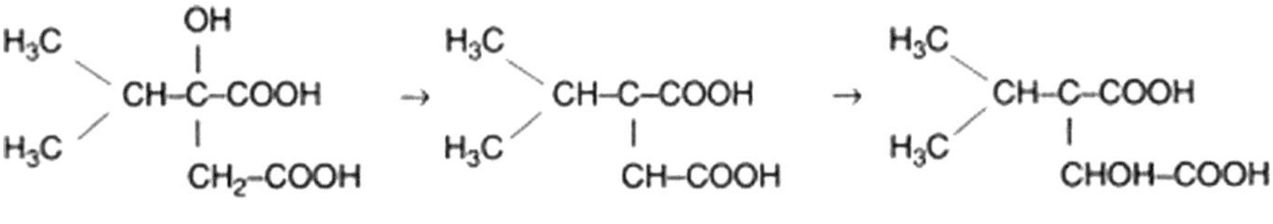

β-isopropylmalate

Isopropylmalate isomerase, is an unstable protein, which genetic data predict to be composed of two different subunits, respectively specified by the two genes *leuC* and *leuD*. The reaction catalyzed by the isomerase is reminiscent of the reaction catalyzed by aconitase in the tricarboxylic acid cycle and may involve the intermediate formation of dimethylcitraconate

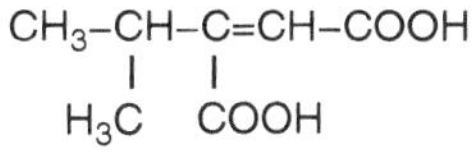

Dimethylcitraconate

An oxidative decarboxylation of β-isopropylmalate, catalyzed by the NAD$^+$-dependent β-isopropylmalate dehydrogenase (*leuB*) leads to the formation of α-ketoisocaproic acid:

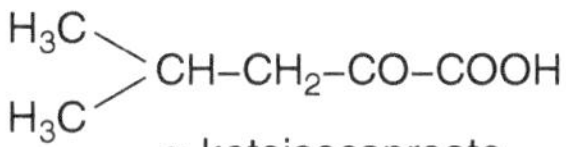

α-ketoisocaproate

The enzyme from *S. typhimurium* has been purified and is a dimer of 35 kDa subunits, requiring a monovalent cation for activity. The comparison of its sequence to that of ten other fungal or bacterial β-isopropylmalate dehydrogenases and to several isopropylmalate isomerases turns up a broadly conserved motif which might delineate the binding site for isopropylmalate, the common ligand of these two enzymes.

Finally, a transamination reaction produces leucine:

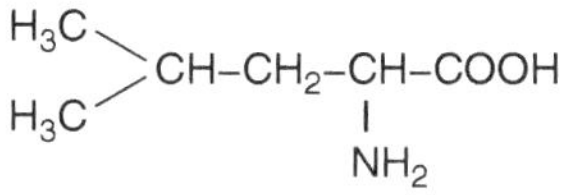

L-leucine

The conversion of α-ketoisocaproate to leucine can be carried out either by the transaminase encoded by *ilvE*, or the aromatic transaminase (*tyrB*).

The *leuABCD* cluster of genes constitutes (in that order) an operon, situated at 2 min on the *E. coli* circular chromosome. The *leu* operon has a single promoter, followed by a leader region and an attenuator (see below).

In a later chapter, we shall see that α-keto-β-methylbutyric acid is also the precursor of the vitamin pantothenic acid.

As in many other cases that become more and more often uncovered, the leucine-forming enzymes display a limited specificity. The pathway has therefore the potential of increasing the chain length not only of α-keto-β-methylvalerate, but also of pyruvate, α-ketobutyrate, and even α-ketoisocaproate through one or several cycles of elongation to yield longer keto acids and their aminated derivatives. Thus, mutants of *Serratia marcescens* have been described which accumulate α-aminobutyrate, norvaline, norleucine and homoleucine. In some cases, where several copies of the leucine genes are present, for example on multicopy plasmids, these endogenously synthesized unnatural amino acid analogs may be incorporated in *E. coli* proteins.

Isoleucine Synthesis from Pyruvate

Isoleucine formation may be entirely disconnected from aspartate in certain organisms. An alternative pathway from pyruvate and acetyl CoA, the citramalate pathway requires fewer ATP equivalents and was probably the ancestral pathway. The pathway from oxobutyrate to isoleucine is the same as that described in Chapter for *Enterobacteriaceae* (see Chap. 26).

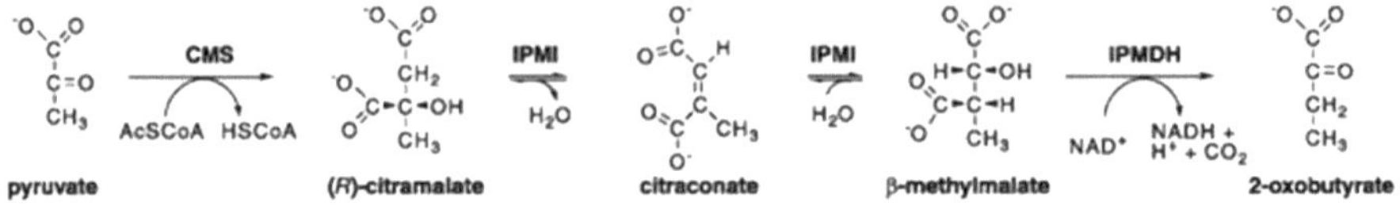

CMS citramalate synthase, *IPMI* isopropylmalate isomerase, *IPMDH* isopropylmalate dehydrogenase

Regulation of Valine, Isoleucine and Leucine Biosynthesis

We have already seen that threonine dehydratase is subject to feedback inhibition by isoleucine, and α–aceto-α-hydroxyacid synthetase by valine. In both cases, the target of allosteric inhibition is the first enzyme specific to the synthesis of the particular amino acid; the branch leading to leucine does not deviate from this rule: α-isopropylmalate synthetase is subject to allosteric feedback by L-leucine in coliform bacteria and in yeast.

The controls by repression in *Enterobacteriaceae* are more complex. It is necessary to remember that each of the last four steps in the biosynthesis of valine and isoleucine are catalyzed by a common enzyme. Repression of their synthesis presents a delicate problem since they bring about the production of two essential metabolites. In addition, the first reaction leading to the biosynthesis of leucine uses an intermediate in the synthesis of valine as a substrate.

The solution of this problem is the following: except for the reducto-isomerase which is induced by α-aceto-α-hydroxyacids, valine, leucine, and isoleucine are all required to repress the synthesis of the other enzymes in the biosynthetic pathway

Table 29.1 Multivalent repression of the synthesis of branched-chain amino acids. The results are expressed as specific activities, μmoles of product formed per mg protein per hour

Limiting amino acid	Threonine deaminase	Dihydroxyacid dehydratase	α-isopropylmalate synthase
None	4	5.8	1.2
Isoleucine	74	30	3.6
Valine	77	37	2.4
Leucine	58	24	12

leading to valine and isoleucine. In contrast, leucine alone is enough to repress the synthesis of the enzymes of its own pathway. Table 29.1, adapted from Freundlich, Burns and Umbarger, shows the effect on several of the enzymes involved of limiting each of the amino acids during growth in a chemostat.

This phenomenon has been called multivalent repression by the authors; we have encountered another case of it when we studied the divalent repression of the synthesis of the enzymes of the threonine operon. Again, the repression occurs through attenuation. Here, the sequence of the leader peptide involved in attenuation is

Met Thr Ala Leu Leu Arg Val Ile Ser Leu Val Val Ile Ser Val Val Val Ile Ile Ile Pro Pro Cys Gly Ala Ala Leu Gly Arg Gly Lys Ala

and is seen to contain several leucine, valine and isoleucine grouped residues at which the uncharged specific tRNAs can cause stalling of the ribosomes translating the RNA, favoring the transcription of the structural genes. (See the Chap. 21).

In contrast, the leader peptide involved in the regulation of the leucine operon, which is also under attenuation control and repressed by leucine alone, contains four tandem leucine residues:

Met Ser His Ile Val Arg Phe Thr Gly Leu Leu Leu Leu Asn Ala Phe Ile Val Arg Gly Arg Pro Val Gly Gly Ile Gln His

Whereas expression of the *ilv* operon in enteric bacteria is controlled primarily by attenuation in response to the intracellular levels of the cognate aminoacyl tRNAs, in the dimorphic bacterium *Caulobacter crescentus*, the expression of *ilvD*, coding for dihydroxyacid dehydratase is positively regulated by a protein coded by a gene (*ilvR*) which auto regulates its own expression.

The following scheme summarizes the reactions involved in the biosynthesis of the amino acids derived from phosphoglyceric acid and pyruvic acid.

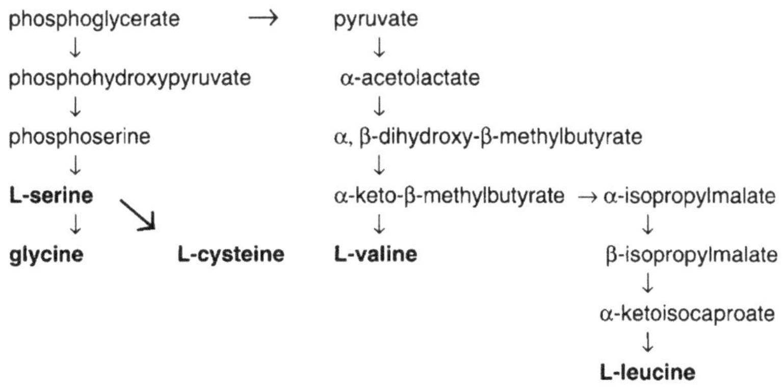

Selected References

Serine Biosynthesis

Bell JK, Pease PJ, Bell JE, Grant GA, Banaszak LJ (2002) Eur J Biochem 269:4176–4184
Chiba Y, Oshima K, Arai H, Ishii M, Igarashi Y (2012) J Biol Chem 287:11937–11941
Hester G, Stark W, Moser M, Kallen J, Markovic-Housley Z, Jansonius JN, Mol J (1999) Biol 286:829–850

Serine Hydroxymethylase

Scarsdale JN, Radaev S, Kazanina G, Schirch V, Wright HT, Mol J (2000) Biol 296:155–168
Schirch L (1982) Adv Enzymol 53:83–112
Stauffer GV (1983) In: Herrmann KM, Somerville RL (eds) Amino acids: biosynthesis and genetic regulation. Addison-Wesley Publishing Company, Inc., Reading, p 103–113

Sulfite Reductase

Siegel LM, Murphy MJ, Kamin H (1973) J Biol Chem 248:251–261
Guillouard I, Auger S, Hullo MF, Chetouani F, Danchin A, Martin-Verstraete I (2002) J Bacteriol 184:4681–4689

Cysteine Synthesis

Agren D, Schnell R, Oehlmann W, Singh M, Schneider G (2008) J Biol Chem 283:31567–31574

Cysteine Biosynthesis

Salsi E, Campanini B, Bettati S, Raboni S, Roderick SL, Cook PF, Mozzarelli A (2010) J Biol Chem 285:12813–12822

Cysteine Synthesis in Methanogens

Helgadóttir S, Rosas-Sandoval G, Söll D, Graham DE (2007) J Bacteriol 189:575–582

Valine and Leucine

Umbarger HE (1987) In: Ingraham JL, Low KB, Magasanik B, Schachter M, Umbarge HE (eds) *Escherichia coli* and *Salmonella typhimurium*, cellular and molecular biology. American Society for Microbiology, Washington, DC, p 352–367

Isoleucine Synthesis from Pyruvate

Drevland RM, Waheed A, Graham DE (2007) J Bacteriol 189:4391–4400

Aspartate-β-decarboxylase

Tate SS, Meister A (1971) Adv Enzymol 35:503–543

Chapter 30
Selenocysteine and Selenoproteins

Outlook

Selenium occurs normally in living things as a highly specific component of certain enzymes and amino acid transfer nucleic acids (tRNAs). In bacteria, biosynthesis of essential selenoenzymes has been shown to be unaffected by wide variations in sulfur levels. The naturally occurring selenoenzymes identified from bacterial sources include glycine reductase, formate dehydrogenases, a hydrogenase, nicotinic acid hydroxylase, xanthine dehydrogenase. The selenoenzyme, glutathione peroxidase, and other selenoproteins of unknown function have been isolated from animals. In certain enzymes, e.g. glycine reductase, formate dehydrogenase, hydrogenase and glutathione peroxidase, the chemical form of selenium has been identified as selenocysteine. A labile, unidentified form of selenium is present in nicotinic acid hydroxylase, and by inference, xanthine dehydrogenase.

Apart from being a constituent of the selenopolypeptides of the FDHs, selenium is also incorporated into modified nucleosides of some tRNAs. The seleno-tRNAs serve as examples of a different type of biological macromolecule that is specifically modified with selenium. The major seleno-tRNAs in *Clostridium sticklandii* and *Escherichia coli* have been identified as glutamate and lysine isoaccepting species. The selenium-modified nucleoside is 5-methyl-aminomethyl-2-selenouridine (mnm5Se2U), which is the chemical analog of 5-methylaminomethyl-2-thiouridine, a previously identified minor base of *E. coli* $tRNA2_{Glu}$. The seleno-$tRNA_{Glu}$ of *C. sticklandii* contains one gram atom of Se per mole of biologically active tRNA. Loss of Se from the modified nucleoside in this tRNA results in concomitant loss of glutamate charging activity suggesting that selenium is essential for interaction of the synthetase and its cognate tRNA.

The formation of many of the enzymes involved in fermentation reactions depends on the availability, the uptake, and the incorporation of a number of metals and metal-containing cofactors. The major ones involved are Fe, Mo, Ni, Co and Se.

G.N. Cohen, *Microbial Biochemistry*,
DOI 10.1007/978-94-017-8908-0_30,

Selenium Binding Protein

In the methanogen *Methanococcus vannielii*, selenium transport in the cell is established by the selenium-binding protein, SeBP. SeBP sequesters selenium during transport, thus regulating the level of free selenium in the cell, and delivers it specifically to the selenophosphate synthase enzyme. In solution, SeBP is an oligomer of 8.8-kDa subunits. It is a symmetric pentamer. The solution structure of SeBP was determined by NMR spectroscopy. Each subunit of SeBP is composed of an α-helix on top of a 4-stranded twisted β-sheet. The stability of the five subunits stems mainly from hydrophobic interactions and supplemented by hydrogen bond interactions. The loop containing Cys59 has been shown to be important for selenium binding, is flexible, and adopts multiple conformations. However, the cysteine accessibility is restricted in the structure, reducing the possibility of the binding of free selenium readily. Therefore, a different selenium precursor or other factors might be needed to facilitate opening of this loop to expose Cys59 for selenium binding.

Enzymes Containing Selenocysteine

Formate Dehydrogenases

The requirement of selenium for the degradation of formate was recognized early but it took more than 30 years before the major features of its metabolism were resolved.

Selenium is incorporated into three formate dehydrogenase (FDH) isoenzymes in the form of a single selenocysteine residue. The identity of two of these isoenzymes was disclosed by labelling with the ^{75}Se radioisotope. Fermentative growth of *E. coli* on glucose resulted in the specific labeling of an 80 kDa selenopolypeptide with the formation of the FDH_H enzyme activity. Growth under nitrate respiratory conditions labeled a 110 kDa selenopolypeptide which is a subunit of FDH_N, an enzyme which delivers the electrons withdrawn from formate to nitrate reductase.

A third FDH (FDH_0) is present in cells grown in the presence of oxygen or anaerobically in the presence of nitrate. It possesses a 110 kDa selenopolypeptide and appears to feed the electrons derived from formate into either the aerobic or the nitrate respiratory chain. Its actual physiological role, however, has yet to be resolved.

Formate dehydrogenase H from *E. coli* contains multiple redox centers, which include a molybdopterin cofactor, an iron-sulfur center, and a selenocysteine residue that is essential for catalytic activity. The addition of formate to the native enzyme induces a signal typical of Mo(V) species, detected by electron paramagnetic resonance (EPR) spectroscopy. Substitution of 77Se for natural isotope abundance Se leads to transformation of this signal, indicating a direct coordination of Se with Mo. Mutant enzyme with cysteine substituted at position 140 for the selenocysteine residue has decreased catalytic activity and exhibits a different EPR signal. The amino acid sequence flanking the selenocysteine residue in formate dehydrogenase H is similar to a conserved sequence found in several other prokaryotic molybdopterin-dependent enzymes. In most of these other enzymes a cysteine residue, or in a few cases a serine or a selenocysteine residue, occurs in the position corresponding to the SeCys-140 of formate dehydrogenase H. At least one of the ligands to Mo should be provided by an amino acid residue of the protein. This ligand could be the Se of a selenocysteine residue, the sulfur of a cysteine residue, or, in the case of a serine residue, oxygen.

The Glycine Reductase Complex

This enzyme complex catalyzes the reductive deamination of glycine, which is coupled to the esterification of orthophosphate resulting in the formation of ATP.

A selenium-containing protein, selenoprotein PA, is an essential component of this complex. The sequence near the selenocysteine (Sec) residue is -Cys-Phe-Val-Sec-Thr-Ala-Ala-Gly-Ala-Met-Asp-Leu-Glu-Asn-Glu-Lys and shows no homology with those of two other selenoproteins, glutathione peroxidase and formate dehydrogenase. An interaction of selenoprotein PA with the thioredoxin system, components of the NADPH-dependent reduction of glycine has been uncovered in *Eubacterium acidaminophilum* and *Clostridium litorale*.

The genes for thioredoxin reductase, thioredoxin and selenoprotein PA from *Eubacterium acidaminophilum* have been cloned, sequenced and characterized. Protein PA was highly similar to those of *Clostridium purinolyticum* and *Clostridium sticklandii* involved in glycine reductase. Thioredoxin reductase and thioredoxin of *E. acidaminophilum* could be successfully expressed in *E. coli*.

The substrate-specific selenoprotein B of glycine reductase (PBglycine) from *Eubacterium acidaminophilum* was purified and characterized. The enzyme consisted of three different subunits with molecular masses of about 22 (α), 25 (β) and 47 kDa (γ), probably in an $\alpha 2\beta 2\gamma$ composition. PBglycine purified from cells grown in the presence of [^{75}Se] selenite was labeled in the 47-kDa subunit. The 22-kDa and 47-kDa subunits both reacted with fluorescein thiosemicarbazide, indicating the presence of a carbonyl compound. This carbonyl residue prevented N-terminal sequencing of the 22-kDa (α) subunit, but it could be removed for Edman degradation by incubation with *o*-phenylenediamine. A DNA fragment was isolated and sequenced which encoded β and α subunits of Pbglycine (*grdE*), followed by a gene encoding selenoprotein A (*grdA2*) and the γ subunit of PBglycine (*grdB2*). The *grdB* gene contained an in-frame UGA codon which confirmed the observed selenium content of γ subunit. Peptide sequence analyses suggest that *grdE* encodes a proprotein which is cleaved into the previously sequenced N-terminal β subunit and the α subunit of PB glycine. Cleavage most probably occurred at an -Asn-Cys- site concomitantly with the generation of the blocking carbonyl moiety from cysteine at the alpha subunit.

The Nicotinic Acid Hydroxylase of Clostridium barkeri

Molybdenum, assayed by atomic absorption spectrometry, copurifies with the selenium-containing nicotinic acid hydroxylase from *Clostridium barkeri*. Fluorescence spectral studies on the enzyme indicate the presence, along with flavin, of another component. The fluorescence spectra of this component obtained after the aerobic denaturation of the nicotinic acid hydroxylase are similar to the fluorescence properties reported for the "pterin-like" cofactor from xanthine oxidase and several other molybdoproteins. Nicotinic acid hydroxylase from *C. barkeri* contains molybdenum, selenium, iron, acid-labile sulfur, and flavin with the occurrence of a "pterin-like" cofactor also a likely component.

The enzyme exhibits a stable Mo (V) electron paramagnetic resonance (EPR) signal ("resting" signal). This signal is correlated with the selenium content and nicotinate hydroxylase activity of the enzyme. Substitution of ^{77}Se for normal selenium isotope abundance results in splitting of the Mo (V) EPR signal of the native protein without affecting the iron signals of the FeS clusters. The Mo (V) EPR signal and nicotinic acid hydroxylase activity of enzyme isolated from cells grown in selenium-deficient medium are barely detectable. In contrast, the EPR signals of the FeS clusters, the electronic absorption spectrum, the NADPH oxidase activity, and the chromatographic behavior are changed little and are typical of active selenium-containing enzyme. An EPR signal indicative of the presence of molybdenum in the selenium-deficient enzyme also is exhibited. From these results it is concluded that a dissociable selenium moiety is coordinated directly with molybdenum in the molybdopterin cofactor and, moreover, that this selenium is essential for nicotinic acid hydroxylase activity.

Whereas selenium is present in FDH in a selenocysteine residue, in nicotinic acid hydroxylase, it occurs in an unidentified labile cofactor. The pterin component has been identified: FDH contains molybdopterin guanine dinucleotide whereas nicotinic acid hydroxylase contains molybdopterin cytosine dinucleotide. The protein which consists of four dissimilar subunits occurs in forms of different molecular masses. There are 5–7 Fe, 1 FAD, and 1 Mo per 160 kDa protein. Mo in the enzyme is bound to a dinucleotide form of molybdopterin and is coordinated with selenium. Mo (V), flavin radical, and two Fe2S2 clusters could be observed with EPR spectroscopy. The Se cofactor which is essential for nicotinic acid hydroxylase activity could be released from the enzyme as a reactive low molecular weight compound by a number of denaturing procedures. Parallel losses of Se and catalytic activity were observed during purification and storage of the enzyme. Addition of sodium selenide or selenophosphate does not restore the catalytic activity of the enzyme. Instead, it is reversibly inactivated by these compounds and also by sulfide. Cyanide, a common inhibitor of Mo-containing hydroxylases, does not affect the catalytic activity. The "as isolated" enzyme exhibits a Mo (V) EPR that was detected at early stages of purification. Nicotinic acid hydroxylase exhibits a high substrate specificity toward electron donor substrates. Its properties differentiate it from known Mo-containing hydroxylases.

Hydrogenases

[NiFeSe] hydrogenases are metalloenzymes that catalyze the reaction $H_2 \leftrightarrow 2H^+ + 2e^-$. They are generally heterodimeric, contain three iron-sulfur clusters in their small subunit and a nickel-iron-containing active site in their large subunit that includes a selenocysteine ligand.

Xanthine Dehydrogenase

A specific dehydrogenase, different from nicotinic acid hydroxylase, was induced during growth of *Eubacterium barkeri* on xanthine. The protein designated as xanthine dehydrogenase was enriched to apparent homogeneity using a three-step purification scheme. It exhibited an NADP-dependent activity. In addition it showed an NADPH-dependent oxidase and diaphorase activity. A molecular mass of 530 kDa was determined for the native enzyme and SDS/PAGE revealed three types of subunits with molecular masses of 17.5, 30 and 81 kDa indicating a dodecameric native structure. Molybdopterin was identified as the molybdenum-complexing cofactor using activity reconstitution experiments and fluorescence measurements. The molecular mass of the cofactor indicated that it is of the dinucleotide type. In an (αβ) heterodimeric xanthine dehysrogenase from *Rhodobacter capsulatus* containing the XdhA and XdhB subunits, the molybdenum cofactor (Moco) was shown to be absent, indicating that dimerization of the (αβ) subunits has to precede Moco insertion. The enzyme contained iron, acid-labile sulfur, molybdenum, tungsten, selenium and FAD at molar ratios of 17.5, 18.4, 2.3, 1.1, 0.95 and 2.8 per mol of native enzyme. Selenium is not attached to the protein in a covalently bound form such as selenocysteine.

Acetoacetyl CoA Thiolase

Clostridium kluyveri grown in the presence of $Na_2{}^{75}SeO_3$ produces a thiolase that copurifies with ^{75}Se. Based on several criteria, the selenium moiety in this protein is selenomethionine. The ^{75}Se-labeled amino acid in acid hydrolysates of the radioactive protein co-chromatographed with authentic selenomethionine on an amino acid analyzer and on TLC plates. Incubation with S-adenosylmethionine synthetase and ATP converted the ^{75}Se-labeled amino acid to a product that was indistinguishable from authentic Se-adenosylselenomethionine. The native selenoenzyme, Mr = 155,000–158,000, is composed of four subunits of Mr 38,000–40,000. Thiolase of similar molecular weight that is less acidic and lacks selenium is also produced by *C. kluyveri*. The factors that control the relative levels of the two enzymes in the cell have not been identified Further studies led to the conclusion that selenium occurs in thiolase adventitiously and is not required for any biological function.

Gene Products Involved in Selenocysteine Biosynthesis and Incorporation

Mutants (previously called *fdh* and now renamed *sel*) had been isolated that were pleiotropically defective in the activity of two FDHs. A reisolation and detailed analysis of several classes of mutants disclosed that they carried a lesion in one of the steps of biosynthesis or incorporation of selenocysteine. The *selC* gene codes for a unique tRNAsec which is charged by seryl-tRNA synthetase. Seryl-tRNAsec is then bound to selenocysteine synthase (the *selA* gene product) via a Schiff base between the amino group of the seryl residue and the carbonyl group of the pyridoxal phosphate group of the enzyme. The elimination of a water molecule yields enzyme-bound dehydroalanyl tRNAsec. Addition of reduced and activated selenium to the double bond results in the liberation of selenocysteyl-tRNAsec from the enzyme. The activated selenium compound is a product of the reaction catalyzed by the *selD* gene product (selenophosphate synthetase). This enzyme cleaves both anhydride bonds of ATP, thereby transferring the γ-phosphate to selenide, which results in selenophosphate synthesis and the release of AMP and the release of AMP and the β-phosphate of ATP.

Selenocysteine Synthase

Selenocysteine synthase is a homodecamer of 50 kDa subunit constituent High-resolution electron microscopy revealed that the decamer is able to bind five seryl tRNAsec molecules. This stoichiometry is confirmed by biochemical analyses.

Selenophosphate Synthetase

Selenophosphate synthetase, the *selD* gene product is a 37 kDa monomer. It possesses a conspicuous Cys17-Gly-Cys19 motif close to the N-terminus of the protein, and it was shown that the cysteyl residue at position 17 is essential for catalytic activity.

The selenophosphate synthetases from several organisms contain a selenocysteine residue in their active site where the *E. coli* enzyme contains a cysteine. The synthesis of these enzymes, therefore, depends on their own reaction product. To analyse how this self-dependence is correlated with the selenium status, e.g. after recovery from severe selenium starvation, the gene for the selenocysteine-containing selenophosphate synthetase from *Haemophilus influenzae* (*selDHI*) was transferred in an *E. coli selD* strain. Gene *selDHI* gave rise to a selenium-containing gene product and also supported – via its activity – the formation of *E. coli* selenoproteins.

Selenocysteine Lyase

Selenocysteine lyase (SCL) is a pyridoxal 5′-phosphate-dependent enzyme that specifically catalyzes the decomposition of L-selenocysteine to L-alanine and elemental selenium. The enzyme (from *Azotobacter vinelandii*, coded by *nifS*) was proposed to function as a selenium delivery protein to selenophosphate synthetase in selenoprotein biosynthesis.

Selenocysteyl tRNA

Selenocysteyl-tRNA sec requires a specialized translation factor SELB, for the decoding process at the ribosome. SELB is homologous in its function to EF-Tu and therefore binds guanine nucleotides and the charged $tRNA_{sec}$. However, in contrast to EF-Tu, it can discriminate between aminoacyl residues. Therefore, seryl-$tRNA_{sec}$ is not a substrate for the protein. SELB also has the capacity to bind to a specific recognition motif which is present in mRNAs coding for prokaryotic selenoproteins and which is located immediately downstream of the UGA directing selenocysteine insertion. It is noteworthy that tRNASec carries the anticodon UCA that matches the UGA termination codon. Thus a termination codon can be read by a native tRNA; It is the longest known tRNA species and has a variable region of 22 nucleotides in which several invariant sequence positions deviate from the consensus.

The flux into and the fidelity of selenocysteine biosynthesis and incorporation are regulated at several levels. First, seryl tRNA synthetase charges $tRNA_{sec}$ with only 1 % of the catalytic efficiency compared to $tRNA_{ser}$, which reflects the very minor amount of serine carbon required for selenocysteine synthesis. Second, $tRNA_{sec}$ possesses structural features which discriminate it from $tRNA_{ser}$ in its reaction with selenocysteine synthase. Finally, SELB binds $tRNA_{sec}$ only when it is charged with selenocysteine: seryl-$tRNA_{sec}$, the precursor, is not recognized.

The role of stoichiometry between mRNA, translation factor SELB and selenocysteyl-tRNA in selenoprotein synthesis was investigated. It was found that overproduction of SelB in an otherwise wild-type genetic background reduced UGA read through to less than 1 %. Concomitant overexpression of *selC* (the gene for selenocysteine-specific $tRNA_{sec}$) completely reversed the inhibition. To gain information on the domain structure of this specialized translation factor, the *selB* genes from two bacteria unrelated to *E. coli* (*Clostridium thermoaceticum* and *Desulfomicrobium baculatum*) were cloned and sequenced. The derived amino acid residue sequences were compared to those of SelB from *E. coli* and *Haemophilus influenzae* and to EF-Tu sequences. The alignment revealed that SelB contains all three domains characterized for EF-Tu. A fourth, C-terminally located domain shows only limited sequence conservation within the four SelB proteins.

A structure-function analysis of SelB from *E. coli* was performed. It showed that a C-terminal 17 kDa subdomain of the translation factor, when expressed separately, specifically binds the mRNA secondary structure. The recognition motif itself could be reduced to a 17 nucleotide minihelix without loss of binding affinity and specificity. A truncated SelB lacking the mRNA binding domain was still able to interact with selenocysteyl-$tRNA_{sec}$. Expression of the mRNA binding domain alone suppressed selenocysteine insertion in vivo by competing with SelB for its binding site at the mRNA. The results indicate that SelB can be considered as an EF-Tu homolog hooked to the mRNA via its C-terminal domain.

Selenocysteine $tRNA_{([Ser]Sec)}$, which decodes specific UGA codons and inserts selenocysteine into designated proteins, contains isopentenyladenosine. Isopentenylation is required for efficient translational decoding of UGA and synthesis of selenoproteins.

The *sel* genes are expressed constitutively in *E. coli*, probably reflecting the need for the synthesis of selenopolypeptides under aerobic (FDH_0), nitrate respiratory (FDH_N and FDH_0) and fermentative conditions (FDH_H).

Insertion Sequences (SECIS Elements)

The incorporation of selenocysteine into proteins is directed by specific UGA codons and mRNA secondary: structures, designated SECIS elements. In bacteria, these elements are positioned within the reading frame of selenoprotein mRNAs immediately downstream of the triplet coding for selenocysteine and they tether a complex of the selenocysteine-specific elongation factor SelB., GTP and selenocysteyl-tRNAsec to the site of UGA decoding.

A SECIS-like structure was identified in the 5′ non-translated region of the *selAB* transcript, encoding selenocysteine synthase and SelB. It specifically binds to SelB and the formation of a SelB-GTP-selenocysteyl-$tRNA_{sec}$ complex on the SECIS-like element represses expression of the downstream gene. This effect is abolished by mutations preventing formation of the complex. The regulating pattern observed correlates with the levels of *sel* gene products. As quaternary complex formation on the SECIS-like element does not influence the transcription rate and only slightly reduces the level of *selAB* mRNA, it is concluded that the structure is involved in regulating translation efficiency, thereby coupling selenocysteine biosynthesis to the availability of the trace element selenium.

The nature of the minimal selenocysteine insertion sequence (SECIS) in *E. coli* has been investigated: a stem-loop structure of 17 nucleotides is necessary on the condition that it is located in a proper distance (11 nucleotides) downstream the UGA codon. SECIS includes the information for an additional function: the prevention of UGA read-through under conditions of selenium deficiency. This information is contained in a short mRNA region consisting of a single C residue adjacent to the UGA on its downstream side, and an additional segment consisting of the six nucleotides immediately upstream from it. These two regions act independently and additively, and probably through different mechanisms. The single C residue acts as itself; the upstream region acts at the level of the two amino acids, arginine and valine, for which it codes. These two codons at the 5′ side of the UGA correspond to the ribosomal E and P sites. The model for the *E. coli* fdh SECIS is presented as a multifunctional RNA structure containing three functional elements. Depending on the availability of selenium, the SECIS enables one of two alternatives for the translational machinery: either selenocysteine incorporation into a polypeptide or termination of the polypeptide chain.

Selenocysteine and Archaea

Selenocysteine is encoded by a UGA codon in all organisms that synthesise selenoproteins. This codon is specified as a selenocysteine codon by an mRNA secondary structure, which is located immediately 3′ of the UGA in the reading frame of selenoprotein genes in Gram-negative bacteria, whereas it is located in the 3′ untranslated region of eukaryotic selenoprotein genes. The location and the structure of a similar mRNA signal in archaea has been investigated. Seven selenoproteins were identified for the archaeon *Methanococcus jannaschii* by labelling with ^{75}Se and by SDS/polyacrylamide electrophoresis. Their size could be correlated with open reading frames possessing internal UGA codons from the total genomic sequence. One of the open reading frames, that of the VhuD subunit of a hydrogenase possesses two UGA codons and appears to code for a selenoprotein with two selenocysteine residues. A strongly conserved mRNA element was identified that is exclusively linked to selenoprotein genes. It is located in the 3′ untranslated region in six of the mRNAs and in the 5′ untranslated region of the fdhA mRNA. This element, which is present in the 3′ non-translated region of two selenoprotein mRNAs from *Methanococcus voltae*, is proposed to act in decoding of the UGA with selenocysteine.

Biochemical Function of the Selenocysteine Residue in Catalysis

The exact biochemical function of the selenocysteine residue in catalysis has not been resolved yet. A cysteine-containing variant of FDH has been constructed genetically and purified. It was found to be catalytically active but with a K_{cat} 2 orders of magnitude less than the Se form. Also, naturally occurring S-forms homologous to the *E. coli* FDHs have been characterized in other organisms. Since they are abundant proteins in these organisms, the selective advantage of having a selenol in the active site might be a means of lowering the amount of enzyme required to fulfil a particular catalytic task.

Selected References

Böck A (2001) Selenium metabolism in bacteria. In: Hatfield DL (ed) Selenium: its molecular biology and role in human health. Kluwer Academic, Dordrecht, pp 7–22
Low SC, Grundner-Culemann E, Harney JW, Berry MJ (2000) EMBO J 19:6882–6890
Lu J, Holmgren A (2009) J Biol Chem 284:723–727
Schumann S, Saggu M, Moller N, Anker SD, Lendzian F, Hildebrandt P, Leimkuhler S (2008) J Biol Chem 283:16602–16611
Stadtman TC (1996) Ann Rev Biochem 65:83–100
Stadtman TC (2002) J Biol Chem 277:49091–49100
Suzuki M, Lee D-Y, Inyamah N, Stadtman TC, Tjandra N (2008) J Biol Chem 283:25936–25943
Thanbichler M, Böck A (2002) EMBO J 24:6925–6934

Chapter 31
Biosynthesis of Aromatic Amino Acids and Its Regulation

Glucose is an aliphatic substance, and so the synthesis of phenylalanine, tyrosine and tryptophan from it presents the biochemist with the problem of the biosynthesis of the aromatic nucleus.

The key to this problem was given 50 years ago by the isolation of polyauxotrophic mutants which required the presence of the three aromatic amino acids in the medium for growth, and in addition *p*-aminobenzoic acid (a precursor of folic acid). The existence of this multiple requirement as a result of a single mutation, immediately suggested that the genetic block consisted in the inability to synthesize a precursor common to all four essential metabolites. After trying 55 cyclic and hydroaromatic derivatives without success, B. D. Davis discovered that shikimic acid could replace the four growth factors. Moreover, growth was proportional to the amount of shikimate added. Other bacterial mutants, in which the lesion lies at a later step in aromatic biosynthesis, accumulate shikimic acid in the medium. It was therefore concluded that shikimic acid was the precursor sought; however, ts identification relied only on chromatographic analysis and on the response given by mutants. Later, the compound was isolated and characterized by the classical methods of organic chemistry.

Shikimic acid has been known since the nineteenth century as a rare constituent of certain plants. We shall now examine how this compound is formed in *E. coli* and how it leads to chorismic acid, from which the different aromatic compounds diverge.

The Common Pathway (Shikimic Pathway)

Formation of Shikimic Acid

A mutant strain which accumulated shikimic acid was grown on glucose labeled at specific positions as sole carbon source. The distribution of carbon atoms in the different carbon atoms of shikimate was then studied. The results showed

G.N. Cohen, *Microbial Biochemistry*,
DOI 10.1007/978-94-017-8908-0_31, © Springer Science+Business Media B.V. 2014

that the carboxyl group and carbon atoms 1 and 2 of shikimic acid were derived from a three-carbon intermediate of glycolysis (C1 to C3, or C4 to C6 of glucose) and that the four remaining carbons could come from a tetrose (C3 to C6 of glucose with some scrambling due to cycling in the hexose monophosphate shunt). This suggested that a seven-carbon intermediate might be formed by the condensation of a C3 derivative with a C4, coming from the pentose phosphate pathway.

A few years later, extracts of *E. coli* were shown to catalyze the conversion of erythrose-4-phosphate (E4P) and phosphoenolpyruvate (PEP) to a seven-carbon sugar derivative which was identified as 3-deoxy-D-arabino-heptulosonate-7-phosphate (DAHP), with the elimination of inorganic phosphate:

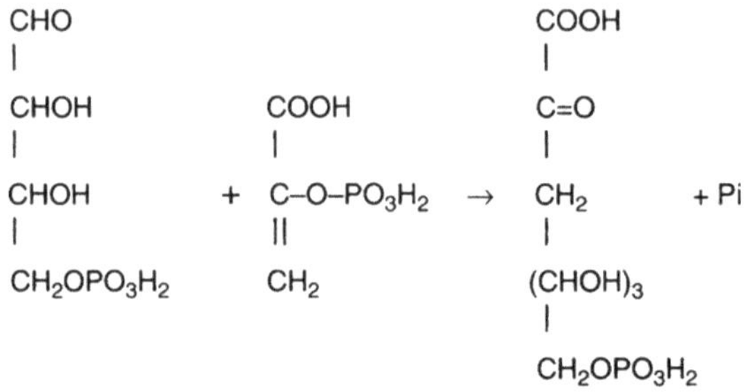

E4P PEP DAHP

As in the case of aspartokinases, a study of the inhibition and repression of 3-deoxy-D-arabino-heptulosonate-7 phosphate synthesis showed that there were at least two isofunctional enzymes catalyzing this reaction. One was inhibited and repressed by phenylalanine; the second was inhibited and repressed by tyrosine. They were respectively called DAHP synthase (phe) and DAHP synthase (tyr). A few years later, a third minor isofunctional enzyme, DAHP synthase (trp), whose activity was inhibited and whose synthesis was repressed by tryptophan was identified.

The three enzymes from *E. coli* and DAHP (tyr) from *S. typhimurium* have been obtained in the pure state, cloned and sequenced.

DAHP (phe) (*aroG*) is a tetramer of identical subunits of 350 residues each, whereas DAHP (tyr) (*aroF*) is a homodimer whose subunits comprise 356 amino acids. The third isoenzyme DAHP(trp) (*aroH*) is also a homodimer whose subunits have a molecular weight of about 40,000. All three enzymes contain one atom of iron per mole of native enzyme. For the two more abundant isofunctional proteins, the reaction mechanism has been analyzed and found to be ordered and sequential, with PEP being the first ligand bound.

The crystal structure of phenylalanine-regulated DAHP-synthase from *E. coli* has been solved. The tetramer consists of two tight dimers. Mutations that reduce feedback inhibition cluster about a cavity near the twofold axis of the tight dimer and are centered approximately 15 Å from the active site, indicating the probable location of a separate inhibitory site.

The barrel is the common protein fold of numerous enzymes and was proposed to be the result of gene duplication and fusion of an ancient half-barrel. The initial enzyme of shikimate biosynthesis possesses the additional feature of feedback regulation. The crystal structure and kinetic studies on chimera and mutant proteins of DAHP synthases from *S. cerevisiae* inhibited by phenylalanine (Aro3p) and by tyrosine (Aro4p) give insight into important regions for regulation in the enzyme: The loop, which is connecting the two half-barrels, and structural elements added to the barrel are prerequisites for regulation and form a cavity on the N-terminal side of the barrel. In the cavity of Aro4p at position 226 is a glycine residue, which is highly conserved in all other tyrosine-regulated DAHP synthases as well. Sequence alignments with phenylalanine-regulated DAHP synthases including Aro3p show a highly conserved serine residue at this position. An exchange of glycine to serine and vice versa leads to a complete change in the regulation pattern. Therefore the evolution of these differently feedback-inhibited isoenzymes required gene duplication and a single mutation within the internal extra element. Numerous additional amino acid substitutions present in the contemporary isofunctional enzymes are irrelevant for regulation and occurred independently.

An Asn-8 to Lys-8 substitution in *E. coli* AroF leads to a tyrosine-insensitive DAHP synthase. This mutant enzyme exhibited similar activities, substrate affinities and positive cooperativity with respect to phospho*enol*pyruvate as the wild-type AroF, but showed decreased thermo stability. An engineered AroF enzyme lacking the seven N-terminal residues also was tyrosine-resistant. These results strongly suggest that the N-terminus of AroF is involved in the molecular interactions occurring in the feedback-inhibition mechanism.

A chimeric protein has been obtained by joining the catalytic domain of the unregulated 3-deoxy-D-arabino-heptulosonate 7-phosphate synthase from *Pyrococcus furiosus* with the regulatory domain of tyrosine- regulated enzyme from *Thermotoga maritima*. This simple gene fusion event on its own is sufficient to confer functional allostery to the unregulated enzyme. The fusion protein shares structural similarities with its regulated parent protein and undergoes an analogous major conformational change in response to the binding of allosteric effector tyrosine to the regulatory domain. These findings help delineate a remarkably facile mechanism for the evolution of modular allostery by domain recruitment.

A single enzyme, 3-dehydroquinate synthase (*aroB*), catalyzes the transformation of DAHP to 3-dehydroquinic acid. The reaction requires NAD^+ and cobaltous ions. The requirement for NAD^+ remained mysterious for a long time, since the reaction does not involve a net transfer of electrons. The problem was finally solved when isotope experiments showed that all the tritium of labeled DAHP is conserved in dehydroquinate. This is due to an oxidation at C-5 by NAD^+, the hydride transfer from the enzyme bound NADH using the same hydrogen atom as was taken from DAHP. The following scheme summarizes these facts.

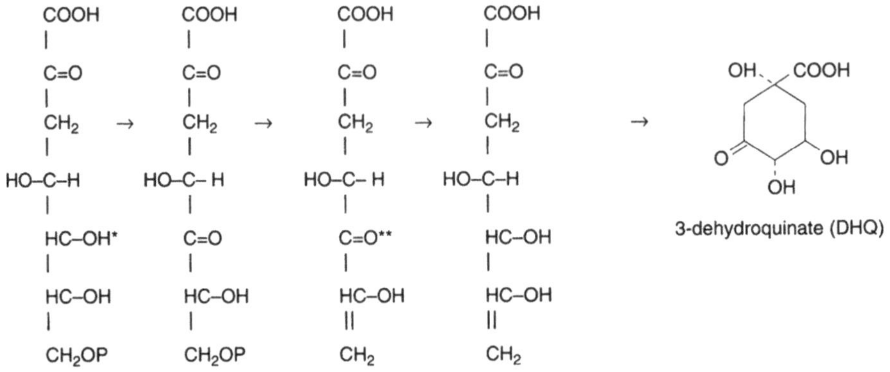

DAHP

3-dehydroquinate (DHQ)

* Site of oxidation by NAD+

** Site of reduction by NADH

Large amounts of DHQ synthase have been obtained by cloning techniques and the enzyme has been obtained in the pure state. It is a monomeric protein of molecular weight ca. 40,000.

Dehydroquinate dehydratase, also called dehydroquinase (*aroD*) removes one molecule of water from DHQ and introduces the first double bond of the aromatic ring. The gene for the enzyme has been cloned and sequenced. The protein has been overexpressed and purified to homogeneity. The native structure of the pure protein is predominantly dimeric, but it exists in equilibrium with monomeric subunits. The enzyme is inactivated by diethylpyrocarbonate; an equilibrium mixture of product and substrate protect against the inactivation which affects one histidine residue, a good candidate for the general base which is postulated to participate in the mechanism of the enzyme. The product of the reaction, which is reversible, is 3-dehydroshikimate.

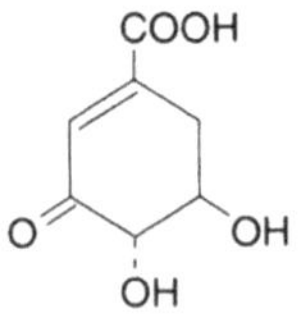

3-dehydroshikimate

This compound is acted upon by shikimate dehydrogenase, formerly known as dehydroshikimate reductase (*aroE*) which reduces it to shikimic acid with the aid of NADPH. The reaction is stereospecific and transfers the hydrogen from the A side of NADPH only. Homogeneous shikimate dehydrogenase has been obtained: it has a molecular weight of 32,000.

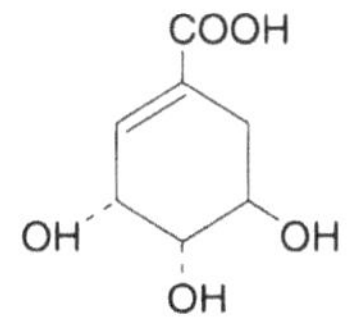

shikimic acid or
3, 4, 5, trihydrocyclohexene-
1-carboxylic acid

It is common to find multiple orthologues of individual enzymes in the shikimate pathway in microorganisms. For instance, the genome of *B. subtilis* carries two versions of dehydroquinate dehydratase (*aroD* and *aroQ*) but a single version of the next two enzymes, *aroE* and *aroK*. In contrast, this pattern is reversed in the genome of *E. coli* K12, which carries just a single orthologue of dehydroquinate dehydratase (*aroD*) but duplicate orthologues for both shikimate dehydrogenase (*aroE* and *ydiB*) and shikimate kinase (*aroK* and *aroL*). The 2.3 Å crystal structure of the *ydiB* orthologue (NAD-specific) shows that the protomer of shikimate-5 dehydrogenase contains two α/β domains connected by two α-helices. The C-terminal domain is a Rossmann fold, to which NAD^+ binds in an elongated fashion. Its binding site contains several unusual features, including a cysteine residue in close apposition to the nicotinamide ring and a clamp over the ribose of the adenosine moiety formed by phenylalanine and lysine residues. The structure explains the specificity for NAD versus NADP in different members of the shikimate dehydrogenase family.

Formation of Chorismic Acid

Shikimate kinase (*aroK*) catalyzes the phosphorylation of shikimic acid by ATP. No mutants were found lacking this activity among all the aromatic auxotrophs tested. The reason resides in the existence of the second shikimate kinase of low affinity for shikimic acid, for which no satisfactory rationale has been found. It may have another major function concerned with some other essential pathway.

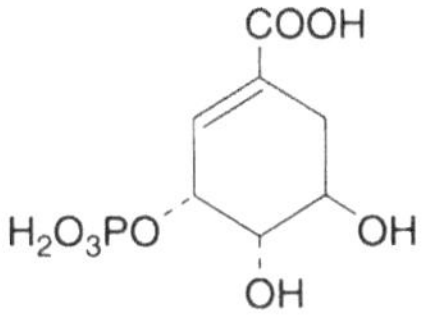

shikimate-3-phosphate

Aromatic auxotrophs blocked in the last step of the common pathway were found to accumulate 5-*enol*pyruvoylshikimate. This is the dephosphorylated form of the actual intermediate, 5-*enol*pyruvoylshikimate-3-phosphate (EPSP) which is synthesized from shikimate-3-phosphate and phosphoenolpyruvate by 5-*enol*pyruvoylshikimate-3-phosphate synthase (*aroA*). The gene has been cloned and overproducing strains have been obtained, from which the pure enzyme has been obtained and studied. It is a monomer of ca. 50,000 Da whose amino acid sequence has been established.

The mechanism is ordered, shikimate-3-phosphate binding first. The reaction catalyzed introduces the three-carbon fragment which is going to become the side chain of phenylalanine and tyrosine, and to be removed in the synthesis of tryptophan.

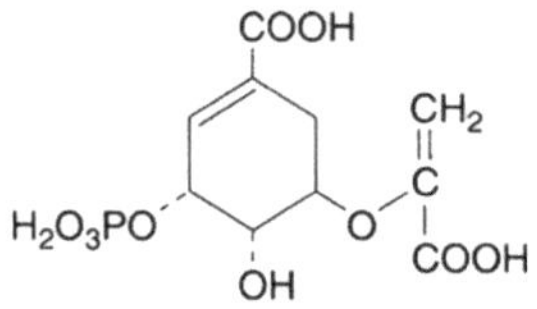

EPSP

The chorismate synthase (*aroC*) introduces a second double bond into the aromatic ring system and removes the phosphate group from EPSP, to yield chorismic acid (from the Greek χορισμοσ, separation) which we shall see, is the common precursor of phenylalanine, tyrosine, tryptophan, *p*-aminobenzoic acid, p-hydroxybenzoic acid and of ubiquinone, menaquinone and enterochelin, and in plants of many alcaloids, cinnamic acid derivatives and lignin.

An intriguing property of chorismate synthase is a requirement for reduced flavin, since the reaction from EPSP to chorismate involves no net oxidation or reduction. How the various chorismate synthases obtain the reduced flavin differs from organism to organism. The enzymes from *Neurospora crassa* and *B. subtilis* are bifunctional, with covalently attached or associated flavin reductases, respectively, that utilize NADPH to produce reduced flavin. In contrast, the chorismate synthases from *E. coli, Pisum sativum* and *Corydalis sempervirens* are monofunctional and must be supplied with exogenous reduced flavin under anaerobic conditions. Single turnover experiments with recombinant enzyme from *E. coli* have shown that a unique, enzyme-bound flavin species accumulates during turnover. It is likely that reduced flavin is directly involved chemically in the conversion of EPSP to chorismate.

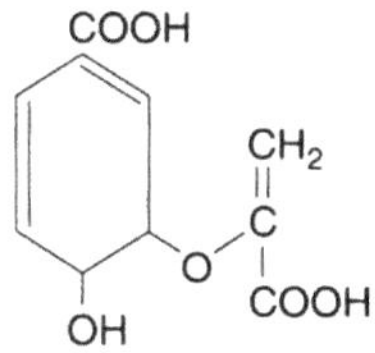

chorismic acid

Physiological Aspects of the Regulation of the Common Pathway

Whereas inhibition by phenylalanine and tyrosine of DAHP synthase (phe) and DAHP synthase (tyr) can be almost total, the inhibition by tryptophan of DAHP (trp) never exceeds 60 %, even at saturation.

The feedback inhibition of the three DAHP synthases has already been discussed. The situation is similar to that we have already encountered in another branched chain, the biosynthesis of amino acids derived from aspartate. Here again, the presence of the three isofunctional enzymes, whose activities and rates of synthesis are affected by individual amino acids allows the modulation of the overall rates of synthesis in response to their availability in the medium or in the cytoplasm.

DAHP synthase (tyr) is repressed not only by tyrosine, but also by very high levels of phenylalanine; DAHP synthase (phe) is repressed by phenylanine, but also by tryptophan, while the synthesis of the third isofunctional enzyme is repressed only by tryptophan.

When *E. coli* grows in minimal medium, DAHP synthase (phe) represents more than 80 % of the total activity; however, under conditions where growth is limited by the aromatic amino acids (minimal medium supplemented with all the amino acids except phenylalanine, tyrosine and tryptophan), DAHP synthase (tyr) is derepressed and becomes the major enzyme. The third enzyme, DAHP synthase (trp) reaches only 5 % of the total activity when fully derepressed by specific tryptophan limitation, perhaps due to an inefficient promoter for the *aroH* gene.

The molecular basis of the modulation of the rates of synthesis of the three isofunctional enzymes will be discussed later in this chapter.

DHQ synthase and dehydratase, shikimate dehydrogenase and *enol*pyruvoylshikimate-3-phosphate synthase are synthesized constitutively, i.e. their level is insensitive to the concentrations of aromatic amino acids in the growth medium. In contrast, shikimate kinase is derepressed up to 10-fold when cells are starved for tyrosine and tryptophan or carry an inactive *tyrR* regulator gene (see below).

Characteristics of the Common Pathway in Several Organisms

With some minor variations, the control pattern of the common pathway is similar to that of *E. coli* in the Enterobacteriaceae examined. *Neurospora crassa* has also the three DAHP synthases.

Yeasts contain two DAHP synthases, one inhibited by phenylalanine, the other by tyrosine.

Bacillus subtilis has a single DAHP synthase. The enzyme is bifunctional and is endowed with chorismate mutase activity. It is composed of four identical subunits of molecular weight 38,500. In some strains, the bifunctional enzyme forms a trifunctional complex with shikimate kinase which, however is a separate protein. The DAHP synthase activity is insensitive to the aromatic amino acids: an efficient inhibition by any one of the three metabolites would be detrimental to the organism. The regulation occurs through its inhibition by chorismate and prephenate.

Thermotoga maritima DAHP synhase is tetrameric, with monomer units comprised of a core catalytic $(\beta/\alpha)_8$ barrel and an N-terminal domain. This enzyme is inhibited strongly by tyrosine and to a lesser extent by phenylalanine. A truncated mutant of lacking the N-terminal domain was catalytically more active and completely insensitive to tyrosine and phenylalanine, consistent with a role for this domain in allosteric inhibition. The structure of this protein was determined to 2.0 Å. In contrast to the wild-type enzyme, this enzyme is dimeric. Wild-type was co-crystallized with tyrosine, and the structure of this complex was determined to a resolution of 2.35 Å. Tyrosine was found to bind at the interface between two regulatory N-terminal domains, formed from diagonally located monomers of the tetramer, revealing a major reorganization of the regulatory domain with respect to the barrel relative to unliganded enzyme. This significant conformational rearrangement observed in the crystal structures was also clearly evident from small angle X-ray scattering measurements recorded in the presence and absence of tyrosine. The closed conformation adopted by the protein on tyrosine binding impedes substrate entry into the neighboring barrel, revealing an unusual tyrosine-controlled gating mechanism for allosteric control of this enzyme.

In *Brevibacterium flavum*, the subunit of the tetrameric enzyme has a molecular weight of 55,000. The enzyme is synergistically inhibited by phenylalanine and tyrosine. However, a second protein is required for chorismate mutase activity. This dimeric (2 × 13,000) protein stimulates DAHP synthase activity. The complex exists as an $\alpha_4\beta_8$ quaternary structure, where chorismate mutase activity is inhibited by phenylalanine and tyrosine, whereas tryptophan activates chorismate mutase but not DAHP synthase.

Whereas *Mycobacterium tuberculosis* DAHP synthase is not inhibited by the addition of Phe, Tyr, or Trp alone, combinations cause significant loss of enzyme activity. In the presence of 200 μm Phe, only 2.4 μm Trp is required to reduce enzymic activity to 50 %. Reaction kinetics were analyzed in the presence of inhibitory concentrations of Trp/Phe or Trp/Tyr. In the absence of inhibitors, the enzyme follows Michaelis-Menten kinetics with respect to substrate erythrose 4-phosphate

(E4P), whereas the addition of inhibitor combinations caused significant homotropic cooperativity with respect to E4P, with Hill coefficients of 3.3 (Trp/Phe) and 2.8 (Trp/Tyr). Structures of DAHP synthases/Trp/Phe/Trp, and/Phe complexes were determined. The/Trp/Phe homotetramer binds four Trp and six Phe molecules. Binding sites for both aromatic amino acids are formed by accessory elements to the core DAHP synthase (β/α)8 barrel that are unique to the type II DAHP synthase family and contribute to the tight dimer and tetramer interfaces. A comparison of the liganded and unliganded structures reveals changes in the interface areas associated with inhibitor binding and a small displacement of the E4P binding loop. These studies uncover a previously unrecognized mode of control for the branched pathways of aromatic amino acid biosynthesis involving synergistic inhibition by specific pairs of pathway end products.

In *N. crassa, Euglena gracilis* and some yeasts, including *S. cerevisiae*, the five central activities of the shikimate pathway are carried by a pentafunctional protein, which catalyzes the conversion of DAHP to 5 *enol*pyruvoylshikimate-3-phosphate without the intermediates ever leaving the protein. The multifunctional protein from *N. crassa* is a homodimer (2×165,000), in which four of the five activities are activated by DAHP.

N. crassa chorismate synthase has been obtained pure as a homodimer (2×55,000). The enzyme has an absolute requirement for reduced FMN and is endowed with an intrinsic FMN: NADPH oxidoreductase activity.

Biosynthesis of Phenylalanine and Tyrosine from Chorismic Acid

It is quite plausible that the first precursor identified in the biosynthesis of these two amino acids, prephenic acid, arises by the rearrangement of chorismic acid as shown below:

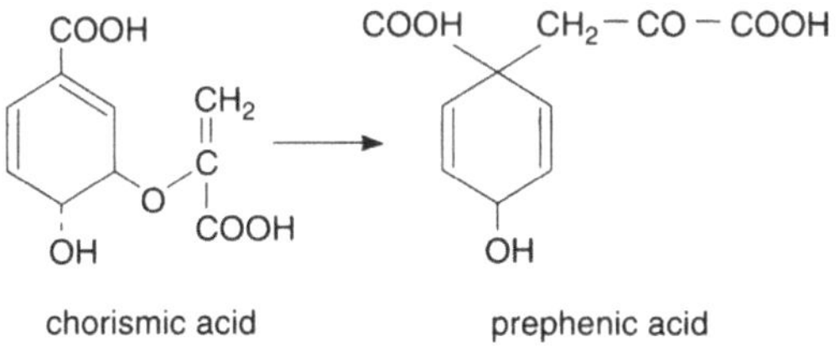

The enzyme catalyzing this reaction has been called chorismate mutase.

During the purification of the *E. coli* enzyme, chromatography of the extracts on DEAE-cellulose yielded two separate peaks of chorismate mutase activity. Prephenate dehydratase was associated with the first peak and prephenate dehydrogenase with the second. Subsequent studies have shown that in each case, both of the activities are carried by a single, bifunctional enzyme.

Chorismate mutase - prephenate dehydratase is the product of the *pheA* gene whereas chorismate mutaseprephenate dehydrogenase is encoded by the *tyrA* gene. The product of the first bifunctional enzyme, which involves a decarboxylation and a dehydration, is phenylpyruvic acid. The second bifunctional enzyme catalyzes also a decarboxylation and in addition a dehydrogenation requiring the presence of NAD^+ as electron acceptor. Its product is *p*-hydroxyphenylpyruvic acid:

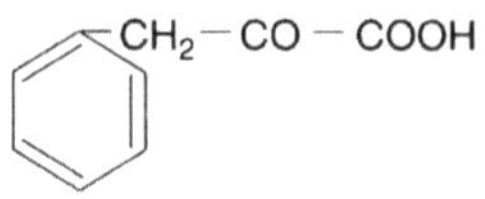

phenylpyruvic acid

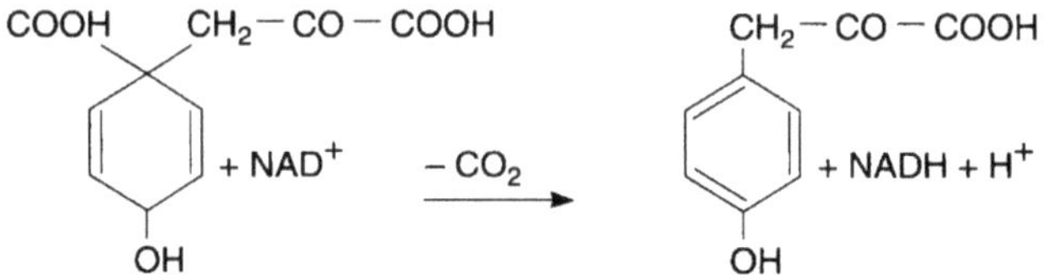

prephenate

p-hydroxyphenylpyruvate

The two enzymes are homodimers with subunit molecular weights of about 40,000.

In the case of chorismate mutase-prephenate dehydratase, a number of studies involving loss of only one of the activities, differential inhibitions and inactivations point to two independent catalytic sites, whereas for chorismate mutase-prephenate dehydrogenase, it appears that the two reactions are catalyzed at closely situated and interacting active sites.

The regulation of chorismate mutase-prephenate dehydratase has been studied with internal titration calorimetry and site-directed mutagenesis: residues important for phenylalanine binding and feedback inhibition have been identified.

Each of the two α-keto acids is the substrate for transamination reactions which yield respectively phenylalanine and tyrosine.

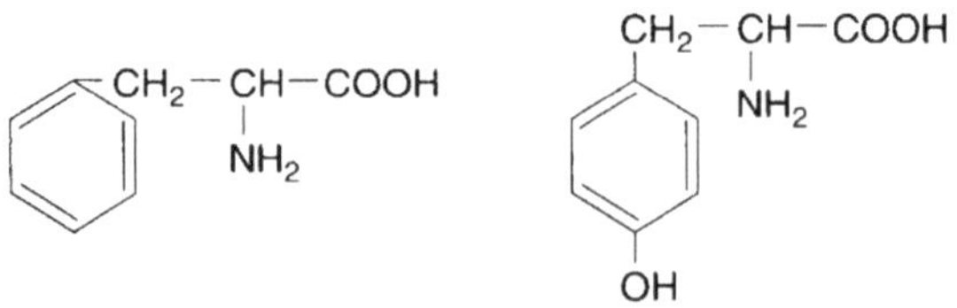

phenylalanine

tyrosine

In both cases, glutamate is the amino donor. The early studies of Rudman and Meister, already referred to several times, showed that there were a number of transaminases with a rather broad specificity in *E. coli*. Phenylalanine and tyrosine can be formed by at least three aminotransferases, the aromatic aminotransferase (*tyrB*) and the aspartate aminotransferase (*aspC*) and the branched-chain aminotransferase specified by *ilvE*. Double mutations in *tyrB* and *aspC* result in a tyrosine requirement, whereas lesions in the three genes lead to a tyrosine and phenylalanine auxotrophy. Based upon relative K_m values of the keto acids for the various enzymes, it seems that under normal physiological conditions the synthesis of the two amino acids is primarily carried out by the aromatic aminotransferase, the product of the *tyrB* gene. The coding region of this gene corresponds to a polypeptide of 397 amino acid residues. The native enzyme is a homodimer, whose activity is repressed by tyrosine and derepressed as a result of a mutation in the *tyrR* regulatory gene.

The tyrR *regulon*

The *tyrR* gene was first identified as being responsible for the resistance of a mutant possessing only the DAHP (tyr) enzyme to the tyrosine analogue 4-aminophenylalanine. Since the analogue acts as a false co-repressor of the synthesis of this enzyme, but is not a feedback inhibitor as tyrosine, derepressed mutants were isolated readily. The mutations were shown to be recessive in diploids and caused the derepressed synthesis of DAHP synthase (tyr) (*aroF*), chorismate mutase-prephenate dehydrogenase (*tyrA*) and of the aromatic aminotransferase (*tyrB*). *tyrR* codes for an aporepressor (513 amino acid residues) which acts on specific operator targets upstream the *aroF* and *tyrA* structural genes, which are part of a single transcription unit.

Later, it was found that *tyrR* controls not only the expression of tyrosine-specific genes, but also that of shikimate kinase, DAHP synthase (phe) and of the TyrR protein itself. In those cases however, the corepressor status is different. Whereas the corepressor is tyrosine in the case of *aroF, tyrA* and *tyrB*, the synthesis of shikimate kinase is repressed by any of the three aromatic amino acids, that of DAHP synthase (phe) by phenylalanine and tryptophan and that of the TyrR protein does not appear to require the presence of an effector molecule.

The TyrR protein has been cloned and sequenced. Its deduced molecular weight is 53,099. The operator loci with which the TyrR protein interacts have been characterized and a consensus "Tyr box" has been identified, existing in one or more copies upstream one or the other structural genes. TyrR is endowed with an ATPase activity which is stimulated by tyrosine. The ATP hydrolyzing domain is located in the central fragment 188–467. Tyrosine in the presence of ATP promotes the hexamerization of this fragment. Although tyrosine-dependent repression of gene transcription by TyrR clearly depends on ligand binding and hexamerization determinants located in the central domain, it would be premature to conclude that transcriptional repression by the TyrR protein is mechanistically connected to the binding and hydrolysis of ATP, in particular since the same hexamerization is provoked by a non-hydrolyzable ATP analogue.

Regulation of the pheA *Gene by Attenuation*

We have just seen that the expression of chorismate mutase-prephenate dehydrogenase is controlled by the TyrR protein combined to tyrosine. In contrast, it appears that the expression of chorismate mutase-prephenate dehydrogenase is mainly regulated by an attenuation mechanism. An analysis of the *E. coli* leader region preceding *pheA* reveals it to have characteristics similar to those reported for the *trp* operon. The leader peptide is 15 amino acid long and contains seven phenylalanine residues.

Met Lys His Ile Pro Phe Phe Phe Ala Phe Phe Phe Thr Phe Pro

Other Organisms. The Arogenate Pathway of Phenylalanine and Tyrosine Biosynthesis

The single activity dimeric chorismate mutase yeast possesses is activated by tryptophan and in some cases, can be negatively regulated by phenylalanine and/or tyrosine, this inhibition being reversed by tryptophan/tyrosine and tryptophan act through the same binding site at the dimer interface. Tyrosine seems to contact the allosteric site by interacting with its phenolic hydroxyl group whereas tryptophan works in an inverse way by opening the allosteric site through the steric size of its side chain.

We have seen that in *B. subtilis*, chorismate mutase and DAHP synthase coexist on a common single protein. This physical arrangement is the molecular basis for the sensitivity of DAHP synthase to inhibition by chorismate and prephenate in a pattern of allosteric control which has been called sequential feedback inhibition.

An alternative pathway of phenylalanine biosynthesis was first discovered in *Pseudomonas aeruginosa*, and an alternative pathway of tyrosine biosynthesis was discovered in cyanobacteria. Transamination in this pathway precedes the dehydration-decarboxylation and the dehydrogenation-decarboxylation step respectively. This pathway uses a prephenate transaminase

prephenate + glutamic acid →

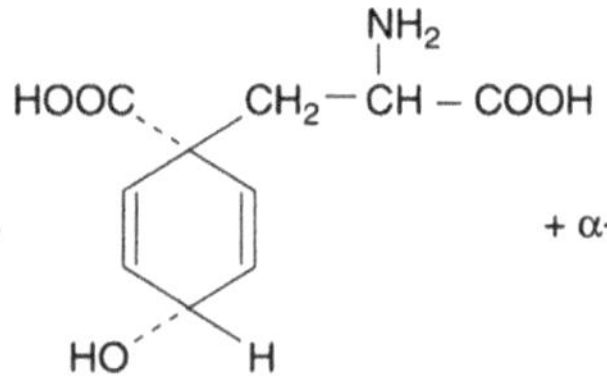

+ α-ketoglutaric acid

arogenate

An arogenate dehydratase leads then to phenylalanine and an arogenate dehydrogenase to tyrosine.

Euglena gracilis synthesizes phenylalanine exclusively by the arogenate pathway. In contrast, activities for both pathways coexist in *Pseudomonas aeruginosa*, where a single dehydrogenase acts upon prephenate and arogenate. In addition, *Pseudomonas* and some other bacterial species can directly hydroxylate phenylalanine to tyrosine.

In *N. crassa*, a single enzyme catalyzes the transamination of both prephenate and *p*-hydroxyphenylpyruvate.

In *Erwinia herbicola*, a monofunctional prephenate dehydrogenase has been obtained by genetic engineering: it retains the dehydrogenase activity but lacks the mutase activity, demonstrating the separability of the two catalytic domains.

In higher plants as in bacteria, different patterns prevail in different species. For example, *Phaseolus* can utilize the two pathways whereas, in *Zea mays*, the arogenate pathway is the only route to phenylalanine synthesis.

Aspartate as a Presursor of Aromatic Amino Acids

Some organisms lack the first two steps of the classical erythrose 4-phosphate pathway for the synthesis of aromatic amino acids. Instead of utilizing E4P and PEP in the initial step, L-aspartate semialdehyde and 6-deoxy-5-ketofructose 1-phosphate (DKFP) are combined to form 2-amino-3,7-dideoxy-D-threo-hept-6-ulosonate under the action of an aldolase called AroA′. A subsequent step, catalyzed by AroB′, generates dehydroquinate, which then merges with the canonical pathway scheme. Thus, in these organisms, aspartokinase is of fundamental importance for the synthesis of aromatic amino acids and other aromatic compounds.

The following figure depicts the steps identified in the methanogen archaeon *Methanococcus maripaludis*. Note that, contrary to many prokaryotes, *p*-aminobenzoic acid does not derive from chorismic acid.

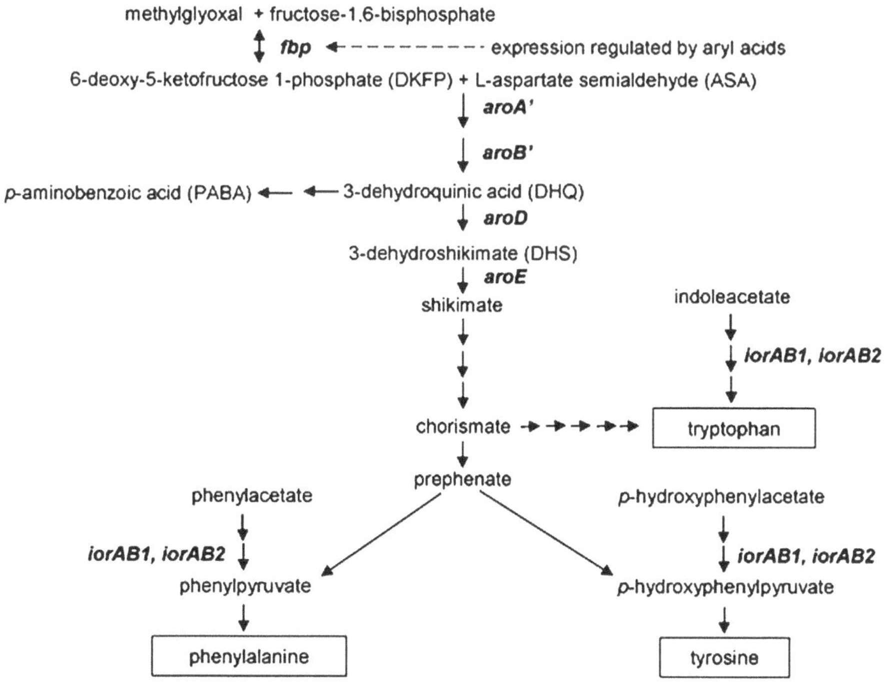

The Biosynthesis of Tryptophan from Chorismic Acid

Early studies showed that some microorganisms requiring tryptophan for growth could also grow if indole or anthranilic acid were added to the culture medium in place of tryptophan. The reactions involved in the biosynthesis of tryptophan as well as the enzymes catalyzing these reactions are now known in detail.

Anthranilate Synthase-Anthranilate Phosphoribosyltransferase

This enzyme, synthesizes anthranilate from chorismic acid and glutamine

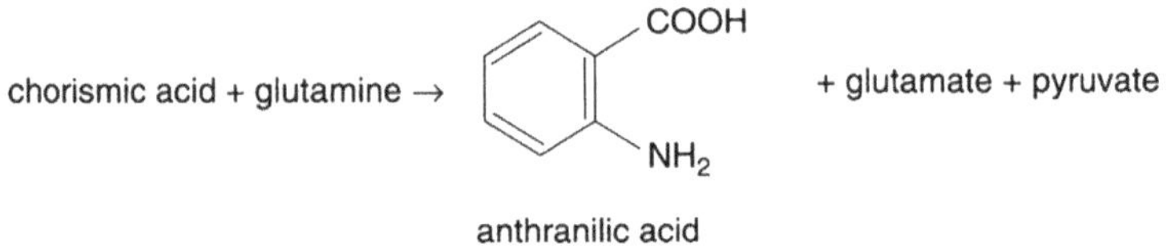

A γ-glutamylenzyme is formed as an intermediate in the reaction and hydroxylamine, which inhibits anthranilate synthase, also inhibits the glutaminase activity.

In *E. coli* and *S. typhimurium*, the enzyme is composed of two molecules each of the polypeptides encoded by the *trpE* and the *trpD* genes, referred to as components I and II respectively. Component I catalyzes the formation of anthranilate from chorismate with ammonia as the amino donor. However, for the reaction to proceed with glutamine as the amino donor, component II is also necessary. Component II, in addition to its glutamine binding site and to its glutamine amidotransferase activity, catalyzes alone the anthranilate phosphoribosyltransferase activity.

Anthranilate + phosphoribosylpyrophosphate → Phosphoribosylanthranilate + PPi

Anthranilate synthase is the target of feedback inhibition in the tryptophan branch of the aromatic pathway. Component I binds tryptophan and since the inhibition is competitive with respect to chorismate, this determines the rate of chorismate channeling toward tryptophan biosynthesis. Component II has the binding site for glutamine (it contains the radioactivity when the enzyme complex is treated with radioactive 6-diazo-5-oxo-norleucine, an analogue of glutamine which binds covalently; this covalent affinity labeling depends on the presence of chorismate and is inhibited by tryptophan).

The enzyme has an $\alpha_2\beta_2$ structure. Component I has 520 amino acids and is devoid of tryptophan, and Component II has 531 amino acids.

Proteolytic digestion of the *S. typhimurium* complex leads to the loss of the phosphoribosyltransferase activity and to a notable loss of molecular weight, resulting from the reduction of the molecular weight of Component II to a 19,000 Da fragment. This small N-terminal piece still carries the glutamine binding site ans still interacts with Component I to yield a functional glutamine-dependent anthranilate synthase activity indistinguishable from that of the native enzyme, but devoid of phosphoribosyltransferase activity, which belongs to the C-terminal piece.

If we screen the different associations between anthranilate synthase (AS) and anthranilate phosphoribosyltransferase (PRT), we find that in pseudomonads, *B. subtilis* and *Serratia marcescens*, glutamine amidotransferase (GAT) is carried by a polypeptide chain which is not fused to PRT (or to AS, for that matter). In these organisms, the conversion of chorismate to phosphoribosylanthranilate is catalyzed by three different proteins.

The sum of the molecular weights of GAT plus PRT is approximately the same, whether the activities reside on the same polypeptide or not.

Phosphoribosylanthranilate Isomerase-Indoleglycerophosphate Synthase

The next two steps are catalyzed in *E. coli* and *S. typhimurium* by a single enzyme, phosphoribosylanthranilate isomerase-indole glycerolphosphate synthase, encoded by the *trpC* gene. Phosphoribosylanthranilate is a ribonucleotide and the isomerase converts it to 1-(o-carboxyphenylamino) 1-deoxyribulose 5-phosphate (CDRP) by an Amadori rearrangement. This compound is then decarboxylated concomitantly with a ring closure to yield indoleglycerol phosphate, the immediate precursor of tryptophan.

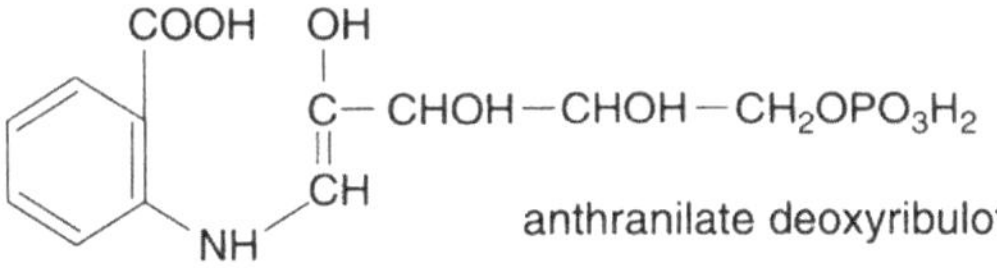

anthranilate deoxyribulotide (CDRP)

CDRP was identified in the non-phosphorylated form as a product accumulating in certain tryptophan mutants of *E. coli* and *Aerobacter aerogenes* starved for tryptophan.

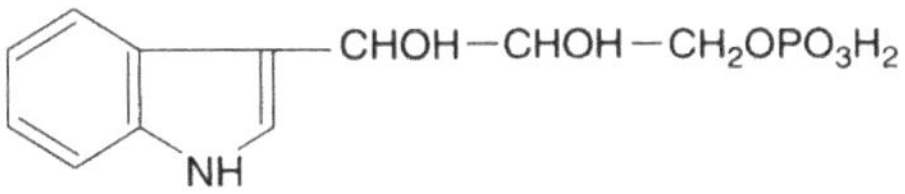

indoleglycerol phosphate (IGP)

The active centers of the two reactions are distinct and do not overlap. The bifunctional enzyme is a single polypeptide of 45,000 Da. Studies carried out with mutants of the enzyme and using limited proteolysis on the wild type enzyme have established that the indoleglycerol phosphate synthase reaction is carried out by the amino terminal half of the polypeptide while the carboxy terminal half forms the isomerase domain. The protein has been crystallized. In some non enteric bacteria and in *Saccharomyces cerevisiae*, the two activities are carried by two distinct polypeptides.

In *Euglena*, the anthranilate phosphoribosyltransferase, the phosphoribosylanthranilate isomerase, the indoleglycerol phosphate synthase activities are associated with the tryptophan synthase activity in a single homodimeric protein (2 × 160,000 Da).

In yeast, the synthesis of anthranilate synthase and indoleglycerol phosphate synthase is under the control of a single genetic locus, but phosphoribosylanthranilate isomerase is a quite separate enzyme.

Tryptophan Synthase

The enzyme from *E. coli* was chosen for the studies which established the collinearity between the nucleotide sequences in DNA and their translation into sequences of amino acids. It has also been an invaluable tool in the early studies on mutagenesis and the genetic code. It presents characteristics which deserve our attention.

In *E. coli*, tryptophan synthase is a tetramer composed of two separate subunits, with an $\alpha_2\beta_2$ quaternary structure. The two subunits can be easily separated by ion exchange chromatography into two α subunits and a β_2 subunit.

Tryptophan synthase of *E. coli* has been an important model for studying subunit structure and function in a multienzyme complex. Although the early studies focused on the genetics and the separate subunits, further work (using limited proteolysis, chemical modification and kinetic studies) has concentrated on the compared properties of the separate and associated subunits and on the changes occurring upon association. The cofactor of the β_2 subunit, pyridoxal phosphate, has served as a useful chromophore probe to study the environment of its active site in the isolated β_2 and when reassociated with α. An extensive X-ray diffraction study has provided an incomparable help to the understanding of the mechanism of tryptophan synthase.

a

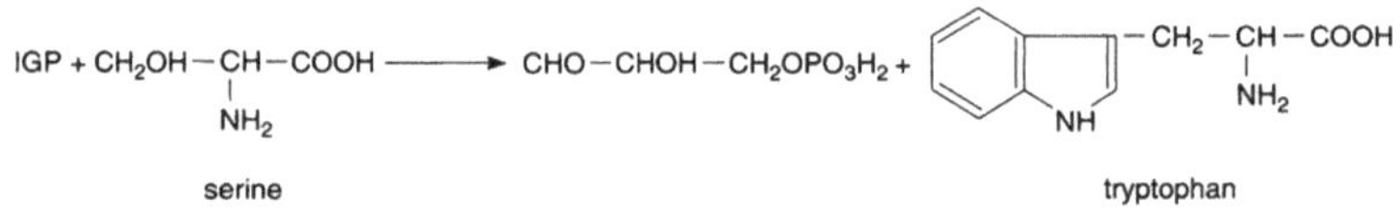

b

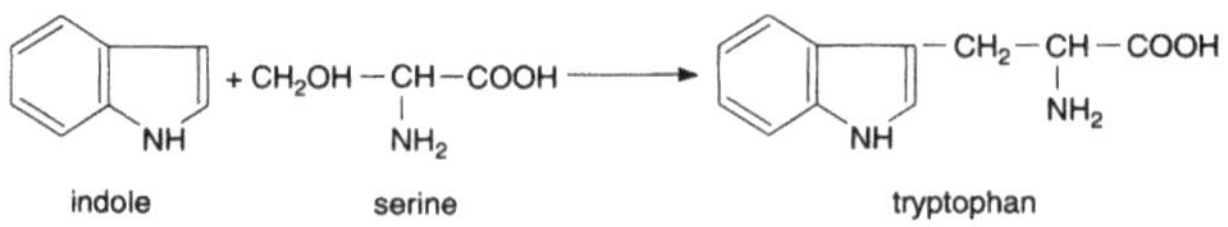

c

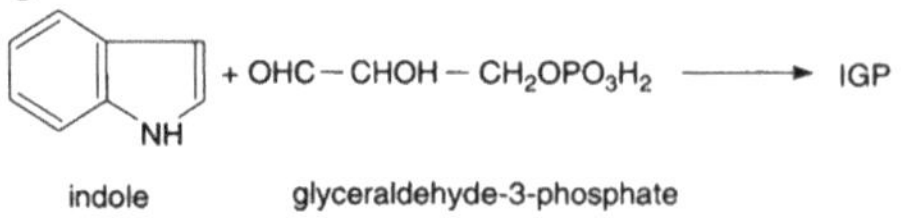

The native enzyme ($\alpha_2\beta_2$) catalyzes the pyridoxal phosphate dependent condensation of indoleglycerol phosphate (IGP) and serine to yield L-tryptophan, and glyceraldehyde-3-phosphate, the physiological reaction (reaction *a*).

The reactions catalyzed by the isolated subunits take place at a much lower rate than the ones carried out by the intact enzyme. It appears that each of the two chains has a specific combining site for indole, specific for the pyridoxal-phosphate dependent reaction *b* in the case of the β_2 chain, and for reaction *c* in the case of the α chain.

The reassociation of the two subunits results in the activation of the two individual reactions and the creation of a catalytic site for reaction (*a*). The active sites of both chains are certainly reorganized in a very special manner in the $\alpha_2\beta_2$ complex, since we know that indoleglycerol phosphate is converted to tryptophan without the free indole participating as an intermediate.

Additional reactions are catalyzed by the isolated subunits: among others, the β_2 subunit can deaminate serine to pyruvate and ammonia (serine dehydratase activity), or condense serine to 2-mercaptoethanol to give S-hydroxyethylcysteine. The existence of these non-physiological reactions has led to an understanding of the mechanism of action of the β_2 subunit: all the reactions in which serine is a substrate occur through a series of Schiff base intermediates formed between enzyme-bound pyridoxal phosphate and serine or derivatives consistent with the general mechanism of β-elimination or β-replacement pyridoxal phosphate-catalyzed reactions.

$$R-CH_2-CH(NH_2)-COOH+H_2O \rightarrow CH_3-CO-COOH+RH+NH_3 \ (\beta\text{-- elimination})$$
$$R-CH_2-CH(NH_2)-COOH+RH \rightarrow R-CH_2-CH(NH_2)-COOH+RH \ (\beta-\text{replacement})$$

The three-dimensional structure of the complex of tryptophan synthase from *S. typhimurium* has been determined by X-ray crystallography at 2.5 Å resolution. The four polypeptide chains are arranged nearly linearly in an αββα order forming a complex 150 Å long. The overall peptide fold of the a subunit, which cleaves indoleglycerol phosphate, reaction (c) is that of an eight-fold α/β barrel. Its active site has been located by difference Fourier analysis of the complex with indolepropanol phosphate, a competitive inhibitor and a close analogue of the natural substrate. The pyridoxal phosphate-dependent β_2 subunit is composed of two domains of approximately equal size, folded into similar helix/sheet/helix structures. The binding site for pyridoxal phosphate lies deep within the interface between the two β subunit domains. The active sites of the two subunits are separated by 25 Å but are directly connected by a tunnel with a diameter matching that of the intermediate substrate indole, facilitating its diffusion from its point of production in the α subunit active site to the site of tryptophan synthesis in the β subunit active site and preventing its escape to the solvent during catalysis. Rapid kinetic studies have confirmed that the tunnel is the preferred route of entry of indole into the β active site in the indole + serine → tryptophan reaction and that indole appears to be channeled between the α and β sites in the overall physiological reaction. The knowledge of the tridimensional structure and of the location of the active site, as well as the availability of site-directed mutagenesis allow a rapid progress in both the detailed catalytic mechanism and the laws governing the reciprocal communications between the α and β subunit in the $\alpha_2\beta_2$ complex. For example, a mutant of tryptophan synthase has been prepared in which one of the residues lining the tunnel β-Cys-170 has been replaced with a bulkier tryptophan residue. Kinetic and structural analyses of the C170W mutant protein by rapid chemical quench methods and x-ray crystallographic analysis show both the transient formation of indole and the obstruction of the tunnel, thus providing direct evidence for the substrate channeling mechanism.

The catalytic activity and substrate channeling of the pyridoxal 5′-phosphate-dependent tryptophan synthase complex is regulated by allosteric interactions that modulate the switching of the enzyme between open, low activity and closed, high activity states during the catalytic cycle. The highly conserved αThr183 residue is located next to the active site and forms part of the subunit interface. The role of the interactions of αThr183 in α site catalysis and allosteric regulation was investigated by analyzing the kinetics and crystal structures of the isosteric mutant Thr183Val. The mutant displays strongly impaired allosteric αβ communication, and the catalytic activity of the reaction is reduced one hundred fold, whereas the β activity is not affected. The structural work establishes that the basis for the missing inter-subunit signaling is the lack of a loop α closure even in the presence of the α subunit ligands, 3-indolyl-D-glycerol 3′-phosphate, or 3-indolylpropanol 3′-phosphate. The structural basis for the reduced α activity has its origins in the missing hydrogen bond between αThr183 and the catalytic residue, αAsp60.

In *N. crassa*, tryptophan synthase is a homodimeric protein of a molecular weight of 150,000. In *S. cerevisiae*, it is also a homodimer of 707 residues per monomer with fused α and β domains. The N-terminal domain (239 residues) is highly homologous to the α subunit of *E. coli* and a C-terminal 389 residue segment is homologous to the β chain of *E. coli*. A 28-residue fragment, with no homology to either subunit, connects the two domains. We have seen that in *Euglena*, tryptophan synthase is part of a multifunctional polypeptide.

The following scheme summarizes the steps identified in the biosynthesis of the aromatic amino acids in *E. coli*.

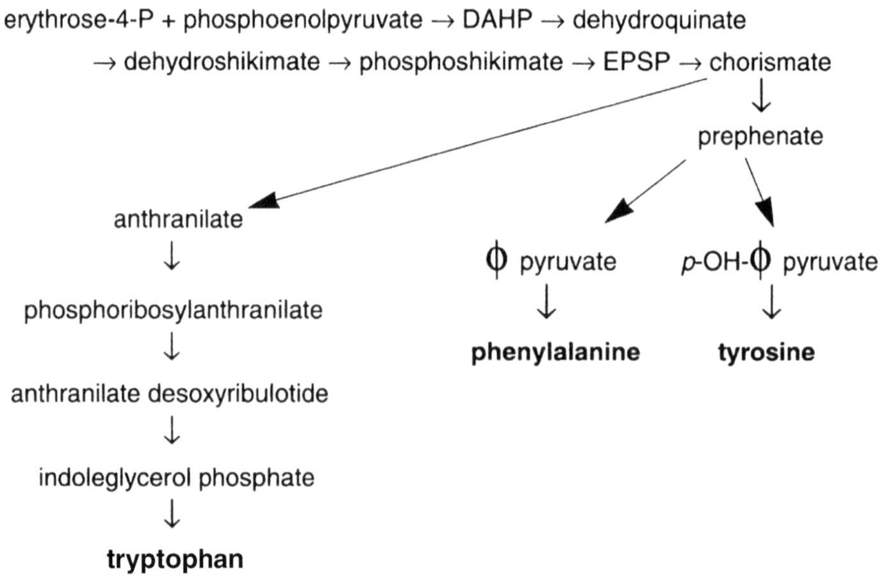

The scheme below summarizes the existing state of our knowledge about feedback inhibition in the biosynthesis of these amino acids in *E. coli*.

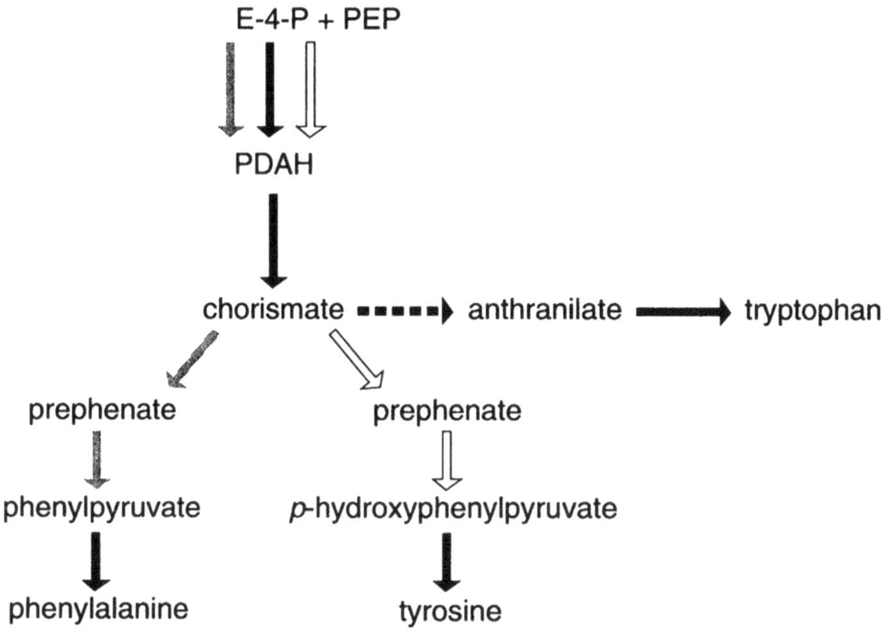

The reactions shown in gray are inhibited by phenylalanine; those in white by tyrosine. The reaction shown by a dotted arrow is inhibited by tryptophan.

In contrast, in *B. subtilis*, there is only one DAHP synthase which is inhibited by chorismate and prephenate:

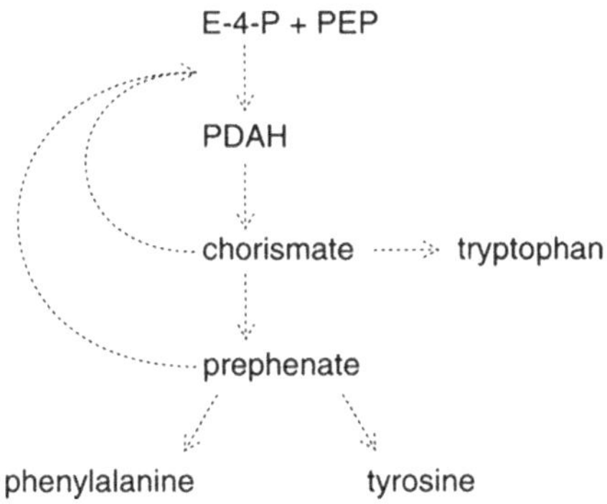

Regulation of Tryptophan Biosynthesis at the Genetic Level. The Tryptophan Repressor

Under physiological conditions, the five enzymes of tryptophan biosynthesis, coded by genes organized into a single operon, are found in roughly equimolar amounts in the cell. When the synthetic rate is altered by physiological means, plots of the activity of one enzyme against the other show coordinate control (see the following chapter for a definition of coordinate repression).

Since the Trp repressor was the first repressor to be characterized in a biosynthetic pathway, it appears useful to describe the rationale of its discovery.

What are the properties to be expected from derepressed mutants? If the synthesis of the enzymes specific for tryptophan biosynthesis are not any more under the control of the size of the intracellular tryptophan pool, one expects to find these enzymes in a greater amount in a derepressed mutant; furthermore, the addition of tryptophan should not reduce appreciably the enzyme content. The greater amount of tryptophan could be correlated with an increased synthesis of tryptophan. This increased synthesis might lead to a resistance to a tryptophan analogue.

The following experiment was carried out by Cohen and Jacob in 1959:

A suspension of *E. coli* is spread onto a solid medium containing 10 mM 5-methyltryptophan, a toxic analogue of tryptophan. Some resistant mutants give rise to colonies, which belong to two categories: (a) those which do not excrete tryptophan; (b) those which excrete tryptophan, as evidenced by a syntrophy test with an indicator strain auxotrophic for tryptophan. Among the excreting bacteria, one finds some which synthesize greater quantities of tryptophan synthase than the wild type bacteria, this greater amount being unaffected by the presence of tryptophan in the medium (Table 31.1).

It was then found that the *trpR* mutation affects in a similar manner all the tryptophan biosynthetic enzymes.

The *trpR* gene has been localized at 0 min on the circular *E. coli* bacterial chromosome, far from the structural genes (28 min) coding for the biosynthetic enzymes, *trpEDCBA* (transcribed in that order, which is also the order of the corresponding biochemical reactions). Its relative location, as well as that of many other repressors with regard of the location of the structural genes they control, is very different of that of the *i* gene vs. the *lac* operon, which may be entirely fortuitous.

Table 31.1 Tryptophan synthase activity of *E. coli* grown with or without tryptophan

	Tryptophan synthase (units)		
	Culture		
Extracts	Without Trp	With Trp	Repression (%)
Wild type	250	66	74
trpR mutant	1,467	1,449	0

The experiment which allows to decide whether *trpR*$^-$ (derepressed) is dominant or recessive with regard to *trpR*$^+$ (wild type) follows: a cross is made between a male Hfr *trpR*$^-$*Sm*^{S} and a female *F*$^-$ *trpR*$^+$*Sm*^{R}; at various times after the injection of the male chromosome, the zygotes are spread onto a solid medium containing streptomycin and 5-methyltryptophan. None of the two parental strains can grow on this medium, the male because they are sensitive to streptomycin (*Sm*^{S}), the females because their growth is inhibited by 5-methyltryptophan. The gene for streptomycin sensitivity is practically never injected, being located at more than 70 min. Only the zygotes having the constitution

$$\frac{trpR^-}{trpR^+\ Sm^R}$$

can give rise to colonies if *trpR*$^-$ is dominant. The recombinants *trpR*$^-$ resistant to streptomycin (*Sm*^{R}) can also give rise to colonies, but they appear much later. The kinetics of appearance of colonies in function of the time of spreading shows that the zygotes are sensitive to 5-methyltryptophan and do not give rise to colonies. In other words, *trpR*$^+$ is dominant over *trpR*$^-$. As in the case of the lac operon, the hypothesis can be advanced that *trpR*$^+$ codes for a cytoplasmic repressor which is not synthesized or synthesized in an inactive form by the *trpR*$^-$ mutants.

One would have guessed that feedback inhibition of anthranilate synthase by tryptophan would prevent tryptophan biosynthesis even at high enzyme levels, but this is clearly not the case.

Most of the physiological and genetic studies on the tryptophan repressor are due to Yanofsky, Crawford and their associates.

The promoter of the *trp* operon is about 200 nucleotides upstream of *trpE*, the first structural gene. The *trp* operator, a 22-base pair target is within this promoter. When this sequence is bound to the repressor, RNA polymerase cannot form a productive complex with the promoter. The binding occurs only in the presence of tryptophan.

In addition of preventing the transcription of the genes of the tryptophan operon, the tryptophan repressor exerts the same effect on the transcription of *aroH*, the gene coding for DAHP (trp), and on its own transcription. Examination of the sequences upstream the *aroH* and *trpR* genes and comparison with the operator of the *trp* operon have allowed the identification of a consensus "Trp box" for the binding of the repressor.

Among the 5-methyltryptophan mutants excreting tryptophan, there was another class, which belonged to the constitutive dominant type. This dominant constitutivity affects in merodiploids all the genes of the tryptophan operon in *cis* position with regard to the mutation, but not those in *trans*. All these mutations were localized in a segment adjacent to the operon, but upstream the first structural gene. They affected the operator.

A Unitary Model for Induction and Repression

These experiments establish the basis for a unitary theory of the mechanism of repression in catabolic and biosynthetic systems: in the case of the *lac* operon, the free repressor blocks transcription and is inactivated upon combination with the inducer. In the biosynthetic systems, the free repressor is inactive and active only in combination with its ligand, the corepressor.

Isolation of the Trp Repressor

Zubay and his colleagues have used the same in vitro protein synthesis system they first used with β-galactosidase, but this time the source of DNA was a segment isolated from a transducing defective bacteriophage carrying a fusion of the lac and trp operons (λ^{-}*.dtrp-lac*) (Fig. 31.1). It is therefore possible to assess the presence of the aporepressor in the *trpR*$^{+}$ strains and its absence in the *trpR*$^{-}$ strains, by measuring the in vitro synthesis of β-galactosidase, which is now under the control of the *trp* promoter and operator. The synthesis of β-galactosidase is progressively reduced by

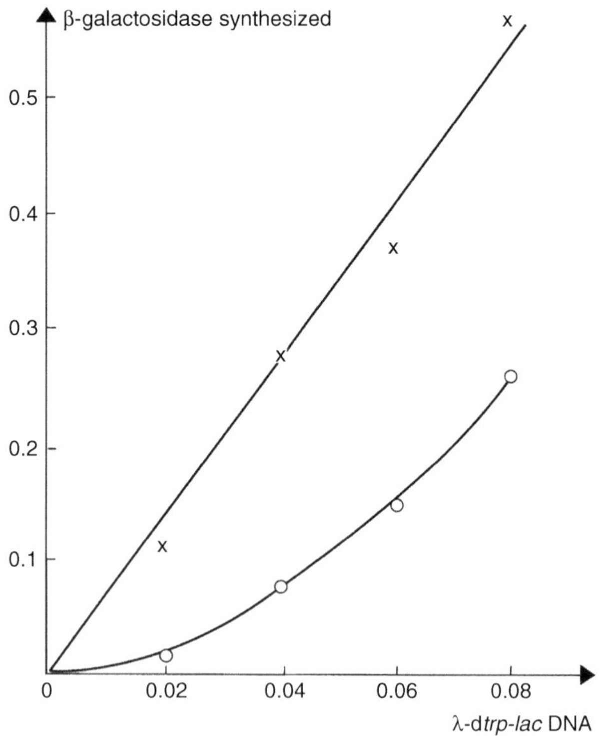

Fig. 31.1 Synthesis of β-galactosidase as a function of the concentration of λdtrp-lac DNA. Experiments performed with *trpR*$^{-}$ supernatants (x) and with the same supernatants (°) where a solution of partially purified Trp repressor (54 μg/ml) was added (From G. Zubay D. E. Morse, W. J. Schrenk and J. H. M. Miller, with permission of Dr Geoffrey Zubay)

increasing the amount of *trpR*⁺ cellular extracts. Such a repression is not observed when β-galactosidase synthesis is programmed by normal *lac* DNA (Fig. 31.2).

This specific and highly specific detection has helped the purification of the Trp repressor, which as many proteins interacting with nucleic acids, is strongly adsorbed by phosphocellulose. It is important to recall that tryptophan is required for β-galactosidase synthesis and that it is always present in this *in vitro* DNA-programmed protein synthesis.

The synthesis of anthranilate synthase and of tryptophan synthase has also been measured directly by an analogous *in vitro* system, but programmed by *trp* DNA instead of *trp-lac* DNA. Anthranilate synthase and the α and β subunits of tryptophan synthase are synthesized in equimolar quantities which suggests that the regulation of translation is similar *in vitro* and *in vivo*. Furthermore, the synthesis of the enzymes follows a temporal sequence corresponding to the relative sequence of the corresponding genes on the DNA.

In order to show the effect of tryptophan, the aporepressor, the synthesis of the specific messenger RNA has been measured. This type of experiment where the synthesis of protein is not measured has shown that tryptophan is indispensable to the effective repression of the expression of the *trp* operon in the presence of the aporepressor.

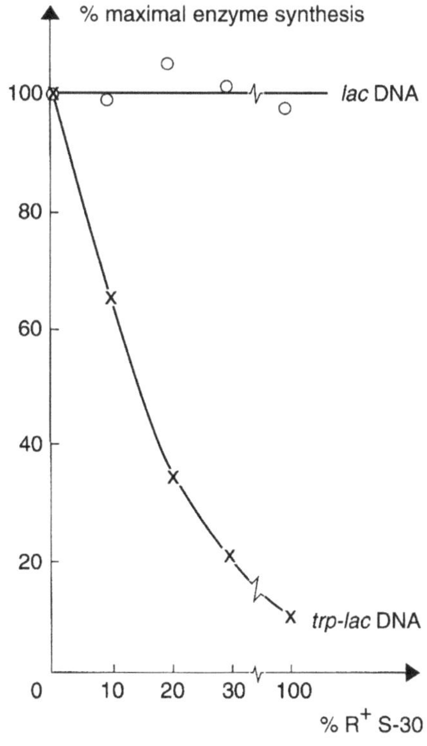

Fig. 31.2 Synthesis of β-galactosidase as a function of the origin of the supernatants. Experiments performed with *lac* DNA or *trp-lac* DNA. The supernatants (S-30) originate from either *trpR*⁺ or *trpR*- strains and they are used either alone, or in variable proportions. Abscissa: percentage of *trpR*⁺ extract (from Cohen and Jacob, 1959)

The *trpR* gene has been cloned and overproduced by bacteria containing genetically engineered plasmids. It codes for a 107-residue polypeptide normally existing as a dimer, whether complexed or not with tryptophan.

In equilibrium binding experiments, the K_D of high affinity binding of Trp repressor to a 43-bp segment containing the natural *trp* promoter operator region was determined to be 2×10^{-10} M. Loss of one or both tryptophan molecules from the complex destabilizes immediately the complex. A heterodimeric aporepressor with a single tryptophan binding site has been constructed: its affinity for operator, when the single site is saturated, is 20–25 fold higher than that of the natural aporepressor, but 20–25 fold lower than that of the natural tryptophan-saturated repressor.

The Trp repressor has been purified and crystallized with and without tryptophan. The ternary complex aporepressor-tryptophan-operator sequence has also been crystallized. The results of their X-ray crystallographic analysis will be presented in the chapter on protein-DNA interactions.

Enterochelin (Enterobactin) Biosynthesis

The Synthesis of 2,3-Dihydroxybenzoic Acid

Enterochelin was mentioned when we considered iron uptake systems. The synthesis of this siderophore is studied in this section because it derives, as the aromatic acids, from chorismic acid. Its synthesis requires six enzymes, one of which (EntB/G) is bifunctional. Figure 31.3 describes its structure and synthesis; starting from chorismate, isochorismate synthase (*entC*) isochorismate pyruvate hydrolase (*entB*) and 2,3 dihydro-2,3-dihydroxybenzoate dehydrogenase (*entA*) lead to 2,3-dihydrobenzoic acid (DHB) the intermediates and the three corresponding enzymes have been purified and characterized. Isochorismate synthase is active as a monomer of 43 kDa and shows strong homology to anthranilate synthase and to PabB (component I of *p*-aminobenzoic synthase, see Chap. 32); two proteins that bind also chorismic acid. Isochorismate being also the precursor of menaquinones (Chap. 33), *ent C* mutants are unable to grow anaerobically on glucose as well as in low iron media. EntB, the isochorismatase is active as a homopentamer of 32.5 kDa subunit. It is bifunctional, participating also in the enterochelin synthetase reaction. The third enzyme, coded by *entA* is an octamer composed of identical subunits of 26,249 Da.

The next reaction

$$3\,\text{DHB} + 3\,\text{L-serine} + 6\,\text{ATP} \rightarrow \text{enterochelin} + 6\,\text{AMP} + 6\,\text{Ppi}$$

is carried out by a multienzyme complex (Ent synthetase) consisting of the *entD, entE, entF*, and *entB/G* gene products, the association of its several polypeptides being transient. Enterobactin contains three amide bonds that are made non

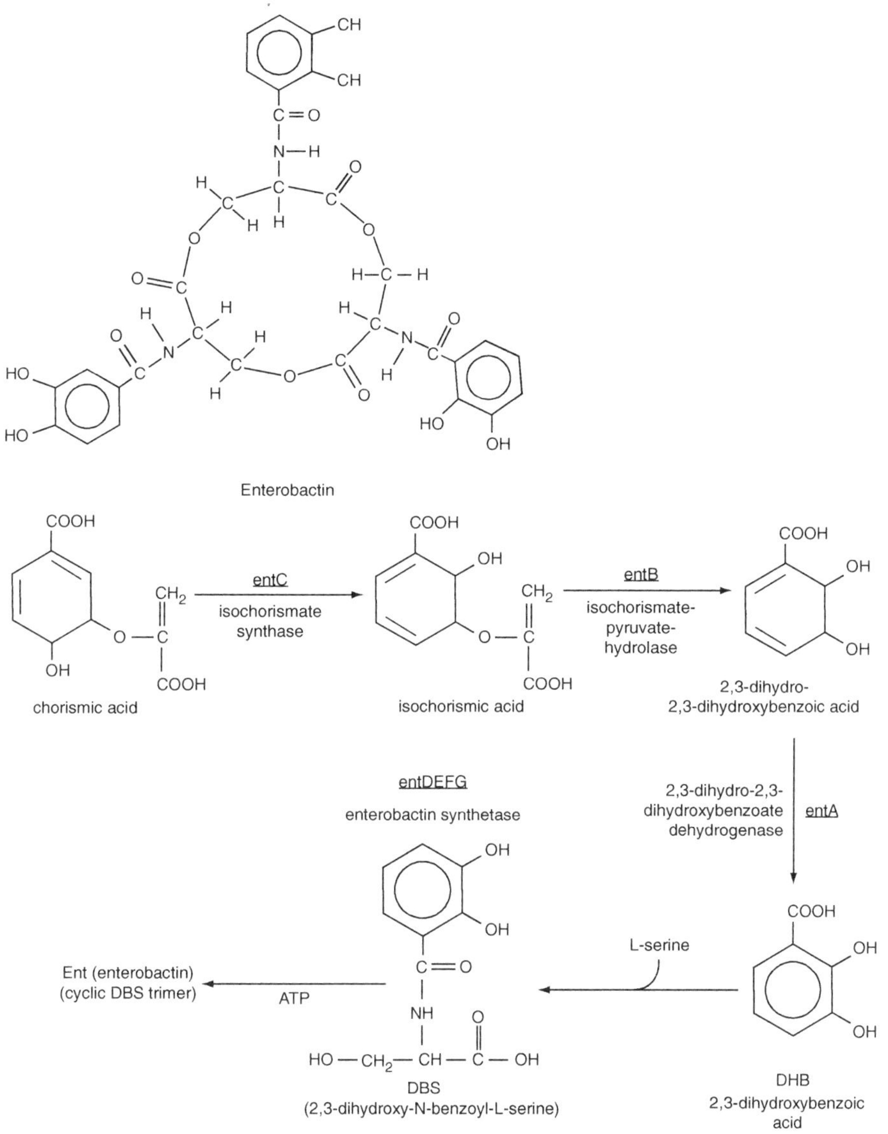

Fig. 31.3 Biosynthesis of enterobactin

ribosomally, by a protein thiotemplate mechanism similar to that of the synthesis of peptide antibiotics. The substrates, DHB and serine, are activated as acyladenylates, bound to a peptide synthetase as thioesters, and sequentially polymerized in the order determined by their synthetase-binding domains. EntF is the peptide synthetase, activating and binding serine via a thioester linkage to bound 4′-phosphopantetheine. EntE adenylate-forming enzyme activating DHB and forming monomeric 2,3-dihydroxy-N-benzoylserine (DBS) by a transamidation with the

EntF-bound serine. Subsequently, DBS units are linked by esterification to yield a cyclic DBS trimer, enterochelin. EntB/G and EntD participate in this process, but no specific activities, have been determined for them as yet.

Selected References

The Common Pathway

Benach J, Lee I, Edstrom W, Kuzin AP, Chiang Y, Acton TB, Montelione GT, Hunt JF (2003) J Biol Chem 278:19176–19182
Cross PJ, Allison TM, Dobson RCJ, Dobson GB, Parker EJ (2013) Proc Natl Acad Sci U S A 110:2111–2116
Cross PJ, Dobson RCJ, Patchett ML, Parker EJ (2011) J Biol Chem 286:10216–10224
Davis BD (1955) Adv Enzymol 16:247–312
Hartmann M, Schneider TR, Pfeil A, Heinrich G, Lipscomb WN, Braus GH (2003) Proc Natl Acad Sci U S A 100:862–867
Herrmann KM, Pittard J (1983) In: Herrmann KM, Somerville R (eds) Amino acids: biosynthesis and genetic regulation. Addison-Wesley Publishing Company, Reading, p 301–350
Jossek R, Bongaerts J, Sprenger GA (2001) FEMS Microbiol Lett 202:145–148
Sprinson DB (1961) Adv Carbohydr Chem 15:235–270
Webby CJ, Jiao W, Hutton RD, Blackmore NJ, Baker HM, Baker EN, Jameson GB, Parker EJ (2010) J Biol Chem 285:30567–30576

Biosynthesis of Phenylalanine and Tyrosine

Christendat D, Turnbull J (1996) Biochemistry 35:4468–4479
Pohnert G, Zhang S, Husain A, Wilson DB, Ganem B (1999) Biochemistry 38:12212–12217

DKFP Pathway. Aspartate as a Precursor of Aromatic Amino Acids

Porat I, Sieprawska-Lupa M, Teng Q, Bohanon FJ, White RH, Whitman WB (2006) Mol Microbiol 62:1117–1131
White RH (2004) Biochemistry 43:7618–7627

TyrR

Dixon MP, Pau RN, Howlett GJ, Dunstan DE, Sawyer WH, Davidson BE (2002) J Biol Chem 277:23186–23192

Tryptophan Synthesis

Hyde CG, Ahmed SA, Padlan EA, Miles EW, Davies DR (1988) J Biol Chem 263:17857–17871
Kulik V, Weyand M, Seidel R, Niks D, Arac D, Dunn MF, Schlichting I (2002) J Mol Biol 324:677–690
Miles EW (1979) Methods Enzymol 49:127–186
Miles EW (1991) Methods Enzymol 64:93–172
Miles EW (2013) J Biol Chem 288:10084–10091
Wilmanns M, Priestle JP, Niermann T, Jansonius JN (1992) J Mol Biol 223:477–507
Yanofsky C (1987) Bioessays 6:133–137

Tryptophan Repressor. Functional Aspects

Cohen GN, Jacob F (1959) Compt Rend 248:3490–3492
Pouwels PH, Van Rotterdam J (1972) Proc Natl Acad Sci USA 69:1786–1790
Rose JK, Yanofsky C (1974) Proc Natl Acad Sci USA 71:3134–3138
Zubay G, Morse DE, Schrenk WJ, Miller JHM (1972) Proc Natl Acad Sci USA 69:1100–1103

Enterochelin

Lingens F, Goebel W, Hoppe-Seylers Z (1965) Physiol Chem 342:1–12
Liu J, Quinn N, Berchtold GA, Walsh CT (1990) Biochemistry 29:1417–1425
Nahlik MS, Brickman TJ, Ozenberger A, McIntosh MA (1989) J Bacteriol 171:184–190
Rusznak F, Liu J, Quinn N, Berchtold GA, Walsh CT (1990) Biochemistry 29:1425–1435
Shaw-Reid CA, Kelleher NL, Losey HC, Gehring AM, Berg C, Walsh CT (1999) Chem Biol 6:385–400

Chapter 32
The Biosynthesis of Histidine and Its Regulation

All the genes of histidine biosynthesis are adjacent and form an operon located at 44 min on the *E. coli* chromosome. The enzyme called phosphoribosyl-phosphate-ATP-pyrophosphorylase (PR-ATP-pyrophosphorylase for short, *hisG*) catalyzes a Mg^{++} dependent reaction between ATP and 5-phosphoribosyl-1-pyrophosphate (PRPP) to give N1-(5′-phosphoribosyl)-ATP:

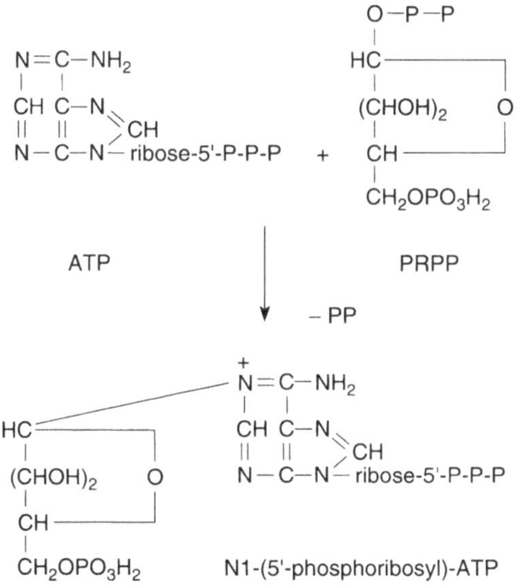

The enzyme has been purified and is a homohexamer under catalytic conditions (subunit molecular weight: 33,000). The enzyme from wild type binds six molecules of histidine and undergoes a conformational change upon binding. It is inhibited non-competitively by histidine with respect to both substrates at physiological histidine concentrations. Other factors may be involved in modulating enzyme

G.N. Cohen, *Microbial Biochemistry*,
DOI 10.1007/978-94-017-8908-0_32,

activity *in vivo*: the *in vitro* inhibition by histidine is enhanced by AMP, ADP and guanosine-5′-diphosphate-3′-diphosphate (ppGpp).

The second intermediate in the biosynthetic chain leading to histidine has been identified as N1-phosphoribosyl-AMP, formed from the preceding compound by PR-ATP-pyrophosphorylase (*hisE*), which removes a pyrophosphate residue. Another enzyme, PR-AMP-1,6-cyclohydrolase (*hisI*), converts this compound to phosphoribosylformiminoaminoimidazolecarboxamide ribotide:

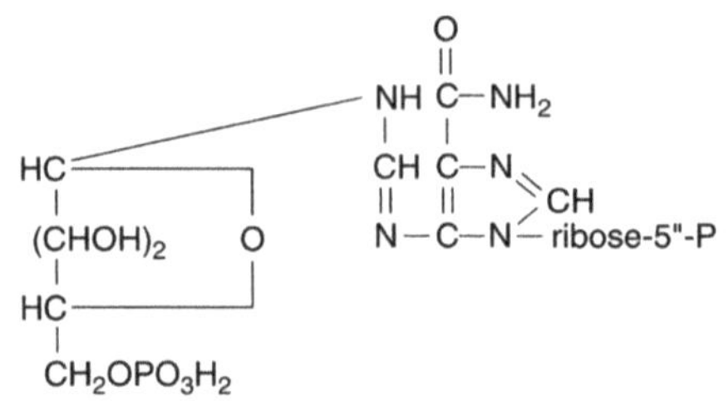

N-(5′-phosphoribosylformimino)-5-amino-4-imidazolecarboxamide-(5″-phosphoriboside)

The examination of the nucleotide sequences of the *E. coli* and *S. typhimurium* *hisE* and *hisI* genes indicates that the two corresponding activities are contained in a single bifunctional protein (subunit MW = 22,800), an observation confirmed by their co-sedimentation as a homodimer.

An isomerase, coded by *hisA*, a monomer of 245 amino acid residues, converts this compound to the corresponding ribuloside.

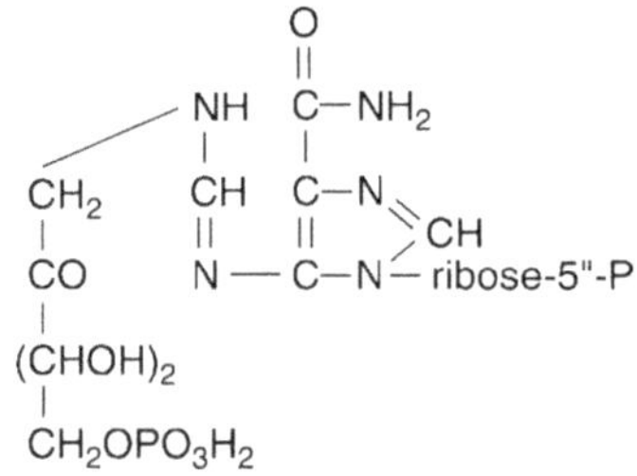

Glutamine provides the amino group for the next reaction, catalyzed by a typical amido transferase (*hisH*) and imidazoleglycerol phosphate is formed by a cyclase (*his F*). The amido transferase is built from two identical subunits of 196 amino acid residues each; the cyclase appears also to be a homodimer (2 × 258 residues).

Two proteins essential for the biosynthesis of histidine in *E. coli* have been overexpressed and purified to apparent homogeneity. The protein encoded by *hisF* has an ammonia-dependent activity that results in the conversion of the biosynthetic intermediate N1-[(5′-phosphoribulosyl) formimino] -5-aminoimidazole-4-carboxamide ribonucleotide (PRFAR) to imidazole glycerol phosphate (IGP) and 5-aminoimidazole-4-carboxamido-1-β-D-ribofuranosyl-5′-monophosphate (AICAR). The second protein, encoded by *hisH* exhibits no detectable catalytic properties with the biosynthetic intermediate PRFAR, glutamine or ammonia. In combination, the proteins are capable of a stoichiometric conversion of glutamine and PRFAR to form AICAR, IGP and glutamate. Neither protein alone is able to mediate a conversion of the nucleotide substrate to a free metabolic intermediate. The HisF and HisH proteins form a 1:1 dimeric complex which constitutes the IGP synthase holoenzyme. In this holoenzyme, glutamine is a 1,000-fold more efficient than ammonia.

In *S. cerevisiae*, the two activities are encoded by a single bifunctional protein, whose N-terminal moiety corresponds to HisH and the C-terminal moiety to HisF. The physical uncoupling of the two activities as observed in *Enterobacteriaceae* is also found in the three eubacteria *Lactococcus lactis*, *Azospirillum brasilense* (a diazotroph), *Streptomyces coelicolor* and in the archaebacterium *Methanobacterium vannielii*.

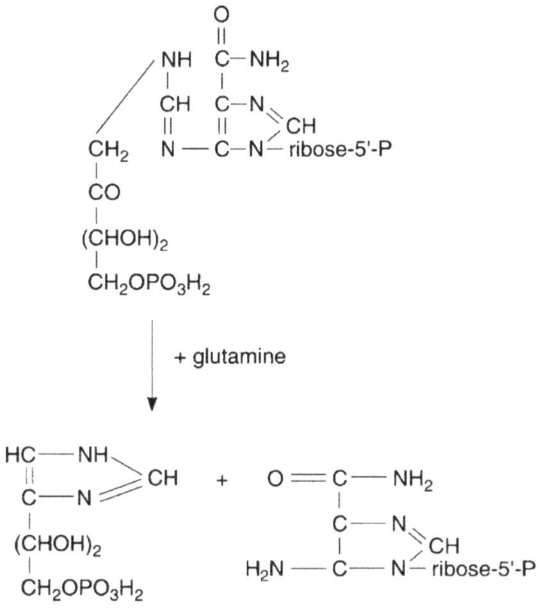

The other product of the reaction, 5-amino-4-imidazolecarboxamide ribonucleotide (AICAR) is itself a precursor of ATP. In this way the purine nucleus of ATP is recycled during the biosynthesis of histidine.

Imidazoleglycerolphosphate dehydratase (*hisB*), a homotetramer, (4×40,000) removes a molecule of water and transforms its substrate into imidazoleacetol phosphate.

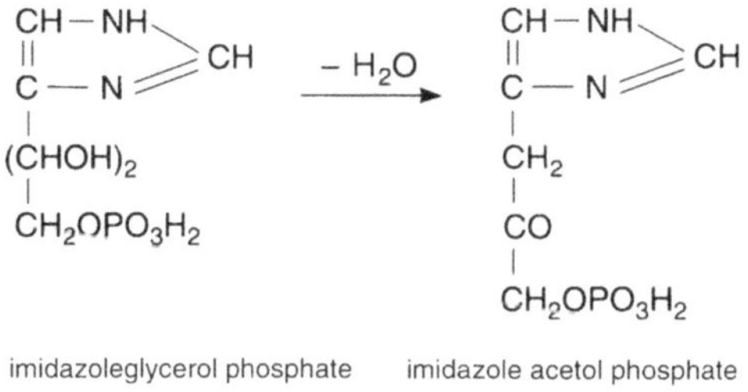

Histidinol phosphate aminotransferase, a homodimer of 2×356 residues (*hisC*) catalyzes a transamination between imidazoleacetol phosphate and glutamate and leads to the production of histidinol phosphate, which is then subject to a specific histidinol phosphatase (which forms a bifunctional enzyme with imidazoleglycerol phosphate dehydratase, both enzymes being coded by *hisB*). The histidinol which is thus formed is then oxidized to histidine by histidinol dehydrogenase (*hisD*), a homodimer of 2×434 amino acid residues.

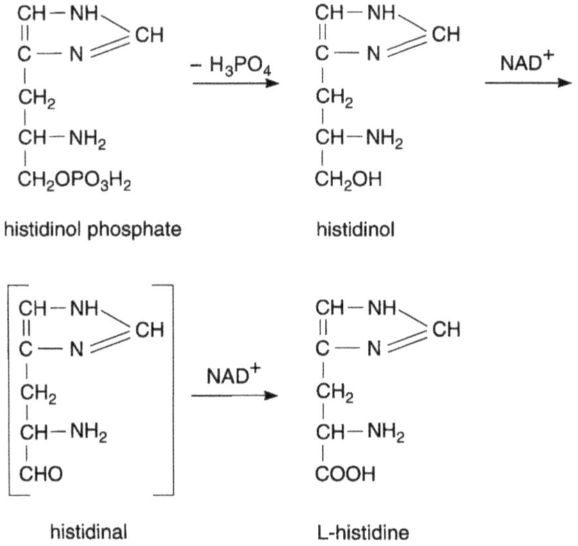

The oxidation of the primary alcohol group of histidinol to the carboxyl group of histidine requires two molecules of NAD^+. In an analogous case of successive reductions of a carboxyl group (aspartic acid to homoserine), two separate enzymes are responsible for the formation of the intermediate aldehyde and for its reduction. Here the aldehyde histidinal is indeed the intermediate, but remains probably enzyme-bound, although there are auxotrophic mutants which can grow on histidinal.

Regulation of Histidine Biosynthesis at the Genetic Level

Table 32.1 represents the histidine operon of *Salmonella typhimurium*. One sees at first glance, that, in contrast to the threonine and the tryptophan operons, the order of the genes on the chromosome (*hisGDCB-HAFIE*) is notably different from that of the reactions catalyzed by the corresponding enzymes.

The question that was addressed by Ames and Garry in 1959 was the following: does the repression of the synthesis of each one of the enzymes of the biosynthetic pathway occur in a coordinate fashion? In other terms, is the variation in the cellular

Table 32.1 The histidine operon of *Salmonella typhimurium*

Gene	G	D	C	B
Reaction	1	10	8	7, 9
Enzyme	PR-ATP pyrophosphorylase	Histidinol dehydrogenase	Transaminase	Dehydratase, phosphatase
Gene	H	A	F	I, E
Reaction	5	4	8	3, 2
Enzyme	Amidotransferase	Isomerase	Cyclase	PR-AMP hydrolase, PR-ATP pyrophosphohydrolase

Note that the order of genes differs from the order of enzyme reactions. *his B* and *his I, E* code for bifunctional enzymes

content of each of these enzymes identical under conditions where the intracellular histidine pool is varied?

To answer their query, Ames and Garry have used mutants in which one or the other of the enzymes of the biosynthetic chain was inactivated. As a result, their growth requires the presence of exogenous histidine in the growth medium. This can be added as histidine itself. The variations of the concentrations of exogenous histidine do not affect the growth rate, but only the final cellular yield. The specific activity of the various enzymes of the pathway is low and identical to those found in the wild type strain growing with or without histidine. This shows that the synthesis of those enzymes is maximally repressed in the wild type. However, in certain mutants, the specific activity rises when histidine becomes limiting and the enzyme extraction is done several hours after growth has stopped. In that case, the enzymes studied are synthesized at the expense of the limiting histidine originating from the slow bacterial protein turnover.

Similarly, histidine can be provided by the addition of substances liberating histidine by slow hydrolysis, carnosine (β-alanylhistidine), histidylhistidine or N-formyl-α-L-histidine, the latter giving the more interesting results.

Figure 32.1 shows the effects of histidine and of various concentrations of formylhistidine on the growth of an histidine auxotroph of *S. typhimurium*.

Figure 32.2a–c show the specific activities of three enzymes of the pathway, compared to a fourth enzyme of the same pathway, histidinol phosphate phosphatase.

It is immediately apparent that the ratio of the specific activities is constant, independently of the auxotroph used and of the growth conditions. Figure 32.2d shows that this relationship does not apply if the relative contents of histidinol phosphate phosphatase and glutamate dehydrogenase, not belonging to the histidine operon, are compared.

The above results have been extended to all the enzymes of the operon. They show that histidine affects equally the synthesis of all the enzymes of the operon. Ames and Garry have named this phenomenon coordinate repression.

This coordinate repression is in keeping with the fact that transcription of the histidine operon leads to the synthesis of a single polycistronic messenger and implies that, at least in the absence of so-called polar mutations, the translation of this messenger leads to equimolar amounts of the different enzymes of the histidine biosynthetic pathway.

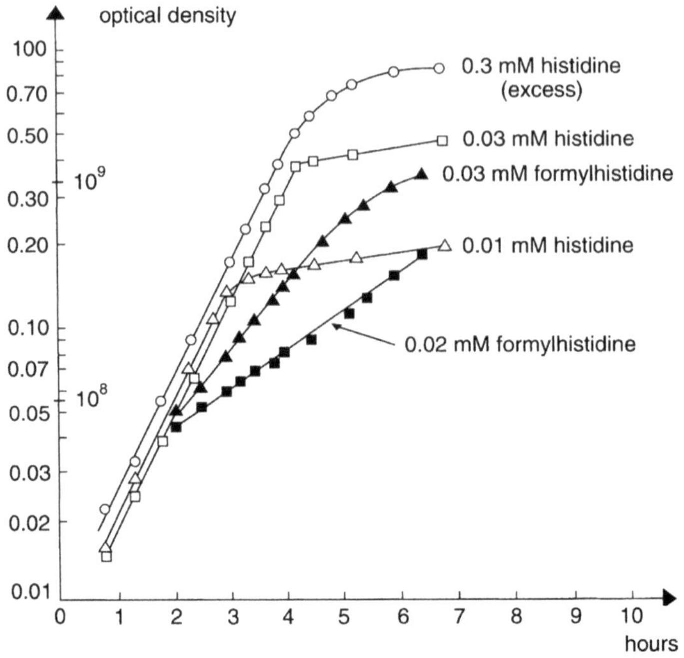

Fig. 32.1 Growth of a mutant requiring histidine for its growth. A 30-fold variation in the exogenous concentration of histidine does not affect the growth rate, but only the yield. In contrast, the growth rate on N-α-formyl-L-histidine depends greatly on its concentration in the medium. The situation is equivalent to an internal chemostat. If the concentration of formylhistidine is lowered to 0.01 mM, there is no growth at all (From B.N. Ames and B.J. Garry, with permission from Dr Bruce Ames)

In 1972, a classic study by Lewis and Ames showed that there was an inverse correlation between the intracellular amount of charged histidinyl-t-RNA, not of free histidine, and the level of the enzymes of the *his* operon in derepressed mutants and a physiologically derepressed strain.

The mutants genetically derepressed have been obtained as in the case of the tryptophan pathway by selecting colonies resistant to the toxic effects of a histidine analogue. The derepressed mutants are distinguished from non derepressed mutants thanks to a peculiar morphology of the colonies. The mutations are grouped in six independent loci of the *S. typhimurium* or of *E. coli* linkage map.

(a) *hisO* was first believed to be an operator locus. Its mutations cause a dominant constitutive phenotype which is only expressed in *cis* in merodiploids. It is thus formally analogous to the lactose and tryptophan operators we have previously studied. The term operator in its formal acceptation does not apply in this particular case and *hisO* mutants are actually promoter mutants, the term operator being reserved to define a repressor binding site. The next classes of derepressed mutants alter either the level, the aminoacylation, the modification or the maturation of histidinyl-tRNA.

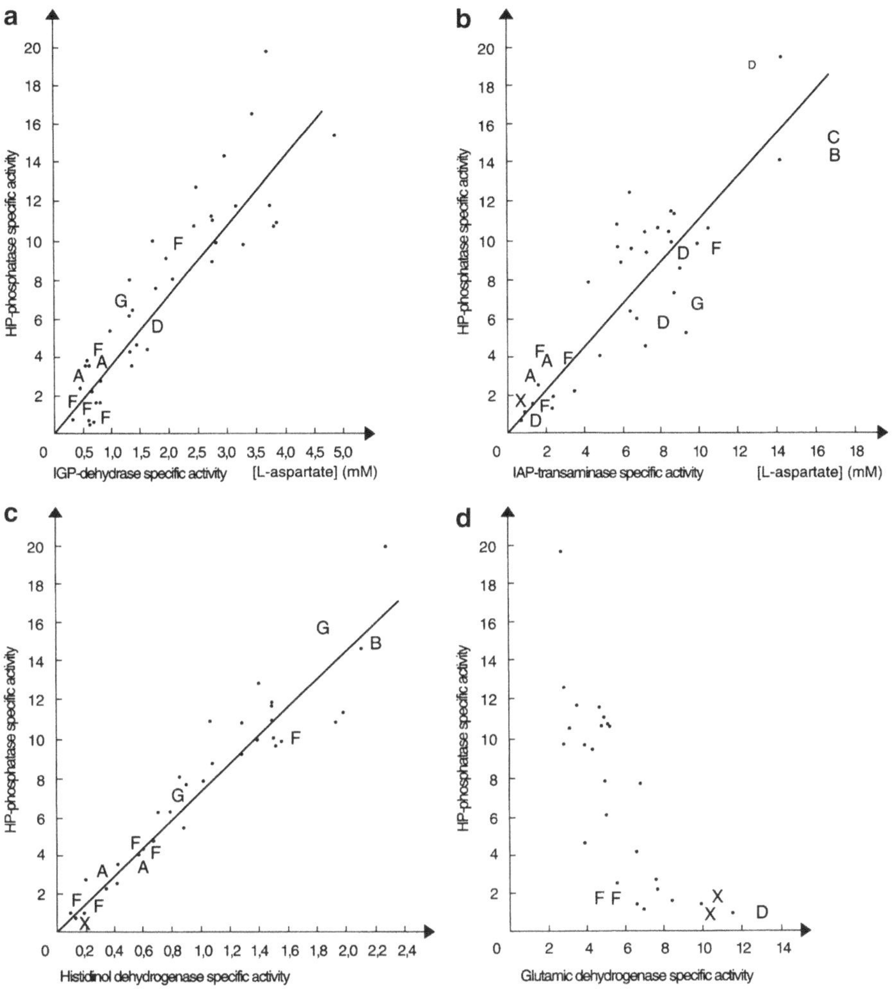

Fig. 32.2 (**a–d**) Specific activities of some enzymes involved in histidine biosynthesis and of glutamate dehydrogenase in various extracts of *S. typhimurium*. Each point represents the specific activities of two of the enzymes in the crude extracts of wild type and of various auxotrophic mutants grown under different conditions. *X* wild type extracts, • extracts of mutants lacking PR-ATP pyrophosphorylase. Extracts of other mutants are symbolized by letters corresponding to their genes (see Fig. 32.1). When a given mutant is lacking one of the four activities examined, the specific activity of the corresponding enzyme is missing from the graph (From B.N. Ames and B.J. Garry, with permission from Dr Bruce Ames)

(b) *hisS* is the structural gene for histidinyl-tRNA synthetase. The gene has been cloned, sequenced and the enzyme purified. It is a homodimer (MW = 2 × 40,000). Mutations in *hisS* cause the synthesis of an enzyme with a considerably lower apparent affinity for histidine than the enzyme from wild type cells. The addition of histidine at a high concentration in the medium results in a normal repression, conforting the idea that charged tRNA is the repression signal.

(c) *hisR* is the gene coding for the single cellular species of histidinyl-tRNA. Mutations in its promoter reduce by about 50 % the cellular content of histidinyl-tRNA without altering its primary structure. The introduction in the bacteria of an episome carrying the wild type *hisR* increases the concentration of this molecular species.
(d) *his U* and *hisW* have been said to be involved in the maturation process of the histidinyl-tRNA, although certain results point to their identity with *gyrB* and *gyrA* respectively, coding for the two subunits of DNA gyrase. The status and explanation of these mutants are still not clear.
(e) *hisT* is the more interesting mutation. The histidinyl-tRNA from *hisT* mutants contain in its anticodon loop two uridine residues which are normally present in the wild type as two pseudouridine residues. Since there are amber mutations of *hisT* which are suppressible, this gene must code for a protein, which has been shown to be pseudouridine synthetase. This enzyme, a monomer of 270 amino acid residues, is specific for the rearrangement of the uridines of the anticodon loop, since a third pseudouridine is normally present in the histidinyl tRNA extracted from the *his T* mutants (it is now clear that pseudouridine is formed by rearrangement of uridine after assembly of the tRNA chain).

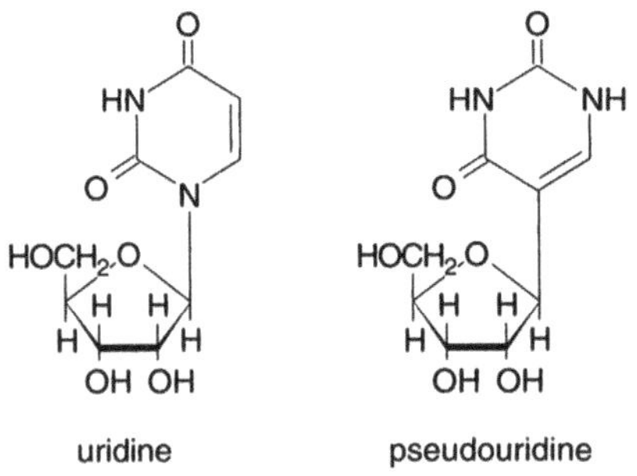

The growth of the *hisT* mutants is practically not affected, indicating that the presence of the two pseudouridines in the anticodon loop is necessary for repression, but dispensable for global protein synthesis. Actually, the histidinyl-t-RNA extracted from the mutants has the same characteristics as wild-type histidinyl-t-RNA in the aminoacylation reaction, be it for reaction rate and extent or affinity for the cognate aminoacyl-t-RNA synthetase.

A word of caution is necessary: although the above studies concluded that the pseudouridine modification is dispensable for cellular growth, they were based on a limited number of experiments with unspecified *hisT* mutants that might have retained residual activity. These conclusions were recently reexamined with strains which were absolutely devoid of pseudouridine synthase. These strains have a uracil requirement which interferes with cell division and lead to the concept that the loss of certain tRNA modifications can actually cause catastrophic effects.

Despite the impact of the repressor-operator model established mainly in the *lac* and *trp* systems, the group led by Bruce Ames considered gene regulatory mechanisms involving charged histidinyl $tRNA^{His}$ in its normal role in protein synthesis and imagined that there was a His-codon-containing RNA "operator" sequence overlapping with the start point of translation of the first structural gene. In the presence of an excess charged histidinyl-tRNAHis, a ribosome would initiate translation of this sequence which would proceed to its termination codon and would somehow interfere with the initiation by ribosomes at the start of the structural genes. Repression would result, and the His codon-containing peptide would be released in the cytoplasm. When histidinyl-$tRNA^{His}$ is limiting, the ribosome would stall at the histidine codons of the operator sequence, allowing the access to the initiation codons of the structural gene. This prophetic theory prefigurates the attenuation model which was beautifully developed by Yanofsky for the tryptophan operon. Actually, in 1974, it was shown that a mutation upstream the structural genes elicits a constitutive *his* operon expression independent of histidinyl-tRNA control.

The attenuation model has been described elsewhere in detail in this volume. Here is the leader peptide sequence of the histidine attenuator, which is made of 16 amino acid residues with seven histidine residues in a row:

Met Thr Arg Val Gln Phe Lys His His His His His His His Pro Asp

Synthesis of Diphthamide, a Modified Histidine, by Archaea

The elongation factor EF-2 of archaea contains diphthamide, the post translationally modified histidine residue found on the eukaryotic EF2 elongation factor (see p.259 of the first edition). Diphthamide is not present in eubacteria.

The first step of its synthesis is the transfer of 3-amino-3-carboxypropyl from S-adenosyl-L-methionine to a histidine residue.

SAM + L-histidine → 2-(3-carboxy-3aminopropyl)-L-histidine

Next, diphthine synthase catalyzes the SAM-dependent trimethylation of this compound to diphthine(2-[3-carboxy-3-(trimethylammonio)propyl] histidine, further amidated by diphthamide synthase to diphthamide (2-[3-carboxyamido-3-(trimethylammonio)propyl]histidine.

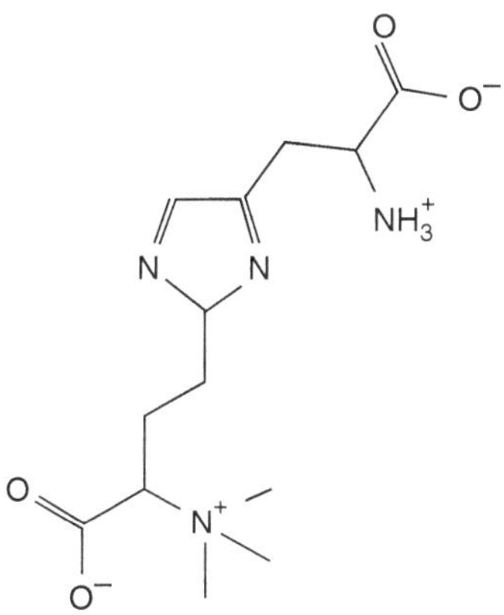

Diphthine

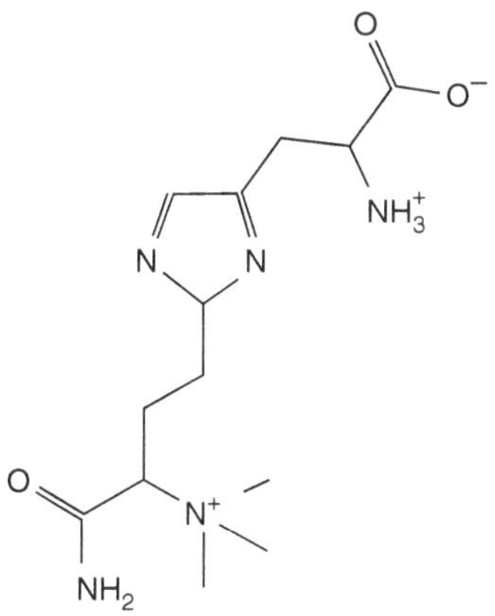

Diphthamide

The homodimeric crystal structures of the archaeal diphthine synthases from *Pyrococcus horikoshii* and *Aeropyrum pernix* have been determined. These structures share essentially the same overall fold as the cobalt-precorrin-4 methyltransferase, confirming that diphthine synthase belongs to the family of methyltransferases.

Selected References

Histidine Biosynthesis and Its Regulation

Ames BN, Garry BJ (1959) Proc Natl Acad Sci USA 45:1453–1461
Artz SW, Holzschu D (1983) In: Herrmann KM, Somerville RL (eds) Amino acids: biosynthesis and genetic regulation. Addison-Wesley Publishing Co., Reading, p 379–404

PR-ATP pyrophosphorylase

Ames BN, Martin RG, Garry BJ (1961) J Biol Chem 236:2019–2026
Martin RG (1963) J Biol Chem 238:257–268

Attenuation

Barnes WM (1978) Proc Natl Acad Sci USA 75:4281–4285

Diphthamide in Archaea

Kishishita S, Shimizu K, Murayama K, Terada T, Shirouzu M, Yokoyama S, Kunishima N (2008) Acta Cryst D64:397–406
Pappenheimer AM Jr, Dunlop PC, Adolph KW, Bodleyn JW (1983) J Bacteriol 153:1342–1347

Chapter 33
The Biosynthesis of Nucleotides

The Biosynthesis of Pyrimidine Nucleotides

Synthesis of 5-Phosphoribosyl-1-Pyrophosphate (PRPP)

The synthesis of PRPP from ATP and ribose-5-phosphate is catalyzed by phosphoribosylpyrophosphate synthetase as follows:

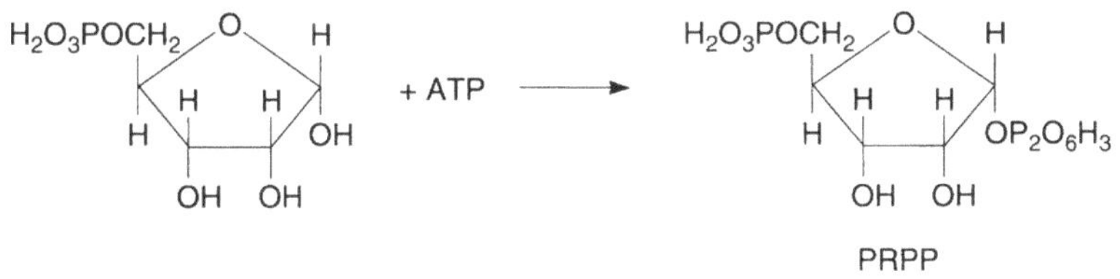

The equilibrium constant favors PRPP synthesis.

PRPP is involved in pyrimidine and purine nucleotide, histidine, tryptophan and pyridine nucleotide syntheses. Ten known enzymes compete for PRPP as substrate, the higher quantitative requirement being for pyrimidine and purine nucleotides.

PRPP synthetase has been purified from *S. typhimurium*, *E. coli*, *B. subtilis*, human erythrocytes and rat liver.

The PRPP synthetase subunit of *E. coli* (encoded by *prs*) has a molecular mass of 31 kDa. Inorganic phosphate is essential for the stability of the enzyme and is required for the activity, although its function in the catalysis remains obscure. Mg^{++} ions are required both as free cations and as ATP chelators. The synthesis of the enzyme is derepressed upon uracil starvation, but the nature of the repressing metabolite remains unknown. The other metabolites deriving from PRPP do not seem to exert an influence on the level of PRPP synthetase.

The intervention of PRPP in the synthesis of pyrimidines occurs at the level of orotidine-5′-phosphate synthesis (see below).

G.N. Cohen, *Microbial Biochemistry*,
DOI 10.1007/978-94-017-8908-0_33,

Synthesis of Carbamylphosphate

In bacteria, the first step in the synthesis of pyrimidines is the synthesis of carbamyl phosphate from ammonia (or an ammonia donor) and CO_2, at the expense of two molecules of ATP:

$$NH_3 + HCO_{3^-} + 2MgATP \rightarrow \underset{\text{Carbamyl phosphate}}{H_2N - COO - PO_3H_2} + 2ADP + Pi$$

This reaction is catalyzed by carbamyl phosphate synthetase, encoded by the genes of the *carAB* operon. The operon is oriented from A to B. The enzyme is a heterodimer the two subunits of which have been cloned and sequenced. The heavy subunit (MW = 117,710), coded for by *carB*, is able to catalyze the reaction with ammonia as the amino donor, whereas the light subunit, coded for by *carA* (MW = 41,270) is actually an amidotransferase which allows the enzyme to function with glutamine (cf. anthranilate synthase for a similar situation). The light subunit binds glutamine and has glutaminase activity.

Carbamyl phosphate synthetase is an allosteric enzyme: both the heavy subunit alone or the complex display a sigmoidal MgATP saturation curve, whereas the kinetics are hyperbolic with respect to glutamine or bicarbonate. The activity of the enzyme is inhibited by UMP and activated by ornithine, IMP, and in the case of the *Salmonella* enzyme by PRPP. All these effectors act by altering the affinity of the enzyme for MgATP.

The mechanism of carbamyl phosphate synthesis by the large subunit is known to proceed through carboxy phosphate. The intermediate is formed by an initial phosphorylation of bicarbonate resulting in an enzyme bound bicarbonate.

$$E + ATP + HCO_{3^-} \rightarrow E.CO_3PO_3^{3^-} + ADP + H^+$$

In the second step, the bound carboxy phosphate undergoes a nucleophilic attack by NH3 yielding enzyme-bound carbamate:

$$E.CO_3PO_3^{3^-} + NH_3 \rightarrow E.CO_2 - NH_2 + Pi$$

Carbamate is phosphorylated by the second molecule of ATP to form carbamyl phosphate in the terminal step:

$$E.CO_2 - NH_2 + ATP \rightarrow E + \text{carbamyl phosphate} + ADP$$

In addition to the synthesis of carbamyl phosphate, the enzyme catalyzes three partial reactions: a bicarbonate-dependent hydrolysis of ATP

$$MgATP + H_20 \xrightarrow{HCO_{3^-}} MgADP + Pi$$

a synthesis of ATP from carbamyl phosphate and ADP

$$\text{Mg ADP} + \text{carbamyl phosphate} \rightarrow \text{MgATP} + CO_2 + NH_3$$

and thc glutaminase activity referred to above, catalyzed by the light subunit.

$$\text{glutamine} + H_2O \rightarrow \text{glutamate} + NH_3$$

There is considerable evidence that each of the MgATP molecules involved in the overall reaction not only act in mechanically discrete steps, but also at separate sites.

Three conserved histidine residues, His243, 781 and 788 are located within the large subunit of carbamyl phosphate synthetase from *E. coli* and were identified by sequence identity comparisons. They were individually mutated to asparagine residues. The mutation at position 243 blocks the formation of the carbamate intermediate from carboxy phosphate and ammonia. The mutant modified at position 781 has an order of magnitude reduction for both carbamyl phosphate and ATP syntheses, consistent with the proposal that the C-terminal portion of the large subunit is primarily involved in the phosphorylation of the carbamate intermediate. The mutation at position 788 synthesizes carbamyl phosphate at a rate comparable to that of the wild type, but ATP synthetase and bicarbonate-dependent ATPase activities of the enzyme are reduced tenfold.

The most common phenotype among the mutants affecting the *carAB* operon is a double requirement for uracil and arginine. Since carbamyl phosphate is also an intermediate in arginine biosynthesis, it follows that mutants which cannot synthesize it require both uracil and arginine. The activation of the enzyme by ornithine is also understandable on a teleonomic basis, since an excess of ornithine would require more carbamyl phosphate to direct the pathway toward arginine.

This enzyme is different from a synthetase, found in vertebrate liver, which uses exclusively ammonia and is activated by N-acetylglutamate, which is required for the active enzyme conformation and does not play a role in CO_2 activation. This enzyme is localized in mitochondria and is used for the synthesis of arginine and urea.

In both *N. crassa* and yeast, there are two carbamyl phosphate synthetases, for pyrimidine and arginine synthesis respectively, designated P and A. The A enzyme from *Neurospora* consists of two different polypeptides, products of unlinked genes. A large subunit is capable of using NH_3 and a small subunit imparts the ability to use glutamine, but has no intrinsic glutaminase activity as opposed to the small subunit of the *S. typhimurium* enzyme. Mutants lacking the small subunit are auxotrophic for arginine. Thus, the glutamine-dependent reaction is the physiological one. Arginine represses the synthesis of both subunits, which is insensitive to pyrimidines. The reverse is true of carbamyl phosphate synthetase P: the mutants devoid of it are pyrimidine auxotrophs.

Synthesis of Cytidine and Uridine Triphosphates

All the *E. coli* genes mentioned in this paragraph have been cloned and sequenced.

Aspartate transcarbamylase (ATCase, *pyrB* and *pyrI*) is an enzyme which carries out the condensation of one molecule of aspartate with a molecule of carbamyl phosphate. The reaction is formally similar to the ornithine transcarbamylase (OTCase) reaction (see Chap. 25):

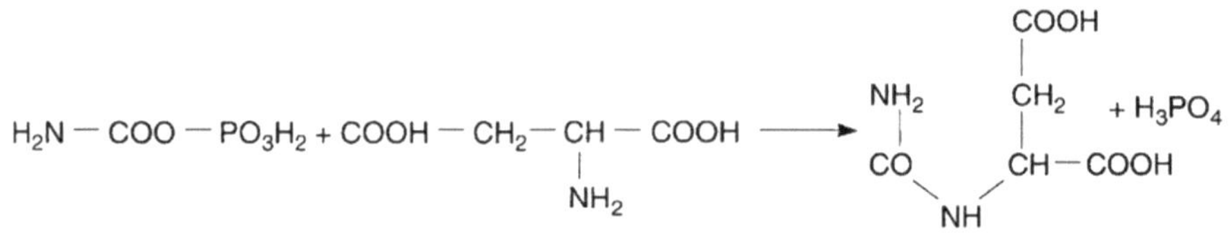

carbamylaspartate
(or ureidosuccinate)

The structure and the regulation of ATCase activity deserve special attention for historical reasons and because it is one of the most thoroughly studied allosteric enzymes. It will be examined in detail in the next paragraph.

Dihydroorotase (*pyrC*), a homodimeric dehydratase (2×38,500) containing one atom of tightly bound zinc per subunit, catalyzes the next step, a reversible cyclization forming the pyrimidine ring:

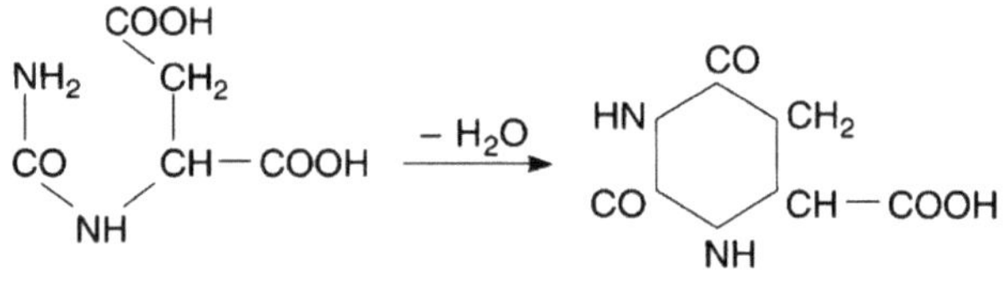

carbamylaspartate dihydroorotate

Dihydroorotate oxidase (*pyrD*) is membrane-bound and linked to the electron transport system of the cell. It consists of two identical flavin-containing subunits of MW 37,000 each and catalyzes the formation of orotic acid, the physiological electron acceptor appearing to be ubiquinone.

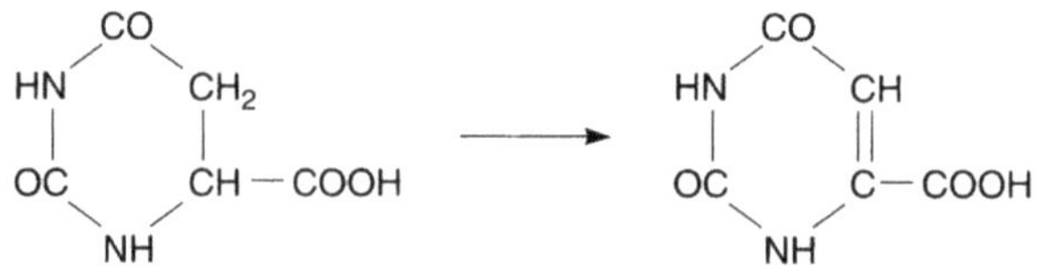

dihydroorotate orotate
(or uracil-4-carboxylate)

In mammalian cells, a multifunctional protein called CAD catalyzes the three first activities of pyrimidine biosynthesis: glutamine-dependent carbamylphosphate synthetase (C), aspartate transcarbamylase (A) and dihydroorotase (D). This protein of 243 kDa is organized into discrete functional domains connected by interdomain linkers. The intermediates carbamylphosphate and carbamylaspartate have been shown to be partially channeled from one active site to the next within the CAD protein.

Phosphoribosyltransferases, proteins we have previously encountered in the synthesis of aromatic amino acids and of histidine, are enzymes also involved in the synthesis of purine, pyrimidine, pyridine nucleotides. They utilize α-D-5-phosphoribosyl-1-pyrophosphate (PRPP) and a nitrogenous base to form a β-N-riboside monophosphate and pyrophosphate. Their importance in nucleotide homeostasis is evidenced by the devastating effects of inherited diseases (hereditary gout, Lesch-Nyhan syndrome, orotic aciduria accompanied by mental retardation) associated to decreased activity and/or stability of these enzymes.

The crystal structure of orotate phosphoribosyl-transferase from *S. typhimurium* has been determined at 2.6 Å resolution. The crystalline enzyme is a dimer, with catalytically important residues from each subunit available to the other subunit, suggesting that oligomerization is necessary for its activity.

The normal pathway of biosynthesis does not involve a direct decarboxylation of orotate to uracil, but a reaction of orotate with 5-phosphoribosyl-1-pyrophosphate (PRPP) catalyzed by orotate phosphoribo-syltransferase (MW = 2 × 23,000, *pyrE*) to produce orotidine-5′ phosphate, anucleoside (OMP).

OMP is converted into uridine-5′-phosphate (UMP) by a specific decarboxylase (MW = 2 × 27,000; *pyrF*).

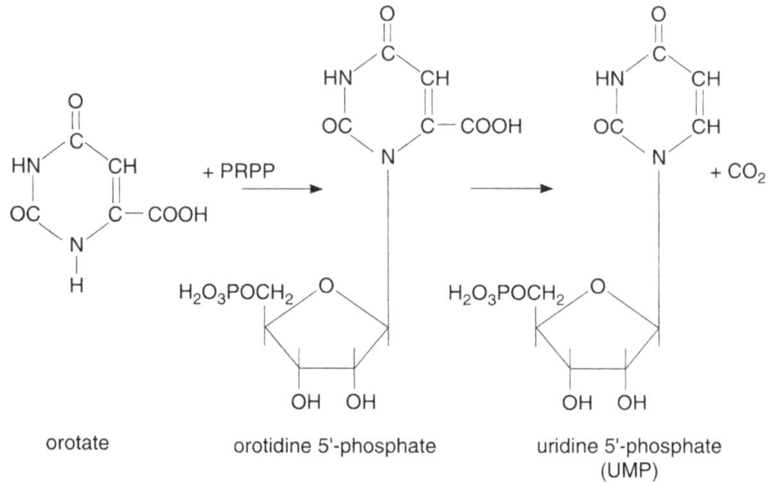

The crystal structure of the UMP complex of orotidine 5′-monophosphate decarboxylase (ODCase) at atomic resolution shows steric competion between the product UMP and the side chain of a catalytic lysine residue. Surface plasmon resonance analysis indicates that UMP binds 5 orders of magnitude more tightly to a mutant in which the interfering side chain has been removed than to wild-type ODCase.

Orotidine 5′-monophosphate decarboxylase (ODCase), encoded by *pyrF* accelerates the decarboxylation of its substrate by 17 orders of magnitude. One argument brought forward against steric/electrostatic repulsion causing substrate distortion at the carboxylate substituent as part of the catalysis has been the weak binding affinity of the decarboxylated product (UMP). The crystal structure of the UMP complex of ODCase at atomic resolution shows steric competition between the product UMP and the side chain of a catalytic lysine residue. Surface plasmon resonance analysis indicates that UMP binds 5 orders of magnitude more tightly to a mutant in which the interfering side chain has been removed than to wild-type ODCase. These results explain the low affinity of UMP and counter a seemingly very strong argument against a contribution of substrate distortion to the catalytic reaction mechanism of ODCase.

A specific kinase, UMP kinase (*pyrH*) phosphorylates the nucleoside monophosphate, UMP, to the diphosphate UDP and nucleoside diphosphokinase (*ndk*), which catalyzes the transfer of the γ-phosphate of any nucleoside-5′-triphosphate to any nucleoside-5′-diphosphate

$$N_1TP + N_2DP \Leftrightarrow N_1DP + N_2TP$$

is responsible for the synthesis of UTP, at the expense of ATP for example. Nucleoside diphosphokinase activity involves a high-energy phosphohistidine intermediate and can be decomposed into two partial reactions:

$$N_1TP + Enz \Leftrightarrow N_2DP + Enz\,P$$

$$=$$

$$Enz\,P + N_2DP \Leftrightarrow E + N_2TP$$

Finally, the second pyrimidine nucleotide, cytidine triphosphate, is formed from UTP by the action of CTP synthetase (*pyrG*).

$$UTP + NH_3 + ATP \rightarrow CTP + ADP + Pi$$

CTP synthetase exists as a tetramer of identical subunits (50 kDa each) under assay conditions, and dissociates to a dimer in absence of UTP and ATP. Glutamine is the preferred amino donor, although like many other amidotransferases we have encountered, ammonia reacts also, albeit with a lower efficiency. The reaction mechanism involves a glutamylation of a reactive cysteine on the enzyme, with the liberation of ammonia which then reacts with UTP. ATP and UTP bind to the enzyme and each ligand induces conformational changes in the protein, which result in tetramerization and also in allosteric activation of the glutamylation reaction.

Direct Utilization of Pyrimidines and of Their Derivatives

In microorganisms, exogenous bases and nucleosides can be utilized by means of numerous enzymes, called the "salvage" enzymes, which enable the cells to utilize preformed bases and nucleosides for nucleotide synthesis.

For example, UMP can be synthesized by uracil phosphoribosyltransferase (*upp*) an enzyme activated by GTP and inhibited by ppGpp and uracil nucleotides. The expression of *upp* is controlled by pyrimidine compounds and the corresponding homotrimeric enzyme (3×23,500 Da) increases considerably upon pyrimidine starvation.

Second, uridine can be synthesized through the action of uridine phosphorylase

$$\text{Uracil} + \text{ribose} - 1 - \text{phosphate} \rightarrow \text{Uridine} + \text{Pi}$$

However, due an unfavorable equilibrium, this reaction can occur only if either uracil or ribose-1-phosphate is present at a high concentration. Uridine can then be processed by uridine kinase.

As we shall see, "salvage" enzymes exist also for the utilization of preformed purines.

Aspartate Transcarbamylase of E. coli

This enzyme has been crystallized from a pyrimidine-requiring mutant of *E. coli* grown under conditions of physiological derepression which enable crude extracts to be obtained with this protein representing more than 10 % of the total soluble proteins. It is to date one of the most studied and best known allosteric enzymes and some of the early results obtained by Gerhardt, Pardee, Schachman and their colleagues have been the foundation for many ideas about metabolic regulation and its mechanism.

The enzyme saturation curve as a function of aspartate concentration is sigmoid, indicating as we have seen several times, that more than one substrate molecule can complex with each molecule of enzyme, and that the binding of the first molecule facilitates the binding of a second.

Cytidine triphosphate, the end product of the biosynthetic pathway starting with the carbamoylation of aspartate, is an inhibitor of the enzyme. The inhibition by CTP is competitive with respect to aspartate and appears as a displacement of the aspartate concentration giving 50 % of the inhibition towards higher values (Fig. 33.1). In other words, the apparent affinity of the enzyme for aspartate decreases in the presence of CTP, but the cooperativity between molecules of substrate is still visible. There is no detectable effect on the value of Vmax.

The maximum inhibition by CTP can reach up to 75 % at saturation of this effector and is increased up to 100 % by synergism between CTP and UTP, the second

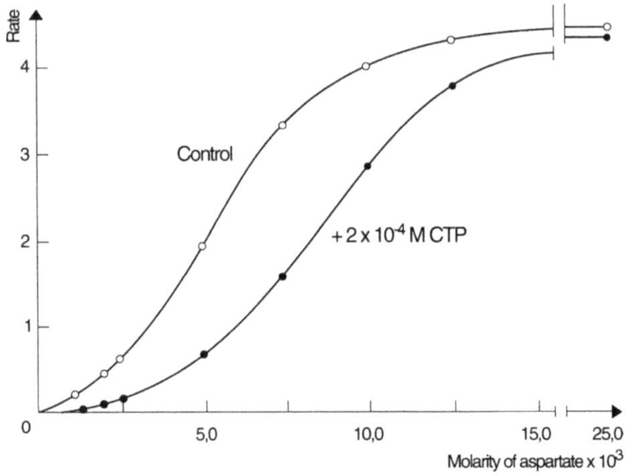

Fig. 33.1 Displacement of the sigmoid saturation curve of aspartate transcarbamylase by CTP. The experiment was carried out with a fixed concentration of carbamyl phosphate and variable aspartate concentrations

pyrimidine nucleotide, which has no effect by itself. By an original method using continuous flow dialysis, England and Hervé have shown that UTP enhances the affinity of CTP 5-fold, and conversely, CTP enhances the affinity for UTP 80-fold. The isolated regulatory subunits (see below) bind the two pyrimidine nucleotides following the same pattern as the entire enzyme. The observations strongly suggest that the synergistic inhibition mechanism relies entirely on interactions between two independent allosteric sites belonging to the same regulatory subunit.

The inhibitory effect of CTP is countered by ATP, an antagonist of ATP. ATP has also an intrinsic activating effect on the uninhibited enzyme; the activation can reach 100 % and is associated with an increase of the apparent affinity of the enzyme for aspartate and a normalization of the saturation curve so as to follow the Michaelis-Henri law. This suggests that ATP induces a conformational change resulting in the loss of cooperative interactions between substrate molecules. The antagonism between ATP and CTP/UTP is assumed to play a role in maintaining a balance between the intracellular pools of purine and pyrimidine nucleotides.

Although the inhibition by CTP is competitive with respect to aspartate, early experiments have established unequivocally that aspartate and CTP combine at different sites on the enzyme surface. This has been shown not only by desensitization of the enzyme towards its allosteric inhibitor, but also by physical separation between different subunits, one of which contains the catalytic site and the other the site with affinity for the allosteric inhibitor.

Physical and chemical measurements have shown that the enzyme has a molecular weight of ca. 310,000. Low concentrations of urea, or treatment by heat or with mercurials, abolishes the sensitivity to CTP, without the catalytic activity being modified. This desensitization is accompanied by major changes in the molecular weight of the enzyme. It can be shown that the action of *p*-mercuribenzoate (*p*MB) dissociates the enzyme into two components which can be separated by centrifugation in a sucrose gradient (Fig. 33.2).

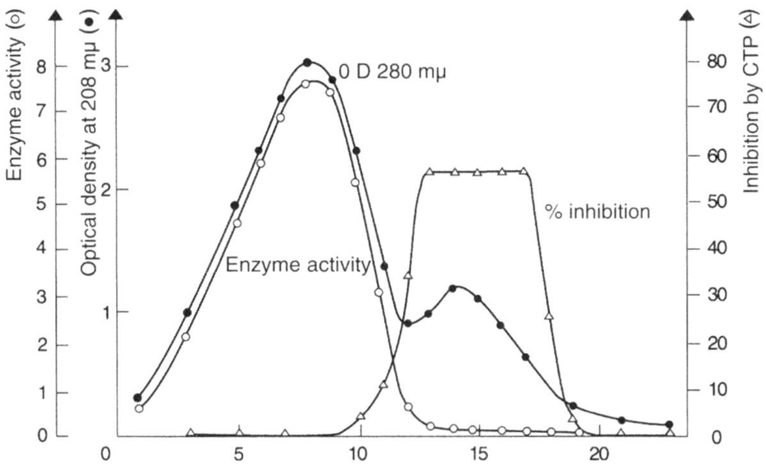

Fig. 33.2 Separation by sucrose gradient centrifugation of the two components of aspartate transcarbamylase. Seven milligrams of *p*-MB treated enzyme were centrifuged at 38,000 rpm for 20 h at 10 °C. The centrifuge tubes were then pierced and fractions of 12 drops collected. The *filled circles* show the protein profile; the *white circles* indicate catalytic activity, measured after addition of 2-mercaptoethanol. The regulatory protein is identified by adding the test fraction to fraction 6, and measuring the resulting activity in the presence and absence of 4×10^{-4} M CTP

Whereas the native enzyme has a sedimentation coefficient of 11.8S, the two components have coefficients respectively 5.8S and 2.8S. The 5.8S fraction has the catalytic activity; it displays a greater specific activity than the native enzyme but differs in that it is completely insensitive to CTP or ATP. It is therefore sufficient for catalytic activity but not for its regulation. Its kinetics as a function of substrate concentration follow strictly the Michaelis law. The addition of 2-mercaptoethanol to the lighter fraction does not cause any catalytic activity to appear, but allows the fraction to regain the ability to bind CTP which the enzyme had lost when treated with the mercurial. The 2.8S component is therefore the carrier of the capacity of the enzyme to be inhibited by CTP, and has no catalytic role.

If 2-mercaptoethanol is added to the unseparated mixture of the two fractions produced by treatment with *p*-mercuribenzoate, nearly all the enzyme is recovered with its native properties of sedimentation and inhibition. The 2.8 component is indispensable for the reaggregation; neither component alone can aggregate upon removal of the mercurial.

Metal ions bound to the regulatory subunit are essential for the reconstitution from isolated subunits; the naturally-occurring metal is zinc with a stoichiometry of six atoms per enzyme molecule. Zinc plays no role in the catalysis, and merely stabilizes the dimeric form R2 of the regulatory subunit and promotes its recombination with the catalytic subunit.

The catalytic subunit, coded by *pyrB* and the regulatory subunit, coded by *pyrI*, have been cloned and sequenced. The catalytic monomer (C) is composed of 311 amino acid residues and the regulatory monomer of 153 residues. Knowing the weight contributions from the 5.8S and the 2.8S components, the calculations show clearly a structure of the type R6C6 for aspartate transcarbamylase. The mercurial agent cleaves the enzyme into two C3 units and three R2 units. X-ray crystallography confirms that the enzyme has a twofold and a threefold symmetry axis.

A 3 % significant ligand-promoted decrease in the sedimentation coefficient was demonstrated, which is in accord with diffusion and X-ray scattering measurements, indicating a swelling of the enzyme. Subsequent crystallographic data obtained by Lipscomb showed that the distance along the threefold symmetry axis increases by approximately 12 Å upon the binding of N-(phosphonacetyl)-L-aspartate (PALA), a bisubstrate analogue which has some of the structural features expected from a transition analogue. In addition to this moving apart of the two catalytic trimers, there is a reorientation of each of them and of the three R dimers.

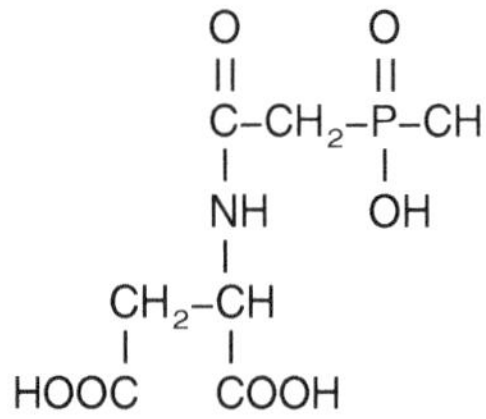

PALA

(In this formula, the structures of aspartate and carbamylphosphate are recognizable).

A quantitative interpretation of the cooperativity of ATCase is dependent upon the knowledge of the energetics of the allosteric transition. Since unliganded ATCase is predominantly in the T conformation and since the R state is detectable only when the enzyme is partially or fully saturated with active site ligands, there is no direct method of determining the T⇔R equilibrium constant for unliganded enzyme.

According to the Monod-Wyman-Changeux model, unliganded ATCase can exist in either of two quaternary conformations, R or T, and the ratio T/R is the allosteric equilibrium constant L. Ligands are assumed to bind reversibly to either form with dissociation constants K_R and K_T. Because the equilibria are linked, the addition of a ligand which will bind preferentially to the R form must promote the T⇔R transition, the shift of the equilibrium caused by the preferential binding of the substrates to the R form being responsible for the homotropic, cooperative, interactions. Effectors binding to the regulatory chains may also show preferential interactions. If, for example, ATP binds more tightly to the R conformation, activation would occur and conversely, preferential binding of CTP to the less active T state would cause inhibition (heterotropic effect). Accordingly, CTP and ATP must affect the T⇔R equilibrium. The model is simple in that ligand affinity is determined by the quaternary state of the molecule, T or R, and not by the number, location, or nature of bound ligands.

Thus, the first models that were proposed to account for the regulatory properties of ATCase postulated that the nucleotides act directly on the transition involved in the homotropic cooperative interactions between the catalytic sites for aspartate binding. Numerous indications have accumulated over the years showing that the case is more complex. The effects of the nucleotides are best explained by local conformational changes, which either alter the affinity of the catalytic sites for aspartate or modulate the stability of the interfaces between regulatory and catalytic chains, thus facilitating (ATP) or hindering (CTP) the transition from T to R state upon substrate binding. X-ray solution scattering experiments and equilibrium isotope exchange kinetics actually show that ATP and CTP do not act directly on the T ⇔ R equilibrium and alter the rate constant for the binding of aspartate to the catalytic sites.

The crystallographic structures of the T and R states, unliganded or liganded with either ATP or CTP have been solved at high resolution by Lipscomb. The protein exhibits several levels of organization. It is composed of two catalytic trimers held together by three regulatory dimers. The catalytic sites lie at the interface between two catalytic chains belonging to the same trimer and involve amino acid side chains belonging to these two catalytic chains. The regulatory sites binding competitively ATP and CTP are located at the extremities of the regulatory dimers.

Each of the two types of chains comprises two domains. One can distinguish a carbamylphosphate and an aspartate binding site in the catalytic chains. Similarly, in the regulatory chains, there is an allosteric domain where the regulatory sites binding ATP and CTP lie, and a zinc domain. This region contains a zinc atom bound to four cysteine residues localized in the C-terminal part of the regulatory chain. These zinc atoms are indispensable to the association between the catalytic and regulatory chains. The effect of mercurials on these cysteines is responsible for the dissociation between the two types of chains which was described above.

The ATP-ligated ATCase is in the T state and the interactions characteristic of this state are not disrupted upon ATP binding to the unliganded enzyme, indicating that, contrary to the simple model prediction, ATP does not promote the T to R transition. The distance between the two catalytic trimers increases only by 0.4 Å upon ATP binding, compared to the 11 Å increase in that distance occurring during the T to R transition.

Although precise structural information for the T and R conformations is of inestimable value, an understanding of the allosteric transition is not yet at hand and requires physico-chemical studies on solutions of wild-type and altered (by chemical modification or by site-directed mutagenesis) enzyme, and an explanation of the results obtained in terms of the structural information available. For example, hybrid ATCase-like molecules have been obtained containing chromophores at well-defined locations and have provided information about the effects of various ligands on the T ⇔ R conversion.

Site-directed mutagenesis may cause slight perturbations in the structure of individual chains at regions remote from the site of amino acid substitutions affecting the T ⇔ R equilibrium. Therefore, the attribution of a change in regulatory properties to the direct effect of a single amino acid replacement may often be an over interpretation. Let us give an example:

The binding site for CTP and ATP is located at the contact region between regulatory chains, and it seemed likely from the crystallographic data that a change in Lys60 would affect nucleotide binding. Indeed, if this Lys is replaced by an histidine, the capacity to bind nucleotides is abolished and the enzyme appears kinetically to be "frozen" in the R state. However, if instead of histidine, Lys-60 is replaced by glutamine or arginine, the enzyme retains both homotropic and heterotropic effects. If Lys60 is substituted by an alanine residue, these effects are altered: ATP activates normally, but the enzyme is now insensitive to CTP. It is thus hazardous and often unwarranted to attribute to a change in a single amino acid residue the properties of this residue in the wild type enzyme.

Other residues are involved in the regulatory process. The regulatory chains are folded in two domains, the effector binding domain and the zinc binding domain, which makes the contacts with the catalytic chains. It follows that the CTP and ATP regulatory signals must necessarily be transmitted through the zinc domain. At the interface between the two domains, Leu32 and Leu76 of the allosteric domain are in contact with Val106 and Leu151 of the zinc domain, forming a hydrophobic pocket in which the side-chain of Tyr77 is inserted. When this tyrosine residue is replaced by a phenylalanine, the effect of ATP on the activity of the enzyme is completely inverted; it becomes an inhibitor instead of an activator, while the inhibition by CTP is maintained. This result shows that the interface between the two domains of the regulatory chain plays an important role in the discrimination between the two regulatory signals.

Site-directed mutagenesis has also been applied to the catalytic chains, taking advantage of the knowledge of the tridimensional structure for the choice of the substitutions. There is an intersubunit interface between a catalytic chain in one C3 trimer and a catalytic chain in the other trimer. This limited contact is observed in the T conformation, but is disrupted in the R conformation. It involves an interaction between Lys164 and Glu239 residues. If these two residues are replaced respectively by a Glu at position 164 and a Lys at position 239, the mutant ATCase exhibits hyperbolic kinetics, no inhibition by CTP or activation by ATP and there is no cooperativity in binding PALA: the unliganded enzyme is in the "frozen" R conformation.

The catalytic trimers of *E. coli* can be transformed to the allosteric form not only by association with the regulatory dimers, but also by site-directed mutagenesis by substituting Arg105 by an alanine residue.

The physiological consequences of disrupting the regulation of ATCase by CTP, however, have been explored: an alternative regulatory strategy that enables precise control of pyrimidine pathway end-product levels, even in the presence of dysregulated biosynthetic flux. The mechanism involves cooperative feedback regulation of the near-terminal pathway enzyme uridine monophosphate kinase, encoded by *pyrH* Such feedback leads to build-up of the pathway intermediate uridine monophosphate, which is in turn degraded by a conserved phosphatase, termed UmpH, with previously unknown physiological function. Such directed overflow metabolism allows homeostasis of uridine triphosphate and cytidine triphosphate levels at the expense of uracil excretion and slower growth during energy limitation. Disruption of the directed overflow regulatory mechanism impairs growth in pyrimidine-rich environments. Thus, pyrimidine homeostasis involves dual regulatory strategies, with classical feedback inhibition enhancing metabolic efficiency and directed overflow metabolism ensuring end-product homeostasis.

The Aspartate Transcarbamylase of Other Organisms

The native *E. coli* ATCase provides the allosteric model for feedback inhibition of the pyrimidine pathway. It shows sigmoidal substrate saturation curves with 100 % activation by ATP, 75 % inhibition by CTP and 100 % inhibition by CTP/UTP synergism. In *Erwinia herbicola*, another member of *Enterobaceriaceae*, the bicistronic operon encodes also a dodecamer holoenzyme $3R_2$-$2C_3$ with 153 residues per R monomer and 311 residues per C monomer as in *E. coli*. However, the sequence diverges by 20 % in the R monomers and 18 % in the C monomers. In spite of conservation of the residues directly implicated in the active site and in the effectors binding sites, the *Erwinia* enzyme has diverged with respect to homotropic and heterotropic responses as exhibited by hyperbolic saturation curves, lack of activation by ATP, and only slight inhibition by CTP. Nonetheless, the UTP/CTP synergistic inhibition approaches 90 % similar to the *E. coli* value and suggests that the *Erwinia* enzyme is a naturally occurring R-state enzyme in contrast to the *E. coli* enzyme, whose T/ R equilibrium approximates an L value of 250.

Other ATCases have been purified; those from *Streptococcus faecalis* and *Bacillus subtilis* do not appear to possess a feedback control mechanism, since their activity is affected neither by ATP or CTP. The *B. subtilis* enzyme is trimeric like the catalytic chain of the *E. coli* enzyme with which its 33,500 Da polypeptide chain shows 35 % homology. Its crystallographic structure is more similar to that of the T rather than that of the R form of the *E. coli* enzyme. By site-directed mutagenesis, if Arg99 (corresponding to Arg105 in *E. coli*) is replaced by an alanine residue, the *B. subtilis* enzyme retains its activity, but the regulatory activity of homotropic substrate cooperativity for aspartate molecules is now observed. Thus, a small number of substitutions could transform a simple metabolic step into a regulatory juncture. It has been proposed that that this possible evolutionary potential could explain in part the natural abundance of oligomeric enzymes. Although the wild type *B. subtilis* enzyme does not show cooperativity for either substrate and is insensitive to both purine and pyrimidine nucleotides, it is developmentally regulated: regulation occurs in the stationery bacterial growth phase, prior to sporulation, where the enzyme is degraded by an ATP-dependent protease.

Aspartate transcarbamylase from *Pseudomonadaceae* consists of six copies of a 36-kDa catalytic chain and six copies of a 45-kDa polypeptide of unknown function. The 45-kDa polypeptide is homologous to dihydroorotase but lacks catalytic activity. *Pseudomonas aeruginosa* aspartate transcarbamylase was overexpressed in *E. coli*. The homogeneous His-tagged protein was isolated in high yield by affinity chromatography and crystallized. Attempts to dissociate the catalytic and pseudo-dihydroorotase (pDHO) subunits or to express catalytic subunits only were unsuccessful suggesting that the pDHO subunits are required for the proper folding and assembly of the complex. As reported previously, the enzyme was inhibited by micromolar concentrations of all nucleotide triphosphates. In the absence of effectors, the aspartate saturation curves were hyperbolic but became strongly sigmoidal in the presence of low concentrations of nucleotide triphosphates. The inhibition was unusual in that only free ATP, not MgATP, inhibits the enzyme. Moreover, kinetic and binding studies with a fluorescent ATP analog suggested that ATP induces a conformational change that interferes with the binding of carbamoyl phosphate but has little effect once carbamoyl phosphate is bound. The peculiar allosteric properties suggest that the enzyme may be a potential target for novel chemotherapeutic agents designed to combat *Pseudomonas* infection.

Regulation of Pyrimidine Nucleotide Synthesis at the Genetic Level

The genes *carAB* and *pyrBI*, which specify the two enzymes made from two different subunits, i.e. carbamylphosphate synthetase and aspartate transcarbamylase are respectively forming each an operon. All the other genes are scattered on the chromosome of *E. coli* or *S. typhimurium*. However, *pyrE* and *pyrF* constitute small bicistronic operons each encompassing proteins of unknown function.

The expression of all six genes is non-coordinately regulated by the availability of pyrimidines, increased by starvation and reduced by addition of pyrimidines.

The synthesis of carbamylphosphate synthetase, which provides the common precursor to uracil and arginine, is subject to a cumulative repression by arginine and a pyrimidine. The effect of arginine is mediated by the arginine repressor-arginine complex.

The nucleotide sequence of the entire *pyrBI* operon has been determined. The genes are transcribed in the order promoter-leader sequence-*pyrB-pyrI*. The structural genes are preceded by a rho-independent terminator, itself preceded by a thymidylate-rich cluster. The existence of these structures have lead to the proposal that there is a tight coupling between the transcribing RNA polymerase and the ribosomes translating a "leader peptide" required for transcription through the attenuator, in a way analogous to the attenuation of transcription observed in the case of the regulation of the biosynthesis of some amino acids. The same mechanism seems to apply to the regulation of the synthesis of orotate phosphoribosyltransferase (*pyr E*).

The genes coding for dihydroorotase and dihydroorotate oxidase, *pyrC* and *pyr D*, do not contain attenuator-like structures, but the leader sequences contain regions of high symmetry which might have some regulatory function.

In the mesophile *Bacillus subtilis* and the thermophile *Bacillus caldolyticus*, the six genetic loci coding for the pyrimidine biosynthesis enzymes are organized into a single operon which includes in addition genes for the salvage pathway enzyme uracil phosphoribosyltransferase and for an uracil permease.

Pyrimidine Homeostasis

The first committed enzyme in the pyrimidine pathway in *E. coli* is allosterically inhibited by cytidine triphosphate. The physiological consequences of disrupting this regulation, however, have not been previously explored. An alternative regulatory strategy that enables precise control of pyrimidine pathway end-product levels, even in the presence of dysregulated biosynthetic flux has recently been identified. The mechanism involves cooperative feedback regulation of the near-terminal pathway enzyme uridine monophosphate kinase (*pyrH*). Such feedback leads to build-up of the pathway intermediate uridine monophosphate, which is in turn degraded by a conserved phosphatase, here termed UmpH, with previously unknown physiological function. Such directed overflow metabolism allows homeostasis of uridine triphosphate and cytidine triphosphate levels at the expense of uracil excretion and slower growth during energy limitation. Disruption of the directed overflow regulatory mechanism impairs growth in pyrimidine-rich environments. Thus, pyrimidine homeostasis involves dual regulatory strategies, with classical feedback inhibition enhancing metabolic efficiency and directed overflow metabolism ensuring end-product homeostasis.

The Biosynthesis of Purine Nucleotides

Biosynthesis of 5-Amino-4-Imidazole Carboxamide Ribonucleotide

Eight successive reactions are necessary to synthesize this compound, the precursor of the purines. Since most of the intermediates in this pathway are not easily available, few thorough enzymological studies have been made.

First, the amide group of glutamine displaces the pyrophosphate group of PRPP to form 5-phosphoribosylamine. This reaction is catalyzed by PRPP aminotransferase (*purF*), a tetrameric enzyme whose subunit has a molecular weight of 53,000. The gene has been cloned and sequenced and in the case of *B. subtilis*, the tridimensional structure of the enzyme has been determined. In both *E. coli* and *B. subtilis*, there are two separate sites for AMP and GMP on each polypeptide chain.

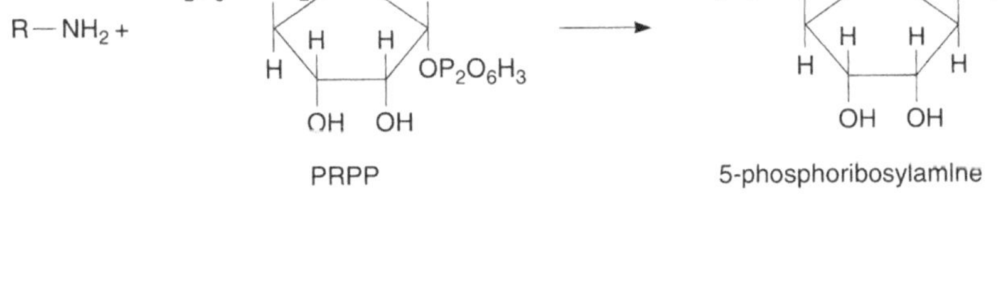

During the reaction, the configuration of the C1 of ribose is inverted from α to β.

As it is the case for many amidotransferases, PRPP amidotransferase can also use ammonia as the amino group donor. Reaction with 6-diazo-5-oxonorleucine, an analogue of glutamine, modifies covalently the N-terminal cysteine of the enzyme which is thus the active site residue required for the glutamine transfer reaction. This enzyme, the first committed step in the biosynthesis of the purine nucleus, is subject to feedback inhibition by AMP and GMP which act synergistically and are competitive inhibitors of PRPP. The hyperbolic PRPP saturation curve becomes sigmoidal in the presence of these allosteric effectors.

The amino acid glycine is added to the phosphoribosylamine molecule in a reaction which uses ATP as the energy source, and which produces 5′-phosphoribosyl-1-glycinamide (glycinamide ribonucleotide, GAR), ADP and Pi. The enzyme catalyzing this reaction, phosphoribosyl glycinamide synthetase is encoded by *purD*.

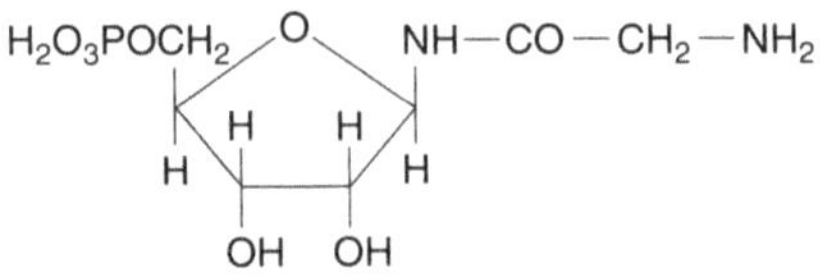

GAR

The free amino group of glycine is formylated by phosphoribosyl formylglycinamide synthetase (also called 10-formyltetrahydrofolate: 5′-phosphoribosylglycinamide formyltransferase, *purL*) to yield a compound whose name is abbreviated to FGAR, the ribonucleotide of formylglycinamide. The formyl donor is N10-formyltetrahydrofolic acid. There is an additional enzyme, GAR transformylase T encoded by *purT* which can utilize formate directly: formate is generated by pyruvate formate lyase during anaerobic growth, whereas an enzyme, encoded by *purU*, provides the major source of formate used in aerobiosis. The purT transformylase has been cloned, sequenced and characterized. It shows no significant sequence homology with the folate-dependent enzyme.

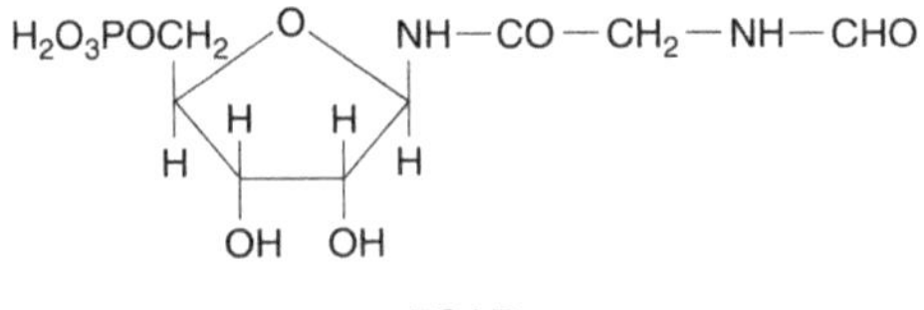

FGAR

The three-dimensional structure of the enzyme has been solved both as an apoenzyme and as a ternary complex with the substrate glycinamide ribonucleotide and a folate inhibitor. The structure is a modified doubly wound sheet with flexibility in the active site; it includes a disordered structure in the apo structure, which is ordered in the ternary complex. This enzyme is a target for anti-cancer therapy and for structure-based drug design.

Another amidotransferase, FGAR amidotransferase (*purG*, MW of the native enzyme = 130,000) transfers in the presence of ATP the amide group of glutamine which becomes the future nitrogen atom 3 of the purine ring, the other products of the reaction being glutamate, ADP and Pi. The product of the reaction which interests us here is FGAM, the ribonucleotide of formylglycinamidine:

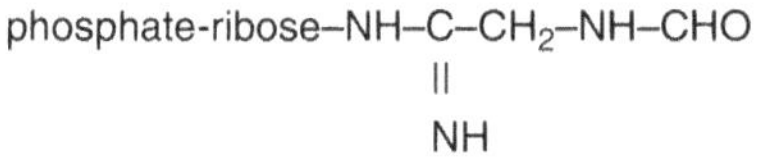

FGAM

An ATP-dependent cyclization completes the formation of the imidazole ring and produces aminoimidazole ribonucleotide (AIR):

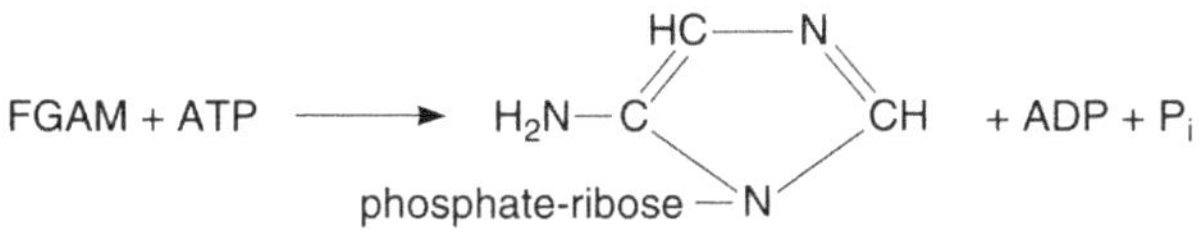

AIR

The gene coding for AIR synthetase (*purI*) has been cloned and sequenced. The protein is a homodimer (2 × 38,000 kDa).

The gene *purE* codes for phosphoribosylaminoimidazole carboxylase which is responsible for the formation of the ribonucleotide of 5-amino-4-imidazolecarboxylic acid (CAIR). Genetic studies dating from 1976 suggested that the enzyme was composed of subunits. Mutations were obtained in the *purE* gene, the other in a gene called *purK*. Both mutations led to auxotrophy for purines. Recent experiments have shown that PurE and PurK are not subunits of the same enzyme but that they catalyze two separate steps in the purine biosynthetic pathway. A new intermediate, a carbamate is formed by PurK from AIR, ATP and bicarbonate whereas PurE is postulated to catalyze a rearrangement of the carbamate to CAIR. The overall reaction results in the introduction of carbon atom 6 of the future purine ring:

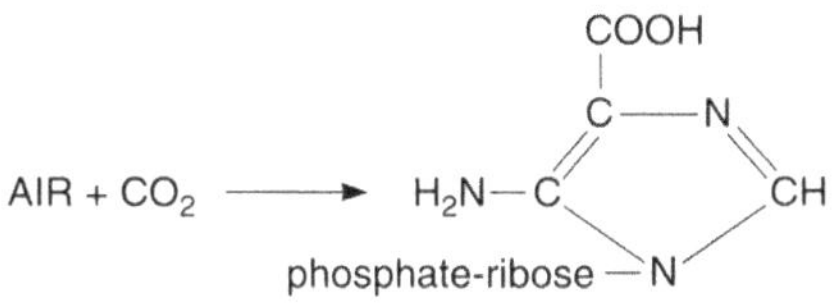

CAIR

This is followed by the addition of a molecule of aspartic acid in the presence of ATP to yield 5-amino-4-imidazole-N-succinylcarboxamide ribonucleotide (SAICAR), a reaction catalyzed by SAICAR synthetase (*purC*), the quaternary structure of which is not known, but which has a subunit mass of 27 kDa.

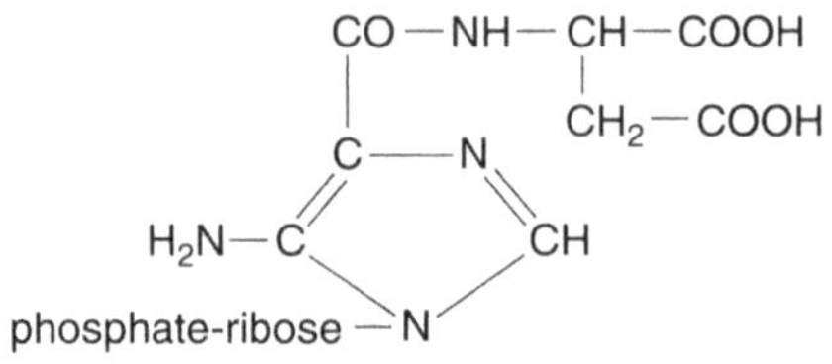

SAICAR

The ribonucleotide of 5-amino-4-imidazolecarboxamide (AICAR) is reached after an elimination of fumaric acid, catalyzed by an adenylosuccinate lyase, which functions also in the synthesis of AMP from IMP (*purB*).

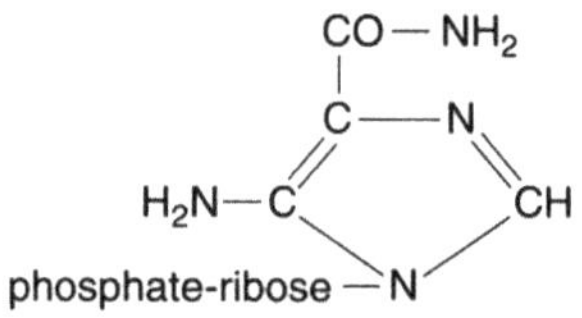

AICAR

Remember that AICAR is also formed during the synthesis of imidazoleglycerolphosphate (histidine biosynthesis).

This is the second time we encounter this curious amination procedure which consists of adding an aspartic acid molecule and then removing a fumaric acid molecule (cf. the synthesis of argininosuccinate, the intermediate between citrulline and arginine). We shall soon see a third example.

This part of the purine biosynthetic pathway is regulated mainly by feedback inhibition of PRPP amidotransferase and by regulation of the expression of the *pur* genes. This expression is repressed when exogenous purines are added to the medium and derepressed upon purine starvation.

Synthesis of Inosinic Acid

Carbon atom 2, the final carbon of the purine ring is introduced by transfer of the formyl group of N^{10}-formyltetrahydrofolate to the 5-amino group of AICAR to yield 5′-phosphoribosyl-4-carboxamide-5-formamidoimidazole (FAICAR), a reaction catalyzed by the product of *purH*.

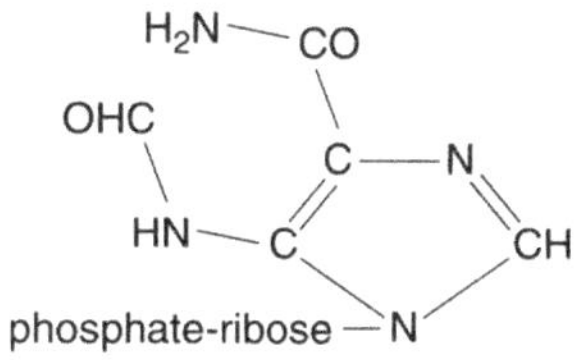

FAICAR

The last step, which gives rise to the completed purine ring, is catalyzed by the product of *purJ*, IMP cyclohydrolase: a molecule of water is eliminated and the second ring is closed in a reaction which does not require ATP:

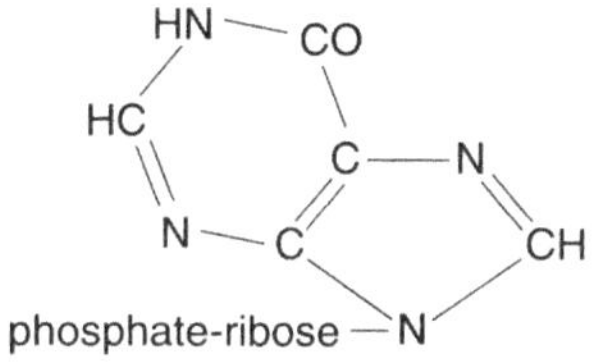

inosinic acid (IMP)

In *E. coli*, it has been shown that the conversion of AICAR into FAICAR, and the subsequent formation of IMP, is catalyzed by a bifunctional enzyme.

Synthesis of Guanylic and Adenylic Acids

Guanylic acid is derived from IMP in two steps. First, IMP is oxidized in position 2 by IMP dehydrogenase, an NAD-dependent enzyme (*guaB*). The *E. coli* enzyme is a homotetramer (4×54,000 Da), containing a cysteine at the IMP binding site. It is inhibited by GMP and ppGpp. The corresponding gene has been cloned and sequenced. The product of the reaction is xanthylic acid (XMP). Xanthylic acid is then aminated at carbon 2 by an amidotransferase, called guanylate synthetase (*guaA*), a homodimer of subunit molecular weight 58,000. As usual, the amidotransferase can use either glutamine or ammonia as the amino donor and requires ATP, which is cleaved to AMP and pyrophosphate; the activity is irreversibly lost when the enzyme is treated with a covalently acting glutamine analogue. Here again, the gene has been cloned and sequenced.

Adenylic acid (AMP) derives also from inosinic acid in two steps. First, aspartate condenses to position 6 of IMP, in the presence of GTP (which is hydrolyzed to GDP and Pi. during the reaction). The enzyme responsible for this reaction is adenylosuccinate synthetase (*purA*). The subunit structure of this enzyme is not known; the molecular weight of its subunit is 56,000. It contains an N-terminal glycine-rich sequence which is homologous to a conserved sequence found in many other guanine nucleotide-binding proteins which appears to play a role in enzyme catalysis A consensus sequence for GTP-binding proteins is present in the eight available sequences of adenylosuccinate synthetase. It contains an aspartate residue (Asp333 in the *E. coli* enzyme). When this aspartate is replaced by an asparagine, the substrate specificity is changed from GTP to xanthosine triphosphate.

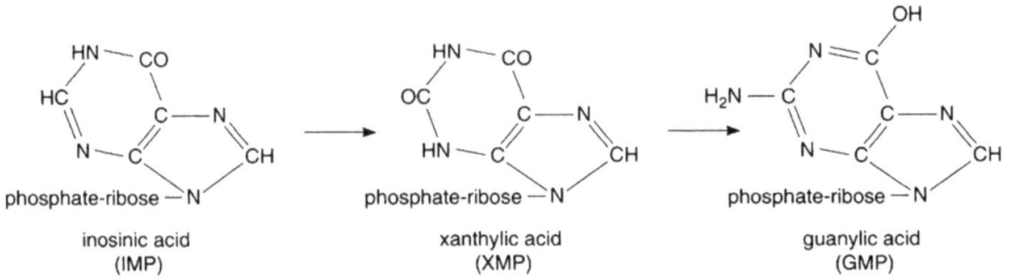

Adenylosuccinate synthetase is inhibited by AMP, GDP and ppGpp.

$$\text{IMP} + \text{aspartate} + \text{GTP} \rightarrow \text{Adenylosuccinate} + \text{GDP} + \text{Pi}$$

The same adenylosuccinate lyase (*purB*), responsible for the transformation of SAICAR to AICAR, catalyzes the removal of a molecule of fumaric acid and adenylic acid (AMP) is produced.

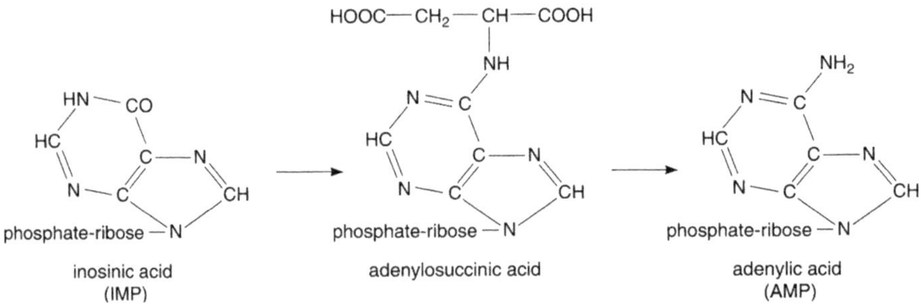

Remarks on the Control of Purine Nucleotide Biosynthesis

Study of the control mechanisms is complicated by the fact that during the biosynthesis of the purine nucleus, a by-product is a precursor of histidine. Also in addition to "salvage" enzymes responsible for the direct formation of nucleotides from preformed bases or nucleosides, hypoxanthine and xanthine, purine bases not found in RNA, can be respectively converted to inosinic and xanthylic acids, precursors of the two "natural" nucleotides.

In addition to PRPP amidotransferase, a certain number of reactions are subject to allosteric inhibition effects. We shall examine only some of them here.

The reactions converting xanthylic acid to guanylic acid, adenylic acid to inosinic acid by deamination, and guanylic acid to inosinic acid, are practically irreversible. In the absence of a rigid control, the three enzymes (xanthosine-5′-phosphate aminase, IMP dehydrogenase and GMP reductase) would theoretically be able to catalyze an irreversible futile cycle of reactions leading to IMP through GMP to IMP again. This cycle would have no result other than the wastage of ATP by the hydrolysis accompanying the amination of XMP. The same difficulty would arise if the conversion of IMP to AMP via adenylosuccinate was coupled with the hydrolytic deamination which transforms AMP to IMP.

These difficulties are resolved by the specific inhibition of IMP dehydrogenase by GMP:

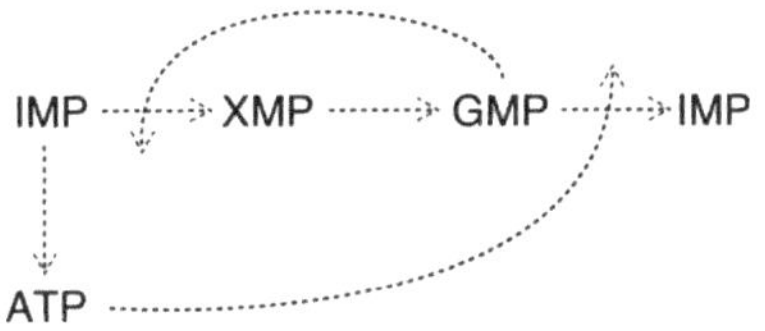

In the same way, adenylosuccinate synthetase in *E. coli* is inhibited by ADP.

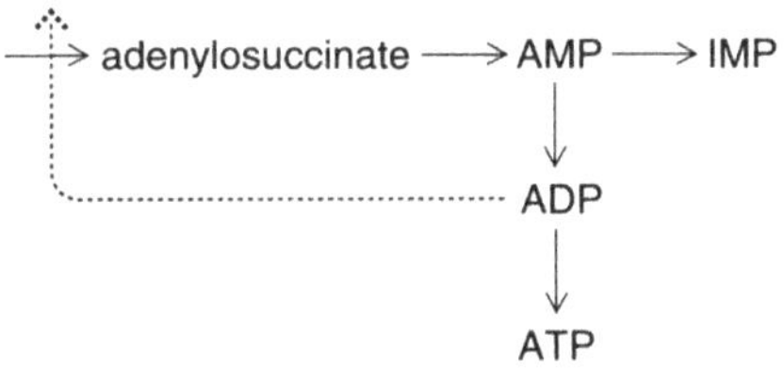

The purine phosphoribosyltransferases catalyze the entry into the cycle of exogenous purine bases, which are not normal intermediates in the synthesis of nucleotides. They catalyze the following type of reaction:

$$\text{Base} + \text{PRPP} \rightarrow \text{Nucleoside monophosphate} + \text{pyrophosphate}$$

E. coli has two of these enzymes, a guanine phosphoribosyltransferase (*gpt*, sequence determined) and an hypoxanthine phosphoribosyltransferase (*hpt*). Both enzymes use guanine and hypoxanthine as substrates, but at different efficiencies, and are inhibited by ppGpp. The product of *gpt* can also salvage xanthine.

In *Bacillus subtilis*, all these reactions are inhibited by a large number of purine nucleotides. From the standpoint of cellular control, it is significant that the nature of the inhibitors and the compared sensitivity of the different reactions can be understood in terms of intracellular equilibria. For example, xanthine phosphoribosyltransferase is especially sensitive to the guanine nucleotides GTP and dGTP: so an excess of either of these prevents the accumulation of XMP, the direct precursor of the guanine nucleotides; incidentally, we just saw that the alternative pathway for the synthesis of XMP from IMP is also inhibited by GMP, making the guanine nucleotides the regulators of every step in their synthesis.

The gene *purR* from *E. coli* encodes a 38 kDa repressor protein which regulates the expression of ten *pur* regulon genes, (which are required for the formation of IMP) and of *guaBA*, which is required for conversion of IMP to GMP. The purine repressor also contributes to the coregulation of at least five additional genes, *pyrC* and *pyrD*, involved in the synthesis of pyrimidine nucleotides, *glyA*, which is required for generation of one-carbon units, *codA* which encodes cytosine deaminase, a salvage enzyme and *purR* itself which it auto-regulates. Hypoxanthine and guanine are the corepressors for binding to a 16-bp conserved operator sequence present in the promoter region or, for the coregulated genes, in the coding regions. The purine repressor is dimeric, it has been overproduced and appears as a complex allosteric protein, having an N-terminal domain of 52 residues for DNA binding and a C-terminal domain of 289 residues for corepressor binding. Binding of the corepressor is accompanied by a conformational change in the corepressor binding domain. The corepressor binding site of PurR appears to be similar to the conserved ligand-binding sites in three periplasmic sugar-binding proteins and in the LacI family of repressors.

The repressor-operator control is independent of the *purB* promoter strength as well as of other 5′ flanking sequences. It appears that the binding of repressor to the *purB* operator inhibits transcription elongation by a steric inhibition "roadblock" mechanism.

From Nucleoside Monophosphates to Nucleoside Diphosphates and Triphosphates

All naturally occurring nucleoside triphosphates can be derived from the mono- and diphosphates by successive phosphorylations catalyzed by nucleotide kinases.

In one organism or the other, enzymes specific for AMP (adenylate kinase, *adk*), GMP, CMP, UMP and dTMP have been resolved and some of them purified to homogeneity. They all use MgATP as the preferred phosphate donor. The first three are active on the deoxynucleoside monophosphates, whereas UMP kinase is specific for UMP; dUMP kinase can use dUMP but not TMP as substrate.

The 2.8 Å crystal structure of *E. coli* UMP kinase bound to its allosteric regulator GTP shows that the GTP-binding site, situated at 15 Å from the UMP-binding site and at 24 Å from the ATP-binding site, is delineated by two contiguous dimers. The overall structure, as compared with those bound to UMP, UDP, or UTP, shows a rearrangement of its quaternary structure: GTP induces an 11° opening of the UMP kinase dimer, resulting in a tighter dimer-dimer interaction. A nucleotide-free UMP kinase dimer has an intermediate opening. Superposition of this structure with that of archaeal UMP kinases, which are also hexamers, shows that a loop appears to hamper any GTP binding in archaeal enzymes. This would explain the absence of activating effect of GTP on this group of UMP kinases. Among GTP-binding residues, the Asp-93 is the most conserved in bacterial UMP kinases. In the previously published structures of *E. coli* UMP kinase, this residue was shown to be involved in hydrogen bonds between the subunits of a dimer. Its substitution by an alanine decreases the cooperativity for UTP binding and suppresses the reversal by GTP of UTP inhibition. This demonstrates that the previously described mutual exclusion of these two nucleotides is mediated by Asp-93.

Adenylate kinase from *E. coli* can be activated by the direct binding of a magnesium ion to the enzyme, in addition to ATP-complexed Mg^{2+}. By systematically varying the concentrations of AMP, ATP, and magnesium in kinetic experiments, it appears that the apparent substrate inhibition of AK, formerly attributed to AMP, was suppressed at low magnesium concentrations and enhanced at high magnesium concentrations. This previously unreported magnesium dependence can be accounted for by a modified random bi-bi model in which Mg^{2+} can bind to AK directly prior to AMP binding.

Once the nucleoside diphosphates obtained, they are phosphorylated to the triphosphates by the non-specific nucleoside diphosphate kinase.

Selected References

Carbamylphosphate Synthetase

Anderson PM, Meister A (1966) Biochemistry 5:3157–3163, 3164–3167
Trotta P, Burt ME, Haschemeyer RH, Meister A (1971) Proc Natl Acad Sci USA 68:2599–2603

PRPP Synthetase

Switzer RL (1973) In: Boyer PD (ed) The enzymes, vol 9, 3rd edn. Academic, New York, pp 607–628

Aspartate Transcarbamylase

Hervé G (1989) In: Hervé G (ed) Allosteric enzymes. CRC Press, Boca Raton, pp 61–79
Jacobson GB, Stark GR (1973) In: Boyer PD (ed) The enzymes, vol 9, 3rd edn. Academic, New York, pp 225–308
Kantrowitz ER, Lipscomb WN (1990) Trends Biochem Sci 15:53–59
Reaves ML, Young BD, Hosios AM, Xu Y-F, Rabinowitz JD (2013) Nature 500:237–241
Schachman HK (1988) J Biol Chem 263:18583–18586

From Carbamylaspartate to UMP

Fujihashi M, Mito K, Pai EF, Miki K (2013a) J Biol Chem 288:9011–9016

CAD Protein

Guy HI, Evans DR (1994) J Biol Chem 269:23806–23816

Nucleoside Monophosphate Kinases

Meyer P, Evrin C, Briozzo P, Joly N, Bârzu O, Gilles A-M (2008) J Biol Chem 283:36011–36018

Pyrimidine Homeostasis

Reaves ML, Young BD, Hosios AM, Xu Y-F, Rabinowitz JD (2013) Nature 500:237–241

Orotidine 5′-Decarboxylase

Fujihashi M, Mito K, Pal EP, Miki K (2013b) J Biol Chem 288:9011–9016

Nucleoside Diphosphokinase

Agarwal RP, Robinson B, Parks RE Jr (1978) Methods Enzymol 51:376–386

PRPP Amidotransferase

Zhou G, Smith JL, Zalkin H (1994) J Biol Chem 269:6784–6789

Adenylate Kinase

Bârzu O, Michelson S (1983) FEBS Lett 153:280–284
Burlacu-Miron S, Gilles AM, Popescu A, Bârzu O, Craescu CT (1999) Eur J Biochem 264:765–774
Tan Y-W, Hanson JA, Yang H (2009) J Biol Chem 284:3306–3313

Chapter 34
The Biosynthesis of Deoxyribonucleotides

Ribonucleotide reductases play a central role in DNA biosynthesis. They catalyze the conversion of NDPs (NTPs) to dNDPs (dNTPs) concomitant with oxidation of thiols within their active sites. A single active site can accomodate reduction of both purine and pyrimidine nucleotides. In addition to the transcriptional and translational controls and to the allosteric regulation of the enzyme activity, the overall regulation may imply the regulation of the production and destruction of auxiliary cofactors and proteins.

The Formation of Deoxyribonucleoside Diphosphates from Ribose Nucleoside Diphosphates

Ribose is not directly reduced to deoxyribose. In *E. coli*, the reduction occurs at the level of nucleoside diphosphates which are reduced to the deoxynucleoside diphosphates. The enzyme catalyzing this reaction is called ribonucleoside diphosphate reductase. The same enzyme acts on all four nucleoside diphosphates. The substrates ADP, GDP and UDP are intermediates in the synthesis of the nucleoside triphosphates, whereas CDP is not, since CTP originates from UTP under the action of CTP synthetase (Chap. 33). CDP must derive from nucleoside diphosphokinase (which catalyzes a reversible reaction) acting on CTP or from CMP produced during phospholipid synthesis and mRNA turnover.

G.N. Cohen, *Microbial Biochemistry*,
DOI 10.1007/978-94-017-8908-0_34, © Springer Science+Business Media B.V. 2014

The Ribosenucleoside Diphosphate (NDP) 'Reductase System of *E. coli*'

Thioredoxin and Thioredoxin Reductase

Early observations established that crude extracts required NADPH for the reduction of ribonucleoside diphosphates. During purification of the system, a preparation was obtained which would no longer catalyze the reduction of CDP in the presence of NAPDH but instead required reduced lipoic acid. Certain other observations then suggested that dihydrolipoate was not the natural hydrogen donor, but a "model substance" and that the true hydrogen donor had been eliminated during the purification. This substance was shown to be a heat-stable protein of low molecular weight, given the name thioredoxin. Thioredoxin can be isolated from *E. coli* in an oxidized disulfide formed referred to as thioredoxin (S_2).

The molecular weight of thioredoxin is 12,000. It consists of a single polypeptide chain of 108 amino acid residues with two cysteine residues at positions 32 and 35 which form the disulfide bond in thioredoxin (S_2). The sequence -Trp-Cys-Gly-Pro-Cys of the active site exposed at the surface of the molecule is significantly conserved in the thioredoxin of most species studied (Fig. 34.1).

The tridimensional structure of S_2 has been determined by X-ray crystallography at 2.8 Å resolution and then refined to 1.68 Å. Figure 34.2 shows that four α-helices surround the central core formed by a five-stranded β-sheet limited by non polar residues. Three strands of the β-sheet are parallel and two are antiparallel. The active site forms a bulge at the surface of the protein.

Conversion by NADPH to reduced thioredoxin, thioredoxin $(SH)_2$, is catalyzed by thioredoxin reductase. The reductase is a flavoprotein which reduces the single cystine of thioredoxin to two cysteine residues. In the presence of catalytic amount of thioredoxin and its reductase, other substances with disulfide bridges are reduced (for example, oxidized lipoate, glutathione, insulin and other proteins). The thioredoxin-thioredoxin reductase system has therefore a wider role than the reduction of ribonucleoside diphosphates. It is thus involved in the oxidoreduction of the disulfide bonds of several proteins, and as such participates in the regulation of the enzymes of the Calvin cycle during photosynthesis according to the light/dark alternance.

Characterization of viable *E. coli* mutants devoid of thioredoxin resulted in the discovery of the glutaredoxin – glutathione system as a hydrogen donor for ribonucleotide diphosphate reductase and has called into question the unique role of thioredoxin. Thioredoxin and glutaredoxin (85 residues) are small proteins containing an active site with redox-active disulfide; they both function in electron transfer via the reversible oxidation of two vicinal -SH groups to a disulfide bridge.

In *E. coli* glutaredoxin, the active site contains a redox-active disulfide/dithiol of the sequence Cys11-Pro12-Tyr13-Cys14. The Cys14→Ser14 glutaredoxin, obtained by site-directed mutagenesis is completely inactive with ribonucleotide reductase, demonstrating that dithiol glutaredoxin is the hydrogen donor for the reductase.

The following figure schematizes the mechanism of reduction of ribonucleoside diphosphates in this complex system:

S*DKIIHL*TDDSFDTDVLKADGAILVDFWA EWCGPCKMIAP

*ILDEIADE*YQGK*LTVAKL* NIDQNPGTAPKYGIRGIP*TLLL*

*F*KNGEVAATKV GALSKGQLKEFLDANLA

Fig. 34.1 Primary structure of *E. coli* thioredoxin. The residues participating to the β-sheets are in *italics*; those pertaining to helical segments are *underlined*. The two cysteines are in *bold* characters. The conserved sequence is WCGPC (residues 31–35)

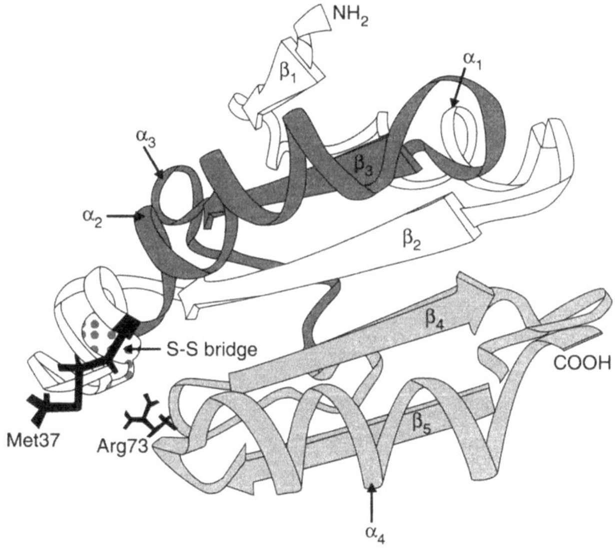

Fig. 34.2 Three-dimensional structure of *E. coli* thioredoxin-S2 from X-ray crystallography at 2.8 Å resolution. The *color code* (1–37; 38–73; and 74–108) refers to fragments generated by selective cleavage at Met37 or Arg73 (From A. Holmgren, with permission of the *Journal of Biological Chemistry*)

$NADPH + H^+$	→	$NADP^+$
	Glutathione reductase	
GSSG	→	2 GSH
Glutaredoxin $(SH)_2$	→	Glutaredoxin (S_2)
	ATP, Mg^{++}	
NDP	→	dNDP
	NDP-reductase system	
Thioredoxin $(SH)_2$	→	Thioredoxin(S_2)
	Thioredoxin reductase	
$NADP^+$	→	$NADPH + H^+$

This scheme emphasizes the fact that thioredoxin (or glutaredoxin) fulfills a catalytic role in the presence of thioredoxin reductase (or glutathione reductase) and excess NADPH. These two systems provide also electrons for the reduction of phosphoadenosine phosphosulfate (see cysteine biosynthesis, Chap. 29) and of methionine sulfoxide.

Recently, a double mutant, lacking both thioredoxin and glutaredoxin has been constructed. Its viability implied the presence of a third, unknown hydrogen donor for ribonucleotide reduction. Two proteins fulfilling this function have been isolated from the mutant. One, Grx2, with a molecular weight of 26,000 was atypical for glutaredoxin whereas the other, Grx3, had a molecular weight of 10,000. The two proteins present the CPYC signature of the glutaredoxin active site. The original glutaredoxin was found to be essential for sulfate reduction, while Grx3 showed activity for ribonucleotide reduction and Grx2 was inactive in both systems.

Ribonucleoside Reductase

This enzyme requires ATP and Mg^{++}.

The ribonucleoside diphosphate reductase from *E. coli* has an $\alpha 2\beta 2$ structure. It contains two different dimeric subunits (B_1, 172 kDa and B_2, 87 kDa), each being composed of two identical protomers, that can be separated during purification. The structural genes of these subunits (*nrdA* and *nrdB*, respectively) have been cloned, sequenced and overexpressed.

Each polypeptide chain of the small subunit B_2 contains a stable radical located on tyrosine 122. The radical lacks one ring electron and thus is an oxidized tyrosine residue. An inactive form of B_2 contains a normal tyrosine 122. In addition to the tyrosyl radical, both the active and inactive form of B_2 contain a binuclear iron center, in which Fe (III) ions are antiferromagnetically coupled to each other by a μ-oxo bridge. Both irons are coordinated to N and O ligands and nuclear magnetic resonance studies indicate that at least one of the ligands is histidine. This center stabilizes the radical and is also involved in its generation, molecular dioxygen being the source of the μ-oxo bridge.

The *Escherichia coli* ribonucleoside reductase (RNR) uses a mechanism of radical propagation by which a cysteine in the active site of the RNR large (α2) subunit is transiently oxidized by a stable tyrosyl radical (Y•) in the RNR small (β2) subunit over a 35-Å pathway of redox-active amino acids: $Y_{122}\bullet \leftrightarrow [W_{48}?] \leftrightarrow Y_{356}$ in β2 to $Y_{731} \leftrightarrow Y_{730} \leftrightarrow C_{439}$ in α2. When 3-aminotyrosine (NH_2Y) is incorporated in place of Y_{730}, a long-lived $NH_2Y_{730}\bullet$ is generated in α2 in the presence of wild-type (wt)-β2, substrate, and effector. This radical intermediate is chemically and kinetically competent to generate dNDPs. $NH_2Y_{730}\bullet$ induces formation of a kinetically stable α2β2 complex. Under conditions that generate $NH_2Y_{730}\bullet$, binding between $Y_{730}NH_2Y$-α2 and wt-β2 is 25-fold tighter ($K_d = 7$ nM) than for wt-α2|wt-β2 and is cooperative; The stabilized species is a compact globular α2β2, consistent with the structure predicted by Uhlin and Eklund's docking model. These results present a structural and biochemical characterization of the active RNR complex "trapped" during turnover, and suggest that stabilization of the α2β2 state may be a regulatory mechanism for protecting the catalytic radical and ensuring the fidelity of its reactivity.

It has been possible to substitute manganese for iron in protein B_2. Its study could provide a model for the native diferrous form of the protein and allows to propose a model for the reduction of the iron center.

Protein B_2 has been crystallized and a refined three-dimensional structure is now available. The large subunit B_1 contains the effector binding sites directing the activity and substrate specificity of the enzyme. The catalytic center is located at the interface between the two subunits, with B_1 contributing the redox active dithiols, in equilibrium with the reduced or oxidized forms of thioredoxin (or glutaredoxin).

The radical is absolutely required for enzyme activity and is thus believed to participate in the activation of the sugar moiety during ribonucleotide reduction. *E. coli* contains an enzyme system that introduces the radical into the protein. This system consists of four proteins and requires reduced pyridine nucleotides, flavin, dithiothreitol and oxygen. Three of these proteins have been obtained in the pure form and characterized as superoxide dismutase, catalase and NAD(P)H: flavin oxidoreductase. The catalase and superoxide dismutase are required to scavenge harmful oxygen radicals formed by the NAD(P)H: flavin oxidoreductase. The fourth protein can be replaced by ferrous ions. This system provides the cell with the ability to regulate the activity of ribonucleotide reductase through the amount of radical present in B_2. Cells devoid, as a result of mutation, of the flavin oxidoreductase are more sensitive to hydroxyurea (see below) during growth, demonstrating the physiological protective function of the protein from the loss of the radical. However, this protein is not essential: *E. coli* cells contain a second tyrosyl radical generating activity, also residing in a flavin reductase, which has been identified as being sulfite reductase (see Chap. 29).

The radical is found buried within the protein. However, it is known to react with a large number of radical scavengers, antioxidants and reductants. This has been the basis for numerous studies on inhibition of ribonucleotide reductase. In particular, hydroxyurea and hydroxylamine are excellent scavengers of the tyrosyl radical and lead to the inactive form of the enzyme. In contrast, the iron center was thought to be chemically inert. Iron is tightly bound to the polypeptide chain, and in order to prepare apoB_2, the apoprotein, extensive dialysis against a strong iron chelator is required. Under anaerobic conditions, substituted hydrazines and hydroxylamines are good reductants of the iron center. The bacterial radical introducing enzyme system has also been shown to function as a protein B_2-diferric reductase and might be the physiological reductant. The formation of ferrous B_2 is a key step during the activation of protein B_2 since the tyrosyl radical can only be generated during reoxidation of the ferrous center by oxygen.

In *E. coli*, the genes for both B_1 and B_2 form a single operon and coordinately transcribed.

Recent genetic evidence involving complementation of *nrd* mutants of *E. coli* suggested the presence *in S. typhimurium* of the genes *nrdE* and *nrdF*, coding for a second ribonucleotide diphosphate reductase of the same type as the one coded by *nrdA* and *nrdB*. These genes are also present in *E. coli* but are, under standard growth conditions, expressed only when present on a plasmid. The amino acid sequences deduced from the corresponding proteins showed a limited identity with other enzymes of the same class, but contained many of their catalytically important residues. The proteins encoded by *nrdE* and *nrdF* (R1E and R2F) are also homodimers. Together they catalyze the reduction of CDP to dCDP, using dithiothreitol or glutaredoxin, but not thioredoxin, as an electron donor. Protein R2F contains an antiferromagnetic coupled dinuclear iron center and a tyrosyl free radical. The *E. coli* and *S. typhimurium* chromosomes have thus maintained the information for a potentially active class I (see below) ribonucleoside diphosphate reductase, whose role in vivo is as yet unknown. Clues to the physiological functions of the new enzyme may come from an analysis of the environmental factors which affect the expression of both types of genes and (or) from a thorough comparison of the allosteric properties of the two enzymes.

Recently, it has been found that the alternative aerobic ribonucleotide reductase of *E. coli*, NrdEF, is a manganese-dependent enzyme that enables cell replication during periods of iron starvation.

Regulation of the Activity of Ribonucleoside Diphosphate Reductase

The large subunit is endowed with a special class of allosteric sites which determine the substrate specificity of the enzyme. Of the four dNTPs, only dCTP does not bind to these sites and its accumulation has therefore no effect on the reductase. As we

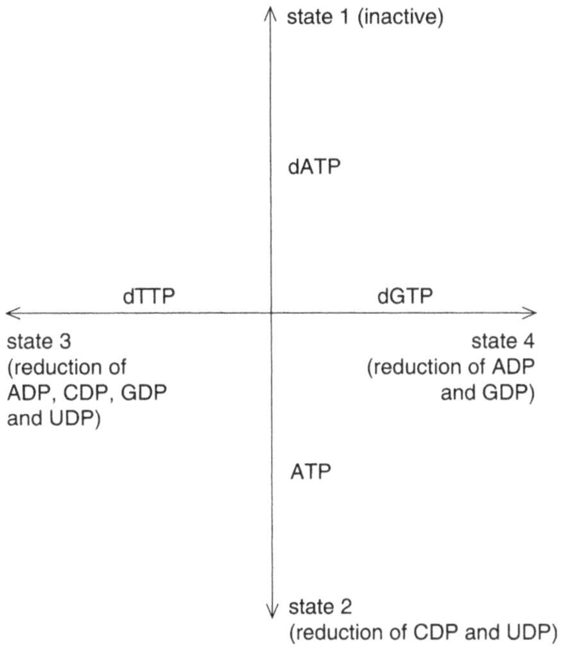

Fig. 34.3 Schematic interpretation of the allosteric effects in the reduction of ribonucleoside diphosphates

shall see, dCTP is an allosteric activator of dCMP deaminase, which provides the dUMP required for dTTP synthesis.

ATP stimulates the reduction of the pyrimidine nucleoside diphosphates, CDP and UDP and has practically no effect on the reduction of purine nucleoside diphosphates; the reduction of the latter is stimulated on the other hand by dTTP and dGTP. dATP strongly inhibit both reductions, its effect being reversed by ATP but not by dTTP.

The current explanation of these interactions, which seem so complicated at first, is as follows: different nucleotides behave as allosteric effectors stabilizing certain states of ribonucleoside diphosphate reductase and so determine the substrate specificity of the enzyme. Thus, ATP stabilizes a state favorable to the reduction of pyrimidine ribonucleotides; dGTP stabilizes a state favorable to the reduction of purine ribonucleotides; dTTP stabilizes a state favorable to both types of reduction, and dATP stabilizes an inactive state. These effects are thought to be one of the foundations of a physiological mechanism for obtaining a balanced supply of the different substrates required for the enzymatic synthesis of DNA (Fig. 34.3).

Figure 34.4 summarizes the different interrelations listed above.

All these studies have been carried out with purified systems. One can surmise however that they are operative in whole cells. This has been shown by experiments where the concentrations of deoxyribonucleotides has been perturbed by the addition of nucleosides or inhibitors, or by mutations in the sites affecting the regulation of the enzyme. The enzyme of certain organisms, like some bacteriophages or herpes viruses lack the allosteric regulation: the loss of this control could be at the

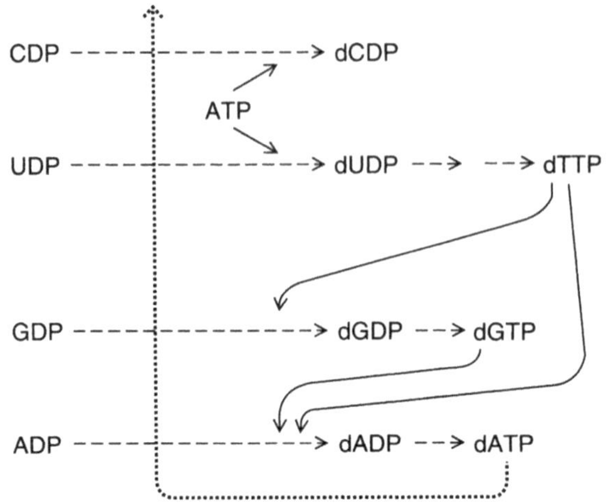

Fig. 34.4 Scheme for the physiological regulation of the synthesis of deoxyribonucleoside diphosphates. The *black arrows* indicate stimulations; the *gray arrow* shows the inhibition by dATP. Synthesis of dCDP and dUDP requires ATP as allosteric effector. Other enzymes transform dUDP to dTTP. DTTP occupies a central position in this scheme: in the presence of ATP, low concentrations of dTTP can cause an additional stimulation of the reduction of pyrimidine ribonucleoside diphosphates, but higher concentrations are inhibitory; this is not shown in the scheme, as the principal effect of dTTP is the stimulation of the synthesis of dGDP. The dGTP formed from this in turn stimulates the formation of dADP. The dADP is transformed to dATP, which is the allosteric inhibitor of the four reduction reactions catalyzed by ribonucleoside diphosphate reductase. The extent of inhibition depends on the concentration of ATP, which reverses the effect of dATP on the reduction of the two pyrimidine ribonucleoside diphosphates (From A. Larsson and P. Reichard, with permission of the *Journal of Biological Chemistry*)

origin of the toxic and mutagenic disequilibria between the different concentrations of deoxyribonucleotides observed during the viral infections.

dCMP Deaminase and Thymidylate Synthase

The above scheme shows that ribonucleoside diphosphate reductase itself, even with its allosteric properties, does not allow the relative reduction of CDP and UDP. It appears that the relative concentrations of dCDP and dTTP is maintained by dCMP deaminase.

dCMP deaminase, also called desoxycytidylate aminohydrolase, catalyzes the deamination of deoxycytidylate (dCMP) to desoxyuridylate (dUMP), thus providing the nucleotide substrate for thymidylate synthase. It has been found in bacteria, sea urchin eggs and in the tissues of warm-blooded animals:

$$dCMP + H_2O \rightarrow dUMP + NH_3$$

It is stimulated by dCTP and inhibited by dTTP. The latter inhibits thus its own production and leads to an accumulation of dCTP, thus restablishing the physiological equilibrium between the two deoxyribonucleotides.

Thymidylate synthase catalyzes the reductive methylation of dUMP into deoxythymidine monophosphate (dTMP) in the sole de novo pathway for this DNA precursor. The cofactor of this reaction is 5–10 methylene tetrahydrofolic acid (CH_2THF), member of a family of compounds encountered whenever a methyl or hydroxymethyl group has to be transferred. It is a dimer of identical subunits of about 35 kDa each, and structural studies have shown that the dimer interface of each monomer is composed of five stranded β-sheets, of which about 29 residues interact with the other monomer. Arg178 and Arg179 are donated from one monomer to the active site of the other, and these residues interact with the phosphate moiety of the substrate dUMP. The enzyme catalyzes the transfer of the methylene group to C5 of dUMP and reduction of the transferred one-carbon unit to a methyl group, utilizing a hydride equivalent provided by tetrahydrofolate. Its tridimensional structure is known: the O4 and N3 of dUMP form hydrogen bonds with asparagine177 of the enzyme. This Asn residue is invariant in all thymidylate synthetases. In the irreversible reaction catalyzed by thymidylate synthase, strong binding of the products would lead to substantial product inhibition. A bound water molecule has been identified in the crystal structure that serves to disfavor binding of the product dTMP. This water molecule is bound to an absolutely conserved tyrosine and is displaced by the C7 methyl group of thymidylate.

If by site-directed mutagenesis, the invariant Asn is replaced by an aspartic acid residue, dCMP is now the preferred substrate (10^6 times the turnover number of that observed with dUMP). Thus Asn177 may be a major determinant of the nucleotide specificity of the enzyme.

In addition to the reaction catalyzed by ThyA, a thymidine kinase dependent salvage of thymidine compounds (*thyK*) from the growth medium can provide the cell with dTMP.

A thymidylate synthase gene from the slime mold *Dictyostelium discioideum* the sequence of which lacks any similarity to *thyA* was used as a bait in manual similarity searches: this gene, named *thyX*, complements the thymidine requirement of an *E. coli thyA* mutant. The corresponding purified protein exhibited a very low activity with methyltetrahydrofolate as the methyl donor. The activity was normal (corresponding to the results of the complementation tests) when FAD and reduced pyridine nucleotides were present. The proposed reaction mechanism is that an electron flow from reduced pyridine nucleotides via flavin pyridine nucleotides is necessary for the activity of ThyX proteins. The nature of the methyl donor remains to be established in the case of ThyX *ThyA* and/or *thyX* genes are present in all completed genome sequences, suggesting that the two proteins are the only thymidylate synthases. The lack of sequence similarity as well as the difference in their enzymatic mechanisms, suggests an independent origin for the two distinct synthases ThyX has a wide phylogenetic distribution, almost exclusively limited to microbial genomes lacking *thyA*.

dUTPase

This enzyme catalyzes the hydrolysis of dUTP to dUMP and PPi and maintains a low concentration of dUTP so that uracil cannot be incorporated into DNA. The dUTPase from *E. coli* is strictly specific for its dUTP substrate, the active site discriminating between nucleotides with respect to the sugar moiety as well as the pyrimidine base. Its tridimensional structure has been determined at a resolution of 1.9 Å. It is a symmetrical homotrimer, not showing the "classical" nucleotide binding domain and whose subunit is composed of 152 amino acid residues.

It has been shown in *S. cerevisiae* that the gene encoding dUTPase is essential for viability. In its absence, cell death occurs from the incorporation of uracil into DNA and the attempted repair of this damage by an excision-repair mechanism. Death can be avoided by providing exogenous thymidylate under conditions where the cell is rendered permeable to this metabolite.

The Ribonucleoside Phosphate Reductase of Other Organisms

The enzyme from *E. coli* is the prototype of the iron-tyrosyl radical enzyme, which is found in mammalian cells, in yeast and in some prokaryotes.

Some other procaryotes e.g. *Brevibacterium ammoniogenes* contain a manganese dependent enzyme, which has an $\alpha\beta2$ structure ($\alpha = 80$ kDa and $\beta = 100$ kDa) instead of $\alpha_2\beta_2$. It appears to possess a binuclear manganese center and to have also a tyrosyl radical.

Another class of reductases (class II) depend on the presence of cobalamin, and *Lactobacillus leichmanii* is characteristic of this group. The enzyme from *L. leichmanii* has been purified to homogeneity. It is a monomer with a molecular weight of 76,000, the activity of which is completely dependent on the presence of 5′-deoxy-5′-adenosylcobalamin, which functions as a radical generator during catalysis and plays the role of the *E. coli* B_2 subunit. The enzyme uses ribonucleoside triphosphates and not diphosphates as substrates. The reaction mechanism of this reductase involves also paramagnetic species intermediates as part of a radical-dithiol system.

The recent isolation of the mammalian and yeast genes coding for the ribonucleoside phosphate reductase has shown that, in addition to the well-documented allosteric regulation, the synthesis of the enzyme is also tightly regulated at the level of transcription. The mRNAs for both subunits are cell-cycle regulated and in yeast, inducible by DNA damage. Furthermore, yeast encodes a large subunit gene that is expressed only after DNA damage. This regulation probably facilitates DNA replication repair processes.

A Ribonucleotide Triphosphate Reductase Reaction in *E. coli* Grown Under Anaerobic Conditions

An additional class (class III) exists which has been detected in the strict anaerobe archaeon *Methanobacterium thermoautotrophicum* and studied in *E. coli* grown under anaerobic conditions. DNA must be obviously synthesized in the absence of oxygen and the enzyme activity that has been characterized in anaerobic *E. coli* is due to a different protein, not coded for by either the *ndrA* or the *ndrB* gene. This enzyme reduces triphosphates and not diphosphates. Its allosteric regulation is not very different from that of the aerobic enzyme. The major difference between the two reductases is the rapid inactivation of the anaerobic enzyme by air and its moderate sensitivity to hydroxyurea, two properties it shares with the *Methanobacterium* enzyme. Since mutants unable to synthesize cobalamin are not inhibited in their anaerobic growth, it is unlikely that a derivative of vitamin B_{12} is involved as a cofactor, as in *L. leichmanii*.

There is strong evidence that in anaerobic *E. coli*, S-adenosylmethionine (SAM) together with a metal (iron) participates in the generation of the radical required for the reduction of the carbon 2′ of the ribosyl moiety of nucleoside triphosphates. The overall reaction occurs in two steps. The first is an activation of the reductase by an auxiliary protein and a boiled extract of *E. coli*, in addition to SAM, this step requires the presence of NADPH, dithiothreitol, and K^+ ions. In the second step, the activated reductase reduces CTP to dCTP with ATP acting as an allosteric effector. During activation, SAM is reductively cleaved to methionine and 5′-deoxyadenosine.

The gene (*nrdD*) for the anaerobic reductase has been cloned and was found to contain a 2,136-nucleotide coding region, corresponding to 712 amino acid residues, and an FNR binding site 228 base pairs upstream of the initiator ATG. The deduced amino acid sequence shows 72 % identity to a gene of coliphage T4, *sunY*, hitherto of unknown function, suggesting that the virus codes for its own anaerobic reductase. The location of an organic free radical formed during activation of the enzyme is proposed to be on Gly 681, since the pentapeptide RVCGY (678–682) is similar to the C-terminal RVSGY of the anaerobically induced pyruvate formate-lyase, where the glycine residue becomes an organic radical during the activation. Exposure to air of the anaerobic reductase having been anaerobically activated in the presence of SAM results in the removal of the 31 amino acid-long sequence downstream of Gly 681.

The apparent molecular weight of the purified anaerobic reductase is ca. 150,000, as estimated by gel exclusion chromatography. Electrophoresis under denaturing conditions separates the protein into an α β band, which seems to be the reductase proper and an α band.

The anaerobic reductase is of special interest from the point of view of evolution and may have been a prerequisite for the transition of an RNA world to a DNA world, at a time when the geochemists tell us that oxygen was scarce. An anaerobic ribonucleotide reductase would then have been an absolute requirement.

The Synthesis of Deoxyribonucleoside Triphosphates from the Diphosphates

The phosphorylation of deoxynucleoside diphosphates to the triphosphates is accomplished by the non-specific nucleoside diphosphate kinase. This NDP kinase has been cloned and sequenced from a number of species, including the prokaryote *Myxococcus xanthus*, the lower eukaryote *Dictyostelium discoideum*, *Drosophila* and mammals showing that this family of enzymes is highly conserved. The X-ray structure of the recombinant NDP kinase from *Dictyostelium* shows that it is a symmetrical hexamer made of identical 17 kDa subunit arranged as three vertical dimers (or three horizontal trimers). The crystallographic analysis of the enzyme with bound ATP and dTDP demonstrates that each subunit carries one nucleotide binding site: the nucleotide is bound of the histidine participating in catalysis.

The origin of dTTP deserves special attention: dCDP is transformed to dCMP by the reversible dCMP kinase. dCMP is then deaminated by dCMP deaminase and dUMP is formed; dUMP is methylated to dTMP by dTMP synthase, an enzyme utilizing 5,10-methylene tetrahydrofolate as the one-carbon donor (*thyA, vide supra*). Thymidylate kinase phosphorylates dTMP to dTDP and nucleoside diphosphokinase catalyzes the synthesis of dTTP.

The sigmoid form of the substrate saturation curves for dCMP deaminases obtained from a great variety of sources, ranging from chicken to bacteria infected with bacteriophage T4, as well as their sensitivity to allosteric inhibition, indicates that dCMP deaminase is an important point of regulation in deoxynucleotide syntheses. The enzymes of every source are inhibited by dTTP and activated by dCTP (hydroxymethyl dCTP for the phage enzyme). When dCMP deaminase is lost by mutation, the cells undergo a decreased DNA synthesis and growth, which is normalized by the addition of pyrimidine deoxynucleosides to the medium (another example of salvage).

Organization of DNA Precursor Synthesis in Eukaryotic Cells

To date, this organization remains unresolved. The existence of a multienzyme complex synthesizing DNA precursors and incorporating them directly into replication forks within the nucleus has been suggested. Countering this hypothesis several studies have characterized thymidylate synthase and ribonucleotide reductase, constituents of the proposed complex as cytoplasmic proteins. Subcellular fractionation of yeast has shown thymidylate synthase is associated with purified nuclei and immunofluorescence analysis suggests that it is situated at the nuclear periphery.

Selected References

Thioredoxin and Glutaredoxin

Holmgren A (1989) J Biol Chem 264:13963–13966
Pille S (1992) Ann Institut Pasteur/Actualités 3:34–50

Ribonucleoside Diphosphate and Triphosphate Reductases

Fontecave M, Nivière V, Mulliez E (1992) Ann Institut Pasteur/Actualités 3:159–165
Gerez C, Fontecave M (1992) Biochemistry 31:780–786
Jordan A, Pontis E, Atta M, Krook M, Gibert I, Barbé J, Reichard P (1994) Proc Natl Acad Sci USA 91:12892–12896
Martin JE, Imlay JA (2011) Mol Microbiol 80:319–334
Minnihan EC, Ando N, Brignole EJ, Olshansky L, Chittuluru J, Asturias FJ, Drennan CL, Nocera DG, Stubbe J (2013) Proc Natl Acad Sci USA 110:3835–3840
Reichard P (1987) Biochemistry 26:3245–3248
Reichard P (1988) Ann Rev Biochem 57:349–374
Rofougaran R, Crona M, Vodnala M, Sjoberg B-M, Hofer A (2008) J Biol Chem 283:35310–35318
Stubbe J (1990) Adv Enzymol 63:349–419
Sun X, Harder J, Krook M, Jörnvall H, Sjöberg B-M, Reichard P (1993) Proc Natl Acad Sci USA 90:577–581
Thelander L, Reichard P (1979) Ann Rev Biochem 48:133–158
Uhlin U, Eklund H (1994) Nature 370:533–539

Thymidylate Kinases

Myllikallio H, Lipowski G, Leduc D, Filee J, Forterre P, Liebl U (2002) Science 297:105–107
Neuhard J, Kelln A (1996) In: Neidhardt FC (ed) *Escherichia coli* and *Salmonella*. American Society for Microbiology, Washington, DC, pp 580–599

Nucleoside Diphosphate Kinase

Dumas C, Lascu I, Moréra S, Glaser P, Fourme R, Wallet V, Lacombe M-L, Véron M, Janin J (1992) EMBO J 11:3203–3208

Tepper AD, Dammann H, Bominaar AA, Véron M (1994) J Biol Chem 269:32175–32180

Chapter 35
Biosynthesis of Some Water-Soluble Vitamins and of Their Coenzyme Forms

Biosynthesis of Thiamin and Cocarboxylase

Thiamin, also known as vitamin B1 can be considered made up of two parts, one being 2-methyl-6-amino-5 hydroxymethyl pyrimidine ("pyrimidine" for the simplicity of presentation) and the other 4-methyl-5-hydroxyethylthiazole ("thiazole").

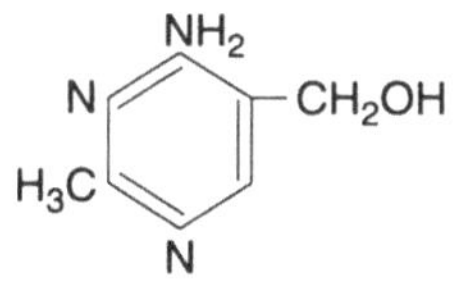

"pyrimidine"

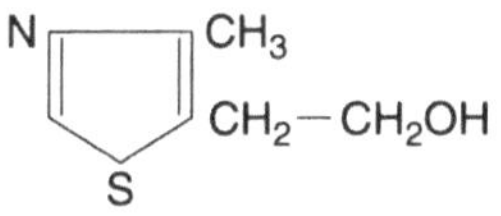

"thiazole"

In *S. typhimurium*, the first five steps in purine biosynthesis also serve as the first steps of the pyrimidine moiety of thiamin (Fig. 35.1).

The condensation of the two parts of the thiamin molecule have been well characterized in yeast and bacteria. In naturally thiamin requiring microorganisms or in auxotrophic mutants, three groups can be distinguished:

1. organisms requiring pyrimidine only and therefore able to synthesize the thiazole moiety;
2. mutants requiring thiazole and therefore able to synthesize the pyrimidine moiety;
3. finally, those requiring preformed thiamin for growth and unable to condense the two parts when they are provided.

The study of these organisms justifies the conclusion that both parts are synthesized separately and then connected. Experiments carried out with cell-free extracts support the following scheme to account for the synthesis of thiamin from its component parts:

G.N. Cohen, *Microbial Biochemistry*,
DOI 10.1007/978-94-017-8908-0_35, © Springer Science+Business Media B.V. 2014

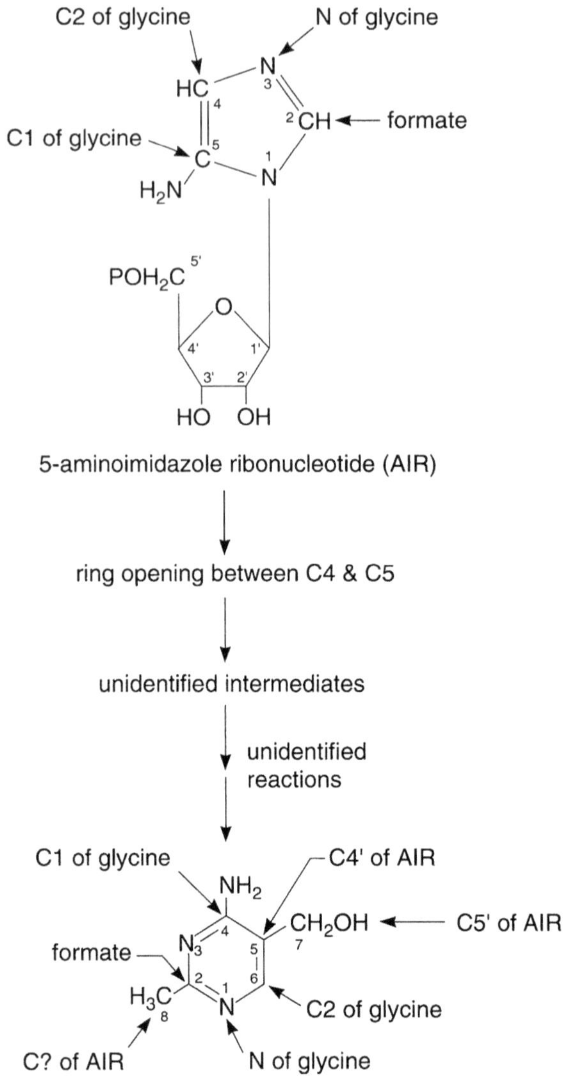

Fig. 35.1 Biosynthesis of the pyrimidine component of thiamin (From G. M. Brown and J. M. Williamson, with permission of the American Society for Microbiology)

$$\text{Pyrimidine} + \text{ATP} \rightarrow \text{Pyrimidine pyrophosphate}$$
$$\text{Thiazole} + \text{ATP} \rightarrow \text{Thiazole monophosphate}$$
$$\text{Pyrimidine pyrophosphate} + \text{Thiazole monophosphate} \rightarrow$$
$$\text{thiamine phosphate} + \text{pyrophosphate}$$
$$\text{Thiamine phosphate} + H_2O \rightarrow \text{Thiamine} + \text{Pi}$$

After several decades of hard and painful work, collaboration between chemists, enzymologists, and structural biologists has finally produced a clear understanding of how prokaryotes produce thiamin. The X-ray crystal structures, biochemical functions and mechanisms of nearly every enzyme in this pathway have been determined. Studies on the formation of the thiazole, the pyrimidine, and thiamin itself, along with a variety of required kinases, are an important contribution to the vast array of biochemical knowledge in the field of vitamin biosynthesis.

The biosynthesis of thiamin pyrophosphate (cocarboxylase, TPP) in prokaryotes is summarized below. The thiazole heterocycle is formed by the convergence of three separate pathways. First, the condensation of glyceraldehyde 3-phosphate and pyruvate, catalyzed by 1-deoxy-D-xylulose 5-phosphate synthase (Dxs), gives 1-deoxy-D-xylulose 5-phosphate (DXP). Next, the sulfur carrier protein ThiS-COO- is converted to its carboxy-terminal thiocarboxylate in reactions catalyzed by ThiF, ThiI, and NifS (ThiF and IscS in *B. subtilis*). Finally, tyrosine (glycine in *B. subtilis*) is converted to dehydroglycine by ThiH (ThiO in B. subtilis). Thiazole synthase (ThiG) catalyzes the complex condensation of ThiS-COSH, dehydroglycine, and DXP to give a thiazole tautomer, which is then aromatized to carboxythiazole phosphate by TenI (*B. subtilis*). Thiazole synthase in *Escherichia coli* is an αβ heterodimer of ThiG and ThiH. ThiH is a tyrosine lyase that cleaves the Cα–Cβ bond of tyrosine, generating *p*-cresol as a by-product, to form dehydroglycine. This reactive intermediate acts as one of three substrates for the thiazole cyclization reaction catalyzed by ThiG. ThiH is a radical *S*-adenosylmethionine (SAM) enzyme that utilizes a [4Fe-4S] cluster to reductively cleave SAM, forming methionine and a 5′-deoxyadenosyl radical. Analysis of the time-dependent formation of the reaction products 5′-deoxyadenosine (DOA) and *p*-cresol has demonstrated catalytic behavior of the tyrosine lyase. The kinetics of product formation showed a pre-steady state burst phase, and the involvement of DOA in product inhibition was identified by the addition of 5′-methylthioadenosine/*S*-adenosylhomocysteine nucleosidase to activity assays. This hydrolyzed the DOA and changed the rate-determining step but, in addition, substantially increased the uncoupled turnover of AdoMet.

Hydroxymethyl pyrimidine phosphate (HMP-P) is formed by a complicated rearrangement reaction of 5-aminoimidazole ribotide (AIR, a molecule we have already encountred in the biosynthesis of the purine ring) catalyzed by ThiC. ThiC (4-amino-5-hydroxymethyl-2-methylpyrimidine phosphate synthase) is a radical *S*-adenosylmethionine enzyme that uses a $[4Fe\text{-}4S]^{+}$ cluster to reductively cleave AdoMet to methionine and a 5′-deoxyadenosyl radical that initiates catalysis. In bacteria, ThiC converts the purine intermediate 5-aminoimidazole ribotide to 4-amino-5-hydroxymethyl-2-methylpyrimidine phosphate, an intermediate of thiamine pyrophosphate biosynthesis. ThiC activity was improved by in situ removal of product 5′-deoxyadenosine. The activity was inhibited by AdoMet metabolites *S*-adenosylhomocysteine, adenosine, 5′-deoxyadenosine, *S*-methyl-5′-thioadenosine, methionine, and homocysteine. Neither adenosine nor *S*-methyl-5′-thioadenosine had been shown to inhibit radical AdoMet enzymes, suggesting that ThiC is distinct from other family members. ThiD then generates hydroxymethyl pyrimidine pyrophosphate. The coupling of the two heterocycles and decarboxylation, catalyzed by thiamin phosphate synthase (ThiE), gives thiamin phosphate. A final phosphorylation, catalyzed by ThiL, completes the biosynthesis of TPP, the biologically active form of the cofactor, which functions in several enzymatic reactions in which it serves as a transient intermediate of the carboxyl group.

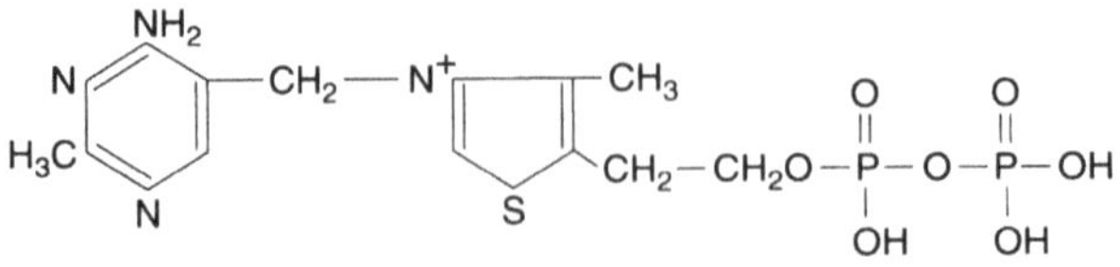

thiamine diphosphate or cocarboxylase

Control of Thiamin Biosynthesis

The regulation of gene expression of most thiamin biosynthetic proteins is controlled at the transcriptional level. Most of the *E. coli* thi genes, are regulated by the presence of thiamin pyrophosphate through interactions with the THI-box riboswitch on mRNA encoding for each operon. The genes *thiI*, *thiL*, *dxs*, and *iscS* are the only thiamin biosynthetic enzymes not regulated by a riboswitch and are not clustered with other thiamin biosynthetic genes. In the case of *thiI*, *dxs*, and *iscS*, this could be because these genes are required for other pathways.

The THI-box consists of a 5′ untranslated region that forms a thiamin-binding site. In gram-positive bacteria, binding of thiamin induces the formation of a Rho-independent transcriptional terminator. In gram-negative bacteria, binding of thiamin masks the Shine-Dalgarno sequence, which is required for the initiation of translation. The thiamin-binding domain of the riboswitch is 1,000-fold more specific for thiamin pyrophosphate than for thiamin monophosphate, with Kd values of 0.1 and 100 μM, respectively, thereby ensuring that only the active form of the cofactor inhibits translation. The fact that thiamin pyrophosphate binds to an mRNA molecule presents an interesting problem in that both molecules carry negative charges on their phosphate groups. The recent crystal structure of a THI-box RNA complexed with thiamin pyrophosphate sheds light on how the binding occurs.

The pathway of thiamin synthesis differs from most pathways we have studied so far, in that whereas the latter can be represented in the form of linear or divergent sequences, the thiamin pathway consists of two converging biosynthetic pathways each of which synthesizes products which do not normally accumulate. In addition, the amounts of the compounds synthesized are several orders of magnitude below those of the amino acids and nucleotides.

The control mechanism found in *S. typhimurium* is summarized in the following figure.

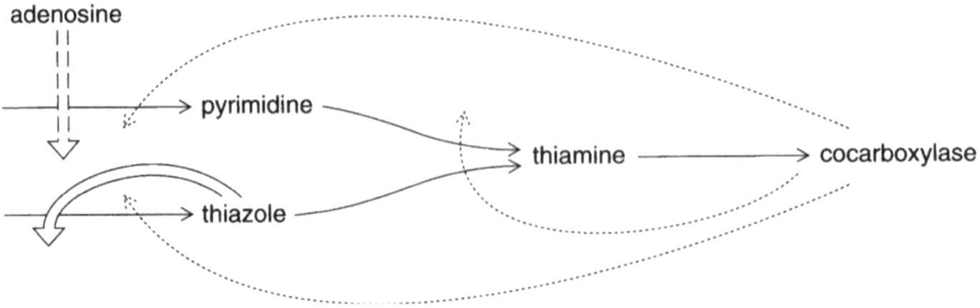

The gray arrows interrupting the biosynthetic reactions represent repressive controls, while the while arrow represents control by feedback inhibition of enzyme activity. The final product, cocarboxylase, can be seen to repress the synthesis of the enzymes involved in both the initial biosyntheses as well as that of the enzyme which condenses the pyrimidine and thiazole. In addition, thiazole exerts an inhibitory effect on an enzyme (not identified as yet) involved in its biosynthesis.

An interesting effect of adenosine is found in this system: preincubation of the organism with this compound depletes its reserve of intracellular thiamin, so causing an enormous derepression of the synthesis of thiamin.

A conserved RNA structure (*thi* box) is involved in regulation of thiamin. This 38-base sequence is highly conserved in the 5′ regions of thiamin biosynthetic and transport genes of Gram-positive and Gram-negative organisms biosynthetic gene expression in bacteria. In *Rhizobium etli*, a deletion analysis of *thiC-lacZ* fusions revealed an unexpected relationship between the degree of repression shown by the deleted derivatives and the length of the *thiC* sequences present in the transcript. Three regions were found to be important for regulation: (i) the *thi* box sequence, which is absolutely necessary for high-level expression of *thiC*; (ii) the region immediately upstream to the translation start codon of *thiC*, which can be folded into a stem-loop structure that would mask the Shine-Dalgarno sequence; and (iii) the proximal part of the coding region of *thiC*, which was shown to contain a putative Rho-independent terminator. A comparative phylogenetic analysis revealed a possible folding of the *thi* box sequence into a hairpin structure composed of a hairpin loop, two helices, and an interior loop. Thiamin regulation of gene expression involves therefore a complex posttranscriptional mechanism, some aspects of which are reminiscent of the attenuation mechanism.

Biosynthesis of Riboflavin

Some lower fungi, like *Eremothecium ashbyii* or *Ashbya gossipyii* synthesize so much riboflavin that the compound crystallizes in the growth medium. It is obvious that such organisms are the tools of choice for studying the synthesis of riboflavin (which is also known as vitamin B2).

Many experiments have shown that addition of purines to the growth medium stimulates the synthesis of riboflavin. Experiments carried out with compounds labelled in different carbon atoms suggest that the purine carbon skeleton, with the exception of carbon 8, is incorporated as such into the isoalloxazine ring of riboflavin. This points to 4,5-diaminouracil or a ribityl derivative of the same, as a possible intermediate; furthermore, this compound has been found to accumulate in a mutant of *Aspergillus nidulans* which requires riboflavin.

It has been since proposed and demonstrated that GTP (1) is actually transformed to N-phosphoribosyl-diaminouracil (2) by a specific cyclohydrolase and that this compound is then deaminated and reduced to 5-amino-2,6-dioxy-4-(5′-phosphoribitylamino) pyrimidine (3).

The deamination and reduction are catalyzed by a single protein, coded by *ribG*. In the N-terminal deaminase domain of *Bacillus subtilis* RibG, structure-based mutational analyses demonstrated that Glu[51] and Lys[79] are essential for the deaminase activity. In the C-terminal reductase domain, the complex structure with the substrate at 2.56-Å resolution unexpectedly showed a ribitylimino intermediate bound at the active site, and hence suggested that the ribosyl reduction occurs through a Schiff base pathway. Lys[151] seems to have evolved to ensure specific recognition of the deaminase product rather than the substrate. Glu[290], instead of the previously proposed Asp[199], would seem to assist in the proton transfer in the reduction reaction.

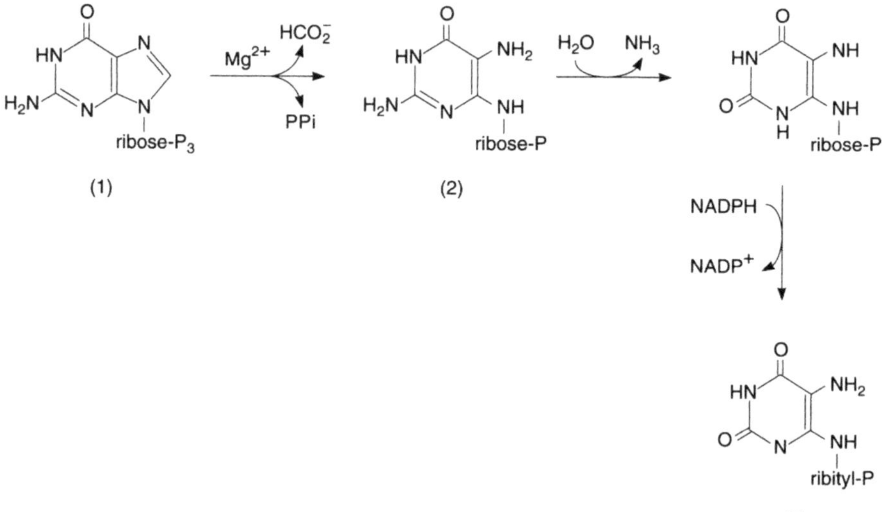

(1) (2) (3)

The next precursor of riboflavin to have been identified with certainty is 6,7-dimethyl-8-ribityllumazine which is synthesized from the previous compound and from carbon atoms 1, 2, 3 and 5 of a pentose. The conversion of the lumazine to riboflavin by riboflavin synthase had been previously demonstrated.

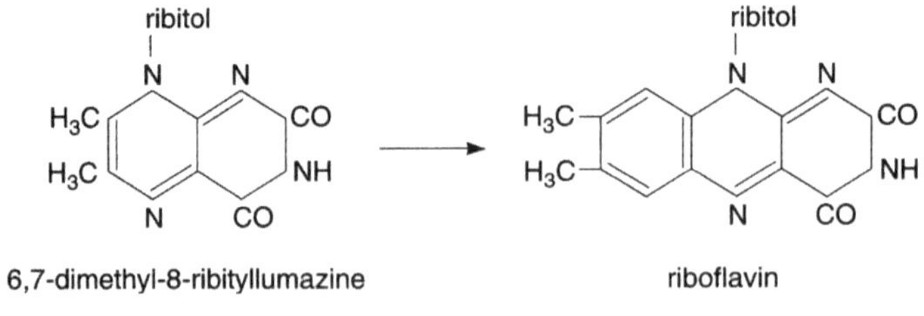

Riboflavin synthase catalyzes in a single enzymatic step the conversion of two molecules of 6,7-dimethyl-8-ribityllumazine to one molecule of riboflavin and one molecule of a ribitylpyrimidine compound. One molecule of the lumazine derivative acts as a donor of a 4-carbon unit and the other is the acceptor.

The structure of the amino-terminal domain of *E. coli* riboflavin synthase has been determined by NMR spectroscopy with riboflavin as a bound ligand. It is functional as a 75 kDa homotrimer, each subunit of which consists of two domains which share very similar sequences and structures. The N-terminal domain (97 residues) forms a 20 kDa homodimer in solution which binds riboflavin with high affinity. The structure features a six-stranded antiparallel β-barrel with a Greek-key fold, both ends of which are closed by an α-helix. One riboflavin molecule is bound per monomer in a site at one end of the barrel which is comprised of elements of both monomers. The structure and ligand binding are similar to that of the FAD binding domains of ferredoxin reductase family proteins. The structure provides insights into the structure of the whole enzyme, the organisation of the functional trimer and the mechanism of riboflavin synthesis. Cys48 from the N-terminal domain is identified as the free cysteine implicated in a nucleophilic role in the synthesis mechanism, while His 102 from the C-terminal domains is also likely to play a key role. Both are invariant in all known riboflavin synthase sequences.

In *B. subtilis*, the lumazine synthase and the riboflavin synthase form a single complex of 60 subunits forming a capsid with icosahedral symmetry enclosing a trimer of α subunits in the central core. The native complex ($\alpha 3\beta 60$) has been crystallized and the structure of the icosahedral capsid has been determined at a resolution of 3.3 Å. The routes for entry of substrates and exit of products are not clear yet.

In different animal and plant tissues and in bacteria, the existence of a flavokinase which catalyzes the phosphorylation of riboflavin at the terminal position of the ribitol chain has been shown:

$$\text{Riboflavin} + \text{ATP} \rightarrow \text{Riboflavin-5-phosphate} + \text{ADP}$$

Riboflavin-5′-phosphate is also called flavin mononucleotide, abbreviated to FMN.

An enzyme has been purified from yeast which catalyzes the formation of flavin adenine dinucleotide (FAD) from FMN and ATP:

$$\text{FMN} + \text{ATP} \rightarrow \text{FAD} + \text{pyrophosphate}$$

In bacteria, riboflavin phosphorylation and subsequent conversion of FMN into FAD are carried out by FAD synthetase, a single bifunctional enzyme. Both reactions require ATP and Mg^{2+}. The N-terminal domain of FAD synthetase appears to be responsible for the adenylyltransferase activity, whereas the C-terminal domain would be in charge of the kinase activity.

The synthesis of flavins is not adjusted as precisely to the physiological needs of bacteria as the synthesis of major metabolites, as is shown by the following results, obtained with *E. coli*:

1. Flavins are produced in large excess during exponential growth; the amount excreted can reach as much as eight times the intracellular concentration.
2. The synthesis is not linked closely to growth and can continue a long time after cessation of growth in *E. coli*; conversely, the growth of lactic acid bacteria which require riboflavin can continue for quite a long time after they have been deprived of the growth factor. It can be shown that flavins have no feedback inhibitory effect

on their synthesis; through control by repression, the amounts of the enzymes necessary for the synthesis of flavins do not vary by more than a factor of 2.

Bacteria would probably not benefit by very strict control of flavins and certain other coenzymes. The wastage of matter and energy involved in the uncontrolled synthesis of coenzymes is very slight indeed. To avoid the small excretion observed (small in absolute terms), it would be necessary to have more rigid control: i.e. lower concentrations of the end product would be required to inhibit its own synthesis (by repression or allosteric inhibition). Lower intracellular concentrations would then be able to affect the synthesis of macromolecules in which enzymes with flavin coenzymes are involved at different points. To give an order of magnitude, the internal concentration of free flavins in *E. coli* is already lower than 4×10^{-6} M.

One case has been analyzed in some detail in *B. subtilis*, where the genes coding for the enzymes catalyzing the reactions from guanosine-5′-triphosphate to riboflavin are organized in a single operon. Expression of the operon is under the control of the *ribC* gene, coding for an aporepressor whose effectors are riboflavin, FMN and FAD. Several *cis*-dominant mutations have been obtained in a region, called *ribO*, which could be the repressor binding site. These mutations, like the *ribC* mutations are expressed as riboflavin overproducers, resulting in accumulation in the growth medium (the highest concentration observed is of the order of 20 mg/l).

Biosynthesis of Nicotinamide, NAD⁺ and NADP⁺

The details of the biosynthesis of the pyridine nucleotides in procaryotes have been mainly worked out with *E. coli*. Some 40 years ago, it was shown that glycerol and a dicarboxylic acid could be the precursors of the nicotinamide moiety of the pyridine nucleotides.

In fact, the actual precursors are iminoaspartate and dihydroxyacetone phosphate (DHAP).

In the first committed step of nicotinamide biosynthesis, the iminoaspartate is formed by aspartate oxidase, encoded by *nadB*.

$$\text{L -aspartate} + \text{FAD} \rightarrow \text{Iminoaspartate} + \text{FADH}_2$$

The reduced form of FAD donates its electrons to oxygen. Iminoaspartate is an unstable intermediate which would spontaneously give rise to oxaloacetate and ammonia in the absence of conversion to quinolinic acid, the next intermediate in the pathway. Quinolinate synthase, coded for by *nadA* condenses iminoaspartate and DHAP, according to

$$\text{iminoaspartate} + \text{DHAP} \rightarrow \text{quinolinate} + \text{Pi} + 2\ H_2O$$

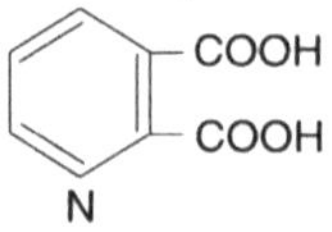

quinolinic acid

nadC codes for a quinolinate phosphoribosyltransferase which catalyzes the reaction of quinolinic acid with phosphoribosylpyrophosphate to yield nicotinic acid mononucleotide (deamido-NAD^+)

$$\text{Quinolinic acid} + \text{PRPP} \xrightarrow{Mg^{++}} \text{deamido} - NAD^+ + \text{PPi} + CO_2$$

NAD^+ synthetase (*nadE*) adds an amide group to the nicotinic acid moiety of deamido-NAD^+.

$$\text{Deamido-}NAD^+ + \text{ATP} + NH_3 \xrightarrow{Mg^{++}} NAD^+ + \text{PPi} + \text{AMP}$$

which completes the synthesis of NAD^+. It seems that an additional subunit is necessary to produce glutamine-dependent NAD^+ synthetase activity.

The synthesis of $NADP^+$ is catalyzed by NAD^+ kinase which transfers a phosphate group from ATP to carbon 2 of the ribose moiety adjacent to adenine in NAD^+. The structural formulas of NAD^+ and $NADP^+$ are set out below.

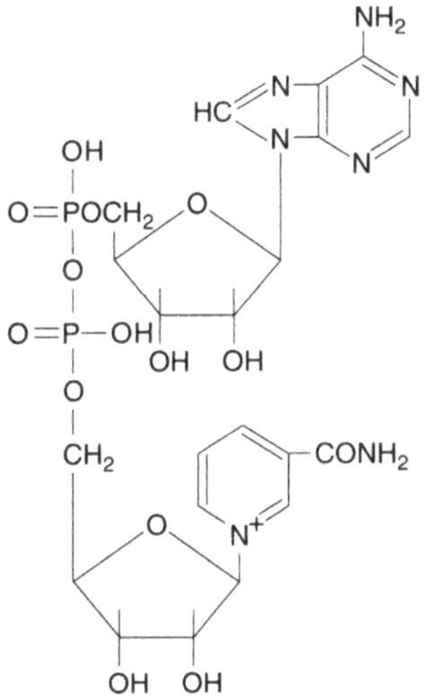

nicotinamide adenine dinucleotide (NAD^+)

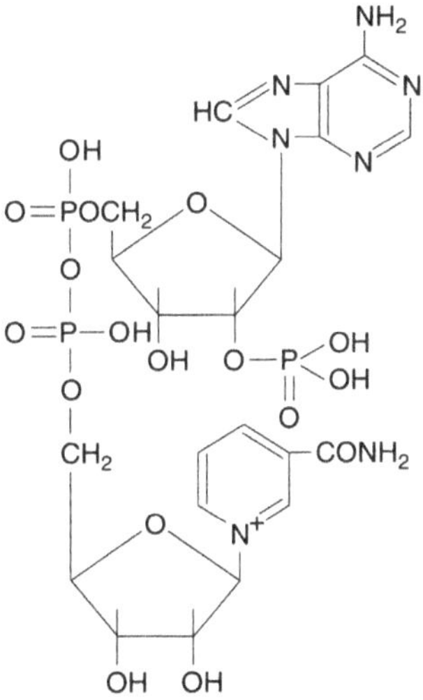

nicotinamide adenine dinucleotide phosphate ($NADP^+$)

Bacteria which, as a result of mutation, or naturally occurring ones often require nicotinic acid or nicotinamide for their growth. These compounds are transformed through specific phosphoribosyl transferases which in the case of nicotinic acid yield deamido-NAD^+ (*pncB*) and in the case of nicotinamide, nicotinamide mononucleotide which is the substrate of an adenylyltransferase converting it to NAD^+.

The pyridine nucleotide cycle is a network of salvage and recycling routes maintaining homeostasis of NAD(P) cofactor pool in the cell. Nicotinamide mononucleotide (NMN) deamidase), one of the key enzymes of the bacterial pyridine nucleotide cycle, was originally described in Enterobacteria, but the corresponding gene eluded identification for over 30 years. A genomics-based reconstruction of NAD metabolism across hundreds of bacterial species suggested that NMN deamidase reaction is the only possible way of nicotinamide salvage in the marine bacterium *Shewanella oneidensis*. This prediction was verified via purification of native NMN deamidase from *S. oneidensis* followed by the identification of the respective gene, termed *pncC*. Enzymatic characterization of the PncC protein, as well as phenotype analysis of deletion mutants, confirmed its proposed biochemical and physiological function in *S. oneidensis*. A comparative analysis at the level of sequence and three-dimensional structure, which is available for one of the PncC family member, shows no homology with any previously described amidohydrolases.

It is interesting to note that deamido-NAD^+ reacts with 5,6-dimethylbenzimidazole to yield nicotinic acid and α-ribazole, one of the components of cobalamin (see under biosynthesis of vitamin B_{12}).

In animals and fungi, the precursor of nicotinic acid is tryptophan. Tryptophan pyrrolase, a heme enzyme, catalyzes the opening of the pyrrole ring by an oxygenation reaction which adds molecular oxygen with the production of N-formylkynurenine:

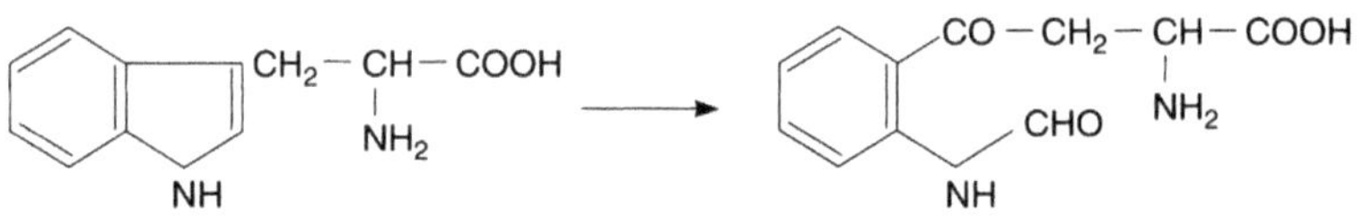

L-tryptophan N-formylkynurenine

The compound N-formylkynurenine loses its formyl group by the action of a specific formamidase and the kynurenin obtained is hydroxylated by kynurenine 3-hydroxylase, the oxygen atom of the hydroxyl group deriving from molecular oxygen and not from water. For an unknown reason, this hydroxylase requires NADPH to function.

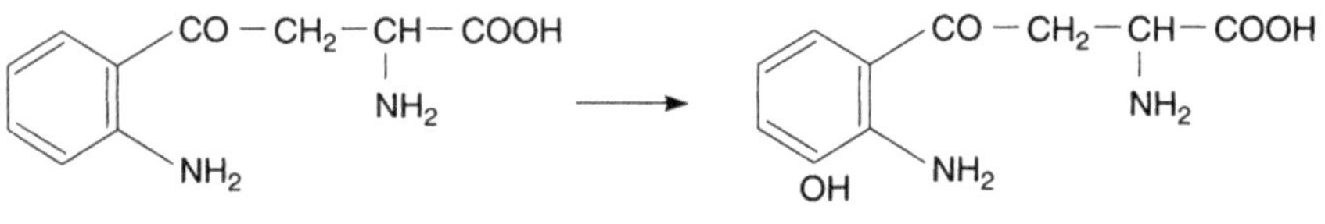

kynurenine 3-hydroxykynurenine

Kynureninase is an enzyme with pyridoxal phosphate as cofactor; it hydrolyzes 3-hydroxykynurenine to 3-hydroxyanthranilic acid and alanine. The benzene ring of 3-hydroxyanthranilic acid is then opened by the action of an oxygenase which has been highly purified from beef liver and is activated by ferrous ions and sulfhydryl compounds. The oxidation product is 1-amino-4-formyl-1,3-butadiene-1,2-dicarboxylic acid:

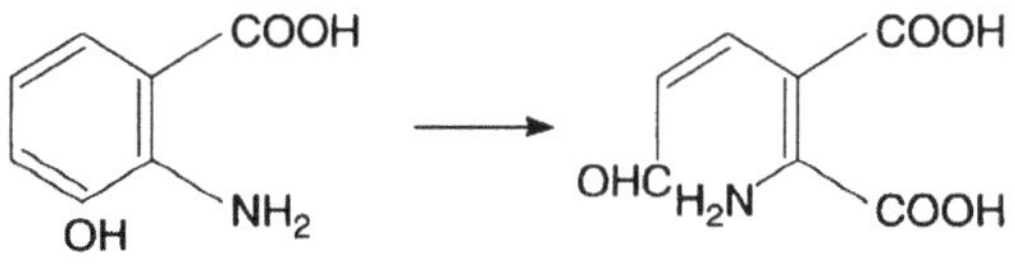

3-hydroxyanthranilic acid   1-amino-4-formyl-1,3-butadiene-1,2-dicarboxylic acid

Specific enzymes carry out the cyclization of this compound to form quinolinic acid, which can either be decarboxylated to nicotinic acid or react with PRPP to yield deamido-NAD$^+$ which is then subjected to the same reactions as those which have been described above for *E. coli.*

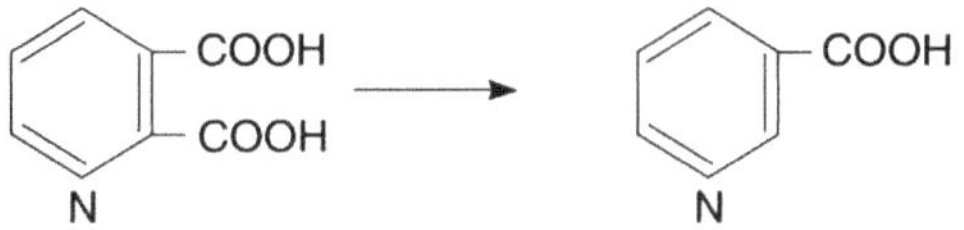

quinolinic acid   nicotinic acid

Regulation of the Biosynthesis of Nicotinamide and Its Derivatives

The *nad* genes are scattered on the *E. coli* and *S. typhimurium* chromosomes and thus do not correspond to a single unit of transcription (operon).

Aspartate oxidase (*nadB*) is under the repressive control of the product of the *nadR* gene with NAD as the probable corepressor molecule. Moreover, L-aspartate oxidase is inhibited by the end-products NAD and NADH, but not by NADP and NADPH. Thus, aspartate oxidase is, both at the transcriptional and activity level, the primary site of regulation of the pathway. In addition to *nadB*, the *nadA* gene, specifying quinolinate synthetase is also under the repressive control of *nadR*.

NAD$^+$ and the ADP-Ribosylation of Proteins

The secreted form of diphtheria toxin is a single polypeptide chain containing two disulfide bridges and no free sulfhydryls. Trypsin cleaves the intact toxin at an arginine residue located within one of the disulfide loops and generates two large fragments A and B (Mr = 21,000 and 39,000 kD, respectively) still linked by a disulfide bridge. This nicked form is still toxic but devoid of enzymatic activity. After reduction, fragment A is now endowed with ADP-ribosyltransferase activity; the role of fragment B is to facilitate attachment and entry of fragment A into the eukaryotic target of the toxin.

Table 35.1 Modification caused by ADP-ribosylation

Toxin	Modified amino acid
Cholera	Arginine
Pertussis	Cysteine
Botulinum	Asparagine
Diphtheria	Diphthamide[a]

[a]An unusual amino acid residue of the human elongation factor EF2,formed by posttranslational modification of a histidine residue

The donor of the ADP-ribosyl moiety is NAD^+, according to

$$\text{N - ribose - P - P - ribose - A + Protein} \rightarrow \text{Protein - ribose - P - P - ribose - A + N + H}^+$$
$$(\text{N = nicotinamide; A = adenine})$$

In the case of diphtheria toxin, the protein target is the eukaryotic elongation factor EF2 and the effect of the toxin is inhibition of protein synthesis.

After diphtheria toxin, other cases of protein ADP-ribosylation have been found in other toxins which play a role in disease. The cholera toxin is an ADP-ribosyltransferase that ADP-ribosylates an arginine residue of G proteins, causing massive fluid secretion from the lining of the small intestine and resulting in life-threatening diarrhea. *Clostridium perfringens* iota-toxin mono-ADP ribosylates Arg177 of actin in eukaryotes, leading to cytoskeletal disorganization and cell death. Several ADP-ribosylating toxins have been cloned and some of the crystal structures have been solved. The toxins have significant similarities in structure with some identity in the amino acids which participate in the catalytic site of ADP-ribosylation. They differ however in the amino acids which they modify (Table 35.1).

ADP-ribosylation is a reversible modification of proteins. We have seen that an ADP-ribosylation cycle regulates dinitrogen reductase in *Rhodospirillum rubrum*. (As in the case of the reversible covalent adenylylation of glutamine synthetase, the inactivating and the reactivating reaction are different.)

Biosynthesis of Para-aminobenzoic Acid, of Folic Acid and Its Derivatives

Examination of the structural formula of folic acid shows that this compound is constituted of three elements, *p*-aminobenzoic acid (*p*-AB), 2-amino-4-hydroxypteridine and glutamic acid. The combination of pteridine and *p*-AB through the methylene bridge, numbered 9 on the formula, is called pteroic acid. Chorismic acid, the common precursor of the three aromatic amino acids, is also the precursor of *p*-AB.

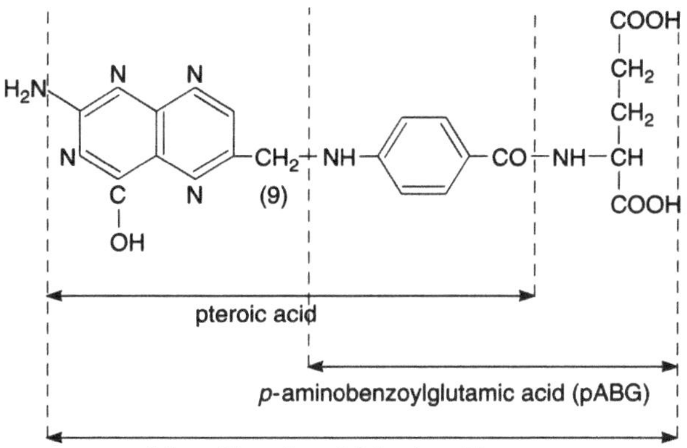

As anthranilate synthase, *p*-aminobenzoate synthase uses glutamine and chorismate as substrates. It produces also glutamate and pyruvate but instead of anthranilate (*o*-aminobenzoate), its specific product is *p*-aminobenzoate. The two aromatic products differ only by the disposition of the amino substituent on the benzene ring.

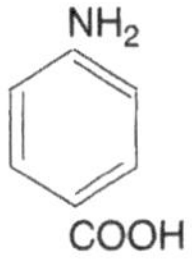

p-aminobenzoic acid

p-Aminobenzoate (pAB) is one of seven aromatic products derived from chorismate in *E. coli*. It is a precurseur of folic acid. It is synthesized by enzymes that catalyze two reactions analogous to the two reactions of anthranilate synthase. ADC (4-amino-4-deoxychorismate) synthase is encoded by *pabA* and *pabB*, and is responsible for the amination of chorismate at position 4, with the concomitant loss of the hydroxyl group. *pabB* encodes the chorismate amination subunit and *pabA* the glutamine amidotransferase. ADC lyase, encoded by *pabC*, catalyzes the elimination of pyruvate from ADC and the aromatization of the resulting product. It is a homodimer of 29.7 kDa subunits and contains a pyridoxal phosphate cofactor.

Prephenic acid and *p*-hydroxybenzoic acid are inactive as substrates for the cell-free system which synthesizes p-AB from chorismate.

Folic acid is a member of a great family of natural pteridines. Before any enzymological work had been done on its biosynthesis, indications had accumulated according to which one purine was a precursor of pteridines. These indications came from experiments done with whole organisms where the incorporation in pteridines of various 14C possible precursors was followed. It was thus observed that guanosine or a nucleotide of guanine was the precursor of the pteridine part of folic acid and that during its incorporation, carbon 8 of guanine was lost, as in the case of the biosynthesis of riboflavin. These observations led to propose a biosynthetic pathway that had already been put forward for the synthesis of the insect pteridine pigments. These hypotheses have now been confirmed by an extensive enzymological work.

GTP is opened by a cyclohydrolase which catalyzes four successive reactions, including the opening of the 5-membered heterocycle of GTP, the loss of formate (corresponding to carbon 8 of guanine), an Amadori rearrangement and a ring closure, the final product being dihydroneopterin triphosphate.

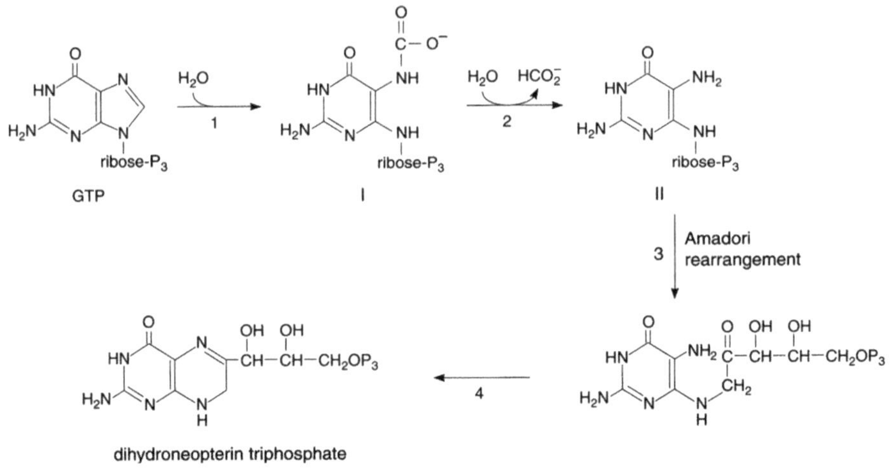

The GTP cyclohydrolase has been purified to homogeneity. It has a molecular weight of 210,000 and is composed of eight identical subunits.

A specific enzyme removes the two distal phosphoryl groups of dihydroneopterin triphosphate, the last phosphate being removed by a non-specific phosphomonoesterase. Thus, dihydroneopterin is obtained.

The next step is catalyzed by dihydroneopterin aldolase (DHNA) and leads to 2-amino-4-oxy-6-hydroxymethyl-7,8-dihydropteridine, abbreviated as dihydropterin-CH_2OH.

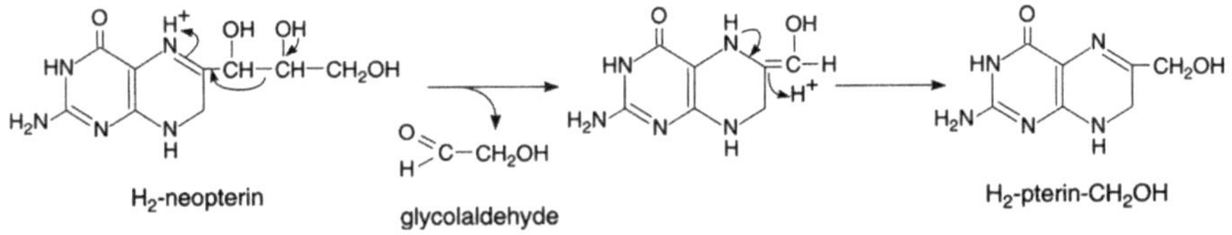

The next enzyme (hydroxymethyldihydropterin pyrophosphokinase, PPPK) utilizes ATP in the presence of Mg^{++} to synthesize the pyrophosphoryl ester dihydropterin-CH2OPP. The enzyme has been purified to homogeneity from *E. coli* and is a monomer of molecular weight 25,000, whose sequence of the 28 first amino acids has been determined. In *Streptococcus pneumoniae*, DHNA and PPPK activities are carried by a bifunctional protein. Dihydropterin-CH2OPP reacts then with *p*-AB with the elimination of pyrophosphate and the formation of dihydropteroate, a reaction catalyzed by dihydropteroate synthase in the presence of Mg^{++}. The enzyme from *E. coli* has been purified to homogeneity, has an apparent molecular mass of 52 kDa and is a dimer. Although this enzyme is capable of using *p-aminobenzoylglutamate* (*p*-ABG) instead of *p*-AB, there are many facts to indicate that *p*-ABG is not a biosynthetic intermediate.

Dihydrofolate synthase couples dihydropteroic acid and glutamic acid in the presence of ATP to produce dihydrofolate.

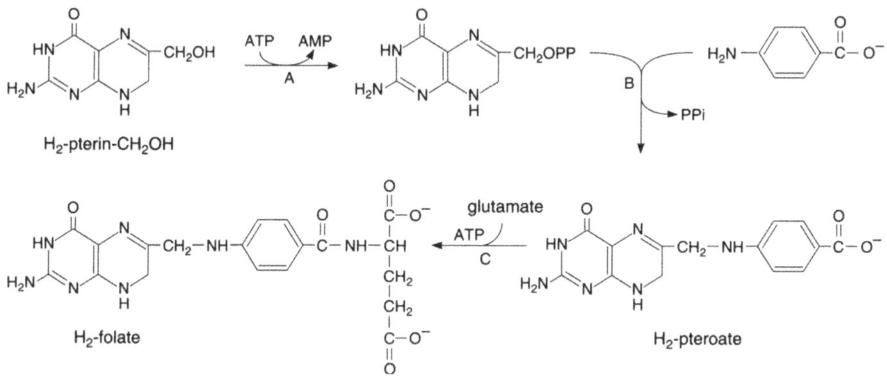

Folic acid is not an intermediate in the synthesis of tetrahydrofolic acid. It is isolated in the oxidized form from natural sources, because of the ease of oxidation of the reduced species. Fortunately for the economy of the cell, there are enzymes to reduce folic acid, the dihydrofolate and tetrahydrofolate dehydrogenases, which use NADH or NADPH.

The tetrahydrofolate dehydrogenase is also called dihydrofolate reductase: it is a monomeric protein of about 180 residues. High-resolution crystal structures have been reported for the enzyme from *E. coli* and *Lactobacillus casei*, as well as for the enzymes from chicken, mouse and man. In each of these structures, the amino and carboxyl termini are located at the protein surface and in adjacent positions not expected on a purely random basis. The structure of several binary complexes with synthetic antifolate drugs or with $NADP^+$, and of a ternary complex with folate and $NADP^+$ has also been determined, so that the catalytic mechanism is known to considerable stereochemical detail.

Other derivatives of folic acid are known: we have already encountered N^5-methyltetrahydrofolic acid and N^5, N^{10}-methylenetetrahydrofolic acid, in connection with the biosynthesis of glycine and methionine where these compounds act as carriers of one-carbon units. Also known are other kinds of tetrahydrofolate derivatives, such as the one used in the transfer of formimino groups (–CH = NH), or such as the 4-formyl, 10-formyl, 10-hydroxymethyl, 5,10-methenyl.

Other naturally occurring forms are the triglutamyl and heptaglutamyl forms of folic acid. In these, the glutamyl groups are linked through their γ-carboxyl groups.

Sulfonamides, which were known since the pioneer work of D. D. Woods in 1940 to be antagonists for the growth of *p*-AB requiring organisms, are competitive inhibitors of *p*-AB in the reaction catalyzed by dihydropteroate synthase.

Biosynthesis of Vitamin B6 Pyridoxine, and of Its Derivatives, Pyridoxal, Pyridoxamine and Pyridoxal Phosphate

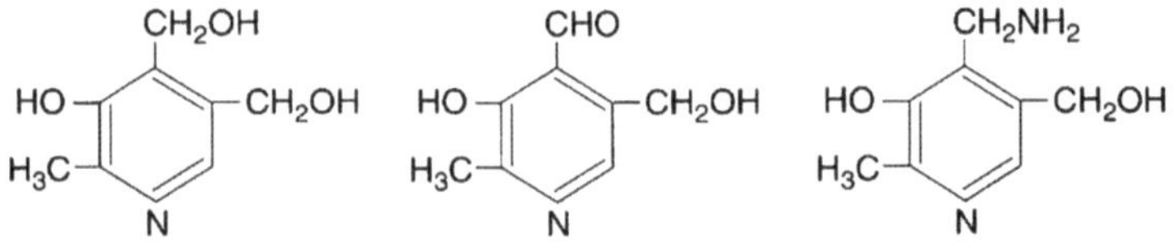

As all these three forms of vitamin B6 are equally active for the growth of vitamin B6-requiring organisms, it is to be expected that they are readily interconvertible. Pyridoxal phosphokinase, which catalyzes the reaction

$$\text{Pyridoxal} + \text{ATP} \rightarrow \text{Pyridoxal phosphate} + \text{ADP}$$

also catalyzes the phosphorylation of pyridoxine and pyridoxamine. Pyridoxine phosphate and pyridoxamine phosphate are transformed to pyridoxal phosphate respectively by a specific transaminase and oxidase.

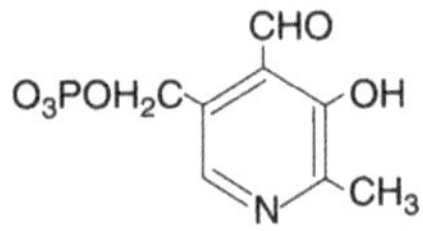

pyridoxal phosphate

The study of the biosynthesis of this group of compounds had not yielded until recently unambiguous data as to its obligatory precursors. This is partly due to the small amounts of pyridoxine derivatives in the cell, but mainly because pyridoxine-less mutants are not always mutants that lack a gene involved in pyridoxine biosynthesis. As an example, it suffices to mention a mutant which grows on either pyridoxine or threonine, due to the fact that its threonine synthase, a pyridoxal phosphate dependent enzyme, has a greatly diminished affinity for its cofactor, and not to the fact that threonine is a precursor of pyridoxine. The literature contains many such examples. Approximately 250 genuine mutants (*pdx*) have been genetically analyzed and fall into five unlinked complementation groups on the *E. coli* chromosome. Their study did not allow until recently to reach a definite conclusion.

Pyridoxine is synthesized in *E. coli* and many bacteria by a branched pathway that leads from the precursor D-1-deoxyxylulose and 4-hydroxy-L-threonine. The pathway leading to D-1-deoxyxylulose results from the condensation of pyruvate and glyceraldehyde 3-phosphate. A specific pathway leading from D-erythrose-4-phosphate, the precursor of aromatic amino acids, (see Chap. 31) to 4-hydroxy-L-threonine has been proposed. Some of the predictions based on this pathway have been confirmed: genetic evidence showed that the products of the *pdxB* and *serC* (coding for phosphoserine aminotransferase) must mediate steps in this branch of the pyridoxine biosynthetic pathway. Enzymological experiments confirmed that *pdxB* encodes a 3-hydroxyacid dehydrogenase that uses 4-phosphoerythronate as its substrate. Furthermore, the immediate precursor, 4-hydroxy-L-threonine meets the pyridoxine requirement of *pdxB* mutants.

A *tktA tktB* double mutant lacking the two transketolases (responsible for sedoheptulose-7-P synthesis) of *E. coli* requires pyridoxine as well as aromatic amino acids for growth. The pyridoxine requirement can be met by either glycolaldehyde or 4-hydroxy-L-threonine. All these results are summarized by the following pathway:

a) pyruvate + glyceraldehyde 3-phosphate $\xrightarrow{tktA}$ deoxyxylulose 5-phosphate

b) xylulose 5-P + ribose 5-P5 $\xrightarrow[tktB]{}$ glyceradehyde-3-P + sedoheptulose 7P

Sedoheptulose 7-P + glyceraldehyde 3-P $\xrightarrow[\text{or } talB]{talA}$ erythrose 4-P + fructose 6-P

D-erythrose-4-P + NAD^+ $\xrightarrow{gapB}$ 4-phosphoerythronate + NADH

4-phosphoerythronate+NAD
↓ *pdxB*
3-hydroxy-4-phosphohydroxy-α-ketobutyrate+NADH

3-hydroxy-4-phosphohydroxy-α-ketobutyrate + glutamate $\xrightarrow{serC(pdxC)}$ 4-phospho-hydroxy-L- threonine + -ketoglutarate

4-phospho-hydroxy-L-threonine + NAD $\xrightarrow{pdxA}$ 1-amino-propan2-one-3-phosphate + NADH + CO_2

c) 1-amino-propan-2-one-3-P + D-1-deoxyxylulose-5-P $\xrightarrow{pdxJ}$ pyridoxine-5-P

pyridoxine-5′-P + O_2 $\xrightarrow{pdxH}$ pyridoxal-5-P + H_2O_2

Pyridoxal 5′-phosphate has another biosynthetic route, which does not coexist in any organism with the above one. Two proteins, neither of which shows similarity with any of the *E. coli* genes, known as PdxS and PdxT, together form a PLP synthase in plants, fungi, archaea, and some eubacteria. PLP synthase is a heteromeric glutamine amidotransferase in which PdxT produces ammonia from glutamine and PdxS combines ammonia with ribulose 5-phosphate, glyceraldehyde-3-phosphate, to form PLP. The structure of the PLP synthase complex (PdxS)PdxT) from *Thermotoga maritima* has been solved at 2.9 Å resolution. This complex consists of a core of 12 PdXS monomers with 12 non interacting monomers attached to the core (Fig. 35.2).

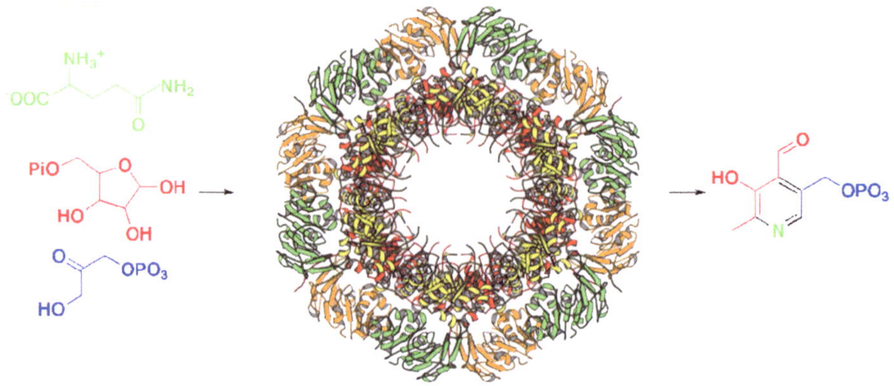

Fig. 35.2 *Thermotoga maritima* PLP synthase (With permission of the American Chemical Society and from Dr. Steven Ealick)

Biosynthesis of Biotin, Biotin CO_2, and Biocytin

Biotin can be thought of formally as a thiophen ring with an n-pentanoyl side-chain and a urea moiety:

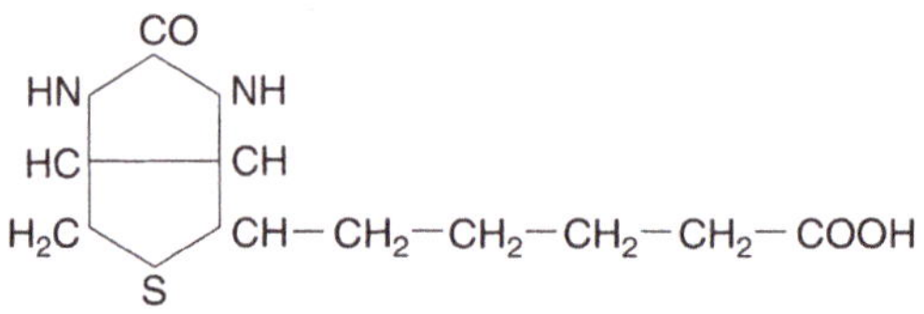

biotin

Studies carried out some 60 years ago had shown that certain biotin requiring mycobacteria could utilize pimelic acid in its place:

$$\mathrm{HOOC-CH_2-CH_2-CH_2-CH_2-CH_2-COOH}$$
$$\text{pimelic acid}$$

In *Escherichia coli*, pimelate is made by bypassing the strict specificity of the fatty acid synthetic pathway. BioC methylates the free carboxyl of a malonyl thioester, which replaces the usual acetyl thioester primer. This atypical primer is transformed to pimeloyl-acyl carrier protein (ACP) methyl ester by two cycles of fatty acid synthesis. The question is, what stops this product from undergoing further elongation? Although BioH readily cleaves this product in vitro, the enzyme is non-specific, which made assignment of its physiological substrate problematical, especially because another enzyme, BioF, could also perform this gatekeeping function. Use of the structure of a 2.05-Å resolution cocrystal structure of a complex of BioH with pimeloyl-ACP methyl ester demonstrates that BioH is the gatekeeper and that its physiological substrate is pimeloyl-ACP methyl ester. The pimeloyl CoA synthase from *B. sphaericus* has been purified to homogeneity, after expression of the corresponding gene in *E. coli*. The synthesis occurs with a concomitant hydrolysis of ATP to AMP. The enzyme is a homodimer with a 28,000 Mr subunit. The next step implies a condensation of pimeloyl CoA with L-alanine, catalyzed by a pyridoxal phosphate enzyme called for short KAP-synthetase:

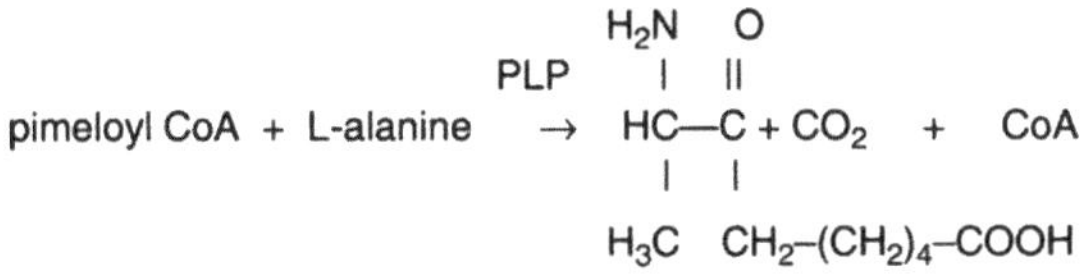

7-keto-8-aminopelargonate (7-KAP)

7-KAP had been observed in the culture media of various microorganisms and biochemically characterized as an open chain structure devoid of sulfur, before having been chemically identified. Its synthesis was demonstrated in cell-free extracts of an *E. coli* mutant auxotrophic for biotin and it was established that the sole reactants were pimeloyl Coa, L-alanine and pyridoxal phosphate (PLP). The molecular weight of the purified enzyme has been obtained by filtration studies and is of 45,000. Genetic complementation studies indicate that it is composed of two or more identical subunits.

The next reaction is a transaminase having as substrates 7-KAP and S-adenosylmethionine (SAM) and leading to 7,8-diaminopelargonic acid (DAPA).

$$\text{7 - KAP} + \text{SAM} \xrightarrow{\text{PLP}} \text{DAPA} + \text{S - adenosyl - 2 - oxo - 4 methylbutyrate}$$

It has been suggested by du Vigneaud as early as 1942 that DAPA could be a precursor of biotin; he had observed that this compound had an activity equivalent to 20 % of that of biotin for an auxotrophic yeast requiring the vitamin for growth. Pontecorvo had made similar observations with *Aspergillus nidulans*. But these findings did not retain the attention and DAPA was accepted as an obligatory intermediate only when it was isolated from the culture medium of a biotin-requiring auxotroph. In the first in vitro experiments, it was established that, in addition to 7-KAP and PLP, methionine was required as the donor of the amino group. However, the methionine requirement was due to the test and it was later shown that SAM was the true amino donor. DAPA synthetase has been purified to near homogeneity and the kinetics of the reaction has been thoroughly analyzed. Its average molecular weight is 94,000 and it is composed of two apparently identical subunits as shown by genetic studies which show intragenic complementation among the *bioA* mutants.

Another precursor of biotin had been discovered in the early 1940s, called dethiobiotin,

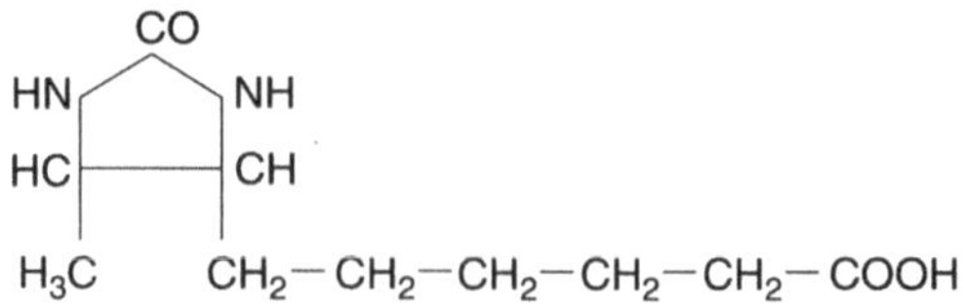

dethiobiotin

where the pimelic acid can be recognized. The precursor-product relationship of DAPA and dethiobiotin has been established through the genetic analysis of a great number of *E. coli* mutants auxotrophic for biotin. However, it was only later that this relationship was established on a biochemical basis when dethiobiotin synthetase, responsible for the transformation, was shown to require ATP and bicarbonate in addition to DAPA. It was found that if ^{14}C bicarbonate was used, the radioactivity was found in the carbonyl of the ureido group of dethiobiotin. A plausible mechanism for the reaction has been proposed:

$$\text{DAPA} + \text{ATP} + CO_2 \rightarrow \text{Dethiobiotin} + \text{ADP} + \text{Pi}$$

Since du Vigneaud and his colleagues demonstrated in 1942 that dethiobiotin was as efficient as biotin to allow the growth of *S. cerevisiae*, a great number of studies were made in vain on the postulated reaction

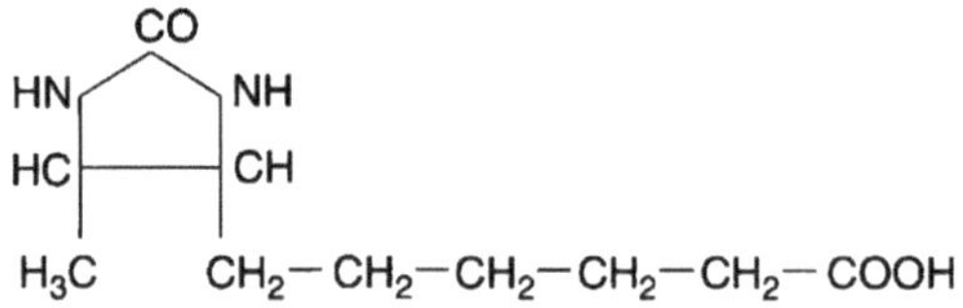

dethiobiotin → biotin

in whole cells of various organisms, due to the absence of a cell-free system able to carry the reaction. The mechanism of incorporation of sulfur was still an open question until 1992, when Ifuku and co-workers observed for the first time the reaction in crude cell-free extracts of *E. coli* overexpressing the *bioB* gene. The *bioB* gene product has been purified in 1994. It runs on native electrophoresis gels with a mobility corresponding to a protein of 82 kD, suggesting that it is a homodimer since the sequence of *bioB* predicts a polypeptide chain of 39 kD. Solutions of the protein are red and have an absorbance spectrum characteristic of proteins with one (2Fe-2S) cluster per monomer. It converts dethiobiotin to biotin in the presence of NADPH, S-adenosylmethionine(SAM), Fe^{2+} or Fe^{3+} ions and additional unidentified factors from the *E. coli* crude extracts. The sulfur donor is not SAM, but probably cysteine or a derivative. The absolute requirement for SAM, for a Fe-S cluster, for NADPH and for FAD suggests that biotin synthase belongs to a family of enzymes which uses SAM as a source of deoxyadenosyl radical with the possible relay of a protein radical, as in the case of anaerobic ribonucleotide reductase, pyruvate-formate lyase or lysine 2,3 aminomutase. All these enzymes catalyze an homolytic cleavage of a C–H bond.

Acetyl CoA carboxylase and other carboxylases contain biotin. The activation of CO_2 necessary for the reaction results from the carboxylation of biotin to form an energy-rich compound, biotin ~ CO_2.

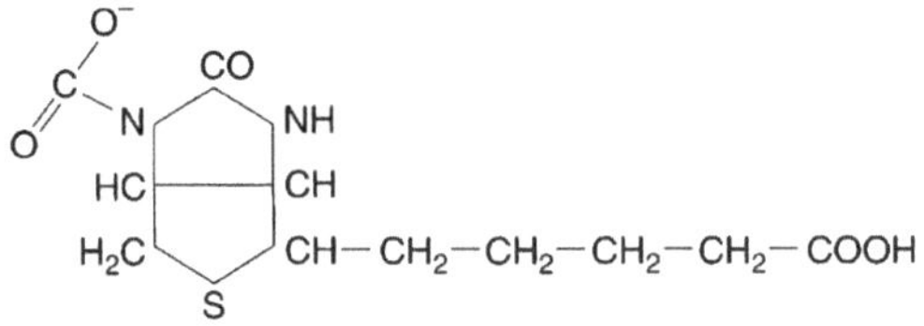

biotin~CO_2

The main difference between the biosynthetic pathways of Gram-negative bacteria and Gram-positive bacteria is the first step. Gram-positive bacteria generate pimeloyl-CoA from pimelate, in a reaction catalyzed by the bioW and bioI gene products, while in Gram-negative bacteria pimeloyl-CoA is synthesized from an unknown precursor through the *bioC* and *bioH* gene products in a reaction that is not yet understood.

A combined form of biotin has been known for a long time, called biocytin (ε-N-biotinyllysine).

Using radioactive biotin, propionylapocarboxylase and ATP, the radioactive holoenzyme can be synthesized by a specific enzyme. Radioactive biocytin has been isolated after proteolytic digestion of the holoenzyme. We can then depict a carboxylated holocarboxylase as follows:

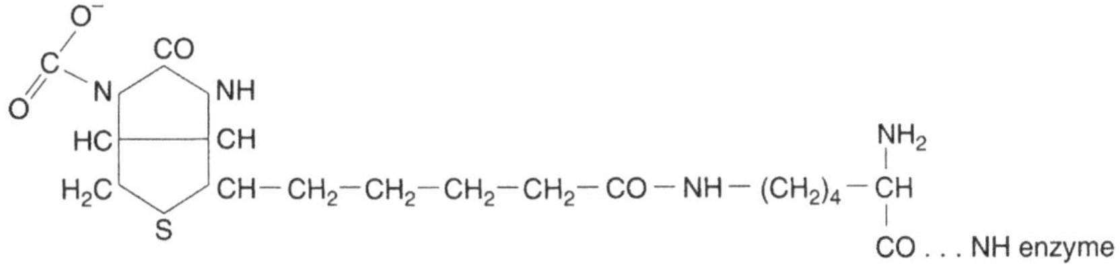

a carboxylated holocarboxylase

The biotinylation is catalyzed by holoenzyme synthetases, which attach a biotin molecule to a specific lysine residue on the apoenzymes.

The sequence flanking the biotin binding site is highly conserved among biotin-dependent enzymes. This conservation may be related to the extensive cross-species activity showed by holoenzyme synthetases: for example, several bacterial apocarboxylases are biotinylated by mammalian holocarboxylase synthetase and vice-versa. The sequence requirements for the biotinylation of human propionylCoA carboxylase by *E. coli* holoenzyme synthetase have been established.

The reaction synthesizing holocarboxylase involves the intermediate formation of biotinyladenylate.

Biotin-dependent carboxylases are widely distributed in nature and have important functions in many cellular processes. These enzymes share a conserved biotin carboxylase (BC) component, which catalyzes the ATP-dependent carboxylation of biotin using bicarbonate as the donor. Despite the availability of a large amount of biochemical and structural information on BC, the molecular basis for its catalysis is currently still poorly understood. The crystal structure at 2.0 Å resolution of wild-type *Escherichia coli* BC in complex with its substrates biotin, bicarbonate, and Mg-ADP has been determined. The structure suggests that Glu^{296} is the general base that extracts the proton from bicarbonate, and Arg^{338} is the residue that stabilizes the enolate biotin intermediate in the carboxylation reaction. The biotin domain of BC is positioned closer to the active site, leading to a 2-Å shift in the bound position of the adenine nucleotide and bringing it near the bicarbonate for catalysis. One of the oxygen atoms of bicarbonate is located in the correct position to initiate the nucleophilic attack on ATP to form the carboxyphosphate intermediate. This oxygen is also located close to the N1′ atom of biotin, providing strong evidence that the phosphate group, derived from decomposition of carboxyphosphate, is the general base that extracts the proton on this N1′ atom. These structural observations are supported by mutagenesis and kinetic studies. Overall, this first structure of BC in complex with substrates offers unprecedented insights into the molecular mechanism for the catalysis by this family of enzymes.

The Biotin Operon and Its Repressor

In *E. coli*, all the biotin genes (except *bioH*) are grouped in a single operon which is located at 17 min on the chromosome of *E. coli*. The genes of the operon *bioA, BFCD* are transcribed divergently from a single regulatory region located between the *bioA* and the *bioB* genes. Transcription in both directions is co-repressed by biotinyl-5′-adenylate and the biotin repressor. In certain other species, the genes are organized in multiple operons, sometimes at different locations in the chromosome. The following table shows the respective function of these genes:

bioH	Steps in pimeloyl CoA biosynthesis
bioF	7-keto-8-aminopelargonic acid synthetase
bioA	7,8-diaminopelargonic acid synthetase
bioD	Dethiobiotin synthetase
bioB	Biotin synthase (introduction of S)

The product of the *bioC* gene converts the free carboxyl group of a malonyl-thioester to its methyl ester by transfer of a methyl group from S-adenosyl-L-methionine (SAM). It allows to synthesize pimeloyl-ACP via the fatty acid synthetic pathway:

$$\text{S-adenosyl-L-methionine} + \text{malonyl-[acyl-carrier protein]} \rightarrow$$
$$\text{S-adenosyl-L-homocysteine} + \text{malonyl-[acyl-carrier protein] methyl ester}$$

All the biotin genes are coordinately repressed when biotin in excess of 1 ng/ml is added to the growth medium.

The repressor is a very interesting bifunctional protein of 321 amino acid residues, which acts at two different levels: in addition to its repressor function, it is endowed with acetyl CoA carboxylase biotin holoenzyme synthetase activity. This protein activates biotin to biotinyl-5′-adenylate and transfers the biotin to acceptor proteins. As soon as these proteins are totally biotinylated, biotinyl-5′-adenylate accumulates and serves as the corepressor of the biotin operon. Thus, we have here the case of an enzyme synthesizing its own co-repressor, a unique property so far among DNA-binding proteins. Mutations in the corresponding gene (*birA*) inactivate the repressor function partially or totally and alter also the enzymatic function. This repressor has recently been crystallized. It binds to a 40-bp symmetric biotin operator site to prevent transcription of the biotin biosynthetic genes. The structure of the repressor is highly asymmetric and consists of three domains. The N-terminal domain is mostly α-helical, contains a helix-turn-helix binding motif and is loosely connected to the remainder of the molecule. The central domain consists of a seven-stranded mixed β-sheet with α-helices covering one face. The other side of the sheet is largely exposed to the solvent and contains the enzyme active site. The C-terminal domain comprises a six-stranded antiparallel β-sheet sandwich. The location of biotin is consistent with mutations affecting the enzyme activity

Two molecules of the monomeric repressor bind cooperatively to one molecule of operator in the nanomolar range of concentration. The data support a model where two molecules of repressor bind with the two operator half sites, and form a dimer only when they bind. Since the complex between repressor and DNA has not been crystallized yet, details of its structure remain an open question.

Biosynthesis of Lipoic Acid

Lipoic acid is a growth factor for *Streptococcus faecalis* and *Butyribacterium rettgeri*.

The chemistry and structure of lipoic acid have been elucidated quite early. It is an 8-carbon un-branched fatty acid (octanoic acid) that contains a terminal 6-8 disulfide (dithiolane) functional group:

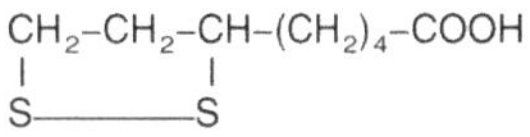

Lipoic or 6,8-thioctic acid

In spite of this information, the detailed enzymic mechanism of its biosynthesis has been delayed due to the minute amounts present in the cells and to an inability to use mutant techniques, although auxotrophs have been obtained in *E. coli* and *S. typhimurium*. Most of our knowledge comes from studies made with putative precursors labeled with radioactive or stable isotopes. When octanoic acid is fed to *E. coli*, it appears to be incorporated as a unit. The present scanty data support the following biosynthetic pathway:

$$\text{Acetyl CoA} \rightarrow \text{octanoic acid} \rightarrow \text{8-mercaptooctanoic acid} \rightarrow \text{thioctic acid}$$

Sulfur is directly inserted at the C-6 and C-8 positions with inversion of configuration at the prochiral center. Cysteine appears to serve as the source of sulfur, although it has not been determined whether the two sulfur atoms are inserted simultaneously or sequentially.

Although a complete pathway of lipoic acid metabolism has been established in *Escherichia coli*, lipoic acid metabolism in other bacteria is more complex and incompletely understood. *Listeria monocytogenes* has been shown to utilize two lipoate-protein ligases for lipoic acid scavenging, whereas only one of the ligases can function in utilization of host-derived lipoic acid-modified peptides. Lipoic acid scavenging requires not only ligation of lipoic acid but also a lipoyl relay pathway in which an amidotransferase transfers lipoyl groups to the enzyme complexes that require the cofactor for activity. Evidence is provided for a new lipoamidase activity that could allow utilization of lipoyl peptides by lipoate-protein ligase. These data support a model of an expanded, three-enzyme pathway for lipoic acid scavenging that seems widespread in the Firmicutes phylum of bacteria.

Lipoic acid and biotin share many common features.

Both biosynthetic pathways involve enzymes of fatty acid biosynthesis. Both function only when covalently linked through formation of an amide linkage to an ε-amino group of lysine of their respective apoenzymes (see the section on biotin). The disulfide group of lipoic acid participates in the acyl transfer reactions of the pyruvate oxidase and α-ketoglutarate dehydrogenase multienzyme complexes and is thus involved in the formation of acetate and succinate.

Lipoic acid analogs have been synthesized where one or both atoms of sulfur have been replaced by selenium. Replacement of either the C-6 or the C-8 sulfur atom with selenium resulted in lipoic acid derivatives with apparently unaltered biological properties. In contrast, simultaneous replacement of both sulfur atoms gave an analog (selenolipoic acid) that inhibited growth of wild-type *E. coli* when present in minimal medium at a concentration of 1.7×10^{-7} M. The inhibition was reversed by either excess lipoic acid or acetate plus succinate. Labeling experiments with ^{75}Se selenolipoic acid showed that it was efficiently incorporated into the α-ketoacid dehydrogenase complexes of growing cells. Spontaneous mutants resistant to selenolipoic acid were isolated and mapped. Their analysis reveals respectively a gene encoding a lipoate-protein ligase and a gene of unknown function in the biosynthesis of lipoic acid.

Biosynthesis of Pantothenate and Coenzyme A

The Synthesis of Pantothenic Acid

Mutants of *E. coli* unable to synthesize pantothenic acid fall into three classes: those which can synthesize pantoic acid, but require β-alanine; those which can synthesize β-alanine, but need pantoic acid for their growth; and finally those which are able to synthesize the two moieties but are unable to join them. The mutants from the latter group require preformed pantothenate for their growth.

β-alanine is formed by decarboxylation of L-aspartic acid, under the action of a specific decarboxylase coded by *panD*:

$$HOOC-CH_2-CH(NH_2)-COOH \rightarrow \underset{\beta\text{-alanine}}{HOOC-CH_2-CH_2-NH_2} + CO_2$$

Pantoic acid derives from α-ketoisovaleric acid, which is also a precursor of valine (see p. 200). An hydroxymethyl group is added to this compound by an enzyme, coded by *panB*, ketopantoate hydroxymethyltransferase or ketopantoaldolase, which depends on the presence of tetrahydrofolate for its activity. The enzyme has been purified to homogeneity. Its product, called ketopantoate for simplicity, is α-keto-β,β-dimethyl-γ-hydroxybutyric acid, which can be utilized for their growth by certain mutants auxotrophic for pantoate.

Ketopantoate is then reduced to pantoate by a reductase, coded by the yet unmapped *panE*, utilizing NADPH as the electron donor.

The *ilvC* gene product, involved in the biosynthesis of valine and isoleucine, is also capableof catalyzing the reduction of ketopantoate. Thus, *panE* mutants do not require pantoate when the *ilvC* gene is abundantly expressed, but when *ilvC* expression is low, pan E mutants require pantoate for growth. The *panE* gene has never been mapped with precision, but the enzyme has been purified.

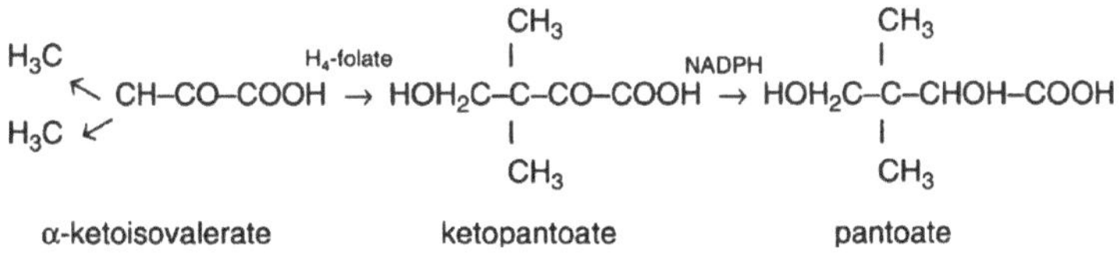

Pantothenate synthetase, coded by *panC*, then couples β-alanine and pantoate to yield pantothenate, in the presence of both a divalent (Mg^{++} or Mn^{++}) and a monovalent (K^+ or NH_4^+) cation. The enzyme has been purified to homogeneity.

$$
\begin{array}{c}
CH_3 \\
| \\
HOH_2C{-}C{-}CHOH{-}CONH{-}CH_2{-}CH_2{-}COOH \\
| \\
CH_3
\end{array}
$$

pantothenic acid

The activity of the enzyme is ATP-dependent. In addition to pantothenate, the other products of the reaction are AMP and pyrophosphate. This reaction has a certain historical importance, since it is the first instance where a net synthesis of a peptide bond has been obtained in vitro and where an enzyme substrate complex has been clearly demonstrated. This reaction can be decomposed as follows:

$$\text{enzyme} + \text{ATP} + \text{pantoate} \rightarrow \text{enzyme - pantoyladenylate} + \text{Ppi} \quad (1)$$
$$\text{enzyme - pantoyladenylate} + \beta\text{ - alanine} \rightarrow \text{pantothenate} + \text{AMP} + \text{enzyme} \quad (2)$$

The Synthesis of Coenzyme A from Pantothenic Acid

The first reaction of the synthesis of Coenzyme A (CoA) is a phosphorylation of the hydroxymethyl group of pantothenate by pantothenate kinase. The phosphopantothenate thus formed is then coupled to cysteine by phosphopantothenylcysteine synthetase; the reaction requires ATP. A specific decarboxylase leads to phosphopantetheine, where the cysteine residue of the substrate is replaced by a mercaptoethylamine group:

$$
\begin{array}{c}
CH_3 \\
| \\
H_2O_3P{-}OH_2C{-}C{-}CHOH{-}CONH{-}CH_2{-}CH_2{-}CONH{-}CH_2{-}CH_2SH \\
| \\
CH_3
\end{array}
$$

4'-phosphopantetheine

The structural gene for *E. coli* homodimeric pantothenate kinase (*coa A*) has been cloned and sequenced. Its polypeptide product has a molecular weight of 35,400. Strains containing the *coaA* gene on a multicopy plasmid express 76 times more pantothenate kinase activity, but their steady-state level of coenzyme A and acyl carrier protein remain unchanged or only slightly increased suggesting that the critical factor controlling the intracellular CoA concentration is regulation of pantothenate kinase activity by feedback inhibition. Actually, it has been shown that the regulation of the enzyme activity involves the competitive binding of CoA to the ATP site, which blocks ATP binding at one site and prevents positive cooperative ATP binding to the second site on the dimer.

It seems that other bacteria, such as *Acetobacter suboxydans* and *Lactobacillus helveticus* can carry pantetheine synthesis without prior phosphorylation of pantothenate; in these organisms, the first phosphorylation occurs at the pantetheine level.

An adenylyltransferase catalyzes the transfer of an adenylyl group from ATP to phosphopantetheine, the products of the reaction being pyrophosphate and dephosphocoenzyme A, to which a phosphate group is added to one of the ribose hydroxyls by dephosphocoenzyme A kinase, thus completing the synthesis of coenzyme A

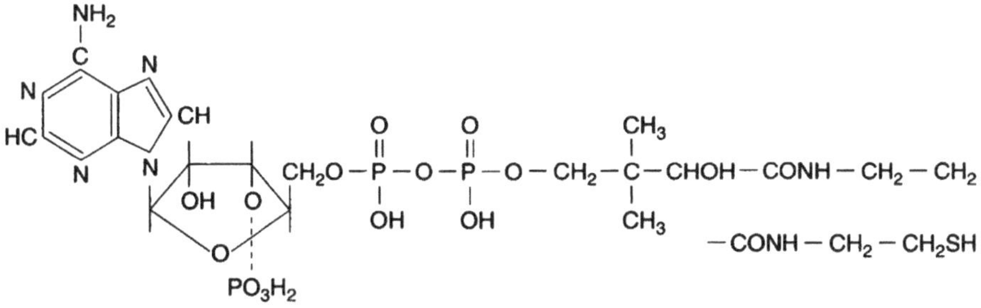

coenzyme A

None of the enzymes catalyzing the series of reactions leading from pantothenate to coenzyme A has been purified to a significant extent. Although many mutants unable to synthesize pantothenate have been isolated, there is only one report of a mutant blocked in the conversion of pantothenate to Coenzyme A. It is a temperature-sensitive mutant of *S. typhimurium* that cannot phosphorylate pantothenate at the nonpermissive temperature. Its genetic locus is at 89 min, close to that of *panD*.

It is the free sulfhydryl group of CoA which reacts with the acyl groups (acetyl, succinyl, malonyl, etc.) in numerous syntheses, in particular in fatty acid biosynthesis and in the tricarboxylic acid cycle.

The Acyl Carrier Protein

This protein has been considered under the chapter dealing with fatty acid biosynthesis.

The Biosynthesis of Inositol

If one compares the structures of glucopyranose and inositol, the hypothesis is readily made that inositol is derived from glucose by cyclization, accompanied by a change of configuration at C-5. Synthesis by animals seems to be limited and this is the reason why inositol has often been classified among the vitamins:

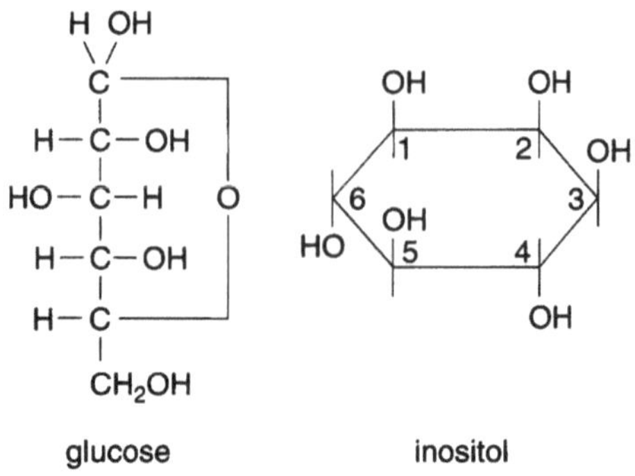

An enzyme system, myo-inositol-1-phosphate synthase, which converts glucose-6-phosphate to myo-inositol-1-phosphate has been clearly demonstrated in yeast. This system has an absolute requirement for NAD^+. In a second step, a phosphatase which requires Mg^{++} ions hydrolyzes inositol phosphate. The same results have been obtained with testicular extracts from mammals.

The Degradation of Myo-inositol by **Bacillus subtilis**

The *iolABCDEFGHIJ* operon of *Bacillus subtilis* is responsible for myo-inositol catabolism involving multiple and stepwise reactions. Previous studies demonstrated that IolG and IolE are the enzymes for the first and second reactions, namely dehydrogenation of myo-inositol to give 2-keto-myo-inositol and the subsequent dehydration to 3D-(3,5/4)-trihydroxycyclohexane-1,2-dione. The third reaction is now shown to be the hydrolysis of 3D-(3,5/4)-trihydroxycyclohexane-1,2-dione catalyzed by IolD to yield 5-deoxy-D-glucuronic acid. The fourth reaction was the isomerization of 5-deoxy-D-glucuronic acid by IolB to produce 2-deoxy-5-keto-D-gluconic acid. Next, in the fifth reaction 2-deoxy-5-keto-D-gluconic acid was phosphorylated by IolC kinase to yield 2-deoxy-5-keto-D-gluconic acid 6-phosphate. IolR is known as the repressor controlling transcription of the *iol* operon. In this reaction, 2-deoxy-5-keto-D-gluconic acid 6-phosphate appeared to be the intermediate acting as inducer by antagonizing DNA binding of IolR. Finally, IolJ turned out to be the specific aldolase for the sixth reaction, the cleavage of 2-deoxy-5-keto-d-gluconic acid 6-phosphate into dihydroxyacetone phosphate and malonic semialdehyde. The former is a known glycolytic intermediate, and the latter was previously shown to be converted to acetyl-CoA and CO_2 by a reaction catalyzed by IolA. The net result of the inositol catabolic pathway in B. subtilis is then the conversion of myo-inositol to an equimolar mixture of dihydroxyacetone phosphate, acetyl-CoA, and CO_2.

Biosynthesis of Pyrroloquinoline Quinone

Pyrroloquinoline quinone (PQQ) is the cofactor of numerous dehydrogenases located in the cytoplasmic membrane or in the periplasm of Gram-negative aerobic bacteria (two known exceptions are the methanol dehydrogenases of *Clostridium thermoautotrophicum* and *Nocardia*, two Gram-positive organisms). The role of these quinoproteins is to catalyze the primary oxidation step of non-phosphorylated substrates such as alcohols, aldehydes and aldoses. *E. coli* possesses all the genetic information required for PQQ biosynthesis and most of the responsible genes have been identified.

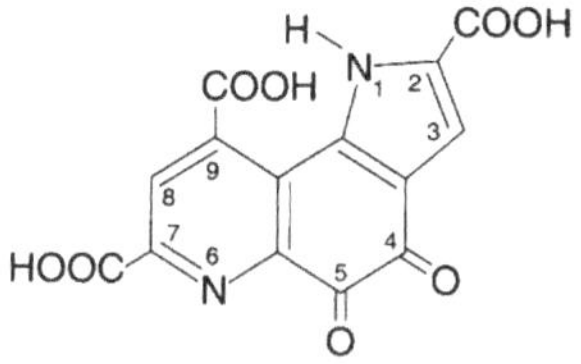

pyrroloquinoline quinone

The two putative precursors of PQQ are glutamate and tyrosine. This result has been obtained by specific ^{13}C labeling of the various possible precursors followed by the analysis of the mass and nuclear magnetic resonance spectra of the products obtained. The part of the PQQ molecule containing N-6, C-7, 8, 9 and the two carboxyls C-7′ and C-9′ derive directly from glutamate; the rest of the molecule originates from tyrosine. An hypothetical assembly mechanism has been proposed (Fig. 35.3), but the enzymatic steps between the two precursors and PQQ are actually unknown. Cross-feeding experiments between different categories of *Methylobacterium organophilum* and the efforts by various laboratories to isolate intermediates led to no results.

The genetic analysis of PQQ biosynthesis rests on the observations of the restored function of biochemical activities (glucose dehydrogenase from *Acinetobacter calcoaceticus* and methanol dehydrogenase of *Methylobacterium organophilum*) rather than on the direct estimation of PQQ. A 5,085 bp DNA fragment from *Acinetobacter* complements all the known *pqq* mutants. This fragment contains contiguous open reading frames (ORF) coding respectively for 29.7, 10.8 and 43.6 kDa polypeptides. Following immediately the last ORF, there is an ORF not involved in PQQ biosynthesis and then a small open reading frame coding for a 24 amino acid polypeptide whose integrity is essential to the synthesis of PQQ.

In the methylamine dehydrogenase from *Thiobacillus versutus* and several other bacteria where the cofactor was believed to be PQQ, it was found that it was actually tryptophyl tryptophanquinone (TTQ, Fig. 35.4). The diheme enzyme MauG uses

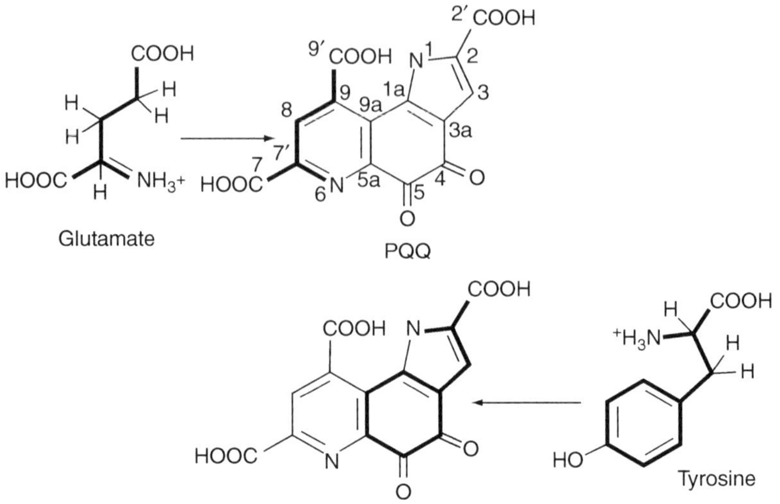

Fig. 35.3 Proposed assembly of the two precursors of PQQ

Fig. 35.4 Tryptophyl tryptophanquinone cofactor of methylamine dehydrogenase from *T. versutus*. The precursor tryptophans are at positions 57 and 108 in the amino acid sequence derived from the gene sequence of *T. versutus* methylamine dehydrogenase (From E.G. Huizinga, B.A.M. van Zanten, J.A. Duine, J.A. Jongejean, F. Huitema, K.S. Wilson and W.G.J. Hol, with permission of Biochemistry)

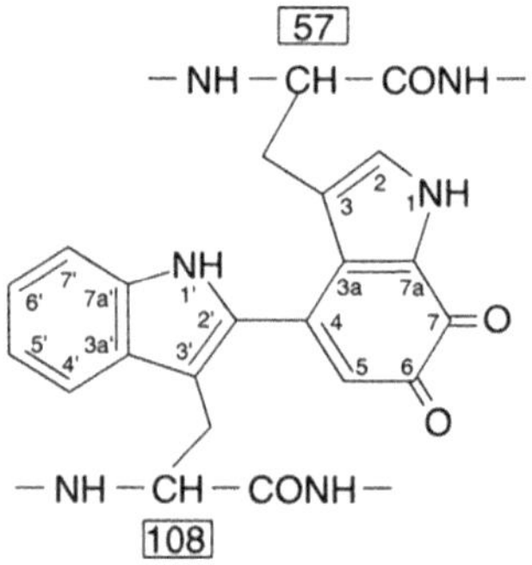

Trp radical chemistry to catalyze formation of theTrp-derived tryptophyl tryptophanquinone cofactor on its substrate protein, methylamine dehydrogenase. The unusual six-electron oxidation that results in tryptophyl tryptopanquinone formation occurs in three discrete two-electron catalytic steps. The exact order of these oxidation steps in the processive six-electron biosynthetic reaction have been determined, and reaction intermediates are structurally characterized. The intermediates observed in crystal structures are also verified in solution using mass spectrometry. Furthermore, an unprecedented Trp-derived diradical species on methylamine dehydrogenase, which is an intermediate in the first two-electron step, has been characterized using high-frequency and -field electron paramagnetic resonance spectroscopy and UV-visible absorbance spectroscopy. In *Paracoccus denitrificans*, the physiological electron acceptor is a 11.5 kD cupredoxin, amicyanin which transfers its electrons to a cytochrome composed of 155 amino acid residues and one heme group. This cytochrome is induced during growth on methylamine or methanol. The electrons are ultimately donated to membrane-bound cytochrome oxidase. The structure of the crystalline ternary complex between methylamine dehydrogenase, amicyanine and the inducible cytochrome has been determined at 2.4 Å resolution (*Science*, 1994, 264, 86–90).

Recently, a second enzyme was shown to depend on tryptophyl tryptophanquinone: the aromatic amine dehydrogenase from *Alcaligenes faecalis*, which catalyzes the oxidative deamination of aromatic amines such as tyramine and dopamine. This enzyme is structurally similar to methylamine dehydrogenase.

Selected References

Thiamin

Baker LJ, Dorocke JA, Harris RA, Timm DE (2001) Structure (Camb) 9:539–546
Challand MR, Martins FT, Roach PL (2010) J Biol Chem 285:5240–5248
Jurgenson CT, Begley TP, Ealick SE (2009) Ann Rev Biochem 78:569–603
Miranda-Rios J, Navarro M, Soberon M (2001) Proc Natl Acad Sci USA 98:9736–9741
Palmer LD, Downs DM (2013) J Biol Chem 288:30693–30699

Riboflavin

Chen S-C, Lin Y-H, Yu H-C, Liaw S-H (2009) J Biol Chem 284:1725–1731
Frago S, Velazquez-Campoy A, Medina M, Biol J (2009) J Biol Chem 284:6610–6619
Zurlinden A, Schweingruber ME (1994) J Bacteriol 176:6631–6635

Pyridine Nucleotide Cycle

Foster JW (1981) J Bacteriol 145:1002–1009
Galeazzi L, Bocci P, Amici A, Brunetti L, Ruggieri S, Romine M, Reed S, Osterman AL, Rodionov DA, Sorci L, Raffaelli N (2011) J Biol Chem 286:40365–40375

Pyridoxal Phosphate

Belitsky BR (2004) J Bacteriol 186:1191–1196
Raschle T, Amrhein N, Fitzpatrick TB (2005) J Biol Chem 280:32291–32300
Raschle T, Speziga D, Kress W, Moccand C, Gehrig P, Amrhein N, Weber-Ban E, Fitzpatrick TB (2009) J Biol Chem 284:7706–7718

Zein F, Zhang Y, Kang Y-N, Burns K, Begley TP, Ealick SE (2006) Biochemistry 45:14609–14620

The Degradation of Myo-inositol by Bacillus subtilis

Yoshida K, Yamaguchi M, Morinaga T, Kinehara M, Ikeuchi M, Ashida H, Fujita Y (2008) J Biol Chem 283:10415–10424

Biotin

Agarwal V, Lin S, Lukk T, Nair SK, Cronan JE (2012) Proc Natl Acad Sci USA 109: 17406–17411
Chou C-Y, Yu LPC, Tong L (2009) J Biol Chem 284:11690–11697

Lipoic Acid

Christensen QH, Hagar JA, O'Riordan MXD, Cronan JE (2011) J Biol Chem 286:31447–31456
Keeney KM, Stuckey JA, O'Riordan MX (2007) Mol Microbiol 66:758–770

Tryptophyl Tryptophanquinone

Lee S, Shin S, Li X, Davidson VL (2009) Biochemistry 48:2442–2447
Luki ET, Liu F, Krzystek J, Shin S, Jensen LM, Davidson VL, Wilmot CM, Liu A (2013) Proc Natl Acad Sci USA 110:4569–4573

Chapter 36
Biosynthesis of Carotene, Vitamin A, Sterols, Ubiquinones and Menaquinones

Synthesis of the Common Precursor

The enzyme β-ketothiolase condenses two molecules of acetyl CoA to form acetoacetyl CoA (see p. 63).

In a manner analogous to the condensation of acetyl CoA with oxaloacetate to give citrate, β-hydroxy-β-methylglutaryl CoA synthase catalyzes the condensation of acetoacetyl CoA with a new molecule of acetyl CoA.

$$CH_3\text{–}CO\text{–}CH_2\text{–}COSCoA + CH_3COSCoA + H_2O \rightarrow$$

```
        OH
        |
CH3–C–CH2–COSCoA + CoASH
        |
        CH2-COOH
```

β-hydroxy-β-methylglutarate

The true intermediate in this reaction appears to be acetoacetyl-ACP. A specific reductase reduces β-hydroxy-β-methylglutarate to mevalonic acid, with the aid of two molecules of NADH (or NADPH).

```
        OH
        |
CH3–C–CH2–CH2OH
        |
        CH2–COOH
```

mevalonic acid

G.N. Cohen, *Microbial Biochemistry*,
DOI 10.1007/978-94-017-8908-0_36,

Mevalonate kinase catalyzes the phosphorylation of mevalonic acid at position 5:

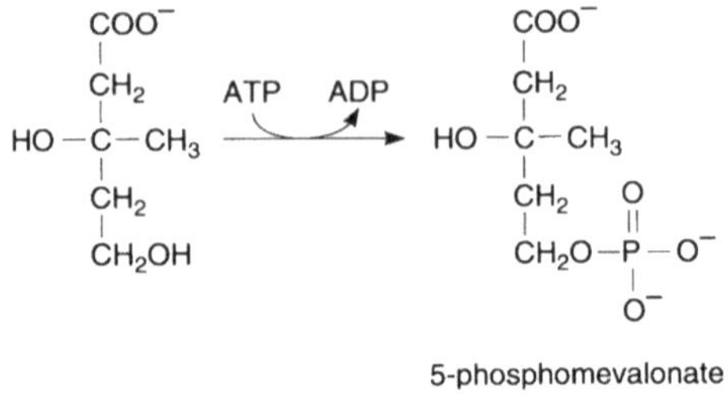

A different kinase adds a second phosphoryl group to this molecule:

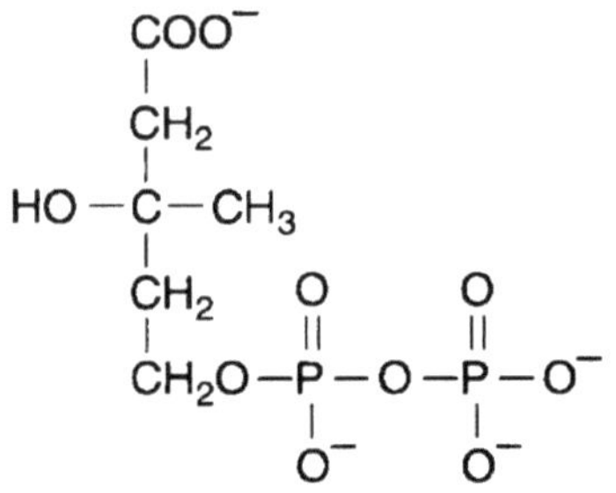

5-pyrophosphomevalonate

A specific anhydrodecarboxylase catalyzes the simultaneous removal of a molecule of CO_2 and the formation of a double bond; the product is an intermediate common to the biosynthesis of carotenoids and sterols, isopentenylpyrophosphate. Information on the determinants of specificity and reactivity of the *Trypanosoma brucei* and *Staphylococcus aureus* mevalonate diphosphate decarboxylases has been obtained by the study of their crystal structures.

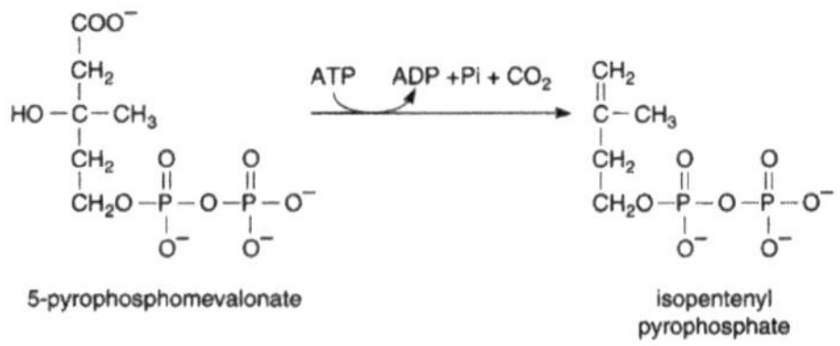

An isomerase is an essential step early in the isoprenoid biosynthetic pathway. It provides the potent electrophile dimethylallyl pyrophosphate (or prenyl pyrophosphate), required for the subsequent prenyl transfer reactions. The pathway is ubiquitous: more than 20,000 naturally occurring isoprenoid metabolites have been identified. End products of the pathway include carotenoids, sterols and ubiquinones. In addition, prenylation is an essential posttranslational modification for many proteins that play key roles in membrane structures, signal transduction across membranes and transport between organelles.

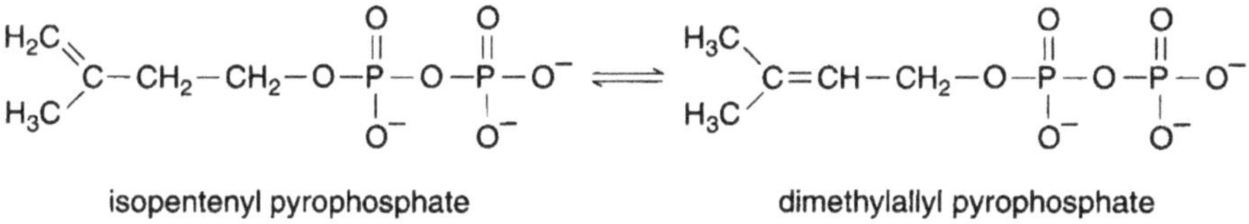

Although these early steps of isoprenoid biosynthesis have been well established in eukaryotes and some archaebacteria, little is known about their mechanism in eubacteria. Some results are indeed absolutely contradictory with the classical scheme through β-hydroxy-β-methylglutarylCoA followed by phosphorylation and decarboxylation to finally yield isopentenylpyrophosphate.

The Non-mevalonate Pathway of Isoprenoid Precursor (Dimethylallyl Pyrophosphate) Biosynthesis

For many years, it was assumed that the mevalonate pathway was the sole route to isopentenyl pyrophosphate and dimethylallyl pyrophosphate. Another route, called the non-mevalonate pathway, to these molecules has been uncovered and studied in the last decade. It consists of eight reactions, catalyzed by nine enzymes, most of which have been structurally characterized. The enzymes of the non-mevalonate pathway for isoprenoid biosynthesis are attractive targets for the development of novel drugs against malaria and tuberculosis. This pathway is used exclusively by the corresponding pathogens, but not by humans. Therefore, it represents a potential target for antimicrobial drugs.

This pathway starts with the synthesis of 1-deoxy-5-xylulose phosphate(DOXP), catalyzed by the eponymous synthase (*dxs*) which condenses pyruvate and glyceraldehyde 3-phosphate. This reaction uses thiamine pyrophosphate as cofactor. The subunit of the synthase is formed of three domains, each similar to equivalent domains in transketolase, which catalyzes a similar reaction.

Pyruvate + D-glyceraldehyde-3-phosphate + H+ → 1-deoxy-D-xylulose 5-phosphate + CO2

The next step is catalyzed by deoxyxylulose 5-phosphate reductoisomerase (*ispC*) which converts DOXP to 2C-methyl-D-erythritol 4-phosphate (*MEP*).

1-deoxy-D-xylulose 5-phosphate + NADP → 2C-methyl-D-erythritol 4-phosphate + NADPH

This enzyme is a target for the antimicrobial fosmidomycin which inhibits it by coordinating an indispensable divalent metal and mimicking the business end of DOXP.

MEP reacts with cytidine triphosphate to produce 4-diphosphocytidylyl-2C-methyl-D-erythritol (*CDP-ME*) and release pyrophosphate in a reaction catalyzed by a specific cytidyltransferase (*ispD*).

2C-methyl-D-erythritol 4-phosphate + CTP → 4-diphosphocytidylyl-2C-methyl-D-erythritol + PPi

An ATP dependent 4-diphosphocytidylyl-2C-erythritol kinase (*ispE*) produces 4-diphosphocytidylyl-2C-methyl-D-erythritol-2-phosphate.

4-diphosphocytidylyl-2C-methyl-D-erythritol+ATP → 4-diphosphocytidylyl-2C-methyl-D-erythritol-2-phosphate + ADP

The crystal structures of *Aquifex aeolicus* IspE nicely complement the one from *E. coli* IspE for use in structure-based design, namely by providing invaluable structural information for the design of inhibitors targeting IspE from *Mycobacterium tuberculosis* and *Plasmodium falciparum*. Similar to the enzymes from these pathogens, *A. aeolicus* IspE directs the OH group of a tyrosine residue into a pocket in the active site. In the *E. coli* enzyme, on the other hand, this pocket is lined by phenylalanine and has a more pronounced hydrophobic character.

CMP is abstracted from the product of IspE by 2-C-methyl-D-erythritol-2,4-cyclodiphosphate synthase (*ispF*), a trimeric protein where a zinc atom is coordinated to a pair of histidine residues, an aspartate and either a water or phosphate if substrate is present. In addition, the enzyme requires a magnesium atom which is positioned between the a and b phosphates.

4-diphosphocytidylyl-2C-methyl-D-erythritol-2-phosphate→ 2C-methyl-D-erythritol 2,4-cyclodiphosphate + CMP

IspG transforms 2-C-methyl-D-erythritol-2,4-cyclodiphosphate to1-hydroxy-2-methyl-2-butenyl-4-diphosphate in a two-electron reduction. IspH then catalyzes production of isopentenyl pyrophosphate and some dimethylallylpyrophosphate. IspG and IspH both contain an iron-sulfur cluster, which explain their oxygen sensitivity. Under anaerobic conditions, catalysis is supported by a combination of NADPH, flavodoxin and flavodoxin reductase.

Two isopentenyl pyrophosphate isomerases, IDI-1 and IDI-2 maintain the supply of dimethylallyl pyrophosphate. The structure of the IDI-1 *E. coli* enzyme has been determined. It is a compact a/b protein. Its flexible N-terminal becomes structured in the presence of Mg^{++}, helping to make the active site in a deep polar cavity. The IDI-II subunit displays the triose phosphate isomerase type a/b8 fold and oligomerizes to a cage-like octamer. The contribution of each of the two isomerases remains unclear.

IspD and IspF are transcriptionally coupled; in some bacteria, as in *Campylobacter jejuni* they are fused in a bifunctional protein. IspDF is unusual because it catalyzes non-consecutive steps. This reminds of the case of the two bifunctional proteins catalyzing in *E. coli* the aspartokinase and homoserine dehydrogenase reactions (see Chap. 26).

A complex comprising three IspD and IspE dimers together with two IspF trimers creates an assembly localizing18 catalytic centers. This leads to the hypothesis of an efficient catalysis of three reactions at the core of isoprenoid precursor biosynthesis, in spite of the absence of enhanced catalytic rates on complex formation and of substrate channeling.

Synthesis of β-Carotene, Carotenoids and Vitamin A

Carotenoids constitute one of the most widespread classes of yellow, orange and red natural pigments found in bacteria, fungi, photosynthetic organisms and also in eukaryotes. It is estimated that 108 tons of carotenoids are synthesized per year, composed of some 600 structurally distinct compounds, often modified by various oxygen-containing functional groups to produce cyclic or acyclic xanthophylls. The majority of them have 40 carbon atoms and are thus members of the tetraterpene family. Since they arise biogenetically from isoprene units, they have the methyl branches typical of isoprenoid compounds. They share the common biosynthetic pathway with related compounds until the condensation of two molecules of geranylgeranylpyrophosphate forming the first carotenoid, phytoene. The degree of conjugation and the isomerization state of the backbone polyene chromophore determine the absorption properties of each carotenoid. Compounds with at least seven conjugated double bonds, such as ζ-carotene, absorb visible light. Some carotenoids occur naturally not only as all-*trans* isomers, but also as *cis* isomers.

Synthesis of the Carotenoids

In the structural formula of carotene reproduced below, the isoprene skeleton of isopentenylpyrophosphate is recognizable:

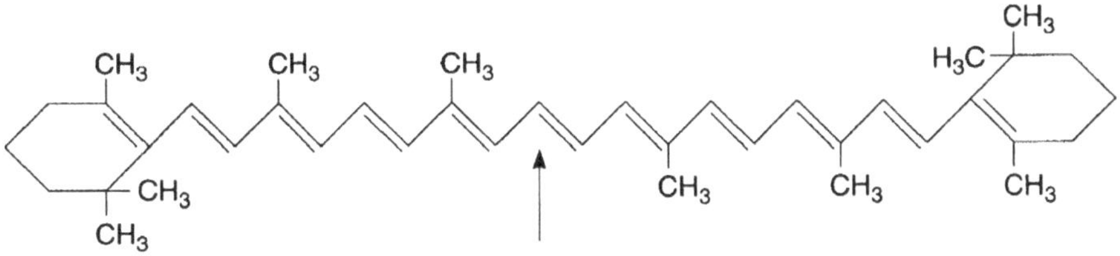

β-carotene $C_{40}H_{56}$

Eleven conjugated *trans* double bonds, four methyl branches, and two terminal β-ionone rings. The α and γ isomers differ at one end.

The isoprenoid nature and the symmetry of the carotene skeleton indicate that it is formed from eight isopentenyl units condensed in the following way:

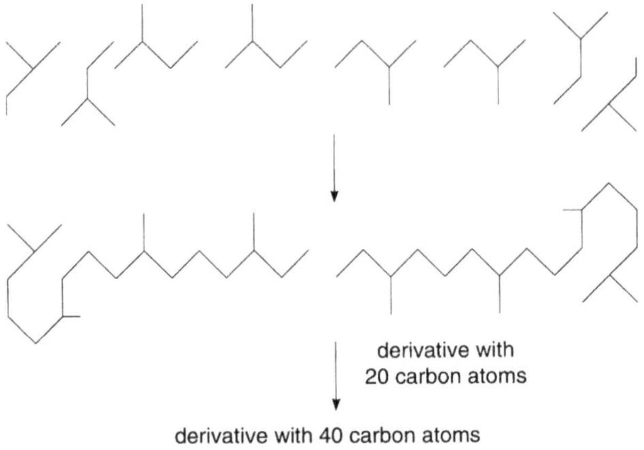

There are many experiments showing that bacterial extracts can produce phytoene from mevalonate, pyrophosphomevalonate and isopentenylpyrophosphate.

From the latter substrate, the synthesis does not require ATP and occurs equally well under aerobic and anaerobic conditions. It can be concluded from various experiments carried out with different organisms that the synthesis of C40 derivatives occurs in steps involving the following intermediates: dimethylallylpyrophosphate → geranylpyrophosphate (C_{10}) → farnesylpyrophosphate (C_{15}) → geranylgeranylpyrophosphate (C_{20}), followed by a condensation of two C_{20} derivatives to give prephytoene pyrophosphate, which is dephosphorylated to phytoene by phytoene synthase:

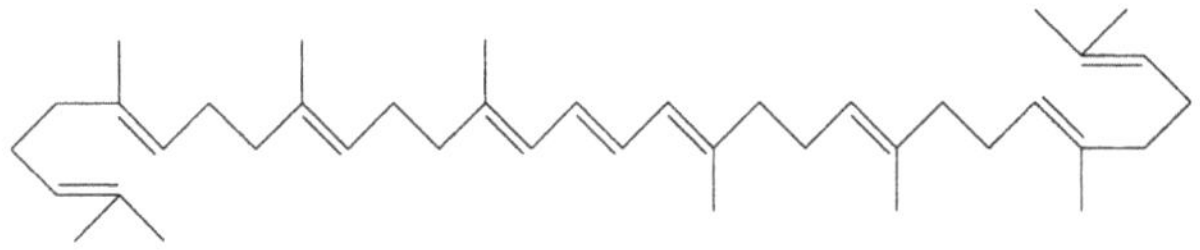

phytoene

Phytoene contains nine double bonds, three of them conjugated, six branch methyl groups and two terminal isopropylidene groups.

In plants, the prephytoene pyrophosphate synthase and phytoene synthase activities are carried by a bifunctional protein. It has been shown in tomatoes that there are two isofunctional phytoene synthases differentially expressed: one of them is induced by fruit ripening, the second being present in all tissues. The phytoene synthase of *Neurospora crassa* has been cloned and sequenced.

Phytoene is the first carotene in carotenoid biosynthesis. Depending on the diversity of the end products of the carotenogenic pathways, a variable number of double bonds is introduced into this colorless hydrocarbon. Anoxygenic photosynthetic bacteria, non-photosynthetic bacteria, and fungi desaturate phytoene either three or four times to yield neurosporene or lycopene, respectively. In contrast, oxygenic photosynthetic organisms (cyanobacteria, algae and higher plants) convert phytoene to lycopene via ζ-carotene in two distinct sets of reactions. At the level of neurosporene or lycopene, the carotenoid biosynthetic pathways of different organisms diverge to generate the extraordinary diversity of carotenoids found in nature.

Desaturation of phytoene proceeds by a dehydrogenase-electron transferase mechanism. The enzyme involved, phytoene desaturase, is feedback-inhibited by subsequent carotenoids in various organisms: the phytoene desaturase is the limiting step in the pathway and is a prime target site for herbicides.

Genes coding for the phytoene desaturases have been sequenced from *Rhodobacter capsulatus*, *N. crassa*, *Erwinia uredovora*, and the cyanobacterium *Synechococcus*. The three first enzymes only share conserved amino acid sequences. The end products of phytoene desaturation found in the various species are different: in the three bacteria: only *Erwinia* and *Synechococcus* produce lycopene, whereas *Rhodobacter* does not. In order to analyze the events, *E. coli* has been transformed by a plasmid of *Erwinia* carrying the genes coding for all the enzymes necessary for the synthesis of 15-*cis* phytoene, the carotenoid precursor, together with a plasmid carrying the desaturase genes of either *Erwinia*, *Rhodobacter* or *Synechococcus*. The products were then analyzed. The *Rh. capsulatus* desaturase formed *trans*- neurosporene, which three additional double bonds, and two *cis-isomers*. The intermediates with only one additional double bond (phytofluene) and two additional double bonds (three isomers of ζ-carotene) have also been detected. The *Synechococcus* desaturase forms two *cis* - isomers of ζ-carotene and small amounts of *trans* - ζ-carotene. Since this cyanobacterium contains lycopene, it needs an additional ζ-carotene desaturase to form this carotenoid which possesses four double bonds in addition to those of phytoene. As for the *Erwinia* desaturase, it forms *trans*-lycopene from 15-*cis* phytoene with small amounts of the *cis* - isomer without the need of an additional enzyme.

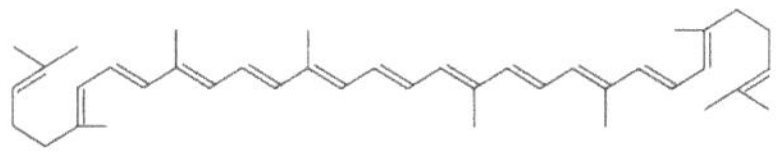

lycopene

Phytofluene contains ten double bonds, five of them conjugated, six branch methyl groups and two terminal isopropylidene groups. As in phytoene, the central double bond between C-15 and C-16 has *cis* configuration.

Phytoene desaturase is universally a membrane-integrated system. Its purification has been hindered by the practical difficulties associated with assaying the enzyme and the rapid loss of activity upon separation from the membrane. Using molecular biological techniques, the gene coding for the desaturase in *Erwinia uredovora* has been overexpressed in *E. coli*: the enzyme however was sequestrated into inclusion bodies requiring urea treatment for solubilization. Purification was finally achieved by standard biochemical techniques. FAD is involved in desaturation, wheras NAD^+ and $NADP^+$ are inhibitory. This contrasts with the desaturase from *Anacystis* where phytoene desaturation is pyridine nucleotide-dependent. In *Rhodobacter sphaeroides*, the phytoene synthase and desaturase enzymes are encoded by a single operon. The desaturase gene has been overexpressed and shown to catalyze three desaturations of phytoene leading to neurosporene. The activity is ATP-dependent.

All phototrophic organisms contain carotenoids in their photosynthetic apparatus. These pigments have a dual role: first, they are auxiliary absorbers of light energy which they transfer to chlorophyll. However, photosynthesis cannot take place without chlorophyll, and the carotenoids cannot therefore replace it. The second role of the carotenoids is the protection of chlorophyll against photo-oxidation: a blue-green mutant of *Rhodopseudomonas spheroides* can synthesize colorless phytoene, but is unable to convert it to colored carotenoids. While this mutant can grow normally in the conditions of anaerobic photosynthesis, it is rapidly killed by simultaneous exposure to light and oxygen. Death is accompanied by the destruction of bacteriochlorophyll. Further experiments have shown that this destruction is a secondary phenomenon and that the mortality curves are the same if the experiment is carried out at low temperature without destruction of chlorophyll. The mechanism of the protection conferred by carotenoids against photo-oxidation requires further work.

In mammals, the cleavage products of several dietary carotenoids fulfill essential roles in nutrition (vitamin A), vision (retinal) and development (retinoic acid). The metabolism of certain cyclic epoxyxanthophylls yields abscisic acid, an important hormone in higher plants. Carotenoids and derivatives provide pigmentation to many birds, fish and crustaceans.

Regulation of Carotenoid Synthesis

Carotenoid synthesis is regulated in different organisms by different endogenous and exogenous factors. Blue light is among the most effective activators of carotogenesis in non-photosynthetic microorganisms. Accumulating evidence points to the possibility of blue light affecting the expression of the carotenoid genes at the transcriptional level. This is certainly the case for the non-photosynthetic gram-negative bacterium *Myxococcus xanthus* and in the fungus *N. crassa*. In the latter case, blue light has been shown to activate specifically the expression of two carotenoid genes, one of them being phytoene synthase, where the level of the specific mRNA increases over 30-fold in illuminated mycelia. In *M. xanthus*, blue light has been shown to stimulate two unlinked genes of carotenoid synthesis. Chemical analyses of mutant strains suggest the existence of a cluster of structural genes being transcribed from a single, light-inducible promoter under the control of three regulatory genes.

Synthesis of Vitamin A

All the carotenes, which are abundant in green plants and fruits, have provitamin A activity. Conversion of carotene to the vitamin occurs mainly in the small intestine, but other organs and serum have a limited conversion activity. The activity is absent in invertebrates. β-carotene is oxidatively cleaved (at the point indicated by the arrow in the formula p. 529) by the intestinal mucosa into two molecules of retinal (an aldehyde), whereas α- and γ-carotene yield only one molecule of retinal. Alcohol dehydrogenase reduces then retinal to all-*trans* retinol (the vitamin proper), which is then esterified (mainly as palmitate) and stored in the liver.

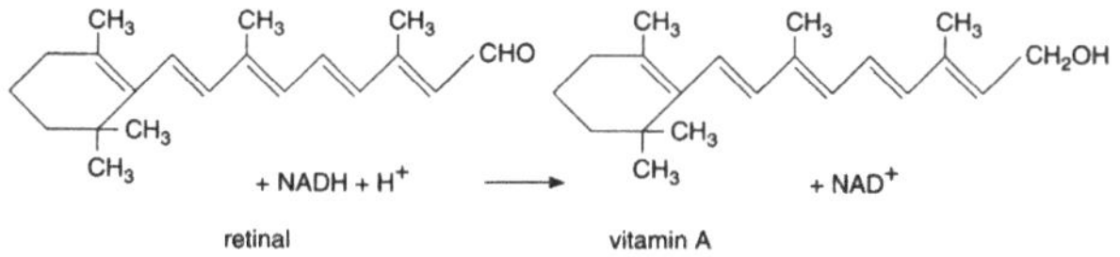

Retinal is the substrate of an oxidase which transforms it to retinoic acid, which can stimulate growth in vitamin A deficient animals, promote cell differentiation and suppress carcinogenesis.

Synthesis of Sterols

The regulation of cholesterol synthesis is a complex biological process, and its understanding is important for the determination in the biochemical function of cholesterol in normal and pathological conditions.

The pathway of synthesis of sterols is identical with the synthesis of carotenoids as far as the stage of farnesylpyrophosphate, the C15 derivative.

Rat farnesylpyrophosphate synthase catalyzes the sequential condensation of isopentenylpyrophosphate with dimethylallylpyrophosphate and then with the resultant geranylpyrophosphate and then with the resultant geranylpyrophosphate. Farnesyl pyrophosphate (FPP) is utilized in the synthesis of squalene, farnesylated proteins, dolichols, coenzyme Q, geranylgeranylpyrophosphate and some other important molecules. The sequence of FPP synthase contains two aspartate rich motifs, involved in catalysis, which have homology with at least 9 prenyl transfer enzymes that utilize an allylic prenylpyrophosphate as one of the substrates. The crystal structure of avian recombinant enzyme, the first three-dimensional structure for any prenyltransferase, has been determined to 2.6 Å resolution. The enzyme exhibits a novel fold composed entirely of α-helices joined by connecting loops. Ten core helices, are arranged around a large central cavity. The two aspartate-rich sequences mentioned above are found on opposite walls of this cavity, with the aspartate side-chains approximately 12 Å apart and facing each other.

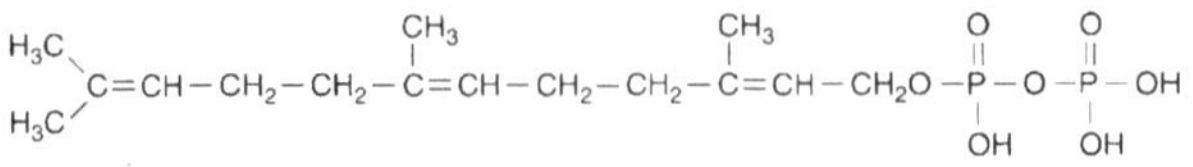

farnesyl pyrophosphate

In a particulate system isolated from yeast and in mammalian liver microsomes, cholesterol is produced from *trans*-farnesyldiphosphate (FPP) with squalene being the first intermediate in this pathway. The formation of squalene is catalyzed by squalene synthase (or farnesyl pyrophosphate: farnesyl pyrophosphate farnesyl-transferase) in two separate steps. Two molecules of FPP condense to form the intermediate presqualene pyrophosphate which is then reductively rearranged to form squalene, the reductant being a reduced pyridine nucleotide. The rat liver gene coding for a truncated, active form of squalene synthase has been cloned and expressed in *E. coli*, and sequenced. The yeast enzyme has been purified to homogeneity and is a single polypeptide of Mr 47,000.

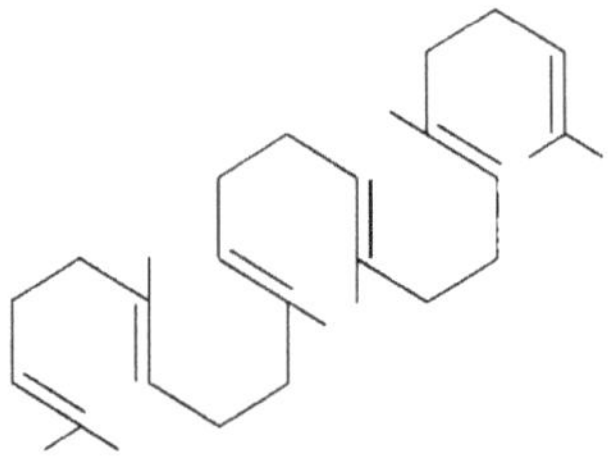

squalene

The highly branched nature of the pathway and the central role of FPP in isoprenoid metabolism makes squalene synthase an attractive target for inhibiting cholesterol biosynthesis without disrupting the flow of FPP into non-sterol metabolites.

The conversion of squalene to cholesterol involves first a cyclization requiring molecular oxygen and NADPH: squalene epoxide (2,3 oxidosqualene) is thus formed, which is then cyclized to lanosterol by a membrane-bound cyclase (Mr = 78,000) which has been purified from rat liver. The *S. cerevisiae* gene coding for this cyclase, lanosterol synthase, has been cloned, characterized and overexpressed. The gene contains a 2,196 base pair open reading frame capable of encoding a 83,000 kDa protein. Lanosterol is then converted to cholesterol in mammals and to ergosterol in plants and yeasts. The different steps involved are outside the scope of this book (some of them are the target of fungicidal antibiotics: ergosterol depletion interferes with the structure of fungal plasma membrane). Suffice it to say that at least 15 successive reactions are involved in the conversion of squalene to cholesterol.

It has been known for more than 50 years that the addition of cholesterol to the diet of laboratory animals reduces the rate of synthesis of cholesterol by liver.

This reduction occurs at the level of β-hydroxy-β-methylglutaryl reductase synthesis and activity: dietary cholesterol suppresses the synthesis of the reductase at the gene transcription level in the liver and is a feedback inhibitor of the enzyme.

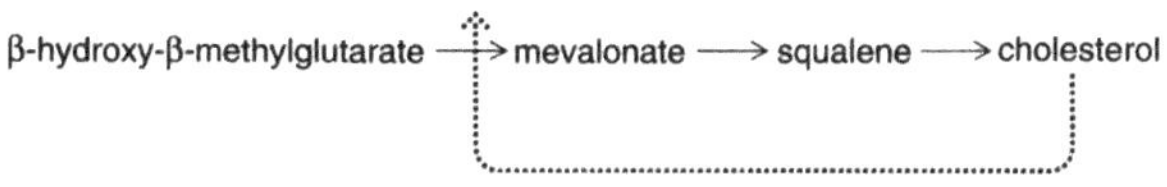

It should be emphasized that the inhibited reaction is the first practically irreversible reaction (committed) in the synthesis of sterols. This feedback inhibition provides a striking example of the absence of structural similarity between an allosteric inhibitor and the substrates of the reaction which it inhibits.

The Biosynthesis of Ubiquinones and Menaquinones

E. coli, together with other Gram-negative facultative aerobic bacteria, contains isoprenoid quinones of both the benzene and the naphthalene series. The major benzoquinone is ubiquinone-8 (with the figure referring to the number of prenyl units present in the side chain), whilst the naphthoquinones are either menaquinone (*me*thylated *na*phthoquinones) or demethylmenaquinone.

Ubiquinone and menaquinone both comprise a nucleus derived from chorismate, a prenyl side chain derived from prenylpyrophosphate and a nuclear methyl group derived from S-adenosylmethionine. In each case, a benzenoid aromatic acid is used as the framework on which the rest of the molecule is constructed.

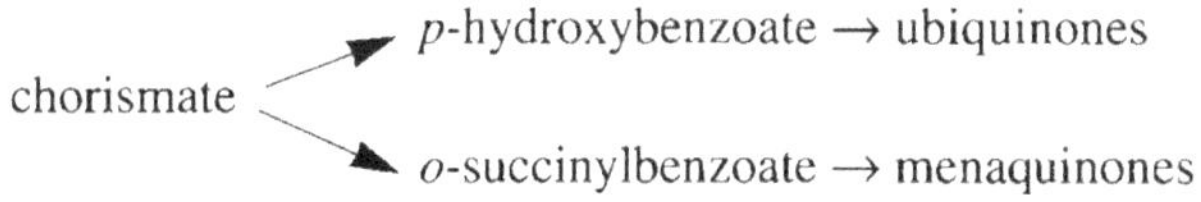

The precursor of benzoquinones is *p*-hydroxybenzoic acid, which itself derives from chorismic acid. The enzyme catalyzing the transformation is called chorismate (pyruvate) lyase and is coded by the *ubiC* gene. The prenylation reaction, under control of the *ubiA* gene, situated in an operon structure together with *ubiC* is catalyzed by *p*-hydroxybenzoate polyprenyltransferase, requires a polyprenylpyrophosphate and results in the release of pyrophosphate. The prenylpyrophosphate, with *n* isoprene units is synthesized through the reaction of dimethylallylpyrophosphate and $(n - 1)$ molecules of isopentenylpyrophosphate. The next step (*ubiD*) is a decarboxylation of the prenylated compound to the corresponding 2-polyprenylphenol. Then other hydroxylation reactions, utilizing molecular oxygen in aerobiosis or the oxygen atom of water under anaerobic conditions, and methylation reactions (where all the methyl groups are provided by S-adenosylmethionine) occur.

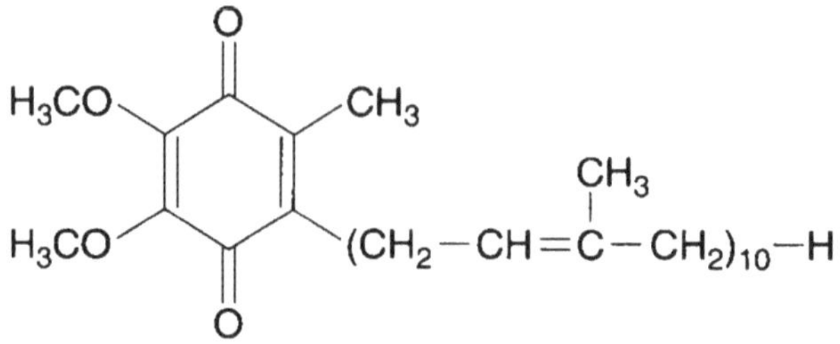

ubiquinone-10

A mutant of *E. coli* unable to grow aerobically on non-fermentable substrates, but which grows anaerobically on fumarate or nitrate, has been isolated. The oxygen consumption by its cell-free extracts is low relative to that of an isogenic control strain, but was restored by adding ubiquinone-1 to the extracts. Transformation with a cloned fragment of chromosomal DNA restored both the cellular quinone content and the ability to grow on succinate. The gene (*ubiG*) responsible was identified to code for the enzyme 2-octaprenyl-3-methyl-5-hydroxy-6-methoxy-1,4-benzoquinone (OMHMB) methyltransferase and catalyzes the terminal step in the biosynthesis of ubiquinone-8. The sequence of the gene and that of the corresponding protein contain a motif which is found in other S-adenosylmethionine-dependent methyl-transferases.

The menaquinone called vitamin K_2 plays an essential role in several electron transport systems by serving as the major electron carrier during anaerobic growth in *E. coli* and mediates also electron flow between dehydrogenases and cytochromes in *Bacillus subtilis*, a strict aerobe. Menaquinone is synthesized from a *men*-specific branch of the shikimate pathway, beginning with the isomerization of chorismate to isochorismate. The isochorismate synthase encoded by the *menF* gene has been overexpressed and purified to homogeneity. The purified enzyme had a relative Mr of 48,000 as determined by sodium dodecyl sulfate-polyacrylamide gel electrophoresis. The native Mr, as determined by gel filtration chromatography, was 98,000, thus establishing that the native enzyme is a homodimer. The enzyme showed a requirement for Mg^{2+} for maximal activity. The product of the *menCD* genes are necessary for the conversion of isochorismate to *o*-succinylbenzoate (OSB) a stable intermediate in the menaquinone biosynthetic pathway, which involves at least seven identified enzymes. The OSB synthase system catalyzes the synthesis of OSB from isochorismic acid and α-ketoglutaric acid in the presence of thiamine pyrophosphate (TPP) and Mg^{++}. The succinyl moiety of OSB is generated by carbon atoms 2 to 5 of α-ketoglutarate, which is decarboxylated by α-ketoglutarate decarboxylase (KDC), the adduct between the resulting succinic semialdehyde and TPP being the source of the succinyl CoA which condenses with isochorismic acid to yield OSB.

$$\alpha\text{-ketoglutarate} + \text{TPP} \rightarrow CO_2 + \text{COOH–}(CH_2)_2\text{–C–(OH)–TPP}$$

"Active" succinic semialdehyde

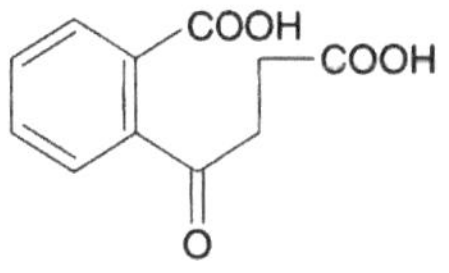

OSB

KDC activity is carried out by an enzyme distinct from the ketoglutarate dehydrogenase complex and has been found to be part of a bifunctional protein, the other activity being 2-succinyl-6-hydroxy-2,4-cyclohexadiene-1-carboxylic acid (SHCHC) synthase. This bifunctional protein is encoded by the *menD* gene. The prearomatic SHCHC is subsequently aromatized with the loss of a hydroxyl group to OSB by the OSB synthase.

Another identified step is catalyzed by naphthoate synthase (coded by *menB*), and forms the bicyclic ring system by catalyzing the conversion of o-succinylbenzoyl coenzyme A to 1,4-dihydroxy-2-naphthoic acid (DHNA), the precursor of demethylmenaquinone (DMK) and menaquinone (MK).

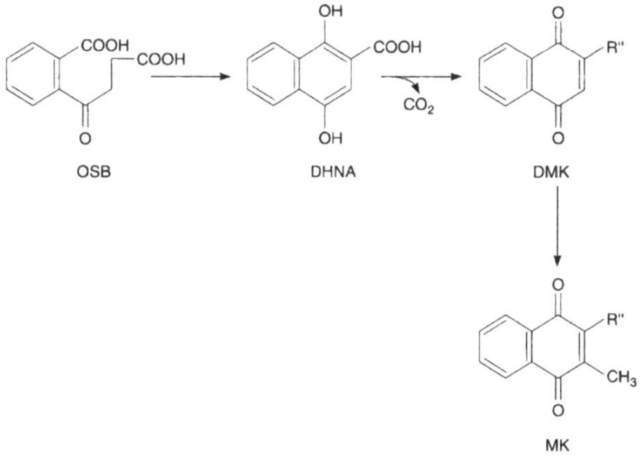

The synthesis of the coenzyme A derivative is due to *o*-succinylbenzoyl coenzyme A synthetase (coded by *menE*) which catalyzes the following reaction:

$$\text{OSB} + \text{CoA} + \text{ATP} \rightarrow \text{OSB}\quad\text{CoA} + \text{AMP} + \text{PPi}$$

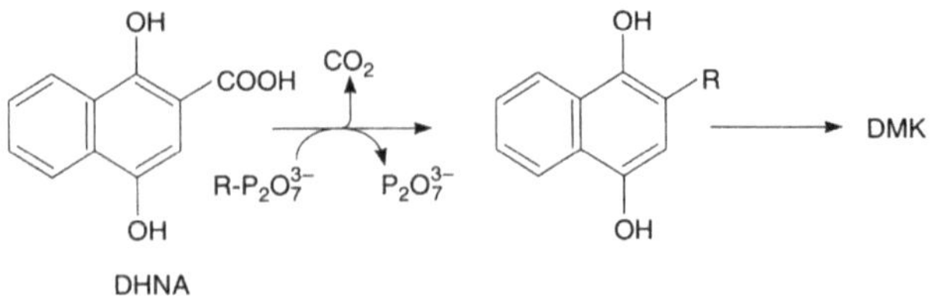

Fig. 36.1 Prenylation of 1,4 dihydroxy-2-naphthoate. *R* prenyl

The overall conversion DHNA → DMK reaction requires at least two stages: removal of the DHNA carboxyl as CO_2, the attachment of the isoprenoid residue (R). A single enzyme, coded by *menA*, apparently catalyzes the two reactions.

The side chain lengths of the menaquinones are probably determined by the availability of the isoprenyl pyrophosphate substrate(s) within the membranes. The prenylation occurs at the level of DHNA (Fig. 36.1).

With the exception of *menA*, all the other identified genes encoding the menaquinone biosynthetic enzymes are clustered at 48.5 min on the *E. coli* chromosome. These genes have been cloned and the sequence of two of them, including *menB* has been reported.

It was reported that various *Streptomyces* species lack orthologs of the *menB–menF* genes, suggesting that an entirely unique biosynthetic pathway is operative in these organisms. Interestingly, these genes are also absent in the pathogenic bacteria *C. jejuni* and *H. pylori*, even though these organisms are known to biosynthesize menaquinone. Isotopic feeding studies confirmed that a unique pathway is employed in *Streptomyces*, and 1,4-dihydroxy-6-naphthoate (DHN) was implicated as an intermediate (see Fig. 36.2). This compound was demonstrated to recover the growth of a menaquinone auxotroph of *Streptomyces coelicolor*, providing strong evidence for this hypothesis. In 2008, Dairi and co-workers reported the discovery of the alternate menaquinone biosynthetic pathway, or the futalosine pathway, and identified several of the genes in either *S. coelicolor* or *Thermus thermophilus*. They found that MqnA is involved in the early steps of futalosine biosynthesis, presumably utilizing chorismate, phosphoenolpyruvate, and inosine as precursors (Fig. 36.2). The enzyme futalosine hydrolase (MqnB or futalosine nucleosidase) then cleaves the hypoxanthine ring from futalosine to generate dehypoxanthinylfutalosine (DHFL). Finally, the enzymes MqnC and MqnD convert DHFL into 1,4-dihydroxy-6-naphthoate. Although the pathway was largely elucidated by generating *S. coelicolor* knock-out strains and isolating the intermediates that accumulated, in the case of futalosine hydrolase and MqnD, in vitro enzyme activity was demonstrated using recombinant *T. thermophilus* enzymes.

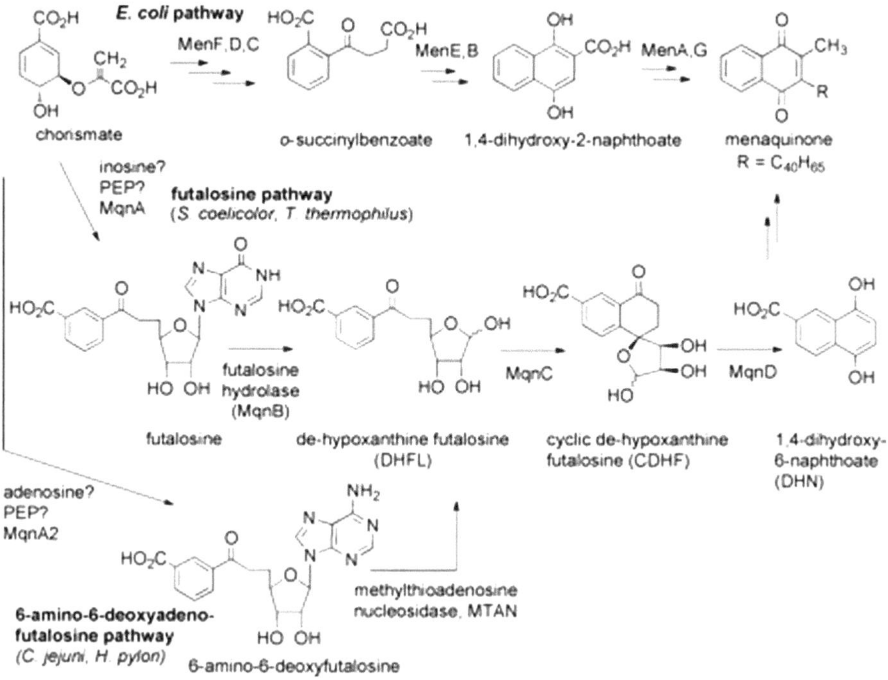

Fig. 36.2 Pathways for menaquinone biosynthesis in bacteria. The *upper pathway* shows the biosynthesis of menaquinone in *E. coli*. The *middle pathway* shows the futalosine pathway employed by *S. coelicolor* and *T. thermophilus*. The *lower pathway* shows the modified futalosine pathway employed by *C. jejuni* and *H. pylori*. *PEP* phosphoenolpyruvate

Selected References

Consult the articles by Bentley and Meganathan, Brown and Williamson, Dempsey, Eisenberg: Ingraham JL, Low KB, Magasanik B, Schachter M, Umbarger HE (eds) (1987) Escherichia coli and Salmonella typhimurium, cellular and molecular biology. American Society for Microbiology, Washington, DC, pp 512–550

Mevalonate and Non-mevalonate Pathways

Byres F, Alphey MS, Smith TK, Hunter WN (2007) J Mol Biol 371:540–553

Crane CM, Hirsch AK, Alphey MS, Sgraja T, Lauw S, Illarionova V, Rohdich F, Eisenreich W, Hunter WN, Bacher A, Diederich F (2008) Chem Med Chem 3:91–101

Hunter WN (2007) J Biol Chem 282:21573–21577

Biosynthesis of Water-Soluble Vitamins

Brown GM (1963) Ann Rev Biochem 32:419–462

Pyridoxine

Zhao G, Winkler ME (1994) J Bacteriol 176:6134–6138

Carotenoids

Armstrong GA (1994) J Bacteriol 176:4795–4802

Menaquinones

Bentley R, Meganathan R (1982) Microbiol Rev 46:241–280

Hiratsuka T, Furihata K, Ishikawa J, Yamashita H, Itoh N, Seto H, Dairi T (2008) Science 321:1670–1673

Li X, Apel D, Gaynor EC, Tanner ME (2011) J Biol Chem 286:19392–19398

Meganathan R, Kwong O (2009) Chapter 3.6.3.3. Biosynthesis of menaquinone and ubiquinone. In: Böck A, Curtiss R III, Kaper JB, Karp PD, Neidhardt FC, Nyström T, Slauch JM, Squires CL, Ussery D (eds) EcoSal Escherichia coli and Salmonella: cellular and molecular biology. ASM Press, Washington, DC. http://www.ecosal.org

Palanappian C, Sharma V, Hudspeth MES, Meganathan R (1992) J Bacteriol 174:8111–8118

Chapter 37
Biosynthesis of the Tetrapyrrole Ring System

The pathways for the biosynthesis of chlorophyll and respiratory pigments of the heme type make use of the same series of reactions up to the protoporphyrin stage. Judicious use of isotopes, of cell preparations catalyzing some of the reactions involved and degradation techniques specially devised for the task, have allowed the brilliant elucidation of the pathway by several groups of workers, among which David Shemin occupies the main place.

Tetrapyrroles are nearly ubiquitous in nature as participants in a wide variety of biological reactions that are central to life, such as electron transfer, gas binding, and one-carbon metabolism. Because of their diverse colors – for example the greens of plant chlorophylls, reds of blood, and blues and browns of avian eggs – tetrapyrroles have been called the "pigments of life". The chemical diversity of cyclic tetrapyrroles owes much to their ability to coordinate a variety of redox active metals in a fashion that allows for fine-tuning of midpoint potentials, increasing or decreasing reactivity of the metal center, and providing protection against undesirable side reactions. One finds magnesium in chlorophyll, iron in hemes, nickel in factor F_{430} (see Chap. 13) and cobalt in cyano-cobalamin (vitamin B_{12}) (see Chap. 38). Although hemes and chlorophyls are synthesized by both prokaryotes and eukaryotes, factor F_{430}, which is involved in methanogenesis, is produced only by some archaea, and cobalamin synthesis is found only in bacteria and archaea.

Synthesis of Protoporphyrin

δ-Aminolevulinic acid (ALA) is the central precursor of porphyrin biosynthesis. Eight molecules of ALA provide all the carbon and nitrogen atoms for the porphyrin skeleton of heme and chlorophyll. In some bacteria and in the mitochondria of yeast, birds and mammals, ALA is formed along the so-called "Shemin route" by the condensation of glycine and succinyl CoA, catalyzed by ALA synthase.

G.N. Cohen, *Microbial Biochemistry*,

DOI 10.1007/978-94-017-8908-0_37,

$$\mathrm{COOH-(CH_2)_2-CO-SCoA+NH_2-CH_2-COOH}$$
$$\mathrm{\rightarrow COOH-(CH_2)_2-CO-CH_2-NH_2+CO_2+CoA}$$
ALA

The enzyme catalyzing this reaction is a homodimer and requires pyridoxal phosphate as a cofactor.

In contrast, in a variety of organisms including *E. coli*, *B. subtilis*, photosynthetic bacteria, cyanobacteria, methylotrophs and methanogenic archaebacteria and in the chloroplasts of higher plants and algae, δ-aminolevulinic acid is formed in a two-step reaction utilizing the five-carbon skeleton of glutamic acid. Glutamate is first activated by an aminoacyl bond to t-RNA^{Glu} (UUC) by a glutamyl-tRNA synthetase step, undistinguishable from that occurring in protein synthesis. An NADPH-dependent glutamyl-tRNA reductase (*hemA*, coding for a 45 kD protein) reduces then the tRNA activated glutamic acid to glutamate 1-semialdehyde. Glutamate1-semialdehyde-2,1 -aminomutase (*hemL*) catalyzes a shift in the five-carbon skeleton to produce ALA:

$$\mathrm{HOOC-(CH_2)_2CH(NH_2)-CHO \rightarrow HOOC-(CH_2)_2-CO-CH_2-NH_2}$$
ALA

This is an interesting transformation, since it is one of the few reactions where a tRNA is utilized in a biosynthetic process other than protein biosynthesis.

ALA dehydratase (*hemB*) catalyzes a reaction between two molecules of δ-aminolevulinic acid to form the pyrrole derivative, porphobilinogen.

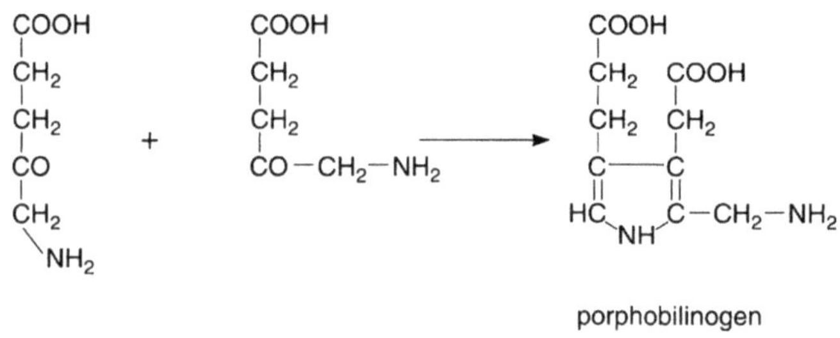

ALA dehydratase is composed of eight identical subunits and requires at least four zinc atoms. The first ALA molecule binds to the enzyme through a lysine group and ends up at one side of the final pyrrole, therefore forming the propionate side chain moiety, whereas the acetate side chain occupies the position near the reactive aminomethyl group.

The assembly of the next intermediate, uroporphyrinogen III, from porphobilinogen requires two enzymes, porphobilinogen deaminase, whose sequence of 313 residues has been deduced in *E. coli* from the nucleotide sequence of the corresponding gene (*hemC*) and uroporphyrinogen III synthase (also called cosynthase, *hemD*). Porphobilinogen deaminase (which should actually be called hydroxymethylbilane synthase, its product being the substrate for the cosynthase) first catalyzes the tetramerization of porphobilinogen to yield the linear 1-hydroxymethylbilane or pre-uroporphyrinogen III, which in the presence of the cosynthase is transformed by ring D inversion and cyclization to uroporphyrinogen III (in ring D, the relative positions of the propionate and acetate side chains are inversed, compared to preuroporphyrinogen; the mechanism of this inversion poses a huge enigma to the organic chemists, and remains to be solved).

The overall stoichiometry of the reaction is as follows:

$$4\ \text{porphobilinogen} \rightarrow \text{uroporphyrinogen III} + 4\,NH_3$$

The mechanism of hydroxymethylbilane synthase has been extensively analyzed. This enzyme contains a unique nucleophile prosthetic group, a dipyrromethane assembled from two molecules of the substrate porphobilinogen linked together. The enzyme itself is responsible for the assembly and attachment of its own cofactor to a conserved cysteine residue (Cys 242). Substitution of this cysteine with serine results in a protein which has no catalytic activity while a similar mutation of Cys99 to serine does not affect enzymatic activity. The sequences of the yeast, *Euglena gracilis* and human enzymes are also known and there is a conserved cysteine residue in all four proteins corresponding to the *E. coli* Cys242 residue. This suggests that the mechanism of action of the enzyme from the other sources is similar to that of the *E. coli* enzyme.

The dipyrromethane cofactor acts as a primer onto which the four substrate molecules are attached sequentially. The enzyme catalyzes the deamination of each substrate, followed by carbon-carbon bond formation and deprotonation. Pre-uroporphyrinogen synthesis proceeds by attaching initially the first substrate molecule (ring A) to the free α position of the cofactor to form a stable enzyme-intermediate complex ES. Subsequent ordered addition of further substrate molecules B, C and D then yield ES2, ES3 and ES4. After the fourth substrate has been added, the enzyme catalyzes the release of pre-uroporphyrinogen by a hydrolysis reaction (Fig. 37.1).

Evidence suggests that hydroxymethylbilane synthase catalyzes the repetitive deamination and condensation reactions at a single catalytic site. Hence, it must also possess the structural capability to manipulate the cofactor and the growing polymer through the catalytic machinery. This is reminiscent of the mobile 4′-phosphopantetheine arm encountered in the study of fatty acid biosynthesis. The recent crystallization of hydroxymethylbilane synthase from *E. coli* throws further light on this extraordinary enzyme. Its three-domain structure has been defined by X-ray analysis at 1.9 Å resolution. The two first domains structurally resemble the transferrins and the periplasmic binding proteins, while the dipyrromethane cofactor is covalently linked to the third domain and bound by extensive salt bridges and hydrogen bonds within the cleft between domains 1 and 2, at a position corresponding to the binding sites for small molecule ligands in the analogous proteins. The tridimensional structure as well as results from site-directed mutagenesis are compatible for a single catalytic site. Interdomain flexibility may aid elongation of the polypyrrole product in the active site cleft of the enzyme. The mechanisms of the tetramerization reaction and cofactor assembly have been ably analyzed using

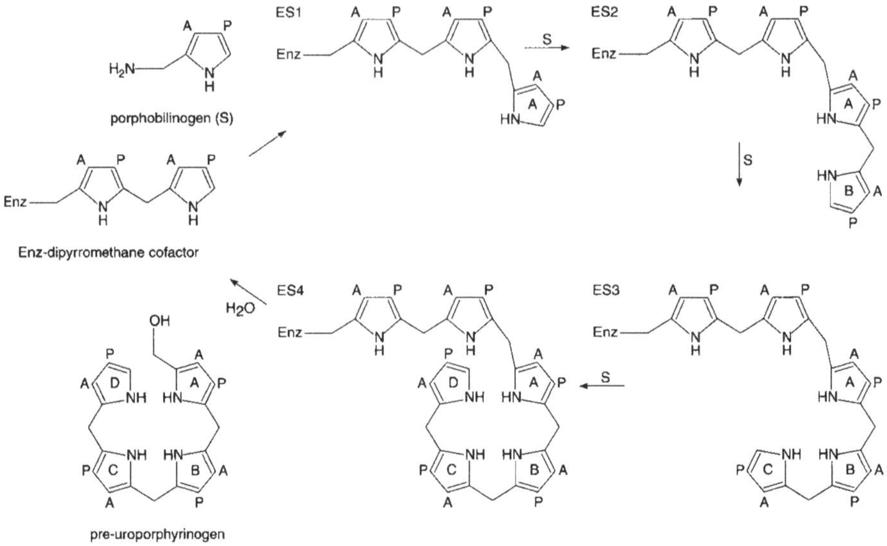

Fig. 37.1 The reaction catalyzed by hydroxymethylbilane synthase. The holoenzyme (*E*) is composed of the apoenzyme (*Enz*) linked to a resident covalently linked prosthetic group called the dipyrromethane cofactor. The four molecules of substrate (*s*) labeled *A*, *B*, *C* and *D* are added enzyme-bound cofactor of the holoenzyme (E) to give the intermediate complexes *ES1*, *ES2*, *ES3* and *ES4*. The release of the product pre-uroporphyrinogen occurs by hydrolysis leaving the dipyrromethane cofactor still linked to the enzyme. *A* acetate side chain, *P* propionate side chain (From P.M. Jordan et al., with permission of the *Journal of Molecular Biology*)

the known tridimensional structure and site-directed mutagenesis of a highly conserved aspartate residue.

hemD, the gene for the uroporphyrinogen III synthase proper (the cosynthase) has been cloned and sequence and the correponding protein overproduced.

After the synthesis of uroporphyrinogen III is completed, the first divergence in the pathway occurs. Decarboxylation commits the pathway towards protoporphyrin IX, heme and chlorophyll. On the other hand C-methylation of uroporphyrinogen III, directs the route towards the synthesis of siroheme (the cofactor of sulfite reductase, see Chap. 29), of factor F_{430} (a nickel-containing cofactor of methanogenesis, see Chap. 13) and of vitamin B_{12} (see Chap. 38).

Uroporphyrinogen III → Protoporphyrinogen →
Heme (Fe) and chlorophyll (Mg)
↓
Precorrin-2 → Vitamin B_{12}
↓
Siroheme (Fe) and Factor F_{430} (Ni)

Uroporphyrinogen III is a tetrapyrrole in which the acetyl residue of the porphobilinogen molecule remains intact. Decarboxylation by urodecarboxylase producing coproporphyrinogen III (*hemE*), in which these acetyls have become methyls, occurs in steps, since porphyrin intermediates with seven, six or five carboxyls have been detected.

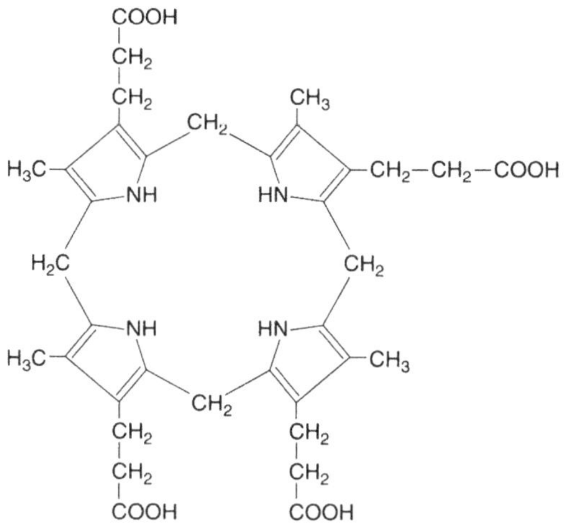

coproporphyrinogen III

An enzyme system, coprodecarboxylase (*hemF*), which in ox liver is firmly bound to mitochondria; carries out two oxidative decarboxylation reactions, the effect of which is to transform two of the propionyl residues of coproporphyrinogen III to vinyl groups, forming protoporphyrinogen IX which is oxidized to protoporphyrin by protoporphyrinogen oxidase (*hemG*).

$$\text{Coproporphyrinogen III} \xrightarrow[-4\,H]{-2\,CO_2} \text{Protoporphyrinogen} \xrightarrow{-6\,H} \text{Protoporphyrin IX}$$

The enzyme from *B. subtilis* is a peripheral membrane protein. It can be recovered in the soluble fraction by salt treatment, in contrast to all previously reported eukaryotic or prokaryotic protoporphyrinogen oxidases, which are membrane-bound. The only bacterial protoporphyrinogen oxidase purified so far is that of *Desulfovibrio gigas*, a multimeric protein of 148 kDa dissociating into three peptides of 57, 18.5 and 12 kDa under reducing conditions.

In addition to *hemF*, a second coproporphyrinogen oxidase, which is oxygen-independent, is encoded by *hemN*.

The porphyrinogens are colorless substances. The color of the porphyrins is due to dehydrogenations which set up a system of conjugated double bonds, a chromophore with strong absorption bands in the near ultraviolet and visible regions of the spectrum.

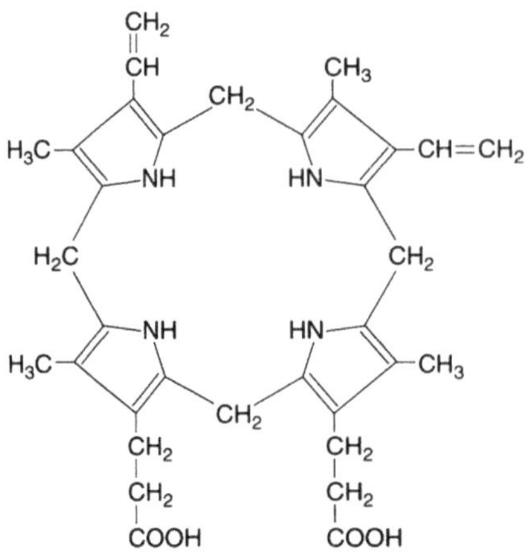

protoporphyrinogen

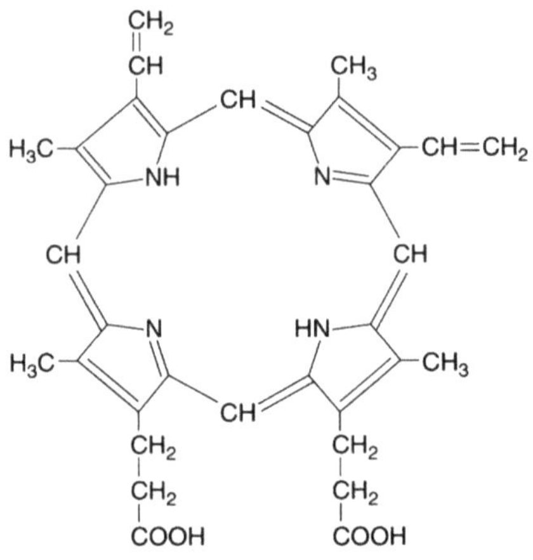

protoporphyrin IX

The Roman numerals following the names of different porphyrinogens and porphyrins are to distinguish the different possible isomers. For example, the methyl, vinyl and propionyl chains in protoporphyrin can be arranged to form fifteen different isomers. In fact, a single series of isomers exist in nature, the protoporphyrin IX series, which we shall now on refer to as protoporphyrin. Other isomers have been isolated from natural sources, but they have no physiological role, and arise from disorders in tetrapyrrole metabolism, or from spontaneous chemical reactions among the precursors of tetrapyrroles.

Synthesis of Heme from Protoporphyrin

The metal atom is incorporated into heme at the protoporphyrin stage. In microorganisms, like *Hemophilus influenzae*, the growth requirement for heme can be satisfied by protoporphyrin. Although the ferrous ion spontaneously forms coordination complexes with protoporphyrin, mutants of *Staphylococcus aureus*, requiring hemin, are found which cannot grow on protoporphyrin. This suggested that the insertion of the iron atom was due to an enzyme, ferrochelatase. Mutants lacking this enzyme have also been described in *E. coli* (*hemH*).

A (2Fe-2S) cluster has been detected in mammalian ferrochelatase; no change in the cluster oxidation state was observed during catalysis.

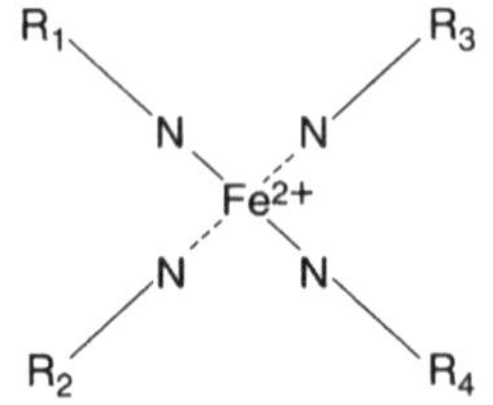

heme or ferrous protoporphyrin

The control of the biosynthesis of the tetrapyrrole ring system occurs at different levels.

Firstly, an iron deficiency in the bacterial growth medium stimulates a considerable build-up of porphyrins exceeding, in *Rhodopseudomonas spheroides* for example, one hundred times the amount of all tetrapyrrole derivatives (bacteriochlorophyll + heme derivatives) normally synthesized in non iron-deficient media.

This phenomenon has been analyzed in detail with cell-free extracts, and it has been shown that the ALA-synthase from this microorganism is very sensitive to hemin, the effect being appreciable even at a concentration of 10^{-7} M. This inhibition is non-competitive with respect to both substrates, succinyl CoA and glycine. Interpretation of the results obtained in vivo is as follows: the control of tetrapyrrole biosynthesis in *R. spheroides* occurs by feedback inhibition of ALA synthase by hemin, the formation of hemin itself depending on the concentration of iron in the medium.

The addition of ALA to cultures of *R. spheroides* also causes a severe repression of the synthesis of ALA synthase and ALA dehydratase. It has been suggested that in liver, heme regulates ALA synthase not by altering the rate of transcription of the coding gene, but by decreasing the half-life of the corresponding mRNA.

δ-Aminolevulinate dehydratase and ferrochelatase are also feedback inhibited by heme.

R. spheroides, in which the regulation of heme synthesis has been the most carefully studied, is a photosynthetic organism which contains another tetrapyrrole, bacteriochlorophyll. This obviously adds a new parameter in the regulation of the synthesis of tetrapyrroles by the photosynthetic organisms.

Heme Biosynthesis in Archaea

Despite the fact that heme biosynthesis has been investigated for a long time, there are still various fundamental open questions. One concerns heme biosynthesis in archaea. Archaea have been shown to possess hemes and heme-dependent proteins Furthermore, essential metabolic pathways such as methanogenesis in methanogenic archaea are dependent on tetrapyrrole cofactors, such as F430 Nevertheless, the genes necessary for the late steps of heme biosynthesis have not been identified. Biochemical and genomic information indicate that heme biosynthesis starts from 5-aminolevulinic acid, which is synthesized from glutamyl-tRNA, and proceeds via porphobilinogen, preuroporphyrinogen and uroporphyrinogen III as intermediates. At this stage the pathway appears to deviate from that known to be operative in eukarya and most bacteria. Most genomes of heme-synthesizing archaea lack genes with homology to *hemE*, *hemF*, *hemN*, *hemG*, *hemY* and *hemH* which encode the enzymes catalyzing the successive conversion of uroporphyrinogen

III via coproporphyrinogen III, protoporphyrinogen IX and protoporphyrin IX to protoheme. The methanogenic archaeon *Methanosarcina barkeri* synthesizes protoheme via precorrin 2, which is formed from uroporphyrinogen III in two consecutive methylation reactions utilizing S-adenosyl-l-methionine as methyl donor. The existence of this pathway, previously exclusively found in the sulfate-reducing δ-proteobacterium *Desulfovibrio vulgaris*, was demonstrated for *M. barkeri* via the incorporation of two methyl groups from methionine into protoheme However, the corresponding enzymes have yet to be identified.

Synthesis of Chlorophyll from Protoporphyrin

The biosynthesis of chlorophyll and haem pigments is a fundamental aspect of photosynthesis, and has been the subject of biochemical and molecular genetics studies for many years. The photosynthetic bacteria are an ideal model, since they are capable of chemoheterotrophic growth in the dark. Then, the synthesis of bacteriochlorophyll can be chemically or genetically disrupted and the properties of non-photosynthetic mutants examined in aerobic cultures grown in the dark. As a result of these studies, it has been possible to identify the series of intermediates in the biosynthesis if bacteriochlorophyll.

Tetrapyrrole biosynthesis proceeds from δ-aminolevulinic acid to protoporphyrin IX. At this point, insertion of Fe^{++} or Mg^{++} yields either heme or Mg-protoporphyrin IX, respectively; in the latter case, subsequent reactions encoded by special genes result in the eventual production of bacteriochlorophyll. Thus, the chelation of Fe^{++} or Mg^{++} is assumed to be the decisive step in this sequence, and an important control point, since it is here that the relative levels of the two tetrapyrroles may be determined. These levels vary widely depending on whether the cells are growing chemoheterotrophically with plentiful aeration, when most of the flux is directed toward heme biosynthesis, or whether they are growing under limited oxygen where the flux is largely directed towards bacteriochlorophyll. In the latter case, total tetrapyrrole production increases up to 100-fold, mainly because of the synthesis of high levels of ALA. Another regulatory problem posed by the versatility of *Rhodobacter sphaeroides* is the consequence of anaerobic photosynthetic growth. The decarboxylation of coproporphyrinogen III to form protoporphyrinogen requires molecular oxygen in animals, and in aerobically grown *R. spheroides*; it appears that anaerobically grown cells need to employ a different enzyme which exhibits requirements for Mg^{++}, ATP, methionine and NAD^+ or $NADP^+$ for activity.

All the intermediates between protoporphyrin IX and bacteriochlorophyll contain magnesium. A mutant of *Chlorella* accumulates the most simple of these compounds, magnesium protoporphyrin, whose structure, except for the metal is identical with that of heme. A gene has been characterized in the plant *Antirrhinum majus* which shows a high degree of homology to the products of two bacterial genes, *bchH* from *Rhodobacter capsulatus* which codes for an enzyme necessary for one of the first steps unique to the synthesis of bacteriochlorophyll, and *cobN* from *Pseudomonas denitrificans* (see Chap. 38) which is necessary for the chelation of cobalt during vitamin B_{12} biosynthesis. It has been proposed that the gene from *Antirrhinum majus* codes for the chelation of magnesium by protoporphyrin IX.

We know little about the mechanism of insertion of the magnesium, except that certain mutants accumulate free protoporphyrin and apparently lack the enzyme required for the incorporation of the metal. Another compound which accumulates in another mutant is the monomethyl ester of the preceding one. The chromatophores of *R. spheroides* contain an enzyme called S-adenosyl-L-methionine: magnesium protoporphyrin methyltransferase (*bchM* in *Rh. capsulatus*) which catalyzes the reaction:

Mg-protoporphyrin IX+S-adenosylmethionine
→ Mg-protoporphyrin IX monomethyl ester
+ S-adenosylhomocysteine

Examination of the structure of chlorophyll indicates which of the two residues of propionic acid is esterified. This step is the best studied at the enzymatic level between protoporphyrin and chlorophyll. Another mutant accumulates a third compound which can be derived from the preceding one only by a series of reactions. This is Mg-vinylpheoporphyrin *a5*, where one of the vinyl groups has been reduced to an ethyl group, the monomethylester of the propionic acid residue has been oxidized and the modified chain cyclized to form a substituted cyclopentanone.

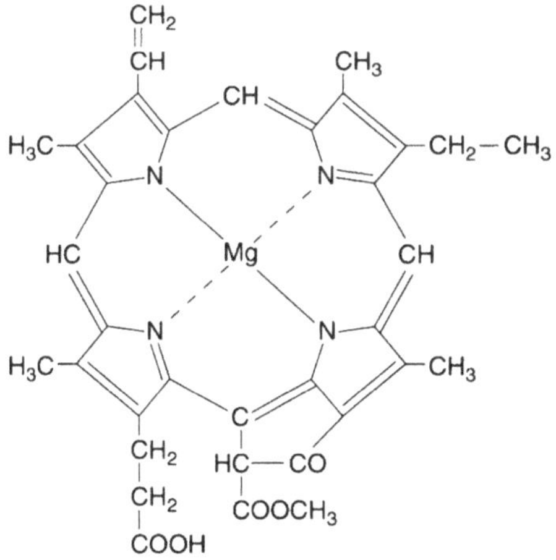

Mg vinylpheoporphyrin a_5
(protochlorophyll)

The reduction of this compound to chlorophyll involves reduction of the tetrapyrrole ring carrying the unchanged propionyl residue and an esterification of this residue by phytol, a C-20 diterpene alcohol. The intermediates in these reactions probably never occur in the free state and take place in higher plants and in some algae in the presence of light only.

The structural formulas of chlorophyll a, bacteriochlorophyll and phytol are found below:

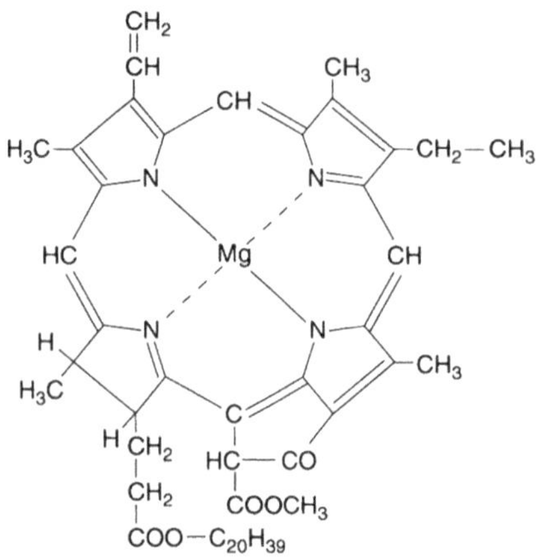

chlorophyll *a*

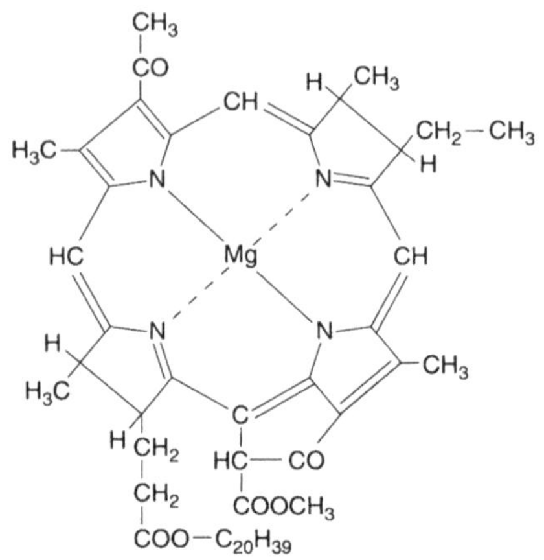

bacteriochlorophyll

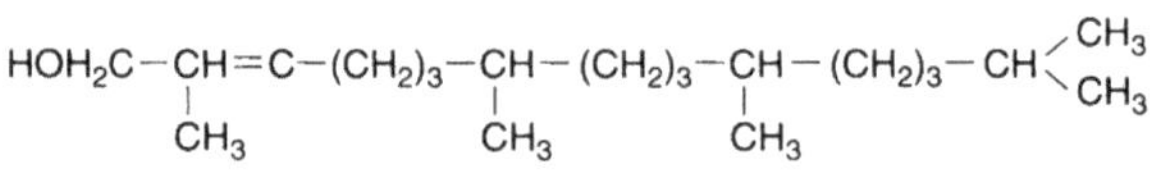

phytol $C_{20}H_{40}$

It was recently discovered that a *bchD* (Mg-chelatase) mutant of *Rhodobacter sphaeroides* produces an alternative bacteriochlorophyll(BChl) in which Mg^{2+} is substituted by Zn^{2+}. Zn-BChl has been found in only one other organism before, the acidophilic *Acidiphilium rubrum*. The *bchD* mutant was found to contain a Zn-protoporphyrin IX pool, analogous to the Mg-protoporphyrin IX pool found in the wild type strain. Inhibition of ferrochelatase with *N*-methylprotoporphyrin IX caused Zn-protoporphyrin IX and Zn-BChl levels to *decline* by 80–90 % in the *bchD* mutant, whereas in the wild type strain, Mg-protoporphyrin IX and Mg-BChl levels *increased* by 170–240 %. Two early metabolites of the Zn-BChl pathway were isolated from the *bchD* mutant and identified as Zn-protoporphyrin IX monomethyl ester and divinyl-Zn-protochlorophyllide.

Biosynthesis of the Phycobilin Chromophores. Chromatic Adaptation

Cyanobacteria and red algae often grow at about one meter under the surface of water where the light absorbed by chlorophyll has already been absorbed by chlorophyll or by organisms living above.

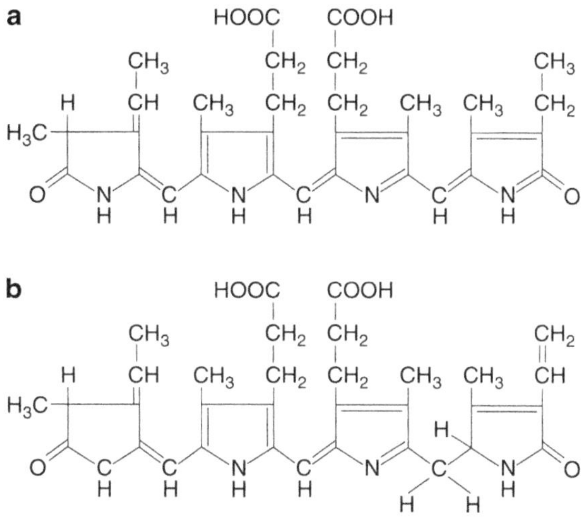

Fig. 37.2 Structures of the chromophores of phycobiliproteins. (**a**) Phycocyanobilin, the chromophore of phycocyanin and allophycocyanin; (**b**) Phycoerythrobilin, the chromophore of phycoerythrin. Both are covalently linked to the proteins with which they are associated

The phycobiliproteins are their major light-harvesting pigments and we shall now examine some of their properties.

They absorb light in a broad region near the middle of the visible spectrum, which reaches the depth at which the organisms thrive, and belong to three different spectral classes. Two blue pigments, allophycocyanin and phycocyanin, which have maxima at relatively long wavelengths, occur universally in cyanobacteria and red algae; the red pigment, phycoerythrin, which absorbs at shorter wavelengths, is formed by some, but not all, members of each group. The phycobiliproteins are contained in granules termed phycobilisomes which occur in regular arrays on the outer faces of the thylacoids, a system of flattened membranous sacs, near the surface region of the membrane. The overall design of the phycobilisome consists of a hemidisk of six rods radiating fanwise from a central core. Phycobilisomes are made of three principal classes of chromophoric water-soluble proteins (phycobiliproteins): allophycocyanin in the core, and phycocyanin and phycoerythrin in the rods, all composed of α and β subunits. All these chains derive from a unique common ancestor, structurally similar and evolutionarily related to the globin family. Open tetrapyrrole chromophores, the phycobilins, are post-translationally grafted to phylogenetically conserved cysteines on both subunits. Phycocyanobilin is the chromophore of phycocyanin and allophycocyanin; phycoerythrobilin is the chromophore of phycoerythrin (Fig. 37.2).

In addition, phycobilisomes contain auxiliary proteins devoid of chromophore, called the linker polypeptides.

The light energy absorbed by these pigments is transferred with very high efficiency to the chlorophyll-containing photosynthetic reaction centers in the thylacoids. The geometrical arrangement of phycobiliproteins in phycobilisomes as well as their spectral properties (overlapping of the spectra of the energy acceptor and of the energy donor) are such that the overall efficiency of energy transfer from one phycobiliprotein to the other and then to the chlorophyll is very high.

Phycocyanobilin arises from the degradation of protoheme via biliverdin IXα. This has been studied in the unicellular red alga, *Cyanidium caldarum*:

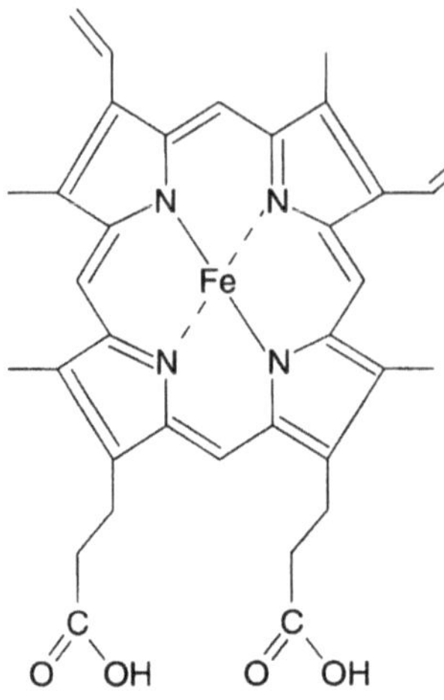

protoheme

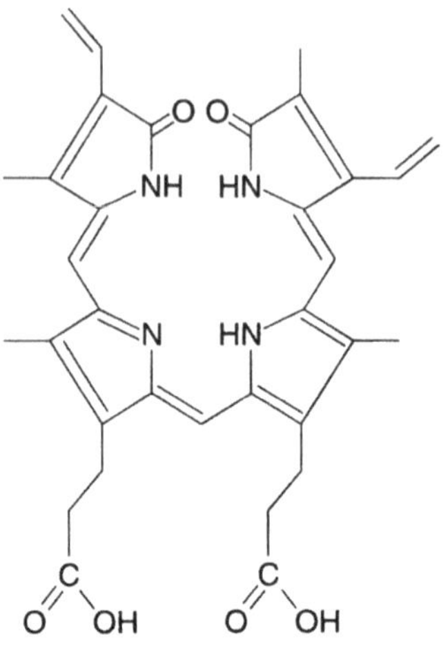

biliverdin IXα

The in vitro transformation of protoheme to biliverdin IXα and of the latter to phycobilins requires NADPH, ferredoxin and ferredoxin-$NADP^+$ reductase, as well as heme oxygenase (required for the opening of the tetrapyrrole ring, and yielding equimolar amounts of biliverdin, CO_2 and iron) and phycobilin formation enzymes. The only role of NADPH in both reactions of phycobilin synthesis is to reduce ferredoxin via ferredoxin-$NADP^+$ reductase: reduced ferredoxin can directly supply the electrons needed to drive both steps in the transformation of protoheme to phycocyanobilin.

In *C. caldarum*, the synthesis of the apoprotein components is induced by light and the phycobilin precursors; ALA, protoporphyrin IX, and protoheme can substitute for light, the regulation being exerted at the level of mRNA synthesis. This has been shown in particular for heme oxygenase.

Studies using interposon mutagenesis with *Synechococcus*, a cyanobacterium, suggest that a phycocyanobilin lyase, composed of two different subunits (coded by the *cpcE* and *cpcF* genes of this organism), is specifically required for chromophorylation of a subunit of apophycocyanin. The phycocyanobilin is covalently attached by a thioether linkage to a given cysteine (α-Cys84) of the apoprotein. The CpcE and CpcF of this organism have been purified. They form an enzymatically active 1:1 complex CpcEF which catalyzes the addition of the bilin to the apoprotein apo-αPC following a simple Michaelis-Menten kinetics. CpcEF catalyzes also the addition of phycoerythrobilin to the apoprotein, but there is a preference for phycocyanobilin both in binding affinity and in the rate of catalysis, sufficient to account for the selective attachment of phycocyanobilin to apo-αPC.

In a natural environment and whatever the geographical location, a great number of physico-chemical and biological factors result in quantitative and qualitative variations of the incident solar radiations. This is due in particular, to the diffusion and absorption properties of the molecules present in the atmosphere, to the circadian rhythms, to the alternance of seasons, to meteorological conditions and to the vegetation. In an aquatic medium, where cyanobacteria thrive, the selective absorption by water molecules of radiation of a wave length higher than 550 nm and the turbidity due to the presence of organic molecules in suspension, modify considerably the transmission spectrum of the incident light. In most cases, beyond a depth of 4 m, the light intensity is greatly diminished and the wave lengths higher than 600 nm are practically absent.

Although it is artificial to dissociate the effect of light intensity from the spectral qualities of the incident light, numerous studies have identified two different responses of the photosynthetic apparatus to the spectral quality of light, namely inverse chromatic adaptation and complementary chromatic adaptation. The first type concerns all photoautotrophic organisms, whereas the second has been shown to occur only in cyanobacteria.

Inverse chromatic adaptation is characterized by a preferential diminution of chlorophyll *a* in cells grown in red light, i.e. in a region of the visible spectrum mostly absorbed by this pigment. Inversely, in yellow or orange light, mainly absorbed by phycobiliproteins, the synthesis of chlorophyll *a* increases.

As early as 1883, Engelmann had reported a correlation between the distribution of cyanobacteria in a water column and their capacity to synthesize pigments complementary of the spectral quality of the incident light. These observations were extended at the beginning of the century to several cyanobacterial species and it was established that the synthesis of the red pigment, phycoerythrin, was predominant when the organisms were grown in green light whereas the synthesis of the blue pigment, phycocyanin, was predominant when the organisms were grown in red light. The interest in this complementary chromatic adaptation was revived in the sixties and the following conclusions may be drawn from the various studies performed:

(a) phycoerythrin and phycocyanin are relatively stable proteins which disappear by dilution during growth, when cells are transferred from red light to green light or vice versa.
(b) when grown in the dark, cells have a pigment composition identical to that of cells grown in red light, i.e. they synthesize only phycocyanin and stop synthesize phycoerythrin.
(c) only those species of cyanobacteria who have the genetic information to synthesize the two types of phycocyanin (of similar spectral properties) can modulate their content in this pigment in answer to the shift from one type of light to the other; in this case, phycocyanin-1 is synthesized whatever the quality of light, while phycocyanin-2 is synthesized only in red light.
(d) the synthesis of the linker proteins is coordinately regulated with the synthesis of the phycobiliproteins with which they are associated in the phycobilisome.
(e) the light signal is perceived by a photoreversible pigment (PRP) analogous to the plant phytochrome, and whose action spectrum shows two maxima at 540 and 640 nm. The two forms of PRP are interconvertible by red or green illuminations of a few minutes. In this purely photochemical process, only the last illumination received determines the type of phycobiliprotein which will be synthesized.

It has been established that the control of pigment synthesis occurs at the transcriptional level and that the induction of the operons leading to the synthesis of the corresponding pigments in green or red light requires protein synthesis.

These studies are presently in full development.

A Type of Chromatic Adaptation Under Conditions of Sulfur Starvation

According to the biotope, the availability of sulfur can considerably vary from about 100 μM in freshwater lakes and 0.2 mM in eutrophic lakes to 27 mM in sea water. In the hot sulfur springs, H_2S concentration can reach 2 mM. Within the same biotope (except in sea water), the sulfur concentrations can vary importantly. Many cyanobacteria have adapted to such ecological niches. In addition to the capacity of certain strains to utilize indifferently H_2O or H_2S, and to be thus able to grow anaerobically or aerobically, cyanobacteria respond in a variety of manners to sulfur starvation.

Three phycocyanin operons have been characterized in a species of *Calothrix*. The operon *cpc1*, which encodes the two phycocyanin 1 subunits, is expressed regardless of the light spectrum, whereas transcription of *cpc2*, which encodes the two phycocyanin 2 subunits and their three associated linker polypeptides is triggered by the red wavelength component. In addition, a third operon, *cpc3* is transcribed at a very low level, if any.

The *cpc3* operon, which remains silent in cells grown under standard conditions, is induced by sulfur deprivation. It includes five open reading frames corresponding to the five genes of the *cpc2* operon. They code for the α and β subunits of a novel phycocyanin and three linker polypeptides, respectively. Cells grown in white light, in a medium containing ample sulfate and therefore brown, turned green when they were inoculated in sulfur-poor medium. Spectral analysis revealed the disappearance of phycoerythrin and a concomitant increase of the phycocyanin absorption peak. The green cells could be indefinitely subcultured in this medium, but this could be reversed if ample sulfur was provided. Messenger RNA hybridization with specific DNA probes showed that the *cpc3* operon was highly expressed upon sulfur limitation while the expression of *cpc1* and *cpc2* was switched off.

The peculiarity of the newly transcribed operon is the absence of codons specifying sulfur-containing amino acids, except for the cysteine codons corresponding to the attachment of the phycobilin chromophores on the α and β subunits of phycocyanins and for the five methionine codons necessary for the initiation of translation of the five genes. In addition the N-terminal methionine residues are eliminated in the mature polypeptides. As a result, the mature proteins coded by the *cpc3* operon are completely devoid of methionine.

Selected References

The Tetrapyrroles

Granick S, Beale SI (1978) Adv Enzymol 46:33–203 (654 references)
Jordan PM (1991) Biosynthesis of tetrapyrroles. In: Bernardi G (ed) New comprehensive biochemistry, vol 19. Elsevier Science Publishers, New York/Amsterdam, 310 pp
Jordan PM, Warren MJ, Mgbeje BIA, Wood SP, Cooper JB, Louie G, Brownlie P, Lambert R, Blundell TL (1992) J Mol Biol 224:269–271
Woodcock SC, Jordan PM (1994) Biochemistry 33:2688–2695

ALA Synthesis

Jahn D, Verkamp E, Söll D (1992) Trends Biochem Sci 17:215–218

Heme Biosynthesis in Archaea

Buchenau B, Kahnt J, Heinemann IU, Jahn D, Thauer RK (2006) J Bacteriol 188:8666–8668

Phycobilins

Biswas A, Boutaghou MN, Alvey RM, Kronfel CM, Cole RB, Bryant DA, Schluchter WM (2012) J Biol Chem 286:35509–35521
Glazer AN (1983) Ann Rev Biochem 52:125–157
Rhie G, Beale SI (1992) J Biol Chem 267:16088–16093
Zhou J, Gasparich GE, Stirewalt VL, de Lorimier R, Bryant DA (1992) J Biol Chem 267:16138–16145

Complementary Chromatic Adaptation

Shukla A, Biswas A, Blot N, Partensky F, Karty JA, Hammad LA, Garczarek L, Gulu A, Schluchter WM, Kehoe DM (2012) Proc Natl Acad Sci U S A 109:20136–20141
Tandeau de Marsac N (1991) Chromatic adaptation by cyanobacteria. In: Bogorad L, Vasil IK (eds) The molecular biology of plastids and mitochondria. Academic, New York

The Effect of Sulfur Starvation on Chromatic Adaptation

Mazel D, Marlière P (1989) Nature 341:245–248

Chapter 38
Biosynthesis of Cobalamins Including Vitamin B_{12}

Vitamin B_{12} can be formally classified with the tetrapyrrole derivatives. It is based on a skeleton called corrin (Fig. 38.1), which differs from the other tetrapyrroles in the absence of a methene bridge between two of the pyrrole rings. In numbering the atoms in this formula, number 20 has been deliberately omitted, in order that all the corrin atoms should correspond to those of the other tetrapyrrole compounds. Some of the biosynthetic intermediates occurring before the formation of hydrogenobyrinic acid are called precorrins.

The structural formula of vitamin B_{12} (or rather of its coenzyme form) is given below (Fig. 38.2).

The formula shows a centrally bound cobalt atom, coordinated on the one hand by four nitrogen atoms from the tetrapyrrole ring and on the other by one of the nitrogens of a benzimidazole derivative. The sixth coordination position is the only known case in biology of an organometallic metal-carbon bond, where the substituent can be a methyl group, or a deoxyadenosyl group (Ado) in the case of the coenzyme forms of vitamin B_{12}. However, the extraction from natural sources usually yields the vitamin in the form of cyanocobalamin, where the substituent is –CN, which can be replaced by -OH (aqua derivative), –Cl, $-NO_2$ or –CNS.

Detailed work has shown that the first steps in the biosynthesis of the vitamin are the same as for hemin and chlorophyll. Glycine, δ-aminolevulinic acid, porphobilinogen and uroporphyrinogen III are readily identified intermediates (see the section on the biosynthesis of the tetrapyrrole ring).

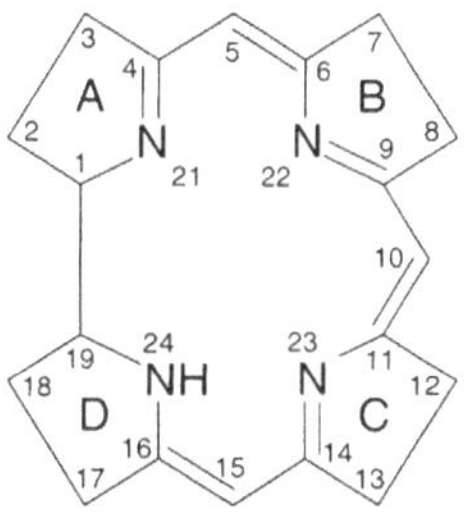

Fig. 38.1 The corrin skeleton

G.N. Cohen, *Microbial Biochemistry*,
DOI 10.1007/978-94-017-8908-0_38, © Springer Science+Business Media B.V. 2014

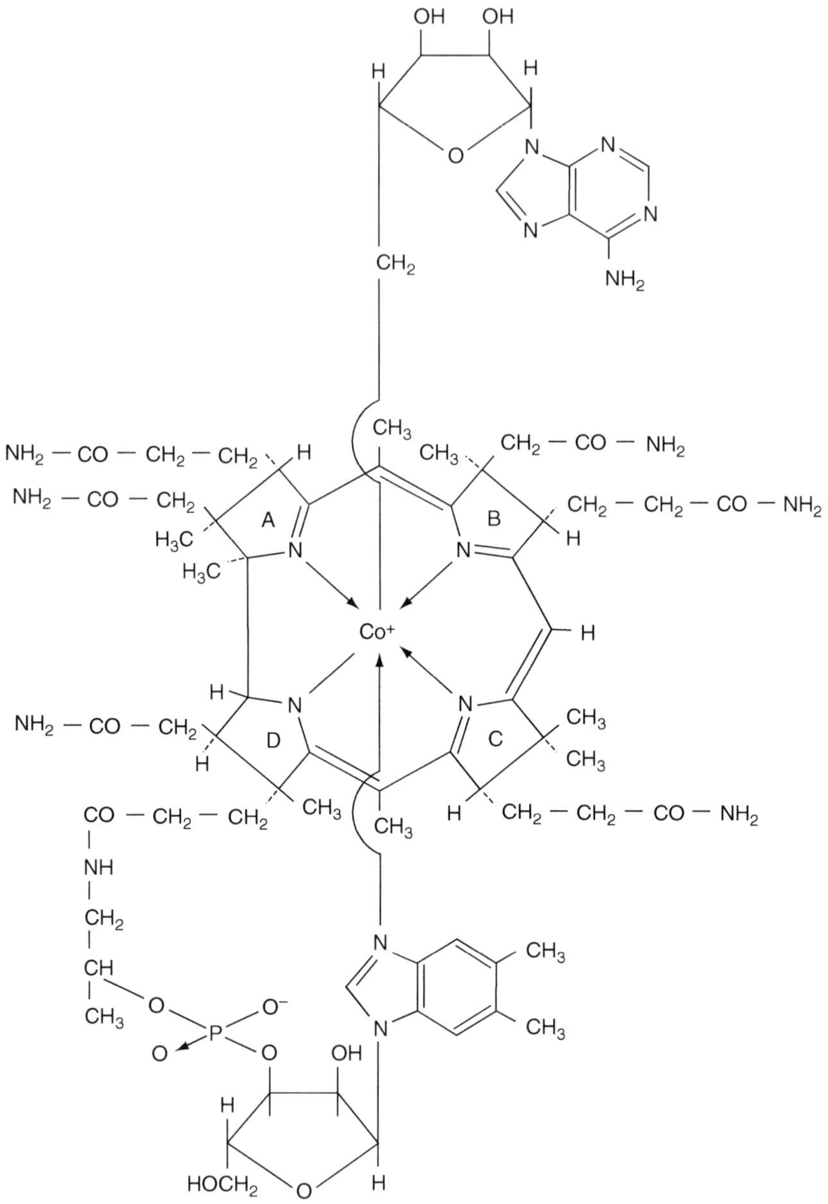

Fig. 38.2 5′-desoxyadenosylcobalamin

Steps for the conversion of uroporphyrinogen III into cobyrinic acid include at least eight C-methylations (the methyl groups coming from S-adenosylmethionine), elimination of C-20, NADPH-dependent reduction of the macrocycle,

Table 38.1 Complementation groups of the genes involved in coenzyme B_{12} biosynthesis in Pseudomonas denitrificans

Complementation group A
Methylation at C-2O, *cobI* (precorrin 2 → precorrin 3)
Methylations at C-17, C-11 and C-1, *cobF* (unidentied enzymes leading from precorrin-3 to precorrin-6x)
Elimination of C-2O, *cobG*
NADPH-dependent reduction of precorrin-6x to precorrin-6y (transfer of a hydride equivalent to C-19), *cobK*
Methylations at C-15 and C-5, decarboxylation at C-12, precorrin- 6y → precorrin -8x, *cobL*
Mutase catalyzing the methyl migration from C11 to C12, precorrin-8x → hydrogenocobyrinic acid, *cob H*
Complementation group B
Insertion of cobalt, cobaltochelatase, *cobST*
Complementation group C
Methylations at C-2 and C-7, uroporphyrinogen III → precorrin 2, *cobA a, c* amidations of hydrogenobyrinic acid, *cobB*
Insertion of cobalt, cobaltochelatase, *cobN*
Cob (I) alamin adenosyltransferase, adenosylation of the *a, c* diamide of cobyrinic acid, *cobO b, d, e*, and *g* amidations, *cobQ*
Enzyme system catalyzing the addition of 1-amino-2-propanol to yield cobinamide, *cobC* and *cobD*
Bifunctional cobinamide kinase-cobinamide phosphate guanylyltransferase, yielding GDP-cobinamide, *cobP*
System possibly involved in the reduction of cobalt leading to Co (I) corrinoids, *cobW*
Complementation group D
Nicotinate mononucleotide: 5,6-dimethylbenzimidazole-phosphoribosyltransferase, *cobU*
Incorporation of 5,6-dimethylbenzimidazole-ribose-5′-phosphate, *cobV*

decarboxylation of the C-12 acetate, methyl migration from C-11 to C-12, and cobalt insertion. In *Pseudomonas denitrificans*, where a thorough genetic analysis has been made, about 20 genes (the *cob* genes) have been cloned, grouped in four genomic loci corresponding to four distinct complementation groups (from A to D, Table 38.1).

Known precorrins are precorrin 3, which carries three methyl groups at carbons 2 and 7 and 20; precorrin 6x, which carries five methyl groups at C-1, C-2, C-7, C-11 and C-17, has a contracted macrocycle, still carries the C-12 acetic acid side chain, and stands at the oxidation level of a dehydrocorrin. The structure of precorrin-6x implies that it has undergone ring contraction and extrusion of C-20, and that it has to be reduced to the oxidation level of a corrin (precorrin- 6y), followed by further methylations at C-5 and C-15, decarboxylation of the acetic residues at C-12 to yield precorrin 8x and methyl migration to C-12, leading to hydrogenobyrinic acid. Cobalamin biosynthesis is considered to require between 20 and 30 steps to transform uroporphyrinogen III into cobalamin.

The ring contraction process that occurs during cobalamin biosynthesis is mediated via the action of two enzymes, CobG and CobJ. The first of these

Fig. 38.3 Synthesis of precorrin-2

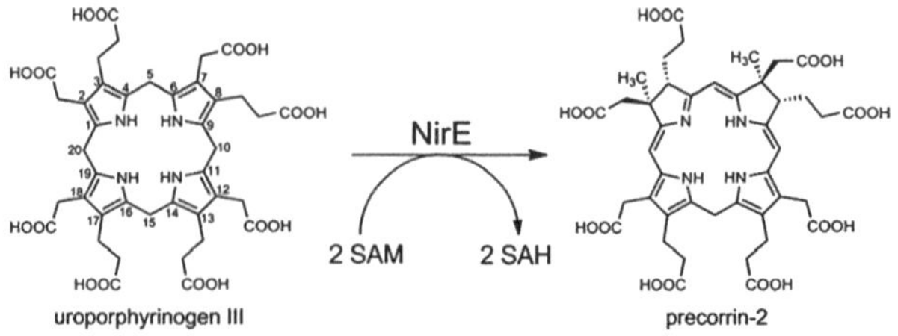

generates a tertiary alcohol at the C-20 position of precorrin-3A by functioning as a monooxygenase, a reaction that also forms a gamma lactone with the acetic acid side chain on ring A. The product, precorrin-3B, is then acted upon by CobJ, which methylates at the C-17 position and promotes ring contraction of the macrocycle by catalyzing a masked pinacol rearrangement. In *Pseudomonas denitrificans* and *Brucella melitensis*, both enzymes contain a [4Fe-4S] center as well as a mononuclear non-heme iron.

A nice parallel between the arrangement of the *cob* genes and the part of the pathway on which the gene products act can be drawn; it is not known whether this correlation results in common regulation of the expression of genes belonging to the same cluster. It is striking that in *S. typhimurium*, the *cob* genes are also organized into three classes related to three parts of the pathway, such a genetic clustering having been also reported in *Bacillus megaterium*.

Only six methyltransferases are required to perform the eight methylation reactions in *Ps. denitrificans*.

The first methylating enzyme, S-adenosylmethionine: uroporphyrinogen III methyltransferase has been purified to homogeneity. This enzyme (NirE) is a homodimer (2 × 30 kDa) and performs the mehylations at C-2 and C-7, leading to precorrin-2 (Fig. 38.3). The next reaction, catalyzed by S-adenosylmethionine: precorrin-2 methyltransferase, also obtained as a homogeneous homodimeric protein (2 × 26 kDa), is a methylation at C-20 on precorrin-2 that generates precorrin-3.

The nature of the reactions catalyzed by enzymes coded by *cobK*, *cobL* and *cobH* have been elucidated. *cobK* codes for a NADPH-dependent precorrin-6x reductase transforming precorrin-6x to the dihydroderivative precorrin-6y. This reductase has been purified to homogeneity and is a monomer of 263 amino acid residues. The product of *cobL* (subunit of 42.9 kDa, oligomeric structure unknown) methylates at C-5 and C-15 and decarboxylates the acetic acid side chain at C-12, yielding precorrin-8x.

A mutase, purified 80-fold, coded by *cobH* transfers a methyl group from C-11 to C-12 and transforms precorrin-8x to hydrogenocobyrinic acid, the cobalt-free counterpart of cobyrinic acid. Figures 38.4 and 38.5 summarize the sequence of events and give the structure of cobyrinic acid.

Fig. 38.4 Biosynthetic sequence from precorrin-6x to hydrogenobyrinic acid in *Pseudomonas denitrificans. SAH* S-adenosylhomocysteine (From D. Thibaut, M. Couder, A. Famechon, L. Debussche, B. Cameron, J. Crouzet and F. Blanche, with permission of the Journal of Bacteriology)

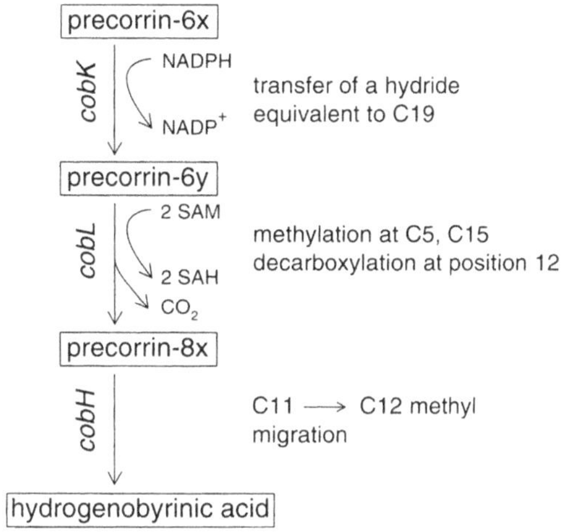

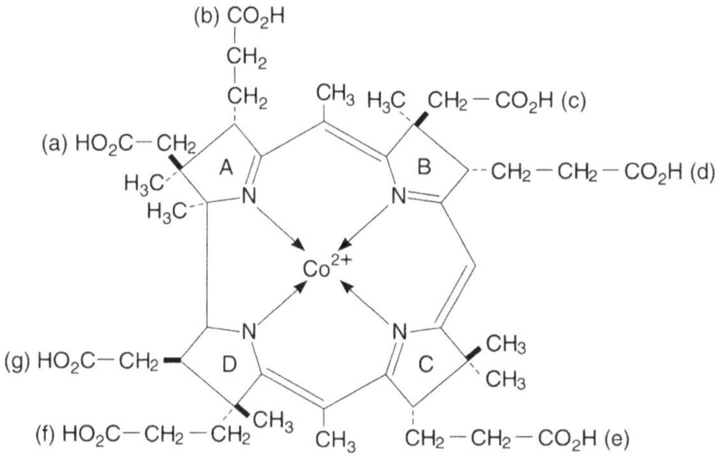

Fig. 38.5 Cobyrinic acid. This acid has first been isolated from cultures of *Propionibacterium shermanii*. If we consider that corrin is in the plane of this page, the *boldface lines* connect substituants above the plane, the *dotted lines* link substituants below the plane to the corrin nucleus. Hydrogenobyrinic acid is the cobalt-free counterpart of cobyrinic acid. The hexaamide corresponding to the six acid groups is called cobyric acid. When 1-amino-2-propanol is added to cobyric acid, cobinamide is obtained

Cobinamide Biosynthesis

The biosynthesis of cobinamide from hydrogenobyrinic acid involves six stepwise amidations of the peripheral carboxyl groups at positions *a*, *b*, *c*, *d*, *e* and *g* to yield the corresponding amides and the attachment of (R) 1-amino-2-propanol at position *f* (Fig. 38.5).

A number of partially amidated corrinoids have been isolated from cultures of microorganisms or transformed into complete corrinoids during feeding experiments to discover the sequence for attachment of unsubstituted amide groups and of aminopropanol.

The major corrinoids existing in nature apparently exist in the coenzyme form. The 5′- deoxy 5′- adenosyl (Ado) axial upper ligand of cobalt is presumably introduced after the completion of the corrin ring, but certainly soon after the cobyrinic acid stage, since this compound has been isolated only in the diaqua form, where the two axial ligands of cobalt are water molecules. Co (I) attacks the 5′ carbon of ATP and displaces the triphosphate group. S-adenosylmethionine synthetase provides the only other biochemical reaction where a nucleophile displaces as a unit the triphosphate group of ATP.

The first amidation activity along the cobalamin pathway from cobyrinic acid has been attributed to an enzyme purified to homogeneity which catalyzes the stepwise amidations to the *c* -monoamide and to the *a, c* - diamide, called cobyrinic acid *a, c* -diamide synthase. This enzyme is encoded by *cobB* (complementation group C).

The single enzyme (encoded by *cobQ*, complementation group C) that catalyzes amidation of 5′-deoxy-5′-adenosyl cobyrinic acid *a, c*- diamide was purified to homogeneity from extracts of a recombinant strain *of Pseudomonas denitrificans*. Its molecular weight has been estimated by gel filtration to be 97,300 and by gel electrophoresis under denaturing conditions of 57,000. Since stepwise Edman degradation provides a single residue at each step of degradation during sixteen degradation steps, this suggests the enzyme is a homodimer. The enzyme acts on the coenzyme form of substrate only and uses glutamine as the amide donor. This cobyric acid synthase is thus a glutamine amidotransferase, and shares the properties of the enzymes of this group: sensitivity to oxidation, requirement for a reducing agent and replaceability of glutamine by ammonia. It catalyzes the four stepwise amidations of carboxyl groups at positions *b, d, e*, and *g* to yield cobyric acid. The specific order in which the four amidations were carried out were not determined although the tri- tetra- and penta-amides were clearly characterized.

A series of studies suggests that cobalt insertion occurs at the level of hydrogenobyrinic acid *a, c* diamide in *Ps. denitrificans*. The supporting evidence is the following: (a) substantial quantities of hydrogenobyrinic acid, hydrogenobyrinic acid *a, c* diamide and cobyrinic acid *a, c* diamide accumulate in cobalamin producing cultures whereas cobyrinic acid has never been detected in these cultures. (b) Hydrogenobyrinic acid is a much better substrate than cobyrinic acid for the cobyrinic acid *a, c* diamide synthase (the enzyme specified by *cobB*). 3) some *cobB* mutants, which are devoid of the *a, c* diamide synthase, do not accumulate cobyrinic acid, still synthesize hydrogenobyrinic acid but not the diamide. The enzyme system catalyzing the insertion of cobalt, cobaltochelatase, is complex and consists of two components, respectively of 140 kDa (coded by *cobN*) and 450 kDa. The latter is composed of two polypeptides coded by *cobS* and *cobT*. Each component is inactive by itself, but the cobaltochelatase activity is reconstituted upon mixing the products of cobN and cobST. The reaction is ATP-dependent and starting from the diamide of hydrogenobyrinic acid, the product has been identified as being the diamide of cobyrinic acid. Insertion of the Ado group takes place at the stage of cobyrinic acid *a, c* diamide in *Ps. denitrificans*, based on two arguments: (1) cobalamin adenosyltransferase purified to homogeneity from this organism, a homodimer of 2×28,000

Mr, does not adenosylate cobyrinic acid, but adenosylates all of the corrinoids where at least positions *a* and *c* are amidated. (2) The *cobO* (complementation group C) mutants, which lack the enzyme, accumulate the diamide in the growth medium.

It has been claimed that in *Propionibacterium shermanii* the incorporation of aminopropanol (deriving from the decarboxylation of threonine) can occur at any of the amidated derivatives from the diamide to the pentaamide. However, in *Ps. denitrificans*, since no partially amidated corrinoids with the aminopropanol have been detected, it is highly probable that an enzyme system (cobC and cobD, complementation group C) inserts this residue at the level of cobyric acid (the hexaamide of cobyrinic acid), to finally yield the cobinamide. In *Salmonella enterica*, a threonine kinase (*pduX*) has recently been found and it is threonine phosphate which is actually the precursor of the aminopropanol.

Cobinamide kinase and cobinamide phosphate guanylyltransferase activities are carried by a bifunctional enzyme.

The biosynthesis of cobalamin from cobinamide is a four-step process in *P. shermani*. The first step is the phosphorylation of the hydroxyl group of the aminopropanol residue of cobinamide, to yield cobinamide phosphate which is then transformed into GDP-cobinamide through the addition of a GMP moiety originating from a molecule of GTP.

$$\text{Cobinamide} + \text{ATP} \rightarrow \text{Cobinamide phosphate} + \text{ADP}$$
$$\text{Cobinamide phosphate} + \text{GTP} \rightarrow \text{GDP - cobinamide} + \text{PPi}$$

In vitro enzymatic synthesis of cobinamide phosphate by cobinamide kinase and of GDP-cobinamide has been demonstrated with crude protein extracts of cobalamin-producing microorganisms (*P. shermanii*, *Propionibacterium arabinosum* and *Clostridium tetanomorphum*) but it is only later that a protein endowed with the two activities has been obtained in the pure state from a recombinant strain of *Ps. denitrificans*. When subjected to Edman degradation, the enzyme showed only one sequence for the 10 first residues at the N-terminus, sequence which was identical to the one inferred from the nucleotide sequence of the corresponding gene, *cobP* (complementation group C). Moreover, the estimated molecular weight of the denatured protein is identical to the one inferred from the DNA sequence.

Labeling experiments showed that the phosphate group of cobinamide phosphate is quantitatively retained as the inner phosphate of GDP-cobinamide during the guanylyltransferase reaction. However, the data are not sufficient to determine whether the bifunctional enzyme possesses one or two catalytic sites.

From GDP-Cobinamide to Cobalamin

The introduction of the nucleotide characteristic of vitamin B12 occurs at the GDP-cobinamide level. This nucleotide contains 5-6-dimethylbenzimidazole as the base. The biosynthetic reactions leading to of base are not known in detail, although in certain *Streptomyces* and *Propionibacteria*, it has been reported that all carbon and nitrogen atoms derive from riboflavin, presumably via an enzyme-catalyzed conversion of FMN involving the action of an oxygenase. Recently, in vitro and in vivo evidence that the BluB protein of the photosynthetic bacterium *Rhodospirillum rubrum* is necessary and sufficient for catalysis of the O_2-dependent conversion of FMNH2 to dimethylnenzimidazole. The product of the reaction was isolated by using reverse-phase high-pressure liquid chromatography, and its identity was established by UV-visible spectroscopy and mass spectrometry. In anaerobic microorganisms, riboflavin is not the precursor of dimethylbenzimidazole and it has been suggested that glycine, as well as methionine and the four carbons of erythrose might contribute to its synthesis.

Nicotinate nucleotide dimethylbenzimidazole phosphoribosyltransferase, first isolated from *P. shermanii* catalyzes the following reaction.

$$\text{5 - 6 dimethylbenzimidazole + nicotinate mononucleotide}$$
$$\downarrow$$
$$1 - \alpha - \text{D - 5 - phosphoribofuranosyl - 5,6 - dimethylbenzimidazole + nicotinate}$$

The characteristic product of the reaction has received the short convenient name of α-ribazole-5′-phosphate. It is coupled to Ado-GDP-cobinamide by an enzyme, cobalamin-5′-phosphate synthase, which requires Mg^{++} for its activity and has been partially purified from *Ps. denitrificans*. It is coded in this organism by the gene *cobV* (complementation group D). The products of the reaction are cobalamin-5′-phosphate and GMP. Finally a phosphatase removes the phosphate group to produce cobalamin, the coenzyme form of vitamin B_{12}. The enzyme is not specific for the coenzyme form or GDP-cobinamide and could use monocyano- or diaquacobinamide equally well. It can use also α-ribazole as a substrate although less well than α-ribazole phosphate. With the nucleoside as substrate, the reaction product is indeed coenzyme B_{12}.

The reaction sequence in the intermediate stages of the coenzyme B_{12} pathway in *Ps. denitrificans* can now be written:

$$\begin{aligned}&\text{Precorrin 2} \to \text{Precorrin3} \to\to \ldots \text{Precorrin 6x} \to \text{Precorrin}\\&\text{6y} \to \text{Precorrin 8x} \to \text{Hydrogenobyrinic acid}\\&\to \text{Hydrogenobyrinic acid a, c - diamide} \to \text{Cob(II) yrinic}\\&\text{acid a, c - diamide} \to \text{Cob(I) yrinic acid a, c - diamide} \to \text{Adenosyl}\\&\text{cobyrinic acid a, c - diamide} \to\to \ldots \text{Adenosylcobalamin}\end{aligned}$$

Only intermediates between precorrin-3 and precorrin-6x remained to be identified to complete the elucidation of the cobalamin pathway in *Ps. denitrificans*.

The anaerobic pathway for the biosynthesis of cobalamin (vitamin B_{12}) has remained poorly characterized because of the sensitivity of the pathway intermediates to oxygen and the low activity of enzymes. One of the major bottlenecks in the anaerobic pathway is the ring contraction step, which has not been observed previously with a purified enzyme system. The Gram-positive aerobic bacterium *Bacillus megaterium* has a complete anaerobic pathway that contains an unusual ring contraction enzyme, $CbiH_{60}$. $CbiH_{60}$ was characterized by electron paramagnetic resonance and shown to contain a [4Fe-4S] center. Assays with purified recombinant $CbiH_{60}$ demonstrate that the enzyme converts both cobalt-precorrin-3 into the ring-contracted product cobalt-precorrin-4 in high yields, with the latter transformation dependent upon DTT and an intact Fe-S center. Furthermore, the ring contraction process was shown not to involve a change in the oxidation state of the central cobalt ion of the macrocycle.

The time of cobalt insertion in cobyrinic acid-producing bacteria such as *P. shermanii* apparently differs from the one just described. It appears that precorrin-3 is synthesized and converted to the cobalt complex, indicating an early chelation.

Methylcobalamin is involved in methionine biosynthesis and in the methylations of tRNAs, whereas 5′-deoxycobalamin is the coenzyme of various isomerization reactions (glutamate mutase, L-methyl-malonylCoA mutase, diol dehydratase) and of ribonucleotide triphosphate reductase in *Lactobacillus leichmanii*. In methane production by methanogenic bacteria, the methylcobalamin contains 5-hydroxybenzimidazole instead of dimethylbenzimidazole.

Vitamin B_{12} coenzymes are also involved in the degradation of L-lysine by some species of Clostridia.

Elucidation of the Anaerobic Pathway for the Corrin Component of Vitamin B_{12}

It has been known for the past 20 years that two pathways exist in nature for the de novo biosynthesis of the coenzyme form of vitamin B12, adenosylcobalamin, representing aerobic and anaerobic routes. In contrast to the aerobic pathway, the anaerobic route has remained enigmatic because many of its intermediates have proven technically challenging to isolate, because of their inherent instability. However, by studying the anaerobic cobalamin biosynthetic pathway in *Bacillus megaterium* and using homologously overproduced enzymes, it has been possible to isolate all of the intermediates between uroporphyrinogen III and cobyrinic acid. Consequently, it has been possible to detail the activities of purified cobinamide biosynthesis (Cbi) proteins CbiF, CbiG, CbiD, CbiJ, CbiET, and CbiC, as well as show the direct in vitro conversion of 5-aminolevulinic acid into cobyrinic acid using a mixture of 14 purified enzymes. This approach has resulted in the isolation of the long sought intermediates, cobalt-precorrin-6A and -6B and cobalt-precorrin-8 (Fig. 38.6).

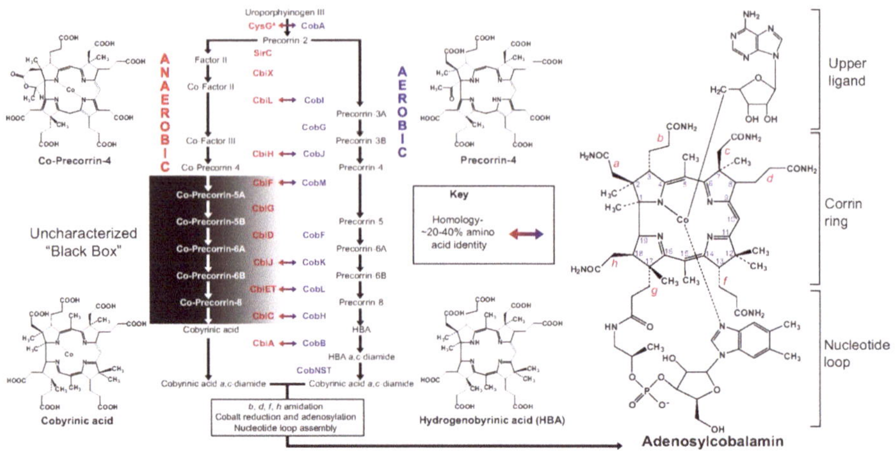

Fig. 38.6 Comparison between aerobic and anaerobic routes of adenosylcobalamin biosynthesis

Selected References

Many references to previous work will be found in the last two chapter.

Crouzet J, Blanche F, Cameron B, Thibaut D, Debussche L (1990) In: Baldwin TO et al (eds) Chemical aspects of enzyme biotechnology. Plenum Press, New York, pp 299–315

Jeter RM, Roth JR (1989) J Bacteriol 169:3189–3198

Moore SJ, Biedendieck R, Lawrence AD, Deery E, Howard MJ, Rigby SEJ, Warren MJ (2013a) J Biol Chem 288:297–305

Schroeder S, Lawrence AD, Biedendieck R, Rose R-S, Deery E, Graham RM, McLean KJ, Munro AW, Rigby SEJ, Warren MJ (2009) J Biol Chem 284:4796–4805

Scott AI, Roessner CA (2007) Pure Appl Chem 79:2179–2188

Storbeck S, Saha S, Krausze J, Klink BU, Heinz DW, Layer G (2011) J Biol Chem 286:26754–26767

Thibaut D, Couder M, Famechon A, Debussche L, Cameron B, Crouzet J, Blanche F (1992) J Bacteriol 174:1043–1049

Threonine Kinase and the Origin of the Aminopropanol Residue

Fan C, Bobik TA (2008) J Biol Chem 283:11322–11329

Fan C, Fromm HJ, Bobik TA (2009) J Biol Chem 284:20240–20248

Origin of the Dimethylbenzimidazole

Gray MJ, Escalante Semerena JC (2007) Proc Natl Acad Sci USA 104:2921–2926

Roth JR, Lawrence JG, Rubenfield M, Kieffer-Higgins S, Church GM (1993) J Bacteriol 175:3303–3316

Anaerobic Synthesis of the Corrin Component of Adenosylcobalamin

Moore SJ, Lawrence AD, Biedendieck R, Deery E, Frank S, Howard MJ, Rigby SEJ, Warren MJ (2013) Proc Natl Acad Sci USA 110:14906–14911

Anaerobic Pathway for the Corrin Component of Cobalamin

Daley HA (2013) Proc Natl Acad Sci USA 110:14823–14824
Moore SJ, Lawrence AD, Biedendieck R, Deery E, Frank S, Howard MJ, Rigby SEJ, Warren MJ (2013c) Proc Natl Acad Sci USA 110:14906–14911

Chapter 39
Interactions Between Proteins and DNA

DNA-Binding Proteins

Transcriptional control is the result of a complex cross-talk between regulatory factors and the basic transcription machinery. In addition to the simple regulation by the mere availability of regulatory factors, transcription has been shown, both in prokaryotes and in eucaryotes, to be modulated in several instances by posttranslational events, such as covalent modification (see the regulation of transcription of glutamine synthetase, Chap. 28), or redox control.

Regulation of transcription occurs largely, both in prokaryotes and in eukaryotes, through the binding of specific proteins, repressors or activators, that bind to specific DNA sequences. The source of the specificity is recognition by some of the amino acid side chains of the pattern of functional groups on the edges of the base pairs in the DNA major groove, mediated by hydrogen bonding. There is also a contribution from the sequence-dependent conformational preferences of the DNA backbone.

Proteins specifically recognizing certain DNA sequences control gene expression by interfering with the transcription process (for example Lac, Trp, and Met repressors, bacteriophage lambda *cro* and λ repressors, the CAP activating protein, etc.). Some other DNA binding proteins play also structural and catalytic roles in other cellular processes (e.g. polymerases and nucleases). How do these proteins recognize a particular base sequence and how do they bind to double-stranded DNA? These questions have become one of the central problems of biochemistry since the beginning of the 1980s and has been studied with physical (crystallography and nuclear magnetic resonance), biochemical and genetic methods.

Ptashne and his colleagues have shown that the Cro repressor binds to six operator sites on the double stranded DNA from phage λ, that they are grouped in two regions, each containing three 17 bp sites whose sequence is similar but not identical. The affinity of the repressor for these sites varies from 1 to 10. The sequence of each site possesses a binary quasi-symmetry and one can deduce by the comparison of all sites a perfectly symmetrical consensus sequence. The repressor monomer is a protein of 66 residues, but in solution, the protein is dimeric and it is this dimer which binds DNA.

G.N. Cohen, *Microbial Biochemistry*,
DOI 10.1007/978-94-017-8908-0_39, © Springer Science+Business Media B.V. 2014

A crystallographic study by Matthews has shown in 1981 that the monomer is composed of three antiparallel β-strands and of three α-helices; the dimer is stabilized by electrostatic interactions between monomers.

The two same groups have studied the λ repressor: the same operator sites as in the case of Cro are recognized, but the affinity of the repressor for these sites varies from 1 to 50, the sites of high affinity for one repressor being sites of low affinity for the other; this difference accounts for the in vivo physiological role of the two proteins. The λ repressor is a dimer of two identical subunits of 236 residues each. The N-terminal domain of the polypeptide (1–92) binds specifically the operator sites and reproduces the effects of the intact repressor on transcription.

On the basis of these observations, a model has been established by Ptashne in 1986 in a very interesting monograph.

Using the *gal* operon and replacing the two spatially separated *gal* operators by *lac* operators, Choy and Adhya have shown that repression of the operon by the tetrameric Lac repressor involves DNA looping and that repression occurs at a step prior to the formation of the first phosphodiester bond of mRNA.

We have seen that the CAP protein intervenes in the positive regulation of several catabolic operons in *E. coli*. When cAMP is present at a high enough concentration, it forms a complex with CAP and this complex binds to DNA at a specific site. A consensus sequence for the different sites has been suggested in 1982, but the individual sites differ considerably from the consensus. It has been shown that the protein is a homodimer whose sequence has been established in 1982. The protein is 209 residues long and is composed of two domains: the C–terminal domain binds to DNA while the N-terminal one binds cAMP and is responsible for the main contacts which lead to the dimer. McKay, Weber and Steitz in 1982 have established the crystallographic structure of the intact dimer bound to cAMP. The C-terminal domain comprises three α-helices and two pairs of short antiparallel β-strands, whereas the contacts necessary to the dimer formation are due to the pairing of two identical helices, each belonging to a different subunit, and to some additional contacts. cAMP occupies part of the interface between the two monomers; although it is entirely buried within the N-terminal domain, it contributes by hydrogen bonds to the stabilization of the dimer.

Before crystals of the complexes between the regulatory proteins and oligodeoxynucleotide corresponding to the putative binding sequences were available, models were constructed to account for the supposed interactions.

At the level of atomic groups, specific recognition refers to the interaction of amino acid side chains and functional groups of base pairs projecting into the solvent. Crystallographic and genetic studies on DNA-binding proteins and their DNA targets revealed numerous hydrogen bonds or hydrophobic contacts between individual bases and mostly amino acids of the helix-turn-helix motif.

In the case of the three proteins, Cro and λ repressors and CAP, a large number of contacts with DNA involve two helices joined by an abrupt turn (helix-turn-helix model). The first helix lies above the large groove of DNA near the sugar-phosphate backbone and the second helix adapts partially or totally in the large groove. The structures of these α-helical units are almost identical in the three proteins: 20–24

Fig. 39.1 The Trp repressor. One subunit is shown in *blue*, the other in *yellow*. The bound tryptophans are shown in *green* and *red*. The two subunits are tightly interlocked (Courtesy of Dr Paul Sigler)

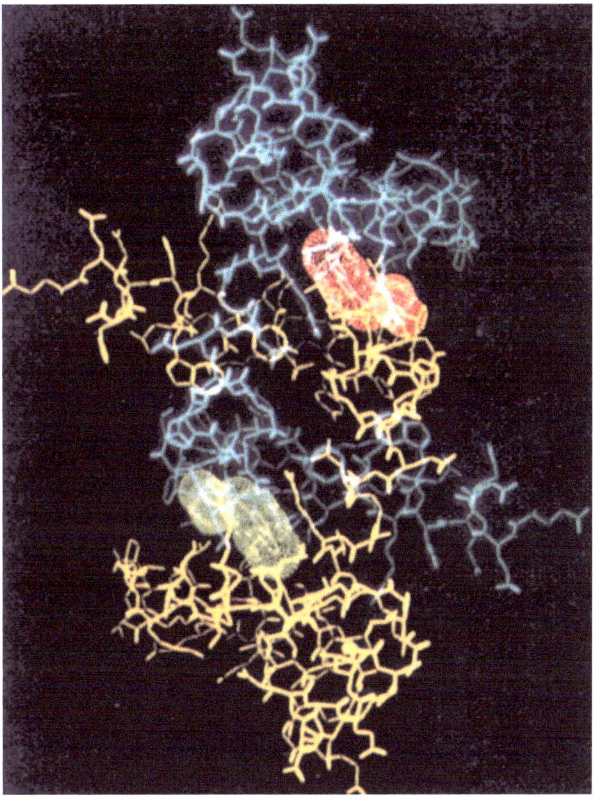

α-carbon atoms are practically superimposable. It appears that the helix-turn-helix motif is specific for DNA binding proteins, since it has not been found yet in other proteins. The three proteins differ in all the other regions: the tertiary foldings are all different; the relative positions of the conserved helices versus the symmetry axis of the dimeric proteins are not identical and the helices in dimers are not as perfectly superimposable as in the monomers; the different orientations of these regions suggest that the different repressors do not dock to DNA in the same manner.

In addition to the homologies in the tertiary structure, sequence homologies are detectable in the totality (repressors from bacteriophages λ, 434 and P22) or only in part of the molecule. Whereas the first case suggests an evolutionary relationship, the homology is restricted in the second case to the helix-turn-helix region in the three proteins whose crystallographic analysis had already been done.

The Trp repressor is a homodimeric protein (2 × 107 residues) consisting of two extensively interlocked subunits (Fig. 39.1). It contains three domains: a central core formed by the N- terminal half of both subunits, and two flexible DNA-reading heads, each formed by the C-terminal half of a subunit. Helices D and E form a helix-turn-helix unit similar to that of the Cro and λ repressors and CAP. The central core serves as a spacer between the reading heads which are 26 Å apart in the aporepressor. X-ray analysis (performed by the late Paul Sigler) of the repressor crystallized with and without tryptophan have revealed that tryptophan induces the formation of a DNA binding surface by increasing the distance between subunits by 8 Å so that they can fit into adjacent major grooves of DNA.

Previous studies had suggested that protein α-helices could fill the large groove of DNA in its B form. From the above results, and from other examples analyzed since, it appears that this is a common method of DNA recognition: an α helix with its side chains has a diameter of about 12 Å, whereas the large groove of DNA is about 12 Å wide and 6–8 Å deep.

Study of the Protein-DNA Complexes

The general rules summarized by Pabo and Sauer in 1984 were the following: the structure which binds the regulatory proteins is B type DNA; the proteins use the symmetry of their subunits to interact with operator sequences which exhibit also a binary quasi-symmetry. Many proteins which interact with DNA are dimers or tetramers and the corresponding targets are quasi-symmetrical. It was therefore probable that these proteins obeyed the interaction model deduced from the crystallographic study of the three above proteins and that they would form symmetrical complexes. The three proteins use an α-helix to establish the majority of the contacts between the protein side chains and the bases of the large groove. The adjacent α-helix contacts the sugar-phosphate backbone of DNA and appears to help the orientation of the first helix. Most contacts appear to occur with a single face of the double helix. However, λ repressor uses a flexible part of the protein to wind around the DNA and to be in contact with both DNA sides. Finally, the recognition of specific base sequences involve hydrogen bonds and van der Waals forces between the protein side chains and the edges of the bases. These specific interactions, added to those with the DNA backbone and to electrostatic interactions, stabilize the DNA-protein complexes. (It should however be recalled that certain proteins, e.g. *E. coli* RNA polymerase are not composed of subunits presenting a symmetry relationship and that the sequences to which they bind do not present a symmetry relationship).

Thus, in 1984, the current models derived from the crystal structure of three proteins were considered as probably correct, but the specialists were aware that model construction, even associated to biochemical and genetic studies, could not give with security a view such as that which could be obtained with real complexes analyzed at high resolution. For example, it is assumed that the operator DNA has a B structure, which is only an approximation. Minor changes in the protein structure, which could occur when the protein binds to DNA, could affect the structure of the complex. It was therefore necessary to study the structure of the binary complexes between repressors and operators, and the ternary complexes formed by the association between operator DNA and activators or repressors (CAP or biosynthetic repressors) in the presence of their specific ligands (cAMP, tryptophan or S-adenosylmethionine, for example).

During 1987 and 1988, the complexes between λ, 434, and cro repressors and their operators have been obtained by Harrison, Pabo, Ptashne and their collaborators. The majority of the above predictions turned out to be correct: the complexes have an order 2 symmetry, the binary axis of the protein coinciding with the binary DNA axis. One admits that the differences in sequence between the two halves of the operator are associated to subtle differences in the protein monomers. DNA appears to be a right handed helix with the B structure. As predicted, a certain deformation of DNA can be observed when the complex is formed: the same 14 bp-operator adopts a different conformation when it binds the Cro or the 434 repressor. In all the three proteins studied, the second helix of the helix-turn-helix motif is situated in the large DNA groove. Experiments with hybrid proteins where each of the helices comes from a different repressor suggest that the geometry of binding to DNA is similar for many repressors.

The observed complexes favor a DNA-protein recognition by contacts between the protein side chains and the base pairs exposed in the large DNA groove, due to hydrogen bonds and van der Waals forces. It is equally clear that DNA adapts itself in order to increase the complementarity between the two types of molecules: multiple interactions with the DNA phosphate backbone contribute to the global affinity of the binding, but it is essentially the contacts between the exposed base pairs and the residues of the reading helix which allow the repressor to discriminate the operator from the other DNA sequences: everything happens as if the protein was directly "reading" the sequential information coded in the DNA groove. A high resolution analysis of a cocrystal of the N-terminal domain of λ repressor and λ operator has revealed multiple hydrogen bonds between two lysines of the recognition helix and guanines of the major groove of the operator DNA.

The Trp repressor seems to have an entirely different mode of interaction; there are no direct contacts established by hydrogen or non polar bonds between the protein and the bases which are known by in vivo experiments to be important for the recognition: these bases are either linked to the protein through intermediate water molecules, or directly exposed to the solvent. Paul Sigler and his colleagues, who described the DNA-Trp operator complex in 1988, think that the Trp repressor provides an example of "indirect reading". In other terms, the operator sequence is recognized indirectly through changes induced in the geometry of the phosphate backbone of DNA which allow then the formation of a stable complex.

These results have raised new questions:

(a) is the distortion of the DNA backbone sufficient to allow the repressor to discriminate between the specific and the non specific sites in the DNA? It is difficult to answer because the deviations between the operator structure and that of a uniform B DNA appear to result from minor local distortions. On the other hand, the difference in affinities of the Trp repressor for its operator and for a random DNA sequence is only 10^4 fold, a figure much smaller than that observed with many other repressors. The formation of any complex of a protein with DNA involves a non negligible electrostatic component, therefore very dependent on the salt concentration. One must therefore be certain that the complex examined by crystallography is identical to the complex studied in solution.

(b) what is the role of water in the recognition process? Sigler and his colleagues think that the hydrogen bonds established between the two macromolecules through the intermediacy of water molecules could contribute to the direct reading of three operator base pairs, but admittedly not to that of two other base pairs known to contribute to the recognition.

The presence of water molecules on the surface of DNA and of the protein adds an additional degree of complexity. Furthermore, tryptophan, the corepressor, in addition to the "allosteric" effect it exerts on the protein conformation, appears to interact directly in conjunction with neighboring residues with certain DNA phosphate residues. A nuclear magnetic resonance study of the Trp repressor-operator complex by Jardetsky questions the possibility that water molecules play a role in the recognition process.

The unique dimeric architecture of the Trp aporepressor built from two highly intertwined chains implies that subunit dimerization is a step on the folding pathway. Despite the complexity of both the architecture and folding kinetics of the Trp repressor, its reversible two-state folding precludes the structural characterization of its folding intermediates. Proteolytic cleavage of each subunit at positions 7 and 71 provides the N and C fragments (8–71 and 72–108, respectively), separating the hydrophobic core into subdomains. Biochemical and biophysical characterization of the purified fragments indicate a native-like dimeric assembly of N fragments and a rather unstructured C fragment. Interestingly, the equimolar mixture of N and C fragments produces an NMR spectrum practically indistinguishable from that of the native protein, indicating a native-like dimeric assembly of TrpR fragments. The results of this proteolytic approach support folding via partially folded dimeric intermediates and illuminates the hierarchy of subdomain folding.

Most of the major families of DNA-binding proteins that have been structurally characterized in procaryotes bind with a α-helix in the major groove. The MetJ repressor, obtained in the homogeneous state by the groups of Greene in USA and Cohen at the Pasteur Institute, has been crystallized in 1988 by the Paris group and by the Phillips group in Leeds. The crystallographic analyses of the aporepressor and of the holorepressor have been carried out by the Leeds group in 1989. As in the case of the Trp repressor, the two subunits are strongly intertwined to form the dimer. S-adenosylmethionine is bound to the dimer at two independent sites, one on each monomer, in a symmetrical manner with regard to the binary symmetry axis of the dimer (Figs. 39.2 and 39.3).

The purine nucleus of SAM inserts in a hydrophobic pocket normally occupied by a phenylalanine residue in the aporepressor, while the methionine moiety lies at the protein surface (Fig. 39.4).

This explains why methionine does not bind to the aporepressor, while S-adenosylhomocysteine binds with about half the affinity of that for SAM. The trivalent sulfur atom, positively charged lies at the C-terminal end of helix B. In contrast to the Trp repressor, there is no major conformational change associated with corepressor binding, and the structures are undistinguishable, except for the absence or presence of SAM. A small conformational change occur at the N-terminus

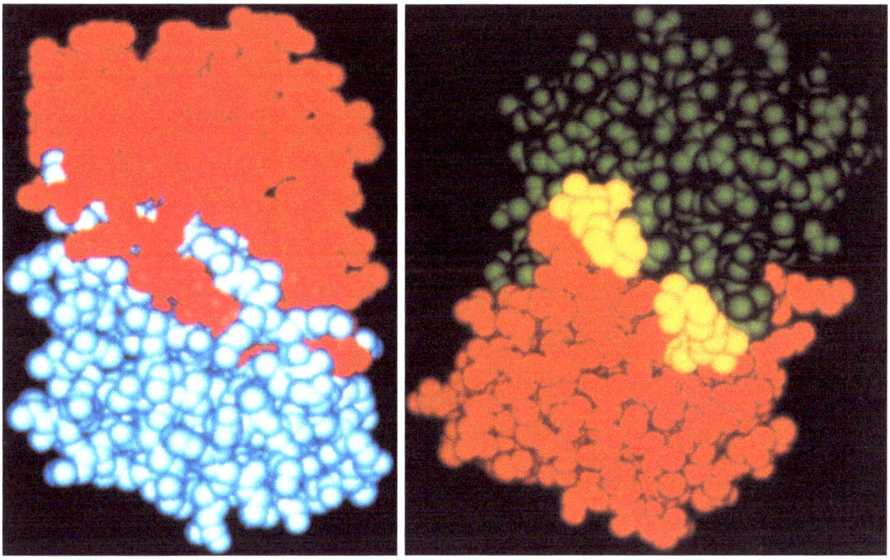

Fig. 39.2 Met repressor space-filling model, with (*right*) and without (*left*) S-adenosylmethionine. One subunit is shown in *blue* or *green*, the other in *red*. The SAM molecule is shown in *yellow* (Courtesy of Dr Simon Phillips)

and at the side chain of Phe65, which occupies the hydrophobic pocket in the absence of corepressor. SAM does not contact DNA in the repressor-operator complex (see below) and its function remains unclear. One possibility is a long-range electrostatic effect based on the presence of the positive charge on the sulfur atom.

Each monomer is composed of three α-helices and of a β-strand, accounting for about 50 % of the sequence, the rest being accounted for by turns or extended chain. In the protein dimer, the β-strands pair to form an antiparallel β-sheet, while the α-helical regions pack against the sheet and against each other to stabilize the dimer. The molecule lacks an helixturn-helix motif which makes it different from all the repressors we have seen so far and from the activator CAP.

The Met repressor has been crystallized with a 19 bp oligonucleotide containing two adjacent 8 bp Met boxes. The analysis of the complex leads to the definition of a new family of prokaryotic regulatory proteins that uses an antiparallel β-sheet for DNA binding (Fig. 39.5).

The methionine repressor binds as two dimeric molecules to two adjacent Met boxes; a dimer binds to each half-site, and each half-site contains a twofold symmetry axis that is coincident with the twofold axis of the β-sheet (Fig. 39.6).

Sequence specificity is achieved by insertion of double-stranded antiparallel protein β-ribbons into the major groove of B-form DNA, with direct hydrogen-bonding between amino acid side chains on the exposed face of the sheet and the base pairs. Lysine23 from each β-strand contacts a guanine, and the neighboring threonine 25 on each strand contacts an adenine. Residues from the N-terminal end of α-helix 2 make backbone contacts. The tetramer is stabilized by interactions between α-helix

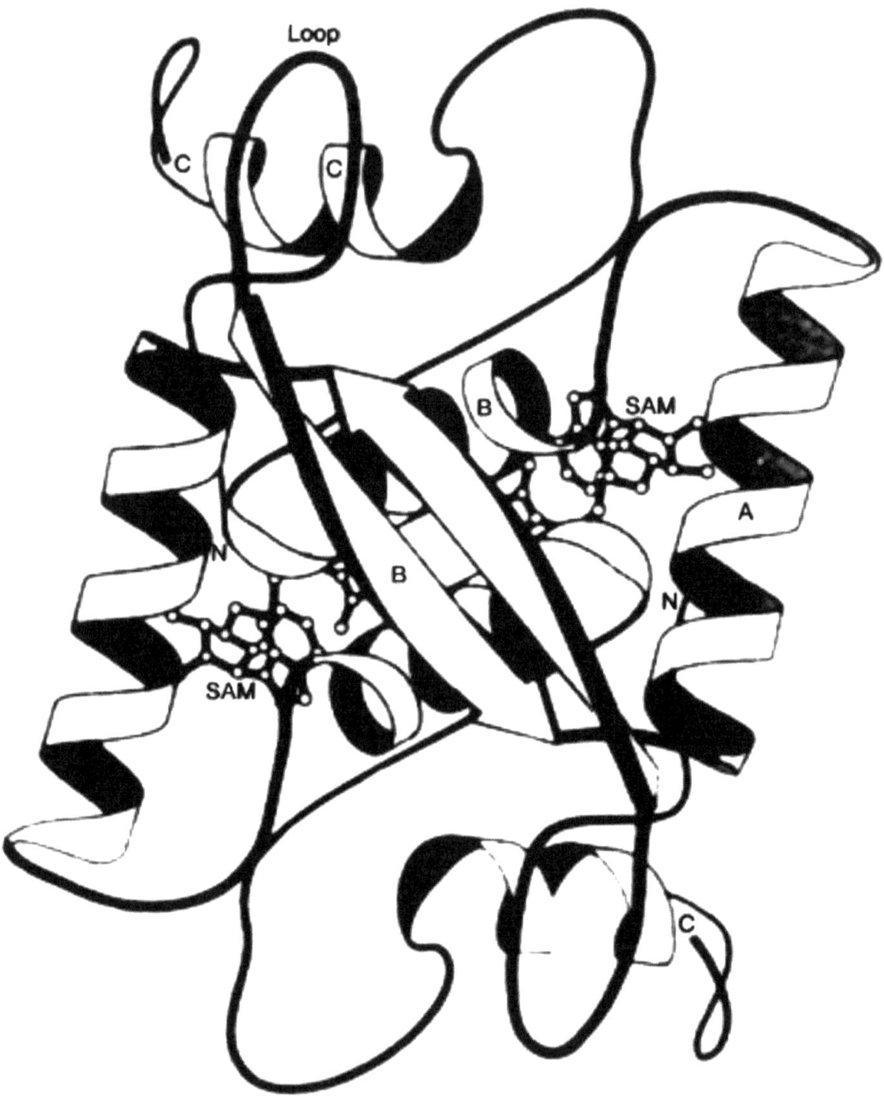

Fig. 39.3 Ribbon representation of the structure of the Met repressor. View along the molecular twofold axis, with the SAM molecules as balls and sticks. One subunit is shaded lightly, with the major elements of the secondary structure labeled and the chain termini marked *N* and *C*. The other subunit is darkly shaded, with termini indicated by *N′* and *C′*. A flexible loop formed by residues 15–20 is also indicated (From I.G. Old, S.E.V. Phillips, P.G. Stockley, and I. Saint-Girons, with permission of *Progress in Biophysics and Molecular Biology*)

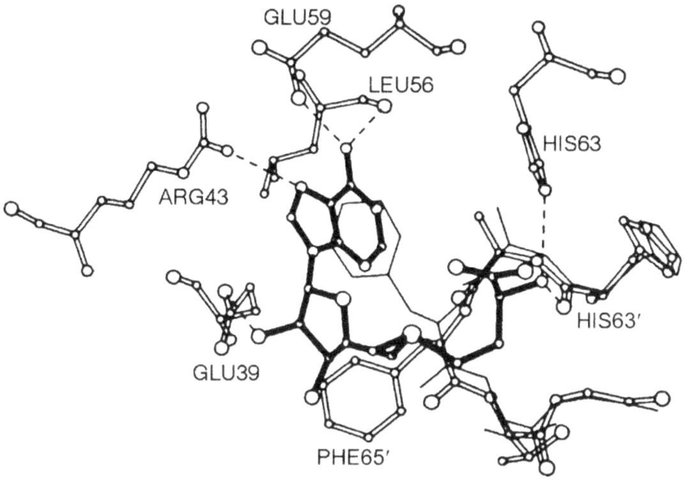

Fig. 39.4 The S-adenosylmethionine binding site. Protein is an open ball and spoke representation for SAM-bound form (=) and as a simple shaded stick representation for the aporepressor form (–), showing the movement of the side chain of Phe 65. The SAM molecule is shown as a *shaded ball* and *spoke* representation. The hydrogen bonds are denoted by *dashed lines*. The two monomers are differentiated by the presence of absence of a prime symbol after residue numbers (Courtesy of Dr. S.E.V. Phillips)

1 on one monomer and α-helix 1 on the other monomer (see p. 165 the sequence of the methionine repressor, which emphasizes the segments involved). The repressor recognizes sequence-dependent distortion or flexibility of the operator phosphate backbone, conferring specificity even for inaccessible base pairs.

Some Other Types of DNA-Binding Proteins

A class of DNA-binding proteins has been identified in eukaryotes which makes use of a binding motif called the "*leucine zipper*". They comprise among others the transcriptional yeast activator GCN4, the oncoproteins *cjun, cfos* and *cmyc* and the C/EBP enhancer. All these proteins contain four or five leucine residues situated exactly seven residues apart in the primary sequence and which can be represented as occurring every two spires from an α-helix; it has been proposed that the leucine residues play an important role in the interdigitation of two α-helices, one from each monomer, this being the structural basis of dimerization. In support of this hypothesis, one notes that GCN4, C/EBP and *cjun* bind as dimers to their protein targets, and that *cjun* and *cfos* can form heterodimers able to bind to DNA. It has been shown further that the leucine zipper repetitive motif can serve for the construction of the heterodimers, but that an adjacent domain, different from the leucine zipper is required for the binding of the heterodimer to DNA. In the case of GCN4, a 60 residues long region containing the leucine zipper is sufficient for dimer formation

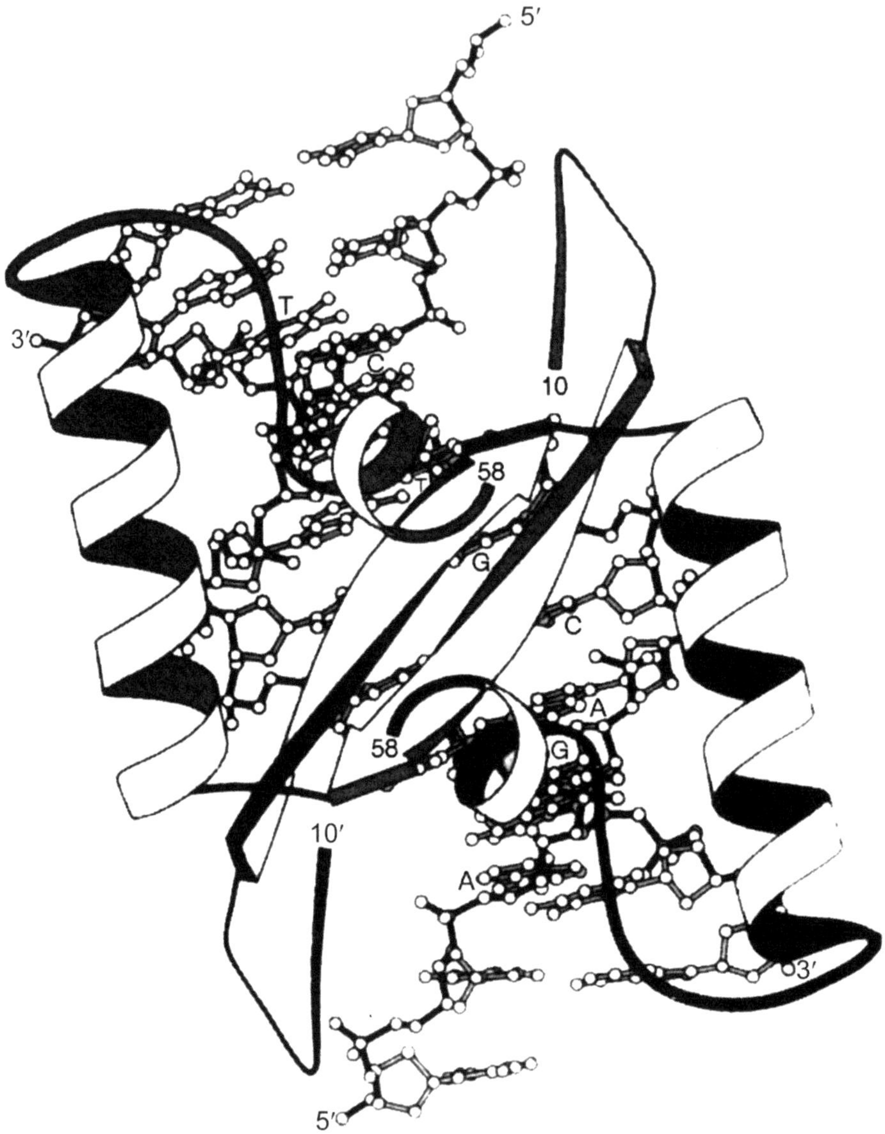

Fig. 39.5 The DNA binding motif, bound to the operator. Viewed as in Fig. 39.3. Residues 1–9 and 59–104, and the SAM molecules have been omitted for clarity. The DNA is shown as *ball* and *stick*, with filled bonds for the sugar-phosphate backbone, and open bonds for the remainder of the ribose rings and the bases. The central region of the DNA fragment shown is a Met-box and the bases have been labeled along one strand. The central twofold axis of the Met-box coincides with the repressor molecular twofold axis. The **β**-strands lie in the central major groove, which faces the reader, while the minor grooves lie to the *upper left* and *lower right* (From I.G. Old, S.E.V. Phillips, P.G. Stockley, and I. Saint-Girons, with permission of *Progress in Biophysics and Molecular Biology*)

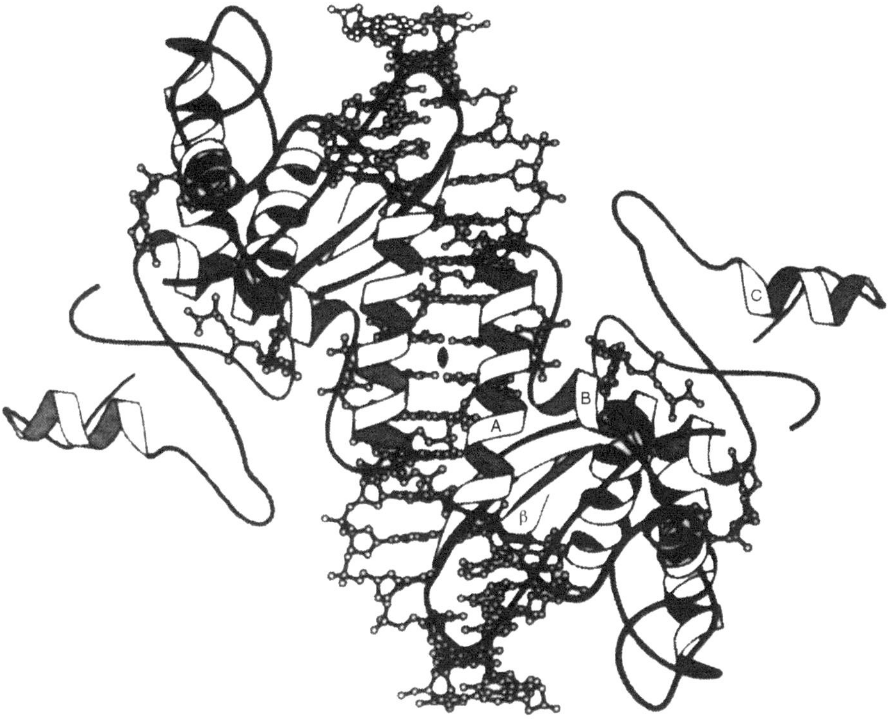

Fig. 39.6 The structure of repressor-operator complex with two repressor dimers, labeled and shaded as in Figs. 39.3 and 39.5. The repressor dimers bind to the same oligonucleotide fragment, one at the *lower right*, the other at the *upper left*. They are related by a twofold central crystallographic axis shown as a black lozenge, which passes through the center of the oligonucleotide between the two Met boxes. The antiparallel **β**-ribbons occupy the major groove of the DNA at *upper left* and *lower right*, while the **α**-helices of adjacent dimers form a long, antiparallel protein-protein contact above the minor groove in the center of the diagram. The corepressor molecules lie on the outer surface of the complex, distant from the DNA. The bends in the DNA helix are not very clear in this view, but slight distortion can be discerned (From I.G. Old, S.E.V. Phillips, P.G. Stockley, and I. Saint-Girons, with permission of *Progress in Biophysics and Molecular Biology*)

and for DNA binding. A synthetic peptide corresponding to this region forms in solution essentially helical stable dimers. A leucine mini-zipper motif has been identified in the C- terminal part of the *lac* repressor (see Fig. 16.7, p. 215) and in the N-terminal -portion of the MetR protein. The leucine heptad repeats of *lac* repressor have the ability to form tetramers.

The prototype of the *zinc finger* structure is the transcription factor TFIIIA for RNA polytmerase III in the frog *Xenopus laevis*.

This protein is characterized by a particular arrangement of cysteine and histidine residues coordinating a zinc atom. The tridimensional structure of a zinc finger has been determined by X-ray diffraction. The zinc atom is coordinated to two cysteine and two histidine residues and the function is to bring closer an α-helix and a β-sheet. The α-helix of each zinc finger interacts with two bases of the DNA large groove. Pabo and Sauer, who have solved the structure of the zinc finger, have shown for the first time that a protein can contain a periodic structure able to recognize a target (the DNA double helix) whose structure is also periodic. The modular structure of the zinc finger proteins allows the recognition of large targets by adapting the number of fingers to the size of the asymmetric target. All the interactions take place on only one strand of DNA, and there are but a few contacts between the deoxyribose phosphate skeleton of DNA and the transcription factor.

There are two types of zinc fingers: (a) the C2H2 type where two cysteine residues face two histidine residues at the base of the finger; the sequence is of the type $hxCx_{2-4}Cx3hx_5hx_2Hx_{3-4}Hx_{2-6}$ (h representing a hydrophobic residue and x any amino acid); the proteins of this type present generally several repeated fingers, three for Sp1, nine for TFIIIA, and up to 38 for other factors.; (b) the C2C2 type where two zinc atoms are coordinated by four cysteine residues, determining two fingers and found in many steroid and thyroid hormone receptors for example. The study by magnetic nuclear resonance of the zinc fingers of the C2C2 type shows that two fingers are necessary for the correct folding of the protein, whereas the folding of a single finger of the C2H2 type can be observed in solution.

Selected References

Pabo CO, Sauer RT (1992) Ann Rev Biochem 61:1053–1095
Ptashne M (1986) A genetic switch. Gene control and phage λ.m128 pp. Blackwell Scientific Publications, Oxford
Roeder RG (1991) Trends Biochem Sci 16:402–408

Trp Repressor. Structural Aspects

Schevitz RW, Otwinowski Z, Joachimiak A, Lawson CL, Sigler PB (1985) Nature 317:782–786
Zhang R-G, Joachimiak A, Lawson CL, Schevitz RW, Otwinowski Z, Sigler PB (1987) Nature 327:591–597

Met Repressor

Phillips SEV, Mainfield I, Parsons I, Davidson BE, Rafferty B, Somers WS, Margarita D, Cohen GN, Saint-Girons I, Stockley PG (1989) Nature 341:711–715
Rafferty JB, Somers WS, Saint-Girons I, Phillips SEV (1989) Nature 341:705–710
Somers WS, Phillips SEV (1992) Nature 359:387–393

Chapter 40
Evolution of Biosynthetic Pathways

Principles of Protein Evolution

The main goal of the student of evolution is to reconstruct past events leading to the structure of contemporary proteins, from which the relationships which they share with a common ancestor can be defined. A single sequence can give rise to different sequences by speciation and divergence or by gene duplication within a single organism followed by divergence. We shall see examples of these two cases in our survey of the evolution of biosynthetic pathways.

Two Theories for the Evolution of Biosynthetic Pathways

Some of the difficulties in understanding the mechanisms whereby multistep metabolic pathways have been constructed is the apparent lack of any selective advantage conferred by individual reactions prior to the establishment of the whole pathway.

In an attempt to solve this paradox, Horowitz proposed in 1945 that evolution proceeded in a retrograde manner, with individual steps being recruited in the reverse direction relative to the final pathway. The last step in the present day pathway was acquired first; the penultimate step next, by mutation in one of the two identical genes following a gene duplication allowing the conservation of the first function in the other gene. And so on. This hypothesis supposed that pristine life took place in a primeval soup, a kind of nutritional paradise in which all the essential metabolites and intermediates were available and that the proposed duplications and mutations were selected one after the other when these metabolites were becoming scarce or exhausted. Twenty years later, Horowitz presented several arguments in favor of his theory: (i) the overlap between product and substrate of the enzymes catalyzing consecutive reactions; (ii) the existence of allosteric inhibition of the first enzyme of a pathway by its end-product which could be regarded as a memory of

G.N. Cohen, *Microbial Biochemistry*,
DOI 10.1007/978-94-017-8908-0_40,

this enzyme for its origin, if one makes the reasonable assumption that the binding sites of the last enzyme and of the first enzyme for this ligand are identical or homologous; (iii) the clustering of some biosynthetic genes in bacterial operons, expected for a duplication not followed by a translocation.

This theory, albeit attractive and although it may apply in some cases, postulates the original presence of metabolic intermediates in general, many of which have very short half-lives.

Martinas Ycas has put forward an entirely different hypothesis, according to which the minimal set of enzymes necessary for the viability of an organism are related by descent from a smaller number of ancestral genes. If true, this poses a problem, since it indicates that earlier organisms contained less than the minimal number of enzymatic components necessary for the present day organisms, in other words that these ancient enzymes were endowed with a broader specificity and were able to carry several indispensable reactions. This hypothesis has been reiterated and extended by Jensen.

The progress of molecular biology and the relative facility of gene sequencing have allowed an experimental investigation of this question. Evolutionary relationships between various enzymes can now be traced back by amino acid sequence comparisons, due to the formidable and ever increasing number of sequences available in the data banks. Amino acid biosynthetic pathways seem especially interesting since, with few exceptions, their different steps are conserved in all organisms where these pathways occur. Therefore, the pathways are probably of very ancient origin and their study potentially applies to the metabolic capacities of the so-called progenote.

We shall now examine some examples, taken from pathways that have been details in preceding chapters of this book.

The Methionine and Cysteine Biosynthetic Pathways

In bacteria, methionine biosynthesis involves a total of nine enzymatic reactions in two convergent pathways from aspartate and serine, respectively. In particular, the sulfur atom of methionine comes from cysteine by a transsulfuration process involving cystathionine and homocysteine. These compounds are the products of two reactions catalyzed by cystathionine γ-synthase and cystathionine β-lyase, which are encoded by the *metB* and *metC* genes, respectively. These pyridoxal phosphate-dependent enzymes both consist of four subunits, each of molecular weight 42,000 and copurify in many chromatographic systems. It has been shown in vitro that cystathionine γ-synthase catalyzes the synthesis of homocysteine from O-succinylhomoserine and H_2S, thus by-passing the cystathionine intermediate.

The understanding of the biochemical similarities shared by these two enzymes came from the comparison of the sequences of the *metB* and *metC* gene products: about one third of the amino acid residues is conserved between the two sequences (Fig. 40.1). This strong similarity, which encompasses the entire length of the two

```
C :  MADKKLDTQLVNAGRSKKYTLGAVNSVIQRASSLVFDSV-EAKKHATRNRANGEL
       +*   *   *   *         *  *   *   *+  *   *  *
B :   MTRKQATIAVRSGLNDDEQYGCVVPPIHLSSTYNFTGFNEPRAHD---------

      FYGRRGTLTHFSLQQAMCELEGGAGCVLFPCGAAAVANSILAFIEQGDHVLMTNT
       + ***   *   +* *  ******* **    *  *+       *+  **  ++
      -YSRRGNPTRDVVQRALAELEGGAGAVLTNTGMSAIHLVTTVFLKPGDLLVAPHD

      AYEPSQDFCSKILSKLGVTTSWF-DPLIGADIVKHLQPNTKIVFLESPGSITMEV
       *  *      * * *    * *      +   *    *+* +***      *
      CYGGSYRLFDS-LAKRGCYRVLFVDQGDEQALRAALAEKPKLVLVESPSNPLLRV

      HDVPAIVAAVRSVVPDAIIMIDNTWAAGVLFKALDFGIDVSIQAATKYLVGHSDA
       *+  *    * *   *+  +***     *    *   * *+ +    **** ****
      VDIAKICHLAREVG--AVSVVDNTFLSPALQNPLALGADLVLHSCTKYLNGHSDV

      MIGTAVCNARCWEQLRENAYL---MGQMVDADTAYITSRGLRTLGVRLRQHHESS
        *  *  *+  + + * *        *     *    *+  ******  *
      VAG--VVIAKDPDVVTELAWWANNIGVTGGAFDSYLLLRGLRTLVPRMELAQRNA

      LKVAEWLAEHPQVARVNHPALPGSKGHEFWKRDFTGSSGLFSFVLKKKLNNEELA
        +    *    * * ++ ** **    ***    *    *     ** *        *
      QAIVKYLQTQPLVKKLYHPSLPENQGHEIAARQQKGFGAMLSFELDGDEQT--LR

      NYLDNFSLFSMAYSWGGYESLILANQPEHIAAIRPQGEIDF--SGTLIRLHIGLE
        *    ***+ *  * ** ****          *    *        * ****+  *+*
      RFLGGLSLFTLAESLGGVESLISHAATMTHAGMAPEARAAAGISETLLRISTGIE

      DDLIADLDAGFARIV
      +******+ **
      EDLIADLENGFRAANKG
```

Fig. 40.1 Comparison of the *E. coli* β-cystathionase (C) and cystathionine γ-synthase (B). The sequences are presented in the one-letter code and have been aligned by introducing gaps (–) in order to maximize identities. *Asterisks* indicate identical residues and + signs indicate accepted replacements (I-L-V, D-E, R-K, T-S) (From J. Belfaiza, C. Parsot, A. Martel, C. Bouthier de la Tour, D. Margarita, G. N. Cohen and I. Saint-Girons)

polypeptides, suggests that these enzymes have a common evolutionary origin. This hypothesis is reinforced by the identification of the lysine residues that binds pyridoxal phosphate in the two enzymes and which are present at the same place in the alignment of the two sequences.

Did the ancestral enzyme perform the two reactions or was it involved in the direct synthesis of homocysteine? The in vitro reaction of the present day cystathionine synthase with O-succinylhomoserine and H_2S supports the second proposal. Moreover, indirect suppressors of *E. coli metC* mutants have been selected (*metQ*) which lead to the synthesis (or the overproduction) of an unidentified sulfur compound which is used by cystathionine synthase to synthesize homocysteine directly. This "new reaction" could represent a throw-back to the ancestral pathway where only one enzyme was responsible for the synthesis of homocysteine from O-succinylhomoserine.

The contemporary cystathionine γ-synthase uses either O-succinylhomoserine or O-acetylhomoserine, depending on the microorganism from which it is extracted. The ancestral enzyme was probably able to utilize both compounds. Thus, there is a striking similarity between the biosynthetic pathways of homocysteine and cysteine, from homoserine and serine, respectively, as already noticed by Ycas (Fig. 40.2).

Cysteine biosynthesis is catalyzed by O-acetylserine sulfhydrylase, the *E. coli cysK* gene product. This is also a pyridoxal phosphate-dependent enzyme, composed of two identical subunits. It was accordingly supposed that this enzyme would belong to the same family as cystathionine γ-synthase and cystathionine β-lyase; when however, the sequence of O-acetylserine sulfhydrylase was established, this appeared not to be the case, except in the vicinity of the pyridoxal phosphate binding site. Rather, the C-terminal moiety of 0-acetylsulfhydrylase was found to be similar to the β-subunit of tryptophan synthase! An evolutionary relationship between *E. coli* enzymes involved in the cysteine and methionine biosynthetic pathways is most likely if we consider these pathways in the yeast *Saccharomyces cerevisiae*. As indicated in Fig. 40.2, there is only one enzyme (encoded by the *MET25* gene) which is involved in the synthesis of homocysteine from O-acetylhomoserine in yeast. This pyridoxal-phosphate-dependent enzyme also catalyses the synthesis of cysteine from O-acetylserine, and is designated as O-acetylhomoserine, O-acetylserine sulfhydrylase. Its nucleotide sequence has been determined and its deduced amino acid sequence has been compared with those of the *E. coli metB* and *metC* gene products. Among about 100 identical residues detected by comparison of pairs of protein sequences, 56 residues are strictly conserved between the three polypeptides. The lysine residues which bind pyridoxal phosphate in cystathionine-γ-synthase and in cystathionine-β-lyase are aligned with a lysine residue in the sequence of O-acetylhomoserine, O-acetylserine sulfhydrylase. The common origin of these three enzymes, which are nowadays involved in the cysteine and/or the methionine biosynthetic pathways, appears to be well-documented. The existence of only one enzyme with broad specificity in *S. cerevisiae* most likely corresponds to a primitive state relative to *E. coli* for which there are at present three different enzymes: the homologous *metB* and *metC* gene products and the *cysK* gene product which derives from another ancestor. In *E. coli*, evolution has selected gene duplications and subsequent mutations leading to specialization in the substrate specificity of the encoded proteins and to different controls of their activity and expression.

In mammals, the transsulfuration pathway runs in the other direction from methionine to cysteine; in spite of this, the mammalian cystathionase (whose substrates are homocysteine and serine) has clearly evolved from the same ancestral gene:

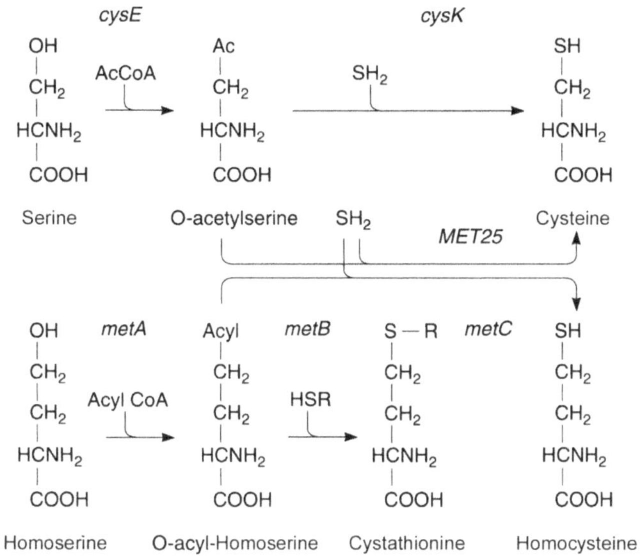

Fig. 40.2 Some of the reactions involved in the cysteine and methionine biosynthetic pathways. The names of the *E. coli* genes encoding the relevant enzymes are indicated above the catalyzed reactions: *cysE* serine acetyltransferase, *cysK* O-acetylserine sulfhydrylase, *metA* homoserine transsuccinylase, *metB* cystathionine γ-synthase, *metC* cystathionine β-lyase (cystathionase). The reactions catalyzed by the *S. cerevisiae* MET25 gene product (O-acetylhomoserine, O-acetylserine sulfhydrylase) are also shown. The radical R represents cysteine, and acyl can be either succinyl or acetyl (From C. Parsot, I. Saint-Girons and G. N. Cohen, with permission of *Microbiological Sciences*)

it shows 39 % identity with the *metB* gene product. In contrast, the mammalian cystathionine synthase shows little similarity with the *metB* and *metC* gene products and appeared to have evolved from the cysteine biosynthetic enzymes.

The Threonine, Isoleucine, Cysteine and Tryptophan Biosynthetic Pathways

Threonine is synthesized from homoserine by a two-step pathway which involves homoserine kinase and threonine synthase. These two enzymes are encoded by the *thrB* and *thrC* genes respectively. Threonine is also the precursor of isoleucine through five enzymatic reactions of which only the first one is specific for the isoleucine pathway, the four latter steps being achieved by enzymes that are also involved in valine biosynthesis. The reaction specific to the isoleucine pathway is catalyzed by threonine dehydratase, encoded by *ilvA*.

From a *B. subtilis ilvA* strain, auxotrophic for isoleucine, suppressor mutations allowing growth on homoserine or threonine have been isolated. These mutations affect the expression or the structure of threonine synthase, the *thrC* gene product.

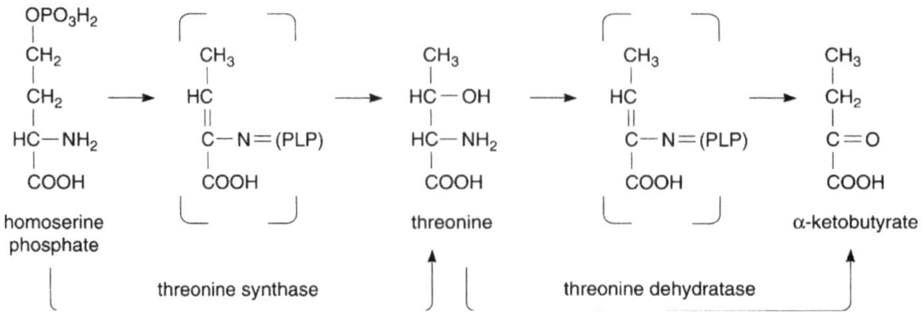

Fig. 40.3 Structure of the substrates and products of the reactions catalyzed by threonine synthase and threonine dehydratase. The structure of the Schiff base between PLP and α-aminocrotonate, the common intermediate of the two reactions is also shown. Inorganic phosphate, ammonia and water, which participate in the reactions, have not been indicated for clarity. PLP stands for pyridoxal phosphate (From C. Parsot, I. Saint-Girons and G. N. Cohen, with permission of *Microbiological Sciences*)

This led to the discovery that threonine synthase is endowed with a low-level threonine dehydratase activity. This activity is sufficient to compensate for the defect in the *ilvA*-encoded threonine dehydratase, once *thrC* expression is derepressed. The second type of mutation gives rise to an altered threonine synthase from which the reaction intermediate, α-aminocrotonate, is eliminated and then spontaneously hydrolyzes to α-ketobutyrate and ammonia. These two compounds actually correspond to the products of the reaction normally catalyzed by threonine dehydratase. As shown in Fig. 40.3, the reaction mechanisms of threonine synthase and threonine dehydratase share the same α-aminocrotonate intermediate.

These observations have led to investigate the sequence of the *B. subtilis thrC* gene. Comparison of *B. subtilis* threonine synthase and *S. cerevisiae* threonine dehydratase revealed 21 % identity and 33 % homology between the two polypeptide sequences. These data constitute strong evidence in favor of a common evolutionary origin for threonine synthase and threonine dehydratase. An important similarity was also detected between the sequences of these two enzymes, as well as that of *E. coli* threonine synthase and of *E. coli* serine dehydratase. All of these enzymes are pyridoxal phosphate-dependent. The lysine residue of D-serine dehydratase identified as covalently bound to pyridoxal phosphate (after reduction of the Schiff base by sodium borohydride) is aligned in Fig. 40.4 with similarly positioned lysine residues lying in conserved regions of the threonine synthases, threonine dehydratase and D-serine dehydratase. Similarity between threonine synthase and threonine dehydratase is further supported by comparison of the available sequences of the products of the *E. coli ilvA* and *tdc* genes which encode the biosynthetic and the catabolic threonine dehydratases.

Three points from these sequence comparisons must be emphasized. Firstly, the homology detected between threonine dehydratase and D-serine dehydratase reveals some interesting perspectives concerning the origin of the enzymes involved in the metabolism of D-amino acids. Secondly, the entire isoleucine biosynthetic pathway

```
THREONINE SYNTHASE                          (thrC , E.coli) :          ..CLELFHGPTLAF K DFGGRFH..   (114)
                                                                                ** *     ** *
THREONINE SYNTHASE                          (thrC , B.subtilis) :      ..VKTEGVNPTGSF K DRGHVLA..   ( 66)
                                                                         ** * * *****    ** *

THREONINE DEHYDRATASE (Biosynthetic)        (ilvA , E.coli) :          ..VKREDRQPVHSF K LRGAYAN..   ( 69)
                                                                         *****  *  **   **** **
THREONINE DEHYDRATASE (Biosynthetic)        (ILV1 , S.cerevisiae) :    ..LKREDLLPVFSF K LRGAYHH..   (116)
                                                                         ********* **   ****  *
THREONINE DEHYDRATASE (Degradative)         (tdc  , E.coli) :          ..LKFENHQRTGSF(K)IRGAFNK..   ( 65)
                                                                         ** * *  ****   ****

TRYPTOPHAN SYNTHASE (Beta chain)            (trpB , E.coli) :          ..LKREDLLHGGAH(K)TNQVLGQ..   ( 94)
                                                                         ******** *         **
TRYPTOPHAN SYNTHASE (Beta chain)            (trpB , B.subtilis) :      ..LKREDLNHTGSH K INNALGQ..   ( 97)
                                                                         ******  ***    *  ***
TRYPTOPHAN SYNTHASE (Beta chain)            (trpB , P.aeruginosa) :    ..LKREELNHTGAH K INNCIGQ..   ( 99)
                                                                         ******  **     *   **
TRYPTOPHAN SYNTHASE                         (TRP5 , S.cerevisiae) :    ..LKREDLNHTGSH K INHALAQ..   (391)
                                                                         ******  ***    *  ***

D-SERINE DEHYDRATASE                        (dsdA , E.coli) :          ..LKKDSHLPSGSI(K)ARGGIYE..   (125)
                                                                         ****  **|***     ****
                                                                                 I
```

Fig. 40.4 Comparison of the amino acid sequences around the pyridoxal phosphate binding site in threonine synthase, threonine dehydratases, and D-serine dehydratase. Only a portion of the sequences is shown. *Asterisks* indicate identical or similar residues present in at least two sets of sequences (accepted replacements: I-L-V-M, D-E, K-R, S-T, G-A). The *arrow* indicates the insertion of an isoleucine residue (I) in the sequence of D-serine dehydratase (between P and S) relative to the other sequences (From C. Parsot, I. Saint-Girons and G. N. Cohen, with permission of *Microbiological Sciences*)

now appears to have arisen from the recruitment of the last enzyme of the threonine biosynthetic pathway and the exploitation of the substrate ambiguity of the enzymes acting in valine biosynthesis. Thirdly, the biosynthetic threonine dehydratases from *E. coli* and *S. cerevisiae* both have a C-terminal fragment of about 150 amino acid residues which is not present in the catabolic threonine dehydratase, in the D-serine dehydratase and in the threonine synthases. This extension could correspond to the acquisition of a regulatory domain by the biosynthetic threonine dehydratases, which are indeed inhibited by isoleucine.

Threonine dehydratase is also endowed with serine dehydratase activity. This is shared by another pyridoxal phosphate-dependent enzyme, namely the β_2 subunit of tryptophan synthase, which catalyzes the last step of the tryptophan biosynthetic pathway: the $\alpha_2 - \beta_2$ heterotetramer of tryptophan synthase catalyzes the hydrolysis of indole-3- glycerol phosphate to glyceraldehyde-3-phosphate and indole, as well as the subsequent condensation of indole onto L-serine to yield L-tryptophan. As we have seen in the chapter dealing with tryptophan biosynthesis, in the absence of indole, the β_2 dimer deaminates L-serine, producing pyruvate and ammonia. Figure 40.5 illustrates the similar mechanism, and Fig. 40.6 the common evolutionary origin of threonine synthase (and therefore of threonine dehydratase) and of the β subunit of tryptophan synthase.

There are numerous identical residues scattered throughout the entire length of the sequences, with the exception of the C-terminal extension of biosynthetic threonine dehydratases discussed above, which is not present in tryptophan synthases from *B. subtilis* or from other organisms. In addition, the lysine residue of the tryptophan synthase β chain identified as bound to pyridoxal phosphate is aligned with the lysine residues of threonine synthases and threonine dehydratases which were proposed to

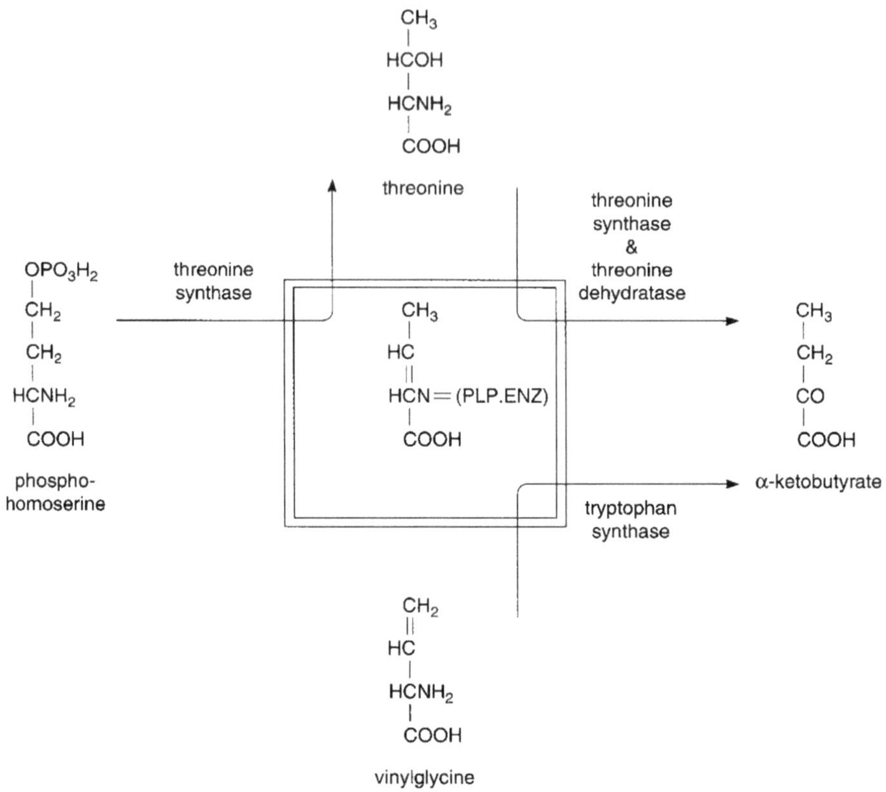

Fig. 40.5 Schematic representation of the substrates and products, and of one of the reaction intermediates in some of the reactions catalysed by threonine synthase and tryptophan synthase. The structure of the pyridoxal phosphate Schiff base of α-aminocrotonate, an intermediate in each of the reactions catalyzed by threonine synthase and dehydratase (see Fig. 40.3) and tryptophan synthase β-chain is shown in the central box (Remember that the β_2 dimer of tryptophan synthase catalyzes the deamination of serine producing pyruvate and ammonia, from enzyme-bound PLP and L-serine, with the formation of an α-aminoacrylate Schiff base intermediate. It can also catalyze the conversion of 2-amino-3-butenoic acid (vinylglycine) to α-ketobutyrate and ammonia via the formation of the α-aminocrotonate Schiff base intermediate (From C. Parsot, with permission of *The EMBO Journal*). For the sake of clarity, the pyridoxal phosphate tautomerization of the double bond from β–γ to α–β carbon atoms involved in some of these reactions, inorganic phosphate, and water, are not presented

bind pyridoxal phosphate. The divergence between threonine and tryptophan synthases is probably very ancient since there are no differences in the biosynthetic pathways of threonine or of tryptophan in eubacteria, lower eukaryotes or plants.

The determination of the nucleotide sequence of most of the genes involved in the tryptophan, threonine, methionine, lysine and isoleucine biosynthetic pathways, and the subsequent comparison of the encoded proteins does not support Horowitz's hypothesis of retrograde evolution in biosynthetic pathways. The results presented above and others are in line with the proposal put forward by Ycas and then by Jensen concerning the substrate ambiguity of primitive enzymes. On the basis of the

```
THREONINE SYNTHASE (B.SUBTILIS )------------------------------MWKGLIHQYKEFLPVTDQTPALTLHEG
                                                              + --+    -+        +     *     *
TRYPTOPHAN SYNTHASE (B.SUBTILIS)--MYPYPNEIGRYGDFGGKFVPETLMQPLDEIQTAFKQIKDDPAFREEYYKLLKDYSG

N.TPLIHLPKLSEQL.GIELHVKTEGVNPTGSFKDRGMVMAVAKAKEEGNDTIMCAST.GNTSAAAAYAARANMKCIVIIPNGKIAFGKL
 *-*    ***+ * *  + ** * ** *** *    +  -  ** *   *-  +  *   * *  **+  --* *++   ** *
RPTALTYADRVTEYLGGAKIYLKREDLNHTGSHKINNALGQALLAKKMGKTKIIAETGAGQHGVAPATVAAKFGFSCTVFMGEEDVARQSL
                                 ↑

AQAVMY..GAEIIAIDGNFD.......DALKIVRSICEKSPIALVNSVNPY......................RIEGQKTAAFEVCEQLGE
    *   ****+-+          +*++     **      +   * *                        ****     -+   *   *
NVFRMKLLGAEVVPVTSGNGTLKDATNEAIRYWVQHCEDHFYMIGSVVGPHPYPQVVREFQKMIGEEAKDQLKRIEGTMPDKVVACVGGGS

APDVLAIPVGNAGNITAYWKGFKEYHEKNGTGLPKMRGFEAEGAAAIVRNEVIENPETIATA.IRIGNPASWDKAVKAAEESNGKI..DEV
    +  -+ * --                 +    - -   **         +      *    * -  *   *         *   - *++  * +
NAMGMFQAFLNEDVELIGAEAAGKGIDTPLHAATISKGTVGVIHGSLTYLIQDEFGQIIEPYSISAGLDYPGIGPEHAYLHKSGRVTYDSI

TDDEILHAYQLIARVEGVFAEPGSCASIAGVLKQVKSGEIPKGSKVVAVLTGNGLKDPNTAVDISEIKPVTLPTDEDSILEYVKGAARV
***+ + *   *+ - ***+     *   +*--+*  *- +  +*  ++  *+* * ** ** +-+ *
TDEEAVDALKLLSEKEGILPAIESAHALAKAFKLAKGMD..RGQLILVCLSGRGDKDVNTLMNVLEEEVKRHV
```

Fig. 40.6 Comparison of *B. subtilis* tryptophan synthase and threonine synthase. The entire amino acid sequences are presented in the one letter code. They have been aligned by introducing gaps to maximize the similarity. *Stars* between the sequences indicate identical residues and + signs indicate conservative replacements: R-K, S-T, D-E, I-L-V-M-F. *Hyphens* indicate that the corresponding residue of tryptophan synthase from *E. coli, S. cerevisiae* and *Pseudomonas aeruginosa* is identical to the residue of *B. subtilis* tryptophan synthase. The ↑ sign indicates the lysine residue of tryptophan synthase which binds pyridoxal phosphate (From C. Parsot, with permission of *The EMBO Journal*)

substrate ambiguity exhibited by some contemporary enzymes, these authors have suggested that ancestral enzymes possessed a broad specificity, allowing them to utilize a wide range of structurally related substrates and thereby yielding small amounts of related products. This process would have provided a biochemically diverse environment in which individual enzyme recruitment would have improved the function of slow, but already existing ambiguous pathways. The specialization of the proteins encoded by *E. coli metB* and *metC* versus the substrate ambiguity exhibited by the *S. cerevisiae MET25* enzyme seems particularly illustrative.

A word of caution is necessary: a similar approach was attempted in the case of enzymes of the arginine pathway, namely N-acetylglutamate kinase and N-acetylglutamate-γ-semialdehyde dehydrogenase, which are functionally analogous to kinases and dehydrogenases of other pathways, which by their successive action also achieve the conversion of a carboxylate into an aldehyde function. To this effect, the corresponding genes of *E. coli, argB* and *argC* have been sequenced and the derived amino acid sequences have been compared with the known sequences of analogous enzymes active in the proline and homoserine biosynthetic pathways and in glycolysis. Figure 40.7 illustrates the series of reactions and the corresponding enzymes considered.

The sequences of the four kinases on the one hand, of the four dehydrogenases on the other, have been compared. In addition, each sequence has been compared to all the protein sequences present in the data banks. No convincing similarity was detected using any of these sequences as a probe (except for short identical stretches along the dehydrogenase polypeptides (obviously the 3-phosphoglyceraldehyde dehydrogenases from other organisms show the expected homology with the homologous *E. coli* enzyme). These negative results contrast with the strong similarities detected when comparing other analogous enzymes operating in different

1) N-acetylglutamate $\xrightarrow{argB}$ N-acetylglutamylphosphate $\xrightarrow{argC}$ N-acetylglutamate semialdehyde

2) glutamate $\xrightarrow{proB}$ glutamylphosphate $\xrightarrow{proA}$ glutamate semialdehyde

3) aspartate $\xrightarrow{lysC}$ aspartylphosphate $\xrightarrow{asd}$ aspartate semialdehyde

4) 3-phosphoglycerate $\xrightarrow{pgk}$ 3-phosphoglyceroylphosphate $\xrightarrow{gap}$ 3-phosphoglyceraldehyde

Fig. 40.7 Series of reactions converting a carboxyl group to an aldehyde group. Only the relevant substrate transformations are shown, with omission of ATP, ADP, NADPH, $NADP^+$, inorganic phosphate and water. For each reaction, the name of the corresponding gene (according to the *E. coli* nomenclature) is indicated above the *arrow*

pathways (aspartate and ornithine transcarbamylases, cystathionine synthase and cystathionase, threonine synthase, threonine dehydratase, D-serine dehydratase and the β-chain of tryptophan synthase). Therefore, in the present case the possibility (distasteful to the author of this book) of convergent evolution should be considered. It remains to be seen whether the tridimensional structure of the two groups of enzymes display similarities indicating homology. In any case, the concept of enzyme recruitment of Ycas and Jensen may be more difficult to prove than suggested by the few supporting examples available.

Although similarities in the reactions catalyzed by enzymes involved in different metabolic pathways must focus attention on their possible evolutionary relationship, no conclusion can be drawn before sequence analyses are completed. Last but not least, it should be kept in mind that even if ancestral organisms possessed a smaller repertoire of enzymes, it does not mean that their metabolism was simpler and led to fewer end products.

The Evolutionary Pathway Leading to the Three Isofunctional Aspartokinases in *Escherichia coli*

The biosynthesis of the amino acids deriving part or all of their carbon atoms from aspartic acid has been thoroughly studied in *Escherichia coli*. The first reaction of the common pathway, leading to diaminopimelate, lysine, methionine, threonine and isoleucine is catalyzed by three distinct isofunctional aspartokinases, differing in the way their synthesis and activities are regulated by the respective concentrations of the end metabolites of the aspartic acid family branched pathway. In addition, two of these aspartokinases are bifunctional and also carry homoserine dehydrogenase activities; these proteins are the tetrameric aspartokinaseI-homoserine dehydrogenaseI and the dimeric aspartokinaseII-homoserine

Fig. 40.8 Comparison between the amino acid sequences of aspartokinase I- homoserine dehydrogenase I and aspartokinase II-homoserine dehydrogenase II of *E. coli*. *Top line*: AKI-HDHI; *bottom line*: AKII-HDHII. Identical residues have been *boxed*. Homologies resulting from a single base change in the DNA sequence are marked by a *black dot*; those corresponding to a conservative amino acid substitution are indicated by a *cross* (From M. M. Zakin, N. Duchange, P. Ferrara and G. N. Cohen, with permission of the *Journal of Biological Chemistry*)

dehydrogenaseII, respectively encoded by the *thrA* and *metL* genes (see Chap. 26). Comparison of their sequences reveal extensive homology which provides unequivocal evidence that the two present genes derive from a common ancestor (Fig. 40.8).

A more detailed examination of each of the protein sequences revealed internal homologies both in the N- and C-terminal parts of the molecules, two segments of about 130 amino acids are repeated in the first one-third of the polypeptide chain, and two segments of about 250 amino acids and covering the COOH-terminal 500 amino acids also present a significant homology. These observations and data obtained from limited proteolysis experiments (see Chap. 26, Fig. 26.6) suggested that the genes coding for the aspartokinase and homoserine dehydrogenase activities may have evolved independently by a process of gene duplication and fusion previous to the appearance of an ancestral fused gene, coding for a bifunctional aspartokinase-homoserine dehydrogenase enzyme, which was itself later duplicated to give rise to the original *thrA* and *metL* genes. As a part of this model represented in Fig. 40.9, it was proposed that the aspartokinase III coding sequence could have been duplicated before the fusion event which gave rise to the common ancestor of *thrA* and *metL*. The main supporting arguments were (i) that it is possible to isolate a proteolytic fragment

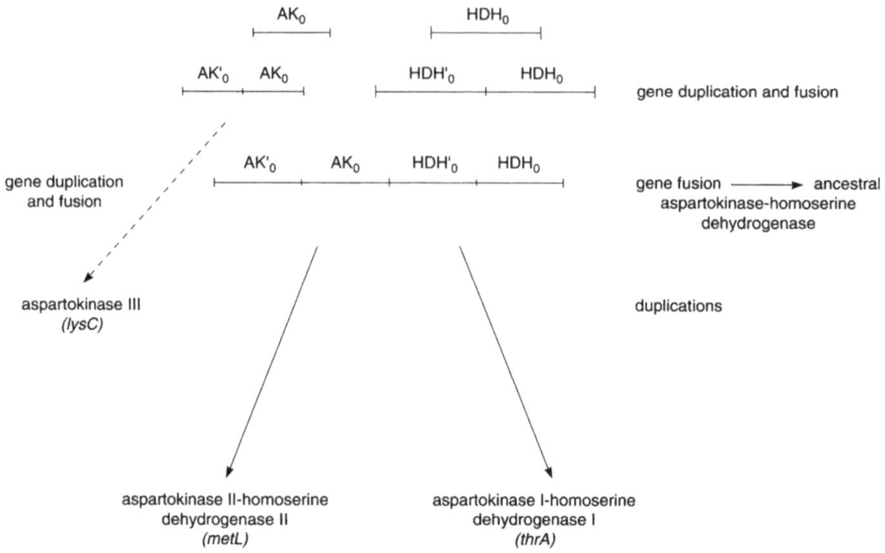

Fig. 40.9 Hypothetical pathway of evolution of *E. coli* aspartokinase-homoserine dehydrogenases. This pathway, proposed in 1984, was proved to be wrong after the sequence of AKIII had been determined. See text (From P. Ferrara, N. Duchange, M. M. Zakin and G. N. Cohen)

of aspartokinase I-homoserine dehydrogenaseI containing 250 N-terminal residues of the protein, endowed with aspartokinase activity and corresponding to the duplicated segments detected in the aspartokinase domain and (ii) that the internal homology in the homoserine dehydrogenase part of the molecule starts at position 330, amino acid sequence 330–565 being homologous to amino acid sequence 570–810 in both aspartokinase-homoserine dehydrogenases I and II. As a consequence, if the aspartokinase III coding sequence had been duplicated before the fusion event, the homology between aspartokinase III and the aspartokinase-homoserine dehydrogenases would not be expected to cover more than the 260–300 N-terminal residues.

The nucleotide sequence of the *lysC* gene, coding for aspartokinase III, was later elucidated and allowed comparison (Fig. 40.10) with the two other proteins and a direct test of the above hypothesis. Three main points must be emphasized from the comparison: (i) the homology between the three proteins extends through the whole aspartokinase sequence (449 residues); (ii) the identities calculated for pairs of proteins show that the aspartokinase III sequence is more similar to the aspartokinase I sequence than to the aspartokinase II and (iii) as previously noticed in the comparison between the aspartokinase segments of the two bifunctional enzymes (Fig. 40.8) and extended by the comparison of the three sequences, the homology is not evenly distributed along the sequences but clearly shows two regions of higher conservation (residues 16–63 and 209–315), the other regions being less homologous with most of the gaps.

The fact that the whole sequence of aspartokinase III is homologous to the N-terminal part of the aspartokinases-homoserine dehydrogenases seems well documented for the N-terminal and the central regions by the long stretches of homology between the three proteins, and for the C-terminal region by the presence of 12

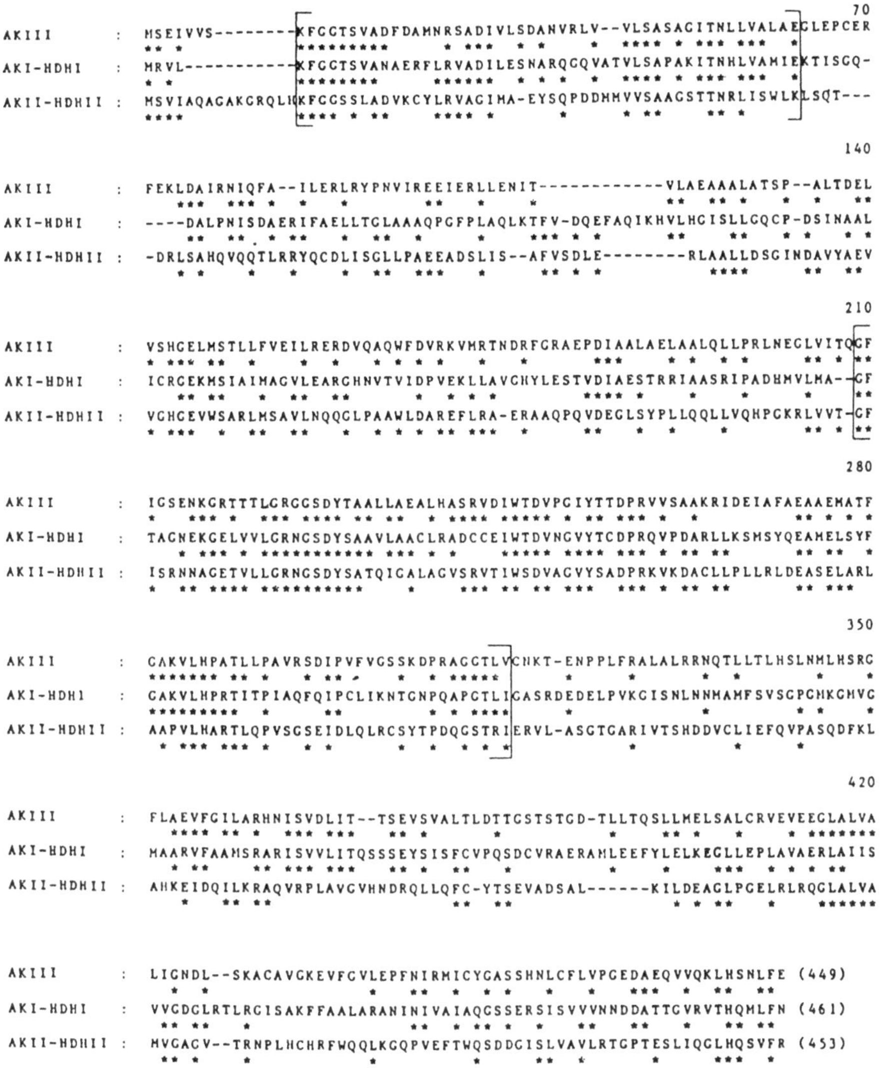

Fig. 40.10 Comparison of the three aspartokinases of *E. coli*. The deduced amino acid sequence of AKIII and part of AKI-HDHI and AKII-HDHII are presented in the one-letter code. Sequences have been aligned in order to maximize homologies by introducing gaps indicated by a *dash* (–). Identical residues are indicated by *stars* under the sequences. *Numbers* refer to the amino acid positions in the alignment and *numbers in parentheses* refer to the position of the last presented amino acid in the original sequences. *Brackets* delimitate the most conserved sequences and are discussed in the text (From M. Cassan, C. Parsot, G.N. Cohen and J-C Patte, with permission of the *Journal of Biological Chemistry*)

identical residues between the last 36 positions of aspartokinase III and the corresponding part of AKI-HDHI. As this last segment is part of the duplicated segment of the dehydrogenase moiety of AKI-HDHI, the homology between the C-terminal part of AKIII and the central region of AKI-HDHI seems to rule out the separation of the ancestor of *lysC* prior to the fusion event between the aspartokinase and dehydrogenase fusion. It is therefore proposed that the separation of the *lysC* ancestral gene has arisen *after* the fusion of the aspartokinase and homoserine dehydrogenase coding sequences and that *lysC* could derive from *thrA*. However, this evolutionary pathway supposes that the mutation rates of the three genes have not been very different and that no genetic conversion occurred, these assumptions being beyond any experimental verification. Whether the duplication event which gave rise to *lysC* encompassed part or all of the homoserine dehydrogenase coding portion of *thrA* is not determinable; there is no homology between the 3′-flanking region of *lysC* and the corresponding region of *thrA*, but even the nucleotide coding sequences show only about 40 % identity, indicating that selective pressure has operated in order to conserve amino acid rather than nucleotide sequences.

The high degree of conservation of the amino acid sequences observed for segments 16–63 and 209–315 probably reflects structural and/or functional constraints. Attempts to utilize these sequences as probes to explore sequences of proteins endowed with kinase activity or concerned with aspartate metabolism failed to detect any significant region of homology that could lead to an educated guess for the localization of the aspartate or ATP-binding sites. On the other hand, data obtained from limited proteolysis on aspartokinase-homoserine dehydrogenases I and II have led to the conclusion that the native polypeptides have a triglobular structure: a N-terminal domain (corresponding to residues 1–245) carries the aspartokinase activity, a central domain (corresponding to residues 250–500) is involved in protein association contacts, and a C-terminal domain (residues 500–820) is endowed with the homoserine dehydrogenase activity. Thus, part of the second very conserved region in the primary sequence (residues 209–315) spans the central domain observed by limited proteolysis (residues 250–500) and may very likely be involved in protein association contacts leading to the polymeric state, dimeric in the case of AKIII and AKII-HDHII and tetrameric in the case of AKI-HDHI.

The Aspartokinase and Homoserine Dehydrogenase Activities of Microorganisms Other Than Enterobacteriaceae

The pattern of isofunctional aspartokinases and homoserine dehydrogenases and probably of bifunctional aspartokinases-homoserine dehydrogenases seems restricted to the enteric bacteria (Chap. 26). What about the sequence of the correponding enzymes in other genera? The beginning of an answer can already be given.

The *Bacillus subtilis* aspartokinase II (Chap. 27) sequence, as determined from the sequence of its gene, is homologous in its entire length (408 amino acid residues) to *E. coli* aspartokinase III. Both *E. coli* and *B. subtilis* monofunctional aspartokinases have thus about 300 amino acid residues in excess of the 250-residue aspartokinase domain that was defined by limited proteolysis on aspartokinase-homoserine dehydrogenases. Therefore, it seemed of particular interest to determine the sequence of an enzyme endowed only with homoserine dehydrogenase activity in order to complete the studies on the structural relationship between aspartokinase-homoserine dehydrogenase, aspartokinase and homoserine dehydrogenase. The obvious choice was that of *Bacillus subtilis* homoserine dehydrogenase, an enzyme which is inhibited by threonine at concentrations of the same order which inhibit homoserine dehydrogenase of *E. coli* AKI-HDHI. The *hom* gene coding for the dehydrogenase in *B. subtilis* is part of a single transcription unit that encompasses the *thrB* and *thrC* genes, coding respectively for homoserine kinase and threonine synthase.

The *hom* gene codes for a polypeptide of 433 residues. Figure 40.11 shows the comparison of its sequence with its counterparts in *E. coli* homoserine dehydrogenases I and II.

The following observations can be made: (1) the similarity between the *B. subtilis* and the *E. coli* homoserine dehydrogenases commences at the exact point where the similarity between *E. coli* and *B. subtilis* aspartokinase ends. (2) some residues are conserved in all three enzymes, thus allowing an hypothesis as to the function of some regions in the homoserine dehydrogenases. (3) the *B. subtilis* enzyme has a C-terminal extension of 108 amino acid residues that has no counterpart in the *E. coli* enzymes.

The beginning of the sequence of *B. subtilis* homoserine dehydrogenase can be aligned with the sequences of AKI-HDHI and AKII-HDHII from residues 462 and 456, respectively. Similarly, the C-terminal end of *B. subtilis* aspartokinase II corresponds to residues 468 and 461 in AKI-HDHI and AKII-HDHII, respectively. Therefore aspartokinases-homoserine dehydrogenases can be viewed as the products of a direct fusion of aspartokinase and homoserine dehydrogenase. One can speculate that, in an ancestral chromosome, genes coding for aspartokinase and homoserine dehydrogenase were in tandem. By the simple loss of a stop codon at the end of the aspartokinase gene, a fused gene coding for a bifunctional protein could have been created. This could have happened after the emergence of the *Enterobacteriaceae*, since bifunctional AK-HDHs have been detected only in these organisms. A comparable artificial fusion between the *S. typhimurium his D* and *hisC* genes has been obtained under appropriate selection conditions, showing that such a scheme is not unreasonable.

The comparison of homoserine dehydrogenases indicates that 52 amino acid residues are strictly conserved between the three sequences (indicated by *asterisks* in Fig. 40.11). Among these highly conserved residues, one finds the motif Gly-X-Gly-X-X-Gly, which is encountered in all known nucleotide binding domains. Around these glycine residues, the small, hydrophobic amino acid residues which are predicted to be of structural importance for the β-α-β folding on this domain are also present at the correct positions in the sequences of the three homoserine dehydrogenases. (This motif is present at similar positions in the homoserine dehydrogenases of *Corynebacterium glutamicum* and *Brevibacterium lactofermentum*). It thus seems likely that this region constitutes the NADPH binding domain of homoserine dehydrogenase.

The entire sequence of *B. subtilis* homoserine dehydrogenase (HDH, *hom*) has been aligned with parts of the sequences of *E. coli* AKI-HDHI and AKII-HDHII (HDHI and HDHII). Symbols = and – indicate identical and similar (I-L-V-M, R-K, D-E, G-A, S-T, F-Y) residues, respectively. *Asterisks* above the alignment indicate positions strictly conserved in the three sequences and arrows refer to the glycine residues of a proposed nucleotide binding domain. The most conserved region is *boxed*. Numbers in

Fig. 40.11 Comparison of *E. coli* and *B. subtilis* homoserine dehydrogenase amino acid sequences

parentheses refer to the position of the relevant residue in the original sequence. From C. Parsot and G. N. Cohen, with permission of the *Journal of Biological Chemistry*.

The most conserved region among the three homoserine dehydrogenases lies between positions 200 and 221 of *B. subtilis* homoserine dehydrogenase, where 12 out of 22 residues are strictly conserved (see the *boxed* region of Fig. 40.11). Interestingly, there are 7 charged residues among these 12 conserved positions, whereas there are only 12 charged residues that are conserved in the full length of the polypeptides. It seems reasonable to speculate that this 22-amino acid region could be involved in substrate binding or catalytic activity. Such a region is also detected by comparison with the homoserine dehydrogenase of *Corynebacterium glutamicum* and *Brevibacterium lactofermentum.*

The comparison between the three enzymes indicates that the *B. subtilis* homoserine dehydrogenase has a 108-amino acid residue C-terminal extension relative to the *E. coli* enzymes. Exploration of protein sequence libraries failed to detect any similarity between this extra segment and any other protein sequence. The following observations suggest that this part of the *B. subtilis* dehydrogenase could constitute a domain involved in the regulation of enzyme activity. First, since this region is not present in the *E. coli* enzymes, it cannot be directly involved in the catalytic activity (even though a truncated polypeptide, lacking the last 80 residues is not endowed with detectable activity, it may be that it is incorrectly folded). Second, numerous lines of evidence indicate that the threonine sensitivity of *E. coli* homoserine dehydrogenase I is intrinsically associated with a threonine-sensitive kinase domain and thus appears to be a consequence of the fusion of homoserine dehydrogenase to aspartokinase. Therefore, the threonine sensitivity of the homoserine dehydrogenase from *B. subtilis* is likely to be achieved by a mechanism different from the mechanism used for regulation of the homoserine dehydrogenase in the *E. coli* bifunctional enzyme. This indirect evidence led the authors of this work to propose that the C-terminal sequence detected in *B. subtilis* homoserine dehydrogenase could constitute a regulatory domain involved in the sensitivity of this enzyme to threonine. This hypothesis provides an attractive scheme for the fusion of two enzymes that do not catalyze two consecutive reactions of a metabolic pathway, namely aspartokinase and homoserine dehydrogenase. One can speculate that the ancestral homoserine dehydrogenase was not regulated and that two different strategies have arisen to ensure regulation of this enzyme activity, i.e. fusion to the C-terminal end of an already regulated aspartokinase or addition of a regulatory domain to the C-terminus of homoserine dehydrogenase. This hypothesis is supported by the fact that the homoserine dehydrogenases of *Corynebacterium glutamicum* and of *Brevibacterium lactofermentum*, also species where separate proteins catalyze threonine-sensitive aspartokinase and homoserine dehydrogenase, possess also a C-terminal extension of about 100 residues at their COOH-terminal, for which there is no counterpart in either of the *E. coli* proteins.

Transmembrane Facilitators

Many transport proteins of bacteria and eukaryotes are thought to possess a common structural motif of 12 transmembrane-spanning α-helical segments, similar to the lactose permease we have examined in some detail. Using statistical methods Marger and Saier have found that five families of proteins which facilitate uniport, symport and antiport comprise a single superfamily. These five families are drug-resistance proteins, prokaryotic and eukaryotic sugar facilitators (among which glucose transporters, galactose, lactose and maltose permeases from various yeasts), bacterial citrate transporters, bacterial sugar-phosphate transporters and finally some disaccharide symporters, including LacY, the product of enterobacterial Lac permease. A total of more than fifty sequences have been compared. The intrafamily and interfamily comparisons show a degree of homology sufficiently high to exclude that it might have occurred by chance. The authors propose that the superfamily arose before the appearance of eukaryotes on earth, the duplication events having occurred some 2–3 billion years ago.

DNA-Binding Regulator Proteins

Due to the great progress in computer-aided search of the data bases, it has been possible to find homologies between bacterial regulators homologous to the Gal and Lac repressors. Each member of the GalR-LacI family has an amino-terminal DNA binding domain and several regions involved in effector recognition and oligomerization in the carboxyl terminal part of the protein. The DNA-binding domains, which can assume the form of a helix-turn-helix, are the most conserved portions of these proteins. This family includes many members belonging to diverse Gram-positive and negative organisms, that regulate a variety of biosynthetic and transport function. A very illuminating alignment has been presented by Weickert and Adhya in 1992 where it is shown that these proteins show a very high degree of similarity

(60 %) through the entire sequence. Since a portion of the operator sequences occupied by these proteins is also conserved, a similar DNA structure may be required for specific recognition of DNA by members of the family. The compilation and alignment presented should simplify the study by mutagenesis of current and new member proteins by identifying, in advance, residues which are more likely involved in the critical functions of DNA binding, inducer interactions and oligomerization. It also may give some hints for protein evolution.

Selected References

Two Books

Loomis WF (1988) Four billion years, an essay on the evolution of genes and organisms. Sinauer Associates, Sunderland, p 286
Mortlock RP (ed) (1992) The evolution of metabolic function. CRC Press, Boca Raton, p 330

Two Different Theories on the Evolution of Biosynthetic Pathways

Horowitz NH (1945) Proc Natl Acad Sci USA 31:153–157
Horowitz NH (1965) In: Bryson V, Vogel HJ (eds) Evolving genes and proteins. Academic, New York, pp 16–23
Jensen RA (1976) Ann Rev Microbiol 30:409–425
Parsot C, Saint Girons I, Cohen GN (1987) Microbiol Sci 4:258–262
Ycas M (1974) J Theoret Biol 44:145–160

Common Origin of Cystathionine-γ-Synthase and Cystathionase

Belfaiza J, Parsot C, Martel A, Bouthier de la Tour C, Margarita D, Cohen GN, Saint Girons I (1986) Proc Natl Acad Sci USA 83:867–871

Common Origin of Threonine Synthase, Threonine Dehydratase, D-Serine Dehydratase, and the B Chain of Tryptophan Synthase

Parsot C (1986) EMBO J 5:3013–3019
Parsot C (1987) Proc Natl Acad Sci USA 84:5207–5210

Comparison of Arg Genes with Homologous and Analogous Enzymes

Parsot C, Boyen A, Cohen GN, Glansdorff N (1988) Gene 68:275–283

Evolution of the E. coli Aspartokinases and Homoserine Dehydrogenases

Cassan M, Parsot C, Cohen GN, Patte J-C (1986) J Biol Chem 261:1052–1057
Ferrara P, Duchange N, Zakin MM, Cohen GN (1984) Proc Natl Acad Sci USA 81:3019–3023
Zakin MM, Duchange N, Ferrara P, Cohen GN (1983) J Biol Chem 258:3028–3031

Structural and Evolutionary Relationships Between E. coli Aspartokinase-Homoserine Dehydrogenases and Monofunctional Homoserine Dehydrogenases

Parsot C, Cohen GN (1988) J Biol Chem 263:14654–14660

Superfamily of Transmembrane Facilitators

Marger MD, Saier MH Jr (1993) Trends Biochem Sci 18:13–20

DNA-Binding Regulator Proteins

Weickert MJ, Adhya S (1992) Biol Chem 267:15869–15874

Index

G.N. Cohen, *Microbial Biochemistry*,
DOI 10.1007/978-94-017-8908-0,

B

E

F

Q

MIX
Papier aus verantwortungsvollen Quellen
Paper from responsible sources
FSC® C105338

If you have any concerns about our products,
you can contact us on
ProductSafety@springernature.com

In case Publisher is established outside the EU,
the EU authorized representative is:
Springer Nature Customer Service Center GmbH
Europaplatz 3, 69115 Heidelberg, Germany

Printed by Libri Plureos GmbH
in Hamburg, Germany